Environmental Systems and Societies

Virginia D'Britto, Öykü Dulun,
Emma Shaw, Joseph Cazabon,
Thierry Torres
Manish Semwal

HODDER Education

Although every effort has been made to ensure that website addresses are correct at time of going to press, Hodder Education cannot be held responsible for the content of any website mentioned in this book. It is sometimes possible to find a relocated web page by typing in the address of the home page for a website in the URL window of your browser.

Hachette UK's policy is to use papers that are natural, renewable and recyclable products and made from wood grown in well-managed forests and other controlled sources. The logging and manufacturing processes are expected to conform to the environmental regulations of the country of origin.

To order, please visit www.hoddereducation.com or contact Customer Service at education@hachette.co.uk / +44 (0)1235 827827.

ISBN: 978-1-0360-0280-0

© Virginia D'Britto, Öykü Dulun, Emma Shaw, Joseph Cazabon and Thierry Torres 2024

First published in 2024 by
Hodder Education,
An Hachette UK Company
Carmelite House
50 Victoria Embankment
London EC4Y 0DZ

www.hoddereducation.com

Impression number   10 9 8 7 6 5 4 3 2 1

Year   2027 2026 2025 2024

All rights reserved. Apart from any use permitted under UK copyright law, no part of this publication may be reproduced or transmitted in any form or by any means, electronic or mechanical, including photocopying and recording, or held within any information storage and retrieval system, without permission in writing from the publisher or under licence from the Copyright Licensing Agency Limited. Further details of such licences (for reprographic reproduction) may be obtained from the Copyright Licensing Agency Limited, www.cla.co.uk

Cover photo: Background image © yauhenka – stock.adobe.com. Anticlockwise from top left: © sweet – stock.adobe.com / © greenbutterfly – stock.adobe.com / © fragrant smell of orchids – stock.adobe.com / © icarmen13 – stock.adobe.com

Illustrations by Integra Software Services Ltd

Typeset in India by Integra Software Services Ltd

Produced by DZS Grafik, Printed in Slovenia

A catalogue record for this title is available from the British Library.

MIX
Paper | Supporting responsible forestry
FSC™ C104740

Contents

Introduction v

How to use this book vi

1 Foundation 1
- **1.1** Perspectives 2
- **1.2** Systems 24
- **1.3** Sustainability 47

2 Ecology 83
- **2.1** Individuals and populations, communities and ecosystems 84
- **2.2** Energy and biomass in ecosystems 119
- **2.3** Biogeochemical cycles 139
- **2.4** Climate, biomes and their influences 163
- **2.5** Zonation, succession and change in ecosystems 178

3 Biodiversity and conservation 191
- **3.1** Biodiversity and evolution 192
- **3.2** Human impact on biodiversity 219
- **3.3** Conservation and regeneration 245

4 Water 267
- **4.1** Water systems 268
- **4.2** Water access, use and scarcity 291
- **4.3** Aquatic food production systems 331
- **4.4** Water pollution 373

5 Land 415
- **5.1** Soil 416
- **5.2** Agriculture and food 456

6 Atmosphere and climate change 513
- **6.1** Introduction to the atmosphere 514
- **6.2** Climate change – causes and impacts 528
- **6.3** Climate change – mitigation and adaptation 562
- **6.4** Stratospheric ozone 594

7 Natural resources 607
- **7.1** Natural resources, uses and management 608
- **7.2** Energy sources, use and management 634
- **7.3** Solid waste 666

8 Human populations and urban systems 693
- **8.1** Human populations 694
- **8.2** Urban systems and urban planning 719
- **8.3** Urban air pollution 740

Glossary 757

Acknowledgements 767

Index 769

Online content
Go to our website www.hoddereducation.com/ib-extras for free access to the following:
- Theme 9 HL extension lenses (environmental law, economics and ethics)
- Answers to review questions and exam-style questions

Introduction

Welcome to *Environmental Systems and Societies (ESS) for the IB Diploma,* designed to meet the criteria of the new International Baccalaureate (IB) Diploma Programme ESS Guide. This coursebook provides complete coverage of the new IB ESS Diploma syllabus, with first teaching from 2024.

Differentiated content for SL and HL students is clearly identified throughout.

Three key concepts comprise the foundation subtopics of Theme 1: Foundation and should be taught at the start of the course:

- Perspectives: The concept of perspectives provides a deeper understanding of worldviews, individual perspectives and their related value systems.
- Systems: Systems theory provides a useful tool for holistic analysis, and gives insight into understanding the mechanics and purpose of human-constructed systems and the function of natural ones
- Sustainability: The concept of sustainability is central to ESS. Resource management issues are pivotal to sustainability, and students' attention is drawn to this throughout the course.

Taken from IB Diploma Programme Environmental systems and societies Guide, page 7

Environmental systems and societies (ESS) aims to empower and equip students to:

1. develop understanding of their own environmental impact, in the broader context of the impact of humanity on the Earth and its biosphere,
2. develop knowledge of diverse perspectives to address issues of sustainability,
3. engage and evaluate the tensions around environmental issues using critical thinking,
4. develop a systems approach to provide a holistic lens for the exploration of environmental issues,
5. be inspired to engage in environmental issues across local and global contexts.

Taken from IB Diploma Programme Environmental systems and societies Guide, page 19

About the authors

Virginia D'Britto is an experienced ESS and Biology teacher and examiner with more than 15 years of classroom experience.

Dr Öykü Dulun is an experienced IBEN member and curriculum specialist. She has taught Biology and TOK for over 10 years. She's an IB examiner and workshop leader for both subjects.

Dr Emma M Shaw has taught environmental courses for over 20 years. She is a highly published researcher and educator with a lifelong passion for ecology and arachnology.

Joseph Cazabon has over 20 years' experience as a teacher of Biology, Environmental Science and ESS in Trinidad and Tobago and the UK. He is co-author of the Caribbean Interactive Science series of text and activity books.

Dr Manish Semwal has been teaching IBDP Environmental systems and Societies and Geography for the past 23 years. He has published more than 14 research papers in various journals.

■ IB advisor

Thierry Torres has taught in IB schools for over 15 years. He teaches DP ESS and Geography at Antwerp International School where he is also IBDP Coordinator.

How to use this book

The following features will help you consolidate and develop your understanding of ESS, through concept-based learning.

Key terms
◆ Definitions appear throughout the margins to provide context and help you understand the language of ESS. There is also a glossary of all the key terms at the end of the book.

Common mistake
To help you avoid falling into regular misunderstandings.

TOK
Links to Theory of Knowledge (TOK) allow you to develop critical thinking skills and deepen your understanding by bringing discussions about the subject beyond the scope of the content of the curriculum.

Link
This feature allows links to be made between different parts of the course, framed using levels of organization and concepts, helping you to form a holistic appreciation of material in the syllabus. Because of the interdisciplinary nature, some context may be required.

Guiding questions
- The key prompts from the ESS Guide to help you view the content through the required conceptual lenses.

SYLLABUS CONTENT
▶ This coursebook follows the exact order of the contents of the IB ESS Diploma syllabus.
▶ Syllabus understandings are introduced naturally throughout each topic.

ATL ACTIVITY
Approaches to learning (ATL), including learning through inquiry, are integral to IB pedagogy. These approaches-to-learning skills activities get you to think about real-world situations.

Concept
Highlighting the links to the three themes that underpin the ESS for the IB Diploma course (perspectives, systems and sustainability).

REAL-WORLD EXAMPLE
An opportunity to apply the topic/concept being explored.

Tool
The Tools features explore the skills and techniques that you require and are integrated into the ESS content to be practised in context. These skills can be assessed through internal and external assessment.

REVIEW QUESTIONS
Formative questions to provide students with the opportunity to test their knowledge.

EXAM-STYLE QUESTIONS
Prepare students for assessment with questions designed to test understanding and knowledge in the format of the exam.

HL lenses
Cover and link to the main content of the HL extension lenses (HL.a, HL.b and HL.c).

Skills are highlighted with this icon. You are expected to be able to show these skills in the examination, so we have explicitly pointed these out when they are mentioned in the Guide.

International mindedness is indicated with this icon. It explores how the exchange of information and ideas across national boundaries has been essential to the progress of knowledge.

The IB learner profile icon indicates material that is particularly useful to help you towards developing the following attributes: to be inquirers, knowledgeable, thinkers, communicators, principled, open-minded, caring, risk-takers, balanced and reflective. When you see the icon, think about what learner profile attribute you might be demonstrating – it could be more than one.

Theme 1
Foundation

1.1 Perspectives

> **Guiding questions**
> - How do different perspectives develop?
> - How do perspectives affect the decisions we make concerning environmental issues?

SYLLABUS CONTENT

This chapter covers the following syllabus content:
- ▶ 1.1.1 A perspective is how a particular situation is viewed and understood by an individual. It is based on a mix of personal and collective assumptions, values and beliefs.
- ▶ 1.1.2 Perspectives are informed and justified by sociocultural norms, scientific understandings, laws, religion, economic conditions, local and global events, and lived experience, among other factors.
- ▶ 1.1.3 Values are qualities or principles that people feel have worth and importance in life.
- ▶ 1.1.4 The values that underpin our perspectives can be seen in our communications with and actions in the wider community. The values held by organizations can be seen through their advertisements, media, policies and actions.
- ▶ 1.1.5 Values surveys can be used to investigate the perspectives shown by a particular social group towards environmental issues.
- ▶ 1.1.6 Worldviews are the lenses shared by groups of people and through which they perceive, make sense of and act within their environment. They shape people's values and perspectives through culture, philosophy, ideology, religion and politics.
- ▶ 1.1.7 An environmental value system is a model that shows the inputs affecting our perspectives and the outputs resulting from our perspectives.
- ▶ 1.1.8 Environmental perspectives (worldviews) can be classified into the broad categories of technocentric, anthropocentric and ecocentric.
- ▶ 1.1.9 Perspectives and the beliefs that underpin them change over time in all societies. This can be influenced by government or NGO campaigns, or through social and demographic change.
- ▶ 1.1.10 The development of the environmental movement has been influenced by individuals, literature, the media, major environmental disasters, international agreements, new technologies and scientific discoveries.

Note: There is no additional higher level content in 1.1.

1.1.1 A perspective is how a particular situation is viewed and understood by an individual, based on assumptions, values and beliefs

> **Concept**
>
> **Perspectives**
>
> According to the Britannica Dictionary, a **perspective** is 'a way of thinking about and understanding something (such as a particular issue or life in general)'.
> To define it further, we need to view issues from a different point of view, one that emerges from a wide range of factors and different circumstances: life experiences, **personal assumptions**, scientific understanding, **economic status**, **ethical beliefs**, and **personal values**.

- **Perspective** – a point of view; a particular way of seeing or considering something.
- **Personal assumptions** – made up of our beliefs and preconceptions, which are based on our previous experiences as well as our cultural background and the education with which we have been provided. For instance, someone who has grown up in a rural area may have a different perspective on the importance of protecting natural resources, as well as being more aware of the source of food, or how necessary it is to drive a car, than someone who has grown up in an urban area.
- **Economic status** – an individual's income and occupation. For example, a person with a high income may have a different perspective on the importance of economic growth than someone with a lower income.
- **Ethical beliefs** – the moral principles and values that an individual holds. For example, a person with strong religious beliefs may have a different perspective on the morality of certain actions than someone without strong religious beliefs.
- **Personal values** – can include beliefs about the importance of family, community and personal responsibility. For example, an individual with a strong sense of community may have a different perspective on the importance of protecting the environment than someone who places a higher emphasis on individual rights.

People should reflect on their own perspectives while they acquire in-depth knowledge on a particular issue. One relevant example is the issue of rising sea levels and the impact of these on several countries' future existence.

> **REAL-WORLD EXAMPLE**
>
> **The Maldives**
>
> The Maldives is an archipelago located in the Indian Ocean, which faces unique environmental and social challenges due to its geographical characteristics.
>
> **Figure 1.1** The Maldives is an archipelago located in the Indian Ocean
>
> In the Maldives, rising sea levels pose a significant threat to the country's very existence. The Maldives is known for its low-lying islands, with the highest point only a few metres above sea level. As global sea levels rise due to climate change, the Maldives is becoming increasingly vulnerable to coastal erosion, flooding and saltwater intrusion into freshwater sources.

People from the Maldives, especially those living in the coastal areas, have a unique perspective on climate change and its consequences. They witness the immediate effects of sea-level rise on their livelihoods, land and homes. Scientists predict that sea level might rise up to 0.9 cm a year. Their perspective is rooted in the fear of losing their homes and way of life due to the encroaching ocean.

On the other hand, people from regions less affected by rising sea levels may have a different perspective. While they may also be concerned about the overall impact of climate change, they might not fully grasp the urgency and personal implications that the Maldivians are facing. The severity of the issue might not resonate as deeply, as it doesn't directly threaten their immediate surroundings.

When individuals from different regions come together to discuss climate change and its impacts, the perspectives from the Maldivians can highlight the urgency of the situation and the need for global cooperation and climate action. At the same time, it can help others empathize and understand the gravity of climate change on a more personal level, rather than just as an abstract concept. This understanding can lead to better-informed discussions, and collaborative efforts to address the environmental and social challenges posed by climate change.

1.1.2 Perspectives are informed and justified by many different factors

Understanding the arguments that support a viewpoint can assist us in critically evaluating that viewpoint, and developing our own informed viewpoint and conclusions.

◉ Common mistake

An argument is not the same as a perspective. A perspective is an individual's unique understanding. A perspective is shaped by the individual's beliefs and experiences, and influences how they perceive information. It shows a broader worldview. By contrast, arguments are statements made to support a personally held perspective or to counter a different one. An argument is a focused presentation of reasons and evidence aimed at supporting a specific viewpoint or countering an opposing one. Arguments arise from perspectives, serving as tools to express and defend a particular stance within a broader outlook.

◆ **Worldview** – a broad and comprehensive framework that shapes people's perceptions and understandings of their surroundings as well as their actions.

◉ TOK

You may remember from Theory of knowledge (TOK) courses that the relativist view of truth is discussed to promote tolerance. However, this may not help us to arrive at a conclusion. According to this view, the truth of a statement is relative to a particular perspective. All views are usually relevant to their contexts, and we would be in a better position to understand them if we try to listen to the different perspectives underlying them.

Link

Three areas where we can observe perspectives that range from relativism to absolutism are religion, politics and history. We should be aware of the national and political interests in the writing of history, for instance.

Local events, such as community gatherings, cultural celebrations, or even local political decisions, can shape our perspectives by exposing us to diverse viewpoints within our immediate surroundings. Attending a town hall meeting or engaging in a community debate allows us to hear different arguments and to gain a deeper understanding of the issues affecting those around us. These experiences can either reinforce our existing perspectives or open our minds to new possibilities.

On the other hand, global events have a far-reaching impact on our perspectives and the arguments we formulate. Events like international conflicts, economic crises, or advancements in science and technology can shift our **worldview** on a broader scale. Access to global news and information allows us to comprehend how interconnected the world is and how decisions made in one part of the world can have repercussions in another.

One such issue is climate change. Local events like extreme weather patterns in our region can directly influence our perspectives on the urgency of environmental action. Simultaneously, global events, such as international climate summits or reports on rising sea levels in distant countries, can provide a broader context and reinforce the need for a collective response to this global challenge.

The more we learn and the deeper our understanding, the more our opinions develop and thus our perspectives widen. Sometimes, our personal perspective can help us empathize with others. At other times, it can prevent us from understanding their position, or even wanting to. Our perspectives and individual knowledge develop our life experiences, however. Our perspectives are shaped by some factors, including our own personal assumptions, scientific understandings, economic conditions, ethical beliefs and personal values. Then we try to justify our positions using arguments based on ethics, logic, religion, politics or pragmatism.

One of the numerous benefits of supporting diversity in all parts of society is that it allows everyone to meet, respect and learn from people who have different views, experiences and viewpoints.

As another point of view, cultural relativism may help us to become more respectful to different cultural and religious perspectives. This is since according to relativism, no one set of cultural values is better than another. This supports the idea that all cultural perspectives are equally valuable and valid.

People all come from different communities and have had varied life experiences. This means that we should improve our knowledge and comprehension of different perspectives in order to develop stronger arguments. We must also become more conscious of our own prejudices, assumptions and perspectives, which are influenced by where and when we live, the communities we belong to, and even the language we use. Cross-checking knowledge claims will undoubtedly help us to obtain actual information and hence grow more confident in our own understanding.

Ethical beliefs are another factor that may shape our perspectives. Ethics refers to moral principles and ideals. Ethical debates may centre on the morality of a certain action or choice. For example, an ethical principle that says all living things have a right to exist would require an argument that deforestation is wrong. Another argument that says renewable energy is a better option than burning fossil fuels would be based on the logic that renewable energy is a more sustainable option.

Arguments that are made using critical thinking would focus on the logical consequences of particular actions or decisions. An argument based on religion would focus on the spiritual ramifications of a particular action or choice. For example, a religious conviction that all living things have a soul may underpin an argument that animals should not be exploited for food.

Pragmatism, as a philosophical approach, relates to utility and practicality. Furthermore, pragmatist arguments would focus on the practical repercussions and usefulness of a specific activity. An argument for recycling, for example, could be based on the pragmatic conclusion that recycling conserves resources and decreases waste.

◆ **Pragmatism** – a philosophical approach that relates to utility and practicality.

1.1.3 Values are qualities or principles that people feel have worth and importance in life

The three types of values are moral, personal and societal values. These values are influenced by our cultural background, religion, family, education and experiences. Values are significant in this sense because they shape how people perceive the world around them. Values can guide us to take actions that align with our values, regardless of the external circumstances. For example, someone who values environmental conservation may choose to purchase a hybrid car despite the fact that it is more expensive.

Values are qualities and ideas that we feel are important in our lives and that might influence how we interact with one another in the community. Individuals who share similar ideals are more likely to form strong bonds. This also encourages collaboration towards common goals.

Furthermore, values might influence our perspectives and choices. A person who values sustainability, for example, may have a different perspective on government rules and policies than someone who values economic growth.

Everyone should be conscious of their own values and how they influence their perceptions, behaviours and decisions. It is equally critical that we understand and appreciate the values of others, even if they differ from our own, in order to build a diverse and globalized society based on effective communication, understanding and respect.

For instance, in the context of environmental ethics and conservation, **intrinsic value** is often associated with the inherent worth of the natural world and its components, such as ecosystems, species or individual organisms. It suggests that these entities have value and deserve moral consideration simply because they exist and have their own inherent characteristics, rights and dignity.

◆ **Intrinsic value** – the value that a thing has in and of itself. Intrinsic value does not depend on whether a thing is useful or beneficial to people.

The recognition of intrinsic value in nature has implications for environmental ethics and conservation practices. It implies that nature should be protected not solely for its instrumental value to humans, but also due to the ethical responsibility to preserve it for its own sake. This perspective underpins conservation efforts aimed at preserving biodiversity, ecosystems and ecological integrity.

1.1.4 The values that underpin our perspectives

In our interconnected world, our values shape not only our individual perspectives but also our communication and actions within the wider community. Moreover, organizations play a significant role in society, and their values can be observed through various channels such as advertisements, media representations, policies and actions. Understanding the underlying values behind our perspectives and those held by organizations is important for navigating the complexities of environmental issues and sustainability.

■ The values that underpin our perspectives can be seen in our communication and actions within the wider community

Values are deeply rooted beliefs that guide our thoughts, decisions and behaviours. They form the foundation of our perspectives on environmental, social and ethical issues. By examining our values, we can gain insight into why we communicate and act in particular ways within the wider community. For instance, by examining our values of equality and justice, our communication and actions may prioritize inclusivity and fairness in addressing environmental challenges. If we value collaboration and empathy, we may engage in open and respectful dialogue, seeking common ground for collective environmental solutions. Alternatively, if we value competition and self-interest, our communication may be more confrontational or centred on personal gain.

In addition to this, our actions often speak louder than words and reveal our true values. Whether it's participating in environmental protests, volunteering for conservation initiatives or making sustainable choices in our daily lives, our actions reflect our commitment to certain values. By observing the actions of individuals, we can gain insights into their environmental consciousness and priorities.

■ The values held by organizations can be seen through advertisements, media, policies and actions

Organizations convey their values through their advertisements and media representations. Advertisements, whether in print, on television or online, reflect the values that companies wish to associate with their products or services. For example, an advertisement promoting a sustainable lifestyle may signal an organization's commitment to environmental responsibility. Similarly, media representations of organizations can shed light on their values by highlighting their environmental initiatives, ethical practices or community engagement.

An organization's values are often embedded in its policies and guidelines. By examining these documents, we can gain an understanding of the organization's stated priorities and principles. For instance, an organization with a strong commitment to sustainability may have policies promoting waste reduction, renewable energy use or ethical sourcing practices. Policies can provide insights into how organizations translate their values into tangible actions.

> ### ● HL.a.12: Environmental law
> Legal and economic strategies for sustainability. Achieving sustainability involves both legal and economic approaches. For instance, laws imposing fines for illegal dumping align with ethical concerns for responsible resource use. Simultaneously, economic strategies attach value to ecosystem services, connecting ethical considerations with economic incentives for sustainable practices.

Like individuals, however, organizations are judged by their actions. Whether it is implementing sustainable practices, engaging in community outreach or supporting environmental causes, an organization's actions reflect its values. For instance, a company that actively supports environmental conservation through donations or partnerships demonstrates a commitment to environmental stewardship. By analysing the actions of organizations, we can evaluate the alignment between their stated values and their real-world impact.

Different values often lead to tensions both between individuals and between organizations. These tensions can arise when the actions taken by one entity are perceived as conflicting with the values of another. For example, in the context of environmental issues, a company that prioritizes economic growth and profitability over environmental concerns might face criticism and tension from environmental advocacy groups or individuals who prioritize conservation and sustainability.

Similarly, within organizations, employees may hold diverse values and beliefs, which can lead to conflicts over the direction the organization should take. For instance, employees who advocate for stronger social responsibility initiatives might clash with those who prioritize cost-cutting measures, leading to internal tensions.

Furthermore, disagreements between organizations can arise due to differences in their core values and priorities. For instance, in the realm of sustainability, organizations focused on immediate profit generation might be at odds with those dedicated to long-term environmental preservation and social welfare. These differing values can lead to competition, conflicting interests or even public disputes between organizations.

In such scenarios, open dialogue and understanding each other's perspectives become crucial in finding common ground and possible areas of collaboration. By acknowledging and respecting the different values held by individuals and organizations, it becomes possible to bridge gaps, address tensions, and work towards collective solutions to environmental and social challenges. Additionally, transparency in actions and decision-making processes can help build trust and foster a more constructive approach to resolving conflicts arising from differing values.

1.1.5 Values surveys for investigating perspectives towards environmental issues

You should be familiar with how to carry out surveys of a particular social group to identify perspectives towards a specific environmental issue, and with how to evaluate the likely impact of these values.

Values surveys are a tool for investigating the perspectives of a specific social group. These surveys seek to identify the values and beliefs that shape a group's perspectives on a specific environmental issue.

A values survey normally begins with the selection of a specific social group. Then a set of survey questions is developed to probe that particular chosen group's values and beliefs, and finally the collection and analysis of data takes place. The survey questions are designed to collect information about how the group views the current environmental issue, as well as what factors impact their opinion.

Environmental Protection vs. Economic Growth

Share of U.S. adults that think the environment should be prioritized over the economy and vice versa

— Environmental protection   — Economic growth

Latest poll surveyed 1,009 U.S. respondents (18+ y/o) March 1-23, 2022
Source: Gallup

statista

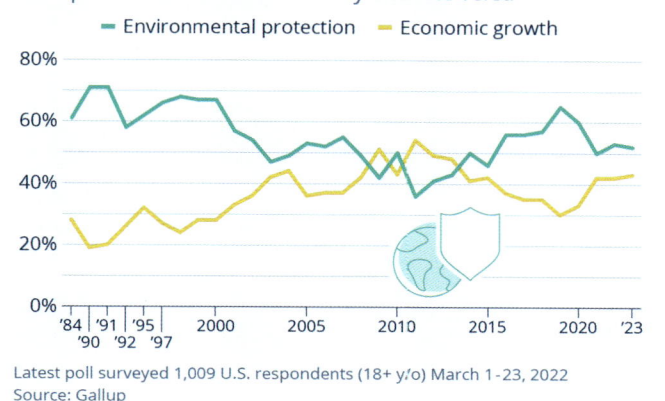

■ **Figure 1.2** Survey results showing prioritization of environmental protection over economic growth (source: Gallup/Statista)

A values survey on a specific environmental issue, such as deforestation, could help to obtain data on the group's beliefs about the necessity of safeguarding natural resources, the moral consequences of clearcutting forests, and the economic rewards of logging.

After collecting the survey data, this can be analysed to identify patterns and trends in the group's perspectives. This can provide insight into how the group's values and beliefs are likely to impact the environmental issue at hand.

For example, a fierce debate in people's minds in the United States is between prioritizing the environment (even at the cost of slowing economic growth) or the economy (even if the environment suffers). This debate has alternated a lot back and forth over the last few decades. According to Gallup survey data in the 1980s and 1990s, the environment was the obvious winner in this moral quandary when seen nationally. This, however, began to alter as the new millennium began.

As the infographic in Figure 1.2 shows, the conflict between the two intensified in the United States as the impacts of the 2008 financial crisis began to bite. In March 2009, the economy had moved to the forefront of most people's attention, with a majority (51 per cent) choosing the economy as a priority over the environment (42 per cent). Between 2015 and 2019, the Earth regained control of American hearts and minds, but the COVID-19 pandemic appears to have altered the balance once more in 2020 and 2021.

Survey results could show that the majority of the respondents prioritize economic growth over environmental protection, implying that they are less likely to support conservation efforts to protect the environment. You should be familiar with the process of conducting values surveys to identify perspectives on environmental issues.

● HL.c.7, HL.c.8, HL.c.9, HL.c.10, HL.c.11: Environmental ethics

Think of moral standing and major ethical approaches as different sets of rules that guide people's choices. For example, some think that being a good person means treating nature kindly, while others think it's more about what good things come out of our actions. By asking questions that relate to these rules, we can learn how they shape attitudes towards environmental issues.

● HL.c.13: Environmental ethics

Environmental and social justice movements. Sometimes, environmental problems are connected to fairness and justice, just like how we treat each other. Understanding this helps us to ask questions that show how people care not only about nature but also about making things fair and equal for everyone.

1.1 Perspectives

Tool 1: Experimental techniques

Questionnaires, surveys, interviews

You need to use a tool like Google Forms or SurveyMonkey, or other polling functions such as on social media, to collect data.

To create and administer surveys, you should be able to:
- identify and justify your choice of an appropriate target audience
- construct relevant open or closed questions with multiple-choice responses or a Likert scale, as appropriate
- choose and justify an appropriate method and size of sample, for example, random, convenience, volunteer or purposive
- show ethical awareness, such as anonymity or consent of respondents over the age of 12
- pilot or trial the survey to gain feedback for modification.

This will assist you in comprehending the perspectives of various social groups and determining how these values are likely to impact specific environmental issues.

To help you understand the process, here's a step-by-step guide to follow:
1. Define the purpose of your research: What particular information would you like to obtain? (Remember that your purpose should also align with the topics and concepts covered in your ESS Course Book.)
2. Create a well-structured survey: Your questions must be written as clearly and concisely as possible. Make sure that your questions are not biased. Avoid leading or loaded questions and cover all relevant aspects of the topic of your concern.
3. Determine the specific group of participants: One factor you should consider is the group's demographic information, such as age and gender, which will depend on the purpose of your research. Make sure that your sample size is appropriate to obtain the desired results.

Tool 2: Technology

Use of digital technology

You then need to choose an appropriate software for data analysis.
1. Administer the survey: Decide a suitable way you can distribute the survey. This could be via an online platform such as Google Forms, SurveyMonkey or social media, or by paper-based questionnaires or face-to-face interviews. Provide clear instructions for respondents, maintain confidentiality if required, and encourage honest and thoughtful responses.
2. Record and organize the data: Data must be collected and recorded in a systematic way. If you are using an online platform, it may collect the data automatically to a spreadsheet. If you are using paper-based surveys or interviews, you will need to manually enter the participants' responses into a spreadsheet or data analysis software.
3. Analyse the data: Once your data collection is completed, you need to apply data analysis methods to draw meaningful conclusions from your survey. Data analysis may include both qualitative analysis (coding and thematic analysis of open-ended responses) and quantitative analysis (statistical measures, such as percentages, averages, correlations). For analysis, you may use Microsoft Excel, SPSS or other online statistical analysis tools.

4. Interpret the data: Examine patterns, trends and relationships within the data. Look for significant findings and connections to your research purpose. You may consider comparing different subgroups or demographic categories to gain a deeper understanding of your survey results.
5. Present your findings: Clearly summarize your findings using appropriate tables, graphs, charts and written explanations. Decide the most effective way to communicate your findings to your intended audience. This can be through a report, presentation or visual display.
6. Reflect and discuss: Reflect on three limitations of your survey. Consider the extent to which the sample or their responses may be biased. Consider the context of relevant theories, concepts and real-world applications while discussing the implications of your findings. Consider the ethical considerations involved in conducting and reporting survey research.

Remember to follow ethical guidelines and obtain any necessary approvals or consent before administering any surveys with human subjects. Additionally, consult your ESS teacher or supervisor for specific guidance and requirements related to survey design and data analysis for your coursework or assessments.

Designing effective values surveys for a specific social group involves crafting questions that capture a range of viewpoints concerning a particular matter and evaluating how these perspectives might influence the issue's dynamics.

For instance, consider a survey aimed at understanding the attitudes of a community towards renewable energy adoption. The survey could include questions that explore different reasons for supporting or opposing renewable energy, such as economic benefits, environmental concerns or concerns about visual impact. By gathering these varied perspectives, the survey assesses how each viewpoint might shape the community's overall stance on renewable energy projects, subsequently impacting decisions related to policy support, investment or development.

This comprehensive understanding of perspectives allows for tailored strategies that address the concerns and motivations of various groups within the community, facilitating more informed and effective approaches to promoting sustainable energy initiatives.

HL.c.2, HL.c.3, HL.c.7, HL.c.8, HL.c.9, HL.c.10, HL.c.11, HL.c.13: Environmental ethics

The concepts of different ethical frameworks, moral standing, major ethical approaches, and environmental and social justice movements give us tools to create questions that reveal what people think about the environment, why they think that way and how their beliefs might impact their choices. This way, we can understand different perspectives on environmental issues and come up with solutions that consider everyone's viewpoints.

ATL ACTIVITY

Consider that you would like to explore how students' unique perspectives relate to their attitudes towards specific environmental or sustainability issues. Design a research project where you create and administer surveys using online collaborative survey tools. Choose an appropriate statistical tool to analyse the collected data. Additionally, consider using behaviour-over-time graphs to visually depict any observed changes in students' lifestyles over the course of the study.

1.1.6 Worldviews shape people's values and perspectives

Worldviews are broad and comprehensive frameworks that shape people's perceptions and understandings of their surroundings as well as their actions. They can also be considered as people's 'lenses' on the world, which are shaped by the cultural, philosophical, ideological, religious and political factors that people use while making sense of the world. Therefore, people's worldviews influence the way they perceive the world and their place in it by shaping their values and perspectives.

A worldview is a set of beliefs, values and assumptions that impact a person's vision of reality and their behaviours. A person with a religious worldview, for example, may believe that the natural world was created by a higher power and, as a result, prioritize environmental conservation based on their religious views.

The interactions of individuals with others are also influenced by their worldviews. People who share similar worldviews have a stronger bond and sense of community with one another. They are also more likely to work together to achieve common goals.

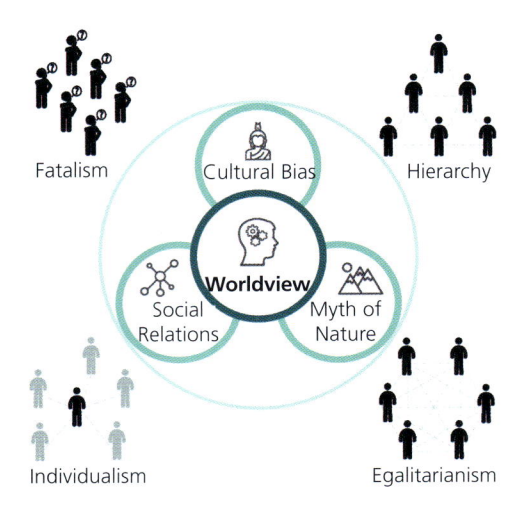

■ **Figure 1.3** Worldview can be shaped by cultural, philosophical, ideological, religious and political factors

● Common mistake

People may not acknowledge or realize that as a result of the development of the internet and social media, one's perspective can be influenced by a far greater variety of worldviews than just those of the local community. Consequently, models that attempt to classify perspectives, though helpful, are invariably inaccurate as individuals often have a complex mix of positions. A person's perspectives and actions, for example, may be shaped differently if they have a combination of religious, environmental and economic worldviews. These worldviews will assist people in effectively navigating and communicating in a diverse and globalized society.

1.1.7 Environmental value systems

A model for understanding the factors that influence our various perspectives on environmental issues is called an environmental value system. It proposes that an individual's perceptions and understanding of environmental issues are shaped by the inputs they receive from various sources, such as the media, education and worldviews. These inputs are then processed by an individual's personal value system, which is comprised of the individual's beliefs, values and assumptions. This process results in outputs like judgements, opinions and actions that are influenced by the individual's value system.

For example, a person who learns about deforestation from a news source that emphasizes the economic benefits of logging may form an opinion that prioritizes economic growth over environmental protection.

■ **Table 1.1** Influence of value systems on decision making: a case on deforestation perspective

Input	Perspective	Output
"Loggers make a significant contribution to the state's economy as they purchase supplies from local businesses." (Source: http://economic-impact-of-ag.com)	Logging as a mean to sustain economic growth.	May form an opinion that prioritizes economic growth over environmental protection.
"The Solomon Islands in the South Pacific are under threat from illegal and unsustainable logging which is destroying its biodiverse rainforests at an alarming rate." (Source: www.globalwitness.org)	Deforestation as a threat to biodiversity.	May develop a viewpoint that prioritizes environmental protection over economic growth.

1.1.8 Classification of environmental perspectives

■ Introduction to technocentric, anthropocentric and ecocentric perspectives

There are many ways to classify our perspectives. These models are useful, but imperfect, as individuals often have a complex range of positions. Technocentrism assumes all environmental issues can be resolved through technology and there can be unlimited economic growth in the **cornucopian view**. Anthropocentrism views humankind as being the central, most important element of existence, and it splits into a wide variety of views. Ecocentrism sees the natural world as having preeminent importance and intrinsic value. It favours small-scale, low-technology lifestyles with restraint in the use of all natural resources.

◆ **Cornucopian view** – the belief that considers that nature is there to be made use of by humanity.

● HL.a.11: Environmental law

Extending legal personhood to natural entities, like granting legal rights to rivers, demonstrates a deepening ethical commitment to environmental protection. This aligns with ecocentric perspectives, where nature holds intrinsic value. Such legal recognition strengthens environmental safeguards beyond anthropocentric viewpoints.

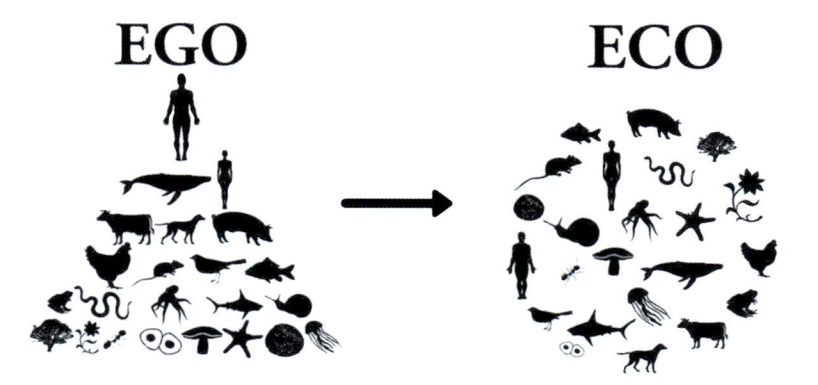

■ **Figure 1.4** The way people perceive human beings in relation to nature can change (diagram by Steffen Lehmann)

The characteristics, values and priorities of environmental perspectives are discussed below with examples and real-world implications.

Ecocentrism

Ecocentrism (values centred on ecology) and technocentrism (values centred on technology) are two competing viewpoints on the potential of human technology to impact, regulate and even preserve the environment. Ecocentrics, particularly 'deep green' ecologists, regard themselves as subject to nature rather than in command of it. They have lost faith in contemporary technology and the bureaucracy that comes with it. Ecocentrics will argue that nature should be respected for its processes and products, and that low-impact technology and self-reliance are preferable to technological domination of nature.

Technocentrism

Technocentrics have unwavering faith in technology and industry, and they believe that people have complete control over nature. Although technocentrics acknowledge the existence of environmental concerns, they do not regard them as problems that can be remedied by reducing industry. Rather, environmental issues are viewed as scientific problems to be solved. Indeed, technocentrics believe that scientific and technical growth is the way forward for both rich and developing countries, as well as the solution to our current environmental concerns.

Conservationist ideas, as well as technology's ability to safeguard nature, should ensure that today's level of life is preserved in the future, but not at the expense of environmental damage.

Anthropocentrism

Anthropocentrism literally means 'being human-centred', but in its most important philosophical form it is the ethical view that only humans have intrinsic value. In contrast, all other beings are only valuable because of their potential to serve humans or because of their instrumental worth (source: L. Goralnik and M.P. Nelson, in *Encyclopedia of Applied Ethics* (Second Edition), 2012).

In some ways, all ethics is anthropocentric, because only humans have the cognitive ability to formulate and perceive moral value. This agency positions humans at the centre of any ethical system we develop, leading some researchers to argue that anthropocentrism is the only logical ethical system available to us. Many other researchers, however, contend that this situation is an ethically uninteresting truth, rather than a limiting element in the type of ethical system we construct to help us discern what is good and evil, right and wrong. We can recognize the limitations of our human lens while making decisions about where we find value in the world.

Because we are moral agents, the same cognitive faculty that allows us to see the world in comparison to ourselves also allows us to treat or value other things with respect or as goals in themselves. Ontological anthropocentrism is the concept of a human-centred universe in which human cognition affects our ethical approach. Ethical anthropocentrism is another form of anthropocentrism that sees humans as the exclusive possessors of inherent value. But not all ethical anthropocentrism is created equal. From this vantage point, one can either view humans in isolation and dismiss nonhuman relationships as unimportant for decision making, which can be referred to as narrow anthropocentrism, or one can understand humans in an ecological context, as embedded in and dependent on a plethora of relationships with other beings and systems, which can be referred to as enlightened, or broad, anthropocentrism.

Environmental ethics discussions frequently centre on ethical anthropocentrism, which attempts to analyse our valuing of the natural world in order to decide how we should act in relation to that world. What do we value in nature (and how do we define nature), why do we value it and how do these values present themselves? In this approach, whether motivated by ethical anthropocentrism or a more inclusive perspective, environmental ethics considerations are crucial to environmental policy and decision making. For, just as ontological anthropocentrism emphasizes the limitations of our experience, anthropomorphism frequently demonstrates the human storyteller's attempt to create sympathetic characters who communicate and participate in relationships in the only way the storyteller fully understands, as a human, even if these characters lives do not reflect ecological reality.

● HL.c.2: Environmental ethics

Imagine that ethical considerations aren't only about how people treat each other, but also about how we treat the environment. This concept helps us understand why some people believe that the environment deserves ethical consideration too, just like humans.

● HL.c.3: Environmental ethics

Think of ethical frameworks as different lenses through which people view the environment. Just like wearing different glasses changes what you see, these frameworks influence how people think about environmental issues. By understanding these frameworks, we can ask questions that capture a wide range of viewpoints.

Similarly, anthropocentric thought is sometimes confused with anthropogenic action, or the consequences of humans on the environment. However, environmentalists may suggest that anthropocentrism is at the foundation of many of today's anthropogenic environmental concerns, such as climate change and widespread pollution. The fundamental issue of Lynn White Jr.'s important work in environmental ethics, *The Historical Roots of Our Ecological Crisis*, which articulates a link between ethics and ecological degradation, is the interaction between religion, science and the environment. According to White, our anthropocentric relationship with the natural world is to blame for our current environmental catastrophe; thus, in order to solve our environmental problems, we must reconsider our worldviews or religious interpretations. "What we do about ecology is determined by our conceptions of the man–nature relationship," writes White (1967, p. 1205). Since then, ethicists have taken on the problem of creating a more inclusive moral community by defining and defending it in a series of nested responses about who and what might matter morally, and why.

Summary

Our environmental worldviews can be divided into three categories: technocentric, anthropocentric and ecocentric. These are not mutually exclusive categories and there are numerous alternate schemes.

There is no doubting that each of these perspectives, in addition to the more traditional environmental perspectives, has merits that must be considered when looking for answers to environmental challenges. The task for society is to reach a consensus of viewpoints through a global debate on what is best not only for humans, but also for the rest of nature.

■ Table 1.2 Our environmental worldviews can be divided into three categories: technocentric, anthropocentric and ecocentric

Perspective	Technocentric	Anthropocentric	Ecocentric
Focus	Technology and its role in solving environmental issues.	Human needs, desires and well-being.	Ecosystem health and the well-being of all organisms.
View of nature	Nature as a resource to be harnessed and controlled.	Nature as a means to fulfil human needs and desires.	Nature as an interconnected web of life.
Human role	Dominant and in control of nature and its resources.	Central and superior to other species.	Interdependent with nature; a part of the ecosystem.
Ethics	Human well-being and progress as primary concerns.	Human-centric ethics and rights.	Preservation and conservation of natural systems.
Environmental problems	Can be solved through technological innovation.	Primarily caused by human actions and can be mitigated.	Result from the disruption of ecological balance.
Solutions	Technological advancements, innovation and engineering.	Regulation and management of natural resources.	Conservation, restoration and sustainable practices.
Sustainability approach	Sustainable development and efficient resource use.	Balancing human needs with environmental protection.	Harmonizing human activities with ecological systems.
Criticisms	Disregard for the inherent value of nature and non-human life.	Lack of consideration for ecological interdependencies.	Anthropocentric bias and overlooking ecosystem health.

◆ **Fracking** – a technique for extracting natural gas from shale rock formations, involving injecting high-pressure water, sand and chemicals underground.

There are many ways to classify our perspectives, and these models are a useful but imperfect way of understanding the complexity of environmental worldviews. We all should also be aware that individuals often have a complex range of positions and that it is important to consider the different perspectives and worldviews when addressing environmental issues.

The values that support our viewpoints can be disclosed in our personal communication with the greater community, just as the values of companies can be revealed in their adverts, media, policies and behaviours. These values can be shaped by personal beliefs, education, culture and experiences.

Individuals who value environmental protection, for example, may express their views in social media posts and conversations with friends and family, whereas organizations that value economic growth may express their views through advertisements and policies that prioritize economic development over environmental protection.

Conflict example

Hydraulic fracturing, or **fracking**, is a specific illustration of how conflicting values can lead to tensions between individuals and organizations.

Some individuals and groups recognize the economic benefits of fracking, such as enhanced energy independence and job development. Other individuals and organizations, on the other side, are concerned about the possible detrimental impacts of fracking on air and water quality, as well as the greenhouse gas emissions involved with the process.

Individuals and organizations that favour fracking and those that oppose it face tensions as a result of this difference over values. For example, a local community that values economic expansion may favour the construction of a fracking well, whereas another community that prioritizes environmental protection may reject it. Furthermore, certain organizations may be in favour of fracking while others are opposed, causing tensions.

This example shows how differing values can produce tensions between individuals and organizations. It can be a good tool for people to appreciate the intricacies of environmental concerns, as well as the significance of considering the diverse views and values that form them. Furthermore, it can also help people comprehend the significance of efficient communication and negotiation in reaching a long-term solution.

There are numerous websites that provide information about fracking. However, remember that bias can appear in various forms, such as cultural, economic, political or ideological. To protect yourself from this bias, it is important to use diverse sources, check the background of the authors, check the facts, and critically evaluate the motives and evidences presented. Some reliable sources for information on this matter are:

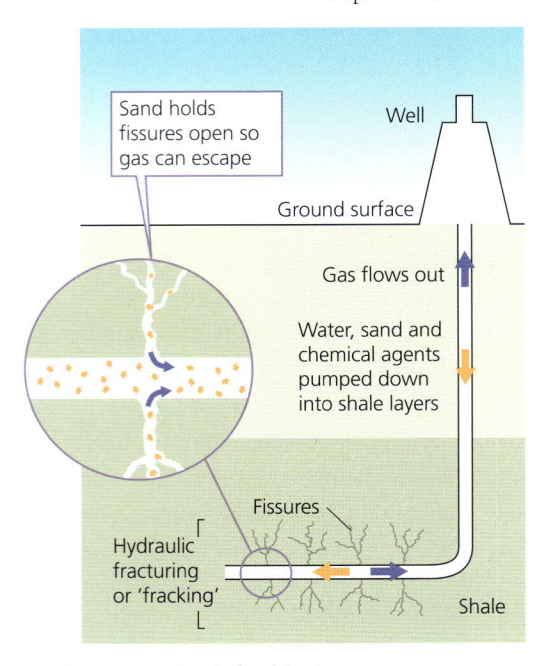

■ Figure 1.5 What is fracking?

■ Figure 1.6 A protest against a fracking development near Southport, UK

1. The Environmental Protection Agency (EPA): Independent executive agency of the US federal government
2. ProPublica: Independent, non-profit news organization
3. Natural Resources Defense Council (NRDC): Environmental advocacy group
4. FracTracker Alliance: Non-profit organization
5. American Association of Petroleum Geologists (AAPG): Professional organization.

Environmental perspectives, which encompass various worldviews, can be understood through different ethical lenses. These lenses, including technocentrism, anthropocentrism and ecocentrism, shape how individuals perceive and interact with the environment. Importantly, these ethical perspectives intersect with legal and economic strategies, creating a comprehensive framework for addressing environmental issues.

> ## HL.a.3, HL.a.4: Environmental law
>
> Environmental law is closely linked to ethical behaviour, as it prevents resource overexploitation and aligns with ethical frameworks like ecocentrism. These laws operate at various levels, from local recycling regulations to international agreements addressing transboundary pollution, all driven by ethical concerns and shared responsibility.

> ## HL.a.5: Environmental law
>
> Environmental constitutionalism exemplifies the integration of ethical considerations into legal frameworks. As environmental rights and obligations find a place in constitutions, ethical values gain legal recognition. Notably, climate change issues are increasingly being addressed through constitutional amendments, reflecting ethical commitments to future generations.
>
> In essence, the integration of these ethical lenses into legal and economic strategies creates a comprehensive approach to environmental issues, fostering responsible behaviour, and promoting long-term ecological and societal well-being.

1.1.9 Changing perspectives and beliefs

◆ **Non-governmental organization (NGO)** – a non-profit organization that is independent of any government. Its main purpose is to address a social, environmental or political issue.

Perspectives and the beliefs that support them vary over time in all communities. Campaigns by government or **non-governmental organizations**, societal and demographic changes, and other factors can all have an impact on this. For instance, increased knowledge and awareness of the effects of climate change may lead to a shift in viewpoints on environmental issues.

By understanding how certain generational shifts have occurred, we may be better able to comprehend value change. For instance, smoking has witnessed considerable changes in attitudes and prevalence during the past few decades, as seen Figure 1.7.

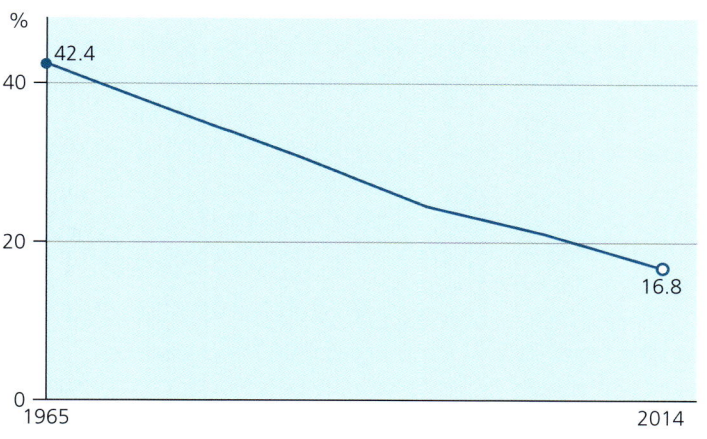

■ **Figure 1.7** Adult smoking rate in the USA has declined since the 1960s

By analysing the attitudes and perspectives of different generations on smoking, we can obtain insight into how these values have changed over time and what may have influenced those changes. For example, we can investigate how the anti-smoking campaigns in the 1960s and 1970s reduced smoking among a generation, while more recently the rise of e-cigarettes and vaping has increased smoking among the younger generation.

A behaviour-over-time graph can help you visualize these changes. The graph could show the percentage of people who smoke plotted over time, revealing changes in attitudes to smoking and the effectiveness of anti-smoking campaigns.

A graph showing the percentage of people who smoke over time, for example, could show that in the 1960s, approximately 42 per cent of adults in the USA smoked, and that this percentage had dropped to approximately 14 per cent by 2020. This graph would depict changes in smoking behaviour and attitudes over time, as well as the impact of anti-smoking campaigns.

1.1 Perspectives

This example shows how values can change over time and how different strategies and campaigns can influence these values. It can also give you a better understanding of how societies' values and priorities change, as well as how to navigate the complex issues that arise in a rapidly changing world.

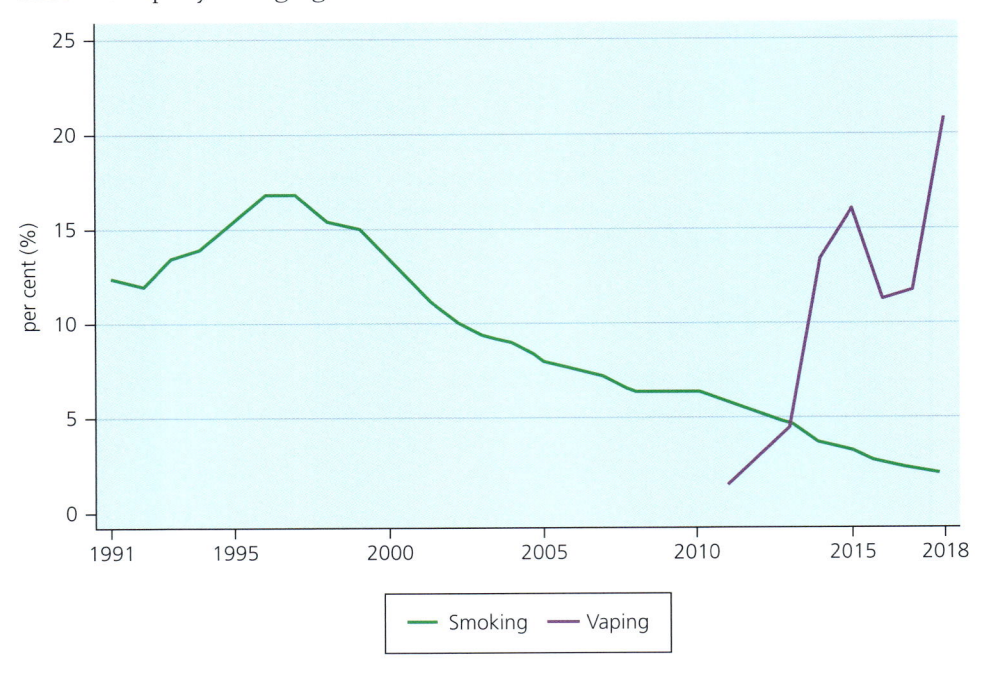

Data sources: Monitoring the Future Study and National Youth Tobacco Survey.

■ **Figure 1.8** Adult smoking rate in US has declined since 1960s

Tool 2: Technology

Use of digital technology

Use the Google Trends website to search for trends over time for an environmental or health issue, such as smoking/anti-smoking campaigns:

1. Open your web browser and go to the Google Trends website (**https://trends.google.com**).
2. In the search bar at the top of the page, enter the keyword or phrase related to the environmental or health issue you want to explore. For example, type 'smoking' or 'anti-smoking'.
3. Once you've entered the keyword, press Enter or click on the magnifying glass icon to perform the search.
4. Google Trends will display the search results related to your keyword. Look for the section titled "Interest over time". This section provides a graph that shows the search interest for the keyword over a specific period.
5. By default, the graph displays search interest over the past 24 hours. You can modify the time range by clicking on the "Past 5 years" drop-down menu and selecting a different timeframe. You can also choose a specific country or region from the "Worldwide" drop-down menu to see regional trends.
6. Examine the graph to analyse the trend of the keyword's search interest over time. Look for any patterns, spikes or fluctuations in the data.
7. Scroll down the page to explore related topics, queries and rising searches. This section provides additional insights into specific subtopics and related search queries associated with your keyword.

Theme 1: Foundation

> **ATL ACTIVITY**
>
> Using the data from Google Trends and your understanding of perspective/worldview concepts, answer the following questions:
> 1. Analyse the trend in search interest for the environmental or health issue you chose (for example, smoking/anti-smoking) over the specified time period. Identify any significant patterns, spikes or fluctuations in the data. What factors might have influenced these trends?
> 2. How do the search interest trends reflect the changing perspectives or worldviews of individuals or society regarding the chosen issue? Discuss any possible correlations between the search interest and societal or environmental factors.
> 3. Compare the search interest trends for different regions or countries. Are there any variations in the patterns? What cultural, social or contextual factors might explain these differences in perspective?
> 4. Based on the insights gained from Google Trends, reflect on the potential implications of the search interest trends for addressing the chosen environmental or health issue. How can this information inform awareness campaigns, policymaking or educational initiatives?
>
> Note: It's important to consider the limitations of Google Trends data, such as the reliance on internet search queries and the potential biases inherent in online search behaviour. Use critical thinking and the integration of multiple sources of information to provide a comprehensive understanding of the chosen environmental or health issue.

Inquiry process

Inquiry 2: Collecting and processing data
1. Download the data displayed in the Google Trends graph and conduct further data analysis or processing.
2. Use a spreadsheet (e.g. Microsoft Excel, Google Sheets) to visualize the data, create additional graphs or calculate statistical measures (for example, averages, growth rates).
3. Explore how the data can provide more in-depth insights into the search interest trends and their implications.

1.1.10 Influences on the environmental movement

Individuals, literature, the media, major environmental disasters, international treaties and technical advances have all had an impact on the evolution of the environmental movement. These factors have changed our understanding of the environment as well as changing societal attitudes and behaviours towards environmental challenges.

Individuals have made major contributions to the growth of the environmental movement. For example, Wawa Gatheru, similar to Swedish activist Greta Thunberg, is a young environmentalist who advocates for environmentally friendly activities in her neighbourhood in Kenya.

1.1 Perspectives

Figure 1.9 Rachel Carson

Environmental issues have also been impacted by literature. Rachel Carson's key work, "Silent Spring", published in 1962, is recognized as a fundamental work in the environmental movement for raising attention about the negative effects of pesticides on wildlife and the environment. James Lovelock's publications on the Gaia hypothesis claim that the Earth is a self-regulating system and provide an ecological viewpoint on the environment. Bill McKibben's *The End of Nature* (1989) was one of the first books to bring the problem of climate change to a wide audience. David Wallace-Wells's *The Uninhabitable Earth* (2019) is a dramatic study of the most recent scientific research on climate change and its repercussions for mankind and the Earth.

Significant environmental disasters have also had an impact on cultural views towards environmental issues. For example, the Minamata tragedy in 1956 was caused by the release of toxic garbage into Japan's Minamata Bay, resulting in widespread pollution and severe health impacts among the local population.

The media has had a large influence on cultural attitudes regarding environmental issues. Al Gore's documentary *An Inconvenient Truth* (2006), for example, raised awareness about climate change and the need for action. Other documentaries, such as *No Impact Man* (2009) and *Breaking Boundaries* (2021), have influenced public opinion on environmental issues.

International treaties have also had an impact on societal attitudes towards environmental challenges. For example, both the 1992 Rio Earth Summit and the 2012 Rio+20 conference helped to increase awareness about environmental challenges and led in the development of international accords to address these issues. United Nations Climate Change Conferences, such as COP 28 in 2023, have also influenced social attitudes towards environmental issues and the establishment of international climate change agreements.

Technological progress has also had an impact on societal attitudes towards environmental challenges. For example, the Green Revolution of the 1950s and 1960s concentrated on raising crop yields via the use of new technologies and methods. This helped to address food security challenges, but it also resulted in higher use of pesticides and fertilizers, which had significant environmental implications. Recent technical developments, including lower energy inputs and enteric fermentation, as well as plant-based meat alternatives, are addressing animal agriculture's environmental impact and supporting more sustainable methods.

As a more specific example, we can investigate changes in behaviours such as smoking, littering, eating meat, or how traditional indigenous lifestyles are being replaced by modern ones. By studying these issues we can gain a better understanding of how perspectives and beliefs change over time and how these changes are influenced by a variety of factors.

ATL ACTIVITY

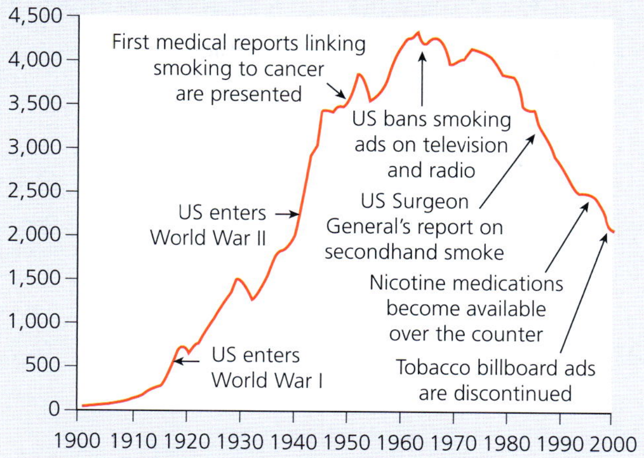

Figure 1.10 Adult per capita cigarette consumption and major smoking and health events, United States, 1900–2000

Annotations on graph:
- US enters World War I
- US enters World War II
- First medical reports linking smoking to cancer are presented
- US bans smoking ads on television and radio
- US Surgeon General's report on secondhand smoke
- Nicotine medications become available over the counter
- Tobacco billboard ads are discontinued

What does the graph in Figure 1.10 tell you about US cigarette consumption between 1900 and 2000? Does this surprise you?

Perspectives and beliefs regarding littering have evolved over time. In the past, littering may have not been seen as a significant concern. However, as environmental awareness grew, people started recognizing the detrimental effects of litter on ecosystems and human health. This shift in perspective has been influenced by several factors. For example, education and awareness campaigns highlighting the harmful effects of plastic pollution in oceans have led to a shift in perspective and a call for responsible waste management. Also, legal measures and regulations against littering, along with fines and penalties, have helped change societal norms and discourage littering behaviour. Strict enforcement and visible consequences have contributed to a change in perspective.

Perspectives and beliefs regarding meat consumption have also undergone changes over time. Historically, meat consumption was seen as a vital part of the human diet and cultural practices, if often a privilege not affordable to all people or at all times. However, several factors have contributed to evolving perspectives on eating meat in Western countries:

- Health concerns: Scientific studies linking excessive meat consumption to health issues, such as heart disease, obesity and certain cancers, have influenced people's perspectives on the health implications of consuming meat. This has led to the rise of vegetarianism, veganism and plant-based diets as alternatives.

- Environmental awareness: The environmental impact of meat production, including deforestation, greenhouse gas emissions and water pollution, has gained significant attention. Increased awareness of these issues has prompted some individuals to reduce or eliminate their meat consumption to mitigate environmental harm.

- Ethical considerations: Concerns about animal welfare and the ethics of raising and slaughtering animals for food have played a role in shaping perspectives on meat consumption. Animal rights movements and documentaries exposing factory-farming practices have raised awareness and led to changes in beliefs about the treatment of animals.

- Cultural shifts: Cultural changes, such as globalization and increased exposure to diverse dietary practices, have contributed to shifts in perspectives on meat consumption. For example, the rise of vegetarian and vegan cuisines and the availability of plant-based meat alternatives reflect changing beliefs and preferences.

1.1 Perspectives

HL.c.13: Environmental ethics

The development of the environmental movement's evolution has been shaped by the convergence of environmental and social justice movements, demonstrating how issues of equity and justice have intertwined with environmental concerns.

HL.a.7: Environmental law

International agreements and protocols have played a significant role in providing a framework for addressing transboundary environmental issues and guiding the movement's direction.

HL.a.8: Environmental law

The impact of UN conferences and international agreements has been instrumental in shaping collective efforts and actions within the movement.

HL.a.9: Environmental law

Institutions created to enforce international agreements have facilitated the implementation of environmentally beneficial measures, contributing to the movement's growth.

HL.a.10: Environmental law

The application of international environmental law in courts has provided legal mechanisms for addressing environmental concerns and driving the movement's progress.

HL.a.11: Environmental law

The concept of granting legal personhood to natural entities has furthered the movement's goals by recognizing the intrinsic value of nature in legal contexts.

HL.a.12: Environmental law

Additionally, the integration of legal and economic strategies for sustainability has contributed to the movement's strategies for fostering a more sustainable relationship between humanity and the environment.

ATL ACTIVITY

You may discover further perspectives by looking at one example of influence from each of the following categories:
- An individual environmental activist, author or the media: for example, Al Gore's documentary *An Inconvenient Truth* (2006), *No Impact Man* (2009), *Breaking Boundaries* (2021).
- An environmental disaster: for example, Minamata disaster (1956), Chernobyl disaster (1986), Fukushima Daiichi nuclear disaster (2011).
- An international agreement: for example, Rio Earth Summit (1992) and Rio+20 (2012), United Nations Climate Change Conferences (e.g. COP 21 and 27).
- Technological developments: for example, the Green Revolution, reduction of energy inputs and enteric fermentation, plant-based meat alternatives.
- Scientific discoveries: for example, pesticide and biocide toxicity, species loss, habitat degradation.

Examples may also be recent, from indigenous cultures or local/global events of your interest.

To summarize, a variety of variables have influenced the evolution of the environmental movement, including individuals, literature, the media, big environmental disasters, international accords and technical breakthroughs. These factors have changed our understanding of the environment as much as societal attitudes and behaviours towards environmental challenges. You must comprehend these impacts in order to develop a deeper awareness of the intricacies of environmental concerns and the need for considering other points of view while tackling these challenges.

> **ATL ACTIVITY**
>
> 1 Explore your perspectives and actions: Engage in debates or discussions with your peers about your own perspectives on various environmental and social issues. Reflect on how these viewpoints might influence the choices you make and your behaviours in relation to these topics. Share insights on the connections between personal beliefs and actions.
> 2 Advocate for change through persuasive materials: Take on the role of an advocate by designing materials that effectively promote a specific environmental or social cause. Use your creativity to develop compelling content that highlights how individual actions can play a pivotal role in driving positive transformations towards a more sustainable society.
> 3 Delve into environmental problem-solving: Participate in discussions that explore the multifaceted approach to addressing environmental challenges. Investigate the roles of politics, intergovernmental organizations (IGOs), non-governmental organizations (NGOs), and even individual efforts through platforms like social media. Consider joining a Model United Nations (MUN) group to collaboratively explore solutions to real-world environmental issues and understand the broader context of problem-solving efforts.

REVIEW QUESTIONS

1 Write an essay with reference to two examples to discuss the factors affecting contrasting environmental values and perspectives.

2 Design your own environmental value system model on the issue of deforestation, using the table below as a template.

Input	Perspective	Output
List some of the sources that shaped your perspective.	State your perspective on the issue of deforestation.	List the actions you would take.

3 Define briefly three environmental perspectives: technocentric, anthropocentric and ecocentric. Give one example for each.

EXAM-STYLE QUESTIONS

1 Discuss, with reference to two contrasting environmental problems, the technocentric belief that technology may provide solutions to environmental problems. [4 marks]

2 Outline the factors that lead to different environmental value systems in contrasting cultures. [4 marks]

1.2 Systems

Guiding question

- How can the systems approach be used to model environmental issues at different levels of complexity and scale?

SYLLABUS CONTENT

This chapter covers the following syllabus content:
- 1.2.1 Systems are sets of interacting or interdependent components.
- 1.2.2 A systems approach is a holistic way of visualizing a complex set of interactions, and it can be applied to ecological or societal situations.
- 1.2.3 In systems diagrams, storages are usually represented as rectangular boxes and flows as arrows, with the direction of each arrow indicating the direction of the flow.
- 1.2.4 Flows are processes that may be either transfers or transformations.
- 1.2.5 Systems can be open or closed.
- 1.2.6 The Earth is a single integrated system encompassing the biosphere, the hydrosphere, the cryosphere, the geosphere, the atmosphere and the anthroposphere.
- 1.2.7 The concept of a system can be applied at a range of scales.
- 1.2.8 Negative feedback loops occur when the output of a process inhibits or reverses the operation of the same process, in such a way as to reduce change. Negative feedback loops are stabilizing as they counteract deviation.
- 1.2.9 As an open system, an ecosystem will normally exist in a stable equilibrium, either in a steady-state equilibrium or in one developing over time (for example, succession), and will be maintained by stabilizing negative feedback loops.
- 1.2.10 Positive feedback loops occur when a disturbance leads to an amplification of that disturbance, destabilizing the system and driving it away from its equilibrium.
- 1.2.11 Positive feedback loops will tend to drive the system towards a tipping point.
- 1.2.12 Tipping points can exist within a system where a small alteration in one component can produce large overall changes, resulting in a shift in equilibrium.
- 1.2.13 A model is a simplified representation of reality; it can be used to understand how a system works, and to predict how it will respond to change.
- 1.2.14 Simplification of a model involves approximation and, therefore, loss of accuracy.
- 1.2.15 Interactions between components in systems can generate emergent properties.
- 1.2.16 The resilience of a system, ecological or social, refers to its tendency to avoid tipping points and maintain stability.
- 1.2.17 Diversity and the size of storages within systems can contribute to their resilience and affect their speed of response to change (time lags).
- 1.2.18 Humans can affect the resilience of systems through reducing these storages and diversity.

Note: There is no additional higher level content in 1.2.

1.2.1 Systems are sets of interacting or interdependent components

A system is a concept used to describe and understand various phenomena in the world, ranging from natural ecosystems to human-made machines and organizations. A system consists of individual elements or components, which can be tangible entities like physical objects, organisms or machinery, or abstract elements like processes, ideas or information.

The components within a system are not isolated but are all connected in some way, such as by physical connections, information exchange, or the influence of one component on another. The interactions have knock-on impacts on the other components or elements they are connected to. These components within a system are organized and structured in such a way that they collaborate to achieve a specific function, purpose or goal. These factors mean that systems can be modelled, manipulated and used for predictions of future issues relating to that system. Simple representations of a system can be presented with a combination of boxes to represent storages of energy or matter, and arrows to show the flow of matter or energy. These arrows show the direction of the flow, either as an input or output of the system.

Solar cell system hybrid type

■ **Figure 1.11** Diagram representing the interconnected systems responsible for providing energy to a house using a mixed solar and grid-power design

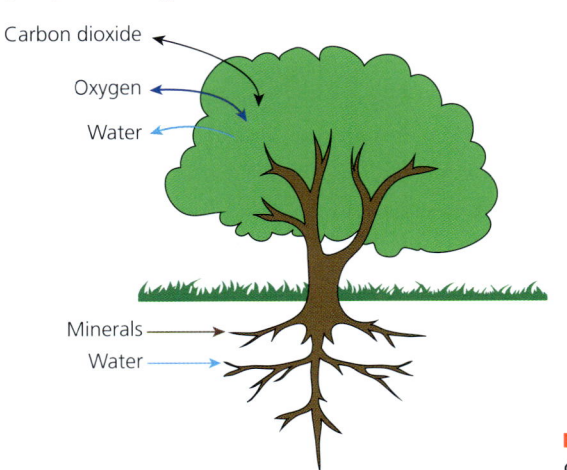

Figure 1.11 represents a number of individual systems that are involved in the development of the power system used within an individual household. It is clear from the diagram that all parts of the system are either directly or indirectly connected to each other. Similarly, the inputs and outputs into plants, such as trees (Figure 1.12), are relatively simple, but vital for the plant to function correctly.

■ **Figure 1.12** A simplified systems diagram showing some of the inputs and outputs for a tree

1.2 Systems

1.2.2 A systems approach is a holistic way of visualizing a complex set of interactions, and it can be applied to ecological or societal situations

◆ **Holistic** – describes how the parts of a system are connected and can only be fully understood by including reference to their part in the whole system.

A systems approach is a comprehensive and **holistic** method for understanding complex interactions and phenomena by viewing them as part of interconnected systems. A crucial aspect of the systems approach is not only understanding each component but also the identification and analysis of interactions and relationships between the different components of a system. These interactions can be both direct and indirect, and they often give rise to feedback loops, where changes in one element influence others. These in turn feed back to affect the original element.

Understanding a system's interactions is key to understanding how the system functions. Reducing systems to their individual parts helps with their individual understanding, however this does not give a clear view of how these systems actually work. Natural and human systems have all evolved to maintain balance to keep them functioning. Equilibrium in systems is important if they are to be self-regulating. Societal systems need to be carefully developed to allow checks and balances that aim to achieve balance.

Systems thinking recognizes that a system's behaviour can exhibit properties that cannot be understood by studying its individual components in isolation. These properties only arise from the interactions and relationships within the system. For example, flood risk can be calculated based on inputs of precipitation upstream from the flood plain, allowing measures to be taken either to reduce flooding or to move residents to minimize damage.

A system, whether it's a natural ecosystem, a machine or a societal organization, can be understood in terms of its storages and flows, which play a crucial role in the exchange of energy and matter within the system. This concept is fundamental in systems thinking and helps us analyse and describe how these systems function.

- Storages represent build-up within a system, where energy or matter is stored over time, like water tanks in a plumbing system. These storages can take various forms depending on the system in question. Within storages there are often a number of different processes taking place that transform the inputs into different outputs. For example, water stored in a lake will be consumed by animals and eliminated as waste, or taken up by plants and transformed into energy through photosynthesis. Figure 1.11 has a storage of energy within the battery system, and Figure 1.12 shows storage of water and nutrients within the tree. Energy and matter can be stored and later released without change, but they can also become part of a process and transformed into a different state of type of energy or matter.

- Flows can be thought of as the processes or interactions that connect storages, like pipes between the water tanks in a plumbing system. They are also responsible for bringing inputs of energy or matter into a system, and transporting outputs of energy or matter out of it. Flows can represent the movement of energy or matter from one location to another (a transfer), or they can represent the change in a substance as a result of processes taking place in the storages (a transformation). Flows can be both transfers and transformations, depending on the process involved. In a lake system, the flow of water into the system can occur through physical flow of water in streams or rivers leading to the lake, or through precipitation. Water leaves this system either via outgoing rivers or streams, or by evaporation transforming water to water vapour in the atmosphere.

Theme 1: Foundation

- Inputs are matter or energy required to make the system function correctly, for example, the inputs required to make the hybrid solar energy system (Figure 1.11) function correctly are solar energy into the solar panels and electrical energy input from the utility grid. A flow in a tree (Figure 1.12) is the flow of water that is transformed in the process of photosynthesis to be converted into glucose for growth of the tree. Outputs are usually waste matter or energy that has been created within the system. These inputs and outputs are crucial for the system's sustainable functioning.

In a systems approach, the concept of boundaries is essential for understanding how a system interacts with its external environment. The system boundary represents the limits of the system in question and separates it from its external environment. The boundary can be physical or conceptual, depending on the nature of the system (Table 1.3).

■ Table 1.3 Simplified breakdown of one aspect of ecological, manmade and societal systems

	Ecological	Manmade	Societal
Example system	Soil system	Hydroelectric dam reservoir	Human population of a city
Inputs	Water, nutrients, animals, seeds, animal waste	Inflow and precipitation	Immigrants and births
Outputs	Water, nutrients, plants	Outflow and evaporation	Emigrants and deaths
Internal flows	Water and nutrients via leaching, flow water into plants	Physical flow	Growth and breeding
Storages	Biomass, soil water, nutrients	Volume of water	Humans
Boundary	Number of plants the water and nutrients can support	The volume of the reservoir at full capacity	The number of people living within the area designated as the city

● **TOK**

As models are simplified representations of real-world systems, how do we decide what to include and what not to include?

Concept

Perspectives

A system can be presented on many different scales, which allows larger systems to be broken down into smaller, more manageable parts, making this a very useful technique to use. The boundary can be applied at different levels to scale the system up or down. For example, the boundary of a human population can be at the scale of an apartment building, housing estate, city, country or continent. The scale of the system will impact the volume and type of inputs and outputs.

1.2.3 In systems diagrams, storages are usually represented as rectangular boxes and flows as arrows, with the direction of each arrow indicating the direction of flow

A system can be broken down into its simplest form to allow it to be studied and these are often represented in simplified diagrams (Figure 1.13). In such diagrams, the flows in and out of the system are represented by an arrow that shows the direction of the flow. Storages are usually represented by a box, which the energy and/or matter flows in to and out of. Systems diagrams can show the scale of different elements of the system by adjusting the size of the storage and the width of the flows. This allows easier interpretation of the impact of the processes on the quantity of matter and energy entering and leaving the system.

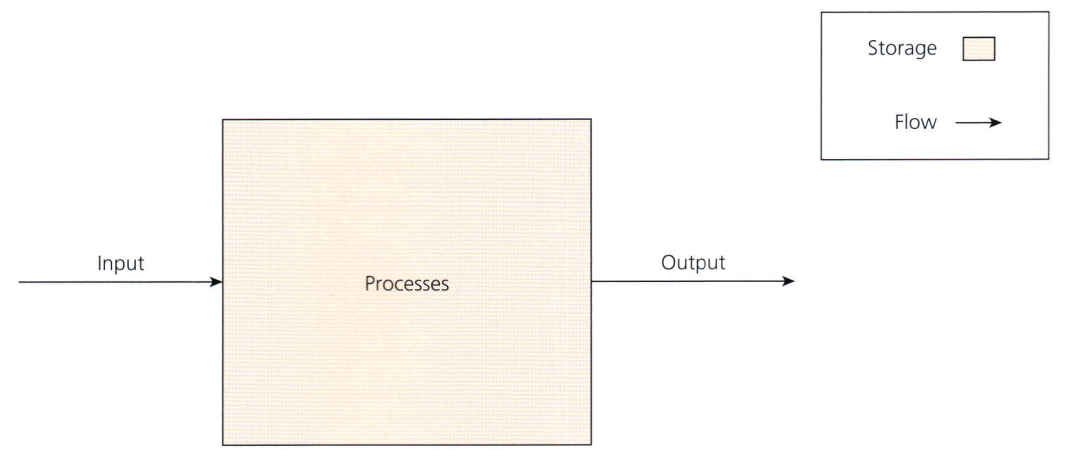

■ **Figure 1.13** A highly simplified system diagram

These visual representations of systems can also help show the complex interactions and relationships within a system in a clear and concise way. They are valuable tools for systems thinking and analysis, and for allowing better understanding of how the system operates and how changes in one part of the system can affect the other parts of the system.

ATL ACTIVITY

Create a self-sustaining community

Microcosms are artificial natural systems that can be used to manipulate natural conditions and to create a self-sustaining system. You can easily create an artificial system such as a terrarium, which is a self-sustaining natural system enclosed within a sealed jar (Figure 1.14).

■ **Figure 1.14** Simple sealed terrariums

Theme 1: Foundation

To create a real ecosystem, all elements of it must work together to support the system as a whole. Each part of the terrarium needs to be carefully considered to ensure that once it is closed, the natural processes will effectively allow the ecosystem to feed itself, clean itself and maintain balance.
- Research the necessary elements for a terrarium to function correctly.
- Collect all the necessary equipment to make either individual or group terrariums.
- Make sure you collect and record data on the inputs to your system – volume, mass, number, etc.
- Build your terrarium and seal it.
- Store your terrarium where there is sunlight, but not in direct sunlight.
- Observe your terrarium at different times of the day. What is happening?
- Create a systems diagram of your system, including the exact sizes and units presented.
- Which transfers and transformations are taking place within the terrarium?
- Create a systems diagram for each process.
- Predict what will happen to the terrarium over time and how the different parts of the system will change, if at all.
- Return to your terrarium at the end of Year 1 and assess how well the microcosm has sustained itself. Is everything you added still alive?
- Explain what has happened in the terrarium during the time it has been sealed. Back up your responses using the data collected before and after the terrarium was built.

1.2.4 Flows are processes that may be either transfers or transformations

Carbon dioxide
Oxygen
Water

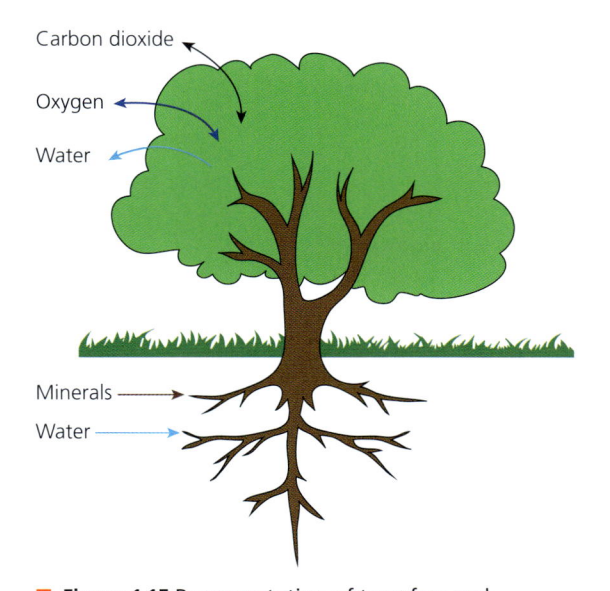

Minerals
Water

■ **Figure 1.15** Representation of transfers and transformations within a tree

Inputs and outputs in a system represent many different processes. Some flows, called transfers, represent the movement of matter or energy from one location to another. These are often physical movements, such as a river bringing water into a lake that will eventually leave the lake via a different river. Transfers are vital in the distribution of resources within an environmental system, affecting the availability of these resources for various organisms and processes.

Other flows represent the transformation of energy or matter into a different state. This can happen as the energy or matter enters the system, while it is within the system or when it exits. For example, water transfers from the soil into the roots of a plant, once it is in the plant some water is transformed into energy through photosynthesis, and some leaves the plant as it is transformed into a gas through evapotranspiration (Figure 1.15). Transformations are critical for environmental systems to maintain balance as they process and recycle resources. These processes are often central to the functioning of an ecosystem.

Concept

Sustainability

Many processes that take place in natural systems are a combination of transfers and transformations, and are balanced by plants. Plants direct more water for photosynthesis during the day, but when sunlight is no longer available, less water will be used in this process and more will leave the plant via evapotranspiration. Plants can close the stomata during the day to limit the loss of water via evapotranspiration. This is one of the ways plants are able to self-regulate to ensure they maximize growth.

■ Table 1.4 Examples of transfers and transformations in different environmental systems

Link	Examples of transfers in environmental systems	Link	Examples of transformations in environmental systems
4.1	Water flowing from a river into an ocean.	2.2.4	Photosynthesis, where plants convert sunlight, water and carbon dioxide into glucose and oxygen.
3.2.9	Animals migrating from one habitat to another.	5.1.4	Decomposition of organic matter by microorganisms, breaking down complex compounds into simpler ones.
6.4, 8.3	Air pollutants being transported by wind from a source area to a downwind region.	6.4, 8.3	The chemical reactions that occur when primary pollutants are converted into secondary pollutants through interaction with parts of the natural system.
5.1	The transfer of nutrients from the soil to plants' roots through absorption.	6.3.9	The conversion of solar energy into heat and kinetic energy to provide a renewable energy source.

1.2.5 Systems can be open or closed

There are two types of system, open and closed, and these have different properties in relation to the movement of energy and matter. An open system allows both energy and matter to be exchanged across the boundary of the system, whereas a closed system has a boundary that prevents matter from entering or leaving, but energy is able to exchange across the boundary (Figure 1.16).

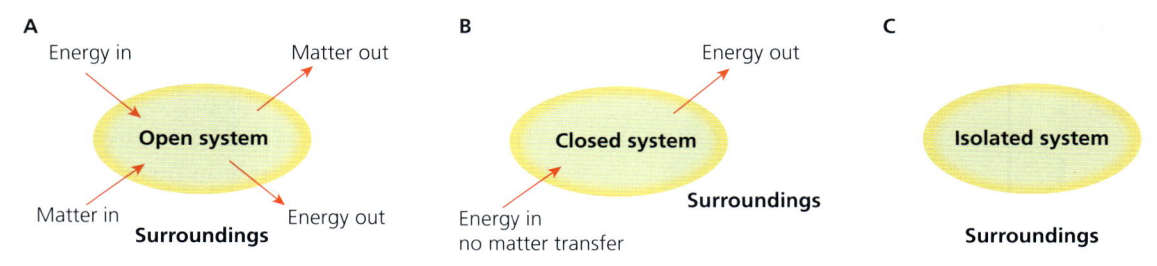

■ Figure 1.16 Systems diagrams showing the flow of energy and matter in open, closed and isolated systems

Most natural systems are open, except for global geochemical cycles that can be considered closed systems. Other closed systems are represented as simulations, such as a terrarium or Biosphere 2. These have a physical barrier stopping matter being added or removed from the system, but the glass enclosure allows energy to be exchanged between the model system and the outside system.

REAL-WORLD EXAMPLE

Biosphere 2

Biosphere 2 is an artificial closed system facility in Arizona, USA that was developed in the late 1980s and was designed to allow the study of ecological systems, biogeochemical cycles, and the interactions between different species and ecosystems within a closed environment. It consists of several interconnected biomes enclosed within a massive glass and steel structure that includes rainforests, savannas, wetlands and even an ocean with a coral reef. The experiments aimed to determine how easily a closed system could support a community, keeping in mind the potential of future space colonies.

In 1991, a group of eight scientists was sealed into the facility for two years to test its ability to support a community of humans. The project faced numerous challenges, including difficulties in managing the closed ecosystem, maintaining proper oxygen and carbon dioxide levels, and addressing unforeseen ecological imbalances that resulted in limited food at times. These challenges provided valuable lessons about the problems of managing closed environments.

■ **Figure 1.17** Biosphere 2 reserve in Arizona, USA, once an experimental closed ecosystem and now an education and tourist facility

1.2.6 The Earth is a single integrated system encompassing the biosphere, the hydrosphere, the cryosphere, the geosphere, the atmosphere and the anthroposphere

The Earth is a complex set of systems that are interconnected and function as a whole system. This means that impacts and changes in one system have direct and indirect impacts on the other cycles. Each of the spheres has its own role in the functioning of the overall system.

- Biosphere: This refers to all living organisms on Earth, including plants, animals, microorganisms and humans. It includes ecosystems and habitats, and the interactions between different species. The biosphere plays a critical role in shaping and influencing the Earth's environment through processes like photosynthesis, respiration and nutrient cycling.
- Hydrosphere: This is made up of all of Earth's water in various forms, including oceans, lakes, rivers, groundwater and even water vapour in the atmosphere. The movement of water through the hydrological cycle affects climate, weather patterns and the distribution of life on Earth.
- Cryosphere: This is the planet's frozen components, such as glaciers, ice caps, ice sheets and polar ice. It plays a vital role in regulating global climate and sea levels, as well as influencing ocean circulation patterns.

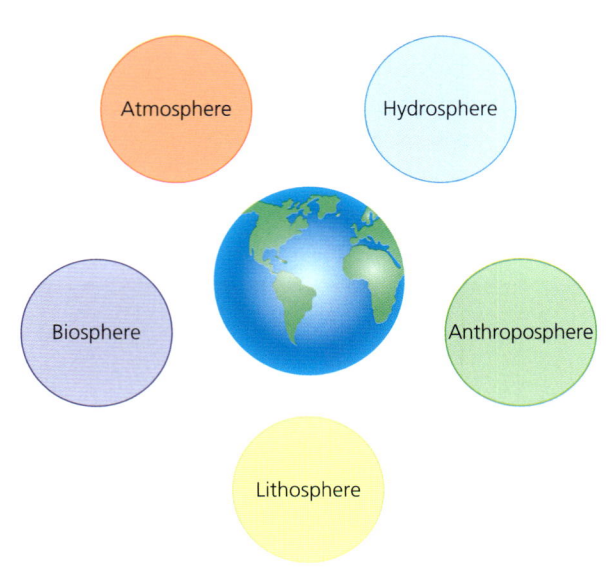

Figure 1.18 Some of the interconnections between the Earth's spheres

- Geosphere: This refers to Earth's solid, rocky components, including the lithosphere (the Earth's crust and upper mantle) and the core. It includes geological processes like plate tectonics, earthquakes and volcanic activity, which influence the Earth's physical structure and landscape.
- Atmosphere: This is made up of the gases surrounding Earth. It contains oxygen, nitrogen, carbon dioxide and other trace gases. The atmosphere is responsible for regulating temperature, weather patterns and climate through processes such as the greenhouse effect and atmospheric circulation.
- Anthroposphere: This is the human presence and impact on the natural spheres through our culture, the built environment and the technologies we have developed. The anthroposphere includes all human activities that impact the natural environment.

Human activities (anthroposphere) can lead to changes in the atmosphere, such as increased greenhouse gas emissions, which affect the climate and biosphere. Melting ice in the cryosphere can contribute to rising sea levels, impacting coastal ecosystems and communities. Geological processes in the geosphere, like volcanic eruptions, can release gases into the atmosphere that influence weather and climate, highlighting the interconnectedness of these systems.

REAL-WORLD EXAMPLE

James Lovelock's Gaia hypothesis

British scientist James Lovelock developed the Gaia hypothesis (also known as the Gaia theory). This proposed that the planet is a self-regulating system, similar to any living organism. This suggests that the Earth's living organisms and their interactions with the environment are what regulates the conditions on Earth. The hypothesis is based around the idea that the environment is naturally trying to find some level of homeostasis (balance) within the systems, and that this is achieved through feedback loops that regulate conditions to maintain some level of balance. The hypothesis emphasizes that the biotic part of a system interacts with the abiotic elements, making adjustments based on the conditions present at a certain time, and the feedback loops keep this balance in the system.

James Lovelock and Andrew Watson developed the DaisyWorld Model to provide a highly simplified visual representation of the way in which the abiotic elements interact with biotic organisms to regulate conditions to maintain them at an optimal level. The model had two different species of Daisy: black daisies that prefer cooler conditions and white daisies that prefer warmer conditions. The black daisies absorb heat and the white daisies reflect heat away from the planet, which is related to the **albedo** of different habitats based on their colour.

◆ **Albedo** – the level of light reflected away from a surface. Dark colours have a low albedo as most of the heat is absorbed, whereas light colours have a high albedo as there is more heat reflected than absorbed.

DaisyWorld began with a moderate climate with a balance of black and white daisies, some absorbing heat and others reflecting heat. If the temperature of the area increased, the number of black daisies would begin to reduce and white daisy numbers would increase. As the white daisy numbers increased, the amount of heat absorbed by the area would decrease, reducing the temperature. The reduced temperature would provide the ideal temperature for the black daisies to be able to return, outcompeting the white daisies. As the black daisies absorb more heat, the planet would begin to heat up, once again making conditions more favourable for the white daisies to start to grow again.

It's important to note that DaisyWorld is a simplified model, but it illustrates the concept that living organisms can contribute to the regulation of environmental conditions.

Since the development of the Gaia hypothesis and the DaisyWorld Model, there have been a number of modifications and extensions to the original ideas. Microbiologist Lynn Margulis joined James Lovelock, putting the ideas into proven scientific concepts through the inclusion of her theories relating to feedback mechanisms present between microbial bacteria, the atmosphere and the planet surface. Margulis also modified the concept of the Earth being a living system to it being an emergent property of the interactions between living organisms. Despite developing the hypothesis into a theory through the addition of scientific evidence, there is still much debate regarding its usefulness. However, the work of both of these scientists continues to drive debate regarding the natural regulation of systems on the planet.

HL.c.7: Environmental ethics

Ecocentrics attribute an intrinsic value to all living organisms on Earth, a perspective that also gives all living things a moral standing. Some ecocentrics extend a moral standing to non-living components of our planet. This concept appeared in the mid-twentieth century with the idea of 'land ethic' in Aldo Leopold's book *A Sand County Almanac*. This concept had a fundamental impact in shaping the environmental ethics movement. It challenges the established anthropocentric concept of nature as an asset for human exploitation and instead promotes an ecocentric perspective that acknowledges the intrinsic value of all components of the land.

The concept of the moral standing of nature is still prevalent today and it continues to fuel the debate on the conservation of our planet's integrity for future generations.

ATL ACTIVITY

Design a Likert survey to discover the views of your peers at school on the moral standing of biotic and abiotic components. You could also ask them about the intrinsic value they place on different components of the Earth.

To design a Likert survey, ensure you ask closed-ended questions and that you include options such as:
- strongly disagree
- disagree
- neither agree nor disagree
- agree
- strongly agree.

Present your results in the format of your choice, which could include graphs to summarize the survey results trends.

1.2.7 The concept of a system can be applied at a range of scales

Systems can be found from a micro to a macro scale, and regardless of the scale they can all be broken down to the same combination of flows, storages, inputs and outputs. A single plant can be an entire system, as can a single ecosystem, a city or the whole planet.

■ Micro scale

Tank bromeliads are a family of plants that do not absorb water and nutrients through their roots, instead they absorb these through the pools that collect in the 'tanks' between the leaves (Figure 1.19). These pools of water develop **detritivorous food webs** (where a combination of microorganisms and bacteria break down detritus such as dead leaves that fall into the pools). These processes break down the nutrients and produce waste that all provide the plant with valuable nutrients. Mosquitos often use these pools to lay eggs and the abundant larvae can attract other species that feed on them.

■ Figure 1.19 Large tank bromeliad plant showing the mini pools of water collected within the plant

◆ **Detritivorous food web** – a food web that is based around the actions of detritivores, which feed on dead and decaying plant material, and are responsible for the breakdown of nutrients and eventual incorporation into the soil.

Link
This Real-world Example is also relevant for Topic 3.1.6 (page 201) and speciation through isolation.

REAL-WORLD EXAMPLE

Rare frog populations in Brazilian bromeliads

The scale of a system for an organism includes all elements that address the niche of that organism, with some organisms spending all parts of their life cycle in a small, enclosed system. One example is *Crossodactylodes itambe*, a rare species of frog that is confined to living in bromeliads on a single mountain top in the Espinhaço Mountain Range of Brazil. This frog is part of a rare genus of frogs that uniquely spend their entire life cycle within the bromeliad plant's aquatic systems. These frogs were found to favour the bromeliads without a large invertebrate population for laying eggs and developing tadpoles (source: Barata, I.M., Silva, E.P. & Griffiths, R.A. (2018) "Predictors of Abundance of a Rare Bromeliad-Dwelling Frog (*Crossodactylodes itambe*) in the Espinhaço Mountain Range of Brazil". *Journal of Herpetology*, 52(3), 321–326. **www.jstor.org/stable/26792529**).

■ Ecosystem scale

Looking at an ecosystem as a system incorporates all the abiotic and biotic elements and interactions taking place: input of nutrients, water and energy; immigration of organisms; predator–prey interactions; mutualistic and parasitic relationships. Most ecosystems have both positive and negative feedback systems at play that regulate or modify the conditions present in the system. Each ecosystem has a specific boundary where conditions change either naturally or artificially, and the vegetation type and conditions change drastically.

■ Global scale

The circulation of atmospheric systems represents a global-scale system. All elements of the system interact, along with inputs from solar radiation and the water cycle. Figure 1.20 shows the fact that the atmosphere circulates in sections, called cells. These cells interact with the temperature, insolation and availability of water. Polar cells circulate cold air around the North and South Poles. As this air reaches areas that are exposed to more intense sunlight, evaporation increases. As the air circulates back to cooler regions, water is released as precipitation. Around the equator, where insolation is at its highest, the Hadley cells circulate warm, moist air that results in regular monsoon seasons, allowing dense tropical rainforests to thrive in these regions.

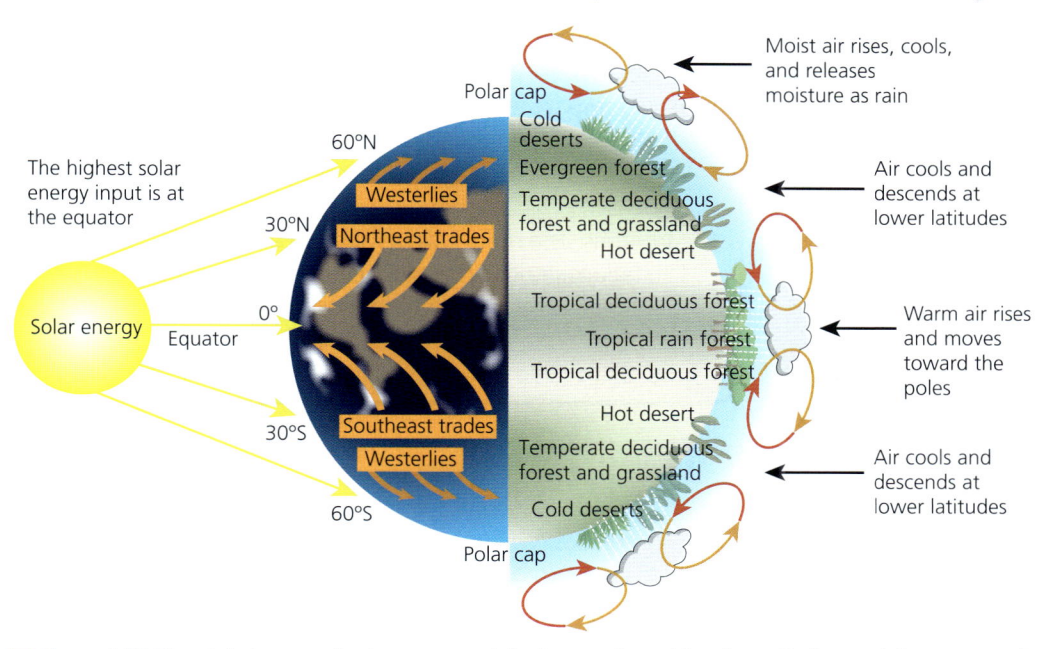

Link

See Theme 6 for greater detail on atmospheric systems.

■ **Figure 1.20** The global atmospheric system and the interaction with solar radiation and the water cycle

Tool 2: Technology

Online simulation models

There are many online simulation models that can help you to understand some of the complex interactions that are part of the environment.

Go to: **https://scied.ucar.edu/interactive/simple-climate-model**.

This is a simplified model that allows you to determine the impact of changing carbon emissions on the amount of carbon dioxide in the atmosphere and on atmospheric temperature.

When you open the model, the carbon emissions are set at 10.5 gigatonnes and the climate sensitivity is set at 3°C. Click on the "Play" button to run the simulation to the year 2100.

1. Write down the year when the temperature crosses the climate-sensitivity boundary, and the final level of atmospheric carbon dioxide and temperature.
2. Adjust the emissions value to determine the following:
   a. What is the upper limit carbon emissions need to be dropped to ensure the temperature does not exceed the recommended temperature limit by 2100?
   b. How low do carbon emissions need to be dropped to stop the rise in temperatures?
   c. How low do carbon emissions need to be dropped to result in a drop in global temperatures by 2100?
3. Evaluate the strengths and limitations of the model.

1.2 Systems

> **Answers**
> 1. The temperature exceeded the recommended temperature limit in 2075. By 2100, the amount of carbon dioxide is 571.87 ppm, and the temperature is 16.20°C.
> 2. a Carbon emissions need to be reduced to at least 7.8 gigatonnes for the temperature to not exceed the limit by 2100.
>   b To stop temperatures increasing, carbon emissions need to be lowered to 1.8 gigatonnes.
>   c To reduce global temperatures by 2100, carbon emissions need to be reduced to at least 1.6 gigatonnes.
> 3. Strengths: The model allows simple calculations and predictions to be made. Limitations: It does not allow carbon emissions to be changed by less than two gigatonnes at a time, and does not take into account any other factors contributing to global temperature increases. Overall, the model allows some level of understanding of the relationship between carbon emissions and global temperatures.

ATL ACTIVITY

Consider the model that you have just used to answer questions about potential changes in carbon emissions. What are the pros and cons of that model? How accurate is it? Does it allow all inputs to be adjusted freely? Is the model able to predict future levels accurately?

Link
There is a link here to Topics 2.4.4 and 2.4.5 (pages 165 and 170) with their detailed description of the global atmospheric circulation and its connections with the distribution of global biomes.

In summary, the global atmospheric circulation system is a dynamic and interconnected system that redistributes heat and air masses across the Earth's surface, leading to the formation of distinct wind patterns and climate zones. Understanding this system is essential for predicting weather, explaining climate phenomena and addressing climate-related challenges.

1.2.8 Negative feedback loops occur when the output of a process inhibits or reverses the operation of the same process, in such a way as to reduce change. Negative feedback loops are stabilizing as they counteract deviation

As demonstrated with the DaisyWorld Model, environmental systems have checks that keep them in balance and stop changes becoming so extreme that they permanently change the system. Negative feedback loops are essential mechanisms in the environment as they help to maintain stability and balance in various natural processes.

◆ **Equilibrium** – balance between inputs and outputs in a system. In an ecosystem, this could be the balance between the number of organisms and their needs.

Negative feedback loops begin with a system or process in **equilibrium**. This state represents the optimum condition that the system aims to maintain. External factors or disturbances can cause the system to move away from this initial state. This can be an increase or decrease in any variable or condition within the system. The organisms in the system respond to an increase in availability of resources and modify their behaviour. For example, in a predator–prey relationship, increased predator numbers due to immigration from another area will result in an increased number of the prey species being consumed. Once the prey abundance drops to a certain level, the predator population will stop growing and eventually might even decrease. This reduces the predator pressure on the prey species and allows its population to recover, which eventually stimulates the predator population to begin to rise again. This cycle repeats over and over as long as there are no other major impacts on the species or their habitat. In negative feedback a system will always return back to the steady state, even if only for a brief amount of time.

Theme 1: Foundation

> **ATL ACTIVITY**
>
> The DaisyWorld Model is an example of negative feedback and can be easily represented as a simple systems model. First, you need to think about what the storages and flows in this system are. In this case, the flows are of heat energy and sunlight that are stored in plants and the planet as heat and biomass. The albedo (reflectivity) of the planet's surface is part of the process that is taking place to stop the planet from getting either too hot or too cold.
>
> ■ **Figure 1.21** Systems diagram of the negative feedback loops that take place in the natural cycle of fluctuations in the Earth's temperature
>
> - Create your own systems diagram of the negative and positive feedback loops described as examples in Topics 1.2.8 and 1.2.10.
> - Remember that arrows show the direct of the flow and boxes are storages.
> - Use the example in Figure 1.21 for guidance.

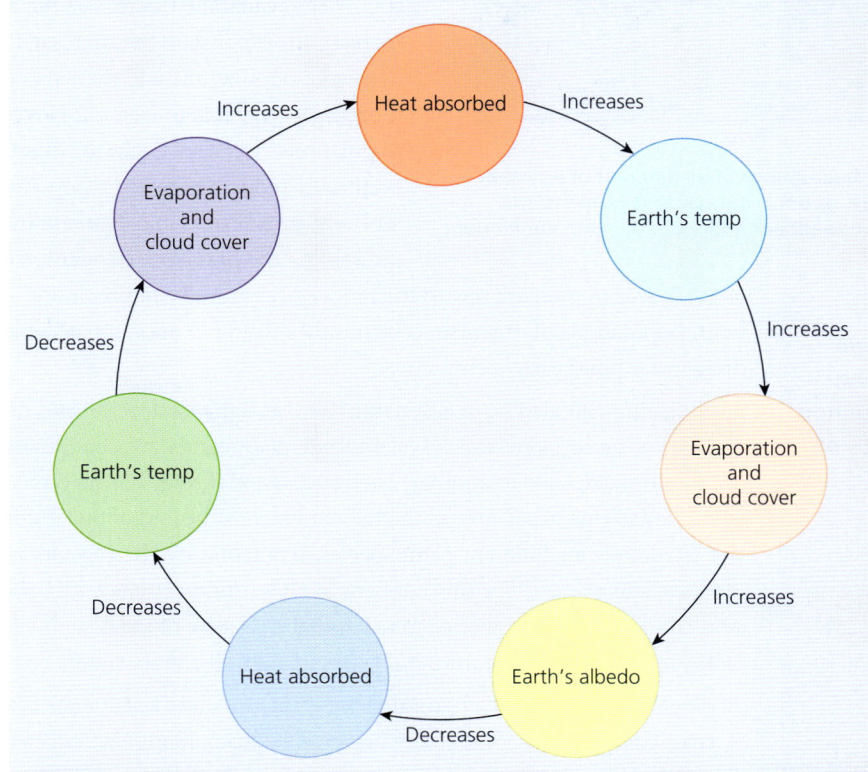

1.2.9 As an open system, an ecosystem will normally exist in a stable equilibrium, either in a steady-state equilibrium or in one developing over time (for example, succession), and will be maintained by stabilizing negative feedback loops

Within the environment, the natural state for the ecosystem is to be constantly striving for balance. This is not a passive state of equilibrium, where there is no overall change in the system, but is a consistent fluctuation around a point of equilibrium (Figure 1.22). These fluctuations are controlled through negative feedback loops, resulting in a system that is able to self-regulate.

In a steady-state equilibrium, an ecosystem maintains relatively stable conditions over time, with certain variables fluctuating around an average level of those variables. Stabilizing negative feedback loops play a crucial role in achieving and maintaining this equilibrium.

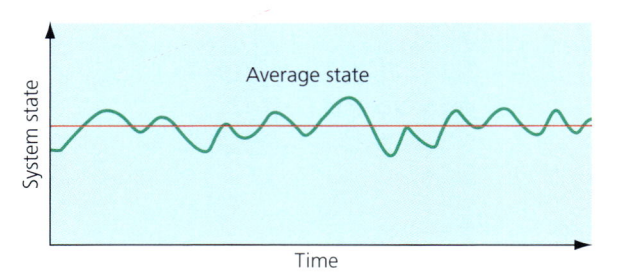

■ **Figure 1.22** The fluctuations of an element of a system around the average, optimal state of that element; this represents negative feedback and a steady-state equilibrium

For example, a freshwater pond ecosystem has a population of algae in the pond that is controlled by the availability of nutrients, by grazing by herbivores, and by competition for resources. If nutrient availability increases, the algae populations will also increase in response to the increased levels of nutrients; the herbivore populations will then also increase as there will be more algae available to feed off. However, as these herbivore populations rise, they begin to consume more of the algae, eventually reducing the algae population. This reduction in food for herbivores results in a decrease in their population size and growth. As the herbivore population continues to decrease, the algae is able to increase due to the reduced pressure from predators. This pattern then repeats, with populations of organisms constantly shifting to maintain a stable equilibrium in the freshwater pond ecosystem.

Link

Topic 2.2.15 (page 127) examines the impact of fluctuation in the food webs of ecosystems.

Equilibrium can also be in a dynamic state as habitats develop and move through the natural process of succession. An ecosystem undergoes significant changes over time as it transitions from one stable state to another. Each successional stage has its own unique plants, animals and conditions. This is because as the habitat reaches the end of one successional stage, the conditions are favourable for different plant species to outcompete the vegetation that was characteristic of the previous stage. Once this happens, a new equilibrium is established, and the plants, animals and conditions slowly change to reach that new state of balance. If left undisturbed, the ecosystem will continuously evolve until it reaches its natural climax state.

Ecological succession is a natural process that involves a sequence of stages with distinct species compositions. Stabilizing negative feedback loops also play a role in these processes. For example, primary succession occurs where there is no previous ecological history, such as a newly formed volcanic island or a glacial retreat zone. The initial pioneer species, often lichens and mosses, colonize and break down rocks to create soil; this alters environmental conditions and the formation of soil eventually allows more complex plants like shrubs and trees to establish themselves. Negative feedback stops any species within each stage from becoming too dominant, and naturally pushes the ecosystem through a series of successional stages until it reaches the final, climax community stage. For instance, as vegetation becomes more established it begins to provide shade, which modifies temperature and moisture conditions under the vegetation and in the soil. This allows different plant species to germinate, grow and thrive. New species modify the soil profile as they die and release their nutrients back to the soil. This gradual development leads to a stable climax community in the long term that is often dominated by large shrubs and trees.

Link

Chapter 2.5 examines the process of primary and secondary succession within ecosystems.

1.2.10 Positive feedback loops occur when a disturbance leads to an amplification of that disturbance, destabilizing the system and driving it away from its equilibrium

Positive feedback loops occur in natural systems when a disturbance triggers a chain of events that amplifies the impact of the initial disturbance. These loops can lead to rapid and sometimes

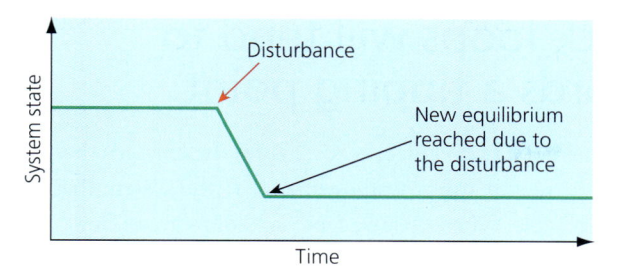

Figure 1.23 The impact of disturbance on the steady state of a system, resulting in the development of a new system; this represents positive feedback

extreme changes in the system that can make the system less stable. Figure 1.23 shows that a disturbance has been so great as to change the level of equilibrium in the system with no real way of returning back to the original equilibrium level.

Positive feedback begins with an initial disturbance or change in the system's state. This disturbance can be external or internal and may result from natural processes, human activities or a random event. This can be on any scale from a tree falling in a forest and creating a clearing, to a hurricane destroying a whole forest or human-led deforestation for agriculture. This change can happen rapidly, resulting in a dramatic change in conditions.

For instance, if a tree falls in a forest and creates a clearing, the level of light and temperature in that immediate area will instantly change. This may be too much for the shade-loving species that are growing underneath the forest canopy. If these plants are not able to tolerate the heat or light, they will die. This will allow species that prefer more light to be able to thrive in this area.

This results in a deviation from the original steady-state equilibrium and the development of a new equilibrium. In some cases these changes continue to amplify as the changes in the system have cascading impacts elsewhere, for instance in ice-albedo positive feedback. As climate change rapidly continues, the overall increase in the temperature of the planet is increasing, resulting in the melting of the polar ice caps. As the albedo of ice is so high, a loss of ice volume and the resulting increase in the oceans (which have a low albedo due to their dark colour) will increase the absorption of heat. This increased absorption of heat accelerates the ice melt, creating a self-reinforcing cycle of warming and ice loss.

The consequences of a positive feedback loop can vary depending on the system and the nature of the disturbance. In some cases, it can lead to sudden and dramatic shifts, such as ecosystem collapses, climate tipping points, or the abrupt release of stored energy.

Common mistake

People often get confused by the positive and negative names of the feedback loops – these are not related to the direction of change. They relate to the relationship with the state of equilibrium.

Negative feedback loops return to the state of equilibrium, regardless of whether we might consider the factor to be moving in a positive or negative direction. For example, an increase in temperature increases the evaporation rate that increases cloud cover, reflecting greater amounts of energy back into the atmosphere, reducing temperatures back down. As temperatures drop, the level of evaporation drops. This leads to reduced cloud cover, which results in the cycle starting again and temperatures increasing. This would be negative feedback.

Positive feedback loops continue moving away from the original state of equilibrium, to create a new equilibrium. This shift in the level of equilibrium can be in either a positive or negative direction. For example, an increase in global temperatures leads to a decrease in the amount of ice at the poles, increasing the depth of the oceans, which results in the planet absorbing more heat and the temperature rising even further. This is an example of positive feedback.

1.2.11 Positive feedback loops will tend to drive the system towards a tipping point

◆ **Tipping point** – the point that represents the critical threshold beyond which an environmental system undergoes rapid and often irreversible changes.

The **tipping point** of a system is the point where it can no longer recover from the changes that are taking place. This point is characterized by a rapid and cascading change in the system that drives it to a new point of equilibrium (Figure 1.24). When positive feedback loops are taking place in a system, some kind of change is required to stop the system reaching its tipping point and changing to become unrecognizable from the ecosystem it previously was.

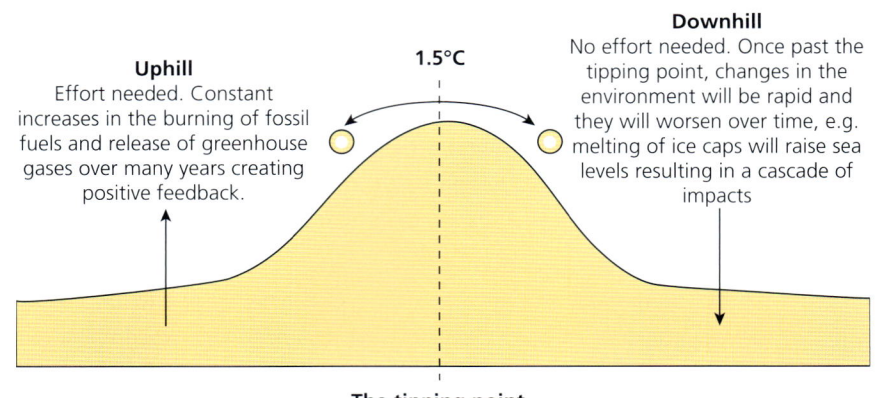

Uphill
Effort needed. Constant increases in the burning of fossil fuels and release of greenhouse gases over many years creating positive feedback.

1.5°C

Downhill
No effort needed. Once past the tipping point, changes in the environment will be rapid and they will worsen over time, e.g. melting of ice caps will raise sea levels resulting in a cascade of impacts

The tipping point
A tipping point in the climate system is thought to be a global temperature rise of 1.5°C. When exceeded, there will be large changes in the state of many parts of the Earth's systems

■ **Figure 1.24** The tipping point of global temperature increases, showing some of the causes and effects

● HL.a.2: Environmental law

It is extremely difficult to stop the rapid changes that can take place as an ecosystem or population of species reaches its tipping point. This often needs to include the development of environmental laws to halt the worsening of situations. Examples of this include:
- The development and implementation of the Montreal Protocol to ban CFC use to reduce the loss of the stratospheric ozone layer that protects the planet from harmful UV rays. See Topic 6.4.8 (page 599) for more details.
- The development of CITES by the IUCN to allow fines and punishment to be associated with illegal trade in protected plant and animal species and their parts. This also links to Topic 3.3.2 (page 246).

1.2.12 Tipping points can exist within a system where a small alteration in one component can produce large overall changes, resulting in a shift in equilibrium

Some of our systems are highly sensitive to changes that are not naturally driven. Human interactions with the environment can often trigger these cascades if the means for a system to repair itself have been removed. Some of the tipping points are related to climate change and its impact on some of the ice sheets around the Arctic.

The Arctic sea ice will be rapidly lost with even a small increase in global temperatures. The loss of ice reduces the Earth's albedo (reflectivity), causing more solar radiation to be absorbed by the ocean, which in turn speeds up the ice melting. This feedback loop can push the Arctic towards an ice-free state during the summer months, which has large impacts on global climate patterns.

The Greenland ice sheet will also rapidly melt with just a small increase in global temperatures. As the ice sheet melts, the Earth's gravitational attraction on nearby ocean water reduces, which could lead to rising sea levels at a faster rate than originally predicted. This could have significant negative impacts on coastal regions and low-lying areas.

There are also a number of ecological systems on the planet that are very easily changed by small modifications in the inputs to the system. For example, coral reef ecosystems are highly sensitive to changes in the ocean temperature and acidification. A slight rise in sea surface temperatures can lead to coral bleaching, causing corals to expel their symbiotic algae. This can disrupt the mutualistic relationship, weaken the corals, and change the system to become algal-dominated, altering the entire ecosystem.

Rainforests are equally vulnerable to change. The Amazon rainforest relies on a delicate balance of rainfall patterns, and yet deforestation and climate change can reduce rainfall, making the region drier. This could push the Amazon towards a tipping point where it transitions from a lush rainforest to a drier, savanna-like ecosystem, with significant consequences for biodiversity and climate.

1.2.13 A model is a simplified representation of reality; it can be used to understand how a system works, and to predict how it will respond to change

Models are used to present complex systems and interactions in a simplified way that allows all parts of the system to be understood and connected. This enables predictions to be made in relation to changes in inputs or processes. Models are key tools in many fields including weather, risk analysis, engineering and conservation, and allow these systems to be studied in a way that is easier to understand and control.

A model simplifies real-world processes to allow elements of the model to be manipulated to provide future predictions based on how that system has responded historically. A model often uses mathematical equations to present predictions that can be used to show the potential impacts of continuing on current terms. For example, climate modelling has allowed predictions to be made regarding the tipping point level in relation to changes in global temperatures. This has allowed decision-makers to propose strategies on how to reduce the risk.

There will be many models referred to, investigated and applied throughout this course.

1.2.14 Simplification of a model involves approximation and, therefore, loss of accuracy

In order to understand complex systems they need to be broken down and simplified, and models provide a way to do this. To make an interaction less complex, certain elements need to be removed to allow an individual process within the system to be understood. However, in doing this the simplified model becomes a less accurate predictor of change. Parts can become distorted in terms of their importance, when in the real world these processes do not happen in isolation of all other processes and pressures.

Each time a model parameter is simplified or removed, the accuracy decreases; over time, however, the accuracy can be verified and calibrated by comparing it to data that are collected from real-world scenarios. This allows the reliability of specific models to be determined. The reliability of a model will also be related to the intention of the model that has been developed. Often a model is created for a specific purpose, such as to support the development of international strategies and agreements, and can therefore be biased.

Despite this limitation of models, technology is developing that is capable of integrating different models to take into account all of the fluctuation and connections between systems. This has allowed current climate model predictions to be made with some level of accuracy, and these are continuously updated as the reality of climate change continues to present new information to keep the models up to date.

> **ATL ACTIVITY**
>
> Climate models are essential tools used by climate scientists and policymakers. Their main purpose is to demonstrate the dynamics of the climate system by using quantitative methodologies and calculations in computer simulations. Inputs can be changed and the outputs examined in order to predict future climate changes.
>
> Evaluate the role of climate models in improving our understanding of climate change.
> 1 Critically examine three different climate models from the following websites:
>   o https://interactive-atlas.ipcc.ch
>   o https://climatereanalyzer.org
>   o https://worldview.earthdata.nasa.gov
>   o https://climexp.knmi.nl/start.cgi
>   o https://scied.ucar.edu/interactive/simple-climate-model
> 2 Identify the strengths and limitations of each of the three models you have chosen.
>   You could outline your findings as a table with the following headings: model, strength, limitations. Alternatively, you could write your evaluation as an extended response, between 300 and 500 words in length. Your written response should include a conclusion giving an overall appraisal or judgement of specific examples and the role played by climate models in deepening our understanding of climate change.

1.2.15 Interactions between components in systems can generate emergent properties

♦ **Emergent properties** – these are unpredictable outcomes from the interactions between different components within a system.

In the real world, parts of systems do not operate in isolation from one another. The reality is that within these systems there are many complex interactions that can be predicted through the knowledge gained from studying simplified models. These **emergent properties** are relatively unpredictable, employing information only from systems in isolation, and therefore these need also to be studied and understood to try to determine potential outcomes.

■ Table 1.5 Examples of emergent properties throughout the book

Chapter link	Emergent properties	Explanation
Chapters 2.1 and 2.2	Ecosystem stability	The interactions among species, nutrient cycling, and predator–prey relationships help to increase the resilience of an ecosystem and its ability to resist disturbances.
Chapter 6.2	Climate patterns	Climate patterns result from interactions between factors like temperature, ocean currents and atmospheric circulation. Phenomena like El Niño and the Indian monsoon are emergent behaviours of these complex interactions.
Chapter 8.2	Urban heat islands	The 'urban heat island' results from the interactions between buildings, concrete surfaces, human activities and weather conditions. This causes urban areas to have higher temperatures than surrounding rural areas.

There are many reasons why these emergent properties occur in natural systems but are not predicted in simulation models:

- Non-linear interactions: Some chemical interactions do not follow a linear pattern, and small changes in one factor may result in a large reaction in another element. These reactions also often have upper limits that we are yet to reach in the real world. This could result in changes that we are not currently able to predict.
- Feedback loops: Our understanding of feedback loops is based on past and current data and conditions. Predictions can be made based on these data, but unpredictable changes in different factors can trigger the development of new interactions and feedback that have yet to be seen.
- Hierarchy of emergence: In some systems changes in one part have cascading effects that can result in amplified impacts. This is clear in food webs when changes in one trophic level result in impacts for both lower and higher trophic levels (see Chapter 2.2).
- Scientific understanding: As the natural world is studied in greater detail, discoveries continue to be made regarding the reality of impacts on the natural systems. Understanding the complex interactions through models that take all elements and interactions into account is the only way to develop a better understanding of these emergent properties.

1.2.16 The resilience of a system, ecological or social, refers to its tendency to avoid tipping points and maintain stability

Resilience is the ability of a system to absorb the impact of disturbance and to recover from disruptions to different parts of the system. Resilience allows the system to maintain an overall balance and stability that stops these disturbances from pushing areas to their tipping point.

Resilient systems are effective at absorbing disturbances, whether they are internal or external. When a system can absorb small changes without undergoing a rapid transition, it is less likely to approach a tipping point. Pioneer communities have less resilience than the older and more developed communities present in climax habitats (see Topic 2.5.8, page 185). The limited complexity within the newly emerging system can easily cause pioneer communities to collapse as a result of small changes in abiotic conditions. As the community develops through successional stages, however, the diversity in all trophic levels increases significantly and small changes impacting a small number of species are absorbed by the remaining unaffected organisms.

As resilience increases, the system becomes more adaptable and likely to evolve and change in reaction to direct impacts on systems. This allows systems to naturally adapt to slow changes in conditions in the systems and allows them to maintain stability throughout. Negative feedback loops also develop over time and act to regulate the system.

In an ecological system resilience is related to the natural feedback loops, the presence of limiting factors such as space and resources that stop natural systems from exceeding these aspects. For example, a wetland habitat is limited by the supply of water and the type of soil present in an area. This stops the wetland from spreading into other areas where the water table may be too low to allow the wetland, so stopping the habitat expanding beyond an area where it can support itself and maintain balance. Adaptability to maintain the natural cycles, such as the water, carbon and nutrient cycles, is vital to maintain stability within these systems.

Social systems operate in similar ways, but the inclusion of humans into a system makes the predictability of resilience more difficult. Laws and social structures are in place to develop balance and equity in resource availability, but this is not achieved in all parts of the world.

◆ **Resilience** – in an environmental context, the inherent ability of a natural system to absorb various disturbances and reorganize while undergoing state changes to maintain critical functions.

> **ATL ACTIVITY**
>
> Apply different forms of media to present a persuasive argument
>
> Visit this website: **https://interconnectedrisks.org**.
>
> Design an infographic to help others understand the concept of tipping points and how human activity contributes to lowering the resilience of systems on Earth. Your design should summarize one of the following examples from the UN's Interconnected Disaster Risks report 2023: accelerating extinctions, groundwater depletion, mountain glacier melting, or unbearable heat. You should use the following headings, which can also be found in the technical report on each tipping point example:
> - Tipping point
> - Causes (referred to as 'Drivers and root causes')
> - Impacts (referred to as 'Current and future impacts')
> - Solutions (referred to as 'The future we want to create').
>
> Include a labelled system diagram or feedback mechanism diagram to illustrate the positive feedback of your chosen tipping point.
>
> Use any infographic tool, such as Piktochart, Google Slides, Google Charts, Infogram, Snappa, Canva.
>
> In order to create an effective and captivating infographic, follow the guidance below:
> - Make it eye-catching by using simple and clear diagrams as well as great pictures.
> - Include numerical data that really sum up the tipping point mechanism.
> - Remember to keep your text simple and always consider your audience.

1.2.17 Diversity and the size of storages within systems can contribute to their resilience and affect their speed of response to change (time lags)

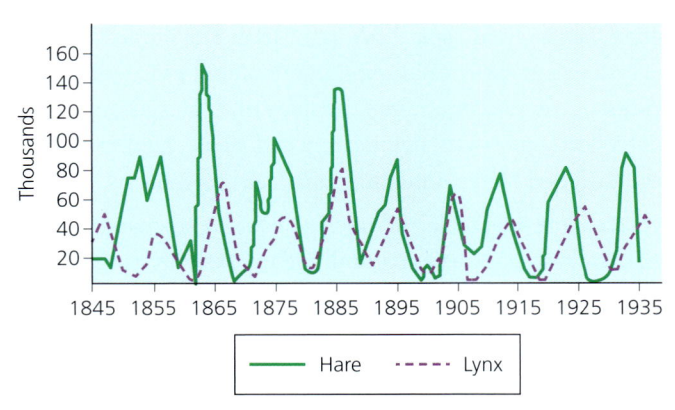

Figure 1.25 Relationship between predator (lynx) and prey (hare) numbers over time, showing the lag effect

As mentioned above, when a system is diverse the natural fluctuations in different parts of the system, such as trophic levels in a food web, do not impact the overall functioning of the system. Small natural changes in a system, such as the death of a tree in a forest due to a lightning strike, mean the loss of carbon storage within the overall forest system will be minimal and therefore it will not cause imbalance in the system. However, this buffering against the impact of change has a threshold whereby if a change is too great, the system can no longer recover from the change that has taken place.

Natural change in ecological systems takes a while to take effect due to the size of the storages within the system. The larger the storages, the greater the resilience and the longer it takes for a change to begin to have a cascading impact. For instance, changes within prey numbers in an ecosystem will not instantly stimulate a growth in the population of predators. Once prey populations have expanded, the predator population will react and increase their breeding in response (Figure 1.25).

The impact of change over time can be easily observed when you look into the impact of removing natural habitats and replacing them with large-scale monoculture agriculture.

This removes all flora and fauna that is not part of the agricultural system to maximize the level of resources available for the crops to grow. Immediate changes in biodiversity can be clearly seen, but over time the levels of nutrients, water and the quality of the soil will also drop as the natural system for returning nutrients to the soil has been removed from the system.

> **REAL-WORLD EXAMPLE**
>
> **North American prairies shift to agriculture**
>
> The vast prairies of North America used to be abundant grassland systems that supported a diverse range of plants and animals. This area is characterized by flat, low-lying land, making it perfect for conversion to agriculture. As the land is flat there is less loss of nutrients or water via leaching, making it ideal for crop growth. As this is an ideal area, the conversion to arable crops has been rapid and extensive. Most of the storages of carbon, energy and nutrients have been rapidly lost to make way for a monoculture area of crops. The resulting biodiversity in the area is dramatically reduced, leaving it with very low resilience and vulnerable to collapse from small changes in conditions.

1.2.18 Humans can affect the resilience of systems through reducing these storages and diversity

The planet's systems are highly vulnerable when change is dramatic and rapid. This characterizes the nature of many anthropogenically driven changes in systems. There are many human activities and actions that impact our natural systems and their resilience. This is particularly the case as the speed of anthropogenic change is too rapid for natural systems to change at the same speed. There are many examples of where anthropogenic activities have significant impacts on reducing the storages and diversity in natural systems.

■ Reduction in system storages

- Deforestation: Through the rapid clearing of large areas of forest, the carbon that is stored in the biomass of the trees and the soil is lost from that system. The removal of these trees also limits the capacity of the system to absorb and store more carbon dioxide. This reduces the resilience of the system, making it more vulnerable to extreme changes in weather.

- Dam construction: Building dams increases the storage of water in one area and redirects the storages in the river and groundwater. Downstream of the dam the storage of water is significantly reduced as the dam controls the release and flow of water. This impacts the resilience of all downstream habitats.

- Overfishing: Continued removal of fish from the oceans and freshwater is happening at an alarming rate that cannot be naturally restocked. This removes the storage of food sources in the fish community and creates an ecological imbalance in the food webs.

■ **Figure 1.26** Monoculture soybean system with no diversity that is maintained by the application of herbicides to eliminate non-crop plant species

- Invasive species: The introduction of invasive species results in a dramatic reduction in the biodiversity of habitats due to their competitive advantage over native species. Reductions in biodiversity reduce resilience as the fewer species at each trophic level, the less the food web system can recover from change.

Human activities that reduce storages and diversity within ecological systems can decrease their resilience and increase vulnerability to disturbances, decrease their adaptive capacity and make it harder for them to recover from shocks. Recognizing the importance of storages and diversity in maintaining resilience is crucial for sustainable management and the protection of natural systems. Efforts to restore and enhance these components can help build more resilient systems in the face of ongoing environmental and societal challenges.

HL.b.3: Environmental economics

According to environmental economics, market failure takes place when the free market has negative impacts on the environment. If goods and services are not allocated effectively by the market, then the resulting outcomes are not in the best interest of society. Several factors can contribute to negative impacts on the environment:

- Externalities: Pollution from industrial activities can be harmful to the environment and public health. However, the financial cost of this is often not taken into account when producing goods. As a consequence, there is an excessive amount of polluting goods and services.
- Tragedy of the commons: In situations where resources are shared, companies or individuals may exploit or damage them, because they do not have to bear the full cost of their actions. This can lead to the depletion and destruction of common resources. This is typically the case for overfishing, water scarcity and deforestation.
- Lack of property rights: Under certain circumstances, property rights over natural resources are not well defined, which might result in overuse and exploitation, such as for groundwater resources and fisheries.

ATL ACTIVITY

Research one example of market failure linked to an environmental disaster. Share your example with the rest of your class. Reflect on the activity and consider the role played by externalities, the tragedy-of-the-commons situation or the lack of property rights for each example.

REVIEW QUESTIONS

1. State the names of the five natural spheres.
2. Describe the three types of systems model.
3. Explain the pros and cons of the original DaisyWorld Model.
4. Explain why simplified models do not take emergent properties into account.
5. Explain the differences between positive and negative feedback mechanisms.

EXAM-STYLE QUESTIONS

1. Discuss the potential impacts of deforestation on the carbon storage present in a rainforest ecosystem. [3 marks]
2. Justify the use of models to create predictions in the event of natural disasters. [7 marks]

1.3 Sustainability

> **Guiding questions**
> - What is sustainability and how can it be measured?
> - To what extent are the challenges of sustainable development also ones of environmental justice?

SYLLABUS CONTENT

This chapter covers the following syllabus content:
- ▶ 1.3.1 Sustainability is a measure of the extent to which practices allow for the long-term viability of a system. The term is generally used to refer to the responsible maintenance of socio-ecological systems such that there is no diminishment of conditions for future generations.
- ▶ 1.3.2 Sustainability comprises environmental, social and economic domains.
- ▶ 1.3.3 Environmental sustainability is the use and management of natural resources in a way that allows replacement of the resources, and recovery and regeneration of ecosystems.
- ▶ 1.3.4 Social sustainability focuses on creating the structures and systems that support human well-being, including health, education, equity, community and other social factors.
- ▶ 1.3.5 Economic sustainability focuses on creating the economic structures and systems to support production and consumption of goods and services that will provide for human needs into the future.
- ▶ 1.3.6 Sustainable development meets the needs of the present without compromising the ability of future generations to meet their own needs. Sustainable development applies the concept of sustainability to our social and economic development.
- ▶ 1.3.7 The unsustainable use of natural resources can lead to ecosystem collapse.
- ▶ 1.3.8 Common indicators of economic development such as gross domestic product (GDP) neglect the value of natural systems and may lead to unsustainable development.
- ▶ 1.3.9 Environmental justice refers to the right of all people to live in a pollution-free environment, and to have equitable access to natural resources, regardless of issues such as race, gender, socioeconomic status or nationality.
- ▶ 1.3.10 Inequalities in income, race, gender and cultural identity within and between different societies lead to disparities in access to water, food and energy.
- ▶ 1.3.11 Sustainability and environmental justice can be applied from the individual scale to the global scale.
- ▶ 1.3.12 Sustainability indicators include quantitative measures of biodiversity, pollution, human population, climate change, material and carbon footprints, and others. These indicators can be applied on a range of scales, from local to global.
- ▶ 1.3.13 The concept of ecological footprints can be used to measure sustainability. If these footprints are greater than the area or resources available to the population, this indicates unsustainability.
- ▶ 1.3.14 The carbon footprint measures the amount of greenhouse gases produced, measured in carbon dioxide equivalents (in tonnes). The water footprint measures water use (in cubic metres per year).
- ▶ 1.3.15 Biocapacity is the capacity of a given biologically productive area to generate an on-going supply of renewable resources and to absorb its resulting wastes.
- ▶ 1.3.16 Citizen science plays a role in monitoring Earth systems and whether resources are being used sustainably.

▶ 1.3.17　There is a range of frameworks and models that support our understanding of sustainability, each with uses and limitations.
▶ 1.3.18　The UN Sustainable Development Goals (SDGs) are a set of social and environmental goals and targets to guide action on sustainability and environmental justice.
▶ 1.3.19　The Planetary Boundaries model describes the nine processes and systems that have regulated the stability and resilience of the Earth system in the Holocene epoch. The model also identifies the limits of human disturbance to those systems, and proposes that crossing those limits increases the risk of abrupt and irreversible changes to Earth systems.
▶ 1.3.20　The doughnut economics model is a framework for creating a regenerative and distributive economy in order to meet the needs of all people within the means of the planet.
▶ 1.3.21　The circular economy is a model that promotes decoupling economic activity from the consumption of finite resources. It has three principles: eliminating waste and pollution, circulating products and materials, and regenerating nature.

Note: There is no additional higher level content in 1.3.

1.3.1 What is sustainability?

Sustainability is an approach that guides us towards a world where every action is a step towards balance, harmony and resilience. It focuses on practices that allow for the long-term viability of a system. It is generally used to refer to the responsible maintenance of socio-ecological systems such that the stability of conditions continues for future generations.

■ **Figure 1.27** Lush green forests of the Amazon

■ **Figure 1.28** Area of illegal deforestation of vegetation native to the Laos forest, Asia

■ **Figure 1.29** Plastic waste on the shores of the Black Sea

All activities taking place around us are part of a bigger system. As we increase the resilience of these systems, the overall stability goes on increasing. Highly diverse systems are able to bounce back from negative effects more easily than less diverse systems. The interconnected web ensures that the system does not completely collapse under adverse conditions.

Theme 1: Foundation

1.3.2 Sustainability comprises environmental, social and economic domains

Sustainability is a holistic concept that comprises three interconnected domains: environmental, social and economic. Sustainable practices and policies put emphasis on their future effects and overall benefit to the ecosystem and society. It must be understood that if we continue to live the way we do today, it will cause irreparable damage to the planet and jeopardize the life of many living organisms.

In order to understand and implement sustainability properly, a few models have been suggested. The first model, recognizes sustainability as being supported by three overlapping areas: environment, society and economy. This is considered a weak model of sustainability because only some parts of the three domains overlap. It does not consider the fact that all systems are part of the environment. An alternate model, the strong model of sustainability (Giddings, 2002), presents the economy as embedded in society, and both society and economy as embedded in the natural environment.

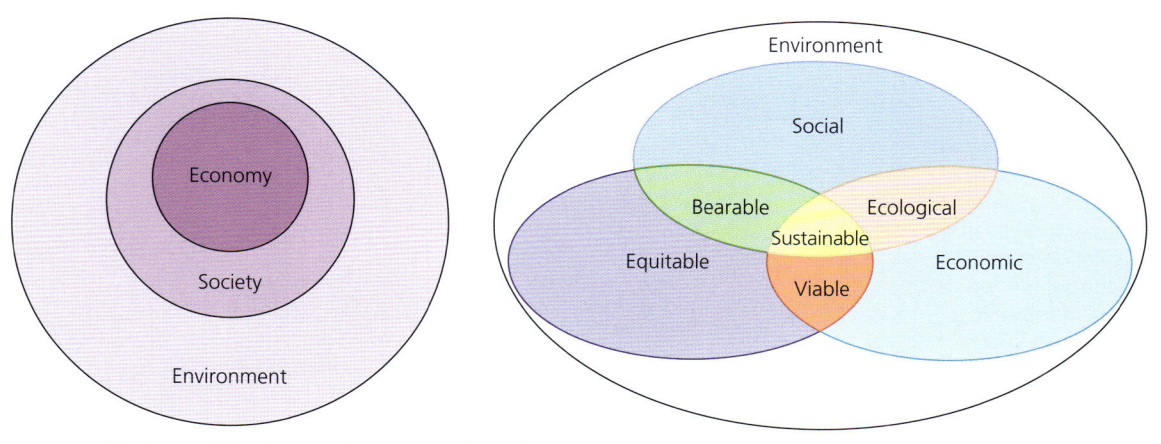

■ **Figure 1.30** Weak and strong models of sustainability

With its ecocentric perspective, the strong model of sustainable development suggests that:

1. In order to sustain human life on this planet with limited resources, it is important to embrace economic models that prioritize sustainable practices.
2. The production and consumption patterns of humans must be rethought so as to ensure the long-term well-being of individuals (both humans and other organisms).
3. Individual and local actions must be connected to the larger global context to ensure that the global issues are addressed adequately and from diverse perspectives.
4. The foundations of human actions must be grounded in ethics that prioritize nature and human well-being over economic benefit.

In the strong model of sustainability the economy is embedded in society, which means that economic decisions and environmental policies must prioritize social well-being and ecological health.

● TOK

In the TOK course you will have come across the concepts of certainty, justification and explanation. Models of sustainability have been used to help stakeholders such as ecologists, policymakers and non-governmental organizations make informed decisions about resource allocation and consumption.

On what basis can we be certain that models are able to accurately detail empirical reality?

1.3.3 Environmental sustainability is the use and management of natural resources in a way that allows replacement of the resources, and recovery and regeneration of ecosystems

Environmental sustainability refers to the conscious and judicious utilization of natural resources while ensuring their long-term availability and the restoration of ecosystems. It involves adopting practices that minimize resource depletion, pollution and habitat destruction. This requires focusing on sustainable land management, using energy and water efficiently, reducing waste generation and conserving biodiversity.

◆ **Natural capital** – the total value of the natural resources in a place (e.g. all the fish in the fishery) that can produce sustainable natural income.

◆ **Natural income** – the yield obtained from the natural resources, and the natural increase in the value of natural capital over time, due to the natural growth of the resource (e.g. the reproduction of fish increases the size of the fish population).

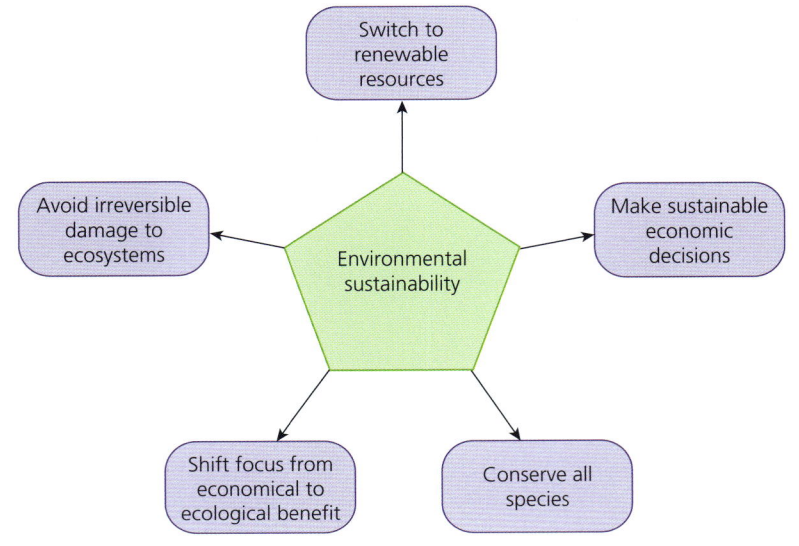

■ **Figure 1.31** Aspects of environmental sustainability

The last 8 years have been the warmest on record

Global land and ocean surface temperature anomalies (degrees Celsius compared to the 20th-century average)

2022: +0.86°C*

Global mean surface temperature 1901–2000
Land 8.5°C Sea 16.1°C Land & sea 13.9°C

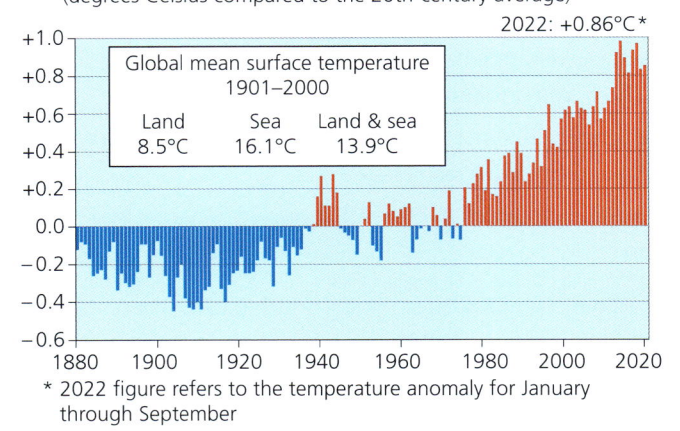

* 2022 figure refers to the temperature anomaly for January through September

■ **Figure 1.32** The average temperature of Earth, 1880–2020 (source: NOAA)

Environmental sustainability establishes the interconnectedness between the environment, society and the economy. Natural resources that can produce sustainable natural income are called **natural capital** and the yield obtained from natural resources is called **natural income**. It is the value that is obtained from the natural resources in the form of goods/products and services.

The varying timescale of the replacement of natural resources allows us to understand the resources that are exhaustible and those which are not. Some natural resources such as solar energy and wind are abundant, while others such as water are fast becoming scarcer. Renewable resources like water and biomass from plants take a short time to replenish, while non-renewable resources like fossil fuels, minerals and uranium take millions of years to be created. Since the sources of these resources are finite, they require careful management. These resources intricately hold the balance of many ecosystems, as will be discussed in the upcoming chapters in this book.

Resource depletion is a major environmental issue that we are dealing with today and this situation is bound to worsen with time. For example, deforestation – the relentless cutting down of forests for timber and land use for urbanization and/or agriculture – reduced the global forest cover by 2.4 per cent between 2000 and 2020. This drastic drop has shown its repercussions in the rising temperatures of the world. Deforestation contributes to almost 10% of the causes of global warming among other anthropogenic activities. The average temperature of the Earth has increased by around 1 degree Celsius since 1880. The act of deforestation is highly unsustainable as the rate of forest removal is greater than the rate of forest renewal. It causes the destruction of habitats across the entire planet. With a more sustainable approach, the management of renewable resources can ensure continued income for a long duration.

◆ **Goods** – commodities that can be sold to gain profits.
◆ **Services** – the direct and indirect contributions ecosystems provide for human well-being and quality of life.

Ecosystems provide a variety of goods and services. Ecosystem **goods** include timber, food, water, air and minerals that can be sold to gain profits. Ecosystem **services** refer to services that are necessary for biodiversity. These include flood protection, air purification, climate stabilization and renewal of soil fertility. Table 1.6 captures some of the important ecosystems around the world and identifies the goods and services they provide.

■ Table 1.6 The goods and services obtained from some important ecosystems

Ecosystem	Forests	Marshland	Coastal brackish	Lakes and ponds	Dry land	Rivers	Oceans and seas	Mountains
Example	Amazon	Okavango Delta	Mangroves of Sundarbans	Lake Superior	Savannah	Nile	Pacific	Kilimanjaro
Freshwater	+	+	+	+		+		+
Food	+	+	+	+	+	+	+	
Fuel	+				+			+
Timber/fibre	+				+			+
Medicinal resources	+		+				+	
Genetic resources	+		+		+		+	
Aesthetic resources	+	+	+	+		+	+	+
Spiritual value				+		+		+
Air purification	+							
Water purification	+	+	+			+		+
Climate stabilization	+		+	+			+	

1.3 Sustainability

> **Link**
>
> A detailed description of natural capitals and natural income can be found in Topic 7.1.3 (page 610). Natural capital serves as the stock of natural resources present on the Earth (Topic 7.1.2, page 609) and the yield of natural capital produces natural income.

It is important to understand and explain the relationship between natural capital, natural income and sustainability, and discuss the value of ecosystem services to society.

Inquiry process

Inquiry 2: Collecting and processing data

Resource mapping project/activity.

Fieldwork

- Create a comprehensive record of environmental goods and services provided by diverse ecosystems.
- To begin with, work in your local area and then compare your findings with others to explore and understand the interconnectedness of species and their contributions to the ecosystems.
- Use cameras, notebooks and laptops to record raw data and qualitative observations.
- Make use of spreadsheets to process data.
- Present your findings graphically.

◆ **Overshoot day** – Earth Overshoot Day is the day in the year by which the annual resource budget of our planet has been exhausted. For a country, it is the day Earth Overshoot Day would be if everyone on the planet consumed at their rate.

Almost two decades ago, humanity overshot the ecologically sustainable level of resource exploitation. Each country has its own **overshoot day**, which represents that day of the year when we have failed to put back into the Earth's systems all that we have taken from it that year. Every year we begin on 1 January and end on 31 December. Ideally, each year we should be able to put back into the system all that we have taken for that entire year by 31 December. But we have been operating on a deficit for decades now and with every passing year, the overshoot day keeps coming closer to the start.

The Earth has a finite amount of resources. If we continue to live according to our current lifestyles, then the resources of our planet may not be sufficient to fulfil our growing demands. Although the development of technology has improved food security, complete reliance on technocentric approaches may not provide sustainable solutions to this issue. An ecocentric approach is required to fully address this problem. The fact that is clear from Figure 1.33 is that we have crossed the limit of overshoot and are still continuing to do the same things, and this is alarming. A closer look at the major countries and their lifestyles will give us a clear picture of how many resources these countries are extracting from the Earth.

Country Overshoot Days 2024

When would Earth Overshoot Day land if the world's population lived like...

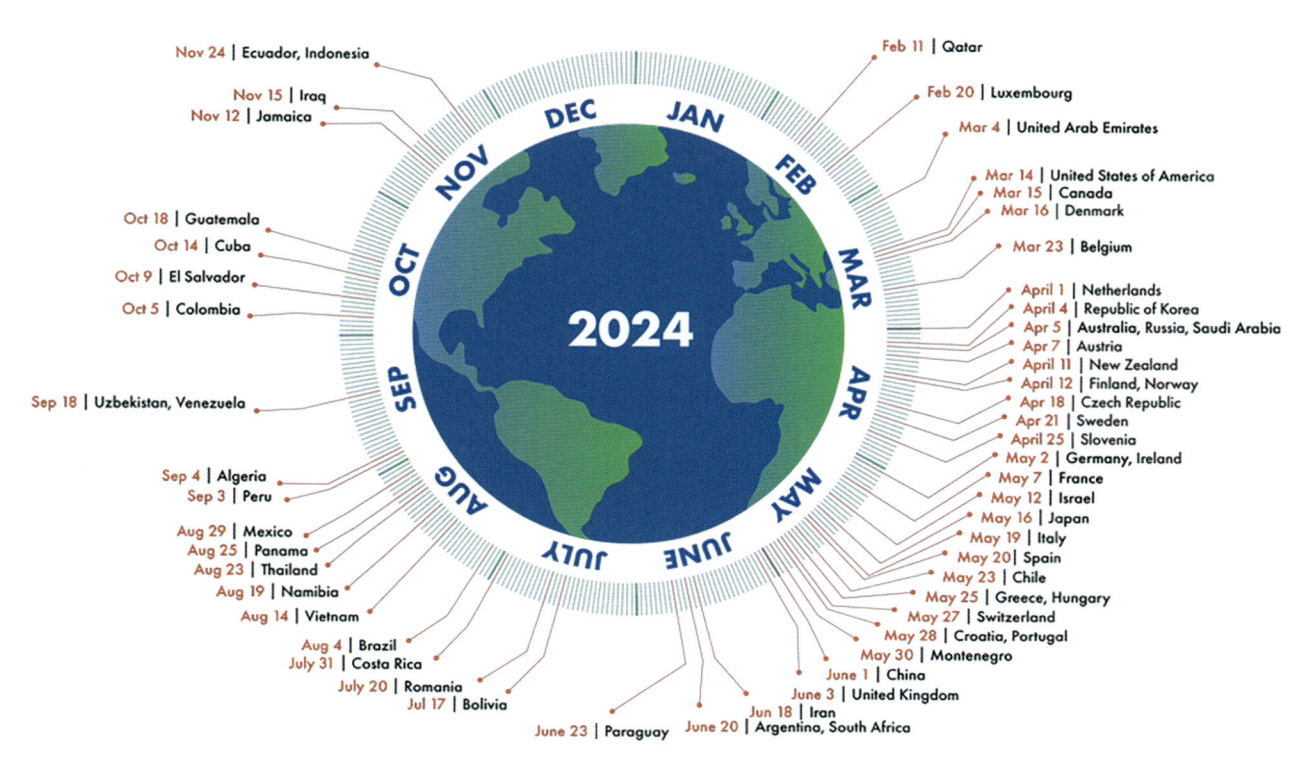

Nov 24 | Ecuador, Indonesia
Nov 15 | Iraq
Nov 12 | Jamaica
Oct 18 | Guatemala
Oct 14 | Cuba
Oct 9 | El Salvador
Oct 5 | Colombia
Sep 18 | Uzbekistan, Venezuela
Sep 4 | Algeria
Sep 3 | Peru
Aug 29 | Mexico
Aug 25 | Panama
Aug 23 | Thailand
Aug 19 | Namibia
Aug 14 | Vietnam
Aug 4 | Brazil
July 31 | Costa Rica
July 20 | Romania
Jul 17 | Bolivia
June 23 | Paraguay
June 20 | Argentina, South Africa
Jun 18 | Iran
June 3 | United Kingdom
June 1 | China
May 30 | Montenegro
May 28 | Croatia, Portugal
May 27 | Switzerland
May 25 | Greece, Hungary
May 23 | Chile
May 20 | Spain
May 19 | Italy
May 16 | Japan
May 12 | Israel
May 7 | France
May 2 | Germany, Ireland
April 25 | Slovenia
Apr 21 | Sweden
Apr 18 | Czech Republic
April 12 | Finland, Norway
April 11 | New Zealand
Apr 7 | Austria
Apr 5 | Australia, Russia, Saudi Arabia
April 4 | Republic of Korea
April 1 | Netherlands
Mar 23 | Belgium
Mar 16 | Denmark
Mar 15 | Canada
Mar 14 | United States of America
Mar 4 | United Arab Emirates
Feb 20 | Luxembourg
Feb 11 | Qatar

For a full list of countries, visit overshootday.org/country-overshoot-days.

Source: National Footprint and Biocapacity Accounts, 2023 Edition
data.footprintnetwork.org

■ **Figure 1.33** Ecological overshoot 2024

How many Earths would we need if the world's population lived like ...

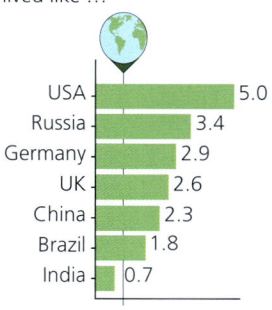

Country	Earths
USA	5.0
Russia	3.4
Germany	2.9
UK	2.6
China	2.3
Brazil	1.8
India	0.7

■ **Figure 1.34** How sustainable are these countries?

◆ **Ecological overshoot** – the point at which human demand exceeds the regenerative capacity of the natural ecosystems. Humanity's ecological footprint exceeds what the planet can regenerate.

Tool 2: Technology

Remote sensing technology

Forests not only provide goods such as timber, food, medicine and fodder, but they also provide ecological services such as purifying the air, climate stabilization and much more. Remote sensing is a tool used to analyse the temporal (time) and spatial (space) changes in land coverage. It serves as a potential tool to study the changing patterns of forest coverage around the world. Human interaction with forests has changed its value over the years. Remote sensing technology allows for accurate imaging of the Earth's surface to show changes at regular intervals.

Figure 1.35 Global forest cover, past and present

Data is collected using different types of platforms like small aircraft, drones for detailed information and satellites for the overall coverage. The technology uses a range of sensors to sense the change in electromagnetic radiation in the form of visible light, infra-red, radar, etc. The sensors create digital imagery of the land in real time. Internet of Trees (**http://internetoftrees.tech**) is an example of how remote sensing technology is used to create an early warning system to protect forests and stop forest fires from spreading.

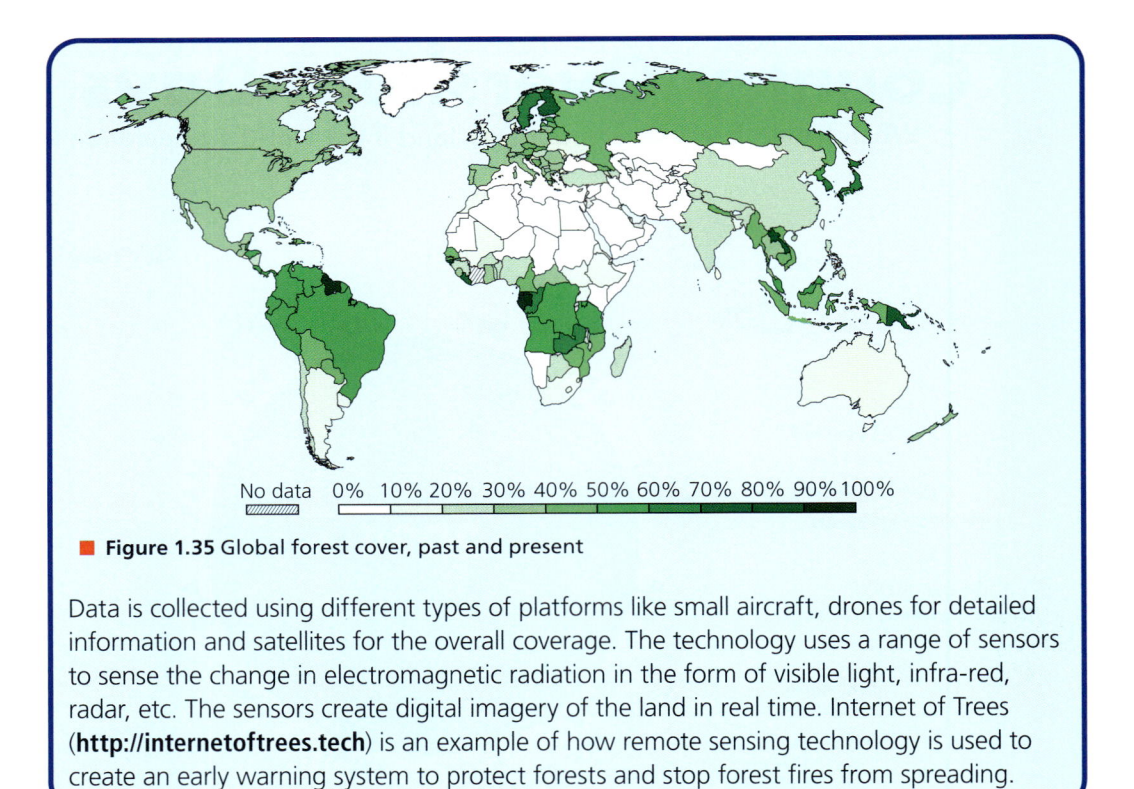

1.3.4 Social sustainability focuses on creating the structures and systems that support human well-being

Social sustainability focuses on how the structure and framework of society and social relationships support equity, justice and human well-being. This includes health, education, community and other social factors. Every individual in a socially sustainable society enjoys a good quality of life. Deep-rooted systemic inequalities owing to gender inequalities and poverty have cost humanity a lot in the past. According to the World Bank, globally the loss in human capital wealth due to gender inequality is estimated at $160.2 trillion. Further, the economic divide in a given society favours a particular section while burdening the others. For example, the landfill areas in cities and towns are usually located on the outskirts of the city. These areas are mostly inhabited by the poorer sectors of society, exposing them to severe health issues like respiratory disorders and exposure to germs, mosquitoes and flies.

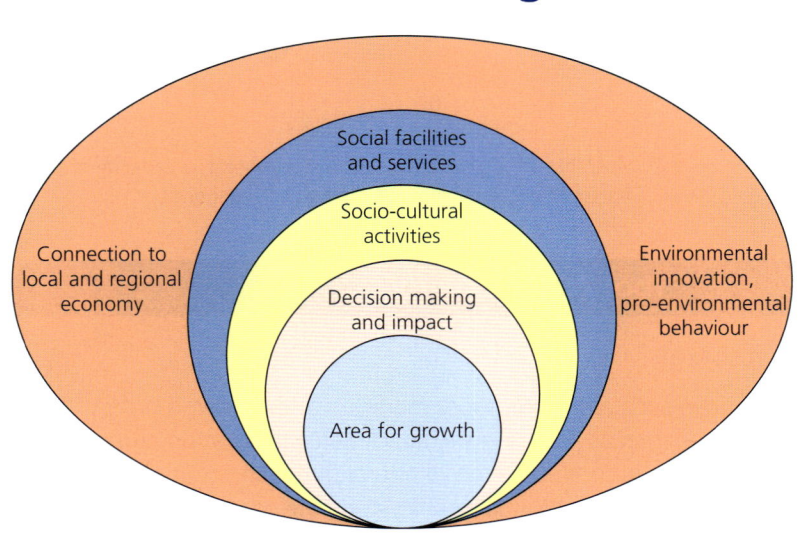

Figure 1.36 Social sustainability

Social sustainability fosters resilient societies, for example in agriculture when farmers come together to share resources during the off-season time (when they are not able to grow crops that

Theme 1: Foundation

have high economic value or they have to leave their fields to replenish their nutrients) to create equity in jobs and also to share profits equitably. This ensures that the entire social structure works together for a common goal and all benefit out of it. Sustainability in this context focuses on the survival of societies and their cultures; it may include consideration of the continued use of a language, belief or spiritual practices in a society.

The importance of cultural sustainability within the context of overall sustainability can be explained with examples such as the preservation and revitalization of indigenous languages. Efforts to document and promote endangered languages, such as the Māori language in New Zealand or the Navajo language in the USA, were successful in safeguarding linguistic diversity and preserving the associated cultural knowledge and heritage. Indigenous communities around the world possess valuable knowledge about sustainable resource management, medicinal plants and ecological wisdom. Integrating and respecting this knowledge within sustainable development initiatives allows for the harmonious coexistence of nature and culture. The preservation of cultural heritage sites, such as historic landmarks, sacred sites and traditional architectural structures, contributes to cultural sustainability. These sites serve as living testaments to a society's history, values and identity. For example, Petra in Jordan is an archaeological site famous for its intricate rock-cut architecture. Preserving this UNESCO **World Heritage site** ensures the safeguarding of Jordan's historical and cultural legacy, attracting visitors from around the world and benefiting local communities.

◆ **World Heritage site** – a site designated by UNESCO as being culturally, historically or scientifically significant.

Link

The UN SDGs are a good example of a framework focused on social, economic and ecological sustainability. See Topic 1.3.18 (page 70) for further details.

> **REAL-WORLD EXAMPLE**
>
> ### The effect of climate change on indigenous populations
>
> Climate change continues to affect people around the world at different levels. In the Arctic region, the Inuit people depend heavily on the environment for resources. They sustain themselves by hunting, fishing and herding. These activities also serve for social and cultural bonding. The effects of climate change have drastically affected these populations as food resources are becoming scarce due to the reduction of ice cover under the influence of global warming. As the ice sheets become thinner and move away from the community, the seals that are the prime source of food for the Inuit also move away. Melting of permafrost is creating lakes and pools and forcing people to shift their homes, causing a change in the social infrastructure.

1.3.5 Economic sustainability

In order to achieve economic sustainability, the relationship between environmental and social elements has to be taken into consideration. The sustained economic health of any economy will depend upon how well it can align its economic development goals with its social and environmental goals.

In terms of resource utilization to meet human needs, there is no economic sustainability without environmental sustainability. Economic activities, such as agriculture, manufacturing and energy production, heavily depend on natural resources like water, minerals, energy sources and ecosystems. These resources are essential for the production of goods and services that meet human needs and drive economic growth. However, if these resources are overexploited, depleted or polluted beyond their regenerative capacity, it jeopardizes not only the environment but also the long-term viability of economic activities. For instance, overfishing can deplete fish stocks, making it unsustainable for the fishing industry in the long run.

Economic sustainability practices aim to ensure that resources are used efficiently, waste is minimized and ecosystems are protected. For example, businesses adopting sustainable practices might reduce energy consumption, use recycled materials or invest in renewable energy sources. These actions not only benefit the environment by reducing ecological strain but can also lead to cost savings and market advantages for businesses.

HL.b.9, HL.b.10: Environmental economics

Ecological economics emphasizes the sustainable use of natural capital and can help bring sustainability. For example, in the context of forestry, ecological economics would advocate for selective logging practices that allow forests to regenerate naturally, preserving biodiversity and maintaining ecosystem services like clean water and carbon sequestration. Similarly, in the agricultural sector, ecological economics promotes organic farming methods that reduce chemical inputs, safeguard soil health, and protect the health of farmworkers and nearby communities.

ATL ACTIVITY

Draft a policy proposal aimed at promoting sustainability within your local community. You can research existing policies, identify areas for improvement and outline their recommendations.

HL.b.15, HL.b.16: Environmental economics

Ecological economists support a slow/no/zero-growth model. In a slow-growth model, the focus is on achieving economic growth at a reduced and sustainable pace. This approach prioritizes the well-being of both present and future generations by emphasizing quality of life, resource efficiency and ecological preservation over rapid expansion.

Bhutan, a small Himalayan kingdom, has a development philosophy of gross national happiness (GNH), instead of GDP. GNH takes into account both economic and environmental factors, as well as social and cultural factors. Bhutan has a number of policies in place to promote sustainable development, such as a requirement that 60 per cent of the country remain forested. It has prioritized the well-being and happiness of its citizens over rapid economic growth. Instead of solely focusing on increasing GDP, Bhutan's GNH model assesses development through nine domains, including psychological well-being, health, education, cultural diversity and ecological diversity.

ATL ACTIVITY

Research about gross national happiness and prepare a 5–6 slide presentation based on your findings. You may want to think about the following:
- The different dimensions of GNH.
- How it is measured.
- What are the challenges in using GNH to measure progress?
- Discuss some of the policies that Bhutan has in place to promote sustainable development.
- How have these policies impacted Bhutan's economy, environment and society?

Several factors govern economic sustainability. The major factors are listed in Figure 1.37:
- efficiency of resource utilization (see Topic 1.3.7, page 58)
- resource allocation and equity in resource distribution (see Topic 1.3.9, page 60)
- green technologies and innovations
- circular economy (see Topic 1.3.21, page 77)
- economic resilience
- ethical consideration in economic decision making (see HL lens c).

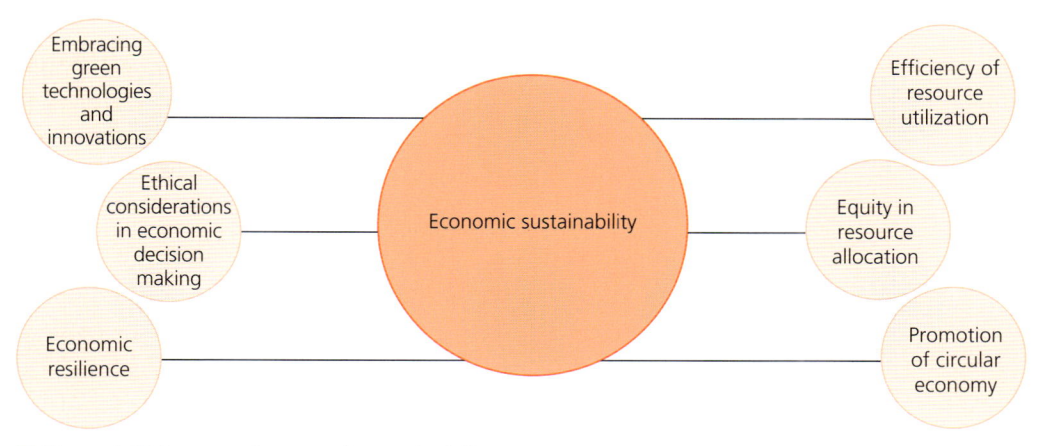

Figure 1.37 Aspects of economic sustainability

To ensure sustainable development, citizens must follow ethical consumption and proper waste management. Relying on the 4Rs (reduce, refuse, reuse and recycle) can be helpful in achieving sustainability.

1.3.6 Sustainable development

The essence of sustainable development is meeting the needs of the present without compromising the ability of future generations to meet their own needs. It applies the concept of sustainability to our social and economic development.

Sustainable development offers a vital framework that guides the progress of human civilization, ensuring a harmonious balance between economic stability, social equity and ecological well-being. It was the groundbreaking Brundtland Report of 1987 that first shed light on the social and economic dimensions of sustainability within the context of sustainable development. The report consists of guiding principles for sustainable development and lays emphasis on the interconnectedness of various aspects of human life and our environment. Sustainable development follows a holistic approach to development, aiming to create harmony between economic growth, social justice and environmental preservation.

The sustainable development model based on the Brundtland Report presents three pillars – social, economic and environmental – with equal weighting and seeks to balance them. However, at that time the model was unable to establish that the reserves of natural resources were one of the major limiting factors because it considered the possibility of economic and social development to occur outside the realms of the environment (nature). A more realistic approach to sustainable development is provided by Giddings' model. Economic decisions cannot be independent of social and environmental aspects. They are based on the utilization of resources that are part of the environment. The end user of the outcome is part of the environment and society. Whatever actions or decisions are made in the realm of economy or society, they all ultimately affect the environment. Often, the decision-makers are a few people whose main focus is rapid economic growth, who are not as concerned about the long-term consequences of their actions. It is therefore necessary to have a system that allows all **stakeholders** to present their views and make informed decisions for the greater benefit of the entire ecosystem.

Sometimes businesses and individuals indulge in **greenwashing** to mislead consumers. This can have a negative impact on sustainable development, such as undermining customer trust, slowing the process of genuine sustainable development, delaying regulatory measures and, above all, having a drastic environmental impact.

◆ **Stakeholder** – an individual, group or organization with an interest in or concern with an issue or a project.

◆ **Greenwashing** – a deceptive practice of conveying a false impression about how environmentally sound products or actions are.

HL.b.4, HL.b.5: Environmental economics

It was reported by BBC News that in 2015, Volkswagen, a leading car manufacturer, was caught cheating on emissions tests for its diesel vehicles. It had installed software that manipulated emissions during testing to appear compliant when, in reality, its cars had been emitting pollutants well above acceptable limits. This damaged VW's reputation and led to significant financial penalties, including fines and settlements.

ATL ACTIVITY

Create a case study of two other companies involved in greenwashing and the consequences.

HL.b.6, HL.b.7, HL.b.8: Environmental economics

All stakeholders must participate in implementing property rights and environmental accounting to avoid overexploitation. Sustainable development can be achieved only when all stakeholders realize the importance of valuation of the natural resources and abide by the conservation protocols. For example, carbon credits and emissions trading. It is essential to involve businesses, governments and environmental organizations in the context of environmental accounting for carbon emissions. Carbon trading systems, such as the European Union Emissions Trading System (EU ETS), allocate emissions allowances to companies. These allowances can be traded, providing economic incentives to reduce emissions. Involving various stakeholders ensures the system's integrity and effectiveness.

ATL ACTIVITY

Think of two ways that you could raise awareness about sustainable consumer choices and their impact on the environment.

TOK

Decisions revolving around how the needs of the current population can be met while ensuring that resources are preserved for future generations are difficult to make.
- On what basis can it be determined whether or not the principle of sustainable development challenges the traditional notions of progress and economic growth?
- How can we know whether or not the paradigm shifts played a role in the evolution of the sustainable development principles?

1.3.7 Unsustainable use of natural resources can lead to ecosystem collapse

Overexploitation of resources can occur due to any of the following reasons:

1. Food: When populations are harvested relentlessly for food, it often leads to overexploitation. One key case is the Newfoundland cod fisheries in Canada. For centuries, the cod population thrived off the coast of Newfoundland, sustaining the livelihoods of local fishing communities. However, relentless overfishing and unsustainable fishing practices led to a catastrophic decline in cod stocks. The once-vibrant ecosystem collapsed, causing immense economic and social consequences. The overfishing practices disturbed the delicate balance of the marine ecosystems, leading to a ripple effect on biodiversity due to thoughtless anthropogenic actions (originating from human activity).

Link

Proper systems embedded in sustainability ensure the availability of resources for a longer period of time (see Chapter 1.2, page 24).

Unsustainable use of resources pushes ecosystems towards the tipping point (see Topic 1.2.12, page 40).

Overfishing and unsustainable fishing practices are covered in Topic 4.3.5 (page 341).

2. Techniques: The methods such as trawlers, dynamite and cyanide used to harvest fish from the natural habitats often lead to more damage than good. The use of dynamite and cyanide for fishing leaves a huge excess of fish that is wasted and left in the water bodies to rot. Trawlers on the other hand scrape the ocean floor for clams, shellfish and oysters, leaving the coral reefs broken and damaged – an entire ecosystem disturbed from its roots.

3. Natural products: The demand for a specific type of natural product can also lead to overexploitation. For example, harvesting mahogany or teak to make furniture and for building purposes.

4. Aesthetic resources: Trading of exotic species of plants and animals for their aesthetic value. For example, the Pramuka bird market in Jakarta is the largest bird market in Southeast Asia and is home to a wide variety of birds, including songbirds, parrots and other exotic birds. Similarly, the Saguaro cacti is removed from the deserts of Arizona, and geckos and chameleons from the forests of Madagascar.

5. Education and research: Some species are sacrificed at the altar of research. The northern leopard frog (*Rana pipiens*) drastically decreased in numbers when it was extensively used for dissections in schools and colleges.

HL.a.1–4, HL.a.6, HL.a.7: Environmental law

International laws provide a framework for resource management. These establish principles and norms that govern the use of shared resources, such as transboundary waters, fisheries and forests. They help to prevent conflict and promote cooperation between states in managing their resources sustainably. They help to prevent overexploitation of economically desirable organisms, which in turn alters habitats. The Convention on International Trade in Endangered Species (CITES) regulates the international trade in endangered species and has been ratified by over 180 countries.

HL.b.3, HL.b.4: Environmental economics

Holding polluters accountable for their actions. The Deepwater Horizon oil spill is an excellent example of an explosion leading to economic and environmental devastation and a six-year-long legal battle. In 2016, the largest environmental damage settlement of $20.8 billion was approved under the US Clean Water Act and Oil Pollution Act in favour of the five US states that border the Gulf of Mexico. Of this money, 80 per cent was directed towards the Gulf Ecosystem Restoration Trust Fund and 20 per cent was dedicated to the Oil Spill Liability Trust.

1.3.8 GDP and green GDP

Economic indicators are measures of the economic performance of a country. They include things like GDP, unemployment rate and inflation rate. Economic indicators are important because they can give us a snapshot of how the economy is doing and can help us identify areas where improvement is needed. Common indicators of economic development include Gross Domestic Product (GDP) per capita, Gross National Product (GNP), per capita income, employment rate and poverty rate. Other indicators include the Human Development Index (HDI) and Foreign Direct Investment (FDI).

However, these indicators are often unable to capture the full picture of progress and sustainability. While GDP measures the monetary value of goods and services produced within a country, it fails to account for the environmental costs or the value of natural systems that underpin the economy. This narrow focus on economic output can lead to unsustainable practices. For instance, countries heavily reliant on fossil fuels may experience high GDP growth but face long-term environmental consequences. GDP as a measure of economic progress can potentially lead to unsustainable development practices. Countries with high population and excessive consumption of natural resources often end up contributing to ecological imbalances. Alternative indicators like GNP account for the value of all the finished goods and services by means of the productions owned by a country's citizens, irrespective of the location of the economic activity. Per capita income gives an indication of the average income of an individual. A good indicator of economic development can be the employment rate, which is the percentage of employed working-age individuals.

Green GDP is a measure of economic growth that considers the environmental costs and benefits of economic activity. It means that economic growth is measured by subtracting the cost of environmental and ecological damage from the GDP. It offers a more comprehensive assessment of economic performance, and one that aligns with sustainability goals.

Green GDP is calculated as follows:

$$\text{Green GDP} = \text{GDP} - \text{Environmental cost}$$

Unfortunately, Green GDP is complex to calculate. There is no one agreed-upon method to measure the exact cost of damage done to the environment. The methods for valuing environmental and ecological costs and benefits are also not standardized. It is a relatively new concept and not widely used by policymakers.

◆ **Per capita GDP** – a measurement of the economic output of a nation per person.

Per capita GDP (or GDP per capita) considers the inequalities in income distribution within a country, providing a more refined understanding of economic well-being in the country. Income distribution is how the income of a country is divided among its citizens. It can have a significant impact on the well-being of the people. Imagine two countries, A and B, with the same GDP. However, the population of country A is 10 million people and that of country B is 100 million. This means the per capita GDP for country A is ten times that of country B. This indicates that an average person in country A has access to more wealth than in country B.

The Nordic countries are a good example of how it is possible to have a strong economy while also ensuring that the benefits of economic growth are widely shared. They have high levels of per capita GDP, but they also have low levels of income inequality. Their focus is on providing a variety of benefits, including healthcare, education, unemployment benefits and retirement benefits, that ensure all citizens have a basic standard of living, regardless of their income.

ATL ACTIVITY

- Select two or more countries and compare their GDP figures. Use **www.ourworldindata.org**.
- Investigate factors that contribute to differences in GDP, such as economic structure, policies or natural resources.
- Make a presentation of your findings.

1.3.9 Environmental justice

Minority ethnic and economically disadvantaged communities often live near environmental hazards, leading to poor health and a diminished quality of life. Environmental justice refers to the right of all people to live in a pollution-free environment, and to have equitable access to natural resources, regardless of race, gender, socioeconomic status, nationality, etc. According to the US Environmental Protection Agency, 'environmental justice is the fair treatment and meaningful involvement of all people regardless of race, colour, national origin or income concerning the development, implementation and enforcement of environmental laws, regulations and policies'. The pervasive nature of environmental justice can be brought to light by examining some local and global examples.

Often, marginalized communities in low-income areas are disproportionately burdened with waste disposal facilities or landfill sites. Communities live in such situations due to economic instability and a lack of political support. Pollutants released from these landfills into the air and water pose great health risks for the nearby residents. Wealthier communities tend to be shielded from such environmental hazards.

REAL-WORLD EXAMPLE

Deepwater Horizon oil spill

The Deepwater Horizon oil spill in the Gulf of Mexico in 2010 is one of the largest oil spills in the history of the USA. On 20 April 2010, a surge of natural gas blasted through the concrete, causing a huge explosion that ripped the oil rig apart. The oil rig eventually capsized and sank on 22 April 2010. The 10,683 m-deep well released millions of barrels of oil into the marine environment. The spill not only impacted marine life and ecosystems but also devastated the livelihoods of coastal communities, particularly those that were reliant on fishing and tourism. The burden of the spill's consequences fell disproportionately on vulnerable communities, exacerbating socioeconomic disparities and depriving individuals of their right to a healthy and sustainable environment.

REAL-WORLD EXAMPLE

Maasai communities

◆ **Pastoralist** – a farmer who breeds animals such as cattle and sheep.

The Maasai people in Kenya and Tanzania have long faced challenges in asserting their land rights. Traditionally, the Maasai have been semi-nomadic **pastoralists**, relying on grazing lands for their livestock. However, their access to ancestral lands has been threatened by encroachment from agriculture, tourism and conservation efforts. In Kenya, land privatization policies and the establishment of national parks and game reserves have led to the displacement of Maasai communities from their traditional lands. These actions have limited their access to grazing areas and water sources, disrupting their traditional way of life and compromising their livelihoods.

Similarly, in Tanzania, the establishment of wildlife conservation areas, such as national parks and hunting reserves, has encroached upon Maasai lands. The creation of these protected areas restricts the Maasai's access to grazing lands and water resources, diminishing their ability to sustain their pastoral way of life and contributing to socioeconomic hardships within the community.

The struggle for Maasai land rights is multifaceted, involving issues of cultural preservation, economic empowerment and environmental sustainability. The Maasai have been advocating for recognition of their land rights, seeking secure tenure and participatory decision-making processes that respect their traditional knowledge and practices. Efforts to address this environmental injustice include advocacy by Maasai-led organizations, collaborations with conservation groups, and engagement with governmental bodies to secure land rights for the Maasai people. There have been successful cases of community-based conservation initiatives that empower the Maasai to actively manage their lands and resources while balancing ecological and economic sustainability.

To promote environmental justice is to amplify marginalized voices and ensure inclusive decision-making processes. By actively engaging and empowering communities affected by environmental injustices, we can work towards equitable solutions that address their specific needs and concerns.

1.3.10 Inequalities in income, race, gender and cultural identity within and between different societies lead to disparities in access to water, food and energy

Socioeconomic disparities, racial inequalities, gender biases and cultural differences can create distinct discrepancies in individuals' access to vital resources, including water, food and energy. Such disparities are evident within societies and can be further exacerbated on a global scale. For instance, disadvantaged communities in developing countries often lack the privilege of clean drinking water, leaving them vulnerable to health hazards and waterborne illnesses. Uneven distribution and availability of food resources contribute to widespread malnutrition and hunger, disproportionately affecting marginalized groups. Certain populations encounter challenges in securing reliable and affordable energy sources, hindering their socioeconomic progress.

In many low-income communities around the world where residents lack the financial means to afford a reliable electricity supply, the situation often results in limited access to modern amenities.

Some holistic efforts that can be taken to address systemic inequalities and ensure equitable access to fundamental resources, regardless of socioeconomic status, race, gender or cultural identity, are:

- investing in renewable energy and energy-efficiency programmes in low-income communities
- providing financial assistance to low-income families
- eliminating discrimination in the housing market
- supporting community-led initiatives.

> **ATL ACTIVITY**
>
> Choose one or two articles that talk about a country's income inequalities, race, gender distribution and cultural identity.
>
> Choose any one aspect, for example, gender distribution.
>
> Write a critical analysis essay on the information presented in those articles. Conduct relevant research to support your arguments.

1.3.11 Sustainability and environmental justice can be applied from the individual scale to the global scale

Very often people confuse 'environmentalism' with 'environmental justice'. The former refers to protection and conservation of wildlife and nature, while the latter serves as a lens through which social justice principles can be incorporated into the realm of fair sustainability. The concepts of sustainability and environmental justice are interconnected and provide a foundation essential for creating a better future for all. Environmental justice operates at a variety of scales, as shown in Figure 1.38.

Theme 1: Foundation

Levels at which environmental justice operates

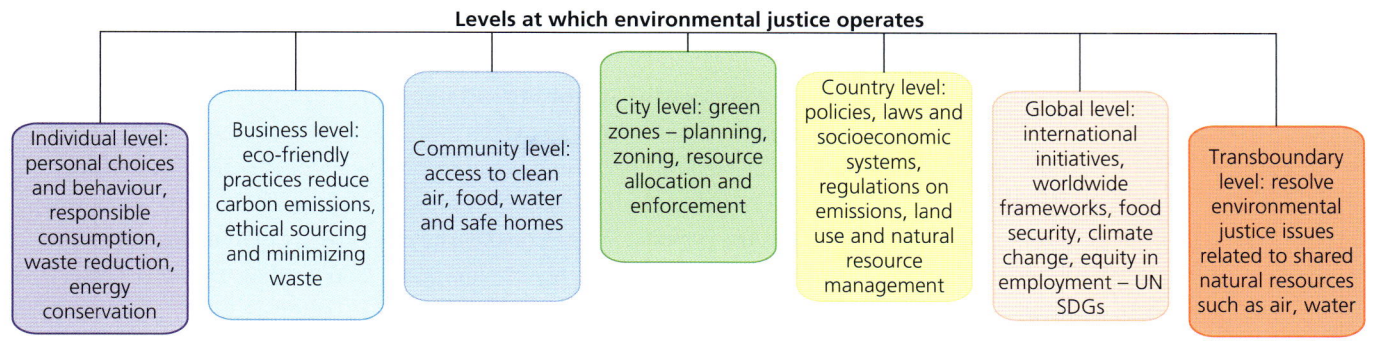

- **Individual level:** personal choices and behaviour, responsible consumption, waste reduction, energy conservation
- **Business level:** eco-friendly practices reduce carbon emissions, ethical sourcing and minimizing waste
- **Community level:** access to clean air, food, water and safe homes
- **City level:** green zones – planning, zoning, resource allocation and enforcement
- **Country level:** policies, laws and socioeconomic systems, regulations on emissions, land use and natural resource management
- **Global level:** international initiatives, worldwide frameworks, food security, climate change, equity in employment – UN SDGs
- **Transboundary level:** resolve environmental justice issues related to shared natural resources such as air, water

■ **Figure 1.38** The multifaceted challenges of environmental justice

Concept

Perspectives

Environmental justice may look different at the individual level compared to the global scale. The daily experiences and interactions with their immediate environment may shape the viewpoints of individuals in a community towards aspects such as access to freshwater, the presence of pollutants and the impact on health. On a global scale, however, individual actions become very minor and the overall perspective of the region comes into focus. Different regions have different perspectives towards the same environmental issues. Vulnerable nations, which are often least responsible for carbon emissions, emphasize the need for adaptation, support and compensation. Meanwhile, developed nations may prioritize mitigation efforts.

1.3.12 Sustainability indicators

Indicators help in making better, informed decisions and effective actions by presenting the data in simplified form to policymakers. Sustainability indicators such as species richness, ecological footprint, air quality index and carbon footprint are measures used to track the progress of sustainable development. They provide valuable information about the environmental, social and economic aspects of sustainability. Sustainability indicators include quantitative measures of biodiversity, pollution, human population, climate change, material and carbon footprints, and others. These indicators can be applied on a range of scales, from local to global. They help policymakers, businesses, organizations and communities evaluate the impact of their actions and make informed decisions.

Inquiry process

Inquiry 2: Collecting and processing data

Inquiry 3: Concluding and evaluating

- Collect secondary data on the air quality index (AQI) for two cities, one from a less economically developed country (LEDC) and the other from a more economically developed country (MEDC) over a range of years (for example 10–15 years).
- You may wish to use the following website: **https://waqi.info**.
- Record the data in a spreadsheet and process the data by performing statistical analyses such as averages, percentage change in AQI, trend analysis correlation between pollution emissions and AQI.
- Represent the processed data graphically.
- Analyse the processed data and provide a valid conclusion.

1.3.13 Ecological footprint

Ecological footprint (EF) is a model used to measure sustainability. It estimates the demands that human populations place on the environment in a given area. It is a hypothetical amount of land that is required to fulfil all the resource needs and waste management for the entire population of that area. If this footprint is greater than the area or resources available to the population, this indicates unsustainability as the population exceeds the carrying capacity of that particular area. The EF depends on various factors such as energy consumption, food production, water usage, transportation and waste generation. It takes into consideration two aspects:

- Biocapacity:
  - Cropland and pastures: The amount of land used to grow food for human consumption and animal feed. It also includes oil crops and rubber.
  - Fisheries: This is calculated from the amount of primary production required to support both marine and freshwater fisheries.
  - Forest: The amount of forest cover required to continuously supply timber, fibre and fuel wood.
- Demand:
  - Built-up land: The amount of land required to build houses and other infrastructure. It also includes hydroelectric reservoir area for human use.

◆ **Global hectare** – a unit that measures the average productivity of the world's biologically productive land and sea areas in a given year.

◆ **Ecological footprint per capita** – the total ecological footprint of a region, divided by the population of that region.

The EF is measured in **global hectares** (gha). It varies significantly from country to country and person to person, reflecting diverse environmental value systems and lifestyles. Less economically developed countries often grapple with limited access to resources and technological advancements, resulting in relatively smaller ecological footprints. However, they face the pressing task of meeting the basic needs of their expanding populations while avoiding further depletion of natural resources. In contrast, more economically developed countries, characterized by higher levels of consumption and resource-intensive lifestyles, tend to have larger ecological footprints. Based on this, the countries with the largest **per capita EF** (resource consumption per person) are Qatar and Luxembourg (Table 1.7).

■ Table 1.7 Ecological footprints per capita, 2022

Country	Ecological footprint per capita (in gha)
Qatar	13.13
Luxembourg	10.96
United States	7.46
Republic of Korea (South Korea)	5.82
Saudi Arabia	5.75
Sweden	4.94
Germany	4.5
China	3.62
Thailand	2.35
Afghanistan	0.80
Rwanda	0.55

> ### ● HL.b.15: Environmental economics
>
> Ecological economics aims at balancing the ecological footprint of a country with its biocapacity, leading to sustainability. Investing in renewable energy, promoting sustainable agriculture, reducing pollution and conserving biodiversity are some of the practices that lead to the balancing of the ecological footprint. People are required to change their consumption patterns to achieve sustainability. For example, Germany is a leader in renewable energy and energy efficiency. The country has also invested heavily in public transportation and other sustainable transportation options. As a result, Germany has been able to reduce its ecological footprint and become a global leader in sustainable development. Promotion of sustainable energy sources not only reduces greenhouse gases but also creates more job opportunities, thus improving the social well-being of the society.

ATL ACTIVITY

Environmental Tax Reform puts a price on the carbon emissions and helps to internalize the environmental cost associated with these emissions.

Research the Environmental Tax Reform using the following document, and prepare a short presentation on the impacts of environmental taxation policies on eco-friendly consumer choices.

Resource: International Monetary Fund, Working Paper, Environmental Tax Reform: Principles from Theory and Practice to Date, Prepared by Dirk Heine, John Norregaard, and Ian W.H. Parry1, July 12. www.imf.org/external/pubs/ft/wp/2012/wp12180.pdf

Tool 2: Technology

Ecological footprint calculator

Research the concept of ecological footprint and how to use the ecological footprint calculator: **www.footprintcalculator.org/home/en**.

- Use the ecological footprint calculator to calculate your own ecological footprint.
- Analyse your results and identify areas where you can reduce your ecological footprint.
- Present comparative data on footprints graphically, using a spreadsheet and graph-plotting software.

1.3.14 Carbon footprint and water footprint

■ The carbon footprint

The carbon footprint is the total amount of greenhouse gases emitted directly or indirectly by an individual, organization or activity. It is calculated by adding up the carbon emissions at every stage of a product's or service's life cycle and is measured in tonnes of carbon dioxide equivalents. In its life cycle, a product or service may emit different types of greenhouse gases, such as carbon dioxide (CO_2), methane (CH_4) and nitrous oxide (N_2O), each of which has its own global warming potential and resulting carbon footprint.

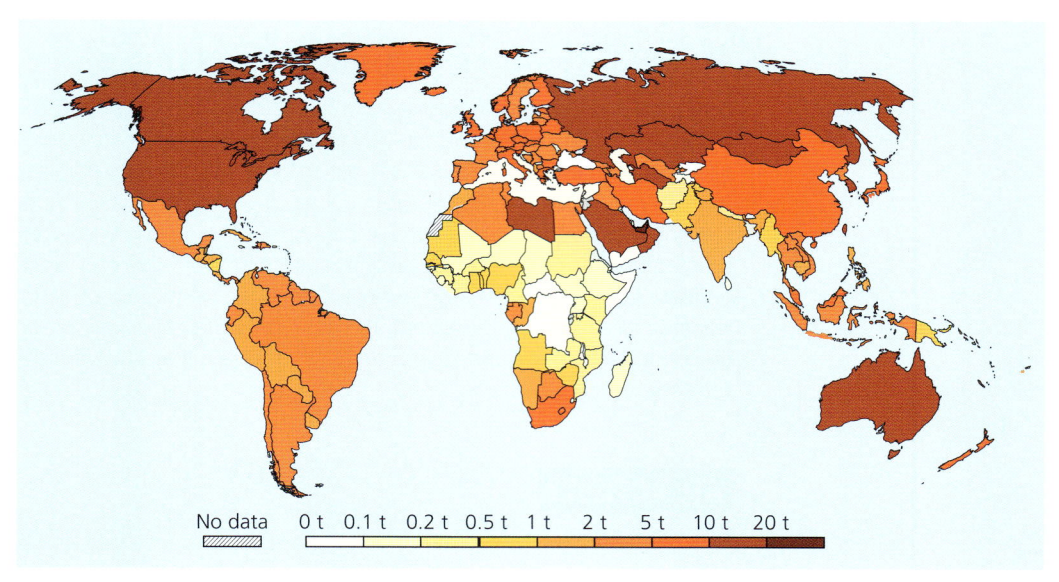

■ **Figure 1.39** Per capita carbon dioxide (CO_2) emissions from fossil fuels and industry

Every individual has a carbon footprint, depending on their choices of food, household appliances and transportation.

Carbon footprint

Food production
- Meat products have a greater per calorie carbon footprint compared to vegetables and crops
- Farmed fish and seafood generate a lower carbon footprint than animal production on land
- Cattle, sheep and other ruminants produce huge amounts of methane
- Transportation of food across borders adds to the carbon footprint

Household
- Fossil fuels vs renewable energy sources
- Centralized home heating and cooling systems
- Energy efficient appliances
- Washing machines and dishwashers – the heating setting on both these machines takes up a lot of energy to heat water for washing clothes and dishes

Transport
- Electric vs petrol and diesel vehicles
- Driving even for short distances
- Keeping vehicles switched on at the signals
- Frequent flying

■ **Figure 1.40** Carbon footprint depending on individual choices (illustration by Virginia D'Britto)

There are three main types of carbon footprint: direct, indirect and embodied.

- Direct emissions are those that are released directly from the source, such as the emissions from a car's exhaust pipe.

- Indirect emissions are those that are released as a result of human activity, but not directly from the source of the activity. For example, the emissions from the power plant that generates the electricity used to power your home are indirect emissions.

- Embodied emissions are those that are associated with the production and transportation of goods and services. For example, the emissions from the production of a car, the transportation of food from the farm to the grocery store, or the construction of a building are all embodied emissions.

You can reduce your carbon footprint through actions such as:
- choosing to walk or cycle more often, especially for short distances
- adopting a plant-based diet or choosing to eat meats that generate lower carbon emissions
- looking at food miles or place of manufacture before buying any food product
- buying local products to reduce the distance that food travels
- switching to renewable sources of energy, such as solar and wind power
- choosing sustainable products that have a low carbon footprint.

The water footprint

In many areas around the world, water scarcity is becoming a serious environmental issue that has arisen due to overexploitation of this resource. A water footprint is a measure of the total volume of freshwater used directly and indirectly to produce a particular product or support a specific human activity. There are three components of the water footprint.

1. Green water: The volume of rainwater consumed by plants during their growth.
2. Blue water: The volume of freshwater from surface and groundwater sources that is used in the production process. It includes water used for irrigation, industrial processes and domestic purposes.
3. Grey water: The volume of freshwater required to dilute pollutants and contaminants produced during the manufacturing or production process to meet water quality standards.

Calculation of water footprint

Let us understand this with the help of an example. Say we are interested in calculating the water footprint of a hamburger. The following data is considered:
- The amount of water used for each step in the production process of the hamburger.
- The amount of water available in the area where the production of the hamburger is taking place.
- The quality of water being used in the production of the hamburger, and the quality of water that is released after the production is completed.

The above data is used to calculate the green, blue and grey water footprints for the hamburger. Water footprint is a sum of the green, blue and grey water footprints. It is expressed in cubic metres per year.

■ **Table 1.8** The water footprint (m³ per ton) of some selected food products from vegetable and animal origin

Food item	Green	Blue	Grey	Total
Vegetables	194	43	85	322
Cereals	1,232	228	184	1,644
Milk	863	86	72	1,021
Pig meat	4,907	459	622	5,988
Bovine meat	14,414	550	451	15,415

You can reduce your water footprint by:
- taking shorter showers
- fixing any leaky faucets or pipes
- installing water-efficient appliances and fixtures
- consuming less meat and more plant-based foods

> **Tool 1: Experimental techniques**
>
> **Questionnaires and surveys**
>
> Calculating your water footprint is possible using some easy calculators available online, such as **https://waterfootprint.org** or **https://footprintcalculator.org**.
>
> 1 Design a questionnaire to study the water consumption patterns of people living in a given area. For example, you may want to study the effect of age group on water consumption.
>   a Frame a relevant and concise research question.
>   b Identify a testable hypothesis.
>   c Identify relevant variables and controls.
>   d Plan for collection of sufficient relevant data.
> 2 Collecting and processing data:
>   a Use an appropriate technological tool, for example, Google Forms, Excel survey or SurveyMonkey®, to design the survey.
>   b Process the collected data using a spreadsheet.
>   c Calculate the means of any values that may be required to answer the research question.
>   d Provide a valid conclusion based on evaluation.

Now that you know what these footprints mean, watch some of these interesting videos that could give you a head start to reduce your environmental footprint:

- The Best Ways to Reduce Your Carbon Footprint | Hot Mess: **https://youtu.be/KdiA12KeSL0**
- How to reduce your environmental footprint | WWF: **https://youtu.be/YbEFJd-fJpQ**

1.3.15 Biocapacity

In recent decades the resource consumption pattern in many countries has shown trends that indicate consumption has exceeded the regeneration capacity of the land (Earth's overshoot day, see Topic 1.3.3, page 50).

Biocapacity refers to the ability of an area to sustain humans by supplying renewable resources and absorbing the resulting wastes. Like environmental footprint (EF), it is also measured in global hectares (gha). It is the biologically productive land and ocean area to provide food, fibre and timber, to accommodate urban infrastructure and absorb excess CO_2. Biocapacity is a direct indication of the current management practices of a given area.

Biocapacity and EF share many criteria, therefore it is possible to compare them for a given region. Ecological deficit is when a country uses more natural resources than its ecosystems can generate. Ecological reserve is when the country's biocapacity is greater than the ecological footprint of its population.

Link

The ability of a land to produce resources and regenerate them depends on the rate of extraction of the resources (Topic 2.2.26, page 135).

In a given land, the producers have a greater ability to sustainably produce resources (Topic 2.2.27, page 136).

Urbanization and overpopulation of urban areas results in decreased biocapacity, which calls for sustainable infrastructure (Topics 8.2.5, 8.2.9, 8.2.11–13, pages 724, 731, 724–741).

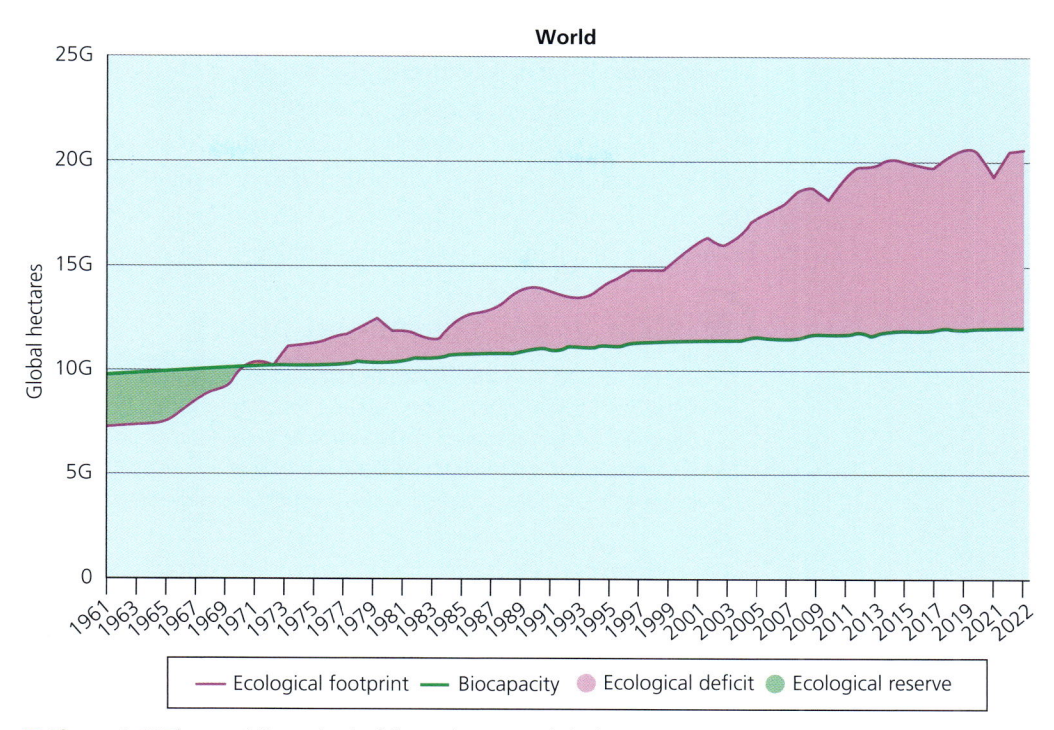

■ **Figure 1.41** The world's ecological footprint exceeds its biocapacity

Different types of lands have different biocapacities to regenerate what people demand from those surfaces. Biocapacity can change from year to year due to climate, management, and also what portions are considered useful inputs to the human economy.

Ecological deficit can lead to improper consumption of resources and waste generation, resulting in environmental degradation, loss of biodiversity and even economic instability.

1.3.16 Citizen science plays a role in monitoring Earth systems and whether resources are being used sustainably

Citizen science is an important tool for monitoring Earth systems and assessing the sustainable utilization of resources. It involves ordinary individuals in scientific research, enabling a deeper understanding of our environment. By actively engaging in data collection and analysis, citizen scientists contribute valuable insights into local challenges and conditions. This localized knowledge is crucial in addressing broader global issues, including climate change.

The information gathered through citizen science initiatives adds depth and relevance to scientific research on environmental systems. It also empowers individuals to actively participate in the larger scientific community and promotes a sense of ownership and responsibility towards the planet. Citizen science fosters a deeper connection between people and their environment, encouraging collective action towards sustainable practices and a more resilient future.

For example, imagine that a group of passionate individuals decided to take environmental monitoring into their own hands in a village situated on the mountains. Equipped with binoculars, field guides and a shared love for nature, they formed a citizen-science initiative to document and track local bird populations. Every weekend, they gathered in parks and nature reserves, observing and recording the species they encountered. Their data not only contributed to important research on bird migration patterns and population dynamics, but also raised awareness about the rich biodiversity in their region. As the word spread, more people joined this initiative, strengthening the sense of community and collective responsibility towards nature. By doing so, these people were able to increase their knowledge about the local ecosystems and also nurture a spirit of stewardship that went beyond the boundaries of their town. Their dedication and commitment serve as an inspiring example of how ordinary individuals can make a meaningful impact on the sustainability of their environment.

HL.c.3: Environmental ethics

Seeing humans as stewards of nature can lead to more ethical decisions. This involves sharing responsibility for caring for and protecting the natural world, for example, monitoring the quality of a local water body. Collection of data such as water-quality parameters (for example, temperature, pH and dissolved oxygen) helps to generate awareness about the water body's cleanliness and overall health, which can help in changing the perspectives of people in the society towards the water source.

ATL ACTIVITY

Write a blog post or article about citizen science to encourage others to participate.

TOK

To what extent can citizen science be used to generate objective evidence, and how can we ensure that it is used responsibly?

Consider the following while answering the above question:
- The different types of citizen-science data and how they are collected.
- The potential sources of bias and error in citizen-science data.
- The methods that can be used to reduce bias and error in citizen-science data.
- The ethical considerations involved in the design, implementation and use of citizen-science projects.

1.3.17 Sustainability frameworks and models

TOK

To what extent can we be certain that models can be used for prediction, given the uncertainty of data used to create them?

We all know that we need to act now and do something to combat climate change. But in the presence of multiple solutions, it becomes very difficult to know which ones would be really helpful and have a long-term effect. Here is where sustainability frameworks and models play an important role. They offer guidance and tools to assess, plan and measure sustainability efforts, enabling organizations and communities to make informed decisions and drive positive change. These models help us simplify and visualize complex concepts, making them more accessible and actionable. However, each model is a simplified version of reality and therefore has its own strengths and limitations.

1.3.18 The UN Sustainable Development Goals model

The UN Sustainable Development Goals (SDGs) are a set of social and environmental objectives that enable people to collaborate in the creation of a more just and sustainable world. They function as a motivation for governments, organizations and communities, as well as individuals, to join forces and strive towards reaching the goals by 2030. The SDGs seek to promote equitable progress that leaves no one behind, by addressing both social and environmental aspects. In order to produce long-lasting positive changes, they offer a framework for collaboration, innovation and meaningful engagement. The SDG model that supports overall development takes into consideration individuals, society and the environment at different levels.

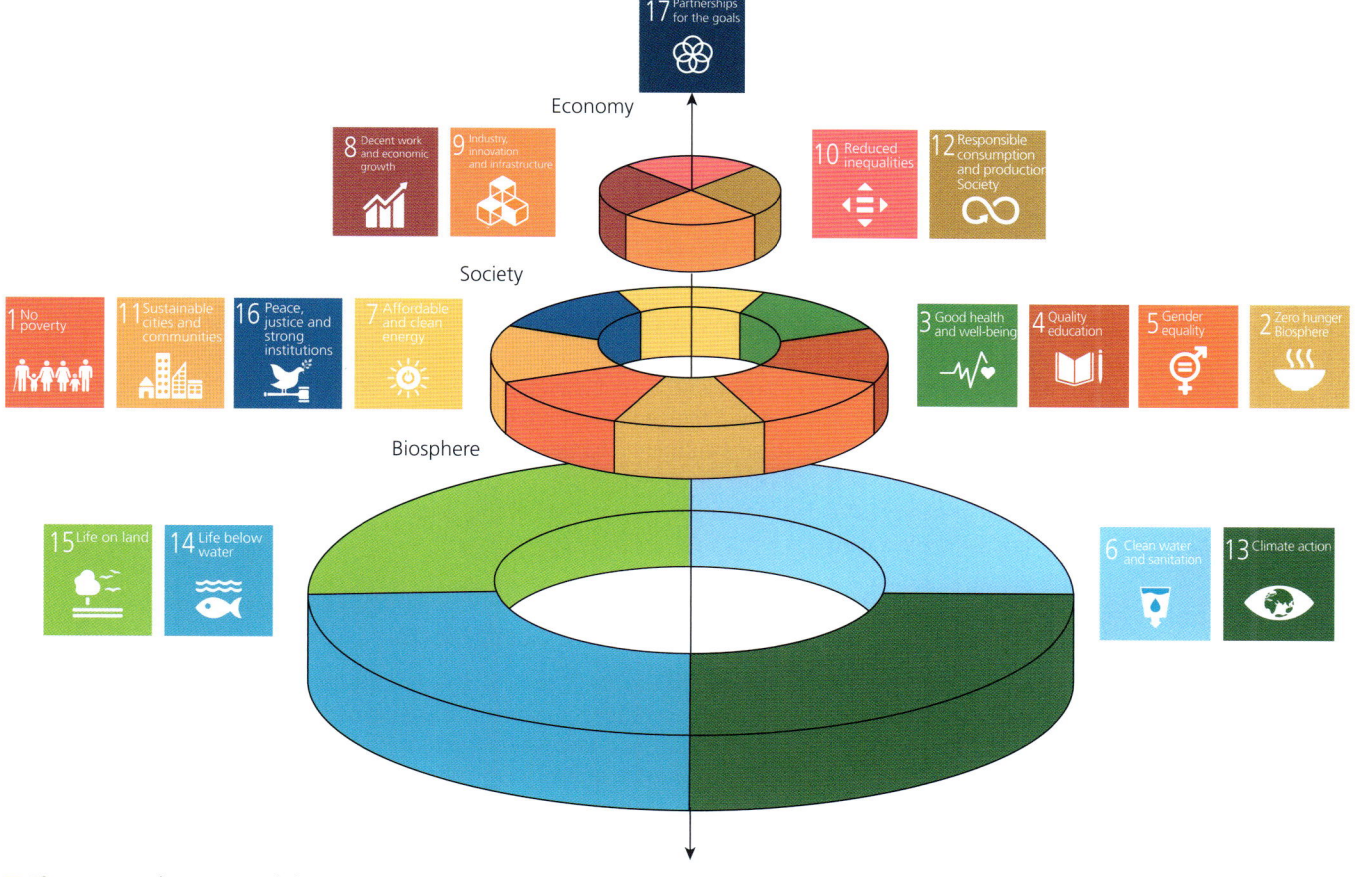

Figure 1.42 The SDG model

The SDGs are a framework of 17 sustainable goals that address the interconnected global challenges faced by humanity, including those related to poverty, education, inequality, climate, environmental degradation and protecting biodiversity, prosperity, and peace and justice. They represent a shared vision for a better future by identifying targets and indicators that focus on establishing sustainability on a global scale. One of the main strengths of the SDG model is that it recognizes the ripple effect of actions and uses it positively. Therefore, the framework ensures that all individuals, irrespective of their background or circumstances, have equal rights to shared resources, such as clean water and sanitation (SDG 6), life below water (SDG 14), life on land (SDG 15) and climate action (SDG 13). These four SDGs form the Biosphere, which supports equity of resources and opportunities in a society, which in turn supports the economy (economic decisions). This commitment to inclusivity and social justice is the foundation of the SDGs. The goals and targets set in the SDGs can be met only by encouraging extensive collaboration and partnership between governments, social bodies and individuals.

1.3 Sustainability

Uses and limitations of the SDG model

Uses	Limitations
• Provides a common framework for countries to develop policies that promote sustainable development. • Mobilizes resources from governments, businesses and civil society to support sustainable development initiatives. • Tracks progress: Provides a set of indicators to track progress on sustainable development and to identify areas where further action is needed. • Raises awareness of the challenges and opportunities of sustainable development. • Inspires people to take action to create a more sustainable future.	• The SDGs are very ambitious and it is not clear if it is possible to achieve all of them by 2030. • There is a lack of funding to support the implementation of the SDGs. • There is a lack of political will from some countries to implement the SDGs. • Some of the SDGs are difficult to measure, making it difficult to track progress. • The SDGs focus on global goals, but they may not be relevant to all countries or local communities.

HL.a.7: Environmental law

International law provides an essential framework for addressing transboundary issues of pollution and resource management. In 1979, Europe, North America, Russia and former Eastern Bloc countries signed the UNECE Convention on Long-Range Transboundary Air Pollution, creating the first international treaty to deal with air pollution on a broad regional basis. The Convention entered into force in 1983, laying down the general principles of international cooperation for air pollution abatement and setting up an institutional framework that has since brought together research and policy. Over the years, the number of substances covered by the Convention and its protocols has been gradually extended, notably to ground-level ozone, persistent organic pollutants, heavy metals and particulate matter.

HL.b.11: Environmental economics

SDG Goal 1 'No poverty' focuses on economic growth, which is the increase in the value of goods and services produced in a country over time, measured as annual percentage change in GDP. Economic growth is one of the most important ways to reduce poverty. As GDP increases, people's incomes tend to rise, and they have more money to spend on essential goods and services. It can lead to job creation, which can provide people with the opportunity to earn an income and improve their standard of living. However, economic growth is not always inclusive, and it can benefit some groups more than others. That is why it is important to ensure that economic growth is sustainable and benefits all members of society. This can be done by investing in education and training, promoting social safety nets and protecting the environment.

1.3.19 The Planetary Boundaries model

◆ **Holocene epoch** – the current geological epoch (era) in which we are living. It began 1,700 years ago after the Pleistocene epoch (ice age) ended.

Human systems interact with the environment, and in certain boundaries they operate at the farthest limits, beyond which the delicate balance that holds life together gets disturbed. There is an urgent need for a novel perspective towards the interrelationship between human and Earth systems. In 2009, a group of 28 esteemed researchers led by Johan Rockström established the nine planetary boundaries. These are the thresholds within which humanity can survive, develop and thrive for generations to come in the **Holocene epoch**. They considered that if these limits were breached, there would be severe consequences for humankind.

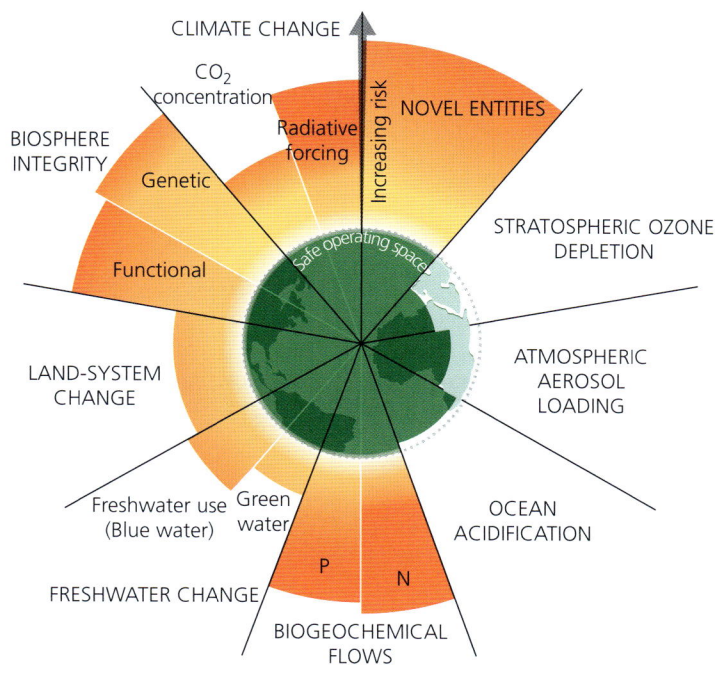

Figure 1.43 The Planetary Boundaries model

The nine planetary boundaries identified are:

1. Climate change: CO_2 concentration, average global temperature
2. Change in biosphere integrity: Biodiversity loss and species extinction
3. Stratospheric ozone depletion: Aerosol concentration
4. Ocean acidification: Carbonate ion concentration in the ocean
5. Biogeochemical flows: Phosphorus and nitrogen cycles
6. Land-system change: Change in forest cover
7. Freshwater use: The amount of freshwater available for animals and plants
8. Atmospheric aerosol loading: Microscopic particles in the atmosphere
9. Introduction of novel entities: Pollution and plastics.

In September 2023, a team of scientists quantified, for the first time, all nine processes that regulate the stability and resilience of the Earth system. These nine boundaries are like tests, with a normal range under which the Earth system would remain healthy. Quantitative data reveals the unhealthy condition of our Earth system.

Figure 1.44 clearly shows that, as of 2023, six of the nine boundaries have already been crossed. The green zone is the safe operating space, the yellow/pale orange zone is the zone of increasing danger and the orange zone is the high-risk zone. Climate change, Land-system change and Freshwater use are in the zone of increasing danger (yellow). Biogeochemical flows, Introduction of novel entities and Change in biosphere integrity are in the high-risk zone (orange or red).

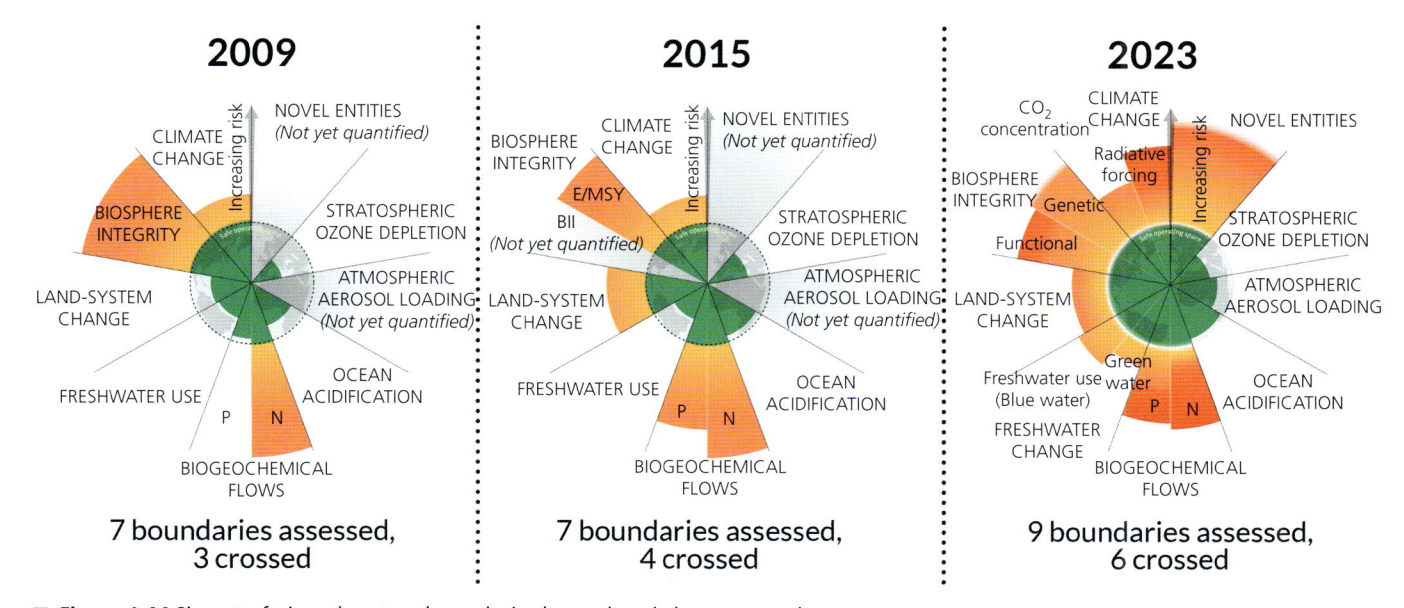

2009
7 boundaries assessed, 3 crossed

2015
7 boundaries assessed, 4 crossed

2023
9 boundaries assessed, 6 crossed

Figure 1.44 Six out of nine planetary boundaries have already been crossed

1.3 Sustainability

■ **Figure 1.45** The Bramble Cay melomys (*Melomys rubicola*) is the first mammal reported to have gone extinct as a direct result of climate change

According to a report by the Stockholm Resilience Centre, the average global temperature has already increased by about 1 degree Celsius since the pre-industrial era, and is on track to exceed 1.5 degrees Celsius by the middle of the century. This is due to human activities such as the burning of fossil fuels, deforestation and intensive agriculture. These actions have also caused a tremendous change in the land-use pattern.

Planetary boundaries have been crossed in freshwater use. Over the recent years, change in freshwater use has affected the ecosystems drastically. Excessive extraction of groundwater and the modification of natural water flows have contributed to habitat degradation, loss of biodiversity and increased water scarcity in certain regions.

Inquiry process

Inquiry 3: Concluding and evaluating

Natural as well as anthropogenic activities have contributed to the increase in concentration of carbon dioxide in the atmosphere. The planetary boundary for climate change has drastically shifted due to this addition.

- Use the data on CO_2 emissions for different countries available on **www.ourworldindata.org**. Interpret the processed data to draw and justify valid conclusions.
- Find relevant information for anthropogenic activities in the chosen countries to support your conclusion.
- Present valid arguments to support your findings.
- Evaluate your methodology of collecting data and the conclusion/s.

■ Uses and limitations of the Planetary Boundaries model

Uses	Limitations
• To assess the impact of human activities on the Earth system. • To set targets for sustainable development. • To guide decision making.	• There is scientific uncertainty. • The scale does not take into account the regional and local variability of environmental systems. • It is a complex system with many interacting parts. • It may not be able to anticipate all of the potential risks of human activities. • There are trade-offs between different planetary boundaries.

● HL.a.7, HL.a.9: Environmental law

Transboundary pollution and resource mismanagement require strict international laws and governing bodies (institutions and organizations). Transboundary pollution is pollution that crosses national borders, affecting the environment and human health in other countries. Resource mismanagement is the overuse or misuse of natural resources, such as water and air, which can also have transboundary impacts. For example, greenhouse gases like CO_2 can cross borders and cause effects in places where they are not being generated.

- The United Nations Framework Convention on Climate Change (UNFCCC) is a treaty that aims to reduce greenhouse gas emissions and prevent dangerous climate change. If implemented properly, this will help reverse the planetary boundary of climate change, which has crossed its threshold.
- The Vienna Convention for the Protection of the Ozone Layer and the Montreal Protocol are treaties that aim to protect the ozone layer and reduce the use of ozone-depleting substances – this aims to influence the planetary boundary of stratospheric ozone depletion.

- The Convention on Biological Diversity (CBD) is a treaty that aims to conserve biological diversity and promote sustainable use of natural resources – this aims to influence the planetary boundary of biosphere integrity.

> **ATL ACTIVITY**
>
> Make a case study of a local and a global example of resource mismanagement and transboundary pollution.

1.3.20 The Doughnut of social and planetary boundaries

British economist Kate Raworth developed the diagram of the Doughnut of social and planetary boundaries in 2012. It is a visual framework for sustainable development that aims to meet the needs of all people within the means of the planet. It is based on two sets of boundaries: a social foundation of essential human needs that everyone should have met, and an ecological ceiling of the planet's limits.

The model is shaped like a doughnut or a lifebelt, with a hole in the middle. The inner circle represents the social foundation: the people who lack access to basic necessities such as food, water, shelter, healthcare, education and equity. The outer circle of the doughnut represents the ecological ceiling: the ecological limits such as climate change, biodiversity loss and pollution, on which life depends and that must not be overexploited. The space between the two circles is the 'doughnut', which represents the safe and just space for humanity to thrive.

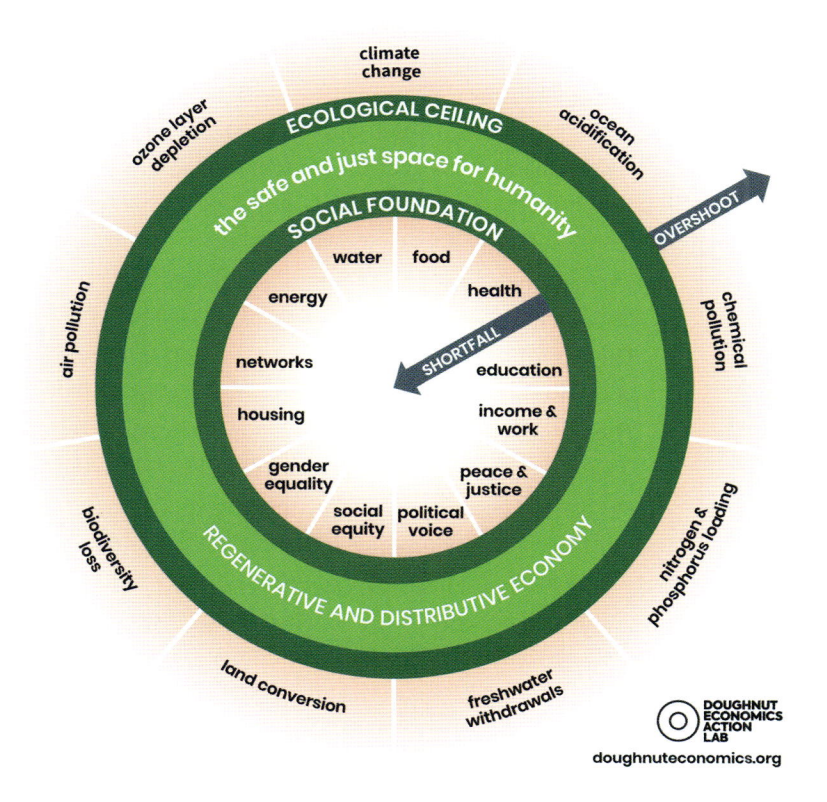

■ **Figure 1.46** The Doughnut of social and planetary boundaries

1.3 Sustainability

The design of the doughnut model takes into consideration aspects that benefit both humans and nature. Such designs are resilient, adaptable, restorative, equitable and inclusive. The social foundation of the model is grounded in the SDGs, while the ecological ceiling is based on the planetary boundaries. There is a small space between which the interplay of these two limits allows sustainable living for humans in harmony with nature. Today, billions of people fall short of the social foundation, while humanity has collectively overshot most of the planetary boundaries. The challenge is to meet the needs of all humans within the limits set by the model. In other words, to pull everyone out of the hole in the doughnut and provide them a safe and just space to live.

REAL-WORLD EXAMPLE

Amsterdam

Amsterdam, the capital city of the Netherlands, has emerged as a leading proponent of the doughnut economics model, spearheading its implementation at the city level. It recognizes that economic growth cannot come at the expense of social well-being and ecological stability.

In 2020, Amsterdam adopted the Doughnut Economy Action Plan to pull itself out of the after-effects of Covid-19. The focus of its approach was the following key areas:
- Investing in renewable energy, such as solar and wind power.
- Reducing waste by promoting recycling and composting.
- Creating new jobs in the circular economy, such as in repair and reuse industries.
- Making public transportation more affordable and accessible.

Investing in green spaces and parks. One of the key elements of Amsterdam's implementation strategy is the integration of the doughnut model into its policymaking processes. The city has established a dedicated team to translate the model's principles into actionable policies and initiatives. They have mapped all material flow from entry to processing to monitor the use of materials and achieve full circular economy by 2050. Through initiatives like the Amsterdam Circular Strategy and the Circular Innovation Programme, a ladder of circularity has been designed for this purpose, showing which processing options are preferable to others. For example:
- The changing use and design of products: Avoiding the use of plastic cups, sharing cars, and producing the same products using fewer raw materials.
- The use phase of products: Policies aimed at prolonging the life cycle as much as possible. Second-hand stores and repair centres play a role here.
- The end of a product's life: Components can be repurposed, while materials can be recycled and, as a last option, incinerated with energy recovery.

Amsterdam is also investing in sustainable mobility and renewable energy. The city aims to become carbon neutral by 2050, emphasizing the importance of transitioning to clean energy sources and promoting electric transportation. Social inclusivity is another critical aspect: the city is committed to addressing inequalities and enhancing well-being for all its residents. Initiatives like Amsterdam Impact, which started in 2015, aim to accelerate the transition to a sustainable and inclusive economy, support entrepreneurs who are working to solve societal challenges, attract and retain talent in Amsterdam, and provide financial and knowledge support for new entrepreneurs and for innovative projects that focus on solving societal challenges and creating a more equitable society.

Amsterdam's commitment to the doughnut economics model extends beyond its borders. The city actively participates in global networks and collaborations to share its experiences and learn from others. By joining initiatives like the Doughnut Economics Action Lab and engaging with cities worldwide, Amsterdam contributes to the collective effort of creating regenerative and distributive economies globally.

■ Uses and limitations of the doughnut economics model

Uses	Limitations
• To identify areas where we are falling short of the social foundation or exceeding the ecological ceiling. • To set targets for sustainable development. • To guide decision making at all levels, from individuals to businesses to governments. • Various scales to support sustainability actions, from countries to businesses and neighbourhoods. • To consider the interconnectedness of social, economic and environmental dimensions.	• The complex framework can be difficult to understand and implement. • It relies on a significant amount of data to measure social and environmental progress, which may be difficult to collect or unavailable for some countries. • It is a relatively new concept, and there is still limited empirical evidence to support its effectiveness.

HL.b.17: Environmental economics

Doughnut economics models can be seen as applications of ecological economics for sustainability. The doughnut model fundamentally attempts to address issues of inequality and injustice. For example, creating jobs for green economy, investing in social safety nets.

ATL ACTIVITY

Create a mind map of all possible applications of ecological economics for sustainability.

HL.c.13: Environmental ethics

The implementation of the doughnut economics model may involve environmental and social justice movements like those that began in the past. Environmental and climate justice movements raise fundamental questions about the use and abuse of the natural world at multiple timescales. For example, environmental justice (see Section 1.3.11), indigenous-led conservations and fossil fuel moratoria, extractivism and 'living well' are some of the key movements.

ATL ACTIVITY

Drawing parallels with historical movements for social and environmental justice, how might the implementation of the doughnut economics model resemble or differ from past movements?

1.3.21 The circular economy model

The circular economy model promotes decoupling economic activity from the consumption of finite resources. It has three principles: eliminating waste and pollution, circulating products and materials, and regenerating nature.

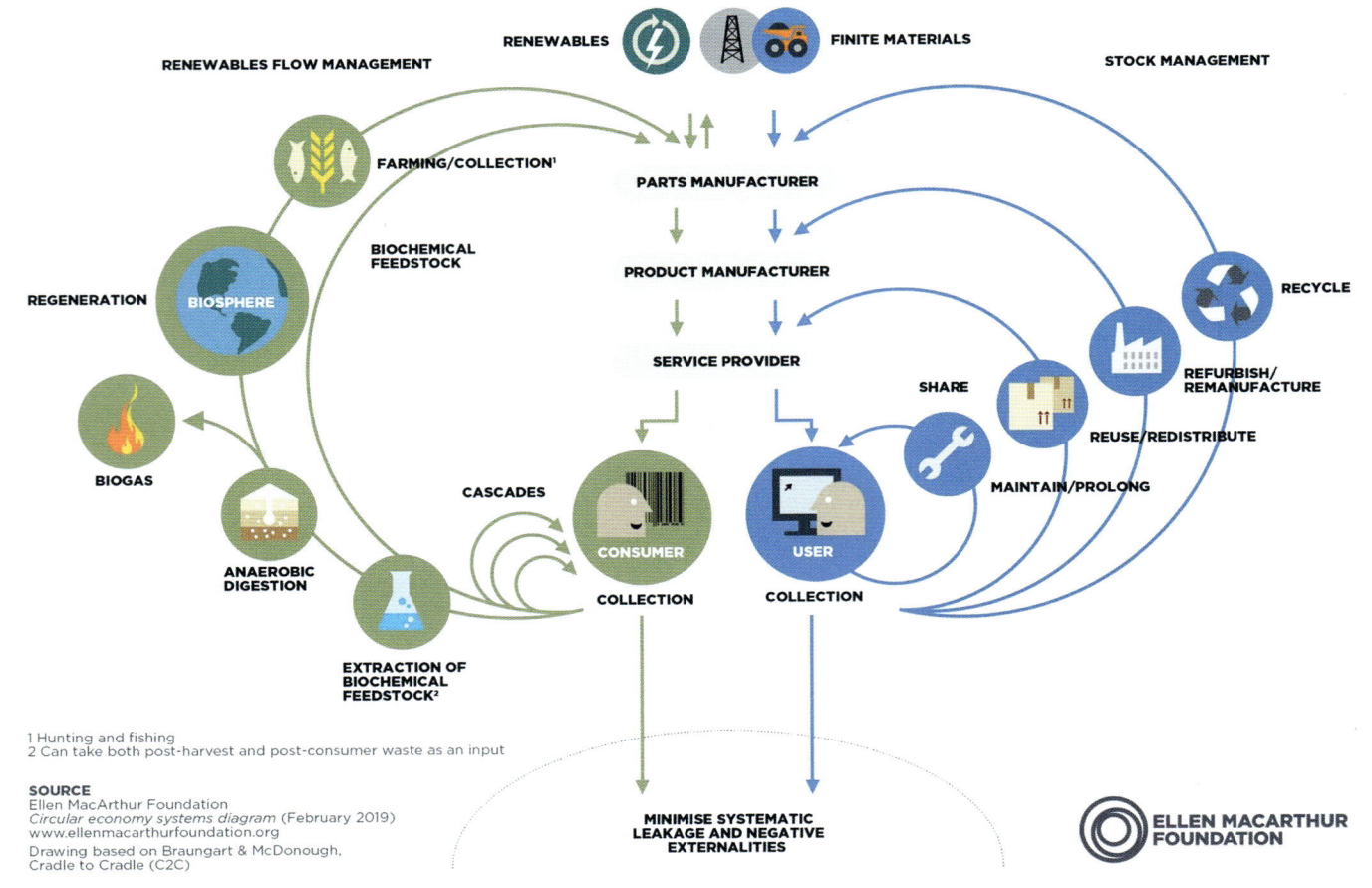

■ Figure 1.47 Butterfly diagram (source: Ellen MacArthur Foundation)

The butterfly diagram in Figure 1.47, developed by the Ellen MacArthur Foundation, visually represents the concept of a circular economy in a captivating and intuitive way. The diagram depicts two wings of a butterfly, each representing a different phase of the circular economy. The wing on the left side represents the biological cycle, emphasizing the importance of regenerating natural systems and designing products with materials that can safely return to the environment, such as through composting or biodegradation. The wing on the right side symbolizes the technical cycle, highlighting the value of retaining products, components and materials in use for as long as possible through strategies like repair, remanufacturing and recycling. The body of the butterfly represents the economy itself, which connects and integrates these two cycles. This visual representation effectively communicates the holistic nature of the circular economy and the need for changes in our production and consumption patterns. The butterfly diagram demonstrates the transformative potential of the circular economy in creating a regenerative and sustainable future.

■ The biological cycle

The left side of the butterfly diagram represents the biological cycle, focusing on materials that can biodegrade and return to the Earth, such as food. Regeneration is at the heart of this cycle, where we actively improve nature instead of degrading it.

- Farming: Regenerative practices in farming, forestry and fisheries can lead to positive outcomes for nature, including healthy soils, increased biodiversity, and improved air and water quality.
- Composting and anaerobic digestion: These processes turn organic waste into compost or biogas, respectively (see Theme 5, page 415).

- Cascades: The biological cycle encourages the use of products and materials already in the economy. This involves repurposing food byproducts for other applications, such as textiles made from orange peels, or animal feed made from leftover human food. When products or materials reach the end of their use, they are returned to the soil.
- Extraction of biochemical feedstock: Biorefineries utilize post-harvest and post-consumer biological materials to produce high-value chemical products. These processes can generate a range of valuable biochemicals and nutraceuticals, contributing to a circular economy.

The technical cycle

The right side of the butterfly diagram represents the technical cycle, focusing on products that are used rather than consumed. The inner loops retain the embedded value of a product by keeping it whole. Sharing, maintaining and reusing are prioritized in these loops, capturing more value and reducing waste.

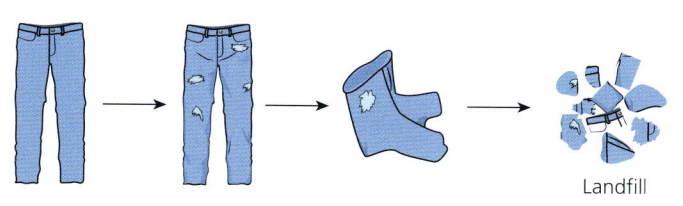

■ Figure 1.48 The linear economy

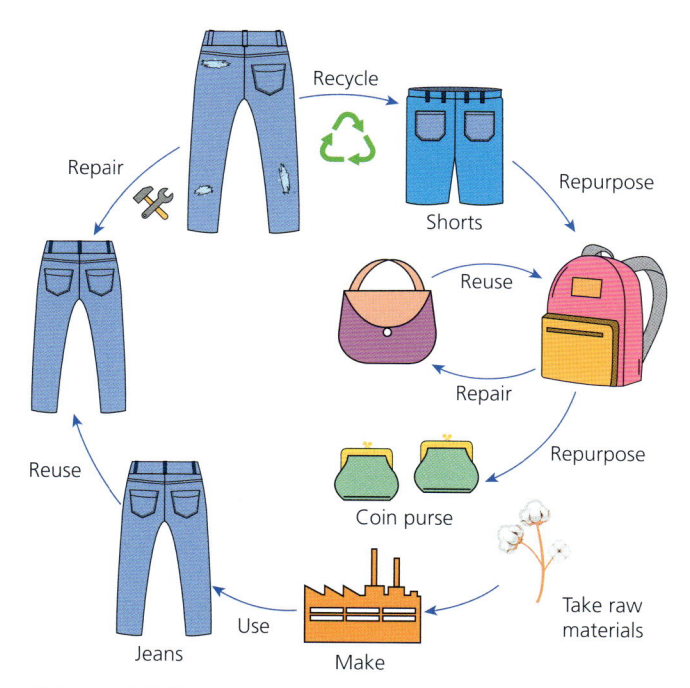

■ Figure 1.49 The circular economy

- Sharing: Sharing products through platforms like tool libraries, shared wardrobes, and car-sharing systems increases utilization and reduces underutilization.
- Maintaining: Proper maintenance of products prolongs their usable life, maximizing their value and guarding against failure or decline.
- Reusing: Reusing products in their original form and for their original purpose reduces the need for new items. Resale platforms for clothing and reusable packaging are examples of successful reuse business models.
- Redistribution: Redirecting products from one market to another ensures their valuable use. Unsold clothing being redistributed to another store is one example.
- Refurbishing: Restoring products to good working order through repair, component replacement or cosmetic improvements maintains their value and extends their lifespan.
- Remanufacturing: When products cannot remain in circulation in their current state, remanufacturing involves re-engineering them to as-new condition. This allows products and components to be used again, representing cost savings.
- Recycling: Recycling is the final step, when a product can no longer be used, refurbished or remanufactured. It transforms the product into basic materials to create new ones. Designing for recycling is crucial, especially for items like single-use packaging.

This model differs from the linear economic model (take–make–waste) in that it focuses on sustainability. This focus is unlike the linear model, which primarily focuses on profitability. In the linear model, resources are extracted, transformed into products, and ultimately discarded as waste after their use. This approach leads to resource depletion, environmental pollution and unsustainable practices. The butterfly model of the circular economy emphasizes the importance of reducing waste, reusing materials and recycling products to maximize resource efficiency.

1.3 Sustainability

The butterfly diagram illustrates this concept by showcasing the interconnectedness of different strategies within the circular economy, such as designing for durability, remanufacturing, sharing platforms and renewable energy sources.

■ Uses and limitations of the circular economy model

Uses	Limitations
• It can help to lessen the environmental impact of economic activity by reducing resource use, pollution and waste. • It can create new jobs and businesses, and it can boost economic growth and social well-being. • It reduces waste by extending the product life cycle through repair, refurbishment and remanufacturing. • It creates a resilient economy that addresses social, economic and environmental challenges.	• Transitioning to a circular economy can be expensive, especially for small and medium-sized businesses. • Knowledge and skill are required for refurbishing and remodelling. • There are technological and design limitations. • The quality of recycled, remodelled or refurbished products may not be good and therefore they may have a short lifespan. • There is a lack of infrastructure. • There are regulatory challenges. • There is difficulty in scaling of the product. • It requires changes to consumer mindset and behaviour.

HL.b.17: Environmental economics

The circular economy involves product stewardship. Responsibility for the sustainable management of a product is attributed to the manufacturer, seller and user.

Manufacturers have the responsibility to:
- use sustainable materials, such as recycled content or materials that are grown in a sustainable way
- use efficient manufacturing processes and renewable energy
- design products for durability and recyclability
- provide consumers with information about the environmental impact of their products.

ATL ACTIVITY

Discuss what responsibilities sellers have in the circular economy.

HL.c.3: Environmental ethics

We have a moral obligation to consume resources responsibly. The circular economy model is based on the principle that we have a moral obligation to protect the environment.
- By reducing our consumption of resources and by recycling and recovering materials, we can help to reduce our impact on the environment and to ensure that future generations have access to the resources they need.
- By designing products and systems for durability and recyclability, we can help to extend the life of natural resources and reduce the amount of waste that we produce.

ATL ACTIVITY

Make a poster or an infographic reflecting the actions that can be taken to promote responsible consumption in a circular economy.

ATL ACTIVITY

CAS project
- Design a creativity, activity, service (CAS) proposal for reducing the carbon footprint of your school or home.
- Keep the following in mind while you write your proposal:
  - It must be a year-long project.
  - You are expected to collaborate with your peers and engage all stakeholders in the process.
- Identify how you will measure the success of your project.
- At the end of one year, if all goes as planned, you may present your work to the school community.

ATL ACTIVITY

Poster presentation
- Research various examples of environmental injustice and inequalities in access to resources.
- Choose any one example and explore it in detail.
- Create a poster to raise awareness about environmental injustice.

ATL ACTIVITY

- Promote the doughnut economics model and/or circular economy strategies to your school community.
- Design a sustainability walk to highlight sustainable options locally.
- Choose one SDG and use it to advocate for a particular issue.

REVIEW QUESTIONS

1. State the constituents of sustainable development.
2. Identify which sustainability model you think is best.
3. State what is meant by environmental justice. Give an example.
4. Outline the role of citizen science in ensuring sustainable use of resources.
5. Describe how the planetary boundaries have changed over time.

EXAM-STYLE QUESTIONS

1. Discuss, with reference to two contrasting environmental problems, the technocentric belief that technology may provide solutions to environmental problems. [7 marks]

2. Outline the factors that lead to different environmental value systems in contrasting cultures. [3 marks]

3. Discuss the potential impacts of deforestation on the carbon storage present in a rainforest ecosystem. [3 marks]

4. Justify the use of models to create predictions in the event of natural disasters. [7 marks]

5. With the help of named examples, explain how embracing the doughnut economics model can lead to a sustainable future. [5 marks]

6. Discuss the importance of social justice and equity in achieving sustainability. [5 marks]

Theme 2
Ecology

2.1 Individuals and populations, communities and ecosystems

> **Guiding question**
>
> ■ How can natural systems be modelled, and can these models be used to predict the effects of human disturbance?

SYLLABUS CONTENT

This chapter covers the following syllabus content:
- ▶ 2.1.1 The biosphere is an ecological system composed of ecosystems, communities, populations and individual organisms.
- ▶ 2.1.2 An individual organism is a member of a species.
- ▶ 2.1.3 Classification of organisms allows for efficient identification and prediction of characteristics.
- ▶ 2.1.4 Taxonomists use a variety of tools to identify organisms.
- ▶ 2.1.5 A population is a group of organisms of the same species living in the same area at the same time, and which are capable of interbreeding.
- ▶ 2.1.6 Factors that determine the distribution of a population can be abiotic or biotic.
- ▶ 2.1.7 Temperature, sunlight, pH, salinity, dissolved oxygen and soil texture are examples of the many abiotic factors that affect species distributions in ecosystems.
- ▶ 2.1.8 A niche describes the particular set of abiotic and biotic conditions and resources upon which an organism or a population depends.
- ▶ 2.1.9 Populations interact in ecosystems by herbivory, predation, parasitism, mutualism, disease and competition, with ecological, behavioural and evolutionary consequences.
- ▶ 2.1.10 Carrying capacity is the average size of a population, determined by competition for limited resources.
- ▶ 2.1.11 Population size is regulated by density-dependent factors and negative feedback mechanisms.
- ▶ 2.1.12 Population growth can be either exponential or limited by carrying capacity.
- ▶ 2.1.13 Limiting factors on the growth of human populations have increasingly been eliminated, resulting in consequences for sustainability of ecosystems.
- ▶ 2.1.14 Carrying capacity cannot be easily assessed for human populations.
- ▶ 2.1.15 Population abundance can be estimated using random sampling, systematic sampling or transect sampling.
- ▶ 2.1.16 Random quadrat sampling can be used to estimate population size for non-mobile organisms.
- ▶ 2.1.17 Capture–mark–release–recapture and the Lincoln index can be used to estimate population size for mobile organisms.
- ▶ 2.1.18 A community is a collection of interacting populations within an ecosystem.
- ▶ 2.1.19 A habitat is the location in which a community, species, population or organism lives.
- ▶ 2.1.20 Ecosystems are open systems in which both energy and matter can enter and exit.
- ▶ 2.1.21 Sustainability is a natural property of ecosystems.
- ▶ 2.1.22 Human activity can lead to tipping points in ecosystem stability.
- ▶ 2.1.23 Keystone species have a role in the sustainability of ecosystems.
- ▶ 2.1.24 The Planetary Boundaries model indicates that changes to biosphere integrity have passed a critical threshold.
- ▶ 2.1.25 To avoid critical tipping points, loss of biosphere integrity needs to be reversed.

HL ONLY

▶ 2.1.26 There are advantages to using a method of classification that illustrates evolutionary relationships in a clade.
▶ 2.1.27 There are difficulties in classifying organisms into the traditional hierarchy of taxa.
▶ 2.1.28 The niche of a species can be defined as fundamental or realized.
▶ 2.1.29 Life cycles vary between species in reproductive behaviour and lifespan.
▶ 2.1.30 Knowledge of species' classifications, niche requirements and life cycles helps us to understand the extent of human impacts upon them.

2.1.1 The biosphere is an ecological system composed of ecosystems, communities, populations and individual organisms

The biosphere is the area of the world where life may be found, and it is made up of many ecological systems such as ecosystems, communities, populations and individual organisms.

- An **ecosystem** is a community of living species and non-living environmental components that interact with one another.
- A community is a collection of distinct populations of organisms that live and interact in the same location at the same time.
- Populations are groupings of individuals from the same species that live in the same place.
- Finally, individual organisms are solitary living organisms that serve as the basic building blocks of the biosphere, as shown in the biological classification pyramid in Figure 2.1.

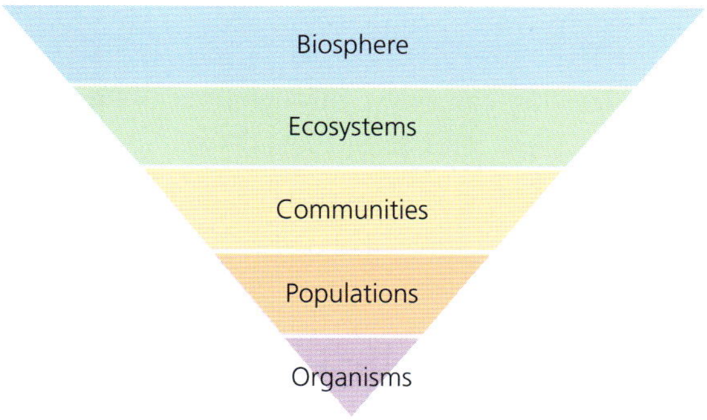

◆ **Ecosystem** – an ecological unit that includes all the organisms living in a particular area (the community) and the physical and chemical components with which they interact.

The biosphere is a complex and linked system in which each component is critical to the ecosystem's balance.

Individual species, for example, form populations that interact with other populations to form communities, which in turn interact with non-living ecosystem components such as water, air and soil to produce ecosystems. These ecosystems can be big or small, and they can be found in a variety of environments including forests, oceans, deserts and grasslands.

Human activities such as deforestation and pollution, and the resulting climate change, can have a negative impact on the biosphere and its components, resulting in ecosystem damage and biodiversity loss. As a result, understanding the biosphere and its components is critical for managing and conserving it for future generations.

■ **Figure 2.1** Biological classification pyramid

2.1.2 An individual organism is a member of a species

A species is made up of individual organisms. According to the biological terminology, a species is a group of living organisms that may interbreed and produce fertile offspring. This indicates that members of the same species are able to mate and have offspring that can reproduce and have fertile offspring.

2.1 Individuals and populations, communities and ecosystems

Figure 2.2 Two members of the same species from the genus *Iris*

There are many different types of organisms in nature, ranging from small microorganisms, such as *E. coli* bacteria, to enormous whales. Each species is distinct and has traits that set it apart from other species. Physical qualities such as size, form and colour, as well as behavioural traits such as mating habits and social interactions, can all be considered as distinguishing characteristics from other species.

Understanding the notion of species is critical in ecology because it serves as the foundation for understanding how various organisms interact with one another and with their surroundings. Species, for example, can be grouped into communities, which are made up of interacting populations. Ecologists can learn about how ecosystems work and how they might be managed and protected by researching communities and species interactions.

It's also worth noting that species are not static entities; they can evolve through time as a result of natural selection, **genetic drift**, and other reasons (see Topic 3.1.3, page 197).

Comprehending the terminology of species and the processes that drive their evolution is critical to comprehending the diversity of life on Earth and how it has evolved over time.

◆ **Genetic drift** – a change in the gene pool of a small population.

2.1.3 Classification of organisms allows for efficient identification and prediction of characteristics

There are believed to be approximately 8.7 million species on Earth. It would be practically impossible to identify and study this enormous number of species without a classification system.

The Linnaean taxonomy, devised by Carl Linnaeus in the eighteenth century, is the most widely used classification system in biology. Each species is given a unique binomial name that consists of two parts: the genus and the species. The genus name begins with a capital letter, while the species name begins with a lowercase letter. Both the genus and species names should be italicized or underlined.

Species in the same genus are thought to have evolved from a common ancestor and they share some morphological and genetic traits. The lion (*Panthera leo*) and the tiger (*Panthera tigris*), for example, are both members of the Panthera genus and share many physical and genetic characteristics, such as their capacity to roar.

Scientists can also predict specific characteristics of a species based on its classification. For example, if a new species is discovered that belongs to the same genus as another species that is known to be carnivorous, the new species is likely to be carnivorous as well.

The dichotomous key is an important tool for species identification. This key is a tool that aids in the identification of unknown organisms by offering a series of options between two traits. The user proceeds through the options until they arrive at the correct identification. Dichotomous keys are commonly employed in ecology and other branches of biology and can be utilized for both plants and animals.

> **Tool 1: Experimental techniques**
>
> **Use of dichotomous keys to identify organisms**
>
> Use the below dichotomous key to identify the insects.
>
> 1a Wings are covered by an exoskeleton – go to step 2
> 1b Wings are not covered by an exoskeleton – go to step 3
>
> 2a Body has a round shape – *ladybird*
> 2b Body has an elongated shape – *beetle*
>
> 3a Wings point outward from the body – *dragonfly*
> 3b Wings point toward the rear of the body - *bee*
>
> ■ **Figure 2.3** Dichotomous key for insects

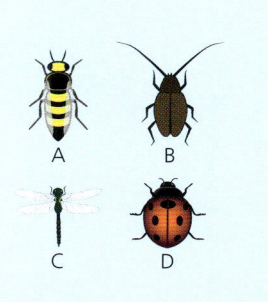

2.1.4 Taxonomists use a variety of tools to identify organisms

Identifying an organism means determining its species. Taxonomists use various tools to do this. Here are some examples:

1. Dichotomous keys: These are tools that allow users to identify an organism by answering a series of questions. Each question presents two choices, and the user must choose the option that best describes the organism they are trying to identify. Based on the user's answers, the key leads them to the correct identification. One example of a dichotomous key is shown in Figure 2.3; this key helps identify insects.

2. Comparison with specimens in reference collections by expert taxonomists: Expert taxonomists compare an unknown specimen to similar specimens in a reference collection. This helps them to identify the unknown specimen by comparing it to known specimens. The Natural History Museum in London, United Kingdom, has a collection of over 80 million specimens, including many examples of insects, birds and plants. The museum's taxonomists use these specimens to identify unknown organisms and add new species to the museum's collection. You can learn more about the museum's collection on its website.

3. DNA surveys: These involve sequencing a small piece of DNA from an organism and comparing it to known sequences in a database. This can help identify an unknown organism, especially if it is difficult to distinguish by physical characteristics alone. The Barcode of Life project is a global initiative to create a DNA barcode database for all species on Earth. This database allows researchers to identify unknown organisms by comparing their DNA barcode to known sequences in the database. You can learn more about the Barcode of Life project here: **www.boldsystems.org/index.php/TaxBrowser_Home**.

These tools are just a few examples of the many methods that taxonomists use to identify and classify organisms. By using these tools, taxonomists can accurately identify organisms and add them to the existing knowledge of species.

2.1.5 A population is a group of organisms of the same species that coexist in the same region and are capable of interbreeding

A population is a group of organisms of the same species that coexist in the same region and are capable of interbreeding. This means that the individual organisms in a group of the same species can mate and have fertile offspring. A population might be as tiny as a handful of individuals or as huge as millions of them.

Populations are essential ecological units to study because they interact with their environment and other populations in a variety of ways. The size of a population, for example, can affect the availability of resources such as food and water, as well as the amount of competition for those resources. Variables that influence population survival and reproduction, such as predation or sickness, also have an impact on populations.

Populations are studied in a variety of ways by scientists in order to better understand their dynamics and interactions with the environment. To acquire information about population size, growth rate and other characteristics, they may utilize technologies such as population surveys, mark-and-recapture studies and population modelling.

Overall, knowing populations is a crucial part of ecology that helps us better grasp the natural world's intricacies.

2.1.6 Factors that determine the distribution of a population can be abiotic or biotic

The term 'range' is used by ecologists to characterize the geographic dispersion of a species' population. A population's range is determined by a number of factors, including both biotic and abiotic influences.

Abiotic elements include environmental, physical and chemical characteristics such as temperature, rainfall, pH, **salinity** and soil type. For example, the polar bear's range is restricted to locations with sea ice, whereas the saguaro cactus' range is restricted to areas with a precise mix of temperature, precipitation and soil conditions.

Competition, predation and mutualism (see Topic 2.1.9, page 91) are examples of biotic elements that occur between living species within an ecosystem. The presence of the predatory red-tailed hawk, for example, which preys on squirrels, limits the distribution of eastern grey squirrels in North America.

◆ **Salinity** – a measure of how much salt is dissolved in water (the dissolved NaCl content of water), usually measured in parts per thousand (‰).

Understanding the factors that influence population distribution is critical for conservation and management initiatives. We can build methods to maintain and restore a population's habitat by understanding what factors limit its range.

2.1.7 Abiotic factors that affect species distributions in ecosystems

◆ **Abiotic factor** – an ecological property associated with the non-living part of an environment, such as temperature, pH and humidity.

Abiotic factors can have a direct impact on the physiology of an organism, and can also influence interactions between organisms. For example, some organisms are adapted to specific temperature ranges, while others may require a certain level of salinity or pH in their environment.
The availability of sunlight can also influence the distribution of plants, as some species require more light than others to carry out photosynthesis.

Dissolved oxygen (DO) is a critical abiotic factor that significantly influences the distribution of species in aquatic ecosystems. It refers to the amount of oxygen dissolved in water and is affected by various factors, including temperature, water movement and the presence of photosynthetic organisms. Aquatic organisms rely on dissolved oxygen to carry out vital biological processes, such as respiration. Fish and other aquatic animals extract oxygen from the water through their gills, while plants and algae produce oxygen as a byproduct of photosynthesis. Adequate dissolved oxygen levels are crucial for maintaining a healthy and diverse aquatic ecosystem.

Soil texture refers to the relative proportions of different particle sizes – sand, silt and clay – in soil. It has a profound impact on the distribution of terrestrial plant species as it affects their ability to obtain water, nutrients and root anchorage.

Soil texture influences several crucial soil properties, such as water-holding capacity, nutrient retention and drainage. Each type of soil texture has unique characteristics that influence the types of plants that can thrive in a particular area.

1. Sand (coarse soil): Sandy soils have large particles with gaps between them, which results in rapid drainage and poor water retention. As a result, plant species adapted to arid or well-draining conditions, such as cacti and certain grasses, are more likely to be found in sandy environments.
2. Silt (medium soil): Silty soils have smaller particles than sand, and offer better water retention than sandy soils. They can support a wider range of plant species, including those adapted to moderately moist conditions.
3. Clay (fine soil): Clay soils have the smallest particles, which leads to excellent water retention but often poor drainage. Species that can tolerate waterlogged conditions, like some wetland plants, are more common in clay-rich habitats.

In order to better understand the relationship between abiotic factors and species distribution, scientists often measure and quantify these factors in different locations. This can help them to identify patterns and determine which factors are most important for particular species so that they can develop strategies for conservation.

You can use the Inquiry process activity to investigate abiotic factors in your local ecosystem.

Inquiry process

Inquiry 2: Collecting and processing data

Exploring your local ecosystem.

Materials
- Field notebook
- Pencils or pens
- Camera or smartphone with camera
- Tape measure
- Sample containers (optional)
- Field guides or identification keys (optional)
- Temperature probes
- pH meters
- Dissolved oxygen meters
- Salinity meters
- Soil texture analysis kit

Steps
1. Field trip: Visit a local ecosystem such as a forest, river, wetland or park. Remember to take a field notebook to take notes and record observations throughout the trip.
2. Observations: Make detailed observations of the ecosystem around you. You can use your senses (sight, smell, touch and hearing) to identify different organisms and describe your surroundings. You can also use your camera or smartphone to take photos of different organisms and habitats.
3. Measurements: Measure and record abiotic factors such as temperature, humidity and soil pH. You can also use a tape measure to measure the distance between different organisms or habitats. Use the temperature probes, pH meters, dissolved oxygen meters, salinity meters and soil texture analysis kit to measure the abiotic factors at each location. Record your data in a notebook or data sheet. If you would like to collect enough data to process for further analysis, repeat the measurements at least three times.
4. Data analysis: Analyse your data to identify any patterns or correlations between the abiotic factors and the distribution of species within the ecosystem. Back in the classroom, review your field notes and photos to identify different organisms and habitats. You can use field guides or identification keys to identify unfamiliar species. You can also analyse your data to identify patterns in the ecosystem, such as the presence or absence of certain species or changes in abiotic factors over time.
5. Report: Finally, create a report or presentation on your local ecosystem. You can include your observations, measurements and data analysis, as well as any conclusions or recommendations you have for preserving or restoring the ecosystem.

Overall, this activity can help you to develop your observation and data analysis skills while also gaining a deeper understanding of the local ecosystems in your community.

2.1.8 A niche is a particular set of abiotic and biotic conditions and resources upon which an organism or a population depends

The role of a species within an ecosystem, encompassing all of the biotic and abiotic elements that affect its survival and reproduction, is referred to as its ecological niche. Each species has a distinct niche that can be described by numerous factors, such as the type of food it consumes, the temperature and humidity ranges it can tolerate, and the type of environment it inhabits.

Take, for example, the African elephant (*Loxodonta africana*) niche. African elephants are herbivores that eat a variety of plant types, including grasses, leaves and fruits. They require a significant amount of land to maintain their grazing needs, and they can travel considerable distances to find water sources.

African elephants have a wide temperature-tolerance range and can survive in a variety of settings, including hot, parched savannas and humid, dense forests. They also disseminate seeds and create clearings in dense vegetation, which helps to sustain the structure and richness of their habitats.

2.1.9 Populations interact in ecosystems by herbivory, predation, parasitism, mutualism, disease and competition, with ecological, behavioural and evolutionary consequences

Herbivores consume plants through herbivory. Consider the relationship of the monarch butterfly and the milkweed plant. Monarch caterpillars consume milkweed plants, which contain poisonous chemicals. Toxins can be stored in the caterpillars' bodies, making them unappealing to predators. Because it influences the location and abundance of monarch butterflies and milkweed plants, this relationship has ecological implications. Among the behavioural repercussions is that specialized behaviours emerge, such as monarch butterflies' ability to detect and locate milkweed plants. The co-evolution of monarchs and milkweed has evolutionary effects, with each species adapting to the other's defences and counter-defences.

■ **Figure 2.4** Herbivory example: A stripy monarch caterpillar (*Danaus plexippus*) feeds on a butterfly milkweed

When one animal eats (preys on) another, this is known as predation. Consider the lions and zebras on the east and southern African savanna. Lions hunt and kill zebras for food, and the zebras attempt to evade capture. Because it influences the distribution and abundance of lions and zebras, this relationship has ecological implications. Behavioural effects include the emergence of defensive behaviours, such as when zebras form herds. The emergence of features that improve predation or defence, such as camouflage in lions and stripes in zebras, are examples of evolutionary consequences.

A parasitic relationship occurs when one organism benefits at the expense of another. Consider the interplay of tapeworms and humans. Tapeworms are parasites that dwell in some humans' intestines and feed on nutrition. This interaction has ecological implications since it can have an impact on the health and survival of the human host. To prevent infection, behavioural implications include the emergence of defensive behaviours such as washing hands and cooking meat. The evolution of characteristics that improve parasite transmission or host resistance is one example of an evolutionary consequence.

Mutualism is a connection that benefits both organisms. Consider the interaction of bees and flowers. Bees take nectar from flowers and utilize it to make honey, pollinating the flowers in the process. Because it influences the range and number of bees and flowers, this relationship has ecological implications. Among the behavioural repercussions is the emergence of specialized activities, such as bees' ability to detect and locate flowers. The co-evolution of bees and flowers has evolutionary repercussions, with one species adjusting to the needs and preferences of the other.

Disease is defined as a relationship in which a pathogen (such as a virus, bacterium or fungus) infects and harms a host. Consider the influenza virus's interaction with people. The influenza virus can cause severe or even fatal respiratory illness in humans. This link has ecological implications since it can have an impact on the health and survival of human communities. Preventive actions such as immunization and hand washing are examples of behavioural consequences.

The evolution of characteristics that improve pathogen transmission or host resistance is one example of an evolutionary consequence. Disease can have ecological, behavioural and evolutionary effects. Diseases have the ability to move through communities and alter their dynamics. When a disease spreads across a population, for example, it might cause a decline in population size, which can have an impact on the entire ecosystem. In some situations, the sickness may act as an evolutionary pressure, favouring those who are more resistant to it.

The spread of white-nose syndrome in bats is one example of a disease interaction. White-nose syndrome is caused by a fungus that grows on bats' noses, wings and ears, forcing them to emerge from hibernation too soon and burn through their fat reserves before the insects they feed on are accessible. This has resulted in a major drop in bat numbers, which has the potential to have a domino impact on the ecosystem. Fewer bats, for example, may result in an increase in bug populations, which can affect plant development and the survival of other organisms that rely on those plants.

ATL ACTIVITY

Models, such as predator–prey relationships, can be used to demonstrate feeding dynamics in the context of these population interactions. For example:
- You can build a simple food web model (see page 124) to show the relationships between different species in an ecosystem. You can create a food web model by using Vensim: **www.vensim.com/documentation/tutorial.html**.
- You can also utilize mathematical models, such as the Lotka-Volterra model (see **https://mathworld.wolfram.com/Lotka-VolterraEquations.html**) to investigate the dynamics of population interactions. This model represents the interactions between predator and prey populations across time.

Competition is defined as a relationship in which two or more organisms compete for the same resource, affecting each other's growth, survival or reproduction. Consider the interaction of two bird species that feed on the same sort of insect. The birds struggle for food supplies, which might impact population growth and distribution. This interaction has ecological implications since it has the potential to alter the makeup and diversity of the community. An example of evolution of specialized behaviours would be the capacity of birds to locate and capture their favourite prey. Evolutionary changes include the evolution of features that increase or decrease competitive capacity.

Biotic factor – an ecological property associated with the living part of an environment, such as competition or predation.

2.1.10 Carrying capacity

The carrying capacity of an ecosystem refers to the maximum population number that it can support based on its available resources. These resources can be biotic or abiotic elements that influence population growth and survival. Abiotic elements that can limit carrying capacity include water availability, temperature and sunlight. A desert ecosystem, for example, may have limited water resources that can only support a certain number of individuals of a specific species.

Biotic factors that can limit carrying capacity include predation, sickness, and competition for resources such as food and space. A population of herbivorous animals, for example, may have a carrying capacity dictated by the availability of food, but a predator population may have a carrying capacity determined by the number of prey species available.

It is crucial to note that carrying capacity is not a set value and can change when the ecosystem changes, such as due to resource variations, natural disasters, and human activities such as deforestation and pollution.

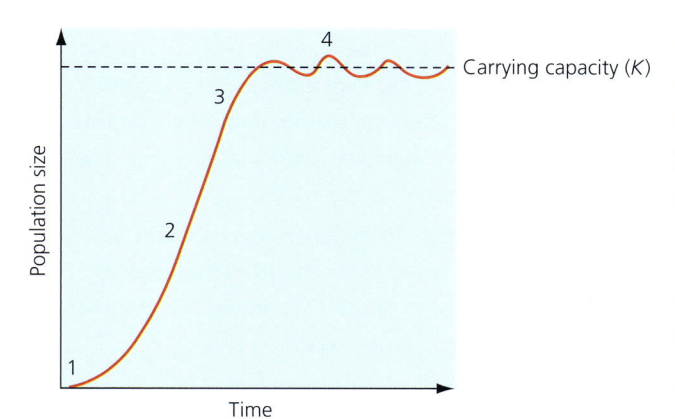

■ **Figure 2.5** Population growth curve controlled by limiting factors

2.1.11 Population size is regulated by density-dependent factors and negative feedback mechanisms

Population size is an important feature of an ecosystem since it determines a species' impact on its environment. Let us look at the factors that influence population size.

Population density-dependent factors tend to be biotic factors that influence population growth rates and are affected by population size. Competition for resources, predation and disease are examples of density-dependent issues. These forces become more intense as population density grows, limiting population growth and restricting population size. These density-dependent factors may cause a logistic pattern of growth as shown in Figure 2.6, where you can see the population size becomes constant at the carrying capacity.

Although density-dependent factors have a considerable impact on population size, they tend to restrict the population at around the carrying capacity. In dense populations, there is an increased danger of predation and infection spread, in addition to competition for limited resources. These are examples of negative feedback, which helps a population return to equilibrium.

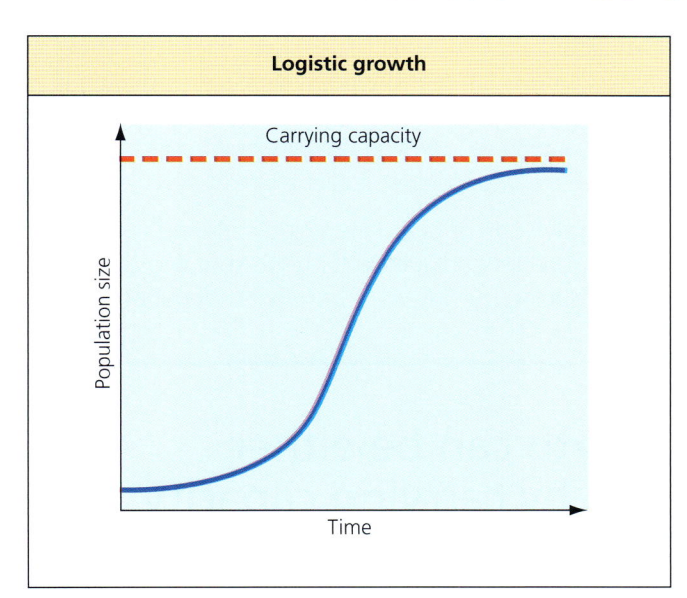

■ **Figure 2.6** There are environmental limits to population growth

2.1 Individuals and populations, communities and ecosystems

◆ **Negative feedback** – a regulatory mechanism in population management where an increase in a particular factor, such as population density, leads to a response that reduces or counteracts that increase, thereby maintaining balance in the system.

Negative feedback mechanisms are processes that minimize or counteract system changes. These mechanisms govern population size and keep it around the carrying capacity in the setting of population dynamics.

The greater risk of predation in dense groups is an example of negative feedback in population management. The likelihood of a predator discovering and devouring prey increases with population size. This increased predation pressure reduces population size, bringing it back to a level where predation pressure is lower.

Pathogen transfer in dense populations is another example of negative feedback in population regulation. Pathogens are more likely to transfer between individuals when population density increases, leading to disease outbreaks. These disease outbreaks have the potential to diminish population size, bringing it to a level where disease propagation is less likely.

Abiotic factors that affect population growth rates but are not affected by population size are known as density-independent factors. Natural disasters, temperature and precipitation are examples of density-independent factors. These factors have a considerable impact on population size, but they do not control it in terms of carrying capacity.

Tool 2: Technology

Computer modelling and simulations to generate data

Materials
- Computer or mobile device
- Internet connection
- Population Dynamics Simulator: https://phet.colorado.edu/en/contributions/view/6715

Steps
1. Open the Population Dynamics Simulator.
2. Explore the simulator and identify the factors that regulate population size, including density-dependent and density-independent factors.
3. Simulate different scenarios, such as changing the carrying capacity or introducing a predator to the ecosystem, and observe the impact on population size.

This simulation allows you to use your knowledge of population dynamics and the elements that influence population size. You can obtain a better understanding of the topic by modelling various scenarios and seeing the influence of negative feedback mechanisms on population size.

2.1.12 Population growth can be either exponential or limited by carrying capacity

Population growth is a fundamental part of ecology, and knowing population growth patterns is critical to understanding ecosystem dynamics. We shall investigate the many patterns of population increase and the elements that impact them.

When a population has unlimited resources and no limiting factors, exponential expansion happens. In this scenario, the population grows at an increasing rate throughout time.
A J-shaped curve is the growth curve of an exponentially rising population, exhibiting a rapid and continuous increase in the number of individuals. However, exponential development is not sustainable in the long term since resources become scarce.

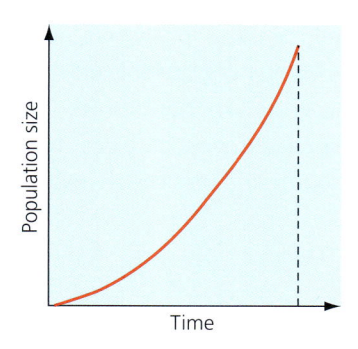

Figure 2.7 A J-shaped curve is the growth curve of an exponentially rising population, exhibiting a rapid and continuous increase in the number of individuals

When density-dependent limiting factors, such as resource competition or increased predation, come into play, population growth is restrained, and the growth curve takes the shape of an S. The S-curve represents a time of rapid growth followed by a period of slowing growth as the population approaches the carrying capacity of the ecosystem. The carrying capacity is the maximum number of organisms that the ecosystem can sustainably support.

The reindeer population on St. Matthew Island in Alaska, northwest USA, is an example of an S-curve population growth trend. A small population of reindeer was transplanted to the island in the early 1940s, and with an abundance of nutrients and no natural predators, this population surged tremendously. However, as the reindeer population grew, the island's scarce resources became depleted. This increased competition for food, resulting in a slower pace of population expansion.

The population eventually reached a point where it surpassed the island's carrying capacity, resulting in a rapid fall in the reindeer population, a phenomenon described as a 'boom and bust' cycle. When the population surpasses the carrying capacity, scarcity of resources and other limiting factors induce a rapid fall in population size, which typically results in a population crash or collapse. This cycle is the result of surpassing the environment's sustainable boundaries, and it serves as a warning example of the need to know carrying capacity and population dynamics.

It is critical to examine both the quantity of individuals and the rates of change through time when evaluating population growth curves. Scientists can acquire insights into population dynamics, the impact of limiting constraints, and the sustainability of ecosystems by researching population growth trends.

2.1.13 Limiting factors on the growth of human populations

Over the past centuries, human populations have seen remarkable growth and expansion, accompanied by significant technological developments and environmental shifts. As a result, human activities have had a significant impact on ecosystems and resource availability. The removal of limiting forces that formerly regulated population increase, such as natural predators and resource competition, has had a number of ramifications for ecosystem sustainability.

Natural predator extinction, whether through deliberate eradication or habitat loss, has upset ecological balance in numerous areas. Without predators to offer checks and balances, many animal populations have grown exponentially, resulting in greater competition for resources and the depletion of food sources.

This imbalance can lead to the decrease or extinction of other species, affecting an ecosystem's total biodiversity and ecological stability.

Humans have been able to circumvent some of the historical constraints on resource availability thanks to technological advancements. Our species has been able to boost resource production and extraction on a global scale thanks to advancements in agriculture, transportation and industry. However, these gains have frequently come at the expense of the environment. Increased use of fossil fuels, deforestation, pollution and habitat destruction have all contributed to environmental degradation, significantly compromising ecosystems' ability to support biodiversity and deliver resources in the long term.

The degradation of the environment, together with the removal of limiting forces, offers enormous difficulties to ecosystem sustainability. Overexploitation of resources, habitat degradation and pollution can all upset ecosystems' delicate balance, resulting in biodiversity loss, diminished resilience and even ecosystem collapse. These changes have repercussions throughout the entire web of life, affecting not only other species but also the overall functioning and services offered by ecosystems.

2.1.14 Carrying capacity cannot be easily accessed for human populations

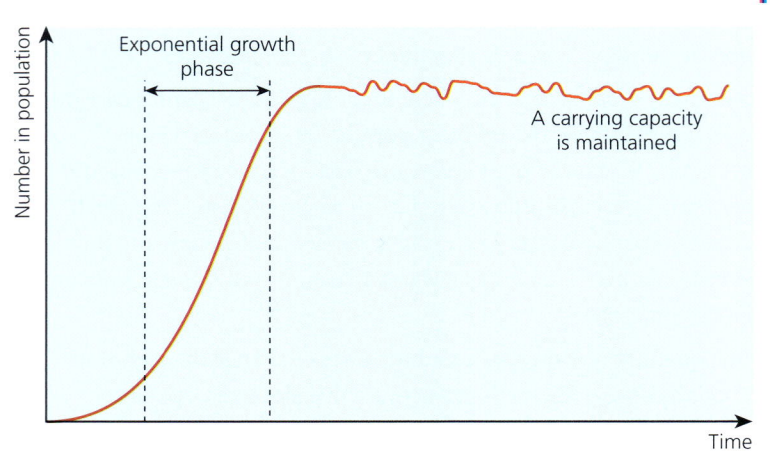

Figure 2.8 Population growth and carrying capacity

Other species' populations tend to establish equilibrium within their environments. However, human populations have demonstrated an extraordinary ability to extend and alter their niche through technological advances and changes in consumption patterns. Humans have built complex resource-extraction techniques, infrastructure and systems, increasing their ability to exploit resources beyond what would be naturally achievable. Because of this mobility and adaptability, human populations have been able to acquire resources from a variety of locations, minimizing the constraints imposed by a given ecosystem's carrying capacity.

Furthermore, technological improvements are fuelling the extension of the human niche, enabling the exploitation of previously inaccessible resources and the development of new ways to meet human requirements. Changes in consumption habits, population expansion and rising resource demand have made carrying capacity estimations for human populations difficult. This is due to the continually changing human habitat, as well as different degrees of resource use and environmental impact. Because of these difficulties, estimates of human population carrying capacity are frequently challenged.

Evaluating the ecological impact of human activities is one technique to calculate the current carrying capacity. An ecological footprint is the amount of land and resources needed to sustain a population's consumption, while also absorbing its emissions. It estimates the environmental impact of human activities and can be used to determine a population's sustainability within a specific ecosystem.

The ecological footprint of a given population is calculated by taking into account elements such as energy consumption, food production, water usage, waste generation and carbon emissions. By examining these criteria, researchers can estimate the ecological resources required to support the existing human population and the accompanying environmental implications.

It is crucial to remember, however, that assessing carrying capacity based on ecological footprint has limits. The calculation is influenced by a number of assumptions and models, and it may not fully describe the complexities of human–environment interactions. Furthermore, carrying capacity is not a constant value as it changes over time as technology, social institutions and environmental factors change.

The concept of human population carrying capacity raises serious concerns regarding sustainability and the need for appropriate resource management. As human populations continue to rise and consume resources at unprecedented rates, it is critical to recognize the planet's ecological constraints and aim for sustainable practices that preserve ecosystem integrity and assure the well-being of current and future generations.

While determining carrying capacity for human populations is difficult due to humans' diverse and changing biological niche, the concept remains essential in the context of sustainability. Understanding the ecological footprint and the environmental repercussions of human activities can provide insights into the constraints and challenges of supporting human populations within ecosystems.

HL.b.9: Ecological economics

In ecological economics, the economy is seen as a subsystem of Earth's larger biosphere, highlighting the interconnectedness of human systems with ecological systems. The concept of carrying capacity, which refers to the maximum population that a certain environment can sustainably support, falls within the realm of ecological economics. Assessing carrying capacity is a complex task, as it involves understanding the balance between resource availability and consumption, the impact of human activities on ecosystems and the ecological limits of a given environment.

Ecological economics recognizes the challenges and complexities inherent in determining the carrying capacity for human populations. Ecological economists emphasize the need to consider the sustainable use of natural resources to ensure that human populations can thrive within the planet's ecological boundaries. By acknowledging the difficulty in assessing carrying capacity, we can understand the importance of ecological economics' holistic approach to addressing the intricate relationship between human societies and the environment.

2.1.15 Population abundance can be estimated using random sampling, systematic sampling or transect sampling

The sampling method used is determined by a variety of criteria and considerations, such as the characteristics of the population, the size of the study region and the research objectives.
Let's look at each of the following sample strategies and why they were chosen:

1. Random sampling: This is the process of randomly picking individuals from a population. This strategy ensures that each organism has an equal chance of being included in the sample, resulting in a representative sample of the population. When the population is homogeneous and researchers are evenly distributed over the study region, random sampling is usually used. It is especially effective for analysing huge populations or species with a very even distribution of organisms. Random sampling reduces bias by decreasing the chances of generalizing a population.

   Random quadrat sampling can be used to estimate population size for non-mobile organisms. Percentage cover is an estimate of the area covered by the plant or animal in a specific frame size (quadrat). The number of occurrences divided by the number of possible occurrences equals the percentage frequency. For instance, if a plant appears in 5 of 100 squares in a grid quadrat, the percentage frequency is 5 per cent.

2. **Systematic sampling:** This entails picking individuals at predefined intervals from a starting location. When the population demonstrates some degree of regularity or pattern, this strategy is appropriate. For example, if a plant or animal is clumped or aggregated, systematic sampling can give a representative sample by picking organisms from each cluster. When there is a known pattern in the population distribution, systematic sampling is more efficient than random sampling since it ensures coverage of diverse locations or clusters.

3. **Transect sampling:** In this method, data is collected along a pre-determined line or transect in the study region. This method is appropriate for analysing transition in the population distribution. To capture differences in population abundance or species composition, transects can be deployed across habitats or environmental gradients. In ecological research, transect sampling is frequently employed to measure variations in population density or species richness along an environmental gradient, such as from shoreline to woodland, or from low to high elevations.

2.1.16 Random quadrat sampling can be used to estimate population size for non-mobile organisms

The sampling method used needs to be determined by the specific research objectives as well as the characteristics of the population and study location.

- Random sampling offers a representative and unbiased measure of population abundance, making it appropriate for homogeneous populations or large-scale investigations.
- Systematic sampling is helpful when there is a regular pattern or clustering in the population distribution, since it allows for efficient coverage of diverse areas.
- Transect sampling can be used to analyse population changes along environmental gradients or along linear features.

When choosing a sampling method, researchers should carefully evaluate the location and timescales of the study, the resources available, and how precisely they would like to estimate the population. To provide accurate and trustworthy estimates of population abundance, appropriate sampling techniques must be used, which can contribute to a better knowledge of ecological dynamics and influence conservation and management initiatives.

Tool 1: Experimental techniques

Quadrat sampling for estimating abundance, population density, percentage coverage and percentage frequency: collecting and processing data

In order to apply the skills of quadrat sampling for estimating abundance, population density, percentage cover and percentage frequency for non-mobile organisms, you can conduct the following activity.

■ **Figure 2.9** Students recording species within a quadrat along a transect of a rocky shore

Materials
1. Quadrats (square frames of known size, e.g. 1 m × 1 m, made of PVC pipes or ropes)
2. Field notebooks and writing instruments
3. Measuring tape or ruler
4. Species identification guides (books, websites or mobile apps)
5. Data sheets or worksheets for recording observations
6. Optional: Cameras for visual documentation

Field setup
1. Select a study area, such as a meadow, forest or shoreline.
2. Lay out transects across the study area, ensuring they cover a representative range of habitats or environmental gradients.
3. Mark the starting points of each transect and measure equal intervals along the transect line.
4. Place the quadrat at each interval along the transect, ensuring consistent positioning and avoiding overlapping.

Sampling process
1. At each quadrat, record the species present, their abundance and other relevant observations (for example, percentage cover).
2. Use species identification guides to correctly identify the organisms.
3. Estimate abundance by counting the number of individuals within the quadrat.
4. Estimate population density by dividing the abundance by the quadrat area (for example, number of individuals per square metre).
5. Estimate percentage cover by visually estimating the proportion of the quadrat area covered by each species.
6. Estimate percentage frequency by calculating the ratio of the number of quadrats where a species occurs to the total number of quadrats sampled.
7. Repeat the process at multiple quadrats along each transect, ensuring a representative sampling effort.

Data analysis
1. Compile the data collected from each quadrat, including species, abundance, population density, percentage cover and percentage frequency.
2. Calculate means, averages and other relevant statistical measures to summarize the data.
3. Analyse the data to identify patterns, relationships and changes in abundance, population density, percentage cover and percentage frequency along the transects.
4. Discuss the findings and draw conclusions about the distribution and ecological characteristics of the non-mobile organisms in the study area.

Reflection and discussion
1. Reflect on the limitations and potential sources of error in the quadrat sampling method.
2. Discuss the significance of abundance, population density, percentage cover and percentage frequency in understanding ecosystem dynamics and monitoring changes over time.
3. Think critically about the implications of your findings and how they relate to ecological concepts and conservation efforts.

2.1.17 Capture–mark–release–recapture and the Lincoln index can be used to estimate population size for mobile organisms

Population size estimate = (M × N) / R

Where M is the number of individuals caught and marked initially, N is the total number of individuals recaptured, and R is the number of marked individuals recaptured.

To apply the skills of capture–mark–release–recapture and use the Lincoln index for population count of mobile organisms, you can conduct the following activity.

> **Tool 1: Experimental techniques**
>
> ### Estimating population size using capture–mark–release–recapture and the Lincoln index
>
> *Materials*
> 1. Animal capture devices appropriate for the target species (for example, Sherman traps, mist nets, pitfall traps). Please check IB's animal policy with your teacher and **do not** harm any organisms. Ensure that all organisms are returned to nature safe and well.
> 2. Marking materials, such as uniquely numbered tags, bands or coloured dyes
> 3. Data sheets or worksheets for recording capture and recapture data
> 4. Field notebooks and writing instruments
> 5. Calculators or devices for performing mathematical calculations
>
> *Field setup*
> 1. Identify the target species and select an appropriate trapping method.
> 2. Set up a trapping area or multiple trap locations within the study area.
> 3. Follow ethical guidelines, and ensure the well-being and safety of the captured organisms.
>
> *Capture–mark–release*
> 1. Capture a sample of individuals from the target population using the chosen trapping method.
> 2. Safely and carefully handle the captured individuals and mark them with unique identifiers (tags, bands or dyes).
> 3. Record the initial number of captured individuals (M) and mark each individual accordingly.
> 4. Release the marked individuals back into their natural habitat.
>
> *Recapture*
> 1. Wait an appropriate amount of time for the marked individuals to disperse and mix with the rest of the population.
> 2. Recapture a sample of individuals using the same trapping method as before.
> 3. Record the total number of individuals recaptured (N) and determine the number of recaptured individuals that are marked (R).
>
> *Data analysis*
> 1. Calculate the estimated population size using the Lincoln index formula:
>
>   Population size estimate = (M × N) / R
>
> 2. Perform the necessary calculations using the captured data and the formula.

3 Discuss the assumptions made when using the Lincoln index, such as the assumptions of random mixing, no birth or death, no emigration or immigration, and equal catchability of individuals.
4 Interpret the population size estimate in the context of the study area and the limitations of the method.

Reflection and discussion
1 Reflect on the accuracy and reliability of the population size estimate obtained through capture–mark–release–recapture and the Lincoln index.
2 Discuss the potential sources of error and bias in the method, such as variations in trap efficiency, marking visibility and assumptions of the model.
3 Explore alternative methods for population estimation and compare their advantages and limitations.

2.1.18 A community is a collection of interacting populations within an ecosystem

A community is a key ecological concept that refers to an ecosystem's interacting populations. It includes a variety of species and their interactions, including competition, predation, mutualism and others. Understanding the concept of community in a small environment allows us to investigate the complex links that exist between species and their ecological dynamics.

A community in a local ecosystem is made up of populations of organisms that coexist and interact within a specified geographic area. This could comprise a variety of plant, animal, fungus and microbe species that live in the ecosystem. Each population within the community has a distinct role in the overall functioning of the ecosystem.

To better understand the concept of community in a particular ecosystem, the following factors must be considered.

■ Species interactions

Interactions between populations shape communities. Competition, predation, herbivory, parasitism, mutualism and commensalism (a connection between two organisms in which one benefits, and the other derives neither benefit nor harm) are some of the numerous sorts of interactions. These interactions are critical in controlling population levels, resource usage and overall ecosystem dynamics.

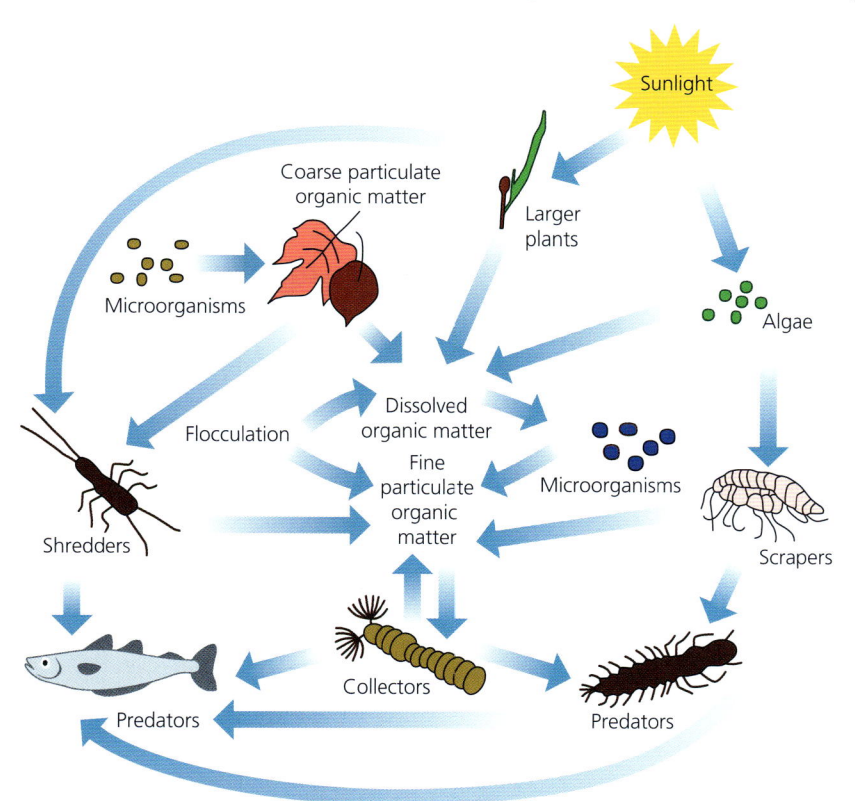

■ **Figure 2.10** Interactions between different species

2.1 Individuals and populations, communities and ecosystems

Energy pyramid

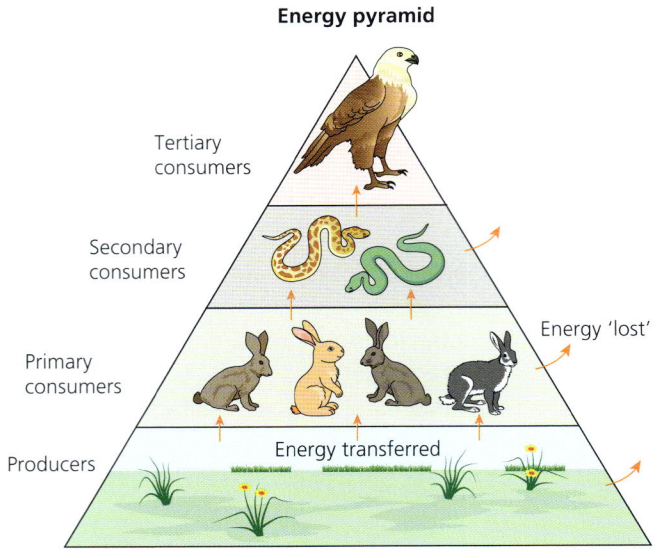

■ Figure 2.11 Trophic levels according to feeding connections

■ Trophic connections

Species are grouped into **trophic** tiers within a population based on their feeding connections. Plants, for example, form the base of the food chain by turning sunlight into energy via photosynthesis. Herbivores eat plants, and carnivores and omnivores eat other animals. Decomposers are essential for breaking down organic debris, recycling nutrients and completing the cycle.

◆ **Trophic** – refers to the different levels in a food chain or web, indicating an organism's position and role in the transfer of energy and nutrients within an ecosystem (see Chapter 2.2).

■ Biodiversity

Different communities have different levels of biodiversity, which refers to the number of species present in an ecosystem. Ecosystem stability, resilience and functionality are all dependent on biodiversity. A diversified community is better prepared to deal with disruptions and environmental changes.

■ Succession

Ecological succession is the process of steady and predictable changes in species composition over time that can occur in communities. Primary succession occurs in places devoid of life, such as newly formed volcanic islands, when species arrive for the first time, whereas secondary succession happens in areas where existing species have been removed by a disturbance and other species move in. Each stage of succession supports a unique set of species, resulting in a dynamic and ever-changing community structure.

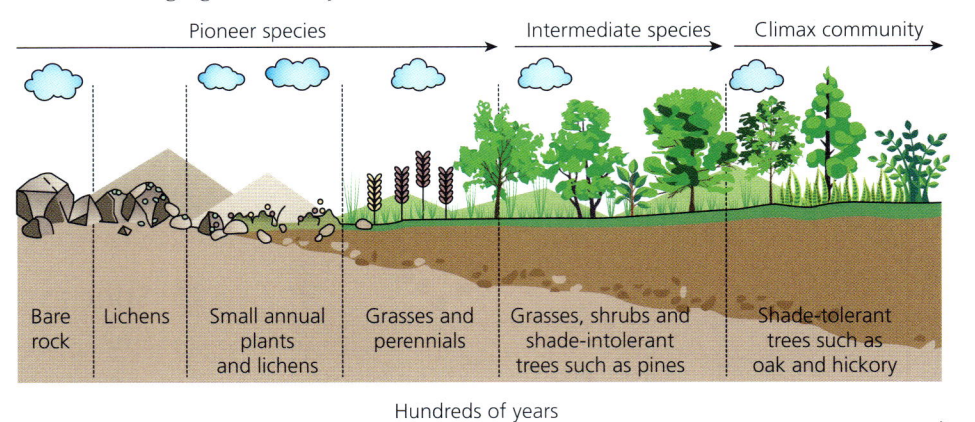

■ Figure 2.12 Primary succession

By studying and understanding the concept of community in a local ecosystem, we gain insights into the complex web of interactions and dependencies that shape ecological systems. This knowledge enables us to appreciate the interconnectedness of species, their roles in ecosystem functioning, and the importance of conservation efforts to maintain healthy and resilient communities.

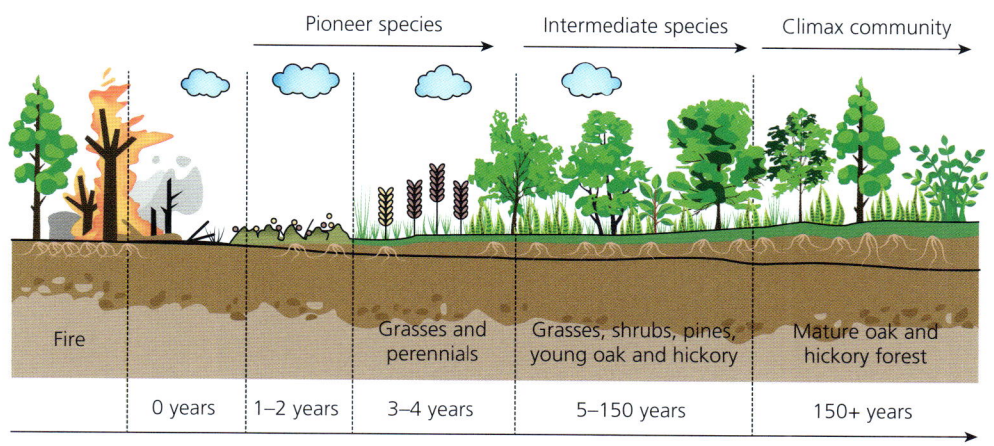

■ **Figure 2.13** Secondary succession

2.1.19 A habitat is the location in which a community, species, population or organism lives

The precise area or environment in which a community, species, population or individual organism lives is referred to as its habitat. It includes the geographical and physical features required for survival, growth and reproduction in the ecosystem. Understanding the notion of habitat in a given ecosystem provides insights into individual species' distinct requirements and their relationship to their environment.

The concept of habitat can be studied in a local ecosystem by addressing the following aspects:

■ Geographical location

A species' habitat encompasses its geographic range or distribution. Latitude, longitude, altitude and specific geographic features can all be used to characterize this. Different species have diverse geographic distributions, ranging from local to worldwide. Understanding the geographical location of a species' habitat allows us to grasp the larger ecological context in which it occurs.

As an example, the African elephant (*Loxodonta africana*) has a wide geographical distribution, occurring in various countries across the African continent. It can be found in habitats ranging from savannas and grasslands in eastern and southern Africa, to dense forests in central and western Africa. Understanding their geographic range helps conservationists implement measures to protect these majestic animals across their vast habitats.

■ Physical characteristics

Habitat also comprises the physical traits and conditions required for the survival of a species. Temperature, sunshine availability, moisture levels, soil composition, topography, and the availability of resources such as food and shelter can all be considered. Each species has unique adaptations and tolerances to various physical circumstances, which influence its habitat requirements.

The saguaro cactus (*Carnegiea gigantea*), as shown in Figure 2.14, is well-adapted to the arid and hot conditions of the Sonoran Desert in North America. It has thick, water-storing stems and shallow but extensive root systems that allow it to survive in the harsh desert environment. Additionally, its ribbed structure helps it expand and contract as it stores and uses water, allowing it to thrive in the specific physical conditions of its habitat.

■ **Figure 2.14** Saguaro cactus (*Carnegiea gigantea*)

Ecosystem type

An essential component of habitat is the type of ecosystem required to provide the environmental conditions required for living. Forests, grasslands, marshes, deserts and aquatic environments all present a unique combination of abiotic and biotic conditions that sustain various groups of organisms. Each ecosystem type has its own set of resources, relationships and ecological processes that determine the habitat's structure and function.

The Bengal tiger (*Panthera tigris tigris*) is adapted to live in a diverse range of ecosystems, including tropical and subtropical rainforests, grasslands and mangrove swamps in the Indian subcontinent. Each of these ecosystems provides unique resources and conditions that support the tiger's survival, ranging from dense cover in the rainforests to open hunting grounds in the grasslands.

Relationships and interactions

Habitat determines not just the physical circumstances but also the interactions and relationships between species. Species within a habitat may compete for resources, create predator–prey partnerships, form mutualistic connections, or engage in other types of ecological interactions. These interactions shape the habitat's structure and dynamics, as well as its distribution and abundance.

The relationship between the clownfish (*Amphiprioninae*) and the sea anemone (*Actiniaria*) is a classic example of mutualism. The clownfish is protected from predators by the stinging cells of the sea anemone, which it inhabits. In return, the clownfish keeps the sea anemone clean and provides it with food scraps. This interdependence influences the distribution and abundance of both species, as they are found together in specific reef habitats where these interactions occur.

■ **Figure 2.15** Interaction between sea anemones and clownfish (orange/black)

2.1.20 Ecosystems are open systems in which both energy and matter can enter and exit

An ecosystem is a complex and dynamic system made up of a population of living organisms that interact with one another and with their physical surroundings.

In a local ecosystem, you can explore this concept by considering the following aspects:

■ Organismal community

An ecosystem is a population of species that interact with one another, such as plants, animals, microbes and fungus. These living organisms can create a variety of connections, including predator–prey interactions, resource rivalry, mutualistic associations and symbiotic relationships. The general structure and dynamics of the ecosystem are influenced by the makeup and variety of the community.

■ Physical setting

Abiotic elements such as sunshine, temperature, precipitation, soil composition, air quality and water availability are all part of the physical environment. These factors influence the distribution and abundance of species through shaping habitat conditions within the ecosystem. Physical factors such as rocks, water bodies and topography also contribute to the definition of the ecosystem's traits.

■ Flow of energy

Energy is a critical component of any ecosystem. It enters the ecosystem largely through sunshine and is transformed into chemical energy by photosynthetic organisms such as plants or algae via the photosynthesis process. Through feeding interactions, this energy is subsequently passed from one organism to another. Energy passes through many trophic levels, beginning with producers (plants), then primary consumers (herbivores), secondary consumers (carnivores) and so on. The flow of energy is critical to the ecosystem's functioning and productivity.

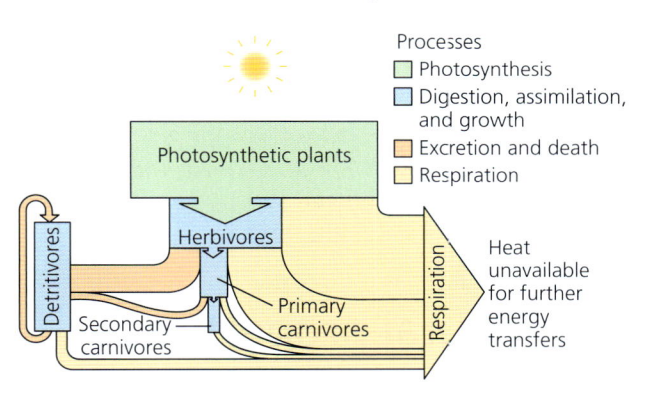

■ **Figure 2.16** Flow of energy in an ecosystem

■ Nutrient cycling

Ecosystems involve the cycling of matter and nutrients, in addition to energy flow. Carbon, nitrogen, phosphorus and water are important nutrients for organism growth and survival. The intake, use and recycling of these elements within the ecosystem is referred to as the nutrient cycle. Bacteria and fungi, for example, serve an important role in breaking down organic matter and returning nutrients to the soil or water, where they can be reused by other species.

2.1 Individuals and populations, communities and ecosystems

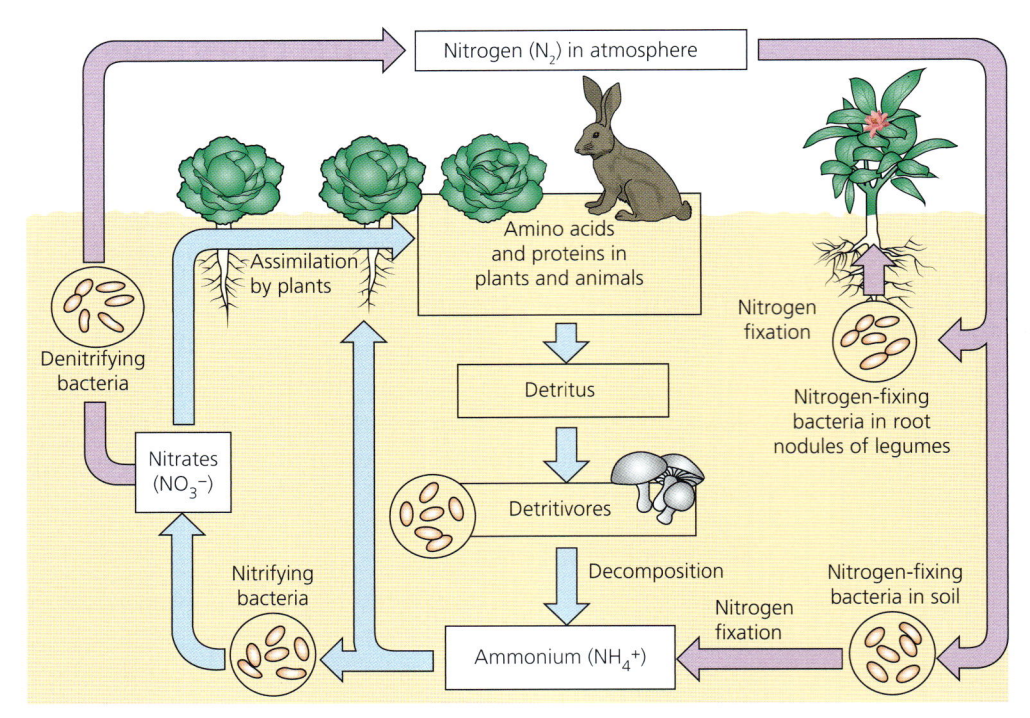

■ **Figure 2.17** The nitrogen cycle (adapted from Spiro and Stigliani, 2003)

Interactions and feedback loops

Complex interactions and feedback loops manage and maintain the functioning of ecosystems. Predator–prey interactions, for example, can influence population sizes, whereas plant–pollinator interactions affect plant reproduction. Feedback loops, such as the one between plants and soil nutrient availability, can have a domino impact on the entire ecosystem. These connections and feedback mechanisms contribute to the ecosystem's stability and resilience.

2.1.21 Sustainability is a natural property of ecosystems

The concept of sustainability is central to the study of ecosystems. It means an ecosystem's ability to preserve its structure, function and productivity over time, while also maintaining the continuity of both biotic and abiotic components. Ecosystems are considered to have a natural property of sustainability since they can balance inputs and outputs in a steady state.

In the context of ecosystems:

- inputs are resources, energy and matter that enters the system
- outputs are waste, energy loss and matter that exits the system

A sustainable ecosystem maintains a balance between these inputs and outputs, ensuring that resources are used efficiently and waste products are recycled or dispersed properly.

Flow diagrams can be used to depict the balance of inputs and outputs in certain ecosystems. They depict the flow of energy and matter via various ecosystem components such as producers, consumers, decomposers and physical processes. These graphics depict the interconnection and resource cycling within the ecosystem.

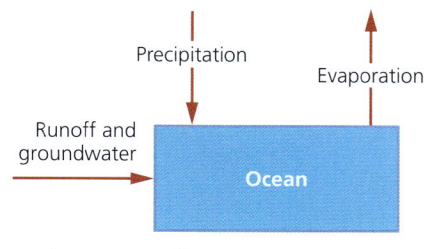

Figure 2.18 A flow diagram of an ocean system, with flows into and out from the storage (the ocean)

In a forest ecosystem, for example, solar energy serves as an input through photosynthesis, by which plants convert sunlight into chemical energy. A fraction of the solar energy captured through photosynthesis is transferred to herbivores when they consume plant material. Carnivores consume herbivores, and the energy moves through the trophic levels. Energy is eventually lost as heat or via metabolic processes and returned to the environment as an output.

Similarly, matter such as nutrients circulates within the ecosystem in a cycle. Plants take nutrients from the earth, which are then transported to herbivores when they devour plants, and then to predators. Decomposers decompose dead organic materials and return nutrients to the soil, completing the cycle.

The fact of some ecosystems, such as tropical rainforests, surviving for millions of years lends credence to the concept of sustainability. These ecosystems have reached a condition of equilibrium in which inputs and outputs have been balanced throughout lengthy periods of time. These ecosystems' complicated web of connections, nutrient cycling and energy movement has allowed them to preserve their integrity and function for extended periods of time. While sustainability is an inherent attribute of ecosystems, human activities have the potential to upset this balance. Deforestation, pollution, habitat destruction and climate change can all have a substantial impact on ecosystem sustainability. Understanding the principles of sustainability and ecosystem dynamics is critical for fostering responsible environmental management and guaranteeing the long-term existence of ecosystems.

2.1.22 Human activity can lead to tipping points in ecosystem stability

Human actions can have a substantial impact on ecosystem stability, sometimes resulting in tipping points that cause the old ecosystem to collapse and a new equilibrium to emerge. A tipping point is a point at which an ecosystem undergoes rapid and irreversible change.

The scenario of deforestation in the Amazon rainforest is one example of a human-induced tipping point. The Amazon rainforest is critical in generating water vapour via a process known as transpiration, in which plants release water vapour into the sky. This water vapour helps to produce clouds, cools the local environment and facilitates precipitation. It generates a self-sustaining cycle that aids in the preservation of the rainforest.

However, extensive deforestation in the Amazon region threatens to destabilize this delicate balance. When huge tracts of forest are destroyed, the transpiration process is considerably curtailed, resulting in less water vapour production. As a result, there is less moisture available in the atmosphere, which leads to a decrease in cloud formation and, as a result, a decrease in rainfall.

This decrease in rainfall has far-reaching implications, affecting not only the local ecology, but also regional and global climate patterns. Reduced precipitation can cause drier conditions, increasing the danger of wildfires and forest damage. It can also have an impact on water resource availability for local residents, agricultural operations and general biodiversity.

The loss of the Amazon rainforest's original equilibrium due to deforestation signals a tipping point. When the tipping point is reached, the ecosystem enters a new phase with different ecological dynamics and maybe reduced biodiversity. The dominance of various plant species suited to drier conditions, altered nutrient-cycling patterns, and changes in general ecosystem structure and function may all contribute to the creation of a new equilibrium.

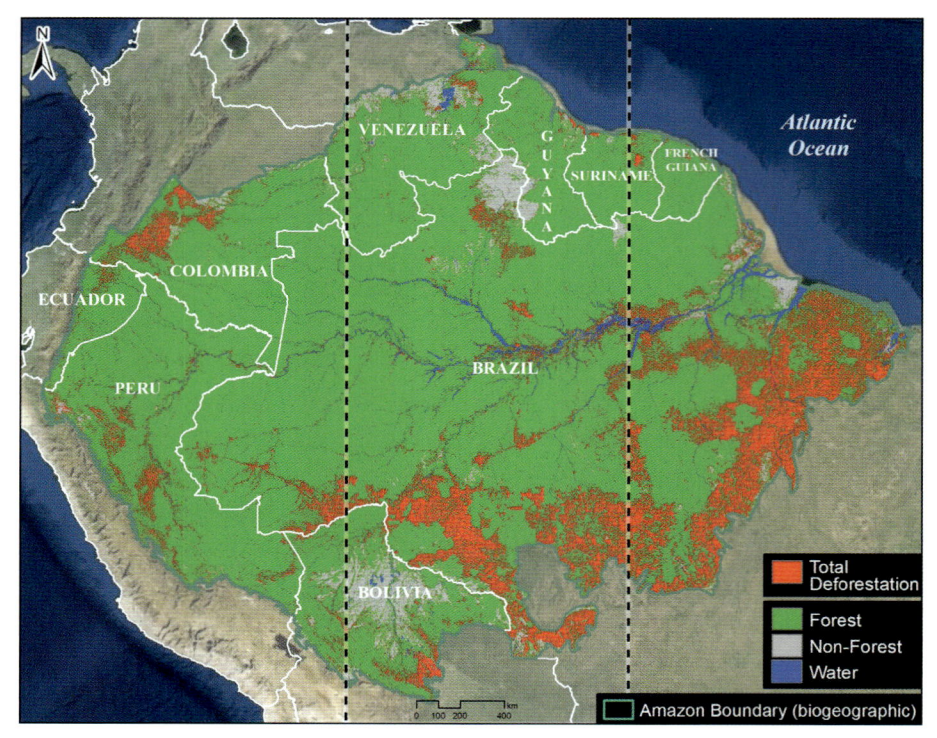

■ **Figure 2.19** Map showing total forest loss in the original Amazon forest biome. An estimated 13.2 per cent has been lost due to deforestation and other causes (data from Amazon Conservation Association and MAAP)

> ### Tool 2: Technology
>
> #### Use GIS maps and data
>
> You can use the GIS maps and data via **https://storymaps.arcgis.com/stories/b6e20235de3a4572bf74775603187784** to see how the changes in land use contribute to deforestation in the state of Rondônia in the southwestern Brazilian Amazon. You can utilize ArcGIS StoryMaps to explore and analyse various aspects of the Amazon rainforest, one of the most ecologically significant regions on the planet. This powerful tool allows you to create interactive and visually engaging narratives, integrating GIS maps and data to study the complex ecosystems and human–environment interactions in the Amazon forest.

The Amazon rainforest highlights how human activities can push ecosystems over their tipping points, resulting in negative feedback loops that magnify the impacts and make restoration challenging. It emphasizes ecological interdependence and fragility, stressing the importance of sustainable practices and conservation efforts to avert irreversible damage.

Understanding the concept of ecosystem stability tipping points is critical for recognizing the potential repercussions of human activities and for steering decision-making processes towards more sustainable practices. We may aim to maintain the stability and resilience of ecosystems by recognizing and reducing possible tipping points, protecting their essential services, and guaranteeing the well-being of both current and future generations.

2.1.23 Keystone species and their role in ecosystem sustainability

Keystone species are critical to the sustainability and balance of ecosystems. These species have a disproportionate impact on the structure and functioning of the community, and their extinction poses a considerable risk of ecosystem collapse.

Let's consider two instances of keystone species to better grasp their significance.

■ Mussels and purple sea stars

Purple sea stars (*Pisaster ochraceus*) serve as a keystone species on the North Pacific coast by controlling mussel numbers. Mussels are fast-growing, competitive filter-feeding organisms that, if left unchecked, can dominate the ecosystem. Purple sea stars, on the other hand, feed on mussels, limiting their population size and preventing them from dominating rocky intertidal areas.

Purple sea stars contribute to the biodiversity and balance of the intertidal community. They create space for other species to develop and prosper by controlling the numbers of mussels, such as barnacles, algae and numerous invertebrates. The wide collection of species contributes to the ecosystem's overall stability and productivity.

■ **Figure 2.20** Purple sea stars can completely change the structure of these communities

When purple sea stars are eliminated from the ecosystem, whether via natural causes or human activity, mussel populations can surge. This results in the extinction of other species, a decrease in biodiversity and changed ecological dynamics. Mussels may come to dominate the ecosystem, resulting in a loss of habitat complexity and a reduction in the availability of resources for other organisms.

■ Elephants and savannah grasslands

African elephants (*Loxodonta africana*) are important keystone species in African savannah grasslands. These huge herbivores eat shrubs and trees, assisting in the management of their growth and the preservation of the open grassland structure. Elephants prevent shrubs and trees from becoming dominant by nibbling on vegetation, fostering grass growth and preserving the savannah habitat.

Elephant grazing creates a mosaic of varying vegetation heights and densities, providing diversified habitats for a range of plant and animal species. Elephants are essential for the survival and reproduction of many species, including grazers, browsers and seed dispersers. Elephants have an impact on nutrient cycling and the allocation of water resources in the ecosystem.

■ **Figure 2.21** *Loxodonta africana* (African bush elephant)

However, when elephant populations are significantly decreased or removed, the savannah grasslands' balance can be thrown off. Without them, bushes and trees can proliferate, changing the grassland into a denser woodland or forest.

This shift in vegetation structure affects the availability of resources and the suitability of the habitat for other species, potentially resulting in a reduction in biodiversity and in ecosystem deterioration.

The examples of purple sea stars and elephants emphasize the critical function of keystone species in ensuring ecosystem sustainability. These organisms have a large impact on community structure, trophic interactions and ecosystem processes. Their abolition can have a domino effect, destabilizing the ecosystem and jeopardizing its overall health and resilience.

◆ **Planetary boundary** – a concept to define the environmental limits within which humans can survive and develop.

◆ **Biosphere integrity** – the overall health and functioning of ecosystems and the biodiversity they support. It is a fundamental component of the Planetary Boundaries model.

2.1.24 The Planetary Boundaries model indicates that changes to biosphere integrity have passed a critical threshold

The model of **planetary boundaries** provides a framework for understanding and measuring the limits of Earth's systems, including the biosphere. The idea of **biosphere integrity**, which refers to the overall health and functioning of ecosystems and the biodiversity they support, is a fundamental component of the model. Unfortunately, human activities have significantly disrupted ecosystems, resulting in a loss of biosphere integrity. As a result, a key threshold has been crossed, as evidenced by changes in species diversity and extinction rates.

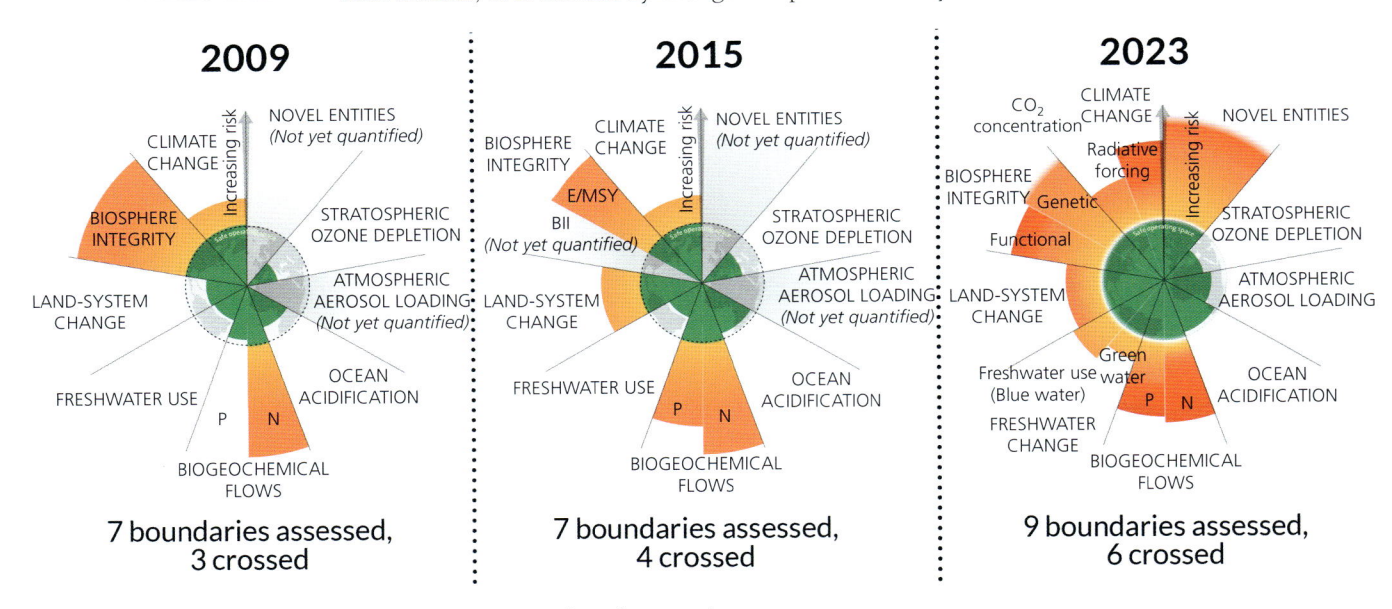

2009 — 7 boundaries assessed, 3 crossed

2015 — 7 boundaries assessed, 4 crossed

2023 — 9 boundaries assessed, 6 crossed

■ **Figure 2.22** The green-shaded polygon represents the safe operating space

The diversity of ecosystems and species is inextricably linked. Ecosystems rely on the presence of various species to function and be resilient. Each species has a distinct role within its environment, contributing to processes such as nutrient cycling, pollination and predation. The interdependence of species ensures the stability and sustainability of ecosystems.

Human actions, such as deforestation, habitat destruction, pollution and climate change, have created catastrophic disruptions to ecosystems globally. These disruptions alter the delicate balance of species interactions and ecological processes, resulting in loss of biodiversity and decreases in ecosystem health. The ability of ecosystems to provide important functions and support varied species reduces as they become increasingly fragmented, degraded or altered.

Examining extinction rates is one method of measuring the influence of human activity on biosphere integrity. Extinction is a natural process, although current rates of species extinction greatly surpass the background rate. This increased extinction rate is mostly due to human-caused factors such as habitat degradation, overexploitation, invasive species introduction and global environmental changes.

Species are undergoing rapid extinction. This is a worrying trend that could impact how ecosystems function, altering ecosystems and jeopardizing their ability to operate properly. The loss of species diversity not only affects our planet's beauty and richness, but it also jeopardizes the stability of ecosystems and the services they provide, such as clean air, fresh water and fertile soil.

To tackle this issue, it is critical to recognize the interconnectivity of ecosystems and species diversity, as well as to take steps to restore and safeguard biosphere integrity. Conservation initiatives should prioritize habitat preservation, habitat-fragmentation reduction, sustainable land use practices and limiting the effects of climate change.

Furthermore, raising knowledge about the value of biodiversity and its role in sustaining ecosystem health is critical for cultivating a sense of stewardship for the Earth and supporting sustainable practices at the individual and societal levels.

We can work towards a more sustainable future by recognizing the importance of biosphere integrity and taking actions to restore and maintain ecosystems. This entails adopting conservation practices, supporting laws that prioritize environmental protection, and cultivating a deep appreciation for the Earth's interrelated web of life. We can work together to reverse the trend of biosphere degradation and maintain a healthy and vibrant planet for future generations.

2.1.25 To avoid critical tipping points, loss of biosphere integrity needs to be reversed

The integrity of ecosystems is critical to the general health and functioning of the Earth's natural systems. We can ensure the continuous provision of niche requirements for species viability by protecting ecosystems. It is therefore vital for humans to carry out activities that attempt to reverse ecosystem damage and preserve species variety in order to avoid hitting critical tipping points and to minimize the loss of biosphere integrity.

Ecosystems are intricate networks of living organisms, their environments and their interactions. They provide a vast range of services that support life on Earth, such as climate management, nutrient cycling, air and water purification, and food and resource availability. When ecosystems are disrupted or damaged, the delicate balance of these services is jeopardized, resulting in negative effects for the environment as well as human well-being.

To reverse the loss of biosphere integrity, it is critical to prioritize ecosystem protection and restoration. This includes the preservation of natural habitats such as forests, wetlands, grasslands and coral reefs, which are home to a varied range of species and provide essential ecosystem services. By preserving intact ecosystems, we assure the availability of suitable habitats and resources to support the continuous survival and reproduction of species.

Ecosystem protection also entails implementing sustainable land- and resource-management practices. This includes preventing habitat damage, pollution and waste, supporting ethical fishing and forestry practices, and implementing sustainable agriculture approaches.

We can reduce the negative consequences on biodiversity and ecosystem health by managing human activities in a way that respects the boundaries and resilience of ecosystems.

Furthermore, maintaining ecosystem integrity necessitates the establishment and implementation of protected areas and conservation legislation. Protected places provide havens for species, allowing them to survive and fulfil their ecological responsibilities in the absence of major human interference. These locations are also useful for researching natural processes and comprehending the dynamics of healthy ecosystems. Conservation policies and regulations establish a legislative framework for long-term resource management and govern decision-making processes in order to prioritize environmental conservation.

It is critical to address the underlying drivers of environmental degradation, such as unsustainable consumption patterns, population increase and climate change, in addition to safeguarding ecosystems. We can reduce the pressures on ecosystems and contribute to their long-term integrity by supporting sustainable lifestyles, minimizing our ecological footprint, and shifting to renewable and low-impact energy sources.

Education and public awareness are critical in reversing the loss of biosphere integrity. We can motivate people around us to become environmental stewards and take action to protect and restore ecosystems in their local communities and beyond by promoting knowledge of the value of ecosystems and species variety.

To summarize, reversing the loss of biosphere integrity necessitates coordinated efforts to protect and restore ecosystems. We can alleviate the negative impacts on biodiversity and strive towards preserving healthy and resilient ecosystems by protecting habitats, implementing sustainable practices, establishing protected areas and tackling the underlying drivers of ecosystem degradation. The integrity of the biosphere is critical not only for the survival of numerous species, but also for the well-being and sustainability of human societies. It is humanity's joint responsibility to act as planet caretakers and ensure a healthy environment for future generations.

(HL) 2.1.26 There are advantages to using a method of classification that illustrates evolutionary relationships in a clade

Cladistic classification is a system for organizing and classifying organisms based on their evolutionary relationships within a clade. A clade is a group of organisms that share a common ancestor. A cladistic method to classification has various advantages for understanding and portraying evolutionary relationships. Here are some of the primary benefits:

1. Reflects phylogenetic evolution: Cladistic classification is based on the idea of phylogeny, which relates to an organism's evolutionary history. Cladistics provides a framework for reflecting the evolutionary relationships and branching patterns across distinct groupings of animals by organizing species into clades based on shared ancestry. This method helps us comprehend the historical relationships and the sequence of evolutionary events that have resulted in the diversity of life on Earth.
2. Objectivity and reproducibility: To determine evolutionary links, cladistics uses objective criteria, notably shared derived characteristics or traits known as synapomorphies. These characteristics are specific to a clade and are inherited from a common ancestor. In cladistic analysis, the application of precise and reproducible criteria helps to reduce subjective biases and allows for independent verification of results. This improves the classification system's scientific rigour and reliability.

3. Predictive power: Cladistic classification gives predictive power by allowing us to make educated guesses about organisms' characteristics and attributes based on their placement within a clade. We can deduce similar evolutionary features and anticipate the presence or absence of specific traits in related species by identifying shared derived traits.
4. Facilitates comparative studies: Because cladistic classification is hierarchical, it enables meaningful comparisons to be made across different taxonomic groups. Cladistics provides a framework for investigating the similarities and differences between species within and between clades by classifying organisms based on shared ancestry. This comparative method contributes to the study of evolutionary processes, to the identification of patterns of diversification, and to the investigation of the functional and ecological implications of diverse features.
5. Incorporates new information: Cladistics is a versatile approach that may add new information into the classification system, such as genetic data. Advances in molecular biology have revealed important information on the genetic relationships between organisms. Cladistic classification can capture a more comprehensive view of evolutionary links and deepen our understanding by merging genetic data with morphological and ecological information.

Finally, using a cladistic method in classification has significant advantages in terms of understanding and portraying evolutionary relationships. Cladistics reflects evolutionary history, gives objectivity and reproducibility, permits predictive inferences, promotes comparative investigations, and can include new knowledge by organizing organisms into clades based on shared ancestry. This method contributes to a better knowledge of the diversity and interdependence of life on Earth, as well as the evolutionary patterns and processes that have moulded the natural world.

2.1.27 There are difficulties in classifying organisms into a traditional taxonomic hierarchy

The classical taxonomic hierarchy, which includes kingdom, phylum, class, order, family, genus and species, has long been used to categorize and organize organisms. It is crucial to note, however, that this hierarchical framework may not always precisely reflect patterns of divergence and evolutionary relationships among animals. Some of the challenges and limits related to classifying organisms into the standard taxonomic ladder are as follows:

1. Evolutionary interactions: The traditional taxonomic order is based mostly on morphological qualities and traits shared by species. However, with the development of molecular biology and genetic-sequencing capabilities, it has become clear that physical similarities alone do not always neatly convey evolutionary relationships. Genetic investigations can reveal unexpected patterns of divergence and relatedness, posing difficulties in classifying organisms.
2. Horizontal gene transfer: Horizontal gene transfer, or the exchange of genetic material between species, complicates organism classification even further. This is a common occurrence in bacteria and other microbes. Because genetic material can be exchanged between distantly related species, the borders between taxonomic groups are blurred, calling into question the old taxonomic system's tight hierarchical structure.
3. Convergent evolution: Convergent evolution occurs when unrelated animals evolve similar features or characteristics due to similar environmental stresses. This can lead to organisms that are not closely related being grouped together based on shared features, even if these traits developed independently. Convergent evolution can make it difficult to appropriately classify species into their appropriate taxonomic divisions.
4. Species concepts: Depending on the species idea utilized, the definition and identification of species can differ. Multiple species ideas, such as biological, morphological and phylogenetic, might result in multiple classifications for the same group of organisms. This variation in species ideas can lead to taxonomic classification discrepancies and conflicts.

5. Taxonomic revisions: Taxonomic revisions and reclassifications are widespread as scientific understanding and methodology progress. New discoveries, such as new species and previously undiscovered evolutionary links, may necessitate changes to the taxonomic hierarchy. Taxonomic revisions can jeopardize the stability and coherence of the hierarchical system since new evidence may compel modifications in organism classification.

6. Hybridization and introgression: Interbreeding between distinct species, known as hybridization, can result in genetic mixing and the transfer of genetic material. As a result, complicated patterns of genetic linkages might emerge that do not neatly correspond with the standard taxonomic system. **Introgression**, the absorption of genetic material from one species into another, complicates even further the classification of organisms into taxonomic groups.

◆ **Introgression** – the transfer of genetic material from one species to the genetic pool of another one.

In summary, while the traditional taxonomic hierarchy has offered a valuable framework for classifying organisms, it is not without challenges and limits. The difficulties of effectively classifying organisms into the traditional hierarchy are exacerbated by evolutionary interactions, horizontal gene transfer, convergent evolution, species conceptions, taxonomic revisions, hybridization and introgression. Alternative techniques and methodologies to solve these issues and give more accurate representations of organismal interactions may be developed as our understanding of genetics and evolutionary processes advances.

TOK

The difficulties of classifying organisms into traditional taxonomic hierarchies raises questions about the limitations of scientific classification systems. We need to explore the challenges of fitting diverse and evolving organisms into rigid hierarchical categories and consider alternative methods, such as molecular techniques or cladistics, that might better represent evolutionary relationships.

2.1.28 The niche of a species can be defined as fundamental or realized

A species' niche refers to the precise collection of conditions and resources on which it relies for survival, development and reproduction within an ecosystem. A niche can be described as fundamental or realized. Let's look at instances of species to better comprehend the notions of fundamental and realized niches.

Fundamental niche

A species' core niche includes all of the environmental circumstances, resources and interactions in which it might theoretically exist and reproduce, in the absence of any limiting factors. Based on physiological adaptations and tolerance limits, it reflects the broadest ecological niche that a species could occupy.

REAL-WORLD EXAMPLE

Joseph Connell's study of barnacle species

In his classic barnacle study, Joseph Connell investigated the essential niches of two barnacle species, *Chthamalus* and *Balanus*. In comparison to *Balanus*, he discovered that *Chthamalus* had a lower tolerance for dehydration and a higher heat sensitivity. As a result, *Chthamalus*' basic niche was limited to the upper **intertidal zone**, but *Balanus* had a broader fundamental niche that extended to the lower intertidal zone (source: **www.ecologycenter.us/population-growth/evidence-for-competition-from-nature-connells-barnacles.html**).

◆ **Intertidal zone** – the area of the shoreline that lies between the high tide and low tide marks.

■ **Figure 2.23** Two barnacle species studied by Joseph Connell: Chtamalus and Balanus

Realized niche

A species' realized niche is the actual set of conditions and resources in which it exists and reproduces, taking into account biotic interactions and other limiting variables like competition, predation and resource availability. Because species confront restrictions and competition in natural environments, the realized niche is frequently narrower than the fundamental niche.

REAL-WORLD EXAMPLE

Brown and green anoles

The brown and green anole lizards from the Caribbean are well-known examples of realized niches. In locations where their ranges overlap, the brown anole prefers lower vegetation levels, while the green anole prefers higher levels. Competitive interactions and resource partitioning form each species' realized niche, resulting in its segregation within the vertical structure of the habitat.

■ **Figure 2.24** Florida's only native anole, the green anole (*Anolis carolinensis*), can change colour from green to brown and vice versa. Despite this ability, they are not true chameleons, which are an entirely different family of lizards. Although green anoles may be brown, brown anoles (*Anolis sagrei*) are never green

It is vital to highlight that the realized niche is the result of evolutionary adaptations and species interactions over time. Biotic interactions, such as competition, predation and mutualism, are critical in defining a species' realized niche and determining its ecological position in the community.

Understanding a species' fundamental and realized niches provides important insights into its ecological needs, adaptations and interactions within an ecosystem. It emphasizes the dynamic nature of species interactions as well as the impact of biotic and abiotic factors on species distributions and cohabitation.

2.1.29 Life cycles vary between species in reproductive behaviour and lifespan

In terms of reproductive behaviour and lifespan, the life cycles of various animals differ significantly. These variances are frequently linked to environmental conditions and successional stages. r-strategist species and K-strategist species are two common categories of species that are adapted to various ecological environments.

r-strategists

r-strategists are species with a reproductive strategy that includes quick colonization of new habitats and high reproductive output. They are also known as opportunistic species. r-strategists have the following characteristics:
- High reproductive rate: r-strategists have a large number of progeny in a short period of time, increasing their chances of survival.
- Limited parental care: r-strategists provide limited parental care since their concentration is on producing a large number of offspring rather than on providing thorough care.
- Short lifespan: These species have shorter lifespans and a high population-turnover rate.

r-strategists are often associated with unstable or unpredictable circumstances, where resources are plentiful but fleeting. To maximize their reproductive success, they take advantage of favourable conditions and rapid resource availability.

Examples of r-strategist species include certain insects, small rodents and many annual plants.

K-strategists

K-strategists are species with a reproductive strategy that is characterized by moderate population increase and efficient resource utilization. They are also known as equilibrium species. K-strategists have the following characteristics:
- Low reproductive rate: K-strategists have fewer offspring but devote more resources to each individual to improve their survival and reproductive success.
- Extensive parental care: These species provide their offspring with extensive parental care, ensuring their survival and growth.
- Longer lifespan: K-strategists have longer lifespans and slower population growth rates because they prioritize long-term population sustainability.

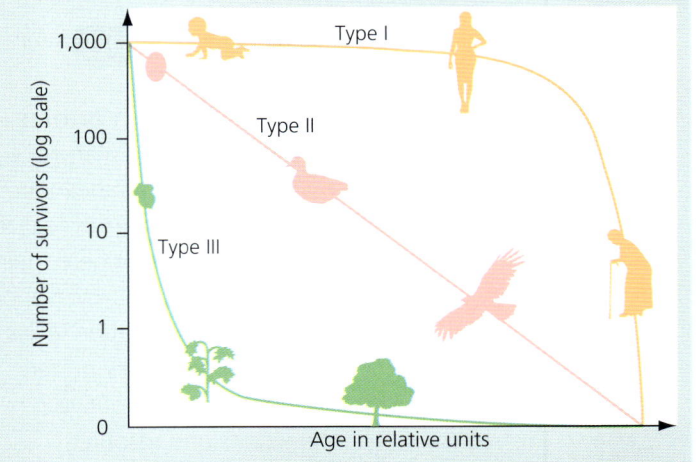

■ **Figure 2.25** This survivorship curve demonstrates the number of individuals in a population that can be expected to survive to any particular age

K-strategists are often associated with stable and predictable contexts with limited resources and high competitiveness. They have characteristics that allow them to compete efficiently for resources and ensure the survival of the species. Examples of K-strategist species include large mammals like elephants and whales, as well as many long-lived perennial plants.

The r/K-selection theory provides a framework for understanding how species adapt to changing environmental conditions and successional stages. While r-strategists are concerned with quick colonization and reproduction in unstable habitats, K-strategists are concerned with long-term survival and successful reproduction in stable situations.

It's crucial to remember that the r/K-selection continuum is a spectrum, and species can exhibit features from either end depending on their ecological situation. Furthermore, the labels r-strategist or K-strategist are not mutually exclusive, but rather describe a general trend in their life history strategies.

■ **Table 2.1** How typical r- and K-strategist populations compare in various aspects

Characteristic	r-strategist	K-strategist
Age of maturity	Early	Late
Reproduction	Early	Late
Level of parental involvement and care	Low, if any	High
Number of offspring	High	Low
Pattern of population growth	J-curve	S-curve
Size of organism	Tend to be smaller	Tend to be larger
Generalist or specialist	Generalist	Specialist

2.1.30 Knowledge of species' classifications, niche requirements and life cycles helps us to understand the extent of human impacts upon them

Understanding the degree of human influence on species requires knowledge of their classifications, niche needs and life cycles. Human activities can have a substantial impact on the natural environment, such as disrupting species' life cycles and ecological connections. Here are some examples of human impacts on species' life cycles, as well as how they are linked:

Disruption of temperature cycles

Climate change, largely driven by human activities such as burning fossil fuels and deforestation, has resulted in global temperature changes. These temperature shifts can have profound effects on the life cycles of various organisms, particularly plants and the animals that rely on them. For example:
- Phenological shifts: Changes in temperature patterns can cause shifts in the timing of critical life cycle events, such as flowering, leaf emergence and migration. If plants flower earlier or later than usual due to warming temperatures, it can disrupt the availability of nectar and pollen for pollinators, affecting their reproduction and survival.
- Trophic mismatch: Climate change can also lead to a mismatch between the life cycles of interacting species. For instance, if the emergence of certain insect species is not synchronized with the peak food availability for their predators or prey, it can disrupt the delicate balance of energy transfer within ecosystems.

Loss of habitat and fragmentation

Human activities, including urbanization, deforestation and land conversion, have led to the loss and fragmentation of natural habitats. These changes can significantly impact species' life cycles by:
- Disrupting breeding and migration: Habitat loss can hinder the breeding and migration patterns of many species. For instance, the destruction of nesting sites for migratory birds can impede their ability to reproduce and complete their annual migration, leading to population declines.
- Altered resource availability: Habitat fragmentation can disrupt the availability of essential resources needed for species' life cycles. Fragmented habitats often have reduced food sources, shelter and suitable breeding grounds, impacting the survival and reproduction of species.

◆ **Bioaccumulation** – the movement of a pollutant from the environment into living organisms, resulting in higher concentrations of the pollutant in organisms than in the environment. Also a process that takes place within an organism, involving the faster absorption and buildup of substance concentration in tissues compared to removal. This accumulation commonly happens through the consumption of contaminated food and through direct absorption from water, often occurring simultaneously.

◆ **Biomagnification** – the increase in concentration of pollutants in organisms at higher trophic levels, typically attributed to the persistence of pollutants that resist natural breakdown processes, leading to their faster transfer up the food chain compared to breakdown or elimination. The highest concentrations are found at the top of the food chain.

Pollution and contamination

Human-generated pollution, including chemicals, toxins and pollutants released into the environment, can have detrimental effects on species' life cycles. Examples include:
- Endocrine disruption: Certain pollutants can interfere with the hormonal systems of organisms, affecting their reproductive cycles, development and behaviour. Endocrine-disrupting chemicals can impact on the reproductive success of fish, amphibians and other aquatic organisms.
- **Bioaccumulation** and **biomagnification**: Some pollutants can accumulate in organisms over time, becoming more concentrated as they move up the food chain. This can lead to disruptions in the life cycles of species at higher trophic levels, affecting their reproduction and overall population dynamics.

ATL ACTIVITY

1. In the world of studying nature, how can we use the skills we've learned in this topic to carefully explore both natural and changed environments, and to compare them with the help of additional information? Discuss what secrets we can uncover.
2. As supporters of saving different kinds of life on Earth, how can we effectively share our concerns about losing biodiversity? Discuss how we can tell powerful stories that inspire action, bringing together science and engaging storytelling.
3. Our active participation in citizen science projects can become a transformative journey by helping people collect data about where different species are and how many there are. Create a project with your classmates to connect what you learned in class with the amazing patterns of the natural world to help people acquire further information about the biodiversity in their neighbourhood.

REVIEW QUESTIONS

1. Describe the term 'carrying capacity' for human populations.
2. Define the term 'niche'.
3. Explain the differences between S- and J-population growth curves.

2.2 Energy and biomass in ecosystems

> **Guiding questions**
> - How can flows of energy and matter through ecosystems be modelled?
> - How do human actions affect the flow of energy and matter, and what is the impact on ecosystems?

SYLLABUS CONTENT

This chapter covers the following syllabus content:
- ▶ 2.2.1 Ecosystems are sustained by supplies of energy and matter.
- ▶ 2.2.2 The first law of thermodynamics states that as energy flows through ecosystems, it can be transformed from one form to another, but cannot be created or destroyed.
- ▶ 2.2.3 Photosynthesis and respiration transform energy and matter in ecosystems.
- ▶ 2.2.4 Photosynthesis is the conversion of light energy to chemical energy in the form of glucose, some of which can be stored as biomass by autotrophs.
- ▶ 2.2.5 Producers form the first trophic level in a food chain.
- ▶ 2.2.6 Respiration releases energy from glucose by converting it into a chemical form that can easily be used in carrying out active processes within living cells.
- ▶ 2.2.7 Some of the chemical energy released during respiration is transformed into heat.
- ▶ 2.2.8 The second law of thermodynamics states that energy transformations in ecosystems are inefficient.
- ▶ 2.2.9 Consumers gain chemical energy from carbon (organic) compounds obtained from other organisms. Consumers have diverse strategies for obtaining energy-containing carbon compounds.
- ▶ 2.2.10 Because producers in ecosystems make their own carbon compounds by photosynthesis, they are at the start of food chains. Consumers obtain carbon compounds from producers or other consumers, so form the subsequent trophic levels.
- ▶ 2.2.11 Carbon compounds and the energy they contain are passed from one organism to the next in a food chain. The stages in a food chain are called trophic levels.
- ▶ 2.2.12 There are losses of energy and organic matter as food is transferred along a food chain.
- ▶ 2.2.13 Gross productivity is the total gain in biomass by an organism. Net productivity is the amount remaining after losses due to respiration.
- ▶ 2.2.14 The number of trophic levels in ecosystems is limited due to energy losses.
- ▶ 2.2.15 Food webs show the complexity of trophic relationships in communities.
- ▶ 2.2.16 Biomass of a trophic level can be measured by collecting and drying samples.
- ▶ 2.2.17 Ecological pyramids are used to represent relative numbers, biomass or energy of trophic levels in an ecosystem.
- ▶ 2.2.18 Pollutants that are non-biodegradable such as polychlorinated biphenyl (PCBs), dichlorodiphenyltrichloroethane (DDT) and mercury cause changes to ecosystems through the processes of bioaccumulation and biomagnification.

- 2.2.19 Non-biodegradable pollutants are absorbed within microplastics, which increases their transmission in the food chain.
- 2.2.20 Human activities such as burning fossil fuels, deforestation, urbanization and agriculture have impacts on flows of energy and transfers of matter in ecosystems.

HL ONLY

- 2.2.21 Autotrophs synthesize carbon compounds from inorganic sources of carbon and other elements. Heterotrophs obtain carbon compounds from other organisms.
- 2.2.22 Photoautotrophs use light as an external energy source in photosynthesis. Chemoautotrophs use exothermic inorganic chemical reactions as an external energy source in chemosynthesis.
- 2.2.23 Primary productivity is the rate of production of biomass using an external energy source and inorganic sources of carbon and other elements.
- 2.2.24 Secondary productivity is the gain in biomass by consumers, using carbon compounds absorbed and assimilated from ingested food.
- 2.2.25 Net primary productivity is the basis for food chains because it is the quantity of carbon compounds sustainably available to primary consumers.
- 2.2.26 Maximum sustainable yields (MSY) are the net primary or net secondary productivity of a system.
- 2.2.27 Sustainable yields are higher for lower trophic levels.
- 2.2.28 Ecological efficiency is the percentage of energy received by one trophic level that is passed on to the next level.
- 2.2.29 The second law of thermodynamics shows how the entropy of a system increases as biomass passes through ecosystems.

2.2.1 Ecosystems are sustained by supplies of energy and matter

The fact that ecosystems exchange matter and energy with their surroundings allows us to think of them as open systems. Matter refers to the physical substances such as nutrients, minerals and organic compounds that cycle through and contribute to the structure and functioning of the ecosystem. Sunlight is the main energy source. Autotrophs are organisms that can produce their own food through photosynthesis or chemosynthesis. As organisms consume one another, this energy is subsequently transferred throughout the ecosystem, resulting in energy flow. At the same time, as organisms interact with their surroundings, matter moves within ecosystems. As plants or other animals are consumed, essential nutrients and components are absorbed by plants from the soil, integrated into their tissues and then transferred to consumers.

2.2.2 The first law of thermodynamics

The flow of energy in an ecosystem is based on the first law of thermodynamics, also referred to as the law of energy conservation. According to this, energy can only be transformed from one form to another within an ecosystem, it cannot be created or destroyed. Within ecosystems, energy is transformed constantly. For instance, the process of photosynthesis, in which autotrophs use sunlight to create organic compounds, transforms solar energy into chemical energy. Consuming these substances allows organisms to transform chemical energy into other forms, such as mechanical energy for movement or thermal energy for body temperature regulation.

2.2.3 Photosynthesis and respiration transform energy and matter in ecosystems

Two fundamental ecological processes, photosynthesis and respiration, are essential for the transformation of both matter and energy. Autotrophs, primarily plants, convert light energy into chemical energy through a process called photosynthesis. In this process, water (H_2O) and carbon dioxide (CO_2) are combined to create glucose ($C_6H_{12}O_6$) and oxygen (O_2).

- Photosynthesis: $6CO_2 + 6H_2O \rightarrow C_6H_{12}O_6 + 6O_2$
- Cellular respiration: $C_6H_{12}O_6 + 6O_2 \rightarrow 6CO_2 + 6H_2O$

The process by which organisms, including plants and animals, release the chemical energy that has been stored within organic compounds is known as respiration. As glucose and oxygen are used up during respiration, carbon dioxide, water and energy are also released. Organisms use this energy for vital functions such as growth, movement, reproduction and other vital processes.

Respiration and photosynthesis are linked processes that control the movement of matter and energy within ecosystems. They are complementary and form a continuous cycle, with respiration releasing energy back into the environment and photosynthesis absorbing energy from it.

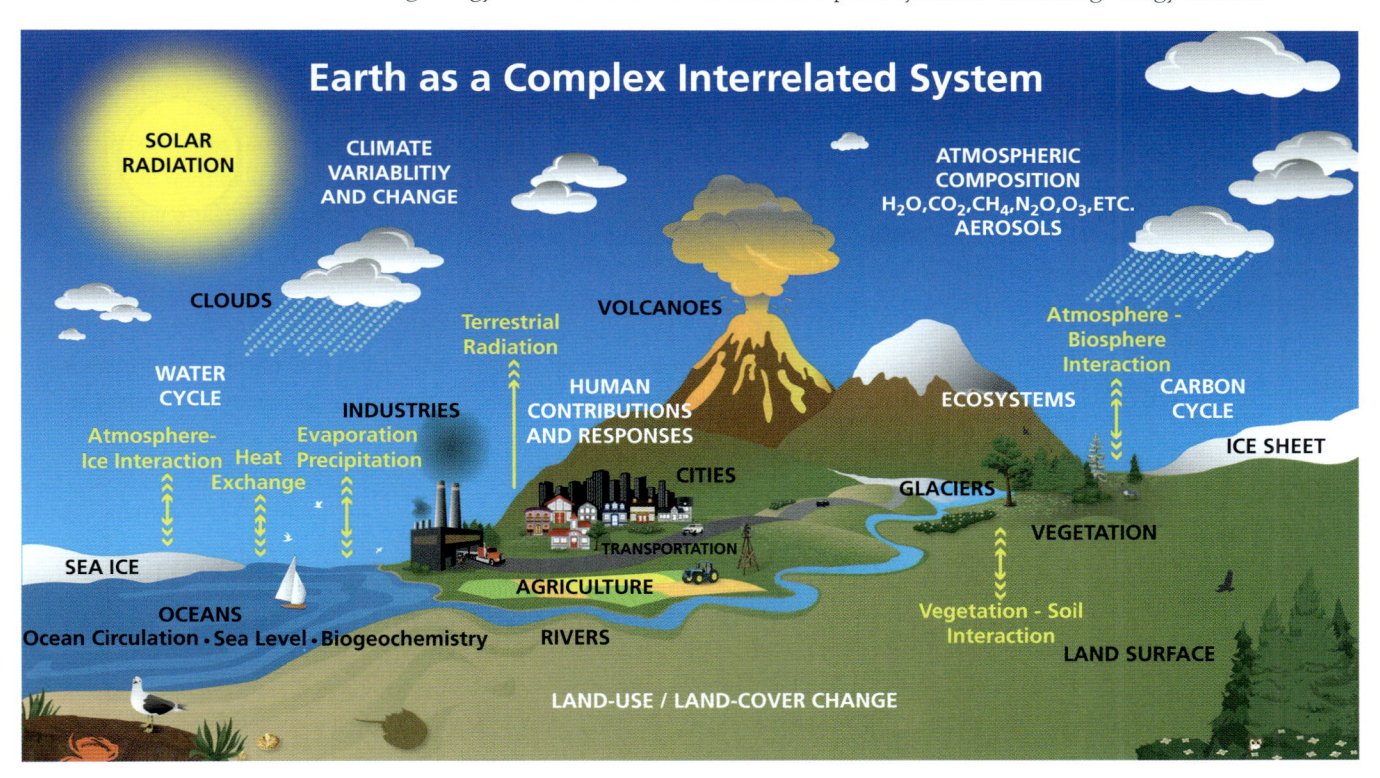

■ Figure 2.26 The matter and energy cycle

2.2.4 Photosynthesis is the conversion of light energy to chemical energy

Autotrophs, including plants, algae and photosynthetic bacteria, perform photosynthesis, which is a crucial process. It entails the transformation of solar light energy into chemical energy in the form of glucose. Autotrophs use pigments, primarily chlorophyll, in photosynthesis to absorb sunlight. The chemical reactions that result in the creation of glucose ($C_6H_{12}O_6$) and oxygen (O_2) from carbon dioxide (CO_2) and water (H_2O) are then powered by this energy.

Autotrophs, as well as other organisms that consume them, use the glucose that is created during photosynthesis as a source of energy. Biomass, which includes the living components of plants and other autotrophs, can be used to store some of the glucose. The maintenance of the autotroph's structure as well as growth and reproduction depend on biomass.

2.2.5 Producers form the first trophic level in a food chain

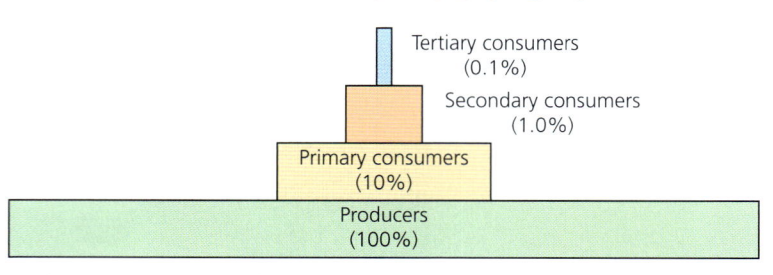

■ **Figure 2.27** Pyramid of energy showing energy loss between each trophic level

Trophic levels in a food chain represent various positions that organisms can occupy depending on their feeding relationships. Producers, which are typically plants, algae and photosynthetic bacteria, form the first trophic level. Due to their ability to transform sunlight energy into sugar (chemical energy) to produce their own food through photosynthesis, they are also known as autotrophs (see Topic 2.2.1, page 120).

By converting solar energy into chemical energy, producers play a crucial part in the dynamics of ecosystems. The food chain's subsequent trophic levels are occupied by consumers – organisms that get their energy from feeding on other organisms, meaning this energy is subsequently transferred throughout the ecosystem.

2.2.6 Cell respiration releases energy from glucose that can easily be used

All living things, including plants, animals and microorganisms, engage in respiration, which is the opposite of photosynthesis. Respiration releases energy from glucose (and other organic compounds) to fuel cellular functions.

The chemical energy contained in glucose is released during respiration when it is broken down in the presence of oxygen. Adenosine triphosphate (ATP), a chemical compound, is then created using this energy. Within cells, ATP supplies the required energy for a variety of active processes like metabolism, growth and reproduction.

2.2.7 Some of the chemical energy released during respiration is transformed into heat

Although ATP is released as energy during respiration, this process is not entirely efficient. During respiration, some of the chemical energy released is converted to heat. This happens as a result of the imperfect efficiency of the conversion of energy from substrates like carbohydrates into the chemical form of energy used in cells (ATP).

A byproduct of energy transformation is the heat energy produced during respiration. The body temperatures of organisms are maintained in part by this heat. The important thing to remember is that once heat is produced inside a specific organism, it cannot be converted back into chemical energy. The body loses this heat and releases it into the surrounding air and space.

2.2.8 The second law of thermodynamics

The effectiveness and quality of energy transformations are subject to the second law of thermodynamics. According to this, some energy must be converted into a less useful form, usually heat, during energy transformations. The largest energy losses in ecosystems happen during respiration, when a sizable portion of the chemical energy from organic compounds is converted to heat.

According to this law, energy transfers within ecosystems are never entirely efficient. The total amount of energy available for higher trophic levels is constrained by the constant loss of energy as heat as it moves through the trophic levels.

Ecological efficiency is the percentage of energy transferred from one trophic level to the next. It is calculated using the following formula:

$$\text{Ecological efficiency} = \frac{\text{Energy used for growth (new biomass)}}{\text{Energy supplied} \times 100}$$

Link

More information about systems is available in Topics 1.2 (page 24) and 5.2 (page 456).

Understanding the second law will help you to recognize the inherent inefficiency of energy transfers and the constraints imposed on how ecosystems operate and what they can produce.

2.2.9 Consumers gain chemical energy from carbon (organic) compounds obtained from other organisms

Consumers are organisms that obtain their energy by consuming other organisms. They have diverse strategies for obtaining energy-containing carbon compounds. Here are some examples of consumer strategies:

- Herbivores: These feed exclusively on plants or plant-derived materials.
- Detritivores: These feed on dead organic matter (detritus), such as decaying leaves or animal carcasses.
- Predators: These hunt and consume other animals.
- Parasites: These obtain nutrients from a host organism, usually without causing immediate death.

- Saprotrophs: These obtain energy from decomposing organic matter.
- Scavengers: These feed on dead animals that they did not kill.
- Decomposers: These play a key role in the recycling of nutrients by converting organic matter into simpler inorganic substances.

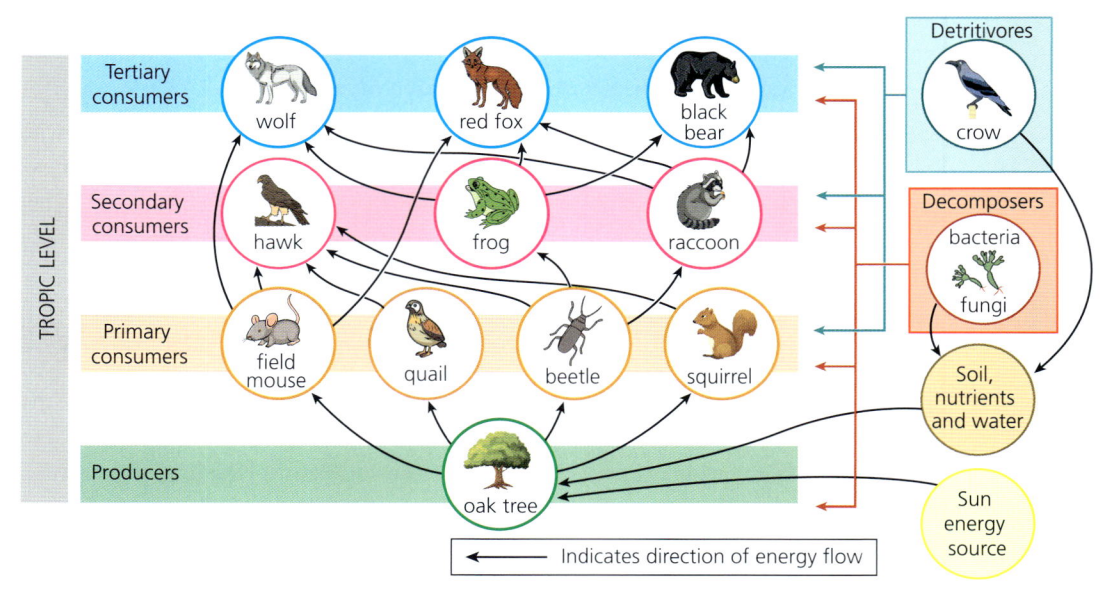

■ Figure 2.28 Partial food web showing trophic levels

2.2.10 Producers are at the start of food chains; consumers obtain carbon compounds from producers or other consumers

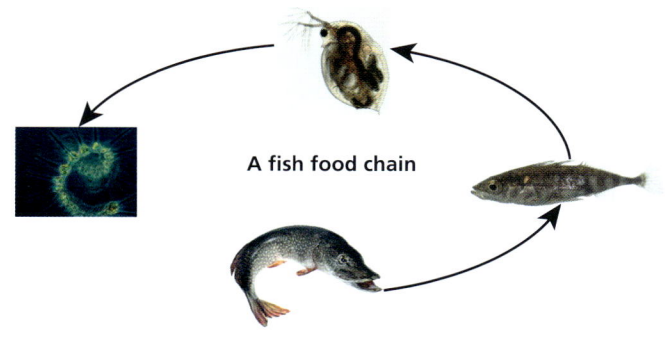

■ Figure 2.29 A food chain

Ecosystems' food chains are built on producers, which include autotrophs like plants, algae and photosynthetic bacteria. They have the unique ability to produce their own carbon compounds through photosynthesis. Plants produce almost all of the food used on Earth.

In a food chain, organic matter moves from primary producers to primary consumers, such as herbivores, then to secondary consumers, such as carnivores or omnivores, and so on. Consumers obtain carbon compounds by consuming producers or other consumers, forming subsequent trophic levels.

The complexity of energy flow and nutrient cycling within ecosystems can be understood by looking at the variety of consumer strategies. In aquatic environments, the smallest plants are phytoplankton, which are eaten by small animal grazers called zooplankton. These in turn are eaten by predators, which may also have predators. Each of these groups of organisms occupies a trophic level, and the size and diversity of each level can affect the whole ecosystem. For example, without predators there may be so many grazers that only a few plants can survive.

> **Tool 4: Systems and models**
>
> ### Create a food chain
> 1. Producers: Grass or sunflower
> 2. Primary consumers: Rabbit or grasshopper
> 3. Secondary consumers: Snake or mouse
> 4. Tertiary consumer: Hawk
>
> Based on the provided data, create a food chain that demonstrates the flow of energy from the producers (grass and sunflower) to the primary consumers (rabbit and grasshopper), then to the secondary consumers (snake and mouse) and finally to the tertiary consumer (hawk). Remember that the arrows in a food chain show the energy flow.

2.2.11 Energy passes through the food chain's trophic levels

Energy-containing carbon compounds are passed from one organism to another in a food chain. In order to fuel their own growth, maintenance and reproduction processes, organisms extract the energy from carbon compounds as they consume food. This energy transfer takes place through the food chain's various trophic levels.

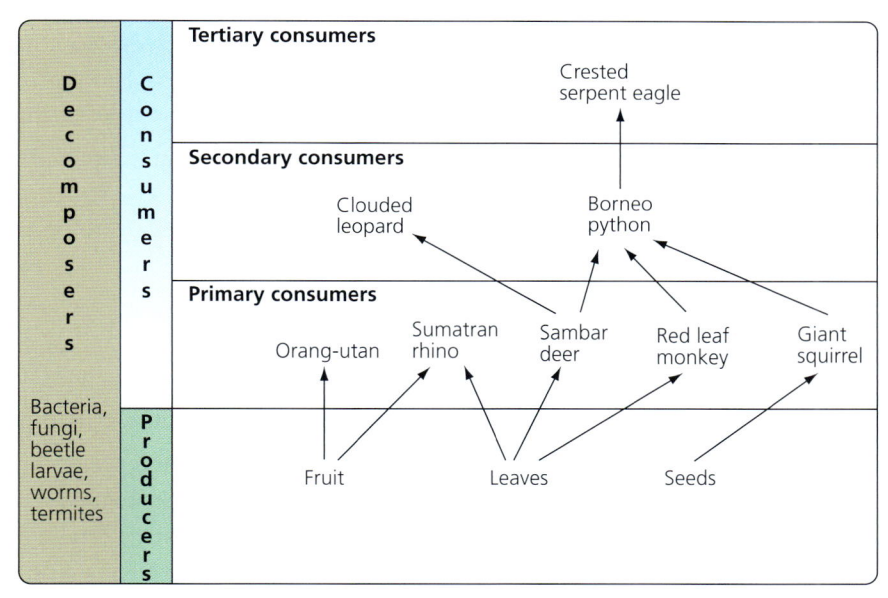

■ Figure 2.30 A rainforest food web from Borneo, showing trophic levels; decomposers feed at each trophic level

The trophic levels in a food chain represent the various positions that organisms occupy depending on their feeding preferences and energy source. Decomposers typically are not part of food chains because they consume a variety of foods. However, taking into account how decomposers transform energy within food webs aids in our comprehension of the significance of nutrient recycling and the flow of energy through various pathways in ecosystems.

2.2 Energy and biomass in ecosystems

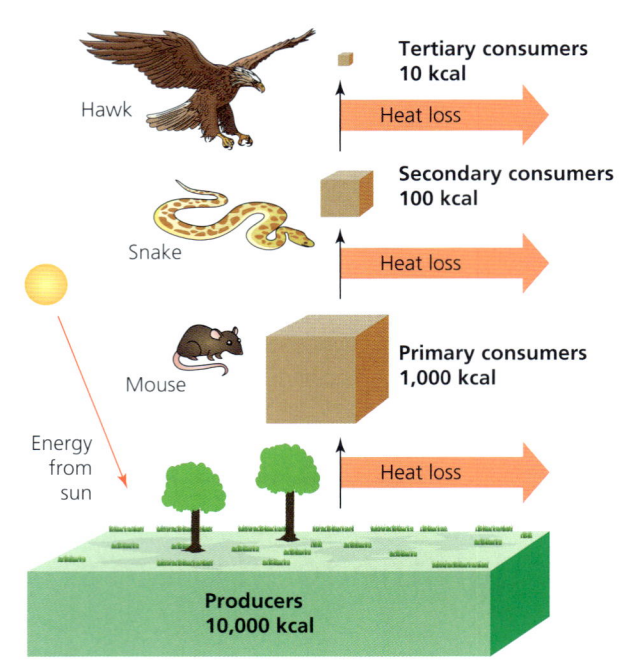

Figure 2.31 Energy flow and trophic levels

2.2.12 There are losses of energy and organic matter as food is transferred along a food chain

As food moves up a food chain, energy and organic matter are inevitably lost from the food chain. At a given trophic level, not all the food that is available is harvested, and of that which is harvested, not all is consumed. Furthermore, not all of what is consumed is absorbed, and not all of what is absorbed is stored as biomass. Respiration causes a significant amount of energy to be lost as heat.

These losses mean there is never a complete transfer of organic matter from one trophic level to the next. These energy losses highlight the inefficiency of energy transfers in ecosystems and the limitations on the amount of energy available to higher trophic levels.

2.2.13 Gross productivity and net productivity

- Gross productivity (GP) is the total increase in biomass that an organism experiences as a result of photosynthesis or other energy-acquiring processes. It measures the amount of energy that is captured and transformed into biomass. However, organisms also need to use energy for their metabolic processes, resulting in energy losses through respiration.
- Net productivity (NP) is the amount of energy remaining after accounting for the losses due to respiration. It represents the energy available for growth, reproduction and storage. Due to the higher energy demands associated with their activities, consumers typically experience greater losses from respiration than producers.
- Net primary productivity (NPP) is the measure of the amount of energy that plants capture and store through photosynthesis, minus the energy they expend during cellular respiration. It represents the surplus energy available for consumption by herbivores and higher trophic levels, making it a key indicator of the ecosystem's capacity to support life.

In an ecological study, the initial step involves determining the dry weight of the food initially presented to the population under examination. As time progresses, researchers collect samples and ascertain the dry weight of the food that remains after a certain number of days have elapsed. The next crucial step entails subtracting the weight of the remaining food from the initially presented quantity, effectively calculating the dry weight of the food that has been consumed. At the same time, researchers collect and assess the dry weight of faeces generated by the population over the designated period. By subtracting the weight of the faeces from the previously calculated food consumed, the amount of food absorbed (the gross productivity) can be determined. This can be expressed as the difference between the food eaten and the loss due to faeces:

Gross productivity = Food eaten − Faecal loss

To refine this measure, the final weight of the population's gross productivity is divided by the number of days encompassing the study. This comprehensive approach provides insights into the efficiency of food utilization and resource allocation within the studied ecosystem.

Equation for net primary productivity (NPP):

$$NPP = GPP - R$$

Where GPP = gross primary productivity and R = respiratory loss.

Understanding the terms 'gross productivity' and 'net productivity' enables us to evaluate the amount of energy required to support organism growth and development as well as any potential effects on ecosystem productivity as a whole.

Link

More information about gross productivity and net productivity is available in Topic 2.5.10 (page 187).

2.2.14 The number of trophic levels in ecosystems is limited due to energy losses

Ecosystems can only have a certain number of trophic levels because energy is gradually lost as it moves up the food chain. Higher trophic level organisms cannot use the energy that is released by respiration and lost as heat by lower trophic level organisms. Along the food chain, there are also additional energy losses due to incomplete consumption and incomplete absorption of consumed energy.

Typically, only around 10 per cent or less of the energy flowing to a trophic level is available to the next level. The length of food chains in ecosystems is constrained by this energy limitation, which also explains why there are typically a finite number of trophic levels. To calculate the efficiency of transfer, calculate the percent of energy that is transferred from the first trophic level to the second trophic level. Divide energy from trophic level one and multiply by 100. This amount is the percentage of energy transferred. Remember to add a per cent sign.

It is important to keep in mind the fact that organisms at higher trophic levels do not necessarily need to consume more food in order to have enough energy. The structure and dynamics of food chains are greatly influenced by the inefficiency of energy transfer and the limited energy supply at higher trophic levels.

2.2.15 Food webs show the relationships within an ecosystem

Compared to simple food chains, food webs provide a more comprehensive representation of trophic relationships within an ecosystem. They illustrate the interconnectedness of various species and the multiple feeding interactions that occur in a community. Arrows in food webs indicate the direction of energy flow and the transfer of biomass between different trophic levels. It's important to note that in a food web, species may feed at multiple trophic levels, reflecting the complexity of their diets and interactions.

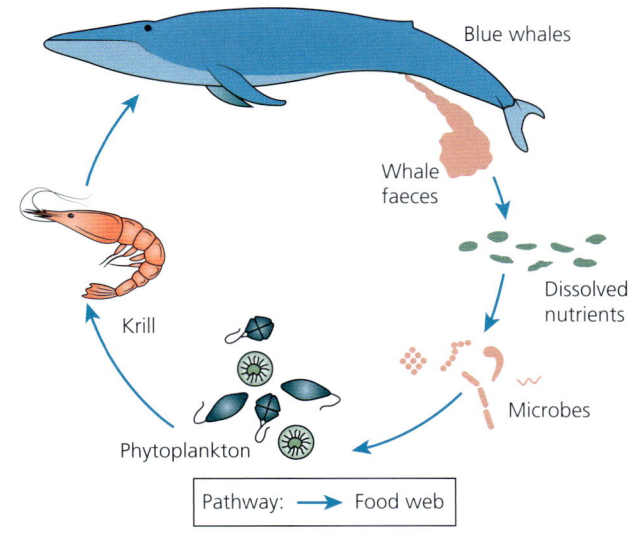

■ **Figure 2.32** A food web

2.2 Energy and biomass in ecosystems

> **Tool 4: Systems and models**
>
> **Construct a food web**
>
> To begin constructing a food web:
> 1. choose the ecosystem or setting you wish to study
> 2. list the primary producers, herbivores, omnivores and carnivores present in that particular habitat (limit to six to eight organisms)
> 3. depict the interconnections and energy transfers among the species using arrows or connectors to show predator–prey relationships.

2.2.16 Biomass of a trophic level can be measured by collecting and drying samples

Biomass refers to the total mass of organic matter in a given population or trophic level. It is an important measure of the energy available within an ecosystem. To determine the biomass of a trophic level, biological samples are collected and dried to remove water content. The dry mass of the samples is considered equal to the mass of organic matter (biomass) since water represents the majority of the inorganic matter in most organisms.

2.2.17 Ecological pyramids

Ecological pyramids are graphical representations used to depict the relative numbers, biomass or energy of organisms at different trophic levels within an ecosystem. They provide a visual representation of the structure and functioning of ecosystems. There are various different types.

- Pyramids of number (also known as pyramids of productivity): These illustrate the number of individuals at each trophic level, with the primary producers forming the base of the pyramid and subsequent trophic levels narrowing as the number of individuals decreases.

- Pyramids of biomass: These represent the total biomass (organic matter) present at each trophic level. The biomass decreases as you move up the pyramid. Figure 2.34 shows an inverted pyramid of biomass.

- Pyramids of energy: These illustrate the flow of energy through different trophic levels. They represent the amount of energy flowing to each trophic level per unit area and per unit time (usually measured in kilojoules per square metre per year) (see Figure 2.31 on page 126).

The shape of ecological pyramids can vary depending on the ecosystem and its dynamics. Factors such as the efficiency of energy transfer, the number of trophic levels and the productivity of the ecosystem influence the shape of the pyramid.

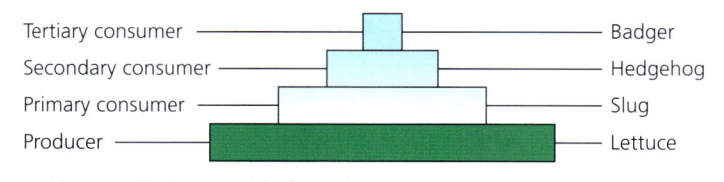

■ Figure 2.33 A pyramid of number

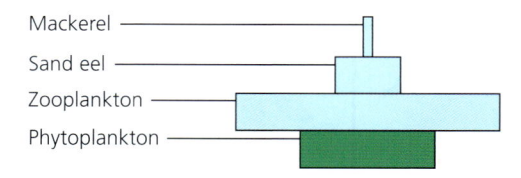

■ Figure 2.34 A pyramid of biomass for a north Atlantic food chain. Changes in feeding patterns and seasonal variations can lead to pyramids of biomass being inverted

Theme 2: Ecology

2.2.18 Non-biodegradable pollutants cause changes to ecosystems through the processes of bioaccumulation and biomagnification

Certain pollutants, such as polychlorinated biphenyl (PCBs), dichlorodiphenyltrichloroethane (DDT) and the metal mercury, are non-biodegradable and have long-lasting effects on ecosystems. These pollutants can cause significant changes and disruptions to the environment through two processes: bioaccumulation and biomagnification.

The term 'bioaccumulation' describes the gradual rise in the concentration of non-biodegradable pollutants in living things or trophic levels. These pollutants build up in the tissues of organisms as they consume food or drink that contains them, frequently in higher concentrations than in the environment. This can eventually have a negative impact on the survival and health of the affected organisms.

On the other hand, biomagnification refers to the accumulation of non-biodegradable pollutants along a food chain. Pollutants are transferred and accumulate in higher concentrations at higher trophic levels when organisms eat other organisms that are contaminated with them. This process takes place because the pollutants are difficult for organisms to break down or get rid of. Therefore, the top predators of the highest trophic levels typically have the highest levels of these pollutants.

> **REAL-WORLD EXAMPLE**
>
> **Minamata Disease**
>
> The discovery of Minamata Disease in Japan in 1956 revealed the devastating impact of mercury pollution on the local population. A nearby factory's chemical waste pipe contaminated Minamata Bay with mercury, which was transformed into toxic methylmercury by bacteria in the water. This methylmercury accumulated in fish muscles, leading to staggering and death in local cats that consumed the fish. Subsequently, people who relied on fish from the bay were also affected, with over 2,000 deaths and many thousands experiencing severe injuries. This incident highlighted the dangers of pollution and the vulnerability of developing foetuses and children to environmental contaminants.
>
> A similar tragedy occurred in Iraq during the 1970s, when seed grain treated with a mercury-based fungicide led to mercury poisoning in the population. These events teach us that pollution cannot be solved through dilution, and emphasize the importance of understanding the harmful effects of environmental contaminants on human health.
>
> *source:* www.healthandenvironment.org/environmental-health/social-context/history/mercury-the-tragedy-of-minamata-disease

2.2.19 Non-biodegradable pollutants are absorbed within microplastics, which increases their transmission in the food chain

Microplastics are tiny plastic flecks with a diameter of less than five millimetres. As a result of plastic pollution, these particles are frequently found in the environment. Non-biodegradable pollutants, including those mentioned earlier, have a tendency to absorb on to the surfaces of microplastics. This absorption increases the transmission of the pollutants in the food chain.

Pollutants containing microplastics are ingested by organisms and are subsequently absorbed by their bodies. Microplastics can affect different trophic levels of the food chain because they can be consumed by a wide variety of organisms, including primary consumers (herbivores), secondary consumers (carnivores) and even decomposers.

2.2.20 Impact of human activities on ecosystems

Human activities have profound impacts on ecosystems, influencing the flows of energy and transfers of matter. Several key activities include:

- Burning fossil fuels: This releases carbon dioxide (CO_2) and other greenhouse gases into the atmosphere, which contributes to climate change and global warming. These changes may disturb the equilibrium of matter and energy in ecosystems, which may affect primary productivity and the availability of resources for organisms.
- Deforestation: Clearing forests for agriculture, logging or urban development leads to the loss of ecosystem biomass. Forests play a crucial role in capturing and storing carbon dioxide, maintaining biodiversity and regulating water cycles. Deforestation disrupts these processes, altering energy flows and matter cycling in ecosystems.
- Urbanization: Urban development involves transforming natural landscapes into built environments, resulting in the loss of habitats and alteration of ecosystems. Urban areas often have reduced biodiversity, limited green spaces and increased impervious surfaces, affecting the availability of resources and altering energy flows within the urban ecosystem.
- Agriculture: Intensive agricultural methods can harm ecosystems, such as the use of fertilizers and pesticides. Aquatic ecosystems may be harmed by excessive nutrient runoff from agricultural fields, which can cause **eutrophication** in water bodies. Pesticides may also have an adverse effect on non-target organisms like pollinators and natural pest controllers, further affecting trophic interactions.

◆ **Eutrophication** – a process of excessive nutrient enrichment, particularly nitrogen and phosphorus, in water bodies, leading to rapid and dense growth of algae and other aquatic plants. This excessive growth can result in oxygen depletion, harming aquatic life and disrupting the ecological balance of the ecosystem.

Inquiry process

Inquiry 2: Collecting and processing data

Analyse the impacts of human activities on energy flows and matter transfers in ecosystems.

Instructions

1. Research and select a specific human activity related to burning fossil fuels, deforestation, urbanization or agriculture.
2. Identify the key ecological impacts of the chosen human activity on energy flows and transfers of matter in ecosystems. Consider both direct and indirect effects.
3. Create a visual representation, such as a flowchart or diagram, to illustrate the ecological consequences of the human activity. Include the affected trophic levels, changes in energy flows, disruptions in matter cycling, and potential consequences for biodiversity and ecosystem functioning.
4. Analyse the short-term and long-term effects of the human activity on the ecosystem's stability and resilience. Consider the potential cascading effects on other trophic levels and the overall ecosystem structure.
5. Present your findings and visual representation to your classmates or teacher. Discuss the implications of human activities for ecosystems and explore possible strategies or solutions to mitigate the negative impacts.

(HL) 2.2.21 Autotrophs and heterotrophs: carbon synthesis and nutrient acquisition

◆ **Synthesize** – to combine or produce something by bringing together different elements or components to form a coherent whole.

Autotrophs are organisms that have the remarkable ability to **synthesize** complex organic molecules, such as carbohydrates, lipids and proteins, from inorganic sources of carbon and other elements. They are also commonly referred to as primary producers because they form the foundation of food chains and provide energy for other organisms within ecosystems.

Autotrophs utilize various mechanisms to convert inorganic substances into organic compounds. The most well-known process is photosynthesis, which harnesses solar energy to synthesize carbon compounds. Within specialized organelles called chloroplasts, autotrophs possess pigments, such as chlorophyll, which capture energy from sunlight. This absorbed energy fuels the chemical reactions that transform carbon dioxide (CO_2) and water (H_2O) into glucose ($C_6H_{12}O_6$) and oxygen (O_2). The glucose serves as a vital energy source for the autotroph, while the released oxygen supports other organisms in their respiratory processes.

In addition to photosynthesis, some autotrophs found in unique environments, such as deep-sea hydrothermal vents and sulfur springs, employ an alternative pathway known as chemosynthesis. These autotrophs derive energy by oxidizing inorganic molecules, such as hydrogen sulphide (H_2S) or methane (CH_4), and utilize this energy to fix carbon and produce organic compounds.

While autotrophs can produce their own carbon compounds, heterotrophs are organisms that cannot synthesize organic molecules from inorganic sources. Instead, they rely on consuming other organisms to obtain the carbon compounds they need for energy and growth. Heterotrophs occupy various trophic levels in food chains and food webs, including herbivores, carnivores and decomposers.

■ **Table 2.2** How different heterotrophs acquire the necessary carbon compounds for their own metabolic processes

Type of heterotroph	How this type acquires the necessary carbon compounds for its own metabolic processes
Herbivore	Primarily consuming autotrophs directly, these organisms feed on plants or algae to obtain the organic compounds synthesized through photosynthesis and acquire the necessary carbon compounds for their own metabolic processes.
Carnivore	In contrast to herbivores, carnivores acquire carbon compounds by feeding on other animals. These secondary consumers obtain energy and nutrients by consuming the tissues of other organisms. Through this process, energy and carbon compounds flow through different trophic levels in the ecosystem.
Decomposer	Decomposers play a critical role in recycling organic matter and releasing carbon compounds back into the environment. Organisms like fungi and bacteria break down dead organic material, including plant and animal remains, returning the carbon compounds to the soil or water where they can be reused by autotrophs.

In summary:
- Autotrophs synthesize carbon compounds from inorganic sources and form the foundation of energy flow in ecosystems through processes like photosynthesis and chemosynthesis.
- Heterotrophs, on the other hand, obtain carbon compounds by consuming other organisms, whether they are autotrophs or other heterotrophs.
- These processes of carbon synthesis and acquisition are fundamental to the functioning of ecosystems and the interdependence of different organisms within them.

2.2 Energy and biomass in ecosystems

2.2.22 Photoautotrophs and chemoautotrophs: Energy sources and ecosystems

Photoautotrophs are a group of autotrophic organisms that utilize light as an external energy source for photosynthesis. They are commonly found in ecosystems where sunlight is available. Photoautotrophs include plants, algae and some bacteria. By harnessing the energy from sunlight, they convert inorganic substances, such as carbon dioxide (CO_2) and water (H_2O), into organic compounds like glucose ($C_6H_{12}O_6$).

During photosynthesis, photoautotrophs capture light energy using specialized pigments, such as chlorophyll. This energy is then used to power the chemical reactions that convert carbon dioxide and water into glucose and oxygen. Photoautotrophs play a crucial role in producing organic matter and releasing oxygen into the environment, which supports the survival of other organisms.

Chemoautotrophs, on the other hand, utilize exothermic inorganic chemical reactions as an external energy source for chemosynthesis. They do not rely on light energy. Instead, they derive energy from the oxidation of inorganic compounds, such as hydrogen sulphide (H_2S), ammonia (NH_3) or methane (CH_4).

Chemoautotrophs are often found in environments where little or no light is available, such as deep-sea hydrothermal vents, volcanic hot springs or caves. In these habitats, chemoautotrophs are the principal source of energy that sustains food webs. They form the basis of ecosystems that are not dependent on sunlight for energy. Chemoautotrophs use the energy generated from inorganic chemical reactions to fix carbon dioxide and synthesize organic compounds, which serve as nutrients for themselves and other organisms in the ecosystem.

Chemoautotrophs play a vital role in supporting diverse ecosystems by providing a source of energy in environments where other forms of autotrophy are limited. These organisms are often associated with unique and extreme habitats, where they form symbiotic relationships with other organisms, such as tube worms or deep-sea mussels.

■ **Table 2.3** Examples of photoautotroph and chemoautotroph nutrition

Nutrition mode	Source of energy	Source of carbon	Examples
Photoautotroph	Light	CO_2	Plants, cyanobacteria, algae
Chemoautotroph	Inorganic chemicals	CO_2	Prokaryotes

2.2.23 Primary productivity and measurement protocols

Primary productivity refers to the rate at which biomass is produced in an ecosystem using an external energy source, such as light or exothermic inorganic chemical reactions. This process also uses inorganic sources of carbon (CO_2) and other necessary elements. It is an essential way to measure how energy moves and how much biomass accumulates in an ecosystem.

The unit commonly used to express primary productivity is kilograms of carbon per square metre of ecosystem per year (kg carbon per m² per year). This unit provides a standardized measure that allows for comparisons across different ecosystems and enables scientists to assess the productivity of various habitats.

Determining primary productivity in ecosystems can be accomplished using different protocols, depending on the specific conditions and research objectives. Two commonly employed approaches are:

1. Laboratory-based protocols: In a laboratory setting, primary productivity estimates can be obtained by examining photosynthesizing samples. This method involves isolating a representative sample of autotrophs, such as algae or plant material, and subjecting them

to controlled conditions that simulate the relevant environmental factors. By measuring the amount of carbon dioxide consumed or oxygen released over a specific period, scientists can calculate the rate of primary productivity.

2 Field-based protocols: These involve measuring changes in biomass over time within a natural ecosystem. For example, in a grassland ecosystem, researchers may establish study plots where the vegetation is carefully monitored. By periodically measuring the biomass of vegetation within these plots, either by harvesting and weighing the plant material or by using non-destructive techniques such as optical sensors or remote sensing, scientists can determine the rate of biomass accumulation and, therefore, the primary productivity of the ecosystem.

Both laboratory-based and field-based protocols have their advantages and limitations.
- Laboratory-based experiments provide controlled conditions that allow for precise measurements of photosynthetic rates.
- Field-based measurements provide insights into ecosystem-level productivity and dynamics.

By combining data from different protocols, scientists can gain a more comprehensive understanding of primary productivity within ecosystems.

2.2.24 Secondary productivity and measurement

Secondary productivity refers to the gain in biomass by consumers in an ecosystem. It measures the rate at which energy and nutrients from ingested food are assimilated and converted into new biomass. This process occurs after primary consumers (herbivores) consume autotrophs (plants or algae), or after secondary consumers (carnivores) consume other organisms.

The units used to express secondary productivity are the same as those used for primary productivity: kilograms of carbon per square metre of ecosystem per year (kg carbon per m^2 per year). This standardized unit allows for easy comparison and assessment of secondary productivity across different ecosystems.

To calculate secondary productivity, it is necessary to consider the amount of carbon obtained from ingested food and assimilated into the consumer's biomass. Faecal waste, which consists of undigested and unabsorbed material, is excluded:

Secondary productivity = Carbon obtained from ingested food − Faecal waste

By subtracting the mass of faecal waste from the ingested food, scientists can determine the net gain in biomass from the assimilated carbon compounds.

2.2.25 Net primary productivity and its importance

◆ **Silviculture** – the practice of managing and controlling the establishment, growth and maintenance of forests and woodlands. It is a branch of forestry that involves applying scientific principles and techniques to ensure the sustained production of various forest resources, such as timber, fuelwood and non-timber forest products, while also considering ecological and environmental objectives.

Net primary productivity (NPP) is a crucial concept in ecology as it forms the basis for food chains within ecosystems. It represents the quantity of carbon compounds that are sustainably available for consumption by primary consumers, such as herbivores. NPP can also be considered as the plant growth that can be harvested sustainably by primary consumers in natural ecosystems, or by farmers and foresters in agricultural and **silvicultural** systems.

NPP measures the surplus energy and organic matter produced by autotrophs (primarily plants) during photosynthesis after accounting for their own metabolic needs. It is the difference between gross primary productivity (GPP), which is the total energy fixed by autotrophs through photosynthesis, and the energy used by autotrophs for their respiration. In essence, NPP represents the energy and organic matter that is available for consumption by primary consumers.

NPP is a critical metric because it determines the amount of energy and biomass that can sustain primary consumers and, subsequently, the entire food chain. Primary consumers rely on the energy-rich carbon compounds produced by autotrophs to meet their metabolic needs and support their growth and reproduction. They serve as a link between autotrophs and higher trophic levels, such as secondary consumers and predators.

In natural ecosystems, NPP sustains the intricate web of interactions within food chains and supports the diversity and stability of ecological communities. It provides the energy and resources necessary for the survival and functioning of primary consumers, and their interactions with other organisms shape the dynamics of ecosystems.

In agricultural and silvicultural systems, NPP assumes significance for farmers and foresters. It represents the harvestable plant growth that can be sustainably utilized for human purposes. Farmers rely on NPP to assess the productivity of crops, while foresters consider NPP to determine the sustainable yield of timber or other forest products.

Understanding and monitoring NPP is essential for managing and conserving ecosystems, as it informs ecological research, conservation efforts and sustainable resource utilization. By quantifying the availability of carbon compounds for primary consumers, we gain insights into the energy flow and productivity of ecosystems, both in their natural state and in human-managed systems.

Tool 1: Experimental techniques

Measure primary and secondary productivity in ecosystems

Use laboratory and field techniques for measuring primary and secondary productivity, and work out gross productivity and net productivity from data.

Measuring primary and secondary productivity in ecosystems is essential for understanding the flow of energy and biomass within food chains. This tool will provide guidance on using laboratory and field techniques to measure primary productivity and to calculate gross primary productivity (GPP) and net primary productivity (NPP) from collected data. It will also cover methods to estimate secondary productivity. Follow the steps below to gain a better understanding of these concepts and calculations.

Step 1: Measuring primary productivity

Laboratory technique:

- Collect samples of photosynthesizing organisms (plants, algae or cyanobacteria).
- Place the samples in a controlled environment, such as an aquatic chamber or a growth chamber, with suitable light and temperature conditions.
- Measure the amount of carbon dioxide consumed or oxygen released over a specific time period.
- Calculate the rate of primary productivity using the measured gas exchange data.

Field technique:

- Select a study plot within the ecosystem of interest.
- Determine the initial biomass of the vegetation within the plot.
- Monitor the vegetation over time, noting any changes in biomass.
- Calculate the difference in biomass between the initial and final measurements.
- Convert the biomass change to carbon using appropriate conversion factors.
- Calculate the rate of primary productivity per unit area using the measured biomass change.

> *Step 2: Calculating gross primary productivity and net primary productivity*
>
> Gross primary productivity (GPP) calculation:
> - Calculate GPP by summing the primary productivity of all photosynthesizing organisms within the ecosystem.
> - If using laboratory data, sum the measured rates of primary productivity for all relevant samples.
> - If using field data, scale up the measured primary productivity per unit area to the entire ecosystem.
>
> Net primary productivity (NPP) calculation:
> - Subtract the energy used by autotrophs during respiration from GPP. This can be estimated by applying appropriate respiration rates or conversion factors.

2.2.26 Maximum sustainable yields in natural and human-managed systems

Maximum sustainable yield (MSY) is a concept used to determine the optimal level of extraction or utilization of resources in natural ecosystems and in agricultural or silvicultural systems. MSY is based on the net primary productivity (NPP) or net secondary productivity (NSP) of a system, representing the amount of energy or biomass that can be sustainably harvested or utilized without depleting the resource base.

In natural ecosystems

In natural ecosystems, MSY refers to the level of harvesting or utilization that can be sustained over the long-term without compromising the ecological integrity of the system. It is based on understanding the ecological dynamics and limits of the ecosystem, ensuring that the removal or utilization of resources does not exceed the rate of natural replenishment. MSY is often calculated based on the NPP or NSP of the ecosystem, considering factors such as population dynamics, reproductive rates and natural resilience.

Achieving MSY in natural ecosystems is crucial for maintaining biodiversity, ecosystem stability and the sustainable provision of ecosystem services. It involves carefully managing the balance between resource extraction and natural regeneration, while considering both the needs of human communities and the preservation of the ecosystem.

In agricultural or silvicultural systems

In agricultural or silvicultural systems, MSY is relevant for optimizing productivity and yield while ensuring the long-term sustainability of the system. It represents the maximum amount of crop or timber production that can be achieved without degrading soil fertility, depleting water resources or causing long-term damage to the ecosystem. MSY is typically based on the NPP or NSP of the cultivated plants or managed forest.

In agricultural systems, achieving MSY involves employing sustainable agricultural practices such as crop rotation, proper soil management and efficient water use to maintain the productivity of the land over time. It also considers factors like pest management, genetic diversity and the ecological balance of the agroecosystem.

In silvicultural systems, MSY focuses on maximizing timber production while ensuring the long-term health and productivity of the forest. It involves practices such as selective harvesting, reforestation, and maintaining a balance between commercial exploitation and ecological conservation.

By applying MSY principles, natural resource managers, farmers and foresters aim to optimize productivity, maintain ecosystem health, and ensure the sustainable use of resources for present and future generations.

2.2.27 Sustainable yields and lower trophic levels in food production

Sustainable yields in food production are more easily achieved when humans consume organisms from lower trophic levels, particularly plant-based foods.

Energy transfer efficiency

Energy transfer efficiency is a crucial concept in ecology. As energy moves through food chains, there is a natural decline in efficiency from one trophic level to the next. This decline means that less energy is accessible to support higher trophic levels. Therefore, by consuming organisms from lower trophic levels, such as plants, humans can tap into a more significant portion of the available energy in the ecosystem. This approach not only enhances energy efficiency but also promotes higher sustainable yields. Importantly, accessing energy lower down the food chain is beneficial because it reduces the strain on resources and contributes to the overall balance of the ecosystem. This strategy aligns with the idea of utilizing resources more effectively and emphasizes the importance of considering energy dynamics in ecological systems.

Trophic level energy loss

At each trophic level, energy is lost through metabolic processes, such as respiration and growth. This again limits the overall energy available to higher trophic levels.

Resource requirements

Higher trophic level organisms generally require greater amounts of resources, including food and water, compared to lower trophic level organisms. By consuming organisms from lower trophic levels, humans can reduce the demand for resources and achieve higher sustainable yields.

Ecological footprint

The ecological footprint of food production is generally lower for plant-based foods. This footprint includes factors such as land use, water consumption, greenhouse gas emissions and habitat destruction. By focusing on lower trophic levels, especially plant-based foods, food production can be more sustainable.

Biodiversity conservation

Consuming organisms from lower trophic levels allows for better preservation of biodiversity. By reducing the demand for higher trophic level species, there is less pressure on these species and their habitats. This, in turn, helps maintain ecosystem balance and protects species diversity.

2.2.28 Ecological efficiency and energy transfer between trophic levels

Ecological efficiency refers to the percentage of energy received by one trophic level that is passed on to the next trophic level in a food chain or food web. It quantifies the efficiency of energy transfer within an ecosystem.

The efficiency of energy transfer is influenced by various factors, including the complexity of the food web, the physiological characteristics of organisms and the availability of resources.

Link

Further links can be found in Topics 1.3 (page 47), 4.3 (page 331) and 5.2 (page 456).

While it is common to estimate the energy transfer efficiency as 10 per cent, it is essential to understand that this value is neither a fixed amount nor a true average across all ecosystems.

In reality, the actual ecological efficiency can range widely depending on the specific ecosystem and the species involved. Ecological efficiency can be higher or lower than 10 per cent based on several factors, such as the efficiency of digestion and assimilation, metabolic rates, trophic interactions and environmental conditions. Some ecosystems or trophic levels may exhibit higher ecological efficiency due to more efficient energy transfer mechanisms, while others may have lower efficiency due to energy losses during metabolic processes or competition for resources.

Calculating the efficiency of energy transfer between trophic levels involves analysing energy flow data and biomass measurements within a food chain or food web. By quantifying the energy content of each trophic level and determining the amount of energy transferred to the next trophic level, the ecological efficiency can be estimated.

> **Tool 3: Mathematics**
>
> Work out the efficiency of energy transfer between trophic levels
>
> To work out the efficiency of energy transfer between trophic levels with given data, follow these steps:
>
> 1. Determine the total energy available at the initial trophic level.
> 2. Measure the energy content of the organisms at the initial trophic level.
> 3. Calculate the energy content of the organisms at the subsequent trophic level.
> 4. Divide the energy content at the subsequent trophic level by the energy content at the initial trophic level.
> 5. Multiply the result by 100 to obtain the ecological efficiency as a percentage.
>
> Remember that the efficiency of energy transfer can vary greatly depending on the specific ecosystem and the organisms involved. It is crucial to gather accurate data and consider the intricacies of the ecosystem when calculating ecological efficiency.

2.2.29 The second law of thermodynamics shows how the entropy of a system increases as biomass passes through ecosystems

The second law of thermodynamics provides insights into the behaviour of energy and matter in ecosystems, particularly in relation to entropy and the flow of biomass. Entropy refers to the amount of disorder within a system, and the second law states that in any energy transfer or transformation, the entropy of the universe (or a closed system) will always increase.

In living systems, such as ecosystems, organisms exhibit a high degree of organization and low entropy. This organization is maintained through various biological processes, including growth, reproduction and the utilization of energy. However, the second law of thermodynamics shows that the overall entropy of the system, including the transfer and transformation of energy and matter, will increase over time.

In the context of ecosystems, the passage of biomass through trophic levels contributes to the increase in entropy. Biomass, representing the organic matter and energy stored within living organisms, undergoes various transformations as it moves through the food chain or food web.

At each trophic level, energy and biomass are transferred from one organism to another through processes such as consumption and digestion. However, with each transfer, some energy is lost as heat due to metabolic processes, and some biomass is expelled as waste. These processes contribute to the increase in entropy within the ecosystem.

Living organisms, through their metabolic activities, consume energy and nutrients to maintain their own organization and low entropy. This is primarily achieved through respiration, which releases energy from organic molecules and results in the net increase in entropy.

The concept of entropy and the second law of thermodynamics highlight the fundamental principles governing energy and matter within ecosystems. They underscore the irreversible nature of energy transformations and the inherent increase in disorder over time. Despite the increase in entropy, living systems are capable of maintaining a degree of organization through ongoing energy acquisition and utilization.

ATL ACTIVITY

Reflect on this section by answering these questions.
1. Through a careful exploration of primary or secondary data, how can we effectively understand the complex outcomes of pollution on an ecosystem, focusing on its cascading effects on food chains? For instance, how can we understand how a sewage overflow disrupts aquatic communities, while also delving into the vital aspects of health, safety and ethical concerns?
2. Using the concepts of the second law of thermodynamics, how can we enthusiastically promote the idea of adopting a diet that benefits both our planet's health and our own? Combine your scientific knowledge with the real-world feasibility of supporting yourselves and the environment within your student community.
3. Taking an active role in the realm of citizen science, how can your participation contribute valuably to ongoing initiatives centred around microplastics? As you delve into this exploratory journey, how can your hands-on involvement add new dimensions to your understanding of the prevalence and impact of microplastics in our environment?

REVIEW QUESTIONS

1. Discuss the three different types of ecological pyramid (pyramids of number, pyramids of biomass and pyramids of energy). How do these pyramids provide insights into the structure and functioning of ecosystems?
2. Describe the role of autotrophs and heterotrophs in energy and matter transfer within ecosystems. Provide examples of different consumer strategies and explain how energy flows through different trophic levels in a food chain.
3. Discuss the limitations on the number of trophic levels in ecosystems and the factors that contribute to energy losses along the food chain.

EXAM-STYLE QUESTIONS

1. With reference to processes occurring within the atmospheric system:
   a. Identify **two** transformations of matter. [2 marks]
   b. Identify **two** transfers of energy. [2 marks]
2. Outline the procedures in a laboratory-based method to find the gross productivity for a population of named aquatic animals in terms of biomass per day. [4 marks]

2.3 Biogeochemical cycles

Guiding question

- How do human activities affect nutrient cycling and what impact does this have on the sustainability of environmental systems?

SYLLABUS CONTENT

This chapter covers the following syllabus content:
- ▶ 2.3.1 Biogeochemical cycles ensure chemical elements continue to be available to living organisms.
- ▶ 2.3.2 Biogeochemical cycles have stores, sinks and sources.
- ▶ 2.3.3 Organisms, crude oil and natural gas contain organic stores of carbon. Inorganic stores can be found in the atmosphere, soils and oceans.
- ▶ 2.3.4 Carbon flows between stores in ecosystems by photosynthesis, feeding, defecation, respiration, death and decomposition.
- ▶ 2.3.5 Carbon sequestration is the process of capturing gaseous and atmospheric carbon dioxide and storing it in a solid or liquid form.
- ▶ 2.3.6 Ecosystems can act as stores, sinks or sources of carbon.
- ▶ 2.3.7 Fossil fuels are stores of carbon with unlimited residence times, that were formed when ecosystems acted as carbon sinks in past eras and become carbon sources when burned.
- ▶ 2.3.8 Agricultural systems can act as carbon stores, sources and sinks, depending on the techniques used.
- ▶ 2.3.9 Carbon dioxide is absorbed into the oceans by dissolving and is released as a gas when it comes out of solution.
- ▶ 2.3.10 Increases in concentrations of dissolved carbon dioxide cause ocean acidification, harming marine animals.
- ▶ 2.3.11 Measures are required to alleviate the effects of human activities on the carbon cycle.

HL ONLY

- ▶ 2.3.12 The lithosphere contains carbon stores in fossil fuels and in rocks such as limestone that contain calcium carbonate.
- ▶ 2.3.13 Reef-building corals and molluscs have hard parts that contain calcium carbonate that can become fossilized in limestone.
- ▶ 2.3.14 In past geological eras, organic matter from partially decomposed plants became fossilized in coal, and partially decomposed marine organisms became fossilized in oil and natural gas held in porous rocks.
- ▶ 2.3.15 Methane is produced from dead organic matter in anaerobic conditions by methanogenic bacteria.
- ▶ 2.3.16 Methane has a residence time of about ten years in the atmosphere and is eventually oxidized to carbon dioxide.
- ▶ 2.3.17 The nitrogen cycle contains organic and inorganic stores.
- ▶ 2.3.18 Bacteria have essential roles in the nitrogen cycle.
- ▶ 2.3.19 Denitrification only happens in anaerobic conditions, such as soils that are waterlogged.
- ▶ 2.3.20 Plants cannot fix nitrogen so atmospheric dinitrogen is unavailable to them unless they form mutualistic associations with nitrogen-fixing bacteria.
- ▶ 2.3.21 Flows in the nitrogen cycle include mineral uptake by producers, photosynthesis, consumption, excretion, death, decomposition and ammonification.

▶ 2.3.22 Human activities such as deforestation, agriculture, aquaculture and urbanization change the nitrogen cycle.
▶ 2.3.23 The Haber process is an industrial process that produces ammonia from nitrogen and hydrogen for use as fertilizer.
▶ 2.3.24 Increases in nitrates in the biosphere from human activities have led to the planetary boundary for the nitrogen cycle being crossed, making irreversible changes to Earth systems likely.
▶ 2.3.25 Global collaboration is needed to address the uncontrolled use of nitrogen in industrial and agricultural processes, and to bring the nitrogen cycle back within planetary boundaries.

2.3.1 Biogeochemical cycles ensure chemical elements continue to be available to living organisms

Biogeochemical cycles regulate the transport and availability of chemical components within ecosystems. The carbon cycle, nitrogen cycle and phosphorus cycle, for example, are biogeochemical cycles and they ensure that the key materials required for life are continuously recycled and made available to living organisms. Elements that cycle through the **biosphere**, **lithosphere**, **atmosphere** and **hydrosphere** include carbon, nitrogen and phosphorus.

◆ **Biosphere** – refers to the region of the Earth that encompasses all living organisms and their interactions with each other and their environment.
◆ **Lithosphere** – the solid outermost shell of the Earth, consisting of the crust and the uppermost part of the mantle.
◆ **Atmosphere** – the layer of gases that surrounds the Earth. It is composed of various gases like nitrogen, oxygen, carbon dioxide and others, and contains variable amounts of water vapour.
◆ **Hydrosphere** – the total sum of the Earth's water in all its forms, including water on the surface (in oceans, seas, lakes, rivers and groundwater) and water vapour in the atmosphere.

Human activity has a huge impact on these cycles, frequently causing imbalances and interruptions. For example, the combustion of fossil fuels emits significant volumes of carbon dioxide into the atmosphere, which contributes to climate change. Deforestation and fertilizer overuse can disrupt the nitrogen and phosphorus cycles, resulting in water pollution and ecological deterioration.

■ Figure 2.35 Combustion of fossil fuels

■ Figure 2.36 Deforestation

■ Figure 2.37 Nitrogen

■ Figure 2.38 Water pollution

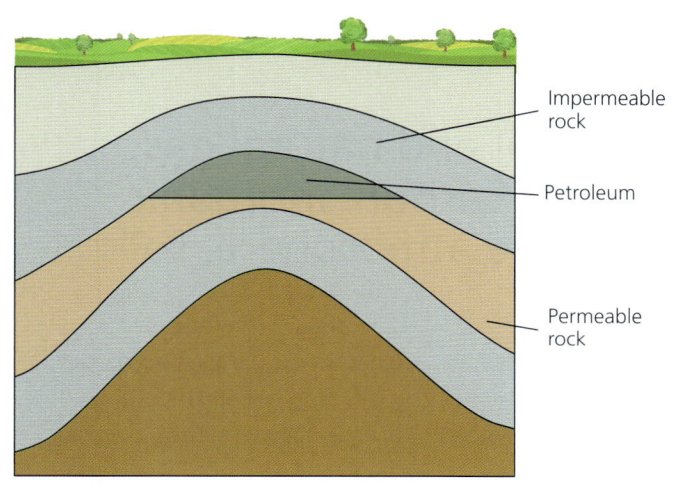

■ Figure 2.39 Petroleum storage in underground rocks

■ Figure 2.40 Forests as sinks

2.3.2 Biogeochemical cycles have stores, sinks and sources

In biogeochemical cycles, elements are stored in different compartments or reservoirs within the Earth's systems. These compartments can be categorized as stores, sinks or sources.

Stores

Stores, also known as storages, are 'reservoirs' in the biogeochemical cycles that are in equilibrium (balance) with their surroundings. They serve as long-term storage facilities for components such as carbon, nitrogen and phosphorus. Fossil fuels, organic material in soils and minerals in rocks are all examples of storage.

Sinks

Sinks represent an element's net accumulation. They absorb more of an element than they release for a short period of time, thereby removing it from circulation. Forests and oceans are examples of carbon dioxide sinks because they absorb more carbon dioxide than they release through respiration.

2.3 Biogeochemical cycles

> **Link**
>
> More about biogeochemical cycles is covered in Chapter 2.3 (page 139).

■ Sources

A net discharge of an element happens within sources. They expel more of the element than they receive, reintroducing it into the cycle. Carbon dioxide is produced through the combustion of fossil fuels and deforestation, which increases its concentration in the atmosphere.

■ **Figure 2.41** Deforestation causes discharge of carbon dioxide from sources

2.3.3 Organisms, crude oil and natural gas contain organic stores of carbon. Inorganic stores can be found in the atmosphere, soils and oceans

Within biogeochemical cycles, carbon exists in both organic and inorganic forms. Living species such as plants, animals and microbes, as well as fossil fuels such as crude oil and natural gas, are organic carbon storage. The atmosphere, soils and oceans are examples of inorganic carbon storage.

When the absorption of carbon is balanced by emission (the same amount goes in as goes out), a store is said to be in equilibrium. This equilibrium provides for a reasonably steady carbon content in the store over time. The average time that a carbon atom spends in a reservoir is referred to as its residence time. Carbon atoms, for example, constantly cycle through activities such as photosynthesis, feeding, respiration, death and degradation in living creatures. By contrast, because fossil fuels need major geological processes in order to develop, their residence period is often measured in hundreds of millions of years.

Human activities such as mining and the use of fossil fuels can upset the equilibrium dynamics of carbon storage. They alter the carbon balance within the Earth's systems and contribute to climate change by rapidly releasing carbon dioxide into the atmosphere.

All types of soil contain organic matter, which contains carbon. This organic matter affects soil properties, including colour, capacity to hold nutrients, nutrient cycling and stability. These in turn impact the ability of the soil to hold water, provide aeration and facilitate workability. Insects and microorganisms decompose plant debris (such as leaves, twigs, dead stems and flowers) into humus. Beyond serving as a nutrient source for plants, carbon compounds in humus interact with substances like iron, aluminium hydroxides and clay minerals. This interaction causes soil particles to bind together, forming stable aggregates. Consequently, this process helps mitigate soil erosion and enhances the effectiveness of root penetration into the soil.

Tool 1: Experimental techniques

Soil organic matter represents a reservoir of carbon. To understand the inorganic storage of carbon better, let's estimate carbon content with the activity below. You might then compare carbon storage between different soil types or land uses.

Estimate carbon content

Organic matter can be estimated from a soil sample. This measurement is different from total organic carbon in that it includes all the elements (for example, hydrogen, oxygen, nitrogen) that are components of organic compounds, not just carbon.

In order to get a measurement of organic carbon per hectare (a useful standardized unit), several calculations need to be done. Use a spreadsheet if possible for this. The stages are set out in Table 2.4.

■ **Table 2.4** Measuring organic carbon per hectare

Stage	Activity	Description
1	Calculate soil bulk density. This varies between soils and it is better to measure it.	Bulk density is the dry weight of a known volume of soil. A soil sample can be taken using a core or metal tube hammered into soil for a given depth. Weigh the sample (in grams). To get a dry soil, heat at 100°C overnight in a large tray to remove all moisture. Weigh again after.
2	Work out the organic-matter content of your soil (%). Unlike peats, which are almost entirely organic, other soils will have much lower organic-matter content.	Use secondary research to determine typical organic carbon contents for soils in your area, e.g. **www.landis.org.uk/services/soilscapes** or colour charts: **https://pubsplus.illinois.edu/products/color-chart-for-estimating-organic-matter-in-mineral-soils-in-illinois**, which will work for grassland soils. If your school laboratory has a muffle furnace and a high-precision balance (measuring grams to at least three decimal places), you may be able to use the loss-on-ignition technique: • Weigh an oven-dry soil sample. • Heat to 500°C for four hours to burn off the organic matter. • Weigh again. Ensure you follow the safety procedures laid down in the risk assessment for this activity by your school laboratory.
3	Calculate organic carbon (%).	Use a conversion factor of 58 per cent of the organic matter.
4	Convert to tonnes per hectare.	This is a useful measure for comparing between different land uses and areas.

Bulk density estimate example

A metal tube 7 cm in diameter (3.5 cm radius), banged into a soil to a depth of 10 cm has a volume of:

$$3.14 \times (3.5 \times 3.5) \times 10 = 385 \text{ cm}^3$$

This soil is weighed and found to be 650 g. It is then dried and reweighed and found to be 500 g.

$$\begin{aligned}\text{Bulk density (BD)} &= \text{Dry soil weight} / \text{Volume} \\ &= 500 / 385 \\ &= 1.3 \text{ g cm}^{-3} \\ &= 1{,}300{,}000 \text{ kg ha}^{-1} \text{ to a soil depth of 10 cm}\end{aligned}$$

2.3 Biogeochemical cycles

> ### Organic-matter conversion to organic carbon example
>
> If a small sample of dry soil has dry weight 10 g and after loss on ignition weighs 9.8 g (note that this could be a much smaller sample), then organic matter is calculated as:
>
> $$0.2 / 10 = 2\% \text{ organic matter}$$
>
> An average conversion factor of 0.58 is commonly used to convert between soil organic carbon and organic matter (although note that this can vary with the type of organic matter, soil type and soil depth):
>
> $$\text{Total organic carbon (\%)} = \text{Organic matter (\%)} \times 0.58$$
>
> Here is the calculation for the amount of carbon in our soil:
>
> $$2 \times 0.58 = 1.16\%$$
>
> ### Carbon estimate per area example
>
> The amount of carbon to 10 cm depth in soil with a carbon value of 1.16 per cent and bulk density of 1.3 g cm^{-3} is:
>
> $$11.6 \text{ (gC kg}^{-1}\text{ soil)} \times 1{,}300{,}000 \text{ (kg soil ha}^{-1}\text{)} = 15.08 \text{ tC ha}^{-1}$$
>
> Note that to do this calculation you need to convert the percentage carbon content to a value in gC kg^{-1} of soil (11.6), and you need to work out the volume of soil to 10 cm depth over 1 ha and then use the bulk density to calculate the mass of the soil (1,300,000 kg).

Link

More about the impacts of carbon is covered in Chapter 6.2 (page 528).

2.3.4 Carbon flows between stores in ecosystems by photosynthesis, feeding, defecation, respiration, death and decomposition

The carbon cycle is the transport of carbon from one ecosystem to another. Photosynthesis, feeding, defecation (releasing excrement), respiration, death and decomposition are all processes that contribute to the movement of carbon.

- Photosynthesis is the process through which plants and some microbes transform atmospheric carbon dioxide into organic molecules, most notably glucose. Carbon is absorbed into plant tissues by photosynthesis, which can later be ingested by other species via feeding.
- Feeding and defecation involve carbon being taken in and given out by an organism.
- The release of carbon dioxide back into the atmosphere as a consequence of cellular respiration occurs in both plants and mammals.
- This carbon dioxide can be taken up again by plants via photosynthesis, completing the cycle. When organisms die, their organic matter decomposes, which is aided by decomposers such as bacteria and fungi. During decomposition, carbon is returned to the soil or atmosphere as carbon dioxide.

2.3.5 Carbon sequestration is the process of capturing gaseous and atmospheric carbon dioxide and storing it in a solid or liquid form

Through the process of photosynthesis, trees and other photosynthetic organisms naturally sequester carbon dioxide. They collect CO_2 from the environment and transform it into organic compounds like sugars and cellulose, which comprise their biomass. This technique helps to remove carbon dioxide from the atmosphere, lowering its concentration and so mitigating climate change.

Organic materials can be buried for long periods of time and undergo geological processes, culminating in the creation of fossil fuels such as coal, oil and natural gas. Planting trees and growing seagrasses can also contribute to the reduction of atmospheric CO_2 levels, which is essential for combating climate change.

Carbon sequestration is critical for minimizing the effects of rising carbon dioxide concentrations in the atmosphere. It entails removing carbon dioxide from the atmosphere or industrial processes and storing it in solid or liquid form for an extended period of time.

However, it's important to note that while tree planting and seagrass restoration are valuable strategies, they are not a sole solution to the carbon emissions problem. They should be complemented by broader efforts to reduce greenhouse gas emissions, transition to renewable energy sources, and implement sustainable land and ocean management practices.

Link

More about the impacts of carbon sequestration is covered in Chapter 6.2 (page 528).

■ **Figure 2.42** A young forest

■ **Figure 2.43** A mature forest

2.3.6 Ecosystems can act as stores, sinks or sources of carbon

Ecosystems play an important role in carbon cycling and can behave as carbon sinks, stores, or sources depending on a variety of conditions. The balance between total carbon inputs and outputs determines the net buildup or release of carbon in an ecosystem. For example:

- In a young forest, where rates of photosynthesis surpass respiration, there is a net intake of carbon dioxide. Such forests operate as sinks, absorbing more carbon dioxide than they emit, helping to reduce atmospheric carbon dioxide levels.
- A mature forest, on the other hand, tends to be in carbon equilibrium, with inputs and outputs balanced. It serves as a carbon store, retaining a reasonably steady carbon supply over time.

■ **Figure 2.44** A recently logged patch of woods on the edge of the White Mountain National Forest in Chatham, New Hampshire

- When ecosystems are disrupted or destroyed, however, they can become carbon sources. Deforestation, for example, causes the release of stored carbon as trees are felled, decayed or burned. Similarly, activities like wildfires and land-use changes can transform ecosystems from carbon sinks to carbon sources.

> **Link**
>
> More about the impacts of carbon stores, sinks, or sources is covered in Chapters 1.2 (page 24), 5.1 (page 416), 5.2 (page 456) and 6.2 (page 528).

2.3.7 Fossil fuels are stores of carbon with unlimited residence times, formed when ecosystems acted as carbon sinks in past eras. They become carbon sources when burned

> **Link**
>
> More about the impacts of fossil fuel use is covered in Chapter 6.2 (page 528).

Coal, oil, and natural gas are massive carbon reservoirs that have accumulated over millions of years. They are made from ancient organic matter, such as plant and animal remains, that has been subjected to extreme heat and pressure throughout geological time.

These are carbon reservoirs with nearly infinite residence durations. When ecosystems served as carbon sinks in the past, they absorbed and stored huge amounts of CO_2 via photosynthesis. This stored carbon eventually became locked in the Earth's crust and turned into coal, oil and natural gas. When fossil fuels are consumed, this stored carbon is released back into the atmosphere as carbon dioxide. This increases atmospheric CO_2 concentrations, which causes climate change and global warming.

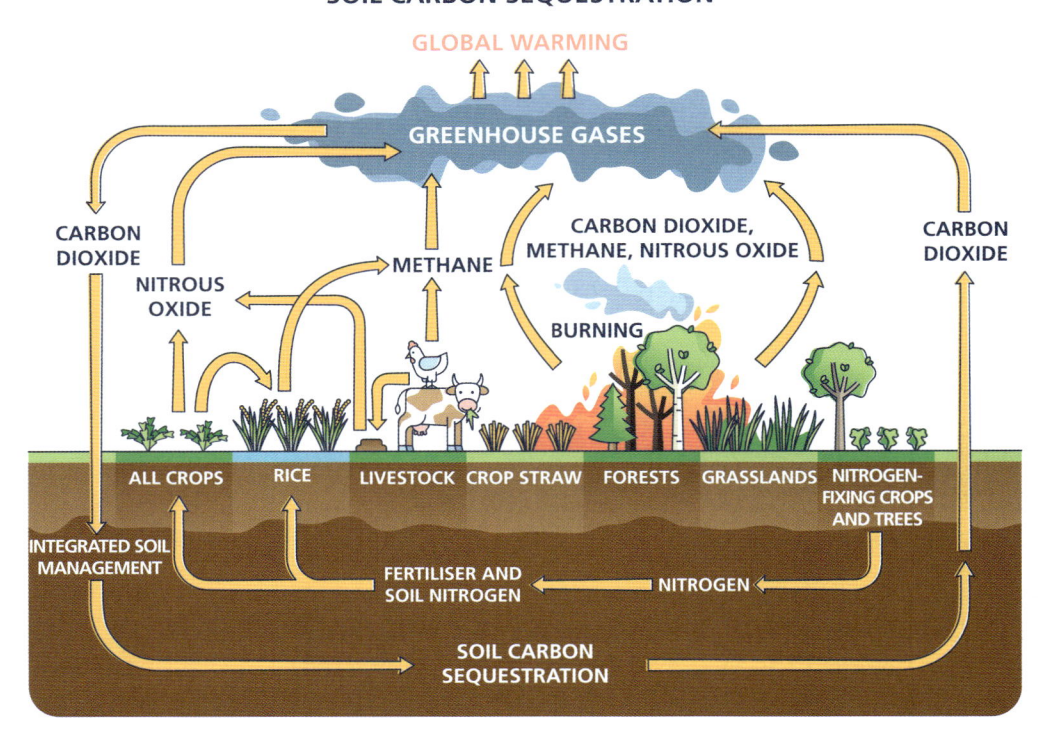

■ **Figure 2.45** The carbon cycle

Theme 2: Ecology

2.3.8 Agricultural systems can act as carbon stores, sources or sinks, depending on the techniques used

Agricultural systems have a large impact on the carbon cycle since various practices can either encourage carbon storage, operate as carbon sources or improve carbon sinks.

Crop rotation, cover cropping and no-till farming are examples of regenerative agricultural practices that promote soil as a carbon sink. These practices increase organic-matter accumulation in the soil, which helps sequester and store carbon dioxide from the atmosphere.

■ Figure 2.46 Crop rotation

■ Figure 2.47 Crimson clover can be used as a cover crop

■ Figure 2.48 Tillage

■ Crop rotation

Crop rotation is a farming practice that involves alternating the types of crops grown in a particular area over time to improve soil health, manage pests, and enhance overall crop yield.

■ Cover cropping

Cover cropping refers to planting specific crops, usually non-commercial, in between main crops to protect and enrich the soil, prevent erosion, and enhance biodiversity. An example of cover cropping is planting a cover crop like clover or rye in a field after the main crop (for example, corn or wheat) has been harvested. The cover crop grows during the off-season, which protects the soil from erosion, improves its structure and adds nutrients. This practice helps to maintain soil health and prepares the land for the next planting season.

■ Tillage

Tillage is the process of preparing soil for planting by ploughing, cultivating or digging to break up the soil, incorporate organic matter and create a suitable seedbed.

■ No-till farming

No-till farming is a conservation method where crops are planted directly into undisturbed soil, without tilling, to reduce soil erosion, improve water retention and maintain soil structure.

Certain agricultural practices can cause the release of stored carbon, thus transforming agricultural systems into carbon sources. This varies based on management practices, environmental conditions and other factors. Draining wetlands, for example, which have large volumes of carbon stores in their soils, may result in the emission of carbon dioxide. Furthermore, practices such as monoculture (growing a single crop) and extensive tillage can hasten the breakdown of organic materials, resulting in carbon dioxide emissions. By implementing sustainable and carbon-conscious agricultural practices, it is possible to minimize carbon emissions and make agricultural systems carbon neutral or even transform them into carbon sinks, where they store more carbon than they release.

The **duration of cropping** and the subsequent uses of produced goods influence agricultural systems' functions as carbon storage, sources or sinks. Longer agricultural rotations or long-term timber cultivation, for example, can improve carbon storage, but fast decomposition of harvested residues can add to carbon emissions.

◆ **Duration of cropping** – how long a particular piece of agricultural land is used for growing crops before it is either rotated to grow a different crop or left empty for a period of time. It is essentially the time period between planting and harvesting the crops in a specific field.

Link

More about agricultural systems is covered in Chapter 5.2 (page 456).

> **ATL ACTIVITY**
>
> Conduct a brief search to define crop rotation, cover cropping and no-till farming, and find examples for each.

HL.b.1: Ecological economics

Agricultural systems play a crucial role in the complex web of economics and the environment. Economics studies how humans produce, distribute and consume goods and services, considering the supply and demand of resources and their impact on market outcomes. While the topic at hand, agricultural systems, might not seem directly tied to market interactions, it's important to recognize that the techniques used in agriculture influence the distribution of a valuable resource: carbon. Just as economics analyses the allocation of resources, agricultural practices determine how carbon is stored, emitted or absorbed within these systems.

HL.b.13: Ecological economics

The relationship between agricultural systems and the carbon cycle serves as a real-world example of how economic principles interact with environmental outcomes. The impact of economic growth on environmental welfare is shown through the effects of agricultural systems on the carbon cycle. Different agricultural practices can affect carbon stores, sources or sinks, paralleling the dual nature of economic growth's impact on the environment. Practices like regenerative agriculture align with sustainable economic growth principles, mitigating climate change, while soil-degrading practices mirror unchecked economic growth's negative consequences. This relationship emphasizes the real-world interplay between economic principles and environmental outcomes, emphasizing the lasting impact of our choices on both economies and the planet's well-being.

2.3.9 Carbon dioxide is absorbed into the oceans by dissolving and is released as a gas when it comes out of solution

By absorbing and releasing carbon dioxide, the seas play an important role in the global carbon cycle. Carbonic acid is formed when atmospheric carbon dioxide dissolves in seawater and combines with the water to generate carbonic acid ($CO_2 + H_2O \rightleftharpoons H_2CO_3$), which is broken down into hydrogen ions and bicarbonate ions.

This process, known as oceanic absorption, enables the oceans to serve as a major carbon sink. The dissolved carbon dioxide remains in the surface waters of the ocean and is transferred to deeper layers by mixing and circulation, depending on the temperature and ocean currents.

However, the use of fossil fuels, in particular, has resulted in a large increase in atmospheric carbon dioxide concentrations. Oceans are currently absorbing more carbon dioxide than they can naturally accommodate at the normal pH. As a result, this extra carbon dioxide disturbs the balance, resulting in ocean acidification.

Link
More about the impacts of carbon absorption is covered in Chapter 6.2 (page 528).

Tool 3: Mathematics

Interpret graphs and data by using two graphs from different sources

Use the two graphs below to discuss the relationship between increased CO_2 concentration and ocean acidification.

Graph 1 below shows the mean seawater pH collected from measurements of pH at the Aloha Station in Hawaii from 1988 to 2022. Graph 2 shows the increase of CO_2 at Mauna Loa since pre-industrial times.

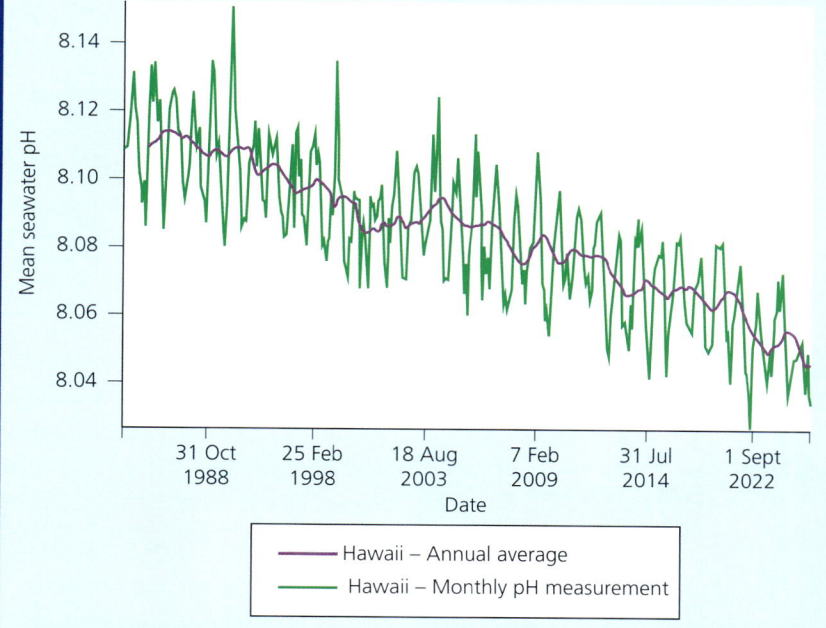

■ **Figure 2.49** Graph 1: Ocean acidification at the Aloha Station from 1988 to 2021 (Source: University of Hawaii)

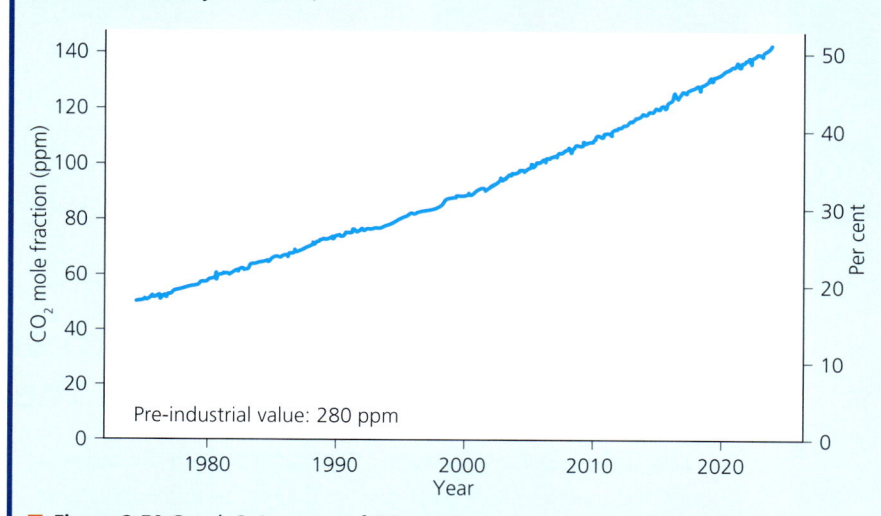

■ **Figure 2.50** Graph 2: Increase of CO_2 at Mauna Loa since pre-industrial times

- Examine the axes, labels, units and any other important information in both graphs.
- Write down your observations and initial insights about the graphs.
- Discuss what trends, patterns or conclusions you can draw from the graphs.
- Pay attention to differences and similarities between the two graphs.

2.3 Biogeochemical cycles

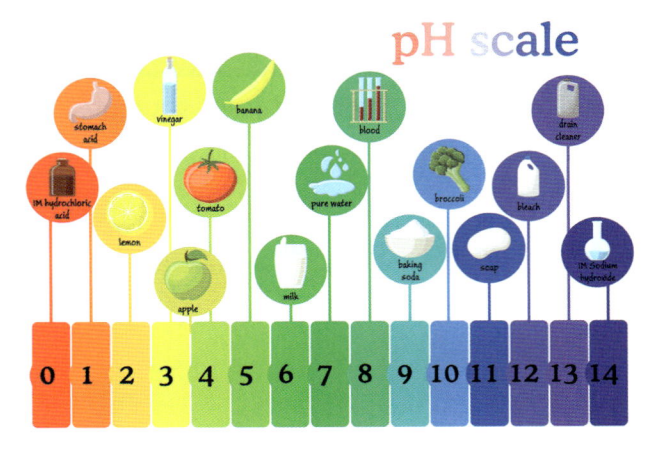

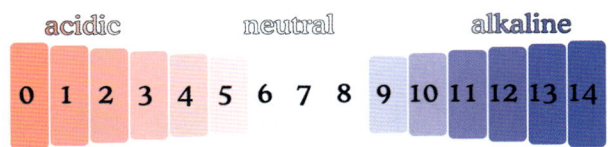

Figure 2.51 pH scale

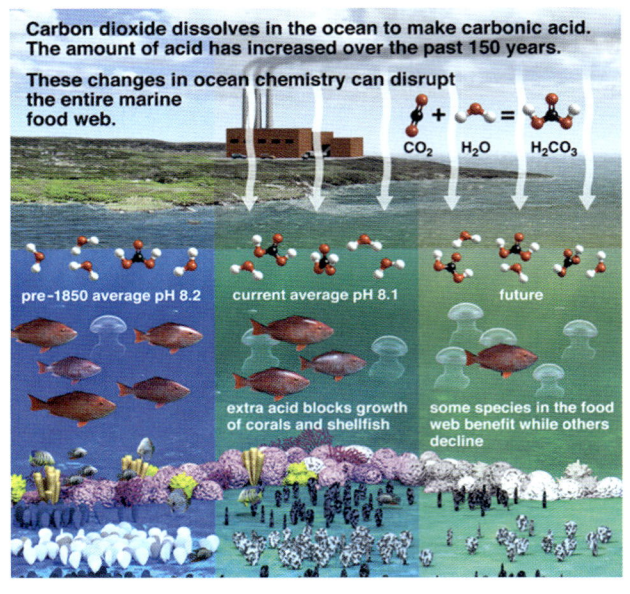

Figure 2.52 Increases in concentrations of dissolved carbon dioxide cause ocean acidification, harming marine animals

2.3.10 Increases in concentrations of dissolved carbon dioxide cause ocean acidification, harming marine animals

As carbon dioxide dissolves in the oceans, it reacts with water to generate carbonic acid, which lowers the pH of the seawater and causes it to become more acidic.

Acidification has a negative impact on many marine creatures, particularly those that rely on calcium carbonate for shell creation or skeletal support. Molluscs such as oysters, clams and snails, as well as coral reefs, rely on calcium carbonate deposition to develop their protective structures. This process of calcium carbonate deposition becomes more difficult in more acidic environments.

Even minor changes in seawater pH can upset the carbonate ion equilibrium, making it difficult for marine species to produce and maintain their shells and skeletons. This can result in slower development rates, weaker structures, and increased vulnerability to predation and other environmental pressures.

Ocean acidification has far-reaching consequences that go beyond individual creatures and can have a domino effect on entire marine ecosystems. Coral reefs, for example, provide a critical habitat for a diverse range of marine animals. If coral growth is hampered by ocean acidification, it can result in biodiversity loss and ecosystem collapse.

Recognizing the repercussions of increasing dissolved carbon dioxide concentrations and ocean acidification highlights the critical need to address the root causes of carbon dioxide emissions and implement actions to protect marine ecosystems.

2.3.11 Measures are required to alleviate the effects of human activities on the carbon cycle

To address the adverse impacts of human activities on the carbon cycle and combat climate change, various measures must now be taken. These measures aim to reduce carbon dioxide emissions, enhance carbon sinks and promote sustainable practices across different sectors.

1. Adoption of low-carbon technologies: Implementing low-carbon technologies, such as renewable energy sources (solar, wind, hydro), energy-efficient systems and electric transportation, can significantly reduce greenhouse gas emissions. These technologies offer alternative energy options that have minimal or no carbon dioxide emissions, and contribute to a more sustainable energy future.

2. Reduction in fossil fuel burning, soil disruption and deforestation: Burning fewer fossil fuels, which is a major source of carbon dioxide emissions, is crucial for mitigating the effects of human activities on the carbon cycle. Additionally, minimizing soil disruption through practices like conservation tillage and promoting sustainable land management can help prevent the release of carbon stored in soils. Furthermore, curbing deforestation and promoting reforestation efforts can enhance carbon sinks by increasing the capacity of forests to absorb carbon dioxide through photosynthesis.

3. Carbon capture through reforestation and artificial sequestration: Reforestation plays a vital role in capturing carbon dioxide from the atmosphere and restoring carbon sinks. Planting trees and restoring degraded ecosystems not only sequesters carbon but also enhances biodiversity and ecosystem services. Additionally, artificial sequestration techniques, such as carbon capture and storage (CCS) technologies, aim to capture carbon dioxide from large point sources, such as power plants and industrial facilities, and store it underground or utilize it in other applications.

REAL-WORLD EXAMPLE

Carbon capture and utilization

Over the past few years, the news headlines have revolved around the escalating global temperatures. Without intervention, temperatures are poised to surge by 1.5°C beyond pre-industrial levels within the upcoming two decades, and a rise of 2°C is also distinctly plausible. To recalibrate our trajectory and prevent overshooting these benchmarks, more potent strategies for mitigating climate change are imperative. One such approach is carbon capture and utilization (CCU).

■ **Figure 2.53** Carbon dioxide is a valuable resource, critical in many industries

Towards the end of 2020, the UK government unveiled an ambitious new objective: to curtail greenhouse gas emissions by 68 per cent over the next decade, in comparison to levels recorded in 1990. The comprehensive ten-point blueprint devised to actualize this goal encompassed investments in CCU infrastructure, with the aim of capturing 10 million metric tonnes of carbon dioxide (the principal contributor to climate change) annually by the decade's end. This translates to mitigating emissions equivalent to what approximately 4 million cars produce in a year.

These measures are crucial for transitioning towards a more sustainable and climate-resilient future.

2.3 Biogeochemical cycles

HL.a.2: Environmental law

Environmental Impact Assessments are a systematic process used to identify and evaluate the potential environmental effects of proposed projects, policies, programmes or activities before they are carried out. The purpose of an EIA is to predict and assess the potential impacts of these actions on the environment and to ensure that decision-makers have accurate and relevant information for making informed choices. An EIA helps in identifying potential environmental risks, suggesting mitigation measures, and providing a basis for regulatory decisions and public input. It is a crucial tool for sustainable development and environmental protection.

HL.b.9: Ecological economics

At its core, ecological economics perceives the biosphere as a delicate system, where the economy is a subset of the Earth's larger biosphere. This lens underscores the importance of recognizing that our social and economic systems are sub-components of the broader ecological framework. The nitrogen cycle, a fundamental process for life on Earth, has come under strain due to uncontrolled human activities. To address this challenge, we must not view economic growth and resource use in isolation but rather as integral parts of the intricate web of life.

The need for global collaboration becomes evident as we aim to restore the balance within the nitrogen cycle and other natural systems. Unilateral actions are insufficient, instead concerted efforts across nations, industries and communities are essential. Collaborative initiatives can harness the expertise, resources and collective determination needed to minimize the environmental impacts caused by nitrogen cycle disruptions.

(HL) 2.3.12 The lithosphere contains carbon stores in fossil fuels and rocks such as limestone that contain calcium carbonate

The lithosphere, the solid outer layer of the Earth, is a fundamental component of the carbon cycle. It has large carbon reserves in the form of fossil fuels and rocks such as limestone ($CaCO_3$). Coal, oil and natural gas are fossil fuels generated from the remains of biological matter such as partially decomposed plants and marine animals that lived millions of years ago. These carbon-rich materials were buried and subjected to tremendous pressure and heat over time, eventually transforming them into fossil fuels.

Limestone is a sedimentary rock consisting mostly of calcium carbonate. It can be created by a variety of processes, including the fossilization of reef-building corals and molluscs with calcium carbonate-containing hard shells. Furthermore, limestone can be formed by both biological (shells) and non-biological (precipitated from water) processes.

In this way, carbon can spend hundreds of millions of years in these lithospheric carbon stores as important long-term carbon reservoirs.

Link
More about the lithospheric carbon stores is covered in Chapter 6.2 (page 528).

2.3.13 Reef-building corals and molluscs contribute to the formation of limestone by depositing calcium carbonate

Corals and molluscs that form reefs play an important part in the production of limestone. These organisms have hard components made mostly of calcium carbonate, such as skeletons or shells. While they are alive, they continuously build and maintain these hard structures through the deposition of calcium carbonate from the surrounding water. However, when these organisms die, their remains, including the calcium carbonate structures, begin to accumulate on the ocean floor or seabed. Layer upon layer of these remains build up and undergo compaction, eventually forming limestone over extended periods of time.

Link
More about the formation of limestone is covered in Chapter 6.2 (page 528).

Limestone, as a sedimentary rock, is the greatest carbon storage in the Earth's systems. It is crucial to remember, however, that not all limestone is generated by the fossilization of animal remains. Limestone can also be generated by a variety of different biological and non-biological processes.

2.3.14 Organic matter from plants and marine organisms became fossilized in coal, oil and natural gas over millions of years

Organic matter was critical in the production of fossil fuels such as coal, oil and natural gas in previous geological ages. Partially degraded plants (such as ancient ferns and mosses) became fossilized in coal, while partially decomposed marine species (like prehistoric algae and microscopic organisms) became fossilized in **porous rocks** containing oil and natural gas.

The intricate transformation of organic matter into major reserves of coal, oil and natural gas unfolded over tens of millions of years. This phenomenon was particularly pronounced during specific geological epochs when conditions were optimal for the development and preservation of these fossil fuels. It's noteworthy that not all oil and gas resulting from decomposed marine species was confined to reservoirs; some found its way into specific rock formations, creating diverse geological features.

> **Porous rock** – rock with interconnected open spaces or pores that allow fluids (such as water, oil or gas) to pass through and be stored within the rock structure.

> **Link**
> More about the production of fossil fuels is covered in Chapter 6.2 (page 528).

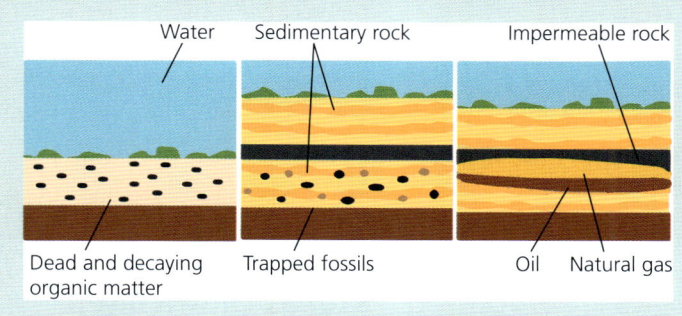

■ **Figure 2.54** Origin of oil and gas (adapted from energyeducation.ca)

2.3.15 Methane is produced from dead organic matter in anaerobic conditions by methanogenic bacteria

In certain environments where oxygen is scarce, such as wetlands, marshes and the digestive systems of some animals, dead organic matter undergoes a unique process leading to the production of methane (CH_4) gas. This process is facilitated by specialized microorganisms known as methanogenic bacteria. As dead organic material accumulates in these anaerobic (oxygen-depleted) conditions, the methanogenic bacteria break down the organic matter through a series of biochemical reactions. Methane gas is generated as a byproduct of this decomposition process. The methane produced in these environments plays a significant role in global greenhouse gas emissions and climate change.

2.3.16 Methane has a residence time of about ten years in the atmosphere and is eventually oxidized to carbon dioxide

Methane, a simple yet impactful hydrocarbon, plays a crucial role in the Earth's atmospheric composition. Methane doesn't stay in the atmosphere for long, only about 10 years, which is shorter than some other greenhouse gases. During this time, however, it is of great significance both environmentally and in terms of climate impact.

Imagine a molecule of methane released into the atmosphere, whether from natural sources like wetlands or livestock, or from human activities such as fossil fuel extraction and agriculture. Once in the atmosphere, over the span of about a decade, various chemical reactions guide methane's fate, eventually leading it towards oxidation into carbon dioxide and water vapour.

2.3 Biogeochemical cycles

These reactions, primarily driven by hydroxyl radicals (OH), initiate the process of methane's breakdown. As methane encounters hydroxyl radicals, it undergoes a series of transformations, eventually forming carbon dioxide (CO_2) as one of the end-products. The conversion from methane to carbon dioxide is vital because, while methane is a potent greenhouse gas in its own right, the impact of carbon dioxide on the climate is more persistent and enduring.

Methane's potency as a greenhouse gas is striking. Pound for pound, it possesses a much greater heat-trapping ability than carbon dioxide over a shorter timeframe. This attribute magnifies its influence on global warming. For instance, over a 20-year period, methane's heat-trapping capacity is more than 80 times greater than that of carbon dioxide. This intense heat-trapping property underscores the urgency of addressing methane emissions as a critical step in mitigating short-term climate change.

The gradual accumulation of methane in the atmosphere could lead to enhanced short-term warming and extreme climate change impacts if the methane emissions are left unchecked. However, the relatively short residence time of methane provides a unique opportunity for mitigation. By implementing strategies to restrict methane emissions, such as improving agricultural practices, minimizing leakage from fossil fuel infrastructure and capturing methane from waste-management processes, we can effectively limit its contribution to climate change.

Recognizing methane's potency as a greenhouse gas prioritizes reducing methane emissions as a pivotal endeavour in the pursuit of a more sustainable and climate-resilient future.

> **Link**
>
> More about the impacts of methane is covered in Chapter 6.2 (page 528).

2.3.17 The nitrogen cycle consists of organic and inorganic stores of nitrogen

The nitrogen cycle is an important biogeochemical cycle in which nitrogen is cycled through various organic and inorganic forms throughout ecosystems. It includes both biological and inorganic nitrogen reserves.

Proteins and other nitrogen-containing carbon compounds found in living organisms such as plants and animals, as well as decaying organic matter, are organic nitrogen reserves in ecosystems. When organisms die, their organic matter enters the nutrient pool and helps to replenish the organic nitrogen reserve.

Inorganic nitrogen storage includes atmospheric nitrogen (N_2) and other nitrogen molecules present in soil and water, such as ammonia (NH_3), nitrites (NO_2^-) and nitrates (NO_3^-). The majority of the nitrogen in the atmosphere is inert and must be transformed into useful chemicals through specialized biological and physical processes.

Understanding the various nitrogen reserves throughout the nitrogen cycle provides insights on nitrogen availability and transformation in ecosystems.

> **Tool 4: Systems and models**
>
> **Construct a systems/flow diagram**
>
> Create a systems diagram of the nitrogen cycle from a given set of data, illustrating the various stores, processes and flows of nitrogen within the cycle.

2.3.18 Bacteria play essential roles in the nitrogen cycle

Bacteria are key players in the nitrogen cycle, performing essential roles in the transformation of nitrogen within ecosystems. Several important processes involve bacterial activity.

1. Nitrogen fixation: Nitrogen-fixing bacteria convert atmospheric nitrogen (N_2) into ammonia (NH_3), making it available for other organisms. These bacteria have the ability to convert nitrogen gas into a form that plants can utilize for their growth and development.
2. Nitrification: Nitrifying bacteria oxidize ammonia (NH_3) into nitrites (NO_2^-) and further into nitrates (NO_3^-). This conversion allows plants to take up nitrogen in the form of nitrates, which is an essential nutrient for their growth.
3. Denitrification: Denitrifying bacteria convert nitrates (NO_3^-) back into atmospheric nitrogen (N_2), completing the nitrogen cycle. This process occurs in anaerobic (oxygen-depleted) environments, such as waterlogged soils, where nitrates are used as an alternative electron acceptor.
4. Decomposition: Bacteria, along with other decomposers, play a crucial role in the breakdown of organic nitrogen compounds in dead organisms, releasing ammonia (NH_3) as a byproduct.

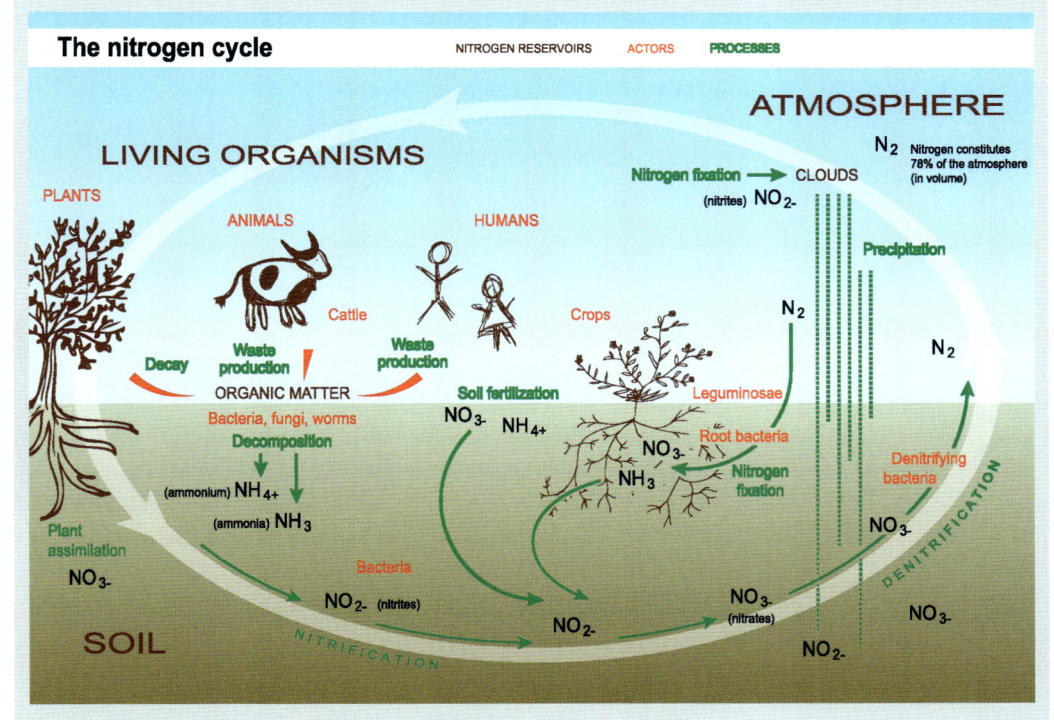

■ Figure 2.55 The nitrogen cycle (source: European Environment Agency)

2.3.19 Denitrification occurs in anaerobic conditions, such as waterlogged soils

Denitrification, the process of turning nitrates (NO_3^-) back into atmospheric nitrogen (N_2), occurs primarily in anaerobic circumstances where oxygen is scarce or non-existent. Waterlogged soils, where oxygen supply is decreased, are one of the key habitats where denitrification occurs.

Bacterial decomposition of organic waste in wet, anaerobic soils consumes available oxygen, generating an environment ideal for denitrifying bacteria. Denitrifying bacteria employ nitrates as an electron acceptor instead of oxygen in these conditions, resulting in the release of nitrogen gas (N_2) back into the atmosphere.

2.3 Biogeochemical cycles

Leaching – the process where nitrates in waterlogged soils are washed away and lost due to the movement of water through the soil.

Furthermore, waterlogged soils can lead to reduced plant growth as nitrates are washed away through denitrification and **leaching**. However, some plant species, like insectivorous plants such as pitcher plants and sundews, have adapted to these conditions by capturing and digesting insects to obtain nitrogen as a nutrient source.

2.3.20 Some plants form mutualistic associations with nitrogen-fixing bacteria to obtain atmospheric nitrogen

Plants are unable to use atmospheric nitrogen (N_2) directly as a nutrition source. To address this limitation, certain plant species have formed mutualistic relationships with nitrogen-fixing bacteria. This means that both the plants and the microbes profit from these partnerships.

Nitrogen-fixing plants, often known as legumes, give a favourable habitat and nutrients to nitrogen-fixing bacteria, which live in specialized structures called nodules on their roots. In exchange, the bacteria transform nitrogen in the air into ammonia (NH_3), which the plant can use for growth and development.

Plants benefit from this mutualistic relationship with nitrogen-fixing bacteria because it gives them a competitive edge in settings where nitrogen is a limiting component for plant growth. These plants can flourish in nitrogen-deficient conditions and outcompete other plant species by absorbing nitrogen from the atmosphere.

Plants that fix nitrogen include legumes like soybeans, clover and peas. These plants serve an important function in providing nitrogen to the soil, increasing fertilizer availability for other plants and contributing to ecosystem biodiversity and productivity.

■ **Figure 2.56** Nitrogen-fixing microbes in legumes

2.3.21 Flows in the nitrogen cycle include mineral uptake by producers, photosynthesis, consumption, excretion, death, decomposition and ammonification

The nitrogen cycle involves various flows or pathways through which nitrogen moves within ecosystems. These flows encompass both biological and geological processes, contributing to the cycling of nitrogen between different forms and reservoirs.

1. Mineral uptake by producers: Producers, such as plants and algae, take up inorganic nitrogen from the environment in the form of nitrates (NO_3^-) and ammonium (NH_4^+) through their root systems. This uptake allows them to incorporate nitrogen into their tissues.
2. Photosynthesis: Producers use nitrogen-containing compounds, such as amino acids, to synthesize (make or produce) proteins and other organic molecules through photosynthesis. This process enables the transfer of nitrogen from inorganic to organic forms within plants and algae.
3. Consumption: Consumers, including herbivores, carnivores and omnivores, obtain nitrogen by consuming producers or other organisms. Through their diet, consumers assimilate organic nitrogen compounds present in the tissues of their food sources.

4. **Excretion:** Animals excrete nitrogenous waste products, such as urea or uric acid, as a byproduct of metabolic processes. These nitrogenous waste products release nitrogen back into the environment, where it can be further utilized in the nitrogen cycle.
5. **Death and decomposition:** When organisms die, decomposers, including bacteria and fungi, break down their organic matter. During decomposition, nitrogen-rich compounds, such as proteins and nucleic acids, are broken down into simpler forms, releasing ammonia (NH_3) as a result.
6. **Ammonification:** Ammonification is the process by which organic nitrogen compounds, including those released during decomposition, are converted into ammonium (NH_4^+) by specific decomposer bacteria. Ammonification replenishes the inorganic nitrogen pool in the soil or water.

■ Figure 2.57 The nitrogen cycle

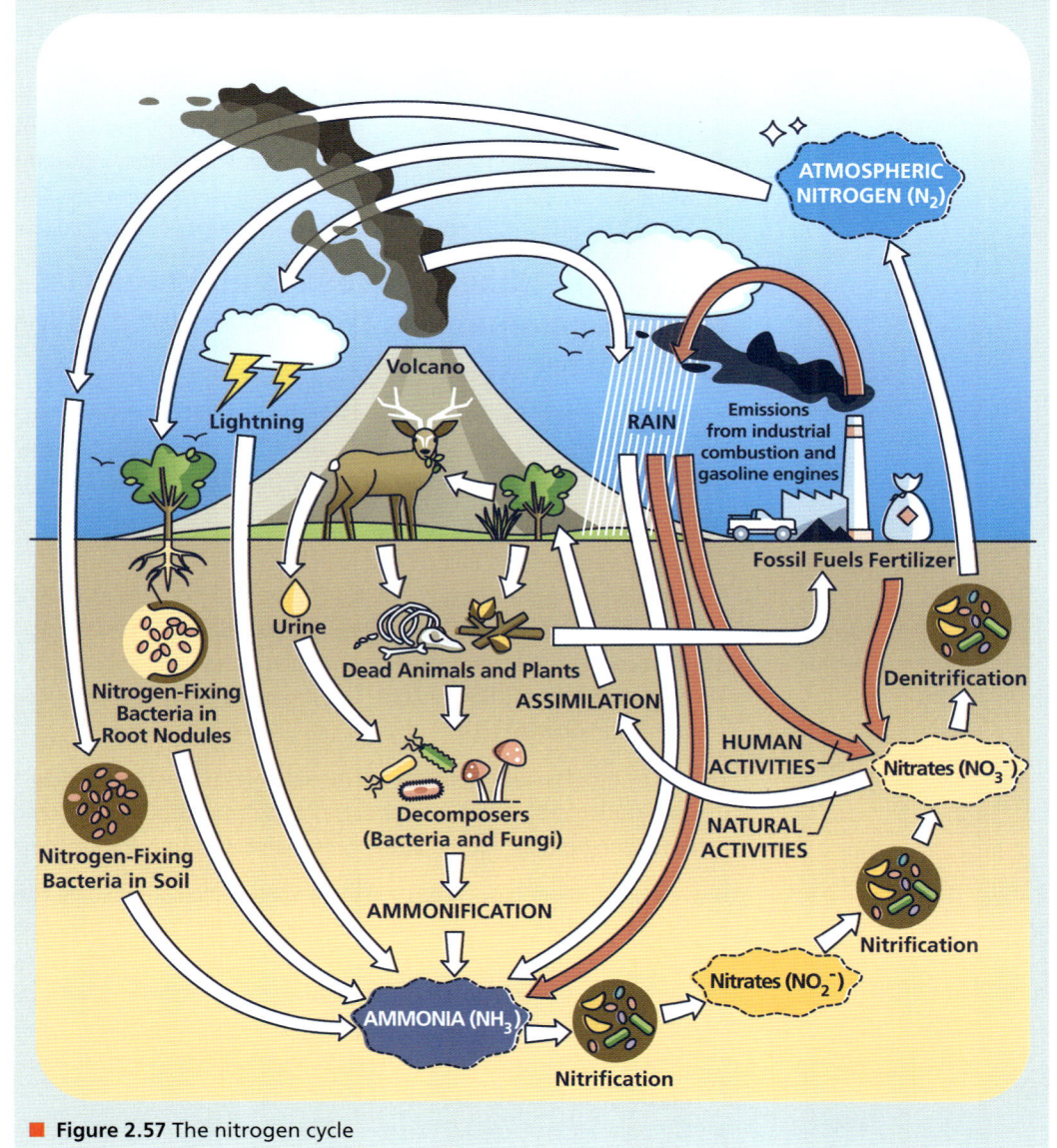

2.3 Biogeochemical cycles

2.3.22 Human activities, such as deforestation, agriculture, aquaculture and urbanization, can significantly impact the nitrogen cycle

Human activities have a profound influence on the nitrogen cycle, leading to significant alterations in the distribution and availability of nitrogen in ecosystems. Some key human activities that affect the nitrogen cycle include:

1. Deforestation: The removal of forests disrupts the natural nitrogen cycle by eliminating nitrogen-fixing plants and reducing organic matter decomposition. This disruption can result in decreased nitrogen inputs to the soil, impacting nutrient availability for other organisms.
2. Agriculture: Agricultural practices, such as the use of synthetic fertilizers, contribute to nitrogen enrichment in ecosystems. Excessive use of fertilizers can lead to nutrient runoff, causing eutrophication (see Topic 2.2.20, page 130) in water bodies and altering aquatic ecosystems.
3. Aquaculture: Intensive fish farming, a form of aquaculture, often involves the use of feed that contains high levels of nitrogen. Uneaten feed and fish waste contribute to nutrient-loading in aquatic systems, affecting water quality and potentially causing ecological imbalances.
4. Urbanization: Urban areas experience increased nitrogen inputs due to human activities such as vehicle emissions, industrial processes and sewage disposal. These inputs can result in elevated nitrogen concentrations in the environment, impacting air and water quality.

2.3.23 The Haber process is an industrial method used to produce ammonia from nitrogen and hydrogen for use as fertilizer

The Haber process, named after the German chemist Fritz Haber, is an industrial process that allows the large-scale production of ammonia (NH_3). This ammonia is primarily used as a key component in fertilizers, which provide essential nutrients, including nitrogen, to support crop growth and increase agricultural yields.

The process involves the following steps:

1. Nitrogen fixation: Nitrogen gas (N_2) is extracted from the atmosphere.
2. Hydrogen production: Hydrogen gas (H_2) is obtained from natural gas or other sources.
3. Catalytic reaction: Nitrogen and hydrogen gases are combined in the presence of a catalyst at high temperature and pressure. This promotes the reaction between nitrogen and hydrogen to form ammonia (NH_3).

The Haber process has revolutionized modern agriculture by providing a means to produce synthetic (human-made) fertilizers on a large scale. It has significantly contributed to increased crop yields and improved food production, supporting the needs of a growing global population.

However, the Haber process has some drawbacks and causes some concerns:

- Energy intensive: The production of ammonia through the Haber process requires high energy input, primarily in the form of fossil fuels. This reliance on non-renewable energy sources contributes to greenhouse gas emissions and environmental impacts.
- Nitrogen losses: Once applied as fertilizer, a significant portion of the nitrogen from synthetic fertilizers can be lost through leaching, volatilization or runoff, leading to environmental pollution and nutrient imbalances in ecosystems.
- Eutrophication: Excessive use of synthetic fertilizers, including ammonia-based fertilizers, can contribute to eutrophication in water bodies. Nutrient runoff from agricultural fields can promote excessive algal growth, deplete oxygen levels and harm aquatic organisms.

> ● **HL.b.9: Ecological economics**
>
> The Haber process is linked to ecological economics, as it involves the industrial production of ammonia, which has economic implications and can impact the biosphere.

2.3.24 Increases in nitrates in the biosphere from human activities have led to the planetary boundary for the nitrogen cycle being crossed, making irreversible changes to Earth systems likely

Human activities have significantly impacted the nitrogen cycle, leading to increased levels of nitrates in the biosphere. This excess nitrogen input can have severe consequences for ecosystems and the environment, contributing to the crossing of the planetary boundary (see Topic 2.1.24, page 110) for the nitrogen cycle.

1. Nitrate accumulation: Human activities, such as the use of synthetic fertilizers in agriculture and the combustion of fossil fuels, release large amounts of nitrogen into the environment. This excessive nitrogen input leads to the accumulation of nitrates (NO_3^-) in soils, bodies of water and the atmosphere.
2. Eutrophication: The accumulation of nitrates in water bodies can result in eutrophication. Excess nitrogen fuels the growth of algae and other aquatic plants, leading to algal blooms. When these blooms die and decompose, oxygen levels in the water decrease, causing harm to fish and other aquatic organisms.
3. Air pollution: Nitrogen oxides (NO_x) released from human activities, such as industrial processes and vehicle emissions, contribute to air pollution. Nitrogen oxides are involved in the formation of smog and can have detrimental effects on human health and the environment.
4. Biodiversity loss: High levels of nitrates can lead to changes in plant species composition and dominance, favouring nitrophilic species over others. This can result in the loss of biodiversity and disrupt the ecological balance in affected ecosystems.

The crossing of the planetary boundary for the nitrogen cycle implies that the current nitrogen inputs and their impacts on the environment are unsustainable. It highlights the urgent need to address the excessive use of nitrogen in industrial and agricultural processes to mitigate further irreversible changes to Earth systems.

2.3.25 Global collaboration is necessary to address the uncontrolled use of nitrogen in industrial and agricultural processes, and to bring the nitrogen cycle back within planetary boundaries

Addressing the challenges associated with the uncontrolled use of nitrogen requires global collaboration and concerted efforts from governments, organizations, scientists and individuals. Several measures can be taken to bring the nitrogen cycle back within planetary boundaries.

1. Sustainable agricultural practices: Promoting these is crucial. This includes optimizing fertilizer application, adopting precision farming techniques, using organic fertilizers, implementing crop rotation and employing agroforestry systems. These practices help reduce nitrogen losses and enhance nutrient use efficiency.
2. Integrated nutrient management: Implementing integrated nutrient-management strategies can help balance nutrient inputs and outputs in agricultural systems. This involves combining organic and inorganic fertilizers, utilizing nitrogen-fixing crops, and managing crop residues effectively to minimize nutrient losses and maintain soil health.

Green infrastructure – eco-infrastructure/nature-based infrastructure involves using natural elements like plants and ecosystems to help with things like water management and environmental sustainability.

3. Wastewater treatment: Implementing efficient wastewater treatment processes helps reduce nitrogen pollution in water bodies. Advanced treatment technologies can remove nitrogen compounds from wastewater before its release, minimizing the impacts on aquatic ecosystems.
4. Sustainable urban planning: Incorporating sustainable urban planning practices can reduce nitrogen emissions from urban areas. This includes promoting public transportation, improving waste-management systems, and implementing **green infrastructure** to enhance nitrogen absorption and filtration.
5. Education and awareness: Raising awareness among individuals, communities and policymakers about the impacts of nitrogen pollution is crucial. Education programmes can foster a sense of responsibility and encourage individuals to adopt sustainable practices in their daily lives.

Global collaboration and coordinated actions are necessary to tackle the uncontrolled use of nitrogen and bring the nitrogen cycle back within planetary boundaries. By implementing sustainable practices, reducing nitrogen pollution and promoting responsible nitrogen management, we can mitigate the negative impacts of excessive nitrogen inputs and safeguard the health of ecosystems and the planet.

HL.b.9: Environmental economics

In our interconnected world, where human activities like deforestation, agriculture, aquaculture and urbanization are altering essential natural processes such as the nitrogen cycle, the principles of ecological economics guide us towards a path of sustainable coexistence with our planet.

HL.b.17: Environmental economics

Embedded within the principles of ecological economics is the concept of the circular economy and the doughnut economics model (see Topics 1.3.21, page 77, and 1.3.20, page 75). These models provide practical frameworks for translating the ideology of sustainability and global collaboration into actionable strategies.

- The circular economy, a manifestation of sustainability, promotes responsible resource management. Just as the nitrogen cycle should function as a closed-loop system within planetary boundaries, so too should our economic practices. Reusing, recycling and regenerating resources mirrors the nitrogen cycle's natural circularity and is integral to addressing its disruption.
- Doughnut economics extends this concept by urging us to operate within the safe operating space between planetary boundaries and social foundations. The nitrogen cycle's changes directly impact both ecological and human systems, exemplifying the need for equilibrium. By adopting the doughnut model, we can ensure that the nitrogen cycle's vital functions are maintained, safeguarding ecosystems and livelihoods.

The challenges posed by changes in the nitrogen cycle highlight the urgency of ecological economics, global collaboration, and sustainable models like the circular economy and doughnut economics. By embracing these principles, we not only restore the delicate balance of the nitrogen cycle but also create a harmonious relationship between human progress and the environment.

Concept

Sustainability

The concept of sustainability emphasizes addressing present demands without sacrificing the capacity of future generations to meet their own needs. It entails managing resources properly, striking a balance between environmental, social and economic issues, and making sure that current actions do not damage the planet's ecosystems or exhaust resources in a way that would be harmful to the welfare of upcoming generations. Sustainable practices promote long-term viability, resilience and equitable outcomes for both the present and future populations by aiming to establish a harmonious balance between human activity and the natural world.

REAL-WORLD EXAMPLE

Global efforts to address nitrogen pollution

One real-life example of global collaboration to address the uncontrolled use of nitrogen and bring the nitrogen cycle back within planetary boundaries is the Global Partnership on Nutrient Management initiative.

The Global Partnership on Nutrient Management is a collaborative effort that brings together governments, international organizations, research institutions and civil society groups from around the world. Its primary goal is to promote sustainable nutrient-management practices, including nitrogen, to ensure food security, protect the environment and mitigate the adverse effects of nutrient pollution.

The rapid utilization of nitrogen and phosphorus lies at the heart of an intricate network of advantages related to development, as well as challenges pertaining to the environment. These elements play a crucial role in the cultivation of crops, and nearly half of the world's food security hinges on the application of nitrogen and phosphorus fertilizers. Nonetheless, surplus nutrients originating from fertilizers, combustion of fossil fuels, and the discharge of wastewater from human activities, livestock farming, aquaculture and industries lead to pollution in the air, water, soil and oceans. This pollution contributes to the decline in biodiversity and aquatic life, depletion of the ozone layer and exacerbation of global warming effects.

These problems are anticipated to escalate as the demand for food and biofuels rises, and as expanding urban populations generate more wastewater. Consequently, there will be an escalating economic toll on countries due to the degradation of ecosystems, particularly in coastal areas, and the subsequent loss of the services and employment opportunities they provide.

The Global Partnership on Nutrient Management (GPNM) was established in response to this pressing 'nutrient challenge'. The GPNM serves as a platform for collaboration between governments, United Nations agencies, scientists and the private sector, aiming to establish a shared agenda. This agenda involves adopting best practices and integrated assessments to ensure that policy decisions and investments can effectively handle nutrient-related problems. Moreover, the GPNM provides a forum where countries and other involved parties can cooperate to deal with nutrient-related matters.

ATL ACTIVITY

1 Design an effective proposal for the adoption of organic fertilizers in your school or community green areas, highlighting their benefits over inorganic alternatives and encouraging sustainable land-care practices.
2 Think about your own school's local community and research to find out if there are any specific instances of environmental justice challenges arising from the exploitation of natural resources for financial gain. Collaborate with your friends to gather data to propose and advocate for equitable solutions and the adoption of sustainable practices.

REVIEW QUESTIONS

1 Outline the different stores and flows in the nitrogen cycle.
2 With reference to the nitrogen cycle, evaluate the impacts of the Haber process on food production systems.
3 To what extent do human activities influence the carbon cycle?
4 What is the role of biogeochemical cycles in ensuring the availability of chemical elements to living organisms?
5 How do ecosystems contribute to the carbon cycle, and what are the different ways in which carbon flows within ecosystems?
6 Why is it important to recognize that fossil fuels were formed when ecosystems acted as carbon sinks in the past and now serve as carbon sources when burned?

EXAM-STYLE QUESTION

1 Even though there is growing global support for ecocentric values, the global consumption of fossil fuels continues to rise each year.

With reference to energy choices in named countries, discuss possible reasons for this situation occurring. [9 marks]

Theme 2: Ecology

2.4 Climate, biomes and their influences

Guiding question

- How can natural systems be modelled, and can these models be used to predict the effects of human disturbance?

SYLLABUS CONTENT

This chapter covers the following syllabus content:

- 2.4.1 'Climate' describes atmospheric conditions over relatively long periods of time, whereas 'weather' describes the conditions in the atmosphere over a short period of time.
- 2.4.2 A biome is a group of comparable ecosystems that have developed in similar climatic conditions, wherever they occur.
- 2.4.3 Abiotic factors are the determinants of terrestrial biome distribution.
- 2.4.4 Biomes can be grouped into various different types, including freshwater, marine, forest, grassland, desert and tundra. Each of these classes has characteristic abiotic limiting factors, productivity and diversity. They may be further classed into many sub-categories, such as temperate forests, tropical rainforests and boreal forests.
- 2.4.5 The tricellular model of atmospheric circulation explains the behaviour of atmospheric systems and the distribution of precipitation and temperature at different latitudes. It also explains how these factors influence the structure and relative productivity of different terrestrial biomes.
- 2.4.6 The oceans absorb solar radiation and ocean currents distribute the resulting heat around the world.
- 2.4.7 Global warming is leading to changing climates and shifts in biomes.

HL ONLY

- 2.4.8 There are three general patterns of climate types that are connected to biome types.
- 2.4.9 The biome predicted by any given temperature and rainfall pattern may not develop in an area because of secondary influences or human interventions.
- 2.4.10 The El Niño–Southern Oscillation (ENSO) is the fluctuation in wind and sea surface temperatures that characterizes conditions in the tropical Pacific Ocean. The two opposite and extreme states are El Niño and La Niña, with transitional and neutral states between the extremes.
- 2.4.11 El Niño is due to a weakening or reversal of the normal east–west (Walker) circulation, which increases surface stratification and decreases upwelling of cold, nutrient-rich water near the coast of northwestern South America. La Niña is due to a strengthening of the Walker circulation and reversal of other effects of El Niño.
- 2.4.12 Tropical cyclones are rapidly circulating storm systems with a low-pressure centre. They originate in the tropics and are characterized by strong winds.
- 2.4.13 Rises in ocean temperatures resulting from global warming are increasing the intensity and frequency of hurricanes and typhoons because warmer water and air have more energy.

2.4.1 Climate and weather

Climate and weather are two of the essential characteristics of the Earth's atmosphere that are used to define ecosystems. Climate refers to the long-term average of various atmospheric conditions, such as temperature, precipitation, humidity, wind patterns, and other characteristics that show the overall atmospheric conditions in a specific place over long periods of time, such as decades or centuries.

Weather, on the other hand, refers to the short-term atmospheric conditions that are frequently experienced across hours, days or weeks. Temperature variations, rainfall events, wind gusts and cloud cover all occur within a specific timeframe. Understanding the difference between climate and weather is critical for understanding how ecosystems adapt to long-term circumstances and respond to short-term fluctuations.

2.4.2 A biome is a group of comparable ecosystems that have developed in similar climatic conditions, wherever they occur

A biome is a collection of related ecosystems that evolved in similar climatic conditions, wherever they appear. Biomes are large-scale ecological zones with diverse climates, vegetation and organisms. Tropical rainforests, temperate forests, deserts, grasslands, tundra, freshwater systems and marine ecosystems are examples of biomes. The concept of biomes enables us to understand how different sections of our planet, despite their geographical distance, can have similar ecological characteristics due to different climatic conditions. The distribution of terrestrial biomes is significantly influenced by factors such as precipitation, temperature and insolation.

Link
More information about the abiotic factors is available in Topic 2.1.6 (page 88).

2.4.3 Abiotic factors are the determinants of terrestrial biome distribution

As an abiotic variable, climate factors such as temperature, rainfall, sunlight, humidity and wind patterns may affect the distribution and structure of the Earth's biomes. Some of the other abiotic variables such as soil composition, topography, nutrient availability and water availability may determine the suitability of a habitat for different species. As a result of all these factors, types of biomes can form in specific climatic zones.

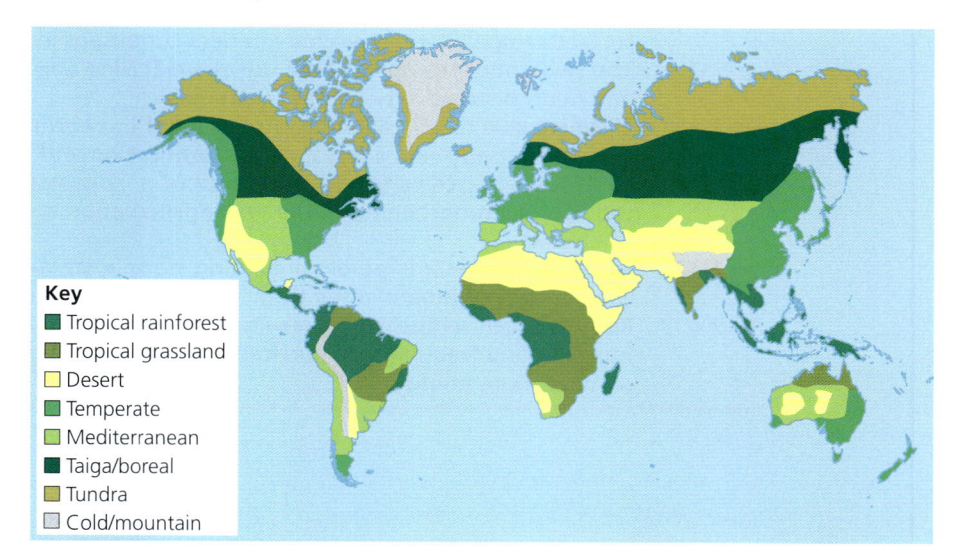

■ Figure 2.58 Global climate zones map

For example, temperate forests experience moderate temperatures and abundant rainfall consistently throughout the year, whereas tropical rainforests have high temperatures and substantial rainfall year-round. Conversely, boreal forests are characterized by cold temperatures and moderate precipitation. These broad biome types can be further subdivided into a plethora of sub-categories based on more particular meteorological, vegetative and ecological characteristics.

> **ATL ACTIVITY**
>
> Using software like Microsoft Excel or Google Sheets, create climate graphs showing annual precipitation and average temperature for different biomes.
> 1. Gather the data: Collect data on annual precipitation and average temperature for the different biomes you are interested in studying. You can find this data from reliable sources like climate research organizations, government agencies or scientific publications.
> 2. Open the spreadsheet software: Open Microsoft Excel or Google Sheets on your computer.
> 3. Enter the data: Create a table with columns for the biome names, years, annual precipitation and average temperatures. Input the data you have collected into the respective cells.
> 4. Create a line graph: To visualize the data, follow these steps:
>   - Excel: a) Select the data in your table. b) Go to the 'Insert' tab and choose the 'Line' chart type. Select the one that displays data over time (usually labelled as 'Line with Markers'). c) A graph will be generated on your spreadsheet.
>   - Google Sheets: a) Select the data in your table. b) Go to the 'Insert' tab and choose 'Chart'. A sidebar will appear. c) In the Chart Editor sidebar, choose the 'Chart type' as 'Line chart'. d) Customize the chart settings as needed, including axis labels and titles.
> 5. Format the graph: Customize the appearance of the graph to make it clear and visually appealing. You can adjust colours, labels, axis scales and add a legend to identify each biome.
> 6. Label the axes: Ensure your graph has clear labels on the x-axis (years) and y-axes (precipitation and temperature). Make sure the units are appropriately labelled as well, as shown in Figure 2.59.
> 7. Title and legend: Add a title to the graph that describes what it's showing (for example, 'Annual precipitation and temperature by biome'). Include a legend if you're comparing multiple biomes.
> 8. Save and share: Save your spreadsheet and consider exporting the graph as an image or PDF if you need to share it in presentations or reports.
>
> ■ Figure 2.59 Example of a climate graph

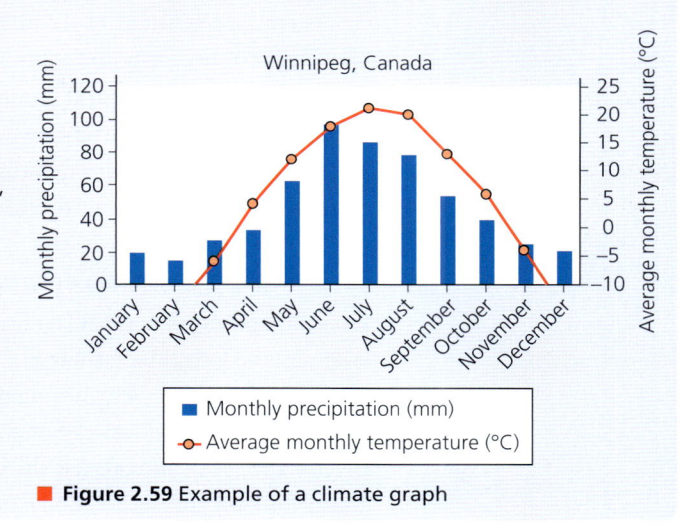

2.4.4 Biomes can be grouped into various different types, including freshwater, marine, forest, grassland, desert and tundra

Each of these groups has characteristic abiotic limiting factors, productivity and diversity. They may be further classed into many sub-categories, such as temperate forests, tropical rainforests and boreal forests.

2.4 Climate, biomes and their influences

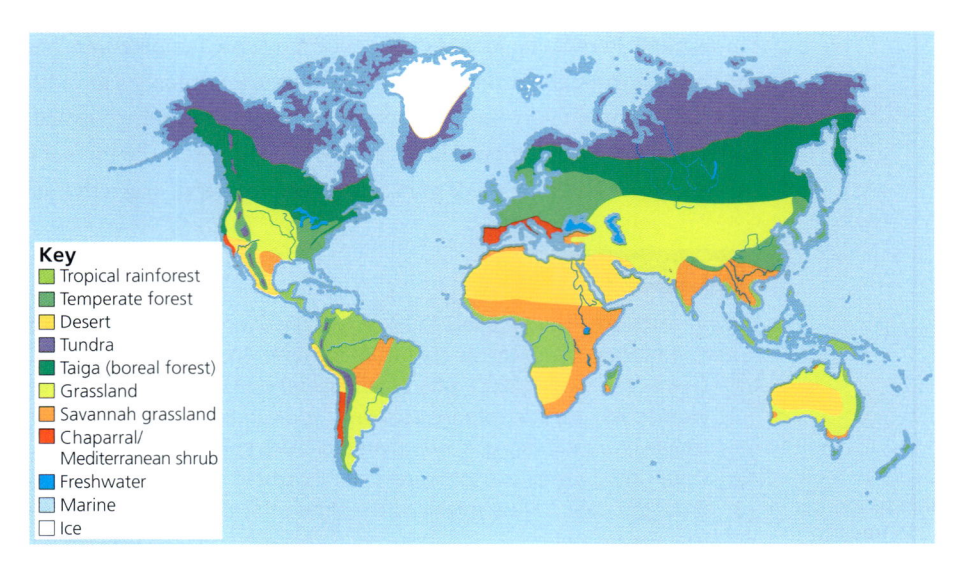

■ Figure 2.60 World biomes map

Key
- Tropical rainforest
- Temperate forest
- Desert
- Tundra
- Taiga (boreal forest)
- Grassland
- Savannah grassland
- Chaparral/Mediterranean shrub
- Freshwater
- Marine
- Ice

■ Tropical rainforests

Tropical rainforests have the most biodiversity of all the biomes and are located in equatorial regions. However, they are under threat because of human activities such as logging and deforestation for agriculture. The typical rainforest vegetation consists of plants with spreading roots and broad leaves that fall off all year round, unlike the trees of deciduous forests that lose their leaves in one season.

The average temperatures range from 20°C to 34°C. Since the temperatures are relatively constant in tropical rainforests, plant growth is year-round rather than being seasonal. In contrast to other ecosystems, a consistent 11–12 hours of daily sunlight provides more solar radiation and therefore more opportunity for primary productivity.

The yearly rainfall in tropical rainforests ranges from 125 to 660 cm, with considerable seasonal variation. Tropical rainforests have wet months in which there can be more than 30 cm of precipitation, as well as dry months in which there may be less than 10 cm of rainfall. However, even the driest month in a tropical rainforest can still have more rain than some other biomes receive in a year, such as deserts. As the yearly temperatures and amount of precipitation support fast plant growth, tropical rainforests have high net primary productivity.

Tropical rainforests' vertical layering of vegetation creates distinct habitats for animals within each layer. On the forest floor, there is a sparse layer of plants and decaying plant matter. A layer of trees rises above this, which is then covered by a closed upper canopy – the uppermost overhead layer of branches and leaves. Some additional trees emerge through this closed upper canopy. Diverse species of animals use the variety of plants and the complex structure of the tropical wet forests for food and shelter. Some organisms live several metres above ground, rarely descending to the forest floor.

■ Hot desert

Hot deserts are located near the tropics of Cancer and Capricorn. The largest hot desert is the Sahara in Africa, which forms the whole width of the continent. Hot deserts' extreme climate creates a challenging environment with very little biodiversity. Few species developed distinct adaptations to survive there. The biotic or living components and the abiotic or non-living components of the hot desert rely on one another – a change in one will lead to a change in the other.

The climate is very hot with summer temperatures exceeding 40°C, although at night the temperatures can drop below 0°C. It is very dry with less than 250 mm of annual rainfall. Hot deserts have two distinct seasons: summer, when the temperature ranges between 35°C and 40°C, and winter, when the temperature ranges between 20°C and 30°C.

Desert soils are very dry, thin, sandy, rocky and generally grey in colour. Rain is soaked up very quickly by the soil. However, due to the lack of rainfall, the surface of the soil may seem crusty. Hot temperatures cause water to be drawn up to the surface of the soil by evaporation. As the water evaporates, salts are left on the surface of the soil.

The distinct characteristics (for example, high temperatures) of hot deserts allow only certain species to survive in such an extreme environment. Plants that show adaptations to allow them to live in hot and dry conditions are called **xerophytic**. They have small leaves to prevent water loss by transpiration by decreasing the surface area. Some plants, such as cactuses, have spines instead of leaves. Some have leaves with a thick, waxy skin on their surface to reduce water loss by transpiration. Their tap roots are long (7–10 metres) to reach deep under the ground to access water supplies. Some plants, known as succulents, store water in their stems, leaves, roots or even fruits.

◆ **Xerophyte** – a plant that is adapted to survive in a habitat that provides limited access to free water, such as a desert or frozen soil.

Boreal forest

The boreal forest, also known as taiga or coniferous forest, is found in most of Canada, Alaska, Russia and northern Europe between 50° and 60° north latitude (see Figure 2.60). Boreal forests can also be found in mountain ranges throughout the Northern Hemisphere above a particular elevation (and below extreme elevations where trees cannot grow). This biome experiences cold, dry winters and brief, cool (moderate temperatures), wet summers. The annual precipitation ranges from 40 cm to 100 cm, and is mostly in the form of snow. Due to the low temperatures, there is little evaporation.

The boreal forest's long and severe winters have resulted in the dominance of cold-tolerant cone-bearing plants. These are evergreen coniferous trees, such as pines, spruce and fir, that keep their needle-shaped leaves all year. Evergreen trees can photosynthesize earlier in the spring than deciduous trees because a needle-like leaf requires less energy from the sun to warm than a wide leaf. In the boreal forest, evergreen trees grow quicker than deciduous trees. Furthermore, soils in boreal forest zones are typically acidic, with little accessible nitrogen. Deciduous trees must create leaves every year, and leaves are nitrogen-rich structures. As a result, in a nitrogen-limiting environment, coniferous trees that retain nitrogen-rich needles may have had a competitive advantage over broad-leafed deciduous trees.

Boreal forests have lower net primary productivity than temperate or tropical wet forests. Boreal forests have a high above-ground biomass because these slow-growing tree species live a long period and collect standing biomass. The diversity of species is lower than in temperate forests and tropical rainforests.

The layered forest structure that is present in tropical rainforests and, to a lesser extent, temperate forests is absent in boreal forests. A boreal forest's structure is frequently composed of merely a tree layer and a ground layer. Because conifer needles degrade more slowly than wide leaves, fewer nutrients are given to the soil to drive plant development.

Tundra

The Arctic tundra is found throughout the Northern Hemisphere's Arctic regions, north of the subarctic boreal woods. Tundra can also be found on mountains at heights above the tree line. The average winter temperature is −34°C, while the average summer temperature ranges from 3°C to 12°C. Plants in the Arctic tundra have a relatively short growing season of 50–60 days. However, there are nearly 24 hours of daylight during this season, meaning plant development is rapid. The yearly precipitation on the Arctic tundra is minimal (15–25 cm), with little variation year to year. And, as in the boreal forests, the cold temperatures generate less evaporation.

Low shrubs, grasses, lichens and small blooming plants are common on the Arctic tundra. There is a lack of species variety, as well as low net primary productivity and above-ground biomass. The soils of the Arctic tundra may remain permanently frozen, a condition known as permafrost. Permafrost prevents roots from penetrating deeply into the soil and slows the breakdown of organic materials, which hinders the release of nutrients from organic matter. Melting permafrost during the brief summer season offers water for a burst of productivity while temperatures and long days allow. The ground of the Arctic tundra might be entirely covered with plants or lichens throughout the growth season.

Temperate forest

Temperate forests dominate the biomes of eastern North America, Western Europe, Eastern Asia, Chile and New Zealand. This biome is widespread throughout the mid-latitudes. Temperatures fluctuate between −30°C and 30°C, with annual lows of below freezing.

Because of these temperatures, temperate woods have distinct growing seasons in the spring, summer and early autumn. Precipitation is very consistent throughout the year, ranging from 75 cm to 150 cm.

This biome is dominated by deciduous trees, with fewer evergreen conifers. Deciduous trees shed their leaves in the autumn and are leafless in the winter. As a result, minimal photosynthesis occurs during the winter dormant period. As the temperatures rise in the spring, fresh leaves emerge.

Because of the dormant period, temperate forests have lower net primary productivity than tropical rainforests. Furthermore, temperate forests have significantly fewer tree species than tropical rainforest biomes.

Temperate forest trees leaf out and shade much of the ground. However, because temperate forest trees do not grow as tall as tropical rainforest trees, more sunlight reaches the ground in this biome than in tropical rainforests.

In comparison to tropical rainforests, temperate forest soils are rich in inorganic and organic nutrients. This is due to the thick layer of leaf litter on forest floors and the reduced nutrient leaching caused by rains. Nutrients are returned to the soil as the leaf litter decays. In addition, leaf litter prevents soil erosion, insulates the ground, and offers habitat for invertebrates and their predators.

> **Link**
> More information on productivity can be found in Topic 2.2.13 (page 126).

Tool 4: Systems and models

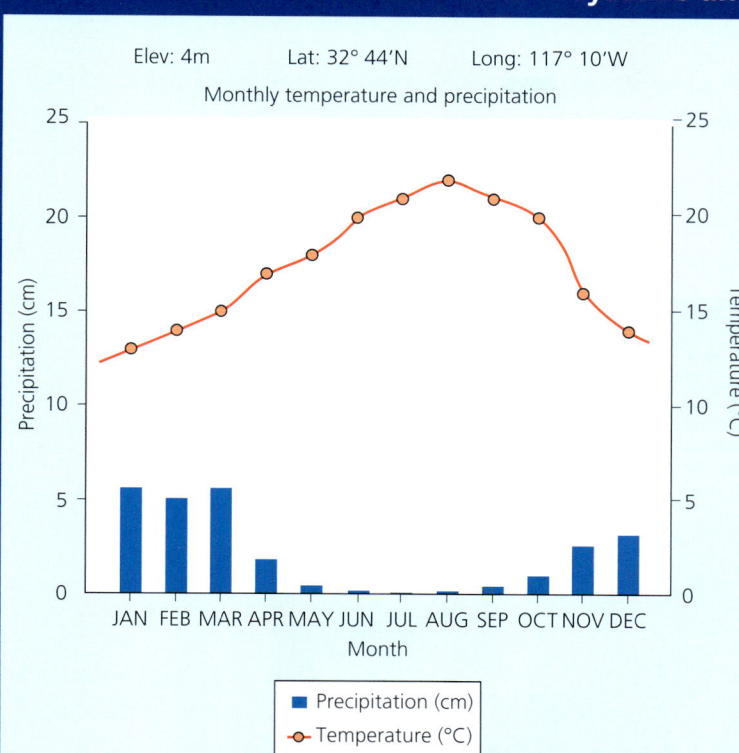

Interpret climate graphs

Climate graphs indicate average temperature and rainfall (precipitation) totals per month on a single graph. For instance, the climate graph in Figure 2.61 shows San Diego's average conditions. It receives most of its rain from November to March, whereas the temperatures do not change much over the year.

● Common mistake

The two vertical axes must be carefully examined so that the temperature axis and rainfall axis are not mixed up.

■ **Figure 2.61** Annual climatograph of San Diego between 1971 and 2000 (source: NCDC Data)

BIOME
TERRESTRIAL ECOSYSTEM WORLD MAP
GEOGRAPHY VECTOR COLLECTION

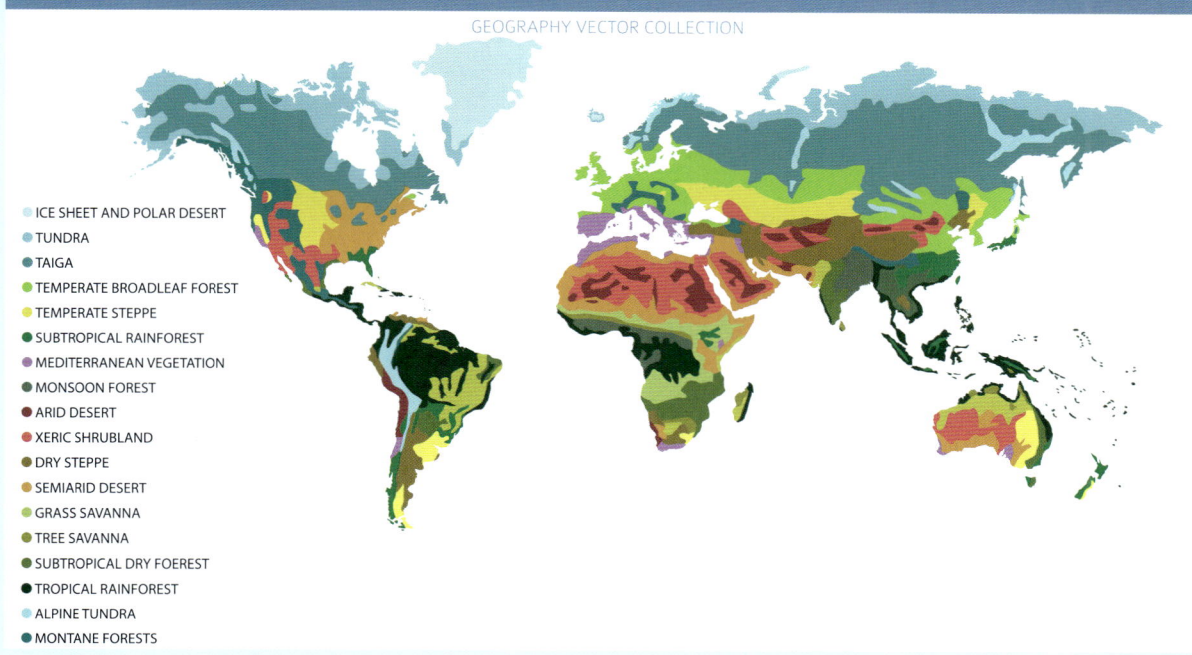

- ICE SHEET AND POLAR DESERT
- TUNDRA
- TAIGA
- TEMPERATE BROADLEAF FOREST
- TEMPERATE STEPPE
- SUBTROPICAL RAINFOREST
- MEDITERRANEAN VEGETATION
- MONSOON FOREST
- ARID DESERT
- XERIC SHRUBLAND
- DRY STEPPE
- SEMIARID DESERT
- GRASS SAVANNA
- TREE SAVANNA
- SUBTROPICAL DRY FOEREST
- TROPICAL RAINFOREST
- ALPINE TUNDRA
- MONTANE FORESTS

■ **Figure 2.62** The world's eight major biomes are distinguished by their characteristic temperature and precipitation amount

Choose a city from **https://drought.unl.edu/Climographs.aspx** and interpret the climate graph to write a brief description of that location's climate in winter, spring, summer and autumn.

2.4 Climate, biomes and their influences

2.4.5 Tricellular model of atmospheric circulation

The sun and latitude play a fundamental role in shaping the atmospheric circulation tricellular model and its implications for global climate patterns. Solar energy, as absorbed and distributed by the Earth's surface, creates temperature differences across latitudes. These variations in temperature drive the movement of air masses within the Earth's atmosphere, resulting in the formation of the Hadley, Ferrel and Polar cells. These cells, each distinct in their location and characteristics, are intricately linked to the distribution of precipitation and temperature, which, in turn, have a profound influence on the structure and productivity of terrestrial biomes.

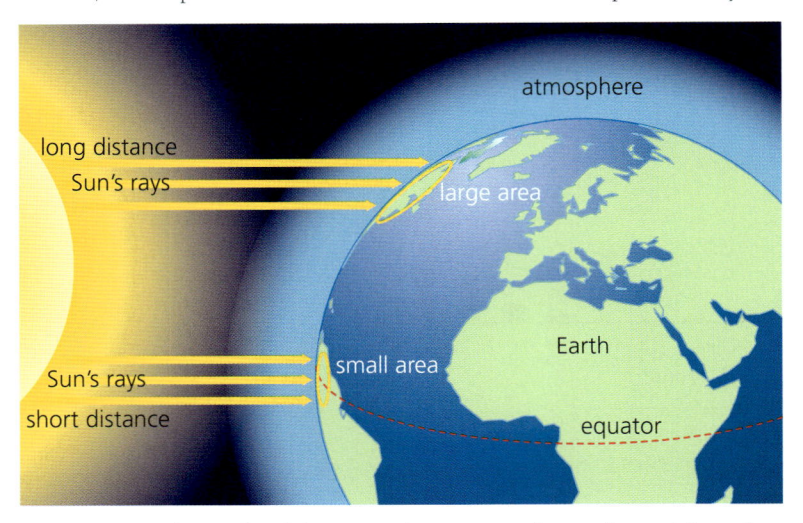

■ **Figure 2.63** The angle of the sun at the equator, the tropics and the poles

- The Hadley cell, positioned near the equator, circulates air around this region due to the intense solar heating. The warm air rises at the equator, creating a low-pressure zone, and then moves poleward at higher altitudes. As it cools and descends in the subtropics, it creates arid conditions and deserts. This temperature variation along the cell's extent contributes to a distinct temperature gradient from the equator to the subtropics.
- The Ferrel cell, existing in the mid-latitudes, is influenced by both the Hadley and Polar cells. Here, the polar air sinks and meets the warm air from the Hadley cell. This creates a belt of prevailing westerlies, leading to changes in temperature and weather patterns. It is characterized by variable weather conditions. The interaction of these cells contributes to the fluctuating temperature gradient typical of the mid-latitudes.
- The Polar cell, positioned around the poles, forms due to the sinking of cold, dense air. As this air mass moves equatorward along Earth's surface, it creates polar easterlies. This frigid air contrasts with the warmer air from lower latitudes, leading to polar fronts and a mix of weather conditions. The temperature gradient in the Polar cell is significant, ranging from the cold polar regions to the temperate mid-latitudes.

By understanding how solar energy interacts with latitude to create these atmospheric cells, you can gain valuable insights into the intricate web of connections that define our planet's climate and ecosystems. The atmospheric circulation tricellular model provides a framework to understand the behaviour of atmospheric systems, distribution of precipitation and temperature at different latitudes. Moreover, it offers an explanation of how these variables collectively influence the structure and relative productivity of various terrestrial biomes.

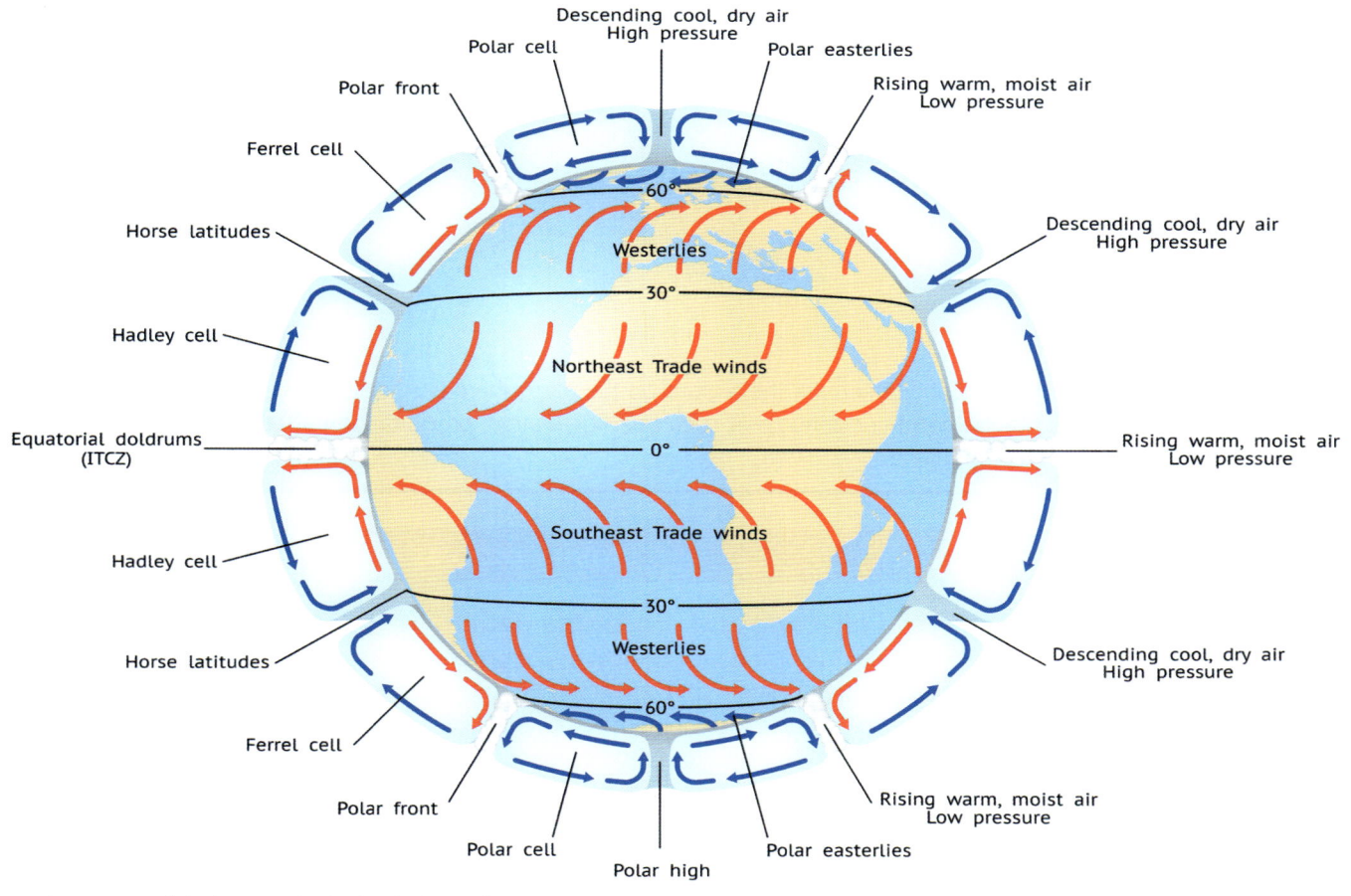

■ Figure 2.64 Global atmospheric circulation cells

> ### Tool 4: Systems and models
>
> #### Use the tricellular model of atmospheric circulation
>
> Use the tricellular model of atmospheric circulation and link it to the planetary distribution of heat and biomes (see Topic 2.4.6, page 172).

2.4 Climate, biomes and their influences

1. Find a blank world map like the one below to label and colour the different cells of the tricellular model on the map. You should use distinct colours to represent the Hadley cell, Ferrel cell and Polar cell.

2. Identify and mark the regions on the map where each cell is and explain the associated atmospheric conditions, such as air circulation, precipitation patterns and temperature gradients.

Inquiry process

Inquiry 2: Collecting and processing data
1. Discuss how the distribution of heat and moisture resulting from the tricellular model influences the formation and characteristics of different biomes around the world.
2. Explore specific examples of biomes found within the latitudinal ranges of each cell and discuss the relationships between climatic conditions and biome types.

2.4.6 The oceans absorb solar radiation, and ocean currents distribute the resulting heat around the world

The seas absorb solar energy primarily through the absorption of sunlight by the ocean surface. The ocean surface directly interacts with sunlight, leading to the warming of the seas. The oceans are extremely important in the management and dispersion of heat on Earth. They absorb a huge quantity of solar radiation from the sun, acting as massive heat sinks. This absorption of solar energy contributes to global temperature regulation by reducing temperature extremes on land.

Furthermore, ocean currents, which are influenced by elements such as wind patterns, temperature differences and the Earth's rotation, aid in the distribution of this absorbed heat around the world. Ocean currents function as conveyor belts, transporting warm or cold water to different places, affecting regional temperatures and the distribution and productivity of marine and coastal ecosystems.

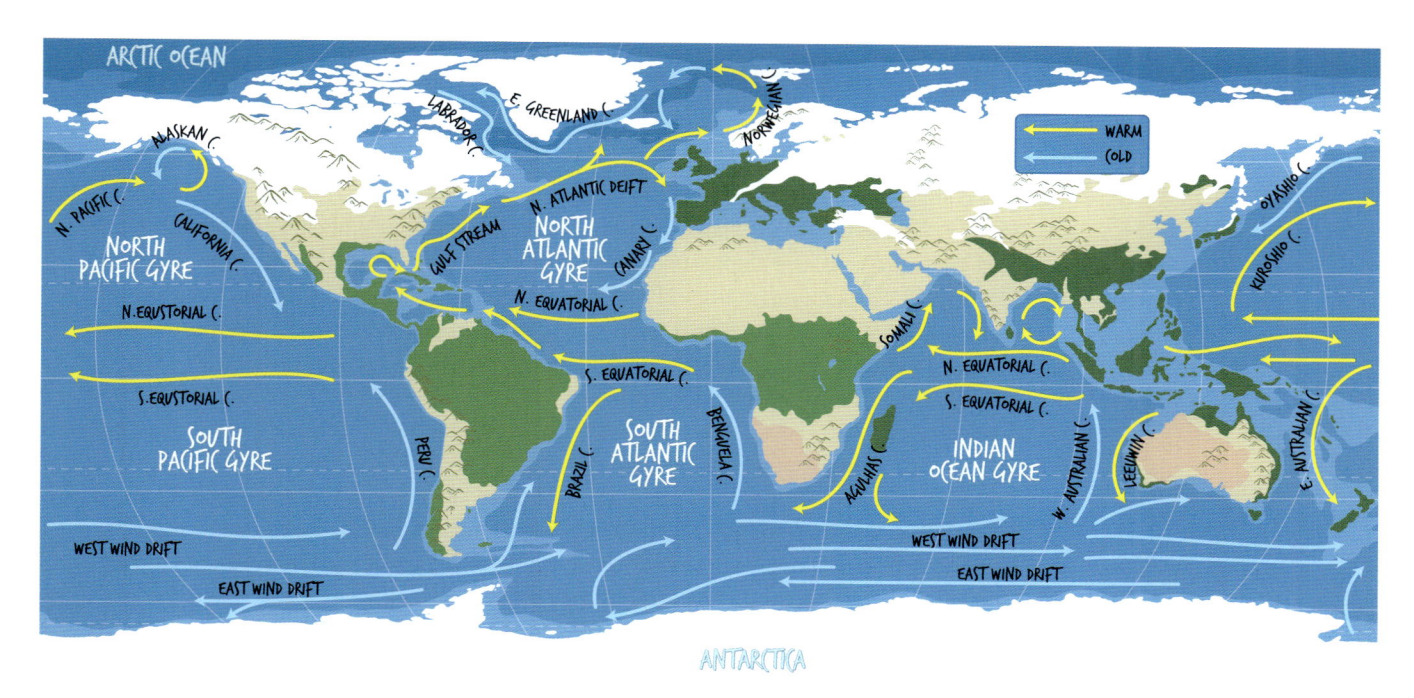

■ **Figure 2.65** Ocean currents distribute the resulting heat around the world

2.4.7 Global warming

Climate change is being driven by global warming, which is principally caused by the accumulation of greenhouse gases in the Earth's atmosphere. Increased concentrations of greenhouse gases, such as carbon dioxide and methane, trap heat within the Earth's atmosphere, causing average temperatures to rise. The change in temperature patterns has an impact on weather systems, precipitation levels and overall climatic conditions around the planet. As a result, biome shifts are occurring, when specific places undergo changes in temperature, rainfall patterns and resource availability.

These changes can have an effect on ecosystem composition and functioning, impacting species distribution, migration patterns and overall ecological balance. They have profound implications for the distribution and composition of biomes on Earth. Some key implications are:

1. Poleward shifts: As temperatures increase, many biomes are shifting their distribution towards higher latitudes. Biomes that were traditionally found closer to the equator are now expanding into regions previously dominated by different biome types.

2. Altitudinal shifts: Similar to poleward shifts, biomes are also shifting to higher altitudes on mountains and hills. As temperatures rise, the suitable climate conditions for certain biomes are found at higher elevations. This can lead to changes in the species composition and ecological dynamics within mountain ecosystems.

3. Ecosystem disruption: The changing climates can disrupt the delicate balance of ecosystems, impacting the interactions between species, their food availability and reproductive patterns. Some species may struggle to adapt to the rapid changes, leading to population declines or even local extinctions.

4. Biodiversity loss: Climate change-induced shifts in biomes can result in a loss of biodiversity. Species that are unable to adapt or migrate to new habitats may face habitat fragmentation, reduced resources and increased competition, which can ultimately lead to biodiversity decline.

(HL) 2.4.8 The three general patterns of climate types

The Earth's climates can be broadly classified into three patterns: tropical, temperate and polar. Each of these patterns has distinct characteristics in terms of temperature, precipitation and general weather conditions.

1. Tropical climates: These are characterized by consistently warm temperatures throughout the year. These areas receive ample sunlight and experience high temperatures, often exceeding 22°C (72°F). Precipitation in tropical climates is abundant, with annual rainfall ranging from around 1,500 to 3,000 mm. These regions usually have distinct wet and dry seasons due to the movement of the Intertropical Convergence Zone (ITCZ). Examples of tropical biomes include rainforests and savannas.
2. Temperate climates: These exhibit noticeable seasonal variations in temperature. Summers are warm to hot, with temperatures often ranging from 20°C to 30°C (68°F to 86°F), while winters are cooler, with temperatures ranging from −3°C to 18°C (26°F to 64°F). Annual precipitation varies widely, ranging from 500 to 2,000 mm. Four distinct seasons are commonly observed, including spring, summer, autumn and winter. Global examples of temperate climates include Mediterranean, humid subtropical and marine west coast climates.
3. Polar climates: These are characterized by extremely cold temperatures, with average temperatures not exceeding 10°C (50°F) even in the warmest months. Winters are long and bitterly cold, with temperatures commonly dropping below −30°C (−22°F). Precipitation is generally low, ranging from 200 to 1,000 mm per year. These regions experience polar night and polar day phenomena, where the sun does not set during certain times of the year and does not rise during others. Examples of polar climates include tundra and ice cap climates.

2.4.9 Predicted biomes may not develop due to outside influences

While temperature and rainfall patterns are important predictors of biome development in a given place, other secondary variables or human interventions can prevent the expected biome from developing. Topography, soil composition, nutrient availability and local disturbances can all have a substantial impact on the actual biome that develops in a given place. Human actions including deforestation, urbanization, pollution and land use changes can also disturb natural ecological processes and modify the intended biome.

Human activities can profoundly influence the development of biomes and alter the natural landscape. For instance:

1. Deforestation: Clearing large areas of forests for agriculture, logging or urban expansion can disrupt existing biomes, leading to habitat loss and changes in biodiversity.
2. Urbanization: Rapid urban growth can replace natural habitats with concrete jungles, affecting local climate, soil and hydrology, and subsequently preventing the development of expected biomes.
3. Pollution: Industrial emissions and pollutants can contaminate soil and water, adversely affecting plant and animal life, and modifying the ecological conditions needed for particular biomes.
4. Land use changes: Converting natural lands into intensive agricultural fields or other human-dominated landscapes can profoundly impact the development of predicted biomes, leading to the establishment of entirely different ecosystems.

2.4.10 The El Niño–Southern Oscillation and La Niña

The El Niño–Southern Oscillation (ENSO) is a climate phenomenon that occurs in the tropical Pacific Ocean and has a global impact on weather patterns. ENSO is defined by changes in wind patterns and sea surface temperatures, which cause changes in atmospheric circulation. El Niño is the warm phase of ENSO, when sea surface temperatures in the central and eastern Pacific rise abnormally.

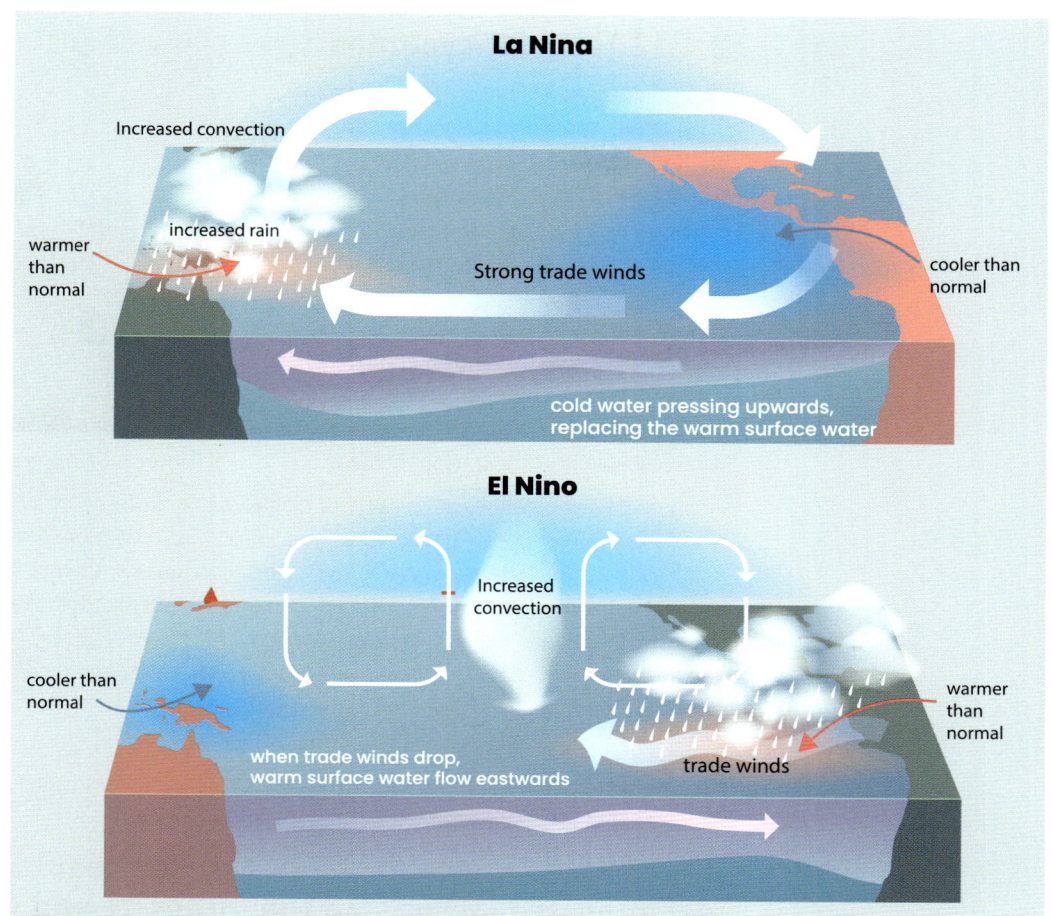

■ **Figure 2.66** El Niño and La Niña represent the warm and cool stages of a repetitive climatic cycle spanning the tropical Pacific Ocean. This phenomenon is known as the El Niño–Southern Oscillation, often abbreviated as 'ENSO'.

In contrast, La Niña depicts the cold phase, with cooler sea surface temperatures in the same location. These opposing phases of ENSO have an impact on global weather patterns, causing changes in precipitation, temperature and storm frequency in diverse places throughout the world. Transitional and neutral phases, depicting times of altering conditions, are also recorded between El Niño and La Niña.

2.4.11 Causes and effects of El Niño and La Niña

El Niño is caused by a weakened or reversed east–west atmospheric circulation pattern (Walker circulation) in the tropical Pacific region. This affects normal wind patterns and ocean currents, which increases surface stratification and reduces upwelling of cold, nutrient-rich water near the coast of northern South America. La Niña is caused by a strengthening of the Walker circulation and a reversal of El Niño's other effects.

Understanding changes in atmospheric and marine circulation patterns can help explain the different effects of El Niño and La Niña on regional climates and the ecological repercussions.

> **ATL ACTIVITY**
>
> Conduct research and prepare a short news section for your school bulletin board to draw attention to the most recent events linked to El Niño.

2.4 Climate, biomes and their influences

2.4.12 Tropical cyclones

Tropical cyclones, often known as hurricanes or typhoons depending where they occur, are fast-moving and extremely intense storm systems that form in tropical areas. They are distinguished by a low-pressure centre, high winds and heavy rainfall. Tropical cyclones form over warm ocean waters, with sea surface temperatures of at least 26.5°C in most cases. As warm, moist air rises and condenses, it releases a vast amount of energy, feeding the cyclonic system and generating strong winds capable of causing major damage.

■ **Figure 2.67** Tropical cyclones

Rising ocean temperatures are a result of global warming caused by increased greenhouse gas emissions. Warmer ocean temperatures give tropical cyclones more energy, increasing their strength and frequency. Because warm water evaporates more quickly, there is more moisture in the atmosphere, which contributes to the production of greater storms. Warmer air can also store more moisture, resulting in higher rainfall associated with tropical cyclones. These climate changes have an impact on the behaviour and characteristics of hurricanes and typhoons, potentially leading to more devastating repercussions in coastal regions.

Understanding the relationship between global warming, ocean temperatures and tropical cyclone strength is critical for assessing future risks and implementing mitigating strategies to moderate their effects.

2.4.13 Rises in ocean temperatures resulting from global warming are increasing the intensity and frequency of hurricanes and typhoons because warmer water and air have more energy

Increased energy in warmer water and air causes the rising of ocean temperatures brought on by global warming. This is turn is boosting both the intensity and frequency of hurricanes and typhoons. This phenomena results from the fundamental idea that warmer temperatures correspond to weather systems with more energy.

The complex interaction between oceanic and atmospheric circumstances is what causes the link between warming oceans and stronger storms. The increase in sea surface temperatures brought on by the ongoing warming of the planet's environment makes it easier for tropical storms to form and grow stronger. Their growth is fuelled by this energy, which makes them stronger and more destructive.

Examples of how increased ocean temperature affects storm activity are becoming more and more clear. Typhoons and hurricanes are occurring more frequently and intensely in areas that formerly experienced relatively reduced vulnerability to such storms. For instance, the frequency of big storms in the Atlantic hurricane season has increased during the past few decades. In addition, the Pacific region has been dealing with stronger typhoons that leave a trail of destruction in their wake. Examples of places where these occur include the Philippines and several areas of Southeast Asia.

Over time, the proportion of Category 4 and 5 hurricanes in the Atlantic Ocean has increased noticeably, according to data from the National Oceanic and Atmospheric Administration (NOAA). In the Western North Pacific, research has shown that powerful typhoons are occurring more frequently. These developments are closely related to the warming of ocean surfaces.

Thus the increasing intensification and frequency of hurricanes and typhoons is driven by the rising ocean temperatures linked to global warming. In order to protect vulnerable people from the rising risks posed by these extreme weather events, this disturbing trend highlights the urgent need for global measures to prevent climate change and its accompanying repercussions.

> **ATL ACTIVITY**
>
> Investigate the impact of climate change on a nearby or regional ecosystem.
> 1 Create a presentation that clarifies the reasons behind the recent changes and the resulting consequences of this transformation.
> 2 Design a project to increase knowledge and gather funds to support communities that are severely affected by intense hurricanes or typhoons.

REVIEW QUESTIONS

1 Outline the two essential characteristics of the Earth's atmosphere that define ecological systems, and how they differ from each other.
2 How do abiotic factors, such as temperature, rainfall and soil composition, influence the distribution and structure of terrestrial biomes?
3 Explain how the tricellular model of atmospheric circulation explains global air movement, precipitation distribution and temperature gradients. How does it influence the formation and characteristics of different terrestrial biomes around the world?
4 Describe the distribution of tropical rainforest and savanna grassland biomes.
5 Compare and contrast the distinct characteristics of these two biomes: tropical rainforest biome and savanna grassland biome.
6 Suggest similarities and differences in net productivity and biodiversity levels between tropical rainforest and savanna grassland biomes.

EXAM-STYLE QUESTIONS

1 Explain how the level of primary productivity of different biomes influences their resilience. [7 marks]
2 Climate can both influence, and be influenced by, terrestrial food production systems.

   To what extent can terrestrial food-production strategies contribute to a sustainable equilibrium in this relationship? [9 marks]
3 With reference to Figure 2.68, outline the effectiveness of mangroves and tropical rainforests in the mitigation of climate change. [2 marks]

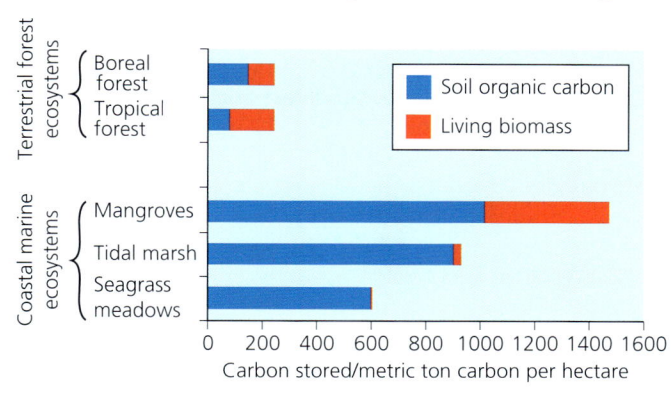

■ **Figure 2.68** Carbon storage in different ecosystems (Source: adapted from thebluecarboninitiative.org)

2.5 Zonation, succession and change in ecosystems

> **Guiding question**
>
> ■ How do ecological systems change over time and over space?

SYLLABUS CONTENT

This chapter covers the following syllabus content:
- ▶ 2.5.1 Zonation refers to changes in community along an environmental gradient.
- ▶ 2.5.2 Transects can be used to measure biotic and abiotic factors along an environmental gradient in order to determine the variables that affect the distribution of species.
- ▶ 2.5.3 Succession is the replacement of one community by another in an area over time due to changes in biotic and abiotic variables.
- ▶ 2.5.4 Each seral community (sere) in a succession causes changes in environmental conditions that allow the next community to replace it through competition, until a stable climax community is reached.
- ▶ 2.5.5 Primary successions happen on newly formed substrata where there is no soil or pre-existing community, such as rock newly formed by volcanism, moraines revealed by retreating glaciers, wind-blown sand or water-borne silt.
- ▶ 2.5.6 Secondary successions happen on bare soil where there has been a pre-existing community, such as a field where agriculture has ceased or a forest after an intense firestorm.
- ▶ 2.5.7 Energy flow, productivity, species diversity, soil depth and nutrient cycling change over time during succession.
- ▶ 2.5.8 An ecosystem's capacity to tolerate disturbances and maintain equilibrium depends on its diversity and resilience.

HL ONLY
- ▶ 2.5.9 The type of community that develops in a succession is influenced by climatic factors, the properties of the local bedrock and soil, and geomorphology, together with fire- and weather-related events that can occur. There can also be top-down influences from primary consumers or higher trophic levels.
- ▶ 2.5.10 Patterns of net and gross productivity change over time in a community undergoing succession.
- ▶ 2.5.11 r- and K-strategist species have reproductive strategies that are better adapted to pioneer and climax communities, respectively.
- ▶ 2.5.12 The concept of a climax community has been challenged and there is uncertainty over what ecosystems would develop naturally were there no human influences.
- ▶ 2.5.13 Human activity can divert and change the progression of succession, leading to a plagioclimax.

Figure 2.69 Rocky shore zonation

2.5.1 Zonation refers to changes in community along an environmental gradient

Zonation is a geographical phenomenon that occurs when environmental circumstances change along a gradient, resulting in separate **populations**. Changes in elevation, latitude, tide level, soil horizons or distance from a water supply are examples of environmental gradients. Zonation occurs as a result of different species' tolerances to specific environmental conditions. As a result, **communities** are structured in a gradient pattern, with different species or communities dominating different zones. For instance, as shown in Figure 2.69, beaches have distinct zones, depending on the water levels. The properties of each zone influence the kinds of marine organisms that live in the habitat.

◆ **Population** – all the members of a single species that live in the same area at the same time (and are capable of interbreeding).

◆ **Community** – all the interacting populations of organisms living in the same area at the same time.

> **REAL-WORLD EXAMPLE**
>
> ### The North Sea coast of the Netherlands
>
> Coastal vegetation along the North Sea coast of the Netherlands exhibits distinct zonation, influenced by various environmental factors from the coastline to the inland. Within this profile, three factors decline rapidly over short distances: salinity, wind velocity and mechanical impact of moving sand particles. The stability of the soil increases as we move inland. This zonation reveals a gradual transition from a vegetation-free zone to an open system with a few plants specially adapted to extreme conditions, such as sand couch and sea rocket, and then to an open grassland dominated by marram grass. Further inland, the landscape transforms into a more enclosed grassland and brushwood, characterized by low shrubs like sea buckthorn, sweet briar, dog rose, honeysuckle and wild privet. Deeper inland, the inner dune wood is found, featuring tall trees such as birch, aspen, oak and sycamore.

2.5.2 Transects can be used to measure biotic and abiotic factors along an environmental gradient in order to determine the variables that affect the distribution of species

The variables affecting the distribution of species can be determined by measuring the biotic and abiotic factors along an environmental gradient with the help of transects. A transect is a line or trail that runs along an environmental gradient to collect data on biotic and abiotic elements at regular intervals for investigating the zonation. Abiotic influences include temperature, humidity, light intensity, pH and soil composition, whereas biotic factors include the existence and abundance of various species. Scientists can find the trends and relationships between these characteristics and species distribution by collecting data along the transect.

Link
Further details of using transects and biotic factors are available in Topic 2.1.15 (page 97).

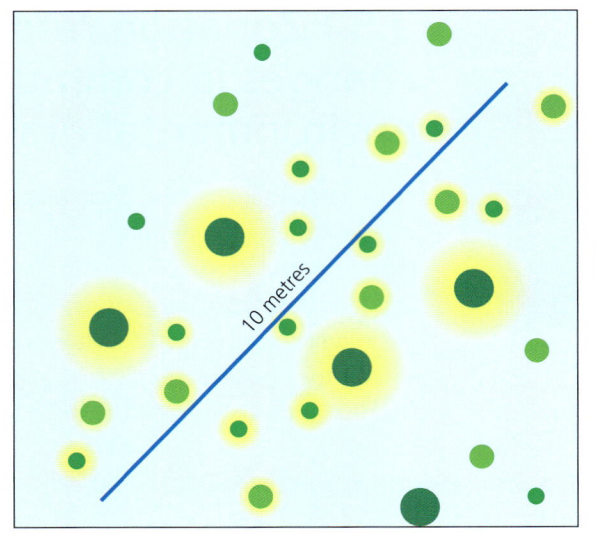

■ **Figure 2.70** A line transect. Organisms are marked with yellow shadow if they are valid to be included in the count as they are within the view of the researcher

> ### Tool 1: Experimental techniques
>
> #### Fieldwork activity: Investigate zonation along an environmental gradient
>
> For carrying out fieldwork you should be able to use transects to measure changes along an abiotic gradient. To show this distribution you can use a kite diagram.
>
> *Materials*
> - Measuring tape or ruler
> - Field notebook or data sheet
> - Sampling quadrat (optional)
> - Digital sensors: such as temperature sensor, pH meter
> - Graph paper or spreadsheet for creating kite diagrams
>
> *Instructions*
> 1. Select an environmental gradient to study, such as a shoreline, mountain slope or forest edge. Ensure that the gradient exhibits clear changes in the environmental factors of your interest.
> 2. Lay out a transect line perpendicular to the gradient, as shown in Figure 2.70. This can be done using a measuring tape or ruler. Mark intervals along the transect line at regular distances (for example, every metre).
> 3. At each interval, record relevant abiotic measurements using appropriate tools. For example, measure temperature, pH, light intensity or any other factors that are relevant to the study.
> 4. If desired, use a sampling quadrat to collect data on the presence and abundance of different species along the transect line. Place the quadrat at each interval and record the species present within it.
> 5. Organize the collected data in tables or figures to analyse the changes in abiotic factors and species distribution along the gradient. Identify any patterns or correlations that emerge.

6 Create kite diagrams, as shown in Figures 2.71 and 2.72, to visually represent the distribution of species along the transect. Each kite diagram should represent a specific interval and include the species found within that interval. The size of each kite should correspond to the abundance of the species.
7 Analyse the kite diagrams and data to draw conclusions about the zonation patterns observed along the environmental gradient. Consider the relationships between abiotic factors and species distribution, and discuss the factors that may be driving these patterns.
8 Reflect on the significance of zonation in ecological communities and its implications for species interactions, biodiversity and ecosystem functioning.

Note: This activity can be adapted based on the specific environmental gradient and factors of interest you want to focus on.

Tool 3: Mathematics

Kite diagrams

There are some key reasons for plotting and visualizing geographical data. Tables of raw data can be difficult to understand and interpret. Graphs and specialist maps can make it easier to highlight patterns and trends in the data, allowing you to understand them better and to spot anomalies. This can also help you determine the reliability of the data.

Kite diagrams are usually used to show changes in the frequency of plant species along a sampling line (see Figure 2.71), often a continuous or interrupted transect. This technique is used in biological sciences to show zonation: changes in a biotic community over a distance due to a changing environmental gradient. Geographers often use these diagrams in ecosystem-survey work, for example, a transect across sand dunes, or to show the changes in species type and frequency associated with impacts such as footpath erosion (see Figure 2.72).

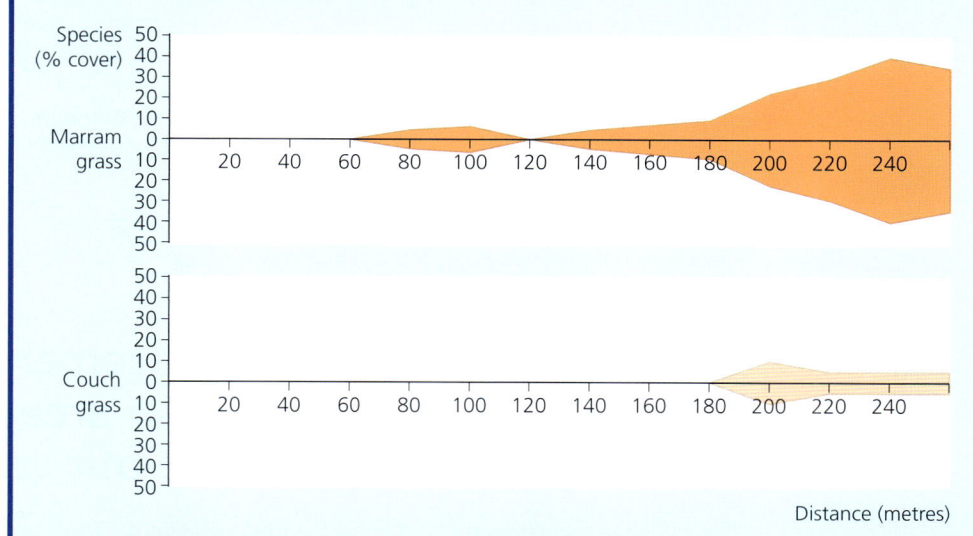

■ **Figure 2.71** An example of a simple two-species kite diagram

2.5 Zonation, succession and change in ecosystems

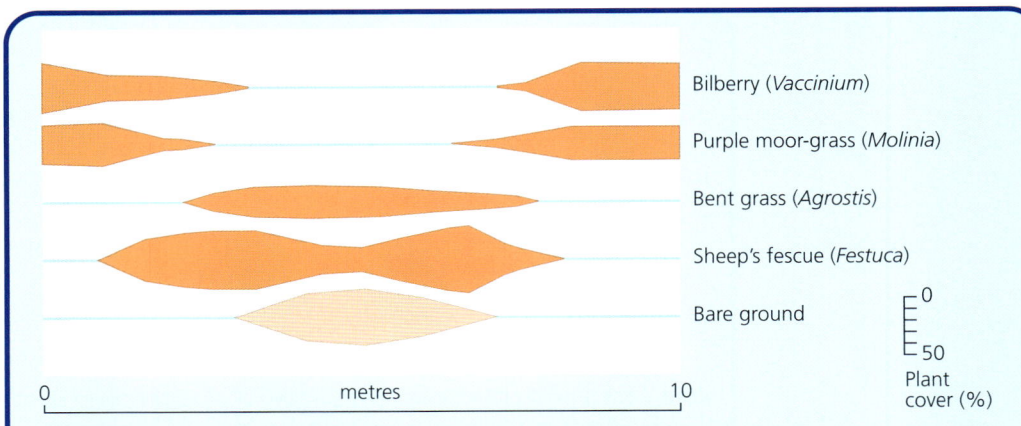

Figure 2.72 A kite diagram showing the change in species (and bare ground) across a footpath. This shows the effect of recreational pressure and trampling on the species type and abundance

How to draw a kite diagram

- Design an appropriate scale line for the horizontal distance of the transect (usually the x-axis).
- Create one row for each type of vegetation cover (there are two in Figure 2.71).
- Each row needs to have the same scale and be wide enough to allow 100 per cent for each type of vegetation.
- At each survey point, plot the value (frequency) on both sides of the x-axis, giving a mirrored effect.
- Join up the points for each row to give the diagram its 'kite' appearance. These are often shaded.

Common mistake

It is important to check you have the following when you are constructing kite diagrams and other types of graphs:

- accurate data representation
- complete data
- correct scaling
- correct and clear labelling
- clear context
- supportive visual elements
- correct interpretation of trends
- appropriate graph types
- graphs
- data precision
- data citations.

To avoid making errors, you should plan carefully, use the right graph type, ensure accuracy and clarity, and include necessary labels and context while citing data sources.

2.5.3 Succession is the replacement of one community by another in an area over time due to changes in biotic and abiotic variables

Succession is the replacement of one community by another in an area over time due to changes in biotic and abiotic variables. This process can take hundreds of years, and pollen records preserved in peat deposits can provide evidence of past successional shifts.

Zonation is *spatial* and refers to the structuring of communities along an environmental gradient, whereas succession is *temporal*: it refers to the sequential replacement of communities over time.

2.5.4 Each seral community (sere) causes changes that allow the next community to replace it

◆ **Seral community** – intermediate stage in ecological succession, representing a transitional group of species in an area undergoing environmental change.

During succession, each **seral community** contributes to the improvement of an area's natural conditions, making it more appropriate for the development of the next community. This evolution occurs as a result of a series of competitive interactions between species. Pioneer species, such as mosses and lichens, can colonize bare rock surfaces and initiate soil formation through breakdown in the early stages of succession. This process of soil formation generates a more conducive environment for the development of larger plants that can outcompete the early colonizers. This cycle continues until a stable climax community is formed, which is often characterized by a diverse array of species.

Initial conditions → Pioneer species → Sequence of plant communities of increasing size → Climax community

| Bare rock and weathering | Lichens | Mosses | Grasses | Grasses, shrubs and young trees | Mature woodland |

Soil formation (Lichens → Mosses → Grasses)

Time →

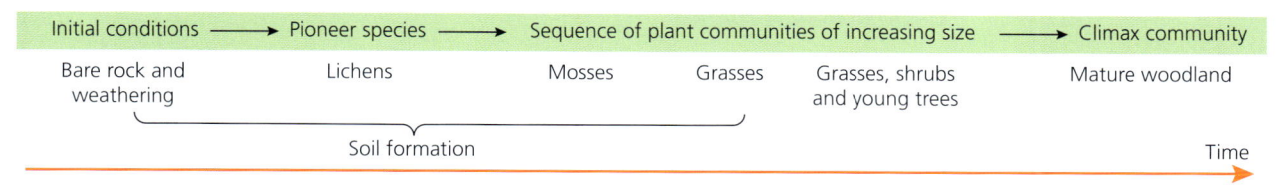

■ **Figure 2.73** The main seral stages during primary succession on bare rock

---

Tool 1: Experimental techniques

Secondary data

Materials
- Mapping software or tools (for example, paper maps or online mapping platforms)
- Secondary data sources (for example, historical maps, aerial photographs, satellite imagery)

Instructions
1. Select a specific area or ecosystem where succession can be observed. This could be a well-documented case study area or a local site with available data.
2. Gather secondary data sources that provide information about the historical changes in the area. These may include historical maps, aerial photographs from different time periods or satellite imagery.
3. Use mapping software or tools to overlay and compare the different data sources. Analyse the changes in vegetation cover, land use or other relevant variables that indicate successional changes over time.
4. Identify the different seral communities (sere) that have been present in the area and their corresponding stages in the successional sequence. For example, distinguish between pioneer, intermediate and climax communities.
5. Map the changes in vegetation or land cover over time, indicating the spatial extent and distribution of each seral community at different time points. Use symbols, colours or other visual representations to differentiate the communities and their stages.
6. Analyse the mapped changes in relation to the biotic and abiotic variables that influence succession. Consider factors such as soil development, nutrient availability, disturbance events or climatic conditions that may have influenced the succession process.
7. Reflect on the implications of succession for ecosystem dynamics, species interactions and ecosystem services. Discuss how human activities can influence or disrupt natural successional processes and the potential consequences of such disruptions.

Note: Depending on the availability of data and resources, the mapping activity can be adjusted to suit the specific case study or local example you choose.

2.5.5 Primary succession

Primary succession occurs in regions where there is no soil or pre-existing community, usually on newly created or exposed substrates. Examples include volcanic rock surfaces, moraines left by receding glaciers, patches of wind-blown sand and water-deposited sediments. The process of succession in such habitats begins with pioneer species that are adapted to harsh circumstances and may colonize these bare substrates. As soil formation occurs and environmental circumstances improve, the pioneer species are gradually replaced by a series of seral communities or stages, as shown in Figure 2.73. This succession results in the formation of a climax community, which is a generally stable and diverse environment.

REAL-WORLD EXAMPLE

The Hawaiian island Kaua'i

Figure 2.74 shows part of the island of Kaua'i, geologically the oldest of the Hawaiian Islands. This island emerged from the sea as a result of volcanic activity around 5 million years ago. After it had cooled down, pioneer species colonized its bare volcanic rock and primary succession began on the island. You can see from the figure that it is now covered by mature forest.

■ **Figure 2.74** An aerial view of the Na Pali coast of Kaua'i, the oldest inhabited Hawaiian island

2.5.6 Secondary succession

Secondary succession occurs when a pre-existing community is disrupted or removed, leaving bare soil behind. Abandoned agricultural fields, or locations ravaged by strong firestorms or landslides are all examples. Unlike primary succession, secondary succession begins with a soil seed bank or the introduction of propagules from nearby places. Propagules are reproductive structures or units of plants that have the potential to develop into new individuals and establish themselves as part of a new community. These structures serve as a means of dispersal, allowing plants to colonize new areas and initiate the process of succession, especially in disturbed or barren environments. Seeds are the most common example of propagules. Seeds contain the genetic material necessary for growth and development, along with some stored nutrients to sustain the new plant initially.

■ **Figure 2.75** Secondary succession of a pine forest after a forest fire in Bastrop, Texas

These propagules can be spread by wind, water or animals. As the process progresses, pioneer species that are well-suited to swiftly colonize damaged environments will colonize the location. The composition of the community evolves throughout time, and distinct seral stages occur until a climax community is established.

2.5.7 Changes over time during succession

Several ecological parameters alter dramatically during succession. Energy flow, productivity, species variety, soil depth and nutrient cycling in ecosystems are all examples of these changes. Table 2.5 explains what changes tend to take place with regards to each of these throughout the process.

■ Table 2.5 Changes to ecological parameters during succession

Ecological parameter	Change throughout the succession process
Energy flow	Energy flow is low in early phases due to limited resources and harsh conditions. It increases in intermediate phases due to greater species diversity, resource utilization and efficient energy capture. It stabilizes or decreases in the climax community due to a balance between resource availability and energy loss through respiration.
Productivity	Productivity is low in early phases due to limited resources and harsh conditions. It increases in intermediate phases due to greater species diversity, resource utilization and efficient energy capture. It stabilizes or decreases in the climax community due to a balance between resource availability and energy loss through respiration.
Species variety	Species variety is limited in pioneer communities with a few opportunistic species. It increases throughout the succession process as more species establish themselves, contributing to ecosystem resilience and stability.
Soil depth	Soil depth increases throughout succession due to the accumulation of organic matter and the breakdown of pioneer species. Deeper soil allows larger plants with deeper root systems to establish themselves. It stabilizes in the climax community.
Nutrient cycling	Nutrient cycling becomes increasingly complex and diversified throughout succession as plant and microbial communities interact and organic matter accumulates. It stabilizes in the climax community.

2.5.8 An ecosystem's capacity to tolerate disturbances and maintain equilibrium depends on its diversity and resilience

An ecosystem's variety and resilience are critical to its ability to endure perturbations and maintain equilibrium. Succession helps to increase the resilience and stability of ecosystems by allowing species to diversify. Ecosystems become more capable of withstanding and recovering from perturbations as species richness and functional diversity grow. This is because a varied population comprises a broader range of features, allowing certain species to endure even in changing conditions.

Human actions, on the other hand, can alter successional processes and impair ecosystem variety and resilience. For example, habitat destruction, fragmentation, pollution or the introduction of alien species might obstruct successional trajectories and result in simplified or degraded ecosystems. These disruptions can reduce an ecosystem's ability to withstand future disturbances, rendering it more vulnerable to permanent changes.

In order to design sustainable management and conservation policies, it is critical to consider the relationships between ecosystem resilience, stability, succession, variety and human activities. Understanding the processes of succession as well as the factors that influence diversity and resilience can assist efforts to protect and restore ecosystems that have been harmed by human-caused disturbances.

(HL) 2.5.9 The type of community that develops in a succession is influenced by climatic factors, the properties of the local bedrock and soil, and geomorphology, together with fire- and weather-related events

A variety of factors influence the type of community that arises during succession.
- Climatic factors such as temperature, precipitation and seasonality all have a role in deciding which species may survive in a given place.
- The local bedrock and soil qualities determine soil development, nutrient availability and pH, which in turn influence the types of plants and creatures that can establish themselves.
- Geomorphology, such as steep slopes and a lack of drainage, can create diverse microhabitats and influence species distribution.
- Disturbances like as fires also influence successional paths. Fire can initiate a reset in succession by eliminating vegetation and supporting the growth of pioneer species that are adapted to post-fire conditions.
- Similarly, major storms or other weather-related events can cause localized perturbations that affect the succession trajectory. Furthermore, top-down influences from primary consumers or higher trophic levels can influence the final community's composition and organization. For example, the reintroduction of wolves into Yellowstone National Park has had a cascade effect on the ecology, affecting herbivore behaviour and hence impacting plant ecosystems.

REAL-WORLD EXAMPLE

Wolves maintain Yellowstone's balance

In the 1920s, government policy let Yellowstone's **apex predator**, the gray wolf, be exterminated, resulting in an ecosystem collapse known as trophic cascade. To restore equilibrium, the conservation community reintroduced the gray wolf in 1995 using the Endangered Species Act. The result has been dramatic.

Without year-round wolf kills to feed on, other scavenger species starved. Without wolves as their principal predator, elk populations soared, leading in catastrophic overgrazing of willows and aspen. Beavers, which rely on these trees for food, housing and dam construction, all but vanished in Yellowstone's northern range. Dams failed, transforming marshy ponds into torrents. There was massive destruction of mature willows and aspens, significant stream erosion, and several plant and animal species were harmed.

In the absence of wolves, the coyote became the apex predator, decimating populations of pronghorn antelope, red fox, rodents and birds. Coyote numbers fell by half as the wolf returned, allowing antelope, rodent, and fox populations to grow.

Elk numbers decreased after the wolf reintroduction in the northern region, whereas beaver colonies increased from one to twelve. Insects, songbirds, fish and amphibians flourish in this environment. Today, biodiversity has been enhanced and scavenger species benefit from regular, wolf-supplied meals.

◆ **Apex predator** – the top-level predator in a food chain or ecosystem, not preyed upon by other organisms.

■ **Figure 2.76** Infographic: Wolves keep Yellowstone in the balance

2.5.10 Patterns of net and gross productivity change over time in a community undergoing succession

Productivity patterns, both gross and net, change over the succession process. Gross productivity is often low in the early stages of succession due to poor starting conditions and a low density of producers, such as pioneer plants. These pioneer plants are often tiny, have limited photosynthetic capability and are adapted to severe circumstances, factors that result in decreased biomass production rates. However, during this stage, the fraction of energy lost by communal respiration is also quite low, resulting in a high net productivity. In other words, despite the low levels of gross productivity, the system is still expanding and accumulating biomass. As succession advances and more species establish themselves the community moves to later stages, when gross output may be higher, particularly in climax communities.

A climax community typically has a varied range of species, including bigger plants with greater photosynthetic ability. As more energy is captured by photosynthesis, this can lead to higher rates of gross productivity. However, when the respiratory demands of a more sophisticated consumer society increase, this higher gross production is offset by increasing energy waste through communal respiration. As a result, net productivity in the climax community approaches zero, suggesting a balance between energy input and output.

2.5.11 r- and K-strategist species have reproductive strategies that are better adapted to pioneer and climax communities, respectively

Different reproductive techniques are used by r-strategist and K-strategist species (see Topic 2.1.29, page 116), which correspond to their adaptations to pioneer and climax ecosystems, respectively. Remember, the ability to produce a large number of offspring distinguishes r-strategists. This reproductive method enables them to quickly colonize new habitats and exploit scarce resources. r-strategists also mature early, have a short lifespan and provide little parental care to their progeny. These characteristics allow them to quickly exploit available resources and widely propagate their genes. Many annual plants and certain small, short-lived animals are examples of r-strategists.

K-strategist species, on the other hand, use a distinct reproductive strategy that is more suited to long-term climax communities. K-strategists generate fewer offspring but devote more resources and parental care to each individual. This technique improves survival and aids the long-term survival of their offspring. K-strategists have a slower reproduction rate, later maturity, longer lifespan and stronger competitive ability. These species are better adapted to climax communities' stable conditions and intensive competition. Long-lived trees, huge mammals and numerous species found in mature forests are examples of K-strategists.

2.5.12 The concept of a climax community

In current ecological thinking, the concept of a climax community, defined as a stable endpoint of succession, has been debated and challenged. Ecologists realize that natural ecosystems are dynamic and subject to a variety of natural and manmade disruptions. As a result, the concept of a fixed peak community has been called into question, and there is growing acceptance that ecosystems can live in alternate stable states.

Different community compositions and structures that can exist under similar environmental conditions are referred to as alternative stable states. These alternative states may emerge as a result of chance events, historical legacies or the influence of critical ecological processes.

The presence or absence of specific species, feedback mechanisms or organism interactions can all alter the trajectory of succession and define the composition of the final community. As a result, a lack of human impacts does not guarantee the establishment of a predictable climax community, because ecosystems have a variety of possible fates.

The wood-pasture ('Vera') hypothesis, for example, questions the accepted notion of climax communities in wood-pasture habitats. According to the hypothesis, substantial grazing pressure from large herbivores, such as wild herbivores or livestock (for example, elephants, horses or deer), maintains a dynamic mosaic of open grassland patches and distributed trees, rather than allowing the system to attain a stable peak state. This hypothesis highlights the importance of primary consumers in creating plant communities and calls into question the idea of a single fixed endpoint.

In summary, there is ambiguity surrounding ecosystem evolution in the absence of human influences, as well as the potential for alternative stable states and dynamic ecological processes to shape community outcomes.

2.5.13 Human activity can divert and change the progression of succession, leading to a plagioclimax

Human activities can drastically affect the succession trajectory, deviating from the natural evolution and resulting in a situation known as a plagioclimax. Plagioclimax refers to a community that has been halted or warped by human intervention, preventing the intended climax community from forming.

The loss of top carnivores from an ecosystem is one example of human-induced plagioclimax (see Real-world Example in Topic 2.5.9, page 186). Top carnivores, such as wolves and large predators, play critical roles in herbivore-population regulation. When apex carnivores are removed from an ecosystem through hunting or other human actions, herbivore populations may grow uncontrollably. Too much herbivore activity can stifle plant development and establishment, affecting successional processes and impeding the transition to a climax community.

Another example is the effects of domesticated livestock grazing. When animals graze heavily on vegetation, it can result in the extinction of some plant species and the promotion of others that are more appealing to livestock or tolerant to grazing. This selective grazing can change the composition and organization of plant communities, causing succession to diverge, and resulting in a plagioclimax state that differs from the expected climax community.

These examples demonstrate how human actions can disrupt natural successional processes, resulting in the formation of alternate community states that diverge from the predicted trajectory. Understanding the effects of human intervention on succession is critical for developing educated conservation and management strategies for protecting or restoring natural ecosystems.

ATL ACTIVITY

Create a digital infographic or poster for your school to explain the research you conducted in one of your fieldworks. Remember to include research question, methodology, data and results.

REVIEW QUESTIONS

1. What is the purpose of using transects in ecological research, and how do they help us understand the factors influencing species distribution?
2. Compare and contrast primary and secondary succession. Provide examples of each and explain how propagules contribute to secondary succession.
3. Explain the concepts of r-strategist and K-strategist species in relation to succession. How do their reproductive strategies relate to pioneer and climax communities?

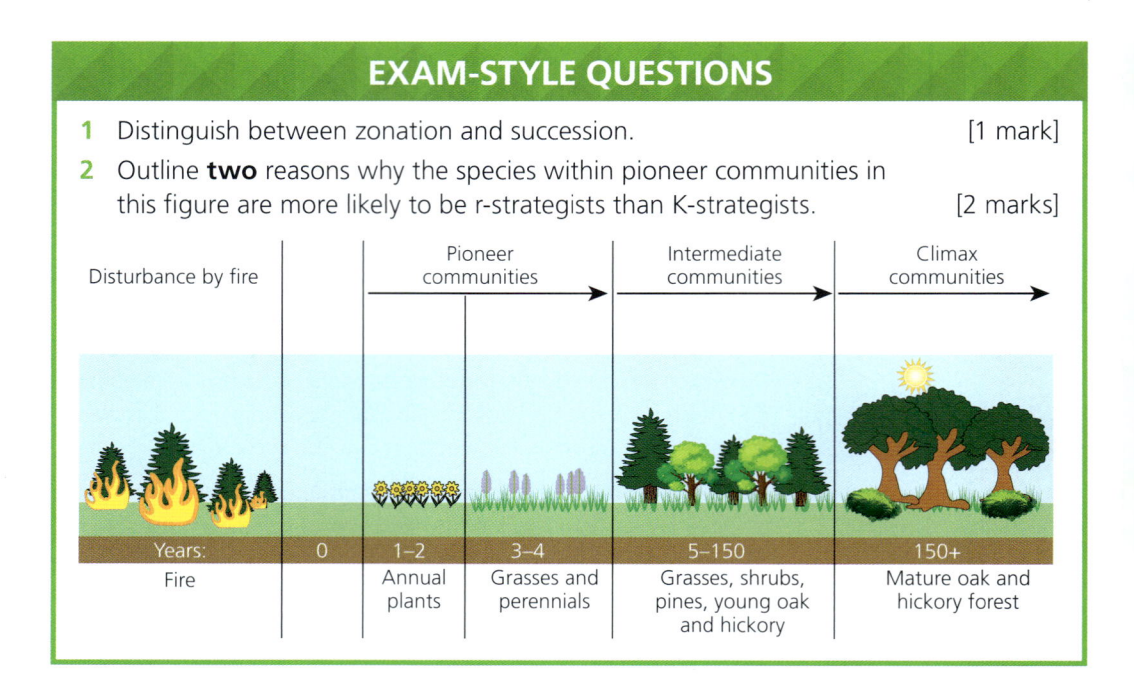

EXAM-STYLE QUESTIONS

1. Distinguish between zonation and succession. [1 mark]
2. Outline **two** reasons why the species within pioneer communities in this figure are more likely to be r-strategists than K-strategists. [2 marks]

Disturbance by fire	Pioneer communities →	Intermediate communities →	Climax communities →
Years: 0	1–2 / 3–4	5–150	150+
Fire	Annual plants / Grasses and perennials	Grasses, shrubs, pines, young oak and hickory	Mature oak and hickory forest

Theme 2: Ecology

Theme 3
Biodiversity and conservation

3.1 Biodiversity and evolution

> **Guiding questions**
> - How can biodiversity be explained and quantified, and why is it important?
> - How does the unsustainable use of natural resources impact biodiversity?

SYLLABUS CONTENT

This chapter covers the following syllabus content:
- ▶ 3.1.1 Biodiversity is the total diversity of living systems. It exists at several levels.
- ▶ 3.1.2 The components of biodiversity contribute to the resilience of ecological systems.
- ▶ 3.1.3 Biodiversity arises from evolutionary processes.
- ▶ 3.1.4 Natural selection is the mechanism driving evolutionary change.
- ▶ 3.1.5 Evolution by natural selection involves variation, overproduction, competition for limited resources, and differences in adaptation that affect rates of survival and reproduction.
- ▶ 3.1.6 Speciation is the generation of new species through evolution.
- ▶ 3.1.7 Species diversity in communities is a product of richness and evenness.
- ▶ 3.1.8 Simpson's reciprocal index is used to provide a quantitative measure of species diversity, allowing different ecosystems to be compared and for changes in a specific ecosystem over time to be monitored.
- ▶ 3.1.9 Knowledge of global and regional biodiversity is needed for the development of effective management strategies to conserve biodiversity.

HL ONLY
- ▶ 3.1.10 Mutation and sexual reproduction increase genetic diversity.
- ▶ 3.1.11 Reproductive isolation can be achieved by geographical separation or, for populations living in the same area, by ecological or behavioural differences.
- ▶ 3.1.12 Biodiversity is spread unevenly across the planet, and certain areas contain a particularly large proportion of species, especially species that are rare and endangered.
- ▶ 3.1.13 Human activities have impacted the selective forces acting on species within ecosystems, resulting in evolutionary change in these species.
- ▶ 3.1.14 Artificial selection reduces genetic diversity and, consequently, species resilience.
- ▶ 3.1.15 Earth history extends over a period of 4.5 billion years. Processes that occur over an extended timescale have led to the evolution of life on Earth.
- ▶ 3.1.16 Earth history is divided up into geological epochs according to the fossil record.
- ▶ 3.1.17 Mass extinctions are followed by rapid rates of speciation due to increased niche availability.
- ▶ 3.1.18 The Anthropocene is a proposed geological epoch characterized by rapid environmental change and species extinction due to human activity.
- ▶ 3.1.19 Human impacts are having a planetary effect, which will be detectable in the geological record.

3.1.1 Biodiversity is the total diversity of living systems, and exists at several levels

◆ **Biodiversity** – the variety and variability of living organisms.

The total diversity of living systems, also known as global **biodiversity**, refers to the variety of all living organisms present on Earth, including plants, animals, fungi, bacteria and microorganisms. Biodiversity encompasses the diversity of ecosystems and habitats. These habitats range from tropical rainforests and coral reefs, to deserts and polar regions. Each different habitat is unique in some way and supports its own specific range of animals, plants and fungi, which results in a unique combination of species and ecological processes.

The diversity of species and the genetic variation within each of those species are also important scales to be considered. High genetic diversity within a species is critical for adaptation and resilience to environmental changes, for disease resistance and for maintaining the overall health of populations. Without this, species are at risk of becoming extinct due to disease or some other environmental change that none of the population is able to adapt to. Cheetahs have one of the lowest genetic diversities of any species, with a predicted diversity of a maximum of 4 per cent, making them highly susceptible to extinction (source: Stephen J. O'Brien *et al.*, Conservation genetics of the cheetah: lessons learned and new opportunities, *Journal of Heredity*, Vol. 108, Issue 6, September 2017, pages 671–77, **https://doi.org/10.1093/jhered/esx047**).

■ Levels of biodiversity

Habitat diversity

Habitat diversity refers to the variety of different habitats within a region. Each habitat provides a unique set of services and functions. Different habitats also support various species specific to them, contributing to overall biodiversity.

Species diversity

Species diversity is the variety of different species within a given area. Each species has its own niche within a habitat and therefore all fulfil different functions within the habitat. Species diversity can also enhance ecosystem productivity and nutrient cycling, resulting in healthier ecosystems.

Genetic diversity

Genetic diversity refers to the variation in genetic information within each species. Populations with higher genetic diversity are better equipped to adapt to changing environmental conditions, diseases and other challenges. A wider range of genetic traits means there is a greater chance of some individuals having advantageous characteristics that can help them survive and reproduce in a changing environment.

■ How diverse organisms are grouped

◆ **Taxonomy** – the science of classifying and organizing in a hierarchical manner.

The diversity of organisms is divided into a hierarchy of **taxonomic** groups that allows species to be grouped with other, similar species. Currently there are thought to be between around 8 million and 30 million different species in the world, with only a fraction of these currently having been discovered and described. The known classification of organisms is divided into Animalia, Plantae, Fungi, Protista (protozoa and algae) and Monera (bacteria). Each kingdom is further split into numerous phyla, which are further split into many different classes. Each species fits into a specific group at each level of classification, and each species has its own individual niche and inhabits its own place in the environment.

Link

More information about the classification of organisms and the use of taxonomy to group organisms is available in Topics 2.1.3 (page 86) and 2.1.4 (page 87).

Currently there are over 2 million species of organisms that have been identified and described, with just over 1 million of those being insects (see Figure 3.1). Around 425,000 species of plants and nearly 6,600 mammals have been described. Despite this, some of these groups are better known than others. The taxonomic group where there is known to be a huge diversity of species that are yet to be described is Insects (see Figure 3.2). Fungi, Arachnids and Algae are other groups where only a small proportion of their projected species numbers have been discovered and described. Conversely, plant and mammal species have the highest proportions of species that are known and described. This means that the biodiversity of organisms is vastly underestimated, highlighting the need to protect current habitats, particularly those that have yet to be fully studied.

Number of described species:

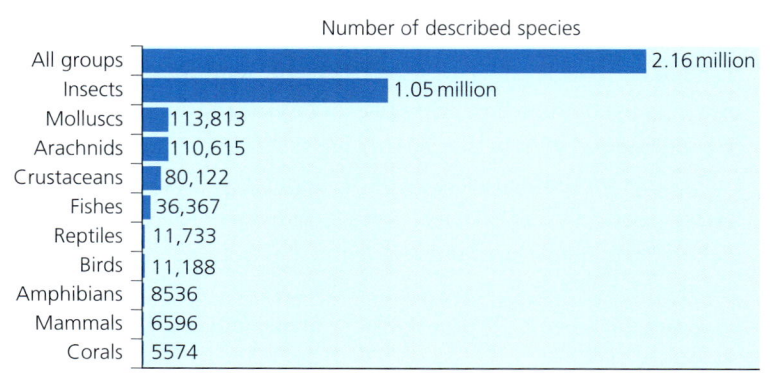

All groups	2.16 million
Insects	1.05 million
Molluscs	113,813
Arachnids	110,615
Crustaceans	80,122
Fishes	36,367
Reptiles	11,733
Birds	11,188
Amphibians	8536
Mammals	6596
Corals	5574

■ **Figure 3.1** The abundance of described species within different taxonomic groups in 2022 (source: IUCN Red List 2022)

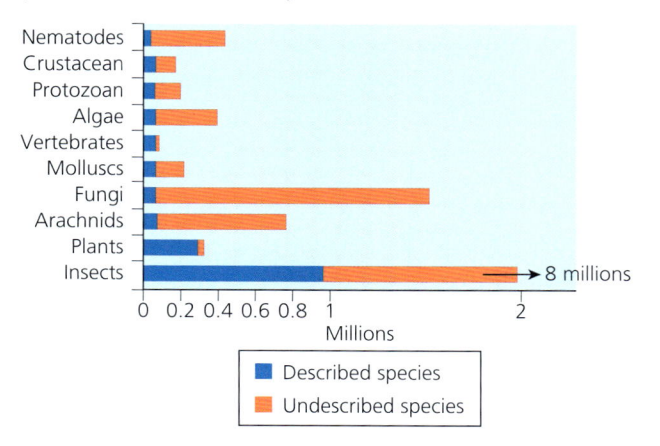

■ **Figure 3.2** Numbers of described and undescribed species of different taxonomic groups in the world

● HL.b.1: Environmental economics

As economies develop, the focus moves towards more technological innovations to help solve some of the global environmental issues. This results in valuable funding being diverted to look at conservation of species, which reduces the money available for taxonomic research. Conservation of species is often funded by museums, universities and societies, such as the Asian Society for Arachnology, which helps funds arachnid research in Asia. However, these societies are reliant on scientists to create opportunities for further research. Most funding must be applied for and this inevitably means that some groups of organisms get less focus than others.

> **ATL ACTIVITY**
>
> Use the internet to research the availability of grants and funding for taxonomic research.
>
> Try to look for general taxonomy grants and then select a specific group of organisms, and search for grants and funding specifically for that group (for example, Arachnids). Try to narrow the search to a single country.
>
> Compare your results with those of others in your class and discuss what you were collectively able to find. You will have found a lot. Would this make it easy to carry out funded research on developing a better understanding of the diversity of organisms on our planet? Consider ways that this could be addressed.

HL.c.4, HL.c.11: Environmental ethics

Canopy fogging has been widely used to collect invertebrates from tree canopies to conduct ecosystem surveys of difficult-to-reach tree canopies. This technique uses harmful pesticides to knock the invertebrates to the forest floor, where they are either passively collected by large funnels, or hand collected from a large sheet positioned under the tree. Organisms are collected and preserved for later identification or classification. This allows scientists to collect large volumes of organisms from all parts of the system, which in a habitat like a rainforest is extremely difficult and time-consuming to do in other ways. This has become a less popular technique over time, but is still used in places.

> **ATL ACTIVITY**
>
> **Think–Pair–Share**
>
> *Think*
>
> Think about the ethics of the canopy-fogging technique and note down your perspectives on these questions:
> - Does every living organism have rights?
> - Is it wrong to kill any organisms, regardless of the possible development of information that might result from it?
> - Is canopy fogging an ethical sampling technique?
>
> Justify why you think the way you do, and remember that there is no right or wrong way of looking at this. Like your environmental values, your ethical values are influenced by factors like your education and your religion (see Chapter 1.1).
>
> *Pair*
>
> Join together with a partner and discuss each other's viewpoints on the ethics of canopy fogging. Remember – there should be no judgement about anyone's perspectives.
>
> *Share*
>
> Share your discussion outcomes with the class. Consider the different perspectives on the ethics of canopy fogging. Has your perspective shifted now you have more information on this? Why or why not?

TOK

How do we know how many species we have, and how do we know how many there are that are unknown?

3.1 Biodiversity and evolution

> ### Concepts
>
> **Perspectives**
>
> We do not always have the same level of information available to us regarding the species that are present and described in the world. This is because taxonomic research is time-consuming and difficult. It often requires long hours in the field collecting specimens and then long hours in a laboratory describing all aspects of the organism if it is a new species. Also, larger organisms have a larger percentage of their total species described, as they are easier to find and classify. Plants also have a larger proportion known as they are not mobile and so are easier to find. Despite the importance of understanding biodiversity, there are many other highly important discoveries that are competing for funding to carry out research.

> ### Concepts
>
> **Systems and models**
>
> The modelling of taxonomic information is called cladistics, and the models are called cladograms. These are highly specialized and different species are grouped together based on common features. These models can be very complex and difficult to interpret, however they provide information regarding potential evolutionary development or different groups of species.

3.1.2 The components of biodiversity contribute to the resilience of ecological systems

Biodiversity refers to the variety and variability of living organisms, so it is important to link biodiversity to resilience. Some of the basic understanding of how ecosystems function is related to the diversity of the food web within the system. As the variety of organisms increases, there are more links within the food web, thus reinforcing its resilience against change. If intra-species diversity is also high then no species should be catastrophically impacted by changes in the abiotic conditions within the ecosystem. The diversity needs to be high at all levels in order to ensure true resilience.

This means that a diverse range of habitats ensures that, if one is compromised, others may continue to provide essential services to other organisms and humans alike. Within a habitat, the higher the variety of species and interactions, the more likely it is to be able to buffer against disturbance or environmental change. If there are numerous different species that serve similar functions, then the environment may be able to compensate for the loss of a species. The most resilient species are those that have the highest genetic diversity.

> ### Concepts
>
> **Sustainability**
>
> In order for an ecosystem to be sustainable, it needs to be able to withstand and change with long-term environmental changes. In order to ensure this, biodiversity needs to be as high as possible for that given system. This way, the loss of more specialized species that cannot adapt easily to change will be buffered by the more generalist species, which are more flexible in the ways they address their needs.

3.1.3 Biodiversity arises from evolutionary processes

♦ **Evolution** – the change in heritable traits passed from generation to generation.

The current level of global biodiversity has taken millions of years to develop to these levels. Biodiversity develops slowly in relation to many circumstances, and the result of this continuous change is continual **evolution**.

As plants have evolved, the varieties of food types and niches have increased, which has resulted in greater diversity of herbivores, which has impacted the variety of carnivores and omnivores. Evolution is the cumulative process of development of inter- and intraspecies diversity. Over time, random genetic mutations will result in individuals within a species having characteristics that may be more favourable to the current environmental conditions. A trait that is related to genetics is heritable, and can be passed down to future generations, making it something that could become an adaptation that could change the species' overall niche over time.

Common mistake

The concept of time in the evolutionary process is a very different timeframe than we are used to. Evolution takes hundreds to thousands of years to fully take place.

Evolution is constantly happening, but there is a perception by some students that it is an ancient process that has resulted in where we are today. This is not the case; evolution is an ongoing, dynamic and neverending process that is necessary for life on Earth to continue.

> **REAL-WORLD EXAMPLE**
>
> **Marine iguanas: Galápagos Islands, Ecuador**
>
> The marine iguanas on the Galápagos Islands have been traced to have the same ancestral line as mainland iguanas, which were brought to the islands. The islands of Galápagos are small and some have limited food availability. It is likely that over time a random genetic mutation allowed iguanas to swim in salt water to reach previously inaccessible food sources. The abundance of the algae they feed on has resulted in the population with this adaptation expanding and developing this ability. Marine iguanas have adapted a nasal gland that collects the excess salt and allows them to 'sneeze' it out. They have also adapted to be able to drop their heart rate from 40 BPM to 10 BPM upon diving to allow them to stay submerged for long periods of time; this could be in response to a reduction in algae abundance closer to the surface as their population increases, therefore requiring individuals to venture deeper in search of abundant food (sources: **https://oceana.org/marine-life/marine-iguana/** and **https://evolutionofplanetearth.com/2018/05/03/to-breathe-or-not-to-breathe-adaptations-of-deep-diving-marine-iguanas/**).

3.1.4 Natural selection is the mechanism driving evolutionary change

♦ **Natural selection** – the change in genetic composition of a species over time as a result of the natural pressures within that species.

The development of the theory of **natural selection** was credited to Charles Darwin in 1859, when he published his theory of evolution in the famous text *On the Origin of Species*. This was based on his ideas about the mechanism that was driving the diversity of organisms he had come across on his expeditions. As he sailed between the Galápagos Islands, he noted that each island had a vastly different habitat, some with sparse volcanic rocky conditions and others with vegetation. He also began to notice that he was finding similar-looking birds on each of these islands, but they had differing beak shapes and sizes (see Figure 3.3). It was clear that each different-shaped beak favoured the food availability on the island it was found on. This was one of the parts of the puzzle that led to the development of the theory of evolution and understanding of natural selection.

Large ground finch
(seeds)

Cactus finch
(cactus fruits and seeds)

Vegetarian finch
(buds)

Woodpecker finch
(insects)

■ **Figure 3.3** Darwin's finches on the Galápagos Islands showing different beak shapes that were adapted to the different food sources available on each island

● Common mistake

Natural selection is not something that is decided, nor is it random. Remember that for every successful adaptation there will have been hundreds, if not thousands, of other random genetic mutations that were not beneficial to the circumstances at the time, and therefore were not passed on. For a trait to become widespread, it has to be something that puts an individual that possesses it at a competitive advantage. This results in the individual being selected as a mate, which allows it to pass those genetic traits on. However, it does not stop there, as for the adaptation to become a real evolutionary change it has to become widespread throughout the population, which takes many generations.

Genetic mutation in the evolutionary process happens randomly, sometimes resulting in positive change and sometimes resulting in organisms that do not survive. Which of these mutations becomes a new trait via natural selection is not random, and is directly related to given circumstances. If the same random genetic mutation appeared in two isolated populations, it might be a positive adaptation in one area but not in the other.

Survival of the fittest is not necessarily related to the most physically 'fit', unless that is the adaptation that gives that species an advantage. It is more related to how an adaptation 'fits' in the current situation.

Concepts

Perspectives

Evolutionary change takes time. How much time exactly is dependent upon many life features of each individual species. The length of time is directly related to life cycle features, such as lifespan, time to maturity, number of offspring and regularity of breeding. Therefore, a mosquito, which takes around 8–10 days to develop from an egg to an adult, will evolve faster than an orangutan, which takes between 10 and 15 years for females to be able to reproduce. This factor significantly impacts a species' ability to adapt to change.

Darwin's theory of evolution can be broken down into a number of different elements: common descent, gradualism, population speciation and natural selection. He hypothesized that we all come from a common ancestor and have slowly, over millions of years, changed in ways that have separated organisms into distinctly different groups. He also theorized that somehow the selection of what was passed to the next generation resulted in a species that was more adapted to the factors present in that particular area. This led to the development of the idea of 'the survival of the fittest', where the genetic variations that provide the best fit to the current situation will pass on their genes and therefore survive. For instance, one of the adaptations for marine iguanas is a wider tail that helps with swimming. This would allow this individual to reach food faster and from further away than others, possibly resulting in that individual becoming larger than the others. This could lead to a competitive advantage in terms of mate selection, and thus lead to the individual's genetics being passed on to further generations, so resulting in more individuals with wider tails.

The theory of natural selection centres around a number of different assumptions, primarily that individuals are variable, and that some of those variations are passed down to offspring. Darwin also stated that more offspring are created than needed and that survival and subsequent reproduction of individuals does not happen randomly.

Although the theories of evolution have been well known for some time, the mechanism of natural selection was not identified until much later, as genetic understanding and testing developed. This was when empirical evidence was found that supported Darwin's theory.

3.1.5 Evolution by natural selection involves variation, overproduction, competition for limited resources, and differences in adaptation that affect rates of survival and reproduction

Variation is the diversity of traits or characteristics within a population of organisms. Individuals within a population may exhibit different traits due to genetic differences or environmental influences. This variation is the foundation upon which natural selection acts. In a population of organisms, some individuals may have a trait that provides a selective advantage in their particular environment, while others may not. The individuals with the trait will be more likely to successfully breed and pass on that trait to their offspring.

Many organisms have the potential to produce more offspring than the environment can support. This leads to a situation of competition, as there are limited resources available to sustain the growing population. Since resources like food, shelter and mates are finite, not all offspring will survive to maturity. The sheer number of offspring produced by organisms means that only a fraction of them will survive and successfully reproduce. This overproduction and limited resource availability results in intense competition among individuals within a population.

Those individuals with advantageous traits that better 'fit' their environment have a higher likelihood of obtaining essential resources, such as food and mates, leading to increased chances of survival and successful reproduction. This means they pass on their genes, including those advantageous traits, to the next generation. Those with less adaptive traits may struggle to acquire these resources, reducing their chances of survival and reproduction. The individuals with advantageous traits have a higher rate of survival and reproduction, and have greater chances of passing on their genes. Over successive generations, the frequency of the advantageous traits increases, leading to the gradual adaptation of the population to its specific environment.

> **Common mistake**
>
> A scientific theory is not just an idea. A theory is created from putting together a system of evidence and information to explain something. For instance, the theory of evolution is supported by many different types of evidence, from fossil records, plate tectonics, common features in different organisms, and the impact of isolation on a population of a species of animal or plant.

> **ATL ACTIVITY**
>
> There are still people who do not believe the theory of evolution. Discuss why this might be with others in the class.

3.1 Biodiversity and evolution

REAL-WORLD EXAMPLE

Coevolution

Sometimes the survival of the fittest can become an evolutionary struggle to 'keep up' with evolutionary changes in some predator–prey relationship. This is also known as **coevolution**. One highly documented example of this evolutionary relationship comes from investigating the relationship between bats and moths. This battle began when bats evolved to use echolocation to detect their prey, often moths, in total darkness. In turn, moths have developed the ability to hear the echolocation clicks from the bat, giving them an early warning system that a predator is nearby. In response, bats have since been reported evolving stealth-like echolocation signals on a different frequency to evade being detected by the moths. Some moth species have already evolved to detect these ultrasonic frequencies.

◆ **Coevolution** – the evolution of two different species in response to changes in the other species.

Tool 2: Technology

Computer simulation data collection

Evolution is something that is very difficult to look at for experimental work, for obvious ethical and time-constraint reasons. There are, however, a number of data simulations that have been developed to allow you to model what would happen in a hypothetical situation. These simulations allow you to manipulate settings to simulate different scenarios and create data that could not be achieved in a real situation. Biology Simulations (**www.biologysimulations.com**), for example, has a number of different simulations.

Whenever you are going to use any kind of simulation or model, you need to be aware of the limitations of the model or simulation you are using, and of the parameters that can and cannot be manipulated.

Go to the link above and select 'Simulations' from the toolbar. Select the option called 'Evolution'.

Here there are a number of simplified simulations investigating different aspects of evolution, including the concepts of natural selection and a simulation that allows 10,000 generations of evolution of a fictional population to be manipulated and tracked.

Both the Natural Selection and Evolution simulations have links to data collection sheets that allow you to test out how different types of populations can evolve over time.

- Select one of the simulations, read the instructions and download the data collection sheets if they are available.
- Run through the simulation. What happened and why?
- Modify the conditions or criteria and rerun the simulation. How has this changed the result and why is the final outcome different?

3.1.6 Speciation is the generation of new species through evolution

Speciation is the process by which new species arise over time from a single ancestral species. It occurs when populations of the same species become genetically distinct and reproductively isolated from each other, preventing interbreeding and gene flow. Speciation is a fundamental consequence of evolution and is driven by various mechanisms. The most common modes of speciation are related to isolation of individuals of a species from others of that species, eventually leading to the different conditions in each area leading them to evolve in different directions, until they are no longer able to breed. This can happen due to a number of different situations involving physical and behavioural isolation.

◆ **Speciation** – the gradual change of individuals of the same species, resulting in distinct species that are no longer able to breed.

Concepts

Perspectives

The scale of speciation depends on the animal or plant. Mobile and large animals can cover large areas, meaning a river might not pose a barrier between individuals. However, for a small rodent the appearance of a river could easily pose a barrier that stops mixing between the two populations.

Speciation due to the development of a barrier to dispersal can easily be recognized when looking at species of aquatic organisms on either side of the Isthmus of Panama. This area of land, which connects North and South America, slowly developed due to deposition of sediment. This has essentially isolated the waters on either side of the isthmus. Over time, this has resulted in genetically different species of crayfish and porkfish in the region (see Figure 3.4). These species were eventually no longer able to mix and began to evolve in line with the conditions on either side of the barrier. Now these are recognized as completely different species that are no longer able to mate and create fertile offspring.

Common mistake

The causes of speciation often happen together and are linked. Geographical isolation might be what drives the eventual behavioural isolation. Species arriving in a new geographical location might have many available resources and niches to occupy, which eventually results in different behaviours. This process takes significant time and can be costly to obtain proof of.

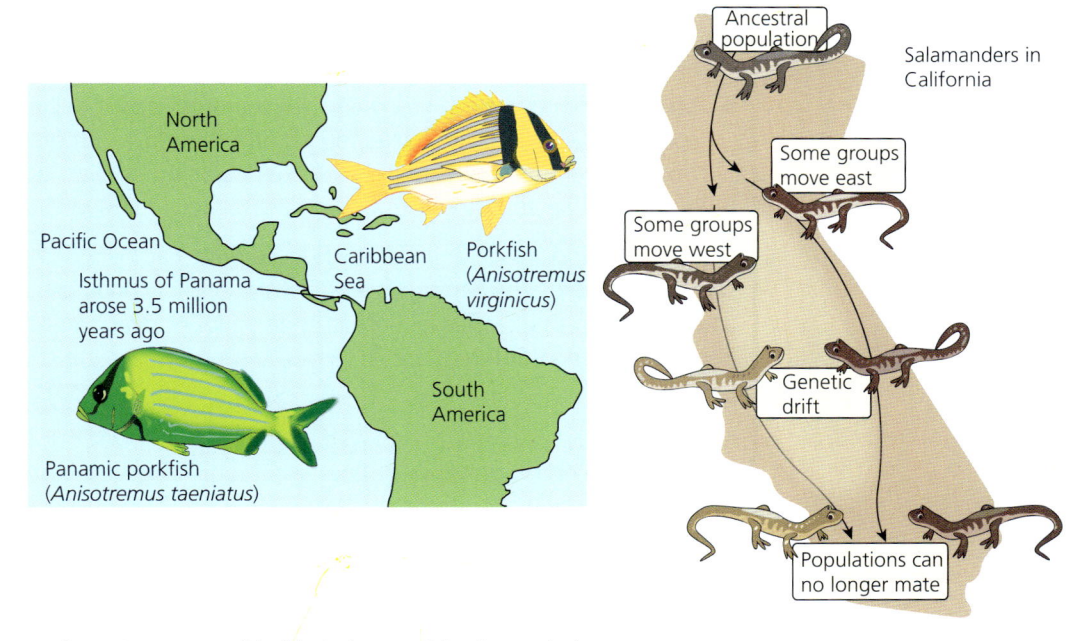

■ Figure 3.4 Geographical isolation resulting in speciation

3.1 Biodiversity and evolution

3.1.7 Species diversity in communities is a product of richness and evenness

Indices of diversity are important to allow it to be measured, compared and used as a tool to determine how well an ecosystem is functioning, also known as ecosystem health. It is not enough to simply look at how many different species there are, as the composition of their populations is also important to give an idea of the resilience of that species within the system.

In order to calculate diversity, two different data points need to be collected:

- The number of different species in an area (richness).
- The number of individuals within each of those species (evenness).

This allows a full picture of the population of organisms that make up that specific community.

- Species richness is important to ensure that there are a number of different organisms at each trophic level and with similar roles in the system.
- Species evenness allows each species to withstand environmental change and have some resistance to disease, as there should be good genetic diversity within that population.

A healthy and resilient ecosystem will typically exhibit both high species richness and high species evenness, reflecting a diverse and well-balanced community.

Tool 1: Experimental techniques

Fieldwork

Data related to diversity are easy measurements to take in the field, but careful planning is needed to ensure you are using the correct techniques for the type of species you are looking at. The scale and mobility of the organism that is being studied are important factors.

- Scale: Scale will help you determine the size of the area that will be assessed, as smaller animals require less space than large animals. Equally, looking at mosses will require a different sample area size than looking at diversity of tree species in a forest.
- Mobility: Plants cannot move and are therefore easier to find and compare, and comparative data can be selected on different days without concerns that conditions might impact the behaviour of the species being studied. Animals, however, can be highly mobile and therefore finding those individuals will be more complex.

Small plants and sessile animals, such as limpets, can be assessed using quadrat sampling to assess percentage cover or frequency. Mobile species of animals can be assessed using mark–release–recapture techniques over a period of time to determine population size in different areas, through using the Lincoln index (see Topic 2.1.17, page 100).

Measures of diversity require species identification, which can be achieved through using local dichotomous keys of flora and fauna. There are also a number of online identification databases and apps that can be used for identification. However, if these are unavailable, it is sufficient to identify organisms as species A or species B. As long as this identification is used throughout the experiment, you will be able to calculate relevant statistics without knowing their specific names.

3.1.8 Simpson's reciprocal index is used to provide a quantitative measure of species diversity, allowing different ecosystems to be compared and for changes in a specific ecosystem over time to be monitored

Diversity can be calculated using one of the indices that have been developed. For ecological work, Simpson's reciprocal index is the most commonly used. This calculation looks at what proportion of the whole community is occupied by each species, giving a value of overall diversity. This allows direct, numerical comparisons to be made, either within a habitat over time or between different sites. It uses the following calculation:

$$D = \frac{N(N-1)}{\Sigma n(n-1)}$$

Where N = the total number of individuals in an area, and n = the total number of each species.

D values range from 1 upwards. The higher the value, the greater the level of diversity within an area, in terms of both richness and evenness.

Interpretation of Simpson's reciprocal index values

- If the index is close to 1, it means there is very little diversity in the community, and one or a few dominant species make up most of the individuals.
- As the index value increases, the community becomes more diverse, with a more even distribution of individuals across different species.
- A value of infinity would theoretically mean an infinite number of species with an equal number of individuals for each species, indicating a perfectly diverse community.

Simpson's reciprocal index is commonly used in ecological studies. However, it is essential to interpret this index in conjunction with other diversity metrics and to consider the context and characteristics of the specific ecosystem being studied.

Inquiry process

Inquiry 2: Collecting and processing data

Simpson's reciprocal index allows the overall diversity of different habitats to be easily compared. Raw data need to be processed to calculate D, which can then be compared statistically between different locations.

Species diversity and species abundance need to be collected using an appropriate method (see Fieldwork feature for Topic 3.1.7, page 202). The results of this will form your raw data that you need to calculate Simpson's reciprocal index:

$$D = \frac{N(N-1)}{\Sigma n(n-1)}$$

If you are comparing two different habitat areas, collection of a quadrat of data is essentially a mini site. Each site should have around 20–30 quadrat samples for sufficient data for analysis. Calculate Simpson's D for each quadrat in each site. Use these values to calculate the mean and standard deviation of Simpson's D for each site. These can be compared graphically using a simple bar chart showing the standard deviation.

A *t*-test can be used on this data to statistically determine if there is a significant difference in the diversity of each site (see Tool 3: Mathematics feature below).

3.1 Biodiversity and evolution

Tool 3: Mathematics

Example: Sampling of plant species in a managed and unmanaged area of grassland

To carry out a *t*-test looking at the difference between diversity in managed and unmanaged grasslands, you would need to first assign the Simpson's values to a relevant group (managed or unmanaged).

Select the 'T-Test Calculator for 2 Independent Means' at **www.socscistatistics.com/tests**. A *t*-test will determine if there is a significant difference between two independent groups of data. To determine the difference, copy the managed data into the 'Treatment 1' box, and the data from the unmanaged habitat into the 'Treatment 2' box.

You then need to select the level of significance required, to determine which value of difference will represent significance. Select '.05' and also select 'Two-tailed' (as you do not know which direction the data will go). Then click 'Calculate T and P Values', and the program will calculate your statistics. These results are presented on a second page and the important statistics from the results page are highlighted in blue.

■ **Table 3.1** Sample data showing Simpson's diversity values in managed and unmanaged grasslands

Quadrat	Simpson's diversity managed	Simpson's diversity unmanaged
1	0.97	4.53
2	1.04	5.46
3	0.93	4.50
4	1.15	4.89
5	0.89	6.87
6	1.21	5.87
7	1.13	4.98
8	0.98	5.36
9	1.04	6.01

The *t*-value is −16.74705. The *p*-value is <0.00001. The result is significant at $p < 0.05$.

Therefore, there is a significant difference in the diversity of the two areas. To determine exactly what the difference is, you can create a graph of the mean and standard deviation of diversity in each area. This will allow a visual representation of the results. In this case, as the diversity values in the unmanaged area are so much higher than in the managed area, an interpretive statement for these results would be that there is significantly higher biodiversity in the unmanaged area when compared to the managed area that was sampled.

3.1.9 Knowledge of global and regional biodiversity is needed for the development of effective management strategies to conserve biodiversity

Global and regional biodiversity-management strategies play a crucial role in effectively conserving biodiversity by addressing conservation challenges at different scales. These strategies often involve coordinated efforts between international organizations, governments, local communities and various stakeholders. Here are some key approaches and elements that contribute to the effectiveness of these conservation efforts.

> **Concepts**
>
> **Perspectives**
>
> In order to be successful, a scaled approach to conservation is needed. Every scale needs to be protected, from global biomes such as tropical rainforests, to individual species such as the cheetah. This means that there also need to be strategies to address each of these levels. Individual species management plans need to be in line with international policy agreements that relate to protecting habitats to maximize the levels of protection for some of our most vulnerable species.

> **Concepts**
>
> **Sustainability**
>
> For a species to be sustainable, it must be able to successfully compete in the race to reproduce. Species are being disadvantaged by the speed at which humans are changing habitats in ways that animals and plants cannot compete against. Humans need to accept their responsibility for both the destruction and preservation of natural systems. This requires systemic changes in behaviour to reduce resource use, waste production and mistreatment of the natural environment.

The UN Sustainable Development Goals (SDGs) created international goals in recognition of the importance of biodiversity with Goals 14 and 15, Life below Water and Life on Land, respectively. These focus efforts globally to improve habitats for different species and to protect species that are negatively impacted by human behaviour. The creation of international agreements and agencies, such as IUCN and CITES, has also improved the power to protect and punish those who break international laws regarding certain species.

International actions also need to be supported by more local policies and actions to ensure that local communities are aware of local issues. Education and awareness help to mobilize action at sites of interest. Local communities can generate interest that could lead to funding to conduct research in relation to the impact on the species in question and on how effective conservation efforts are in addressing the problem.

(HL) 3.1.10 Mutation and sexual reproduction increase genetic diversity

Genetic mutation is a spontaneous, random change in the DNA sequence of an organism's genes. It occurs due to errors during DNA replication, exposure to environmental factors (such as radiation or chemicals) or other natural processes. Mutations introduce new genetic variations into a population, and even small changes in the genetic code can have significant effects on an organism's phenotype (observable traits). Beneficial mutations can lead to advantageous traits that improve an organism's fitness, while neutral or harmful mutations are subject to selection pressures. Advantageous traits improve an individual's chance of mating and passing on that mutation and trait on to their offspring via sexual reproduction. As sexual reproduction involves the combination of genetic material from two parent organisms to produce offspring, it introduces genetic diversity by shuffling and recombining existing genetic material.

Concepts

Perspectives

If a species breeds infrequently then the opportunities for mutations to take place are dramatically reduced. This is also the case with organisms that have few offspring with each birth event. Those species that breed frequently and produce many offspring will statistically have a greater chance of mutation and potential speciation due to the number of times random mutations can take place. For this reason these species – r-selected – are more likely to be able to evolve to withstand the changes in climate over time. K-selected species have few opportunities for random genetic mutations to occur. Equally, K-selected species have long lifespans and therefore any evolutionary changes will take place very slowly.

3.1.11 Reproductive isolation can be achieved by geographical separation or, for populations living in the same area, by ecological or behavioural differences

Geographical isolation

Geographical isolation can take place when a barrier occurs that physically separates individuals of the same species into different groups. This can occur due to tectonic or volcanic activity that results in the dividing of continents or the development of mountains, and prevents individuals from coming into contact with one another.

Some species with widespread distributions can evolve into different species over time due to the different environments across their distribution. Birds offer good examples of this type of isolation as some species can be found all over the planet, but over time these species become different as different pressures impact which of the genetic mutations will result in a trait that is favourable for their specific environment.

Another physical barrier occurs if individuals from a species group become isolated from the group in some way. Again, birds provide good examples when they are blown off course and eventually evolve in relation to the conditions in the new habitat. This can result in different mainland and island species in areas such as the Galápagos Islands. The bird of paradise is a group of birds that has become widely diversified due to this type of gene-flow isolation.

Behavioural isolation

Behavioural isolation occurs when selective pressure modifies specific aspects of a species that are important in terms of mate selection. For instance, many birds use calls, displays and dances to entice a mate; modifications in these behaviours can eventually lead to a reduction in gene flow, with individuals no longer recognizing one another. Others within that group may have evolved to select for recognizing this new mating ritual, eventually leading to a new distinct species that no longer mates with its ancestral species.

Temporal isolation occurs when individuals of a species develop different activities or mating schedules, resulting in individuals no longer coming into contact with one another to mate.

Speciation also sometimes occurs due to environmental pressure, which may drive a few individuals to disperse away from the group in search of food. Genetic material shows that brown bears and polar bears come from a common ancestor. It is thought that the ancestor of the polar bear migrated north in search of food and was cut off by the advancing glaciers of the ice age. This led to adaptations that eventually led to the two bears being two distinctly different species.

The impact of isolation can be easily seen on islands, whether oceanic or continental. Islands tend to have distinct species due to their isolation and limited species pool, resulting in the evolution of unique flora and fauna and high levels of endemism. The length of time the area has been isolated will impact the number of **endemic** species, with older islands having greater numbers.

◆ **Endemic** – a species that is only found in one geographical location and is not naturally found in any other area. These species are usually found in isolated areas.

REAL-WORLD EXAMPLE

Bird of paradise speciation

The group of birds known as the birds of paradise are known for their spectacular array of plumage styles, colours and distinctive mating rituals. However, 20 million years ago only one crow-like ancestor existed. Over millions of years, geographic and behavioural isolation has led to 15 genera (plural of genus) and 39 identified species.

These bird species are found in New Guinea, the surrounding islands, and parts of northern Australia. These species originated from Australia. At the time there were land bridges between New Guinea and Australia and as they drifted apart these populations became isolated. As a non-migratory bird, no gene flow took place once there was an ocean barrier between populations. As the islands around the main island of New Guinea split from one another, more species began forming, resulting in the range of different species observed today.

3.1.12 Biodiversity is spread unevenly across the planet, and certain areas contain a particularly large proportion of species, especially species that are rare and endangered

Biodiversity is directly related to primary productivity because primary producers are at the base of a food web. Therefore, the level of biodiversity partly follows the same patterns as the differences in primary productivity. This is all related to the intensity and duration of sunlight and the availability of water. Therefore, because tropical zones are dominated by regular intense sunlight throughout the year, coupled with intense monsoon seasons, these areas are some of the most biodiverse in the world. There are areas in the world that have particularly high levels of endemism and are at significant risk from destruction, which have been classified as biodiversity hotspots. To qualify as a biodiversity hotspot, an area must have over 1,500 endemic vascular plant species and have lost at least 70 per cent of the primary native populations. Many of these areas are found in tropical regions, but they are not limited to them (see Figure 3.5).

3.1 Biodiversity and evolution

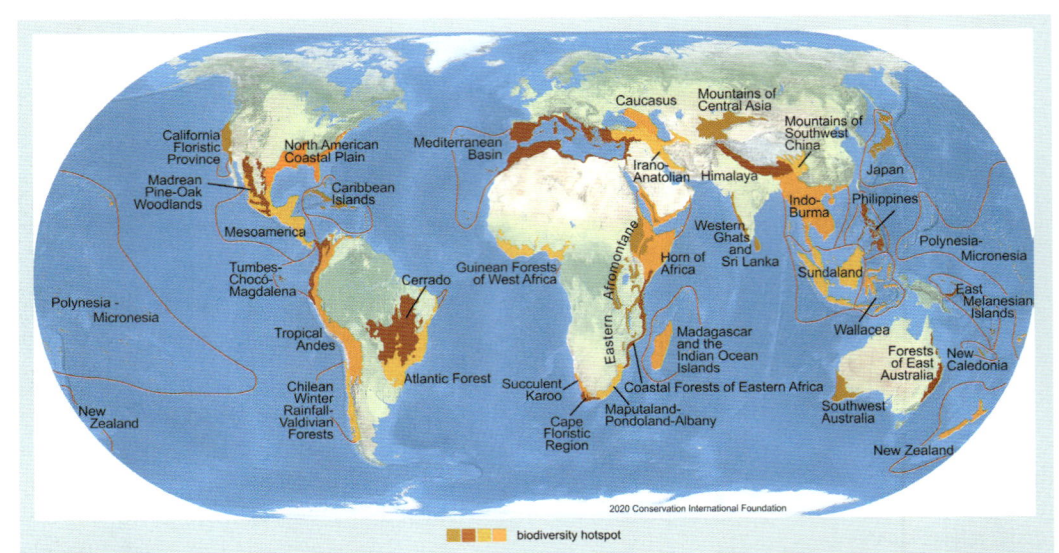

Conservation International (conservation.org) defines 36 biodiversity hotspots – extraordinary places that harbor vast numbers of plant and animal species found nowhere else. All are nearly threatened by habitat loss and degradation, making their conservation crucial to protecting nature for the benefit of all life on Earth.

■ **Figure 3.5** Map of the designated global biodiversity hotspots

Biodiversity hotspots are designated by Conservation International and are able to access funding and grants through the Critical Ecosystem Partnership Fund. The types of habitats vary dramatically, and many of the 36 officially designated areas are located in coastal and island locations, with Japan, New Zealand and Madagascar having the whole country covered under the designation.

HL.a.9: Environmental law

When habitats become designated as having some kind of conservation status, whether as a Ramsar site or a Site of Special Scientific Interest (SSSI), biodiversity hotspots are offered some kind of protection against open and obvious continued destruction.

HL.b.2: Environmental economics

Many biodiversity hotspots are in poor regions where funding is probably not freely available to protect areas and carry out valuable research. With the intervention of international bodies such as the IUCN, however, external resources can be directed to fulfil these needs. Identifying these areas also regularly results in the creation of national parks or protected areas. These generate jobs and income, and eventually may bring tourism into these areas, thereby supporting and strengthening the local economy.

HL.c.7: Environmental ethics

Some of these areas are relatively undiscovered, so the volume of potential new discoveries and cures in these areas is disproportionately high. Therefore, their protection is of utmost importance. In addition, the protection of these habitats can sometimes also directly and indirectly aid the human communities who live in these areas. Conservationists have a responsibility also to consider protecting indigenous communities, traditions and practices as these are often tied into the protection of their habitat.

3.1.13 Human activities have impacted the selective forces acting on species within ecosystems, resulting in evolutionary change in these species

Humans are not only responsible for the loss of many species, but also for the creation of new species via speciation driven by human interaction and intervention in the natural habitat. This can isolate individuals and expose them to environmental stressors that are not put on other populations of that species, eventually leading to new species forming. Human activities such as deforestation, urbanization and agricultural expansion have led to the alteration and fragmentation of many habitats. Species living in these altered environments may experience different selection pressures to those in undisturbed areas, resulting in them selecting for traits that enable them to thrive in these human-modified landscapes. This fragmentation leads to isolation, which is more likely to lead to speciation.

Other human activities that significantly impact species include the climate change that is being experienced as a result of the overuse of fossil fuels. The pollution that is created when fossil fuels are burned is accelerating the rate at which the climate is changing, and is resulting in the increased frequency and severity of extreme weather conditions, such as droughts and flooding. These climate changes are taking place faster than most species on the planet can evolve to keep up, resulting in increased rates of extinction. These changes will result in eventual shifting of biomes and greater species extinction.

> **REAL-WORLD EXAMPLE**
>
> ### Tuskless elephants: Gorongosa National Park, Mozambique
>
> Between 1977 and 1992, Mozambique was engaged in a bloody civil war that resulted in a significant drop in the biodiversity of large mammals in the country. Almost 90 per cent of the elephant population was killed for their valuable tusks, leaving a disproportionately large number of tuskless elephants as they were seen as having no value. In comparison to neighbouring countries, which have a natural rate of 2–4 per cent of elephants being tuskless, around 53 per cent of Gorongosa's adult elephant population are tuskless. During the civil war those elephants that carried genes for having no tusks were at a competitive advantage over tusked individuals, which were regularly killed. Therefore, during this period the population of tuskless elephants expanded, and this is now the dominant trait within the population (source: **www.science.org/content/article/civil-war-drove-these-elephants-lose-their-tusks-through-evolution**).

There are also ways in which humans have purposely changed conditions within an area, resulting in species needing to adapt to these new conditions. Introductions of non-native species as pets or garden plants can lead to serious disruptions in the natural balances within the native habitats. The widespread application of pesticides has also resulted in many changes in species as they adapt to the changing substances that are used to control pest species.

Genetic modification is another way in which humans have artificially created organisms with specific favourable traits. This has been going on for generations as crops are harnessed and improved to feed growing populations. Species can be created to produce higher yields and withstand conditions such as drought. These changes becoming dominant within these populations has isolated those populations from others of the same species.

◆ **Genetic modification** – a technique that takes DNA from one organism and inserts it into another the DNA of another species to change the characteristics of a living organism.

3.1 Biodiversity and evolution

These examples highlight how human activities have altered the selective pressures that shape species' traits and behaviours. Evolutionary changes in response to these selective forces can have both positive and negative consequences for species' survival and interactions with other species, and overall ecosystem dynamics. It's important to recognize and consider these impacts when making decisions about conservation and management of natural systems.

3.1.14 Artificial selection reduces genetic diversity and, consequently, species resilience

Artificial selection, also known as selective breeding, is the intentional breeding of plants or animals for specific desirable traits. While artificial selection can lead to the development of desired traits in domesticated species, it often reduces genetic diversity within populations and, as a result, can compromise species' ability to withstand change. Selective breeding can reduce the genetic diversity of a species since specific beneficial traits are selected for, eliminating the occurrence of other variations within that trait. Most of our current crops and domesticated animals have been through some kind of artificial selection to get to the varieties that we are used to seeing today.

For example, cabbages, cauliflower and broccoli are a few of the crop varieties that we have bred from one plant, in this case the wild mustard plant (*Brassica oleracea*) (see Figure 3.6). Carrots, tomatoes and many other crop species originally had many more varieties than are generally grown today, but over time their beneficial traits were combined to make crops larger, more nutritious and more appealing to the consumer. Dogs all originated from an ancestral wolf species, with the aggression and wild tendencies bred out over many years. This allowed breeds of dog to be developed for specific functions, such as hunting, protection and pulling vehicles. This has also allowed the breeding of new dog breeds to create specific types of pets or even service dogs. While technically all dogs can breed, there are barriers to that happening due to size differences and lack of encounters. Furthermore, owners are able to decide whether or not their dog will be allowed to breed, resulting in a highly lucrative business in breeding dogs.

■ **Figure 3.6** Five of the different vegetable crops artificially bred from the mustard plant (*Brassica oleracea*)

Theme 3: Biodiversity and conservation

Concepts

Perspectives

Organisms have evolved their traits as a result of the pressures that have been placed on them over hundreds, if not thousands, of years. It is therefore possible that the traits we prefer may not be those that are the most important, or the most beneficial, for the species. For instance, snub-nosed dogs such as bulldogs, pugs and boxers were bred to have a shorter snout as it creates a stronger jaw, making these good fighting dogs, and now many snub-nosed breeds are popular for their visual appeal. However, this trait gives these breeds breathing difficulties that can be dangerous for the dog in certain situations.

HL.b.1, HL.b.2, HL.b.6: Environmental economics

Many of our commercial species of crops and livestock are highly profitable as they have been selected either for the quality or quantity of their yield. Breeds or species with lower yields get artificially selected out of the system as they are not as economically viable. As the human population increases, the reliance on these yields also increases. This system does not encourage diversity in crop growth. However, many of the heirloom varieties are coming back to farmers markets and local communities as people look to return to the variety we used to have. Due to the perceived novelty of these varieties, this has become a highly profitable business, with these varieties selling for much higher prices than the common varieties that are grown on an industrial scale.

■ **Figure 3.7** Heirloom carrot varieties

HL.c.4–7, HL.c.12: Environmental ethics

If we consider that all living organisms have intrinsic value, is it ethically correct for us, as humans, to determine what is valuable based on what it provides for us? Aldo Leopold argued that if something led to a greater good or to benefits for the wider good then it was considered to be of value; if it was not of greater benefit, or in fact damaged the balance of the system, then it was considered to have lower value. As much as this might be sustainable, are we right to rank value in that way? There are also many perspectives relating to the value of an organism.

3.1.15 Processes that occur over an extended timescale have led to the evolution of life on Earth

The Earth has been identified as being created around 4.5 billion years ago. Throughout this time the planet has gone through continual sequences of change that have resulted in the continents, habitats and organisms that are present today.

Numerous climate, geological and evolutionary processes have been constantly shifting throughout time. Climatic factors have played a clear role in the development and destruction of habitats and organisms. Geological changes have shaped the continents and landscapes, whether through plate tectonic or volcanic activity. Evolution has resulted in the development of species of plants and animals that specifically fit to the conditions of their habitat at a particular time, resulting in the biodiversity that we are trying to preserve. During the Earth's history there have been a number of major, rapid changes that have been triggered by natural phenomena such as meteor strikes and volcanic events, which resulted in dramatic changes.

Our knowledge of the exact processes that have taken place over time comes from a variety of sources, including the fossil record, carbon dating and information contained in ice core samples.

3.1.16 Earth history is divided up into geological epochs according to the fossil record

Geological time is divided into different periods that relate to the changes in conditions. Most of the names of these periods of time represent the changes that took place during that period. Each 1 billion years is called an eon, and these are further divided into periods of geological time (see Table 3.3). For instance, we are currently in the Phanerozoic (visible life) Eon and the Cenozoic (new life) Era, which is also known as the Age of Mammals.

■ **Table 3.3** Distribution and timescale on the geological timeframe

Eon	Era	Period	Epoch	
Phanerozoic	Cenozoic	Quaternary	Holocene	← Today
			Pleistocene	← 11.8 Ka
		Neogene	Pliocene	
			Miocene	
		Paleogene	Oligocene	
			Eocene	
			Paleocene	← 66 Ma
	Mesozoic	Cretaceous	–	
		Jurassic	–	
		Triassic	–	← 252 Ma
	Paleozoic	Permian	–	
		Carboniferous	–	
		Devonian	–	
		Silurian	–	
		Ordovician	–	
		Cambrian	–	← 541 Ma
Proterozoic	–	–	–	← 2.5 Ga
Archean	–	–	–	← 4 Ga
Hadean	–	–	–	← 4.54 Ga

(Younger ↑ Older ↓)

◆ **Carbon dating** – a technique used to date fossil records that looks at the amount that the Carbon-14 molecules have broken down.

The divisions in time signal major shifts in geological and environmental factors due to a number of different natural disasters. These are visible in the vast fossil record and in climatic records from ice cores and **carbon dating**, showing that these periods were marked with large volumes of extinctions, followed by the evolution of new species.

Fossils have allowed us to trace the evolution of some species and infer information about the environment at the time when the organisms were alive. For instance, life started in a time that had optimum levels of hydrogen, carbon, ammonium and water vapour, which sparked the development of the first organic compounds, such as amino acids. Eventually, organisms developed and the long process of evolution began to create the ancestors of all organisms on the planet.

Concepts

Systems and models

Models have been developed that allow scientists to map evolution through geologic time. For example, cladograms are branching diagrams that show ancestral relationships between organisms. These models are constantly being updated and changed as more information is discovered.

Each division between one epoch and another is not a set period of time, instead it is related to visible changes in the fossil record that represent changes in environmental conditions that caused the variety and diversity of organisms to change. This can be determined through layers of fossils of organisms that are not present in later layers of rock.

One of the indicators in the fossil record is the presence of both aquatic and terrestrial organisms that share similar bone structures, such as the pentadactyl limb that has evolved to adapt to the specific requirements of many species, including a fin in whale species or a wing in a bird or bat (see Figure 3.8). This provides evidence that terrestrial mammals originated from aquatic ancestors.

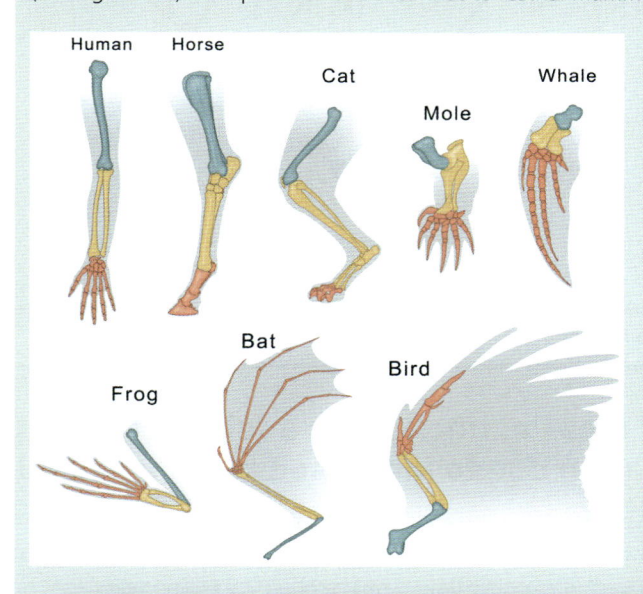

■ **Figure 3.8** The pentadactyl limb evolution in various different types of mammals, showing adaptation to the needs of the animal

Tool 4: Systems and models

Interpret a cladogram

Cladograms are branching diagrams that show ancestral relationships between organisms. The main line of a cladogram shows the passage of time, with current time at the top of the line.

- Each branch on the line is called a clade; this represents the division of a distinct group from the main ancestral pathway.
- Those groups that divide off at the bottom of the cladogram are the most distantly related to the clade at the top of the image. In Figure 3.9, frogs are the most distantly related to mammals.
- Each clade represents a group of organisms with similar features. For example, all marsupials keep their young in a pouch and carry them around to feed and protect them.
- Clades that are next to each other show many similar characteristics. For instance, amphibians and reptiles are similar in many ways, with one of the main differences being that amphibians start their life cycle in the water and develop into land animals, whereas reptiles are born with lungs and do not have the aquatic stage.

3.1 Biodiversity and evolution

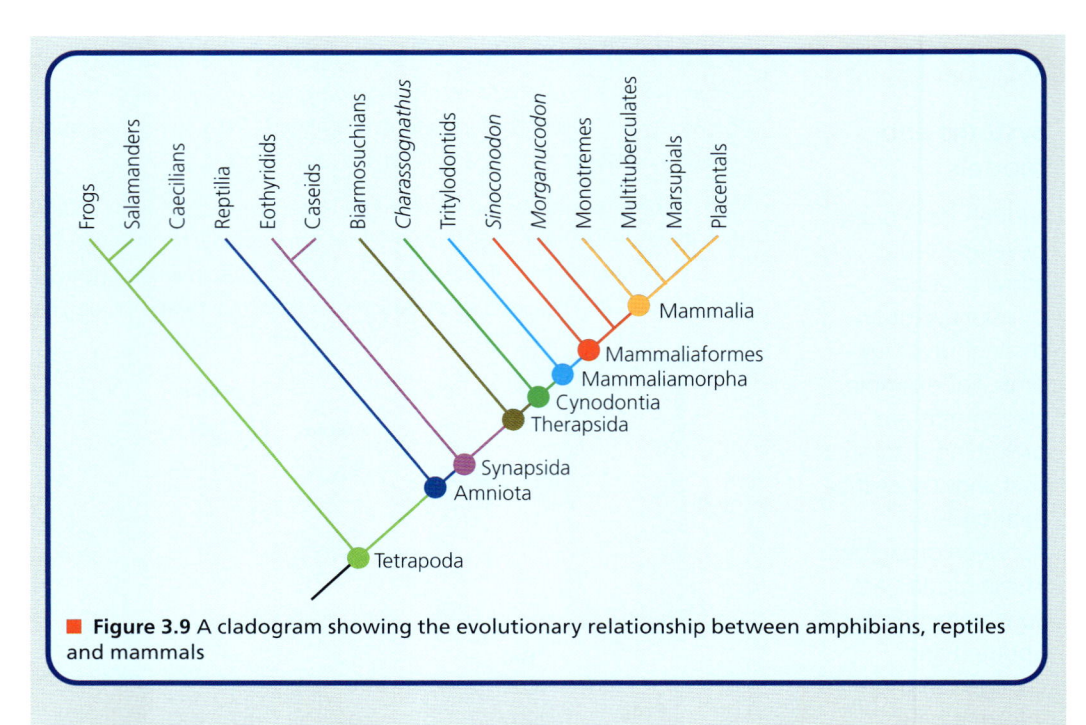

■ **Figure 3.9** A cladogram showing the evolutionary relationship between amphibians, reptiles and mammals

3.1.17 Mass extinctions are followed by rapid rates of speciation due to increased niche availability

There have been five complete, historical mass-extinction events in Earth's history. A mass-extinction event is when there is a dramatic (75 per cent), rapid decrease in the biodiversity of organisms in a short period of geologic time, within around 2.8 million years. Past mass-extinction events have been triggered by major natural disasters and disturbances. It can be hard to determine exactly when the mass-extinction events began and/or ended, and much of the information we have has been derived from the fossil record.

■ **Table 3.4** Causes and impacts of the five historic mass-extinction events (source: National Geographic)

	Ordovician–Silurian	Late Devonian	Permian–Triassic	Triassic–Jurassic	Cretaceous–Paleogene
Timescale (millions of years ago)	440	356	252	201.3	66
Percentage of life extinct	86 per cent of all life; all marine life lost.	75 per cent of all life.	96 per cent of all life.	80 per cent of all life, including most mammals, making way for dinosaurs.	60–76 per cent of all life, including all dinosaur species.
Cause	A global cooling event followed by a global warming event.	Either global cooling or mass algal blooms removed O_2 from the ocean, suffocating most species.	Probably a large volcanic eruption. High levels of CO_2 release, which bacteria fed off, increasing methane levels. This caused acid precipitation, causing oceans to become toxic.	Either a large volcanic event or meteor strike.	Meteor strike on the Yucatan Peninsula, Mexico. Large amounts of ash were ejected into the atmosphere, blocking light and causing global temperatures to drop.

Theme 3: Biodiversity and conservation

● Common mistake

It is hard to get an understanding of the scale and timing of these mass-extinction events. All of these events happened millions of years before humans evolved and therefore we had no influence on them.

Concepts

Perspectives

The fossil record only records animals and plants that had hard carbon-based systems, either internally or externally. Organisms with soft bodies, such as many insects and spiders, cannot be preserved in rock samples. Our knowledge of some of these organisms comes from organisms that have been trapped in tree sap, which has later turned into amber.

■ **Figure 3.10** *Eustaloides setosus Petrunkevitch*, 1942 (family *Araneidae*) in 44–49 Ma (Eocene) Baltic amber

Following each mass-extinction event, a large number of plants and animals evolved to fill the gaps left by those that became extinct. These species were likely adapted to the conditions already, since their ancestors had survived the event that caused the mass extinction. The drop in niches due to changes in conditions, however, meant there may have been greater competition for these niches and for resources. This pressure drove the speciation of organisms and eventually resulted in the evolution of new species that adapted to exploit a different niche. Some species may have become more dominant immediately after a mass-extinction event due to their adaptability and favourable biological conditions, such as breeding rapidly and having many offspring. These species would initially have become more abundant while other species adapted more slowly.

Evidence for these events can be found in fossils and in ice and sediment core samples. Official markers called Global Boundary Stratotype Section and Points (GSSPs or 'golden spikes') have been internationally agreed as representing the beginning of each geological era. These have been discovered in various parts of the world and the golden spike for the lower boundary of the Holocene was found in an ice core sample taken in North Greenland, showing a sharp and marked decrease in hydrogen isotopes.

3.1.18 The Anthropocene is a proposed geological epoch characterized by rapid environmental change and species extinction due to human activity

The term 'Anthropocene' stems from the recognition that human activities have become a dominant force shaping the Earth's environment (*anthropo* meaning 'human' in Greek). There has long been debate as to whether the Earth is currently in a sixth mass-extinction event due to the fact that humans have been altering the planet's land, oceans, atmosphere and biosphere at an unprecedented pace and scale. The start of the Anthropocene is much debated, but its worsening has been linked to the start of the Industrial Revolution, when there was a rapid increase in industrialization, manufacturing, urbanization and the burning of fossil fuels. These activities released substantial amounts of greenhouse gases and other pollutants into the environment. These changes in atmospheric gases could, as has been proven before, result in a dramatic warming of the planet, which coupled with the rates of habitat destruction, pollution, overexploitation and introduction of invasive species, have all contributed to this dramatic loss of biodiversity.

Human activities have disrupted critical biogeochemical cycles, such as the carbon and nitrogen cycles, leading to profound changes in ecosystems and the composition of the Earth's atmosphere and oceans. These changes are creating a rate of environmental change in atmospheric composition, global temperatures and sea level rise that is far faster than any natural geological process. As a result, the rate of biodiversity loss has also continuously accelerated in line with these unpredictable climatic shifts.

Evidence for these disruptions has been found in various locations globally and there has been much debate in the Anthropocene Working Group (AWG) as to which was the real golden spike for the start of the Anthropocene. The first golden spike was identified from Antarctic ice cores and shows a dramatic drop in global CO_2 levels in 1610. This time matches the period when Europeans arrived and settled in North America, resulting in the death of over 50 million members of the indigenous communities from famine, disease epidemics and warfare. This caused vast areas of agricultural land to be abandoned and be reclaimed by natural forests and other vegetation, with a resulting drop of between 7 and 10 ppm of CO_2 in the atmosphere. This change in atmospheric condition caused a noticeable drop in global temperatures at this time, as represented by larger ice deposits during this time.

Another proposed date is 1952, due to the discovery of millions of black, spherical fly ash particles in sediment cores all over the planet. These come from the incomplete combustion of carbon-based fuels such as coal and oil, which has been what has driven the industrialization and development of countries around the world. This shift is clearly visible in many lake cores (see Figure 3.11).

■ **Figure 3.11** Section of a drill core sample taken from the bottom of Crawford Lake in Canada, where the golden spike was discovered (photograph by Dr. R. Tim Patterson)

Another potential golden spike for the beginning of the Anthropocene is the radioactive substance in sediment and ice cores. The first thermonuclear weapons were tested from the early 1950s, leaving large radionuclide deposits in the environment.

> ### Concepts
>
> **Systems and models**
>
> Developments in scientific techniques mean that we now have access to considerable data regarding what happened in the past. C-14 dating has been used to develop a number of possible models for the development of understanding of the Earth's geological history. Some models link data to those from ancient dendrochronological surveys (tree rings), others to data in sediment or ice cores. These data allow modellers to develop predictions about what will happen in the future, based on what has already happened in the past.

3.1.19 Human impacts are having a planetary effect, which will be detectable in the geological record

As the past changes in organisms, climate factors and atmospheric composition are all detectable in the present, it is safe to assume that the changes that are taking place during the Anthropocene will also be detectable in the future. Currently, the impact of our human populations on the planetary boundaries is clear in many pollution events, climate changes and species extinctions. As with previous extinction events, these changes will be clearly visible in the fossil record in the future. Changes in land use will be detected through dramatic change in the soil and sediment created at that time.

Deforestation and other activities that remove large portions of vegetation have a significant impact on the stability of the soil that remains. This soil is often displaced and ends up within streams, rivers and eventually the oceans. This increase in the rate of sedimentation will be clearly visible through cores that can be removed from river and lake sediments.

The movement of animals for livestock reasons, as well as the accidental or intentional introductions of non-native species, means that future fossil records will show species of plants and animals far away from their origin, without clear evidence of natural migration taking place. This current time has also already seen a decrease in around 50 per cent of bird and amphibian species and around 30 per cent of mammal species, which will continue and be clearly evident in fossil records in the future (see Figure 3.12).

Changes in chemicals deposited in the soil and water will also be detectable in future records as the level of chemicals released into the environment increases. Some of these chemicals break down slowly and will still be detectable in the future. As humans continue to develop technologies and substances, their mass uptake and use could leave an everlasting marker for future generations to see.

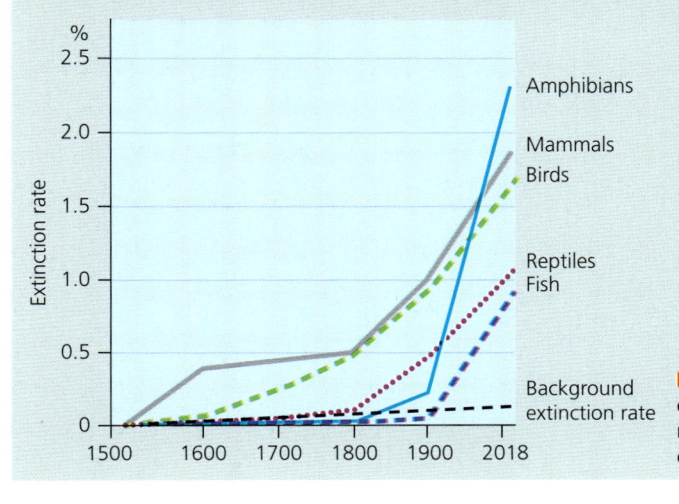

■ **Figure 3.12** Changes in extinction rates over time in relation to the background extinction rate

3.1 Biodiversity and evolution

REVIEW QUESTIONS

1 Identify the three levels of diversity.
2 State the name of the mechanism responsible for evolutionary change.
3 List the steps required for evolution to take place.
4 Identify the most biodiverse site in terms of non-woody plants, based on Simpson's reciprocal values shown below:
   a Oak forest $D = 3.6$
   b Mixed deciduous forest $D = 5.6$
   c Coniferous forest $D = 1.2$.
5 Using the information you learned in Chapter 2 about food webs in ecosystems, identify two reasons for the data presented in Question 4.

EXAM-STYLE QUESTIONS

1 State which two factors increase genetic diversity within a species. [2 marks]
2 Explain how fossils can help to identify the common ancestry of different types of mammals. [4 marks]
3 Using Table 3.4, explain the extinction event that resulted in the greatest loss of species. [6 marks]
4 Explain how evidence is used to reliably determine the start and end of different periods of geological time. [3 marks]
5 Describe how deposits from testing nuclear bombs will be detectable in future core samples. [3 marks]

3.2 Human impact on biodiversity

Guiding question

- What causes biodiversity loss, and how are ecological and societal systems impacted?

SYLLABUS CONTENT

This chapter covers the following syllabus content:
- 3.2.1 Biological diversity is being adversely affected by both direct and indirect influences.
- 3.2.2 Most ecosystems are subject to multiple human impacts.
- 3.2.3 Invasive alien species can reduce local biodiversity by competing for limited resources, predation, and introduction of diseases or parasites.
- 3.2.4 The global conservation status of species is assessed by the International Union for Conservation of Nature (IUCN) and is published as the IUCN Red List. Status is based on the number of individuals, rate of increase or decrease of the population, breeding potential, geographic range and known threats.
- 3.2.5 Assigning a global conservation status publicizes the vulnerability of species and allows governments, non-governmental agencies and individual citizens to select appropriate conservation priorities and management strategies.
- 3.2.6 Investigate three different named species: a species that has become extinct due to human activity; a species that is critically endangered; and a species whose conservation status has been improved by intervention.
- 3.2.7 The tragedy of the commons describes possible outcomes of the shared unrestricted use of a resource, with implications for sustainability and the impacts on biodiversity.

HL ONLY
- 3.2.8 Biodiversity hotspots are under threat from habitat destruction, which could lead to a significant loss of biological diversity, especially in tropical biomes.
- 3.2.9 Key areas that should be prioritized for biodiversity conservation have been identified on the basis of the international importance of their species and habitats.
- 3.2.10 In key biodiversity areas (KBAs), there is conflict between exploitation, sustainable development and conservation.
- 3.2.11 Traditional indigenous approaches to land management can be seen as more sustainable but face challenges from population growth, economic development, climate change, and a lack of governmental support and protection.
- 3.2.12 Environmental justice must be considered when undertaking conservation efforts to address biodiversity loss.
- 3.2.13 The planetary boundary for loss of biosphere integrity indicates that species extinctions have already crossed a critical threshold.

> **Link**
>
> Direct links to human influences on biodiversity can be seen throughout the book, with pollution of both water (Chapter 4.4, page 373) and soil (Topic 5.2.7, page 471) directly impacting plant and animal diversity. Climate change (Chapter 6.2, page 5.2.8), stratospheric ozone depletion (Chapter 6.4, page 594), and urban air pollution (Chapter 8.3, page 740) all have direct and indirect impacts on living systems.
>
> ◆ **Poaching** – the illegal hunting and harvesting of animals.

3.2.1 Biological diversity is being adversely affected by both direct and indirect influences

Biodiversity is impacted in many ways by the actions of humans. Unfortunately, there are more negative impacts than there are positive. Some come from direct actions to the species themselves, while others are related to actions that impact the habitat conditions or features and indirectly impact the diversity of all organisms in the area.

Direct influences on biodiversity

Species are directly impacted when they provide a service for humans. This includes being a suitable food or pet, and poaching of valuable parts of the organism. When an animal or plant becomes a suitable food source for humans, wild populations will begin to be exploited to keep up with demand for the product. Changes in fads and trends can also lead to short-term overexploitation and long-term overharvesting of species.

Some animal species have been over-collected in the wild and bred to develop a trade for that species as a pet. This stretches from tarantulas to species of big cat. Captive breeding does provide a proportion of these animals, but as demand increases faster than these species can breed, the illegal pet trade expands to meet this gap.

Poaching is another direct way that humans impact species numbers. Many species are poached due to the value of certain parts of their body for fashion, medicine/traditional remedies, or to create trophies for hunters. Poaching can involve the illegal capture and smuggling of live animals as well as the slaughter of animals for body parts. Many large, already significantly vulnerable, species are the focus of poaching due to the value of their parts. Elephants and rhinos are poached for their tusks and horns for decoration or medicine. Many big cats are poached for their pelts, which are then used for coats and other elements of the luxury fashion market. Traditional Chinese medicine has also been responsible for the collection of many parts of animals to create cures for ailments. Pangolin scales, sun bear gall bladders and tiger bones, teeth and claws are all thought to have powerful healing properties and this is one of the direct causes of the significant drop in their populations.

Indirect influences on biodiversity

Large-scale global issues such as climate change are impacting species in a number of ways. As conditions change more rapidly than in previous times, evolution is unable to keep pace with issues such as the melting of glaciers and ice at the North and South Poles. This is impacting species such as the polar bear, which already has a limited range and distribution due to its more niche requirements. Melting of ice is driving polar bears closer to expanding human settlements, resulting in death through starvation or shooting due to proximity or the potential threat to human safety.

Pollution is a universal issue around the globe and there are many types of pollution that are created by human activity, in all habitats and parts of our terrestrial (see Topic 5), aquatic (see Topic 4) and atmospheric (see Topic 6) systems, and the species within these areas. Changes in the conditions, or the addition of non-natural materials into these systems, have resulted in many species becoming vulnerable, endangered and even extinct.

> **Concepts**
>
> **Perspectives**
>
> As much as the loss of so many endemic fish species from Lake Victoria is devastating to biodiversity, it also has an impact on the human populations surrounding the enormous lake. Due to the size of the lake and the abundance of fish within it, it has been supporting communities in Kenya, Uganda and Tanzania. The loss of so many fish species is having a devastating impact on food availability for people in the region.

◆ **Invasive species** – harmful, non-native species that are rapidly able to spread and outcompete local species.

More localized issues such as habitat destruction and the introduction of **invasive species** also negatively impact the biodiversity of an area. Habitat destruction frequently occurs to make way for our expanding population and our needs. This displaces some individuals, but the reduction in the size and connectivity of remaining habitats means that they are often unable to support the influx of individuals from a different area. These individuals will likely die without the resources they need to survive.

The introduction of species to a new location and habitat can also create problems for the species already present within those systems. Introductions of alien species can result in them taking over due to the lack of a natural predator. Some examples of species introduction have been catastrophic for the flora and/or fauna of the area. The Stephens Island wren (*Traversia lyalli*), for example, was a small ground-dwelling bird that was once widespread in New Zealand, but the species became endangered through destruction of its nesting habitats due to an increase in urban environments and an increase in predator numbers. By the late 1800s the Stephens Island wren only inhabited one small island off the coast of New Zealand. Following numerous shipwrecks in the area, a lighthouse was constructed on the island and a lighthouse keeper brought his pet cat to the island, which subsequently hunted the flightless bird to extinction in 1898.

● HL.a.6, HL.a.9: Environmental law

There are numerous local, national and international organizations that work together to create a network of knowledge and understanding. This allows laws to be created and internationally enforced. The IUCN CITES list is an example of how multiple levels of monitoring and implementation are required for success.

● HL.b.17: Environmental economics

Due to the nature of development and economic growth, many parts of our environment are beginning to stop being able to buffer the impacts of human development. Doughnut economics can be applied alongside a move towards a more circular system in manufacturing and the provision of goods and services. These models can allow multiple impacts to be addressed at the same time.

● HL.c.4, HL.c.10: Environmental ethics

Many of our natural environments have been overused due to their instrumental value to humans. Humans have used all parts of some systems through the extraction of water, nutrients, food and more. The consequences of these actions are rarely weighed up over the long term, however. The consequences of individual pressures on an environment are difficult to determine, therefore the assessment of multiple pressures and the cumulative effects can be extremely difficult to understand.

3.2.2 Most ecosystems are subject to multiple human impacts

As humans convert more and more space to agricultural land and urban areas, environmental systems become overloaded with nutrients and pollutants that impact them in many different ways. Irrigation and an increase in the non-porous surfaces present in urban areas further disrupts aspects of the nutrients and the water cycle, compounding issues such as temperature increase, input of pollutants and over extraction.

While each individual pollution event or disturbance issue can have a negative impact on the diversity of an area, the combined impact of the many simultaneous issues can result in devastating impacts on all aspects of the natural system. The cumulative effects of different pollutants in the air, water and soil, changes in climate patterns, invasive species introductions and habitat disturbances, can result in the breakdown of the system. Climate change is reducing both habitats and species resilience to the other pressures that are present, making them more vulnerable to extinction.

A holistic approach that considers the interconnectedness of human actions and their effects on ecosystems is crucial for maintaining biodiversity and ecosystem resilience, and the services they provide to humans and the planet.

Inquiry process

Inquiry 1: Exploring and designing

There are two different ways you can design an experiment to investigate the impact of human activity on biodiversity. There are many ways that human activity can impact a habitat, and this might help you decide how to design your experiment.

■ Table 3.5 Ideas for field research to look at biodiversity

Area to be tested	Field collection method	Standardization	Data analysis	Statistical analysis
A habitat at the edge of a human disturbance, e.g. a road, carpark or other manmade structure cutting into a natural habitat	Transects running perpendicular from the edge of the disturbance. Quadrat samples at set distances. Record plant species and percentage cover. Set numerous transects to allow for replication.	The length of the transect and the frequency of quadrats depends on the field situation you have. The general rule is that a longer transect is better, as long as it does not change habitats or come into contact with another disturbance that might impact diversity. The number of transects used to provide reliable data.	The data can be used to calculate Simpson's diversity index along the transect (see Topic 3.1.8, page 203). This will allow you to investigate if there is a relationship between distance from the disturbance and plant biodiversity.	Pearson correlation coefficient or Spearman's rank correlation coefficient (see Topic 8.1.3, page 695). Investigating the relationship between disturbance and its impact on biodiversity.
Comparing two different habitats	Decide on the size of a square area to study in each habitat; this will depend on what organisms you are looking at. For plants, try to have at least a 10 × 10 metre area in each habitat. Record plant species and percentage cover using quadrat sampling.	Use an online random number generator to select 20 random coordinates. Use these in both areas to determine where to place quadrats.	The data can be used to calculate Simpson's diversity index for each quadrat in each habitat.	*t*-test or Mann-Whitney U test Investigating the difference in the diversity in both habitats.

Another technique is to test the impact of a change in management in a habitat. For instance, a grassland that is partly kept mowed and managed, but has unmanaged areas surrounding it. For example, your sports fields if the area surrounding is not kept cut short. In this scenario, a transect would be laid from the unmanaged area, through into the managed area. This would allow a test to be conducted of the impact of management on the plant diversity. This would again consist of using quadrats to determine the species richness and evenness before and after management. This technique would provide data suitable for a paired difference test, such as a paired *t*-test.

3.2.3 Invasive alien species can reduce local biodiversity by competing for limited resources, predation and introduction of diseases or parasites

Each ecosystem has balance at the core of how it functions. This is something that has developed over time, as the area has gone through stages of succession and the species of plants and animals are naturally adapted to those conditions from years of evolution. Non-native species often come into contact with the local flora and fauna, either accidentally or deliberately, and can cause adverse impacts on the local ecosystem balance. Non-native species may not have a natural predator within the new system and may therefore be at an advantage compared to local species. Similarly, if non-native species have a competitive advantage over local species, they might be able to easily outcompete them for resources, effectively taking their place in the food web.

Non-native species may arrive in new habitats by accident, such as attached to imported goods. Spiders are frequently-found passengers within imported fruit consignments, and seeds can be carried in the soil in the tread of someone's shoe. For this reason, certain countries, such as Australia, which has many invasive non-native species impacting its natural ecosystems, have extremely strict import and immigration policies around anything that could bring non-native species into the country.

In certain circumstances, other non-native species are brought into new ecosystems on purpose. Previous lack of understanding about the delicate balance within ecosystems and the potential impacts of introducing unfamiliar species into the balance, has resulted in many non-native species being introduced into a locality for biocontrol purposes, for example, the cane toad in Australia, or for ornamental and aesthetic purposes, such as Japanese knotweed that was sold as a garden variety plant around the world and has rapidly spread and outcompeted many native plants in most places it has been introduced.

TOK
To what extent can humans be considered the ultimate invasive species?

REAL-WORLD EXAMPLE

Cane toad introductions into Australia

The cane toad (*Rhinella marina*, but formerly known as *Bufo marinus*) is a large toad with a native distribution from parts of southern Texas down to the tropics of South America. They can grow to between 15 and 25 cm long, significantly larger than most other amphibian species around the world. Due to their size, these toads are voracious predators and feed on beetles, frogs, other toads and small lizards.

3.2 Human impact on biodiversity

In Australia during the 1930s, sugarcane crops were becoming infested with the cane beetle. These were causing widespread damage to crops in parts of Australia that resulted in significant economic impacts for Australian farmers. Cane toads (given this common name once they were employed to reduce the beetle infestation) were brought into Australia as part of a biological-control management plan.

At first, just over 2,000 captive-bred individuals were released in Queensland. Following success and pressure from the sugarcane manufacturers, thousands more were subsequently introduced. In 2018, there were an estimated 2 billion cane toads in Queensland.

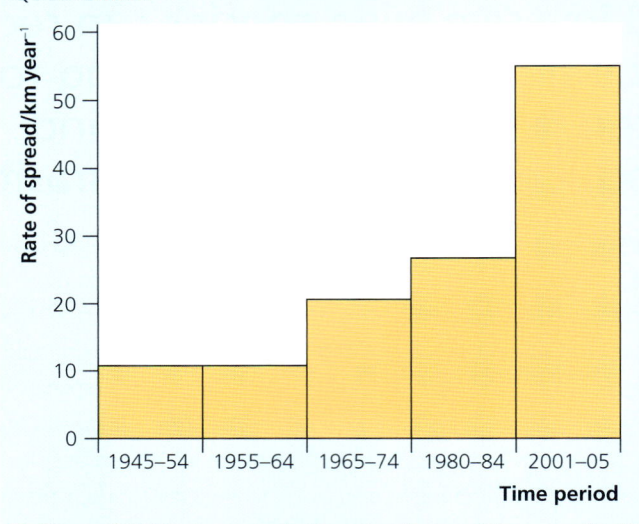

■ **Figure 3.13** The increase in the rate of spread (km year^{-1}) of the cane toad through Australia

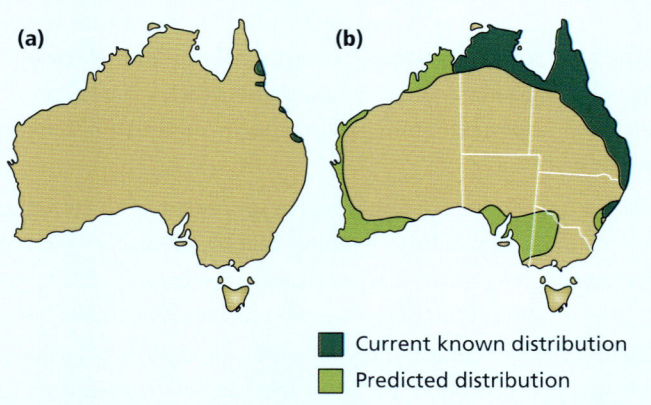

■ **Figure 3.14** (a) Cane toad distribution in 1460, and (b) known and predicted distribution in 2016

Despite the fact that the cane toads consumed the cane beetle, they also ate many other organisms and significantly reduced the availability of food for other, smaller species of frogs, toads and lizards. In addition, the cane toad produces a strong toxin from a gland under its eye that is toxic to many organisms. This has led to the death of many local wild species of animals that try to attack them, as well as domestic pets. This has led to significant changes in the native biodiversity in many parts of Australia (**www.dcceew.gov.au/environment/invasive-species/feral-animals-australia/cane-toads**).

HL.a.1: Environmental law

In 2005, the biological effects, including lethal toxic ingestion, caused by cane toads were listed as a key threatening process under the Environment Protection and Biodiversity Conservation Act 1999 (source: **https://invasives.org.au/our-work/feral-animals/strategy-invasive-species-australia**).

ATL ACTIVITIES

The cane toad used to be called *Bufo marinus*. Why is it now called *Rhinella marina*?

As scientific knowledge and capabilities advance, scientists are discovering new species and mapping out their genetic ancestry. This results in species of plants and animals moving places on mapped phylogenetic trees (see Topic 2.1.27, page 113). This will continue to happen as we discover more species and different techniques.

Although this information does not have any significant impacts on scientific understanding, it can help to more effectively develop correct and targeted conservation efforts for species. It can also suggest possible knowledge, based around other species within the group it has been moved to.

Use the internet to investigate why the cane toad had its scientific name changed from *Bufo marinus* to *Rhinella marina*.

REAL-WORLD EXAMPLE

Cane toad (*Rhinella marina*) introductions into Australia: the solutions

The introduction of cane toads into Australia (see Real-world Example, page 223) rapidly became impossible to control due to the lack of predators and their size. This meant they were able to outcompete local species and to breed rapidly and successfully. The impact of this invasion was significant enough to warrant action to be taken to try to control the population explosion of this species.

Management of cane toads in Australia has not been successful, possibly due to the scale of the problem. Management of this invasive species includes numerous strategies to reduce breeding, tadpole development and adult development. Stopping the spread of the species is one of the key goals of management in an attempt to contain the problem, however this is an extremely difficult solution to action.

In 1999, a plan was put in place to eliminate the threat of cane toads to biodiversity and as a toxic species. This resulted in the species being listed in the Australian Environment Protection and Biodiversity Conservation Act (EPBC Act) for its impacts. This inclusion has allowed the development of codes of practice for dealing with cane toad management, including guidance on harmless euthanasia of the toad to allow residents to be able to deal with populations in their gardens. Toad tracking and reporting by local residents allows the use of fences and other barriers to stop their dispersal and manage the local population. Locals are provided with information for identifying cane toad tadpoles. This allows management of the species before it matures into adulthood, as well as providing identification of adult individuals in relation to other similar species in the areas.

3.2.4 The global conservation status of species is assessed by the International Union for Conservation of Nature and is published as the IUCN Red List. Status is based on the number of individuals, rate of increase or decrease of the population, breeding potential, geographic range and known threats

In response to concerns regarding growing pressure on organisms around the world and the lack of international conservation measures, the International Union for Conservation of Nature (IUCN) was developed in 1948. The primary needs met by the formation of the IUCN were to improve international collaboration, improve awareness regarding the impact of human activities on the loss of species, improve the breadth and depth of scientific knowledge regarding species conservation, and influence laws that promote sustainable practices and conservation at a global level.

Currently, more than 1,400 member organizations and government agencies from over 170 countries work together with the IUCN. It conducts scientific research, manages conservation projects, publishes influential reports, and evaluates the conservation status of species through the Red List of Threatened Species.

The IUCN Red List is a continuously developing assessment process that evaluates the conservation status of known species and assesses their risk of extinction. This information allows conservation efforts to be targeted at specific species, guides scientific research, and provides some of these species with protection against poaching and international trade. A series of categories were created to determine the level of risk of a species becoming extinct.

The seven IUCN Red List Categories are as follows:

- Least Concern (LC): Species that are not currently at risk of extinction. They have a stable population size and distribution.
- Near Threatened (NT): Species that are close to qualifying as Vulnerable but do not yet meet the criteria. They may face future threats if conservation efforts are not implemented.
- Vulnerable (VU): Species that are at high risk of extinction due to declining population size, habitat loss or other threats.
- Endangered (EN): Species that are facing a very high risk of extinction in the wild.
- Critically Endangered (CR): Species that are facing an extremely high risk of extinction in the wild.
- Extinct in the Wild (EW): Species that no longer exist in the wild but are maintained in captivity or as a naturalized population outside their historic range.
- Extinct (EX): Species that no longer exist anywhere in the world.

Determination of the placement of a species is based on a number of different criterion measurements relating to the size of the current populations, how they function, and how well distributed the species is. This allows a level of extinction risk to be determined, allowing species to be placed into the relevant category.

Concepts

Systems and models

The model used to assign species to the IUCN Red List categories has, like all models, some limitations. It is important to reflect on these when you are interpreting information.

One of the biggest limitations is that the classifications are based on known and collected data. Therefore, there are many estimations, inferences and projections that may not be entirely accurate or up to date. Some groups of species are studied more widely than others and therefore information about these will be more reliable. There is a possibility of measurement error resulting in missing essential individuals, causing species to be assigned to the wrong criteria. Finally, interpretation can be confusing for some because a species can have different regional and international statuses. So, a species could be locally Vulnerable but globally Critically Endangered, due to local populations fitting in the Vulnerable category, but due to a lack of global geographical locations it is classified as Critically Endangered.

3.2.5 Assigning a global conservation status publicizes the vulnerability of species and allows governments, non-governmental agencies and individual citizens to select appropriate conservation priorities and management strategies

The IUCN evaluating the conservation status of individual species around the world has also allowed it to prioritize funding for conservation, research and restoration of habitats. Those species that are assigned the conservation status of Extinct in the Wild, for example, the Spix's macaw and the scimitar oryx, are found only in captive populations. Significant research is less likely to take place around these due to the complexity of reintroductions. When species are assigned as Critically Endangered, this allows funding and focus to be allocated to them as a priority, such as the development of captive breeding programmes in zoos. Observations relating to individuals' susceptibility to changes, for example, in conditions or food quality, can be examined in these captive settings, allowing greater understanding of the species itself, which can result in targeted conservation efforts in habitats to limit the effects of observed issues.

Detailed examination of all aspects of a species' global population means that the specific threats to each species can be comprehensively addressed. Species are likely to be under pressure due to more than one factor, resulting in cumulative impacts that can be greater than the sum of individual impacts alone. For instance, if a species is under the threat of poaching for a valuable body part, such as the tusks of an elephant, then reduced population sizes, **habitat fragmentation** and a reduction in food availability will compound the issue and make elephants easier for poachers to find. Therefore, actions relating to the IUCN Red List status recommendations need to be addressed in order to ensure that all stressors impacting that species are considered.

◆ **Habitat fragmentation** – the breaking up of areas of continuous habitat into unconnected, smaller parts.

> **Concepts**
>
> **Perspectives**
>
> Conservation is everyone's responsibility. Governments create laws and regulations that keep organizations accountable for their actions, but that is not where it should stop. Each person also has a role and can influence issues by individual decisions that are made. For instance, by stopping using companies that support unsustainable practices or that promote the use of illegal materials, and by encouraging others to do the same. Take part in citizen science calls for information. For example, in the UK the Royal Society for the Protection of Birds (RSPB) has an annual garden birdwatch that is widely advertised on television networks.

In 1975, the IUCN adopted the Convention on International Trade in Endangered Species of Wild Fauna and Flora (CITES), which was developed to restrict the trade of species and parts of species. By law, CITES has three appendices that are related to the severity of the risk to the species. Appendix I is solely for species threatened with extinction. Appendix II addresses species that are not necessarily at threat of extinction, but for which continued trade could result in potential extinction issues. This relates to species that are poached for body parts and those that are wild caught for breeding for the pet trade. Being caught with species on either appendix list can result in large fines or imprisonment. Appendix III lists species that are protected in at least one country, which has asked other CITES parties for assistance in regulating the trade to prevent unsustainable exploitation.

HL.a.6: Environmental law

It is not only the responsibility of international organizations to create legal protections for habitats and species. Local laws can be effective in local communities that may not understand the implications of international conventions and agreements, and may not even know the international laws exist. If there are local laws prohibiting poaching, illegal logging and collection for the pet trade, and these laws are enforced by local people in the communities, there is a greater chance of changing the behaviour at the source of the problem. These local laws can result in large changes to the availability of organisms and their parts on the international market.

HL.c.4, HL.c.5: Environmental ethics

Intrinsic and instrumental value can impact the level of conservation that is applied to a specific organism. Intrinsic values relate to the organism's natural value as part of an environmental system and its natural balance, whereas instrumental value is more related to the goods and services the organism provides. When looking at instrumental value, sustainability is key to understanding the obligation to preserve systems for future generations.

TOK

Discuss the differences between instrumental and intrinsic values of items in different cultures.

Should sustainability for humans be given more value than the sustainability of natural systems and populations?

3.2.6 Investigate three different named species: a species that has become extinct due to human activity; a species that is critically endangered; and a species whose conservation status has been improved by intervention

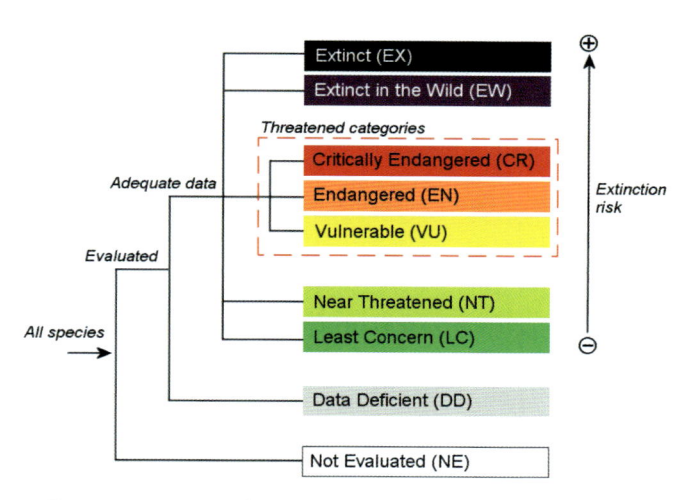

Figure 3.15 IUCN Red List categories showing their related level of extinction risk

Each species on the IUCN Red List has had a comprehensive range of data collected regarding its range, level of fragmentation and population dynamics, among others. Species can be reclassified if necessary. These data are freely available through the IUCN Red List site (**www.iucnredlist.org**), along with information regarding the date of the last evaluation of the species.

It is through this monitoring that species move from the Critically Endangered status to Extinct (or functionally extinct in the case of some species where the only individuals remaining are the same sex). There is disagreement in the real annual rates of extinction, mostly due to the enormity of the data that are still unknown. Conservative estimates are that between 0.01 and 0.1 per cent of species become extinct each year.

Concepts

Perspectives

There are still some parts of the world that have been left untouched and undisturbed by humans. When scientists do gain access to these often difficult-to-access areas, the scale of the unknown becomes very clear. In 2022, scientists made an expedition to the Greater Mekong valley that separates Laos and Thailand, which had never been explored by scientists before. As a result of this study, 380 new species were discovered, including species of snake, gecko, orchid, bat and more. This is not an isolated event and other such large discoveries are commonplace when unexplored habitats are sampled.

REAL-WORLD EXAMPLE

Extinct species

The Pinta Island tortoise (*Chelonoidis abingdonii*) was an iconic tortoise species due to Darwin's discoveries in the Galápagos Islands, off the coast of Ecuador. The tortoise was endemic to Pinta Island and had a modified carapace and elongated neck to allow it to make the most of the vegetation available on the island. Species reduction started early in the 1900s, when the tortoises were captured and eaten by sailors due to their size and ease of capture. The later introduction of goats to the island destroyed much of the favoured vegetation of the species, pushing them to extinction. The species became officially extinct in 2012 when the last known individual, Lonesome George, died.

3.2 Human impact on biodiversity

Critically Endangered species

The Sumatran rhino *(Dicerorhinus sumatrensis)* is the smallest of the rhino species and is currently only found on two Indonesian islands, Sumatra and Borneo. Of those populations still present in the wild, there is a high risk of them not being viable as the breeding success of this species appears to be particularly low. One of the major issues related to the species is its value due to the prevalence of the use of rhino horn in many traditional medicines, as well as the serious destruction and fragmentation of the dense tropical forest it inhabits. The World Wildlife Fund (WWF) has set up the Sumatran Rhino Rescue Alliance to bring together the efforts of all of the conservation organizations to protect the small, localized populations. However, this comes with problems as rhinos often wander out of the protected forest zones, meaning they are no longer under the protection of forest rangers.

Species with an improving status

One of the most iconic species related to conservation, the giant panda (*Ailuropoda melanoleuca*), is a species with an improving status. The famous face of the WWF, this species has been under threat for many decades. The giant panda is a typical specialist species and this has made its conservation difficult. It is a selective grazer that consumes only bamboo. Due to the low caloric density of bamboo, adult individuals need to eat continuously throughout the day. Therefore, any disruption in the bamboo forests they live in will have negative impacts on the species. Human development and subsequent habitat clearance has been by far the greatest challenge impacting this species.

In 2017, after decades of conservation, the giant panda was reclassified by the IUCN from Endangered to Vulnerable after the range of forest and population dynamics were such to remove the species from the Endangered list.

■ Table 3.6 Details of three species' IUCN Red List status and information

Name and status	Pinta Island tortoise Extinct	Sumatran rhino Critically Endangered	Giant panda Improving: Endangered to Vulnerable
Population size	0	30	Over 2,000
Location	0	Fragmented – four locations	Protected reserves in China only
Habitat	Shrubland, Galápagos Islands	Indonesia	China
Main pressures	Hunting, invasive species, reduction in quality of habitat	Poaching, habitat destruction	Habitat destruction, climate change

> **ATL ACTIVITY**
>
> Use the IUCN Red List online search features to find a Critically Endangered mammal species that interests you.
>
> Select the species and extract relevant information from the IUCN page. This should include information about the current population's locations and sizes, and other information.
>
> Research the reasons for the species being Critically Endangered. Identify species-based and anthropogenic factors that are responsible for its conservation status.
>
> Investigate the presence of *in situ* and *ex situ* conservation efforts that are in place, and the reasons for their success or failure.
>
> Using this evidence, make a judgement on whether or not you feel this species can recover.

HL.a.4: Environmental law

Despite numerous laws being in place from local to international levels, there are still parts of the world that do not respect the law as strongly as others. Bribery and corruption can lead to the lack of implementation of laws, allowing poachers and collectors access to organisms regardless of laws. This is one of the reasons that regulations are not a guarantee of successful conservation.

HL.b.6: Environmental economics

The success of conservation programmes is partly related to the economic pressures in the habitat where the species lives and the level of external funding that can be found to help pay for the conservation and protection to take place. Many of the parts of the world with the highest levels of biodiversity and need for conservation are in developing countries that are not financially stable enough to provide funding themselves. This is why it is vital that international organizations can prioritize these locations to ensure sufficient funding is available.

HL.c.4: Environmental ethics

Consider how choices should be made to protect organisms. Is the level of decline enough to give a certain species priority for conservation efforts, or do factors relating to its usefulness to humans or the value of the services or products it provides dictate priority for protection? Or, as some believe, do all living organisms deserve the same level of protection, regardless of impact or specific values?

3.2.7 The tragedy of the commons describes possible outcomes of the shared unrestricted use of a resource, with implications for sustainability and the impacts on biodiversity

The tragedy of the commons is a concept in environmental science and economics that describes a scenario where individuals, acting in their own self-interest, deplete or degrade a shared and limited resource, leading to the depletion of that resource until it is no longer sustainable. The term originates from the use of the 'common' land that was part of early European urban design and that created a relationship between the community and the environment. These areas were designed for common use by residents and no limits were imposed. This inevitably led to some residents using the common resource to a greater degree than others. In response to this situation, both in the past and now, others in the community also increase their use of the resource in an attempt to make things appear more fair. This often continues until the cumulative use of the common resource causes it to become less abundant or even to disappear, ultimately resulting in none of the community having access to that resource (see Figure 3.16).

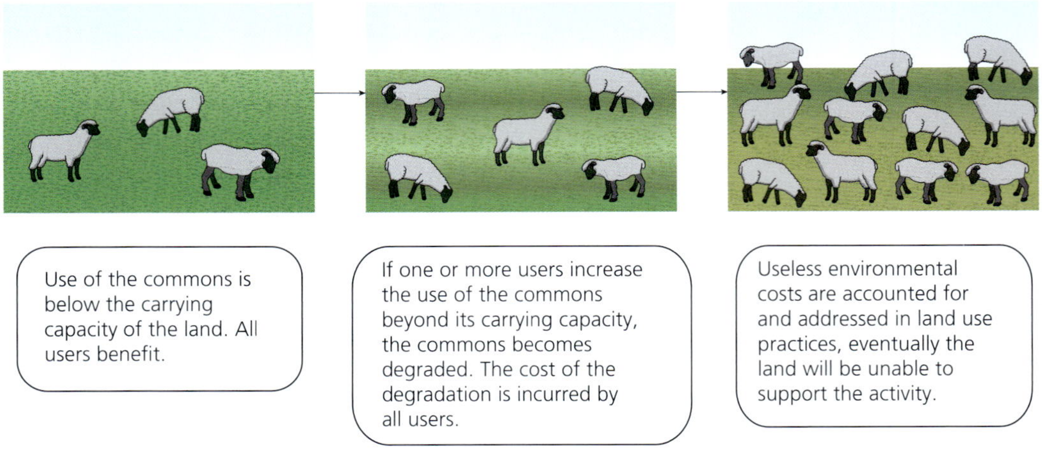

| Use of the commons is below the carrying capacity of the land. All users benefit. | If one or more users increase the use of the commons beyond its carrying capacity, the commons becomes degraded. The cost of the degradation is incurred by all users. | Useless environmental costs are accounted for and addressed in land use practices, eventually the land will be unable to support the activity. |

■ **Figure 3.16** Visual representation of the impact of the unsustainable use of common resources

One of the main issues associated with common resources is that there is a lack of real ownership of the resource, resulting in no regulation, protection or conservation. This also means that determining who is responsible for managing the resource can lead to mismanagement of the resource: successful management requires a careful balance between individual self-interest and the potential shared benefits.

One of the Earth's biggest common resources is the oceans. A small part of the ocean surrounding a country, a set distance from the coast, is owned by that country. However, past this point is what is considered the open ocean, and this is not owned by anyone and is relatively unregulated. This has led to the overexploitation of every aspect of the ocean that humans have a use for. Due to the abundance of fish in the ocean, extraction rates have become unsustainable due to lack of understanding and lack of regulation. Some fish species have been overfished to the point that juveniles are being taken before they are able to mature and reproduce. Waste from around the globe is collected in large parts of the ocean, but no country will take responsibility for mobilizing clean-up efforts, highlighting that lack of ownership also comes with a perceived lack of responsibility for impacts on those resources.

REAL-WORLD EXAMPLE

Grand Banks cod fishing, Newfoundland

Cod fishing off the coast of Newfoundland, Canada is a prime example of the tragedy of the commons and its devastating consequences for both the fishing industry and marine biodiversity. The Grand Banks was historically one of the world's most abundant fishing grounds due to its shallow depths where both cool and warm water currents mix, thus creating ideal conditions for an abundance of nutrients to develop. This area is particularly known for its once abundant cod fisheries. However, over the centuries, unsustainable fishing practices and a lack of effective management led to the collapse of the once-thriving fishery.

The Grand Banks cod fishery operated with little regulation, and fishing practices were driven by short-term profit. Large-scale commercial fishing vessels used advanced technology to catch vast quantities of cod without considering the long-term sustainability of the resource. This resulted in overfishing, where cod stocks were depleted faster than they could naturally replenish.

Since the fishing grounds were open-access (free to anyone), with no property rights or effective regulations, each fishing vessel had the incentive to catch as much cod as possible to maximize its own profits, without considering the collective impact on the fishery's sustainability. As a result, the resource was overexploited, leading to the collapse of cod stocks in the early 1990s. The collapse of the fishery had devastating effects on the livelihoods of fishing communities in Newfoundland. Thousands of people lost their jobs, and the economic balance in many coastal towns and villages was shattered.

In order to address this, Canada implemented strict fishing bans and restrictions, along with efforts to promote sustainable fishing practices and stock rebuilding. Conservation and management measures included moratoriums on cod fishing, the establishment of marine protected areas, and the development of scientific assessments to monitor cod stocks. While some recovery has been observed in recent years, the fishery's full recovery is still a long way from being achieved.

REAL-WORLD EXAMPLE

Plastic pollution in ocean gyres

Ocean currents act to move nutrients, water, organisms and much more around the oceans. These currents move at different speeds in different parts of the oceans and when the current is rapid it will pick up debris and plastic pollution, which it deposits in areas where the ocean currents slow down. These gyres collect large volumes of waste materials that have either accidentally or intentionally ended up in the oceans. As the plastic is washed around it breaks down into smaller particles, resulting in water that is densely packed with small particles of waste. See Topics 4.4.2 for more detail on how ocean gyres are formed and function, and 7.3 for the Great Pacific Garbage Patch.

ATL ACTIVITY

Create a campaign in your school that engages students and staff in plastic collection, recycling and repurposing. Focus on why this is important. Create an article for the school newspaper or social media page, or a presentation for an assembly, about plastic pollution in ocean gyres and how we can all help reduce plastic waste.

If your school is at the coast, organize a service activity with a beach clean-up day to bring the school community together to protect your local environment. Create video diaries, newspaper articles and social media posts showing how much was collected on the clean-up day and how people in the community can reduce their waste. If your school isn't on the coast, organize a community clean-up day to raise awareness that all waste can end up in the ocean.

HL.a.7: Environmental law

Transboundary agreements are required when managing common resources such as the oceans and the atmosphere. This is particularly important as the impacts of actions in one part of the world could easily have rippling effects in wide-reaching parts of the world. The Association of Southeast Asian Nations (ASEAN) has been working to develop agreements addressing the transboundary air pollution that results in significant pollution events on an annual basis.

HL.b.6: Environmental economics

Some products that are considered a common good, such as fish, have monetary value and are therefore more likely to be overexploited and over-harvested due to the potential for financial gain. Elinor Ostrom suggested there should be community management of common resources to monitor usage and allow penalties to be placed on those who overuse the resource, and that users of the resource should all be part of the decision-making processes to make it more fair. This has been put in place in some parts of the world, for instance Iceland has developed a sustainable way of managing its fish stocks through regulating the size of fish that can be caught, designating breeding periods as no-fish periods, and monitoring the number of boats and the size of their catches to ensure the population of fish does not deplete further.

HL.c.9: Environmental ethics

As much as humans would like to act ethically, this is hard to do when livelihoods, incomes and family are at stake. Therefore the concept of consequentialist ethics should be relevant in this situation. That level of comprehension of the wider consequences of overfishing is not something that all communities have, however, or even believe in. Virtue ethics suggests that if overexploitation by one individual was done in order to feed their family, then would this be considered ethical as it is keeping the family alive. This makes the tragedy of the commons a difficult ethical dilemma that can be seen from many angles. Discuss how different stakeholders will be impacted by the overuse of common resources.

ATL ACTIVITY

Consider the concept of overfishing from the view of as many different stakeholders in the issue as possible. This could include an individual fisherman, a fishing company, a Michelin-star restaurant, a scuba-diving tour operator or a customer in a supermarket buying frozen fish. What would each perspective be and what would influence these perspectives? What do you think about the issue? Share your ideas and answers with a partner or the class.

(HL) 3.2.8 Biodiversity hotspots are under threat from habitat destruction, which could lead to a significant loss of biological diversity, especially in tropical biomes

Biodiversity hotspots are areas that harbour an exceptionally high concentration of species, many of which are found nowhere else on Earth. These regions are often characterized by unique ecosystems and a remarkable diversity of plants, animals and other organisms. However, biodiversity hotspots are also particularly vulnerable to various threats, and habitat destruction is one of the most significant factors contributing to the potential loss of biological diversity.

The Mediterranean Basin Biodiversity Hotspot is one of the largest hotspots worldwide. It has an extremely high diversity of plant species, with more than 13,000 endemic to the hotspot. This area is particularly vulnerable due to the increase in urbanization nearby. The area was home to many climax oak, evergreen coniferous and deciduous forests that have been heavily reduced and fragmented as the human population development in the hotspot increased. Now the dominant habitat is marquis, a hard-leaved, shrubby vegetation that is maintained by grazing and fire. As the urban settlements continue to develop and expand, these rural communities and their management practices are being abandoned, leading to the disappearance of the original habitat due to succession.

Many of the hotspots are located in tropical regions, where poverty can be impacting local human communities and increase the vulnerability of their natural habitats. Pressure to sell land for agriculture and mining is common due to the abundance of undisturbed habitats. Deforestation in these areas is also prevalent due to the abundance of large, old trees, many of which are hardwood species and therefore have a high value. This development means that infrastructure must be created to allow access, for example, for logging vehicles, as well as shelter and amenities provided for workers. This can lead to the development of urban communities in these areas. Other regions become vast agricultural areas, as seen in large parts of the deforested Amazon rainforest in Brazil.

HL.a.4: Environmental law

Many of these biodiversity hotspots are protected by international law. However, these laws are sometimes hard to implement due to corruption, lack of local government support and the power of the organizations that are making the deals. The effectiveness of the laws is dependent on the country where the hotspot is located.

HL.b.3, HL.b.7: Environmental economics

In these areas, poverty is often so high that communities and governments are highly vulnerable to accepting offers for the clearance of forests due to the significant financial incentives that are offered. Such countries need to be financially supported so that they do not have to rely on the short-term income from selling their valuable resources in order to develop their economy. International funds are directed into these areas to help them achieve this, mostly due to the fact that they are UNESCO-protected areas.

This does not mean that such activities do not happen, however. Satellite data clearly shows that forest areas are decreasing rapidly in regions such as the Amazon rainforest and parts of Southeast Asia. Protecting biodiversity is a priority, but in some of these areas the economic development of the country and its infrastructure is also a priority. These two things do not necessarily go together well, straining the support that is present and available.

HL.c.4: Environmental ethics

Ethically, it is not considered right to exploit those in developing countries to gain access to their valuable resources, only to remove the resources from that country and make large amounts of money manufacturing goods from those. Much of this pattern can be related to limited education and lack of understanding in some local communities of the long-term implications of selling the rights to their land or resources.

Despite the availability of funding for projects in these hotspots, communities based in the developing countries are at continued risk due to the social issues there. Balancing the need for the protection of valuable natural resources and the development of poorer nations is complicated, as they do not complement each other. Success of conservation efforts is reliant upon local support and engagement, which is difficult when these communities may not have the knowledge to understand the need for conservation of their natural resources. Education, healthcare, infrastructure and employment all need to be addressed to ensure that the conservation efforts are sustainable.

3.2.9 Key areas that should be prioritized for biodiversity conservation have been identified on the basis of the international importance of their species and habitats

Key biodiversity areas (KBAs) are identified using a standardized set of criteria that consider factors such as species diversity, endemism, population sizes and ecological functions. KBAs can include a variety of habitats, from terrestrial and freshwater ecosystems to marine and coastal environments. They help guide conservation efforts by highlighting specific sites that are of utmost importance for biodiversity. They provide a framework for directing resources and attention to areas where conservation actions can have the greatest impact. Many KBAs are strategically located to maintain ecological connectivity between fragmented habitats.

While these areas are not protected by other designations, they are recognized for their importance in a certain life stage or migratory pathway of a specific species. For example, these areas ensure that birds on migration pathways are able to reach their destination, and as such these sites must be able to provide adequate shelter and food to allow the species to arrive at breeding or overwintering sites. The removal of these habitats would result in a significant reduction in the success of migratory birds, due to starvation and exhaustion en route.

KBAs ensure that funding, conservation efforts and recognition are given to areas outside the typical biodiversity hotspot designation. Some KBAs are areas that are essential for the migratory patterns used by birds and animals on an annual basis, while others are sites where ecological interactions might take place that are specific to that region only, and are therefore associated with vulnerable and threatened species and habitats. Many of these sites are within the biodiversity hotspots, but are not limited to them. KBA designation does not come with any protections for that area; it is instead designed to highlight the importance of the area. However, many of these areas are assigned some level of protections due to overlap with international protection relating to World Heritage and **Ramsar** designations.

◆ **Ramsar site** – a wetland designated as having international importance.

REAL-WORLD EXAMPLE

The Critical Ecosystems Partnership Fund

The Critical Ecosystems Partnership Fund (CEPF) was set up in 2000 to protect biodiversity and human communities in biodiversity hotspots around the world. The fund is a hub for developing projects in areas that need help combatting the impacted biodiversity. It helps to empower the communities in these areas to create local conservation leadership. The fund is supported by organizations such as Conservation International, the European Union, the World Bank and others. Currently it has funded over 270 million USD to projects in all of the 36 biodiversity hotspots. These funds ensure that research and conservation are targeted towards the species, habitats and communities in the areas. The fund also provides financial support for projects in key biodiversity areas due to their importance for global biodiversity (source: **www.cepf.net**).

Tool 2: Technology

Designation of an area as a biodiversity hotspot or KBA

Designation of an area as a biodiversity hotspot or KBA is based on specific criteria. This means that there are considerable data collected in these areas. All KBAs have data collected on them that identifies what the specific value of that area is, for example, a breeding site on the migration pathway of a number of bird species. Active monitoring of these areas is required to allow conservation efforts to be quantified and measured. The key biodiversity areas website (**www.keybiodiversityareas.org**) provides a link to the data that are available for every designated area, including the size of the area, the percentage of that habitat that is protected, and the reasons the classification was required in that location. These are freely available and could provide interesting datasets for use in either secondary data-based Internal Assessments or Extended Essay data collection.

REAL-WORLD EXAMPLE

Key biodiversity areas in Canada

Canada has around 482 KBAs developed for the protection of birds. Canada is home to over 450 migratory bird species that rely on these natural environments for breeding and resting along migratory pathways. In addition to the significant migratory bird population, there are many species that are endemic to Canada and associated with some of the threatened habitats. One of the most important KBAs in Canada is Long Point Peninsular and Marshes, close to the Canadian/US border. This is an area of national significance for many aggregations of native birds, as well as being a significant spring and autumn site for migrating waterfowl. The site is also home to six endangered reptiles and amphibians, including Fowler's toad (*Anaxyrus fowleri*), which has 22 per cent of its national population in this area. These toads are particularly vulnerable as they inhabit the sand dunes that are being destroyed by off-road and ATV activities.

HL.a.8: Environmental law

While a KBA designation does not offer specific protections, parts of Canada are protected due to the development of the Migratory Birds Convention Act of 1994. Some areas are protected due to the IUCN-classified species that use that habitat. For instance, Fowler's toad (*Anaxyrus fowleri*) and the spiny softshell turtle (*Apalone spinifera*) are protected species found on Long Point Peninsula, an area which is therefore given protection.

3.2.10 In key biodiversity areas, there is conflict between exploitation, sustainable development and conservation

Many KBAs are located in parts of the world that are under pressure in many areas other than conservation, such as poverty, civil conflict, agriculture and mining pressures. Many of these factors are driven by a need for the country to develop economically, creating a difficult balance between this and maintaining biodiversity. Situations such as this are highly complex and need an individualized approach to be successful. As KBA designation does not come with official protections, it does not have any impact on the other activities happening in these areas.

HL.a.9: Environmental law

According to the IUCN, in Myanmar pressure due to ongoing sporadic civil conflicts within the country makes prioritizing conservation very difficult. This is due to lack of access to important areas and potential danger associated with living in these regions. Contact with community-based conservation organizations has been limited to online communications during the civil coup, and attempts to empower women in local conservation initiatives have also been put on hold as they are opting to stay at home, tend their land and ensure the safety of their family. Some protected areas are also at risk of becoming targets due to the value of the natural resources in the area. Despite some of these areas being protected by international law, the issues arising due to the conflict are making enforcement of these laws almost impossible.

REAL-WORLD EXAMPLE

Indonesia's conservation and the palm oil industry

The cultivation of palms for oil has been a rapidly expanding industry in Indonesia since the 1970s, causing mass destruction of native habitats on the islands. Driven by economics, governments have tried to intervene by placing laws on companies to ensure that they provide profit and benefit to local communities in exchange for growing palm oil on their land. Companies were required to give a portion of the area they occupied to the local community, or to provide them with a certain percentage of the profits from that area of land. Recently there have been protests and civil unrest due to the fact that many of the large palm oil producers have not honoured these legally binding agreements. This leaves locals without any benefit from handing over their land for production. The promise was seen as an incentive to these poor communities, resulting in large areas being given over to palm oil plantations, further fragmenting the native habitats in Indonesia.

Although there is legislation protecting certain areas, this has not prevented illegal palm oil production, which is responsible for the loss of at least 43 per cent of the Tesso Nilo National Park on Sumatra, an established protected habitat for the critically endangered Sumatran tiger (*Panthera tigris sumatrae*) (source: **www.worldwildlife.org/industries/palm-oil**).

3.2.11 Traditional indigenous approaches to land management can be seen as more sustainable but face challenges from population growth, economic development, climate change, and a lack of governmental support and protection

Indigenous communities are the first inhabitants of an area. The communities of their descendants have been living with nature for many generations. Traditional indigenous approaches to land management are often deeply rooted in sustainable practices of hunting, gathering and farming that have evolved over generations. These approaches are closely tied to cultural, spiritual and ecological values, allowing communities to live in harmony with their environments.

These groups are therefore disproportionately impacted by issues that impact the habitats they rely on. Despite indigenous communities accounting for only around 5 per cent of the human population according to the UN, they are stewards for between 20 and 25 per cent of the Earth's surface, which represents around 80 per cent of the planet's biodiversity. Conservationists are looking more closely at the ways indigenous communities have managed to subsist and live in harmony with their environment for generations (see Topic 5.2.18, page 502).

> **REAL-WORLD EXAMPLE**
>
> ### The Baka people of the Democratic Republic of Congo rainforest
>
> The Baka people inhabit their ancestral land in Messok Dja in the rainforest in the Democratic Republic of Congo, central Africa. These people have been living in harmony with the forest for generations and rely on the forest for their food, medicine and all other aspects of life. According to Survival International, the biodiversity of the area is such that the World Wildlife Fund (WWF) is establishing this part of the forest as a conservation zone. This designation comes with regulations whereby boundaries have been placed around parts of the forest so that the Baka people can no longer enter. In addition to the removal of this indigenous tribe from their ancestral land, other parts of the forest are being further removed by continual logging, leaving the Baka people without a livelihood or a safe place to live.

TOK

Should the rights of nature be placed above the rights of indigenous communities?

Global issues such as climate change, population growth and economic development are impacting all communities worldwide. However, indigenous communities have fewer measures in place to protect their already small and vulnerable communities. There is pressure for these communities to begin to engage in economic activity outside their communities, causing aspirations of following the development model of the rest of the world.

HL.a.11: Environmental law

The rights of indigenous communities to retain ownership of their land has long been a global issue as nations develop and want to gain access to these protected and valuable areas of land. Legally, many indigenous communities have been increasingly protected due to international treaties and agreements, and their traditions and practices are used to improve the sustainability of habitat management. However, despite being covered under international law, this is not always the case at a regional or local level.

■ **Table 3.7** Timeline of international meetings that address the rights of indigenous communities (source: Land Portal Foundation)

Year	Event	Rights addressed
1919	United Nations set up the International Labour Organization (ILO).	Improves working conditions and human rights worldwide.
1989	ILO developed the legally binding treaty Convention 169 on Indigenous and Tribal Peoples.	Recognizes and protects indigenous land ownership rights. Ratified by only 22 countries.
1992	The Convention on Biological Diversity was signed at the Rio Earth Summit.	Protects the Earth's plants, animals and humans, and highlights the use of traditional cultural practices to protect diversity. Signed by 193 parties.
2007	United Nations Declaration on the Rights of Indigenous Peoples (UNDRIP).	Recognizes the rights of indigenous peoples in relation to all aspects of life, such as traditions, customs and ceremonies, as well as protection of land with cultural and religious significance. Signed by 143 parties, but is not legally binding.
2022	Stockholm+50 UN general assembly COP 15, Montreal.	Calls for the mainstream implementation of traditional scientific knowledge to tackle climate change and biodiversity loss.

HL.b.7: Environmental economics

Many of these indigenous communities used bartering systems in the past and would exchange produce or services between each other. As the developed world has encroached on their lives, they have been exposed to the trappings that accompany economic wealth, such as electricity and mobile phones. As these communities become more used to seeing these elements of the developed world, it is only natural that those people will want those devices, thus introducing these communities into the monetary world. This can often lead to communities abandoning their traditional practices in favour of faster and more convenient methods of cooking, washing clothes, for example, that increase their need for electricity. This can often come from mainstream power that is likely to be produced through the burning of fossil fuels. As these elements become a greater focus in their lives, communities begin to require an income to cover these costs.

TOK

To what extent is it ethically correct for the developed world to impose limits on developing nations based on their current understanding of the environmental issues they caused?

HL.c.10: Environmental ethics

As the stewards of their ancestral lands, indigenous communities often feel ethically obliged to protect their practices that work in harmony with the environment. However, some would say it is not ethically correct to keep a community living in underdeveloped conditions because the developed world recognizes the mistakes that were made in their own countries and wants to stop that from happening in less developed areas. This practice could be seen as putting the value of the natural environment over the value of these communities, due to hindsight gained regarding the environmental consequences of the development of their own economy.

3.2.12 Environmental justice must be considered when undertaking conservation efforts to address biodiversity loss

Environmental justice emphasizes the fair and equitable distribution of both environmental benefits and the burdens of environmental degradation, ensuring that marginalized and vulnerable communities are not disproportionately affected by conservation initiatives. In the past, indigenous communities have been displaced due to conservation programmes and a lack of education among and legal support for these communities. Therefore, conservation efforts must take into consideration the needs of the indigenous communities as part of the projects.

Displacement and marginalization commonly occur with these communities due to a lack of education and understanding of the developed world. This can lead to these communities being taken advantage of and forced into agreements they may not fully understand. Environmental justice would ensure that these communities are included in decision making related to the management of their land. In order to do this successfully, they need access to free legal support to help them protect the rights to their land.

Environmental justice will also enable indigenous communities to protect and preserve their cultural identity and voice. It is widely recognized that through empowering local communities to be the guardians of their biodiversity, conservation efforts are more successful as local communities also benefit.

REAL-WORLD EXAMPLE

Displacement of the Tanzanian Maasai tribe for a game reserve

The Maasai tribe are the indigenous tribe of northern Tanzania and have lived in this region for hundreds of years, adapting their nomadic grazing practices in response to the climatic pressure of the area. Their land has been fragmented since it was divided between the UK, Germany and Kenya in 1886, resulting in the Maasai losing access to over half of their original grazing land. As agricultural development was taking place, the Maasai were pushed into small, drought-prone areas of their original land.

More recently, there has emerged a new threat to these indigenous lands. The government's focus on wildlife conservation has resulted in the development of large game reserves, where it is illegal for the Maasai to graze their animals or even to collect water. There have been reports of the violent removal of the Maasai tribe from these areas, with settlements being destroyed and burned to force individuals to leave the area (see Figure 3.17). Missionary groups in the Ngorongoro Conservation Area have been assisting in the eviction of these people through the withdrawal of vital healthcare to these communities while they remain in the now-protected reserve area. These factors reduce their access to ancestral lands, and also reduce their culture, community connections and ability to support themselves, leaving them as possibly the most endangered indigenous tribe in Africa (sources: www.hrw.org and www.aciafrica.org).

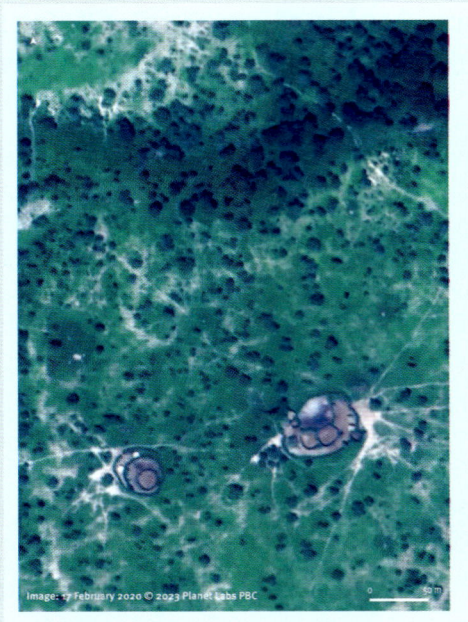

■ **Figure 3.17** Satellite images showing the location of a Maasai tribal compound in 2020 (left image), and a satellite image of the area in 2022 once the boundaries of the new game reserve had been established

Concepts

Perspectives

Ensuring that the communities living within protected areas benefit in some way, either through income from tourism or paid employment, leads to these communities becoming invested in the success of the management of their land. They are also less likely to be tempted to take bribes to access land if they are reliant on the area for their livelihoods (see Chapter 1.1, page 2).

HL.a.11: Environmental law

In order to increase the potential success and strength of environmental protection of indigenous land, legal personhood can be granted to the area to provide it with legal rights against destruction. Legal rights of nature have been developing for over 50 years and these rights were granted to all rivers in Bangladesh in 2019, and to the Komi Memem River and all its tributaries in the Brazilian Amazon in 2023.

HL.c.11: Environmental ethics

Some religions and communities believe that it is wrong to kill a living organism, regardless of the reason. Therefore, many conservation organizations focus on the habitat approach, rather than the single-species approach. However, humans appear to have taken themselves out of this equation in some conservation efforts, and the displacement of communities has frequently taken place to create nature reserves for the protection of biodiversity.

3.2.13 The planetary boundary for loss of biosphere integrity indicates that species extinctions have already crossed a critical threshold

The planetary boundary for loss of biosphere integrity refers to the threshold beyond which human activities cause significant and irreversible harm to ecosystems, leading to biodiversity loss, species extinctions and disruption of ecosystem functions. The amount by which this boundary has been surpassed in different areas is an indicator of how human activities are impacting the stability and resilience of the Earth's biosphere. As we continue to develop, the planet is beginning to get to the point where the damage to some of the essential planetary boundaries has gone past the point of no return – past the tipping point (see Figure 3.18).

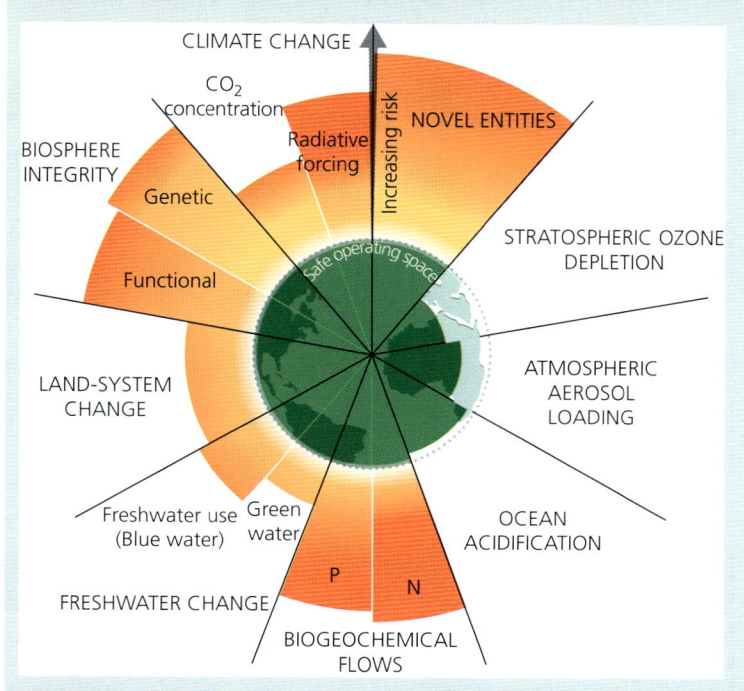

■ **Figure 3.18** Planetary Boundary model showing the status of each boundary; overshoot is shown by the red bars

There is clear evidence that humans have been surpassing the natural extinction rate for some time. The Smithsonian Institute calculated the natural extinction rate, based on the fossil record, as one species per 1 million species, per year. Estimates show that the actual current extinction rates are in fact already 10,000 times over the natural rate of extinction. This enormous overshoot in the extinction rate can be mostly attributed to the impact of human activities on our planet.

As extinction rates of known species increase, there are possibly going to be the same proportion, if not more, of species that have yet to be discovered becoming extinct. The more species that become extinct within a system, the weaker the system becomes. Robust ecosystems can often withstand change because of the diversity within their system. However, if that diversity starts to decrease, the pressure on other species will increase. This will eventually lead to a tipping point and the ecosystem will collapse.

3.2 Human impact on biodiversity

Each species within an ecosystem has a specific role within the system as a whole. This could be as a primary producer that begins the chain of energy transfer between organisms, or as a decomposer that breaks down the dead organisms to return their nutrients back to the system. This means there is a greater risk of individual ecosystem collapse, resulting in a cascade effect due to the interconnected nature of all of our systems, and the potential that this could negatively impact the whole Earth system.

Link

Sustainability is a key factor in trying to reduce the overshoot of the planetary boundaries. The overshoot represents unsustainable impacts that may be hard to reverse or recover from if they move too far away from the point at which our impact on that specific factor is in balance (see Chapter 1.3, page 47).

REVIEW QUESTIONS

1. List two direct and two indirect impacts that human behaviours have on wild plants and animals.
2. Identify which of the following are not factors that make a species prone to extinction: high levels of maternal care; individual offspring per birth; large geographical distribution; aspects of the animal's body used in traditional medicines.
3. Explain one limitation of the IUCN Red List.
4. Describe, using a named example, the development of unsustainable use of resources leading to a tragedy of the commons situation.
5. Identify two factors that can lead to successful species-based conservation.

EXAM-STYLE QUESTIONS

1. Explain why indigenous knowledge is a vital component of conservation plans. [5 marks]
2. Evaluate the pros and cons of an area being designated as a key biodiversity area. [5 marks]
3. Discuss, using named examples, the potential conflicts with conservation that are experienced in many biodiversity hotspots and KBAs. [9 marks]
4. Explain why environmental justice is so important for indigenous communities. [3 marks]
5. Using the three HL lenses, discuss the complexities of an aspect of habitat and/or species conservation. Use named examples to support your answer. [9 marks]

3.3 Conservation and regeneration

Guiding questions

- How can different strategies for conserving and regenerating natural systems be compared?
- How do worldviews affect the choices made in protecting natural systems?

SYLLABUS CONTENT

This chapter covers the following syllabus content:
- ▶ 3.3.1 Arguments for species and habitat preservation can be based on aesthetic, ecological, economic, ethical and social justifications.
- ▶ 3.3.2 Species-based conservation tends to involve *ex situ* strategies, and habitat-based conservation tends to involve *in situ* strategies.
- ▶ 3.3.3 Sometimes a mixed conservation approach is adopted, where both habitat and particular species are considered.
- ▶ 3.3.4 The Convention on Biological Diversity (CBD) is a UN treaty addressing both species-based and habitat-based conservation.
- ▶ 3.3.5 Habitat-conservation strategies protect species by conservation of their natural environment. This may require protection of wild areas or active management.
- ▶ 3.3.6 Effective conservation of biodiversity in nature reserves and national parks depends on an understanding of the biology of target species, and of the effect of the size and shape of conservation areas.
- ▶ 3.3.7 Natural processes in ecosystems can be regenerated by rewilding.
- ▶ 3.3.8 Conservation and regeneration measures can be used to reverse the decline in biodiversity to ensure a safe operating space for humanity within the biodiversity planetary boundary.
- ▶ 3.3.9 Environmental perspectives and value systems can impact the choice of conservation strategies selected by a society.

HL ONLY
- ▶ 3.3.10 Success in conserving and restoring biodiversity by international, governmental and non-governmental organizations depends on their use of media, speed of response, diplomatic constraints, financial resources and political influence.
- ▶ 3.3.11 Positive feedback loops that enhance biodiversity and promote ecosystem equilibrium can be triggered by rewilding and habitat-restoration efforts.
- ▶ 3.3.12 Rewilding projects have both benefits and limitations.
- ▶ 3.3.13 The success of conservation or regeneration measures needs to be assessed.
- ▶ 3.3.14 Ecotourism can increase interdependence of local communities and increase biodiversity by generating income and providing funds for protecting areas, but there can also be negative societal and ecological impacts.

3.3.1 Arguments for species and habitat preservation can be based on aesthetic, ecological, economic, ethical and social justifications

There are multiple reasons why we should preserve our natural habitats and species, all of which are equally valid. This stems from the fundamental right to be alive and thrive, regardless of a habitat's or species' level of importance for some other reason, such as the products it provides for consumption. Everything has value due to its place within the successful functioning of an ecosystem. Some species have other values that can increase the support for preservation (see Table 3.8 for details). Most conservation projects take all these factors into account when developing an action plan. By recognizing and promoting these multidimensional values, conservation efforts are more likely to gain support and be successful in ensuring long-term sustainability.

■ **Table 3.8** Possible examples of aesthetic, ecological, economic, ethical and social reasons for preserving natural habitats and species

Aesthetic	Ecological	Economic	Ethical	Social
Biodiversity and natural landscapes contribute to the beauty and cultural value of the environment. Preserving unique species and habitats allows future generations to appreciate and enjoy the natural world.	Species and habitats play critical roles in ecosystem functioning. Each species contributes to ecological processes such as pollination, nutrient cycling and predator–prey relationships, which maintain ecosystem health and resilience.	Ecosystem services provided by species and habitats have significant economic value. These services include pollination of crops, water purification, carbon sequestration and natural pest control.	Many people believe that all species have an intrinsic value and a right to exist, irrespective of their usefulness to humans. Preserving species and habitats is an ethical responsibility to protect the inherent worth of lifeforms.	Biodiversity and healthy ecosystems contribute to human well-being and quality of life. Access to green spaces and natural areas positively impacts mental and physical health, reducing stress and promoting overall well-being.
Many people find solace, inspiration and recreation in nature. Conserving species and habitats provides opportunities for ecotourism, nature-based recreation and wilderness experiences.	Biodiversity enhances ecosystem stability and reduces the risk of ecosystem collapse. A diverse range of species ensures that ecosystems can better adapt to environmental change and disturbances.	Conservation of natural habitats attracts ecotourism and nature-based recreational activities, generating economic benefits for local communities and economies.	Conservation reflects moral considerations, such as avoiding causing harm to other species, preserving genetic diversity and safeguarding the future of life on Earth, which is also a key element of the Buddhist religion.	Conservation efforts can help address environmental injustices and disparities, ensuring equitable access to natural resources and the benefits derived from biodiversity.

3.3.2 Species-based conservation tends to involve *ex situ* strategies, and habitat-based conservation tends to involve *in situ* strategies

Solutions need to be adapted according to the problem. Therefore, approaches to conservation need to be individualized and based on the specific threats and issues at hand. Furthermore, the strategies needed for conserving a critically endangered species will be distinct from those needed to protect a habitat. Individual species can be taken out of their natural habitat to allow them to

◆ **Ex situ conservation** – conservation that is carried out outside the natural habitat, such as in a zoo, sanctuary or botanical garden.

◆ **In situ conservation** – conservation that is carried out within the habitat of the species.

be protected, preserved and managed more effectively, and this is called **ex situ conservation**. Conservation of a habitat needs to take place within that habitat, and this is known as **in situ conservation** (see Figure 3.19).

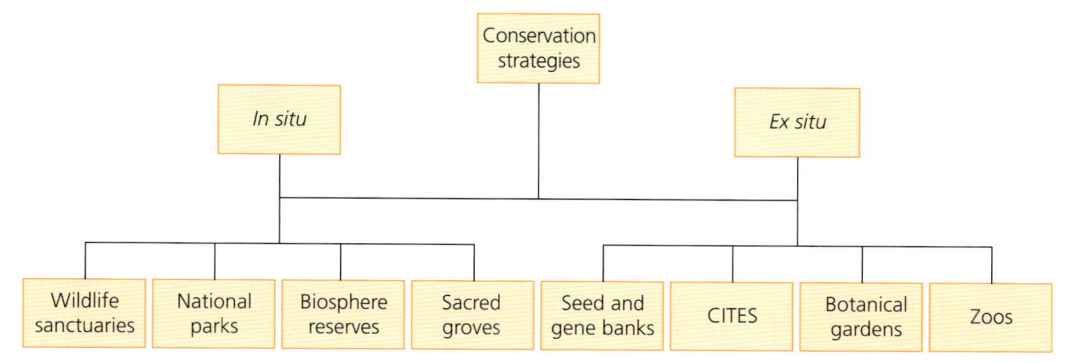

■ **Figure 3.19** The variety of *in situ* and *ex situ* conservation strategies currently widely used

Concepts

Perspectives

Some ancient woodlands have long-term conservation efforts in place because religious beliefs and practices have designated these areas as sacred groves. These are often areas of pristine primary forest that have religious, cultural and spiritual significance for local communities. These groves are protected from any form of exploitation or development due to their religious or cultural significance. Sacred groves can be found in various parts of the world, and their preservation has been instrumental in safeguarding biodiversity and natural habitats. However, due to the isolated nature of these habitats, they are often not included in habitat conservation efforts.

■ Species-based *ex situ* conservation

Many species are under threat because they are experiencing difficulty keeping up with rapid habitat changes or the influx of an invasive species. In such situations it is important to remove some individuals from wild populations to ensure that they do not become catastrophically impacted. This provides valuable protection for those individuals and their genetic material, and also allows scientists to investigate the impact of stressors and conduct captive breeding programmes to help improve wild population numbers. This practice has been common for many years and, as much as zoos originally provided entertainment only, they are now centres for research, breeding and release programmes that are invaluable to the preservation of species.

Zoos are now key providers of funding for *in situ* and *ex situ* research. This not only allows captive populations to be better understood, but also allows conservation to be targeted at the essential requirements for the species. By having a wide variety of animals present in zoos, education and outreach work can help humans to better understand the natural needs of the animals.

Plants are equally at risk of extinction and it is important to maintain a genealogical record and the ability to regrow plant communities if a particular habitat becomes extinct. These are not only a store of valuable information regarding our biological diversity, but they also serve as educational and research hubs that are important to ensure continued and improved understanding of the complexities of biological communities.

REAL-WORLD EXAMPLE

Svalbard global seed vault, Norway

The Svalbard global seed vault in Norway is a global insurance policy where copies of most of the main crop-seed samples are stored. This allows countries to withdraw seeds from the bank in order to redevelop crops after any disastrous natural or human event. The seed bank currently stores over 1 million varieties of seeds from most countries, including some of the indigenous communities around the world. The purpose of this gene-bank collection is to represent food security for the future. It is located in Norway's Arctic territory Svalbard, in a highly protected and isolated location to protect the valuable contents. The local permafrost provides a fail-safe way to maintain the low temperature required for seed preservation. This project was developed by the Crop Trust to ensure that current crop diversity will be available for the future.

The facility was first opened in 2008, and by 2013 almost a third of the world's recorded gene-bank crop-seed samples were stored at Svalbard. After 2015 and the destruction of other seed stores in areas impacted during the Syrian war, a seed withdrawal was made from Svalbard to regenerate Syria's local seed supplies. Increased use of the facility has help to collect enough varieties of seeds to ensure continued diversity of global crops and the ability for these areas to be replanted following a destruction event.

Habitat-based *in situ* conservation

Habitat-based *in situ* conservation aims to protect and preserve intact and functioning natural habitats, rather than focusing on one species. Habitat-based protection considers the dynamics within the habitats and looks to protect all aspects of the abiotic and biotic elements in their natural, balanced state. This involves identifying the pressures on the habitat and addressing these, while protecting and regenerating the current habitat. Many habitats are under threat from multiple factors such as climate change, local political situations, and economic stressors that drive poaching of valuable species. These factors all need to be addressed on an international, national and local level to be successful.

REAL-WORLD EXAMPLE

San Diego Zoo Wildlife Alliance

Ex situ conservation

San Diego Zoo in California is one of the most research-active zoos in the USA. The Wildlife Alliance takes part in research, protection of orphaned or injured animals, captive-breeding programmes, reintroductions and inter-zoo collaborations to ensure genetic variability when mating animals in the zoo. San Diego Zoo is also home to the 'Frozen Zoo®', which is a store of living and non-living genetic material that could be used to help recover extinct species and preserve the genetic variation we have access to.

In situ conservation

San Diego Zoo Wildlife Alliance has a vast range of *in situ* conservation hubs around the world, including tracking and collecting DNA material from tigers in Sumatra, gorillas in African lowland forest, and elephants and rhino in the African savannah. Each of these projects works in collaboration with local organizations that are protecting the habitats and species already. The zoo is able to direct funding towards these efforts, as well as conduct genetic and behavioural research back at the zoo in the USA. The importance of this type of conservation is that it protects the habitat of the threatened species, as well as working to eliminate issues such as illegal poaching and trade in these animals. It also offers legal protection of the land in the form of national parks, wildlife reserves and other protected statuses. This kind of protection means that these areas are usually covered by relatively comprehensive management plans to ensure areas are adequately monitored and protected. Some areas are open to the public, but with restrictions relating to access to parts of the protected habitat. This allows these reserves to generate income through tourism, which can go back to improving the conservation and protection of the area.

REAL-WORLD EXAMPLE

Yellowstone National Park, USA

In situ conservation

Yellowstone was the first area to be designated as a national park in the USA, in 1872. This makes it probably the most comprehensively managed park. Due to the length of time it has been protected, this around 3,500 square mile park is home to many vulnerable species, including bison, wolverines, brown bears and grizzly bears. The diversity of large mammal species shows there is a stable and diverse community of organisms within the park.

Yellowstone has strict regulations in place for visitors, including that all boats brought into the park must be dried to ensure no foreign species are transferred into the park's lakes. Visitors are restricted to staying in specific areas and are made aware of the dangers present from the park's wildlife. The park charges an entrance fee that supports the large body of rangers who continuously patrol and check specific areas. This not only ensures the safety of visitors and animals, but it also creates significant employment in the local area.

3.3.3 Sometimes a mixed conservation approach is adopted, where both habitat and particular species are considered

It is impossible to separate species decline and the conservation of the habitat that is essential to that species. Without protecting the habitat, it is hard for captive breeding programmes to have viable, stable sites that will ensure the reintroduced individuals have a chance at survival. Some areas require a mixed conservation approach to address the multiple issues impacting a species. Many of the Conservation Hubs that San Diego Zoo has developed around the world address both the *in situ* reasons for the species decline and the *ex situ*. Zoo-based work focuses on captive breeding and understanding the needs, behaviours and genetics of the flagship species for that habitat.

In a mixed conservation approach, it is important to select the right way to develop interest in protecting the habitat. Often a specific animal species is chosen to represent that area or be the focus of conservation efforts. These species are often larger species, some selected for their

◆ **Flagship species** – a species that is selected to be the symbol or image associated with a conservation programme. For instance, the giant panda is the flagship for the World Wildlife Fund (WWF).

◆ **Keystone species** – a species that has a disproportionately large impact on a habitat. For instance, the African elephant is a keystone species due to the impact of its presence on all other aspects of the environment.

charismatic features (**flagship species**) and others for their importance in the overall functioning of that ecosystem (**keystone species**). This allows publicity to be generated relating to the specific species, rather than the habitat itself. However, by protecting keystone species, the stability and resilience of the habitat is significantly improved, indirectly improving habitat conditions for all other associated species. These mixed conservation strategies have been proven to be highly successful in the long-term for both the species and the regeneration of its habitat.

REAL-WORLD EXAMPLE

Chengdu Research Base of Giant Panda Breeding

As the flagship species of the WWF, the giant panda has been the focus of much concern and research over the years. Giant pandas have all of the characteristics of a species that is more prone to extinction:

- They are a solitary species that are very territorial and require several square miles of forest to support their individual needs.
- Individuals come together to mate, but females are only able to get pregnant during one short period of time during spring. This, coupled with the size of their territory, makes the chances of a pair coming together at the correct time highly unlikely. The reduction in their numbers and fragmentation of areas due to agriculture and urban development makes this increasingly difficult.
- Single offspring per birth are normal, and it has even been documented that if two offspring are born, one will be abandoned as females are not able to support and care for two juveniles. Panda females will keep the juvenile with them for between 18 months and three years, meaning females are not likely to become pregnant until after this period due to the inability to collect enough food to sustain a juvenile and a pregnancy. Pandas also can take between four and eight years to mature and be able to breed, which slows down the speed at which populations can develop and increase.
- The giant panda has a very specialized diet, eating only bamboo. Bamboo has a very low nutritional value, requiring giant pandas to spend more of the day eating in order to get enough energy to support their large size. Pandas can thrive in forests with abundant bamboo, however these are increasingly being cleared to make way for urban expansion and agriculture. This is causing these areas to be reduced to sizes that can not sustain panda populations.

All of these factors, along with the impact of the rapidly developing and growing human population in China causing the forest that pandas rely on to become damaged and fragmented, make it hard for pandas to gain enough nutrition or to come together to mate.

The Chengdu Research Base in Sichuan Province was founded in 1987 and was designed to create an area for both *in situ* and *ex situ* research to protect the giant panda and its habitat. It has successfully developed a comprehensive programme over the years that addresses the issue with a combination of captive breeding for reintroduction, protection of bamboo habitat in collaboration with the Chinese government, restoration of current habitat areas, and extensive community outreach and education programmes. These efforts have collectively resulted in the reclassification of the giant panda down from the IUCN Endangered classification to Vulnerable, representing a significant success for the species and the research base.

Further to these clear advantages, research has shown that the conservation of the giant panda's habitat has significantly increased the level of carbon sequestered within the forest, thus also supporting the global targets relating to climate change (sources: **http://m.panda.org.cn/en/conservation/2016-05-18/90.html** and **https://doi.org/10.1038/s41598-017-12843-0**).

> **TOK**
>
> To what extent does a spider deserve the same rights as a panda?

> **Concepts**
>
> **Perspectives**
>
> Significant time and money have been invested in saving the giant panda from extinction. While its reclassification as Vulnerable is a considerable win for the panda, how sustainable is this for the other species in that area? There is clear evidence that the habitat conservation in China that focused on the giant panda has indirectly also protected aspects of the bamboo forest system. However, this may have resulted in many of the other large mammals in the region having declined significantly since conservation efforts focused on the panda. Long-term evidence collected from camera traps within the area has shown declines in leopard, wolf and wild dog species since the 1960s.

3.3.4 The Convention on Biological Diversity is a UN treaty addressing both species-based and habitat-based conservation

The Convention on Biological Diversity (CBD) was originally developed as a result of a meeting regarding concerns over the rising rate of anthropogenically influenced extinctions. Collection of signatory countries opened at the Rio Earth Summit in 1992, closing in 1994. The CBD has three main goals: conservation of biodiversity; the sustainable use of the components of biodiversity; and through the 2010 inclusion of the Nagoya Protocol, the fair and equitable sharing of the benefits from genetic resources.

The CBD recognizes the intrinsic role and value of every species within every habitat, and the importance of protecting every organism, since it is biodiversity that makes habitats more resilient and able to withstand change. The protection of interspecies variation is just as important as protecting species or habitat diversity. The CBD also set out to promote the sustainable use of biological resources, ensuring that current and future generations will be able to benefit from the goods and services provided by biodiversity. These goods and services include food, medicine, clean water and ecosystem services.

The 2010 meeting in Nagoya, Japan added a new protocol, the Nagoya Protocol, to the treaty. This emphasized the fair and equitable sharing of the benefits arising from the utilization of genetic resources, particularly access to genetic resources and the sharing of benefits arising from their use. This led to the expansion of establishments like the Svalbard seed vault and the 'Frozen Zoo®' in San Diego Zoo, which are preserving and protecting large volumes of diverse genetic material from the plant and animal species present around the world today.

The convention established a framework for action that includes specific commitments from the parties (countries that have ratified the CBD) to undertake various measures, such as developing conservation action plans and national strategies to promote the understanding of the sustainable use of resources. These guidelines and decisions are discussed at the annual Conference of Parties (COP) meeting. COP 10 resulted in the Nagoya Protocol requiring countries to make their genetic material and localized knowledge available to be shared with other countries. International collaboration and cooperation is encouraged in order to tackle biodiversity issues and so representatives of signed countries regularly meet to discuss unified action against biodiversity loss.

3.3.5 Habitat-conservation strategies protect species by conservation of their natural environment. This may require protection of wild areas or active management

Most species are threatened by more than one single issue, including adjusting to climate change challenges and poaching. The impact of these factors is exacerbated if the habitat the species lives in is also under its own pressures, such as pollution and human encroachment. Therefore, by protecting the habitat, all of the species within it have a greater chance of survival. Designating and protecting areas such as national parks and wildlife reserves involves developing and implementing strategic plans to address and monitor the issues relevant to that habitat. Within these protected areas, human activities are often regulated to minimize disturbances to ensure the preservation of wildlife, the safety of visitors and the development of natural processes.

Encouraging tourism around nature reserves is one way in which these areas can develop and become financially stable. However, this requires careful balance in order to not disrupt the habitat or animals too much or change their natural behaviour.

A key element of habitat management is habitat restoration, as many threatened areas have been increasingly disturbed by human activity for many years. Habitat restoration involves returning the habitat to its natural state, which enhances its capacity to support native species and ecological functions. However, many of the remaining habitats are severely fragmented and some are too small to be viable for some species. In these situations the habitat either needs to be expanded, or it needs to have some connection with other habitat patches. This way there is potential for a flow of genes between populations, reducing the potential drops in genetic diversity that can occur in completely isolated populations of animals. To increase connectivity, ecological corridors can be created and maintained to facilitate movement between the patches and help the population be more resilient against changing conditions.

Other elements that need to be considered when designating an area as a reserve are the proximity to human habitation and how the land surrounding the reserve is being used. This will allow a full picture of the potential issues that are likely to impact on the reserve. For instance, if a protected area is in close proximity to a city, the air and water pollution coming from the city might impact on the habitat. Equally, if the reserve is surrounded by agricultural activities there may be issues with the level of fertilizers and pesticides in the water coming into the reserve. Careful monitoring of water for possible eutrophication would be necessary as part of the management plan for such a reserve.

Areas that are designated for some kind of protection are required to actively manage the habitat, usually to halt the natural progress of succession, as well as to reverse the impact of threats on the species present in that particular stage of succession in that habitat. This requires a good understanding of the successional elements of the habitat and the use of management to halt this change. For instance, an area of protected forest may be at an intermediary stage of succession and have many abnormally large populations of herbivores that are unsustainably changing the floral composition. In that case, active culling of species may need to take place in order to rebalance the herbivore numbers so that the vegetation is not overconsumed.

This requires the continuous active input of management of species to ensure that the balance in the habitat remains constant. This can be difficult to achieve, but there are some areas that have used the need for population control to earn extra income for the nature reserve. This is highly controversial for some, who believe that killing another animal is wrong.

Many reserves employ active management in order to maintain a status quo. South African reserves have been using this model which allows hunting for many years and are able to effectively manage tourism, animal populations and conservation that includes local community involvement. Management of the experience of the visitor is just as important to the reserve as managing the habitat to maintain the current diversity. This management includes maintenance of visitor areas with adequate signage, provision of facilities, safety advice and protection from the wildlife within the reserve.

Some areas require protection against animal invaders as well as poachers, and therefore some protected reserves are island-based and others even employ the use of pest-exclusion fencing. This is a protection that is commonly employed in New Zealand due to the large populations of mammals such as possums and rats that prey on the high number of endemic species of birds and mammals. Although successful at protecting endemic species in these reserves, these reserves have essentially become isolated habitat patches. Not all ecosanctuaries are fenced, and in those that are not a lack of expertise and leadership in effective pest control threatens the effectiveness of these types of sanctuaries.

REAL-WORLD EXAMPLE

Zealandia Te Māra a Tāne, Wellington, New Zealand

Zealandia Te Māra a Tāne is the world's oldest fenced ecosanctuary, and it has a comprehensive 500-year plan to return the area to its native state. The reserve is surrounded by a fence over 8 km long that is designed to keep out introduced mammal species that are not part of the diverse native fauna. Currently, 18 species have been reintroduced to the sanctuary, some of which have not been recorded on the island for over 100 years. Biosecurity checks on people, bags, vehicles, for example, are carried out on visitors to the sanctuary to ensure that no biological material accidentally enters the area.

The effectiveness of the fence has been developed over time in relation to the climbing, jumping and burrowing capabilities of the mammals being excluded. When the fence was established, it was effective at blocking the entrance of these mammals. However, to ensure none of these species were already present within the sanctuary area, a comprehensive eradication plan was needed both during and immediately after the fence was built. Ambitious plans are also looking at replacing some of the current pine with ancient podocarp, a hardwood species that is native to New Zealand. Careful planning has been conducted to ensure that these changes take place on a scale that does not cause shifts in the habitat conditions that would jeopardize the survival of the species. During the process of transforming the habitat, research is being conducted to monitor progress and impacts. Community engagement and education programmes ensure that the importance of these areas is widely recognized and supported.

IUCN's Nature 2030: One nature, one future

The importance of community inclusion in conservation has driven a new initiative launched by the IUCN in 2021. Nature 2030 creates a programme that incorporates humans as a central focus in the management of water, land, oceans and the climate. The project aims to empower communities through education in conservation practices, and includes local community members within all decision-making processes. It aims to reconnect humans and the environment to restore habitats back to a natural balance.

The programme is aligned with the Sustainable Development Goals and aims to aid in the completion of a number of these goals. SDGs 6, 13, 14 and 15 (relating to clean water and sanitation, climate action, life below water, and life on land) have been identified as essential to create healthy ecosystems, which are needed to develop healthy human populations and economies. Empowering the youth of the planet is a key focus, with the recognition that they will be our future managers and leaders. Recognition of developing more sustainable city ecosystems is essential in moving towards a more sustainable future.

> **Inquiry process**
>
> ### Inquiry 3: Concluding and evaluating
>
> When evaluating something, you need to look at both the positive and negative aspects, before making some type of judgement. This can be carried out in relation to experimental methodology and can also be used to look at how effective a management practice has been.
>
> The use of pest-proof fencing in the management of ecosanctuaries in New Zealand has been widely praised and is considered a highly effective conservation strategy. Ecosanctuaries, also known as wildlife sanctuaries or predator-free areas, are enclosed spaces that aim to protect and restore native ecosystems by eradicating or excluding introduced predators and pests. However, there are as many cons as there are pros to the implementation of this practice, as shown in Table 3.9.
>
> ■ **Table 3.9** Comparing the pros and cons of using pest-exclusion fencing around ecosanctuaries in New Zealand
>
> | Pros | Cons |
> |---|---|
> | Predator eradication | High cost to maintain fence |
> | Biodiversity restoration | Limited size |
> | Native ecosystem recovery | Genetic isolation within sanctuary |
> | Facilitate reintroductions into the wild | Easier invasive species management |
> | Easier research and monitoring of species | Ethics of eliminating predators |
> | Tourism and education can generate income | |
>
> When comparing pros and cons, both the number of each and the relative importance of each need to be taken into account to allow a critical evaluation to be made.

3.3.6 Effective conservation of biodiversity in nature reserves and national parks depends on an understanding of the biology of target species, and of the effect of the size and shape of conservation areas

Developing effective management of a protected area requires a detailed understanding of the species that are present and that are being protected within that area. Each species has its own specific niche that it needs to be able to survive, some of which are very specific, and if they are not addressed that species will not thrive.

The size of a protected area is one of the factors that needs to be carefully considered when determining its footprint. A reserve designed to protect a viable population of cheetahs would need to be large enough to accommodate a home range of around 800 km^2, whereas a reserve designed to protect the pangolin does not need to be as extensive as their home range is just over 12 km^2.

Shape, use of buffer zones and transition zones, and connectivity have long been recognized as essential factors required to create an effective protected area. The shape of a reserve is related to the edge effect. Some shapes, such as a circle, have low edge-to-centre ratios, providing a large central area that is not directly linked to the non-reserve habitat.

If the edges of these reserves are hard boundaries surrounded by different ecosystems, then these edges act as transition zones where there is a mix of plant and animal species from both habitats. In addition, some species are specialized to these edge habitats and the conditions present within them.

Since 1971 UNESCO has been developing a network of biosphere reserves around the world. These biosphere reserves are designed to employ all of the relevant known management practices within vulnerable habitats around the world. Each reserve has three levels of zonation: the transition zone; the buffer zone; and the core area (see Figure 3.20). These ensure that the core area is fully protected and can represent a pristine habitat.

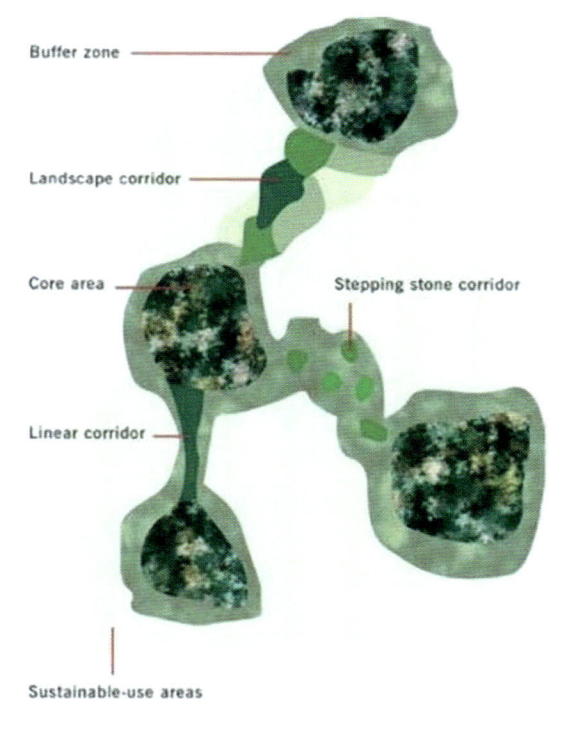

■ **Figure 3.20** Representation of an ecological network designed to link UNESCO biosphere reserve sites

- Transition areas on the edges of protected habitats are recognized as habitats within their own right and are often considered to be not fully one habitat type or another. Therefore, these areas may have a combination of flora and fauna from both of the joining habitats.
- Buffer zones offer protection from poachers as it is more difficult for them to penetrate the core part of the habitat.
- The core area of a protected habitat is therefore buffered against the impact of the outer habitat and represents the real core of the reserve.

3.3 Conservation and regeneration

Alongside the development of individual reserve design, UNESCO recognized the need for connectivity and developed a series of networks that connect these areas together using corridors between the individual reserves. This connectivity is essential to reduce the potential of genetic inbreeding that would ultimately reduce the resilience of a species.

REAL-WORLD EXAMPLE

The Estonian Green Network

In 1983 Estonia was the first country to develop an ecological network, connecting up its fragmented protected habitats. Estonia has high biodiversity due to the lack of overall economic development and the disturbance in the country during the Soviet era. Currently around 50 per cent of the country is covered by the Estonian Green Network, which maintains strict controls over the land use within the network to protect its native habitats and species. Through the development of this network, strategic planning relates to balancing the activities, resource use and development within these areas. Promotion of sustainable behaviours such as recycling and water-use management, and providing recreational opportunities for local residents have also helped to connect the communities to the management of these areas.

3.3.7 Natural processes in ecosystems can be regenerated by rewilding

Over many years, cultivation has changed the natural landscapes in order to provide sufficient food supply for our ever-expanding human populations. The diversity of the natural flora has also been changed and the diversity of wild species has dropped in many regions of the world as a direct result of agriculture. This process has modified many areas beyond recognition, resulting in the loss of many larger mammals that are considered potentially harmful to humans. This has caused scientists around the world to look at the benefits of returning areas to the natural, wild state that was present prior to the adoption of cultivation in that area.

The benefits of rewilding are far reaching, as the restoration of native plants and animals helps to repair the damaged soil, nutrient and water systems and cycles that are part of the abiotic. Repairing this balance allows the developing ecosystem to act as a sink for carbon, helping to address wider global climate change issues. Long-term success relies on developing education initiatives and encouraging locals and tourists to mindfully engage with the environment without creating damage. Rewilding represents a holistic approach to land management that seeks to strike a balance between human needs and the preservation of the natural world, ensuring a sustainable future for both.

Controversially, returning habitats to their wild ancestral roots has meant the reintroduction of some large predator species. Reintroductions of carnivores back into their native habitats has been proven to improve the quality of all levels of the ecosystem, as shown with the successful reintroduction of the gray wolf to Yellowstone National Park, USA (see Real-world Example in Topic 2.5.9, page 186). This has been controversial as this reintroduction was so successful that the gray wolf was removed from the protection of the Endangered Species Act in the USA, resulting a dramatic increase in the volume of people hunting and killing the wolves.

> **REAL-WORLD EXAMPLE**
>
> **Rewilding Hinewai Reserve in New Zealand, 1987**
>
> Hinewai Reserve in New Zealand is a privately owned reserve that was founded in 1987, when 109 hectares of forest was bought for the purpose of restoring the habitat back to its native state. In 2023 the reserve covered 1,250 ha of forest that is under rewilding management. Initial goals included clearing invasive plant species, followed by a minimally invasive management model. The reserve is open to visitors, but there are strict guidelines for visitors to reduce the possibility of fire. The removal of plant specimens, littering and camping are not allowed.
>
> Over the nearly 40 years that the reserve has been developing, the single New Zealander owner has provided the environment with the conditions and help to develop into a biodiverse, native and natural forest system.

> **ATL ACTIVITY**
>
> As a class, host a screening of *Fools and Dreamers: Regenerating a Native Forest*, an award-winning documentary following the journey of Dr Hugh Wilson, the owner of the Hinewai Reserve.
>
> Go to the link **https://happenfilms.com/fools-and-dreamers** and apply to host a screening. This would be a great Earth Day activity.
>
> Create posters for the event and inform students about the importance of biodiversity and repairing our native ecosystems.

Successful rewilding incorporates a number of key factors. These include the reintroduction of top predators back into the community, as their presence creates a trophic cascade that forces the different trophic levels to rebalance to accommodate the top predator. At the other end of the spectrum, the reintroduction of large herbivores such as bison creates patchy disturbance through grazing, thus increasing the overall diversity within the habitat. Restoration of the nutrient cycle and water system flow within a rewilded area is essential for the habitat to maintain balance once the process of rewilding has been completed. This includes allowing dead plants and animals to naturally break down *in situ* and return their valuable nutrients to the soil. Rewilding initiatives must consider the ecological history and current conditions of the area as well as the potential impacts on surrounding human communities, and the needs of both wildlife and people for sustainable coexistence.

3.3.8 Conservation and regeneration measures can be used to reverse the decline in biodiversity to ensure a safe operating space for humanity within the biodiversity planetary boundary

Habitat conservation is a vital component in protecting many of our animal species, but this is not always enough to protect the habitat as it is at that time. As habitats become overused and under-supported, the conditions change to take account of the change in species diversity. These changes reduce the diversity in food webs in the ecosystem to only some of the species interactions, connections and relationships that were present before the habitat was degraded. Therefore, conserving a degraded habitat does not create a long-term solution and it is necessary to focus on regenerating the habitat before conservation measures are implemented.

Before protecting the habitat, it must first be restored to its native, balanced state. Restoration involves limited human interference once invasive species and pollution issues have been dealt with. This allows natural regeneration of the plant species that were once present. This in turn will attract native animals back to the area and many of the natural feedback systems will eventually return. This minimal, hands-off approach allows the natural repair of the water, nutrient and carbon cycles, which would have been damaged through management.

Once the regeneration stage is complete the conservation phase of a project can begin, to allow the ecosystem to be maintained. Once the balance has been regenerated, these habitats should be able to self-regulate to maintain the status quo without much human intervention.

The ultimate goal of any environmental management strategy is to reduce impacts on the nine planetary boundaries so that, as indicated on the doughnut economics model, humans can return to living within the 'safe and just space for humanity'. Biodiversity is one of the planetary boundaries, and careful development of *in situ* and *ex situ* conservation can address what is already present within our ecosystems. This is then coupled with regeneration to restore what was once in that ecosystem. This helps bring back heirloom species of plants and animals to help return the area's biodiversity and stability. If this is employed widely in different ecosystems around the world, there is a possibility that we could slow the rapid decline in biodiversity.

To gain real improvement in biodiversity management, every area should employ a complex suite of management techniques that are tailored to the specific needs of that habitat, in that part of the world, at that time. The type, severity and impact of negative issues vary for each habitat, so complex decision making and implementation of local, national and international requirements is essential for an area to be truly successful.

> **Common mistake**
>
> Students commonly focus on one issue at a time. The real success in conservation is through combining strategies involving many stakeholders and addressing multiple issues at the same time.

3.3.9 Environmental perspectives and value systems can impact the choice of conservation strategies selected by a society

The success of conservation and regeneration is clearly reliant on a diverse approach to address the myriad issues. Legal protection of habitats and species attracts funding that allows both *in situ* and *ex situ* research to be conducted. This allows the development of the management plan over time to enable the protection to be targeted towards specific species-based issues. Through creating a network of protected areas, the long-term health and resilience of populations will improve. Over many years, actively managed reserves are able to create a unique balance of species that they can maintain to protect the specific species adapted to that stage of habitat succession.

How conservation is achieved is very much dependent on the environmental value system of the decision-makers. A country such as Bhutan, which measures progress through Gross National Happiness, would address conservation differently to a country such as Dubai, where economic wealth is the focus of development in the country. Bhutan would be more likely to adopt ecocentric approaches to management, whereas Dubai would be more likely to approach management from an anthropogenic or technocentric viewpoint.

For instance, Dubai has created the Sustainable City, where using innovation it has been able to develop a self-regulating environment that addresses pollution, electricity generation, natural cooling through design and more.

The Constitution of Bhutan mandates there must be at least 60 per cent forest coverage in the country at all times. Current forest coverage is around 71 per cent, contributing significantly to its carbon-negative status. However, as is the case in much of the field of conservation and biodiversity, the reality is that there needs to be a variety of approaches around the world. Without innovation coming from more technocentric-focused nations, developments in genetics and other research would not be prioritized. Nations employing more ecocentric solutions need the knowledge developed from the technocentric discoveries to ensure that *in situ* conservation correctly addresses species and habitat issues. Finally, without any anthropocentric focus these strategies would be less likely to be effective as change often requires changes in human behaviours and actions in order to be sustainable.

(HL) 3.3.10 Success in conserving and restoring biodiversity by international governmental and non-governmental organizations depends on their use of media, speed of response, diplomatic constraints, financial resources and political influence

The success of any conservation or restoration project is dependent on numerous factors. To ensure adequate protection, there needs to be many layers of support. For instance:

- Protection of a habitat: At the local level this involves and empowers the people living within these areas. This provides the local community with work and stability, and means they are less likely to be open to violating protections.
- Governments need to provide financial support and law enforcement for any infringements on the local and international laws: International organizations are often responsible for the development of policy, but not for its enforcement. This is the role of local governments. This is a vital link that ensures legal protections are effective. Local funding can also be redirected to support areas much faster than can be achieved at an international level.
- International organizations must address legal and further financial aid to ensure those countries with large numbers of protected areas do not have to financially support the conservation of all of these areas: This is especially important for the vast majority of biodiversity hotspots that are located in developing countries. International policies and agreements have to be viable and flexible enough to fit in any country's situation.

With this triple layer of support, these habitats and species stand a much better chance of being protected from further damage and depletion. One of the reasons for the importance of having multiple levels of support is the fact that each level has different benefits, which support each other in the protection of the same space. Table 3.10 shows how non-governmental organizations, such as Greenpeace, are essential in creating rapid, widely publicized responses to environmental and social issues that put pressure on governmental organizations at a local and national level. While the NGOs have no legal jurisdiction in terms of protection for species, environments or communities, the pressure they can apply to local governments results in the development of national law, or the inclusion of a new site under current protection strategies.

■ Table 3.10 A comparison of the benefits and limitations of protection at differing levels, from international organizations to NGOs

	International organizations (IOs)	Governmental organizations (GOs)	(International) non-governmental organizations (NGOs and INGOs)
Media	Social media is used by some IOs to advertise and raise awareness, but this is used mostly to present achievements.	GOs make some use of the mainstream media news for official campaigning for support in developing environmental programmes.	NGOs rely on the media to help them relay their message to a wider network. Social media has become a major part of their organizations as they are able to show the problems and the impacts of donations and their work.

	International organizations (IOs)	Governmental organizations (GOs)	(International) non-governmental organizations (NGOs and INGOs)
Response time	Slow response time due to the constraints put on new issues, unless it is a disaster situation. Creation of new international law has to go through many official international meetings, which can take years of planning to become legal.	Medium-scale response time. Evidence of a problem needs to be brought to the attention of national governments, often by NGOs. Governments of different countries will have differing priorities, and environmental issues may not be high on the agenda for all.	Rapid response due to the lack of constraints. This means that NGOs often mobilize aid, raise awareness, and pressure the GOs and IOs to create legal support for environmental issues.
Diplomatic constraints	Membership of organizations is voluntary, so not all countries have to abide by the laws and regulations.	Decisions have to go through many steps before they are officially put in place.	There are no diplomatic constraints. Some NGOs purposely break laws to get media exposure for issues.
Financial resources	Considerable funding is available to be directed at the areas with the greatest needs.	Financial resources depend on the country in question. Some countries' governments have comprehensive funding schemes to help protect the environment, while for others it might not be a priority.	NGOs are often reliant upon donations and fundraising to carry out their work. Financial constraints are common for NGOs. There are grants available from the IOs and GOs, but this takes time.
Political influence	Production of international law that must be upheld by national authorities. These organizations can have a direct impact on policy development and decision making at an international, national and local level.	International laws need to be upheld by national-level governmental organizations. This can be where there are difficulties as some countries are impacted by corruption that may hinder the implementation of strategies developed at the governmental level.	No direct political power, but they use advocacy to influence governments at a local and national level.
Examples	UNEP IUCN	Ministry of Environment, Forest and Climate Change (Bangladesh) Fish and Wildlife Service (United States)	Greenpeace and WWF (NGOs) EcoThailand Foundation (NGO)

For effective conservation, rapid response to emerging threats, such as habitat destruction or species decline, is crucial. Quick action can prevent irreversible damage, minimize species loss, and support ecosystem recovery. This requires a team on the ground who are able to respond to and adapt management to mitigate any local environmental issue that could be negatively impacting a protected area. For instance, bushfires regularly affect large parts of Australia, and the rescue of animals caught in the fires is mobilized by local organizations and communities.

Adequate funding is crucial for supporting conservation initiatives including habitat protection, research, restoration and community engagement. A lack of financial resources can hinder effective strategies, therefore partnerships between governmental, non-governmental and international organizations can leverage diverse expertise and essential resources. If this is backed up by strong political support it leads to the establishment of protected areas, conservation laws, and international agreements that can enhance the efficiency and impact of conservation initiatives.

3.3.11 Positive feedback loops that enhance biodiversity and promote ecosystem equilibrium can be triggered by rewilding and habitat-restoration efforts

Rewilding and habitat restoration aim to restore natural ecological processes, reintroduce key species, and return the ecosystem to a condition where it is able to maintain a healthy natural balance of species. Habitat restoration can be highly complex, especially if the area has been contaminated by industrial processes in the past. This type of habitat restoration needs to address contamination prior to addressing the reintroduction of species to the area. This can involve reintroducing keystone species, predators and herbivores that each play a crucial role in shaping ecosystems. These species can trigger cascading effects through trophic levels. Any introduction of predator populations affects herbivore populations, which in turn impacts plant communities. For instance, the reintroduction of wolves in Yellowstone National Park led to a decrease in elk populations, which allowed vegetation to recover, attracting more birds, insects, and other species that thrive in diverse habitats (see Real-world Example in Topic 2.5.9, page 186). These cascades can lead to more diverse and complex vegetation, which creates more diverse and suitable environments for supporting a wider array of species. This increase in species can lead to increased genetic diversity and various ecological interactions, resulting in a more resilient ecosystem.

> **Concepts**
>
> **Systems and models**
>
> As ecosystem functioning is predictable in terms of energy transfer and cause and effect, it is relatively easy to model. Every system has its limiting factors, such as space, nutrients and water, and so it is easy to predict what will happen as populations change through successional stages. This allows rewilding and restoration to be carefully planned in order to ensure they are successful.

Grey wolf is reintroduced in Yellowstone National Park, USA

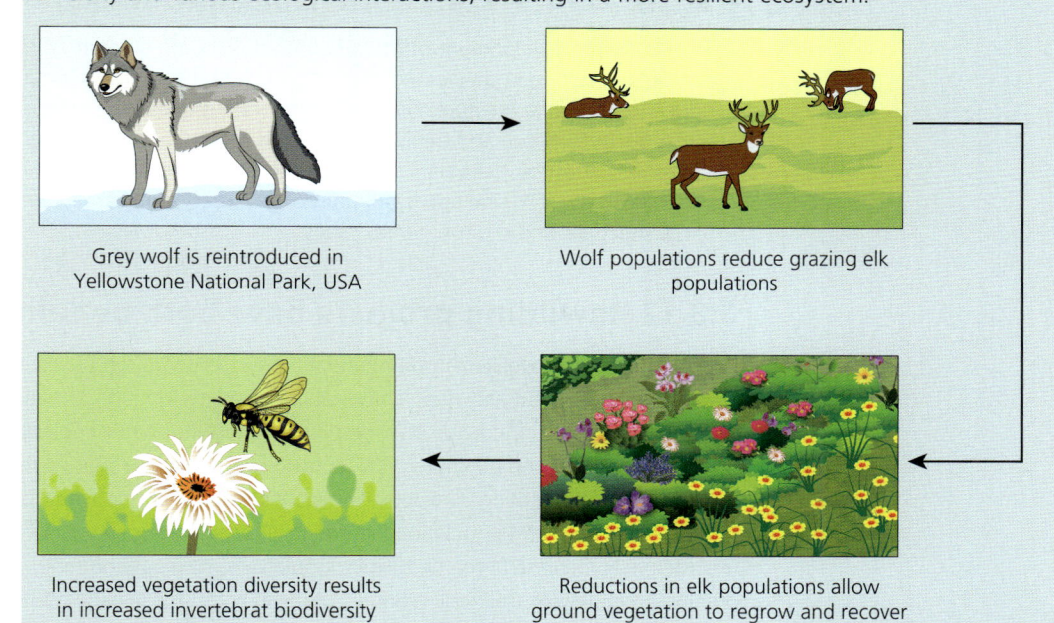

Wolf populations reduce grazing elk populations

Increased vegetation diversity results in increased invertebrat biodiversity

Reductions in elk populations allow ground vegetation to regrow and recover

■ **Figure 3.21** Diagram showing the impact of reintroducing wolves into Yellowstone National Park

Over time, the rewilding or restoration of a habitat improves the soil health and nutrient cycling in the system. This leads to more plant species being able to colonize the area, increasing primary producer diversity and productivity, thus allowing the area to support more robust consumer populations. Rewilded and restored ecosystems tend to be more resilient to the impacts of climate change due to increased diversity and adaptive capacity, allowing these ecosystems to withstand disturbances and maintain equilibrium. By promoting natural interactions and self-regulation within ecosystems, these efforts contribute to enhancing biodiversity, restoring ecosystem functions, and maintaining equilibrium in the face of environmental challenges.

3.3 Conservation and regeneration

HL.a.3–6: Environmental law

Environmental laws take a long time to develop at both national and international levels due to political lobbying and the input of powerful economic stakeholders, which can delay and sway political decisions. This is particularly the case when international law has to be implemented at a national level, where it may not be a priority of the current government, which can reduce its local effectiveness.

> **ATL ACTIVITY**
> - Research the environmental laws of your country. How much focus do they have over other major issues that the government is addressing?
> - Investigate a country that is very different to the country you live in. How does that country prioritize environmental laws and issues?
> - Compare and contrast your two countries with others in your class. Do you see any patterns?

HL.b.13, HL.b.17: Environmental economics

The economic development of many countries is having a negative impact on biodiversity hotspots, resulting in species extinctions and habitat fragmentation in areas that are already under considerable pressure. Many of these areas are also the location of abundant natural resources both above and below the soil. This makes these areas highly vulnerable and brings into question the balance between environmental justice and development to alleviate widespread poverty. The application of doughnut economics will direct attention towards these social shortfalls and environmental issues.

3.3.12 Rewilding projects have both benefits and limitations

Rewilding projects involve restoring natural ecosystems through the reintroduction of native species, restoring habitats, and allowing ecological processes to function as naturally as possible. The level of intervention and support dictate the type of rewilding that is implemented. These projects often lead to increased biodiversity by reintroducing keystone species that have a cascading effect on ecosystems, benefiting all trophic levels. This can lead to increased biodiversity due to the regulation of herbivore populations, ensuring that vegetation levels did not become overgrazed by unsustainable increase in the herbivore populations, as would normally occur without these natural predator populations (see Real-world Example in Topic 2.5.9, page 186).

Restored ecosystems provide essential services such as water purification, carbon sequestration and soil retention, which all lead to greater resilience to disturbance and climate change within the area. For instance, the Knepp Estate Rewilding Project in the UK has improved ecosystem resilience and increased biodiversity through minimal human intervention. Explore their rewilding program at **https://knepp.co.uk**.

However, rewilding can sometimes clash with human activities such as agriculture or livestock grazing, leading to conflicts. The reintroduction of apex predators like wolves can result in concerns about livestock predation, which can affect local communities. Farmers may find themselves losing livestock or crops to these new predators, and some incidences of human attacks have been recorded. In some cases, the original habitat for certain species may no longer exist or may have been significantly altered. This can hinder successful reintroduction efforts. The California condor reintroduction, for example, faced challenges due to habitat loss and lead poisoning. Introducing new species can also lead to unintended consequences, such as the spread of diseases or competition with native species that have adjusted to a habitat without the new species present.

HL.a.2: Environmental law

Rewilding projects are often parts of official schemes developed on a national governmental level and are therefore usually legally protected. This protection is essential to ensure the continued protection and support of ecosystems that have previously disappeared from our landscapes due to human influences.

HL.b.15, HL.b.16: Environmental economics

By rewilding habitats, countries are improving their biocapacity, which helps to balance out the ecological footprint of a country. This can be economically beneficial for the protection of these areas, which in turn not only protect the overall ecosystem but can also lead to increased income through tourism, eventually having the potential to be beneficial to local economies.

HL.c.5, HL.c.10: Environmental ethics

The selection of species for reintroduction may raise ethical questions. For instance, rewilding projects involving captive breeding or cloning might raise concerns about the 'naturalness' of the reintroduced populations. Reintroduction of apex predators into areas where species have evolved without their presence may result in other species becoming endangered. However, the lens of consequentialism could easily argue that restoration of these areas is for the greater good and therefore the impacts on other species can be seen as a justifiable consequence.

ATL ACTIVITY

Discuss your feelings and opinions on the reintroduction of apex predators into a system where they have previously been resident.

3.3.13 The success of conservation or regeneration measures needs to be assessed

The success of conservation or regeneration projects needs to be assessed to ensure that the intended goals are being met, resources are being used effectively, and adaptive management strategies can be implemented if necessary. Assessment provides valuable information for decision making, helps identify areas for improvement, and ensures that conservation efforts are achieving their desired outcomes. Assessments provide a way to quantitatively and qualitatively measure the impact of conservation or regeneration efforts on biodiversity, ecosystem health and other relevant factors, and so to provide objective data to guide decision making. This ensures that future actions are grounded in evidence and best practice. This monitoring should continue throughout the lifetime of the habitat to ensure that these efforts do not stop, with the risk of areas returning to the damage they had previously encountered.

In order to assess the success of the project, before its implementation the development and baseline data of the populations or habitats to be protected need to be determined. Without this careful planning and documentation, it is difficult to determine what is and what isn't working best for the situation, environment and specific area. Once baseline data and clear strategies are in place, the project can be initiated. During the management of such a project, regular monitoring allows for the systematic and regular standardized collection of data that will allow comparisons to be made regarding species diversity, relative abundance, and any other indicators that have been selected to represent levels of success. These data can then be statistically analysed to allow for reflection on how effective a strategy is in that area. Stakeholders and funders can then be regularly updated on progress in the area's conservation.

> **REAL-WORLD EXAMPLE**
>
> ### FORRU-CMU, Chiang Mai, Thailand
>
> The Forest Restoration Unit (FORRU) of Chiang Mai University has been working to restore the forests of northern Thailand since 1994. Projects have been developed that address the restoration of various forest types that have been impacted by clearance for agriculture, tourism development and wild fires. These areas are home to a number of ethnic minority communities, and FORRU has developed many projects that are empowering these communities to be the guardians of these forest areas. This has involved extensive development of education programmes, not only in Thai, but in the local dialects of the communities in order to allow access to all.
>
> The planning and development stages, goals and progress of all projects are openly available via their website, providing clear frameworks for other projects to work with. This project brings together local indigenous communities, academic researchers and funding to develop the programmes (source: www.forru.org).
>
> ■ **Figure 3.22** FORRU's Nursery Officer, Yoi, teaching college instructors from southern Myanmar about various forest restoration activities

● HL.a.2: Environmental law

Areas that are designated as reserves or conservation zones will often fall under one of the many official designations, such as Ramsar sites or IUCN-protected habitats. This means that these areas will have more regulations in place, but they will also be monitored to ensure that goals are being met.

● HL.c.3: Environmental ethics

Many of the ethics behind protecting species relate to the opinion of the individual. What is worthy of protection and what is not can be a contentious issue. Some people operate with the belief that humans are essentially part of the natural system with a specific role as the custodians of the planet. This belief would result in different approaches to conservation than if it were developed by people with the belief that the natural world is there to provide for the human population.

◆ **Ecotourism** – responsible travel to natural areas, which conserves the environment, sustains the well-being of the local people, and involves interpretation and education (source: The International Ecotourism Society, 2015).

3.3.14 Ecotourism can increase interdependence of local communities and increase biodiversity by generating income and providing funds for protecting areas, but there can also be negative societal and ecological impacts

Ecotourism, when planned and managed carefully, has the potential to create a positive cycle of benefits for both local communities and biodiversity conservation. However, it's important to recognize that there can be both positive and negative impacts associated with ecotourism. Costa Rica is one of the countries that has been able to develop a sustainable network of ecotourism that attracts visitors from all over the world (see Figure 3.23).

■ **Figure 3.23** Ecotourism is a key factor in the success of many parts of Costa Rica

Ecotourism provides the core principles of sustainability through tourism, which in turn supports local communities and the environment. Ecotourism brings valuable financial help to these areas without having to apply for any grants for restoration. For successful management of these areas, there needs to be an awareness of the positive and negative impacts, how they interact and sometimes block successful conservation, and how these elements can be managed.

■ **Table 3.11** Comparison of the positive and negative impacts relating to ecotourism

Positive effects	Negative effects
Revenue generated from ecotourism can be invested in conservation efforts, including habitat restoration, anti-poaching initiatives and education programmes.	High tourist numbers can lead to overcrowding and habitat degradation, impacting local ecosystems and wildlife. Trails, infrastructure and waste can negatively affect biodiversity.
Ecotourism provides an alternative source of income for local communities, reducing their reliance on activities that may harm the environment, such as logging or poaching.	Tourist demands for resources such as water, energy and food can strain local ecosystems and put additional stress on natural resources.
The involvement of local communities in decision making and activities leads to a sense of ownership and stewardship over the protected areas.	Excessive tourism can alter local cultures and lead to the commodification of traditional practices, eroding their authenticity.
Ecotourism raises awareness among tourists about the importance of biodiversity and conservation. This education can contribute to a greater understanding of environmental issues.	Overreliance on ecotourism as the primary source of income can make communities vulnerable to fluctuations in tourist numbers, economic downturns or changes in travel trends.
Ecotourism encourages the preservation of local cultures and traditional practices, as these aspects often become attractions for tourists.	If not managed properly, ecotourism can exacerbate inequalities within local communities, as benefits may not be distributed equitably.

Link

Chapter 1.1 (page 2) looks into how our individual values are influenced as we grow up and how these values can shape the way we view certain issues and our approach to their management.

In order to develop without impacting the local communities and environment, sustainable practices can be applied to ease the negative impacts, such as limiting tourist numbers, enforcing responsible behaviour and adopting eco-friendly infrastructure. Community involvement and training is essential to the success of these restoration projects as these communities are continuously interacting with the natural environment. Tourism that provides education and awareness helps not only the environment, but local communities as well. This can be in the form of developing new local skills and crafts to create additional income.

HL.a.6: Environmental law

Success of conservation is reliant on understanding local traditions, systems and behaviours. Therefore, it is vital that laws are drafted at the local level as well as the international level. International laws need to be informed by local strategy development to ensure that the rights and beliefs of communities are taken into account.

HL.b.7: Environmental economics

Placing value on environmental issues or entities is difficult due to the fact that value is perceived differently by different people. Perception of an issue can be influenced by the value system a person developed throughout their lives. Therefore, placing economic value on natural resources is hard to regulate, which makes it difficult to place value on the loss of resources through natural or anthropogenic means.

HL.c.12: Environmental ethics

An ecocentric value system would view the appeal to nature as having significant value and importance, whereas an anthropocentric value system would focus mostly on how the natural environment can support and enhance our lives. Each viewpoint is relevant and valid and this can lead to disagreements in terms of developing management programmes for some protected areas.

REVIEW QUESTIONS

1. State the name for conservation that takes place in a zoo, gene bank or seed bank.
2. Explain why mixed conservation strategies are so successful.
3. Explain the importance of a flagship species.
4. Describe why species unique to New Zealand are particularly vulnerable to extinction.
5. Explain how rewilding is different to not actively managing an area.
6. Explain how rewilding can increase the resilience of food webs.
7. Discuss the need for a mixed support approach, from governmental and non-governmental organizations, when managing protected habitats.
8. Explain the benefit of being an NGO when developing an environmental campaign.
9. Describe the impact of returning an apex predator to a natural habitat following a long period of absence.
10. Use an example to discuss the pros and cons of ecotourism.

EXAM-STYLE QUESTIONS

1. Explain two factors that will make a species more vulnerable to extinction. [2 marks]
2. Determine ways in which IOs, GOs and NGOs could work together to mobilize conservation protection for newly damaged habitats. [5 marks]
3. Using named examples, discuss the conflict with conservation in developing countries and suggest a possible solution. [7 marks]
4. Explain why Australia has such high biodiversity. [5 marks]
5. Discuss the success of fenced eco sanctuaries in New Zealand. [6 marks]

Theme 4
Water

4.1 Water systems

> **Guiding question**
>
> - How do water systems support life on Earth and how do they interact with other systems, such as the carbon cycle?

SYLLABUS CONTENT

This chapter covers the following syllabus content:
- 4.1.1 Movements of water in the hydrosphere are driven by solar radiation and gravity.
- 4.1.2 The global hydrological cycle operates as a system with stores and flows.
- 4.1.3 The main stores in the hydrological cycle are the oceans (96.5 per cent), glaciers and ice caps (1.7 per cent), groundwater (1.7 per cent), surface freshwater (0.02 per cent), atmosphere (0.001 per cent) and organisms (0.0001 per cent).
- 4.1.4 Flows in the hydrological cycle include transpiration, sublimation, evaporation, condensation, advection, precipitation, melting, freezing, surface runoff, infiltration, percolation, streamflow and groundwater flow.
- 4.1.5 Human activities, such as agriculture, deforestation and urbanization, can alter these flows and stores.
- 4.1.6 The steady state of any water body can be demonstrated through flow diagrams of inputs and outputs.

HL ONLY
- 4.1.7 Water has unique physical and chemical properties that support and sustain life.
- 4.1.8 The oceans act as a carbon sink by absorbing carbon dioxide from the atmosphere and sequestering it.
- 4.1.9 Carbon sequestered in oceans over the short term as dissolved carbon dioxide causes ocean acidification; over the longer term, carbon is taken up into living organisms as biomass that accumulates on the seabed.
- 4.1.10 The temperature of water varies with depth, with cold water below and warmer water above. Differences in density restrict mixing between the layers, leading to persistent stratification.
- 4.1.11 Stratification occurs in deeper lakes, coastal areas, enclosed seas and open ocean, with a thermocline forming a transition layer between the warmer mixed layer at the surface and the cooler water below.
- 4.1.12 Global warming and salinity changes have increased the intensity of ocean stratification.
- 4.1.13 Upwellings in oceans and freshwater bodies can bring cold, nutrient-rich waters to the surface.
- 4.1.14 Thermohaline circulation systems are driven by differences in temperature and salinity. The resulting differences in water density drive the ocean conveyor belt, which distributes heat around the world and thus affects climate.

Link
Energy transfers and transformations are covered in Topic 1.2.4 (page 29).

◆ **Precipitation** – any state/phase of water, liquid or solid that falls from the atmosphere to the ground.

◆ **Hydrological cycle** – also known as the water cycle, this is the movement, distribution and transformation of water on Earth. There is a continuous movement of water in its various phases

4.1.1 Movements of water in the hydrosphere are driven by solar radiation and gravity

Approximately 70 per cent the Earth's surface is covered by water. All the water on Earth is collectively called the hydrosphere. This includes the water in the oceans, rivers and lakes, the ice in polar regions, and the water vapour in the atmosphere. Water can move between these different parts of the hydrosphere. For instance, the ice in glaciers can melt, and the water can flow into rivers and eventually the sea. These movements can be explained in terms of energy transfers and transformations.

The two main driving forces behind the movements of water in the hydrosphere are the sun and gravity. Solar radiation provides heat energy, which causes water to evaporate from the oceans and from surface waters such as rivers, lakes and wetlands. This water vapour rises in convection currents in the air, and can be blown by the wind over long distances to places far from where it evaporated. In this way, energy from the sun can cause water to move from low elevations to high elevations. Eventually, when the air cools, the water vapour condenses, and the water returns to the ground as rain, snow or other forms of **precipitation**.

The movement of water from higher areas, such as mountains, to lower areas and eventually to the sea is driven by gravity. Water at the higher elevations possesses gravitational potential energy, which is converted to kinetic energy as it flows downhill. This continuous movement of water in its various phases, is known as the water cycle or **hydrological cycle**.

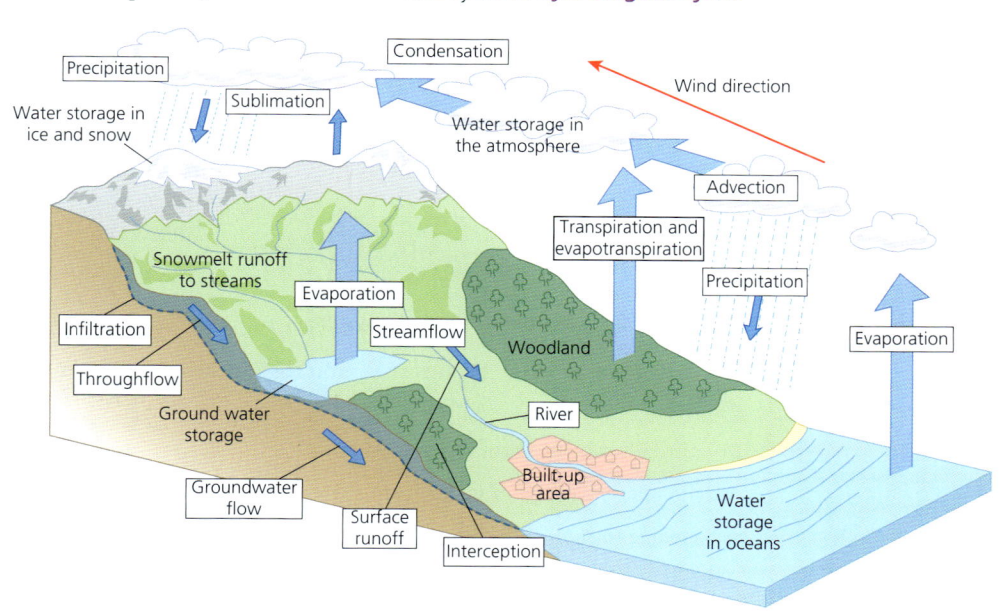

■ Figure 4.1 The hydrological cycle

4.1 Water systems

4.1.2 The global hydrological cycle operates as a system with stores and flows

The hydrological cycle is an example of a biogeochemical cycle. Like the carbon and nitrogen cycles, it functions as a system, with stores and flows. We can examine these stores and flows individually or collectively.

> **Concept**
>
> **Systems and models**
>
> Treating the water cycle as a system allows us to better understand how the various parts of the cycle interact, and how changes to one part of the cycle can affect other parts. This can help us to understand and mitigate the ways in which humans affect the cycle.

> **TOK**
>
> To what extent does the scientific approach of reductionism allow for a holistic view of the natural world? Some scientists think that the scientific method is primarily concerned with seeking to explain the natural world by breaking it down to its constituent parts. For example, the hydrological system can be broken down into sub-systems, such as the drainage basin system. This is a reductionist approach. Others think that a holistic approach is more appropriate, such as for the global hydrological system, as it allows a better understanding of complex systems as a whole.

Consider, for example, a lake with a stream flowing into it, and another flowing out. We can represent this as a simple systems diagram, as shown in Figure 4.2.

Link

Systems, stores, flows and systems diagrams are covered in Chapter 1.2 (page 24).

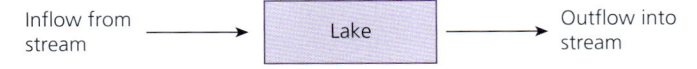

Inflow from stream → Lake → Outflow into stream

■ **Figure 4.2** A lake represented as a system

On closer examination, however, we find that a lake has several other inflows and outflows. These might include inflows such as:

- rainfall or other forms of precipitation
- the overland flow of water from surrounding areas
- sewage inputs from domestic, industrial or agricultural sources.

As well as outflows such as:

- evaporation
- infiltration of water into the groundwater
- abstraction of water for domestic, industrial or agricultural uses.

> **Common mistake**
>
> Ensure you understand that the *global* water cycle can be described as a closed system, because no water enters or leaves the system. Energy, however, enters from the sun and can escape into the atmosphere. Meanwhile, individual *parts* of the water cycle, such as a lake or watershed, are open systems, since water can flow from one part to another.

Theme 4: Water

4.1.3 The main stores in the hydrological cycle are the oceans, glaciers and ice caps, groundwater, surface freshwater, atmosphere and organisms

◆ **Saltwater** – water with a high salinity, the majority found in the oceans, but also in some inland seas.

◆ **Glacial ice** – ice found on land, near the poles, in cold climates or at high altitudes. Gravity causes glaciers to slowly slide over the surface of the land, towards the sea.

◆ **Polar ice caps** – large masses of ice found near the poles. At the North Pole, the ice mostly floats on the Arctic Ocean. At the South Pole, the ice sits on the continent of Antarctica.

◆ **Groundwater** – water found in porous rocks underground.

◆ **Aquifer** – A body of rock that contains groundwater.

◆ **Freshwater** – water with very low salinity, usually found in rivers, lakes, glaciers and ice caps, and groundwater.

The major stores in the hydrological cycle are:

- The oceans (approximately 96.5 per cent of all water on Earth): Comprised of **saltwater**, with an average salinity of approximately 35 parts per thousand (‰) (1 kg of seawater contains approximately 35 g of mineral salts, mostly sodium chloride). This high salt content (normally sodium chloride) means that seawater cannot be used for many human activities, such as farming, industry or household drinking water. Freshwater can be obtained from the sea by the industrial process of desalination, but this is an energy-intensive process, and can therefore be very expensive.

- **Glacial** and **polar ice** (1.7 per cent of all water): This is frozen freshwater, located on high mountains such as the Rockies and Himalayas, on land in the polar regions such as Greenland and Antarctica, and floating on the sea in the Arctic and Southern Oceans. The total amount of ice on Earth fluctuates with the seasons, but is currently declining due to global warming.

- **Groundwater** (1.7 per cent of all water): Some water percolates through rocks on the surface of the Earth, and becomes stored in porous rocks or aquifers. A body of groundwater, stored in porous rock, is called an **aquifer**. Groundwater is generally fresh, but may contain small concentrations of minerals dissolved from the rock. We extract water from the aquifers using wells.

- Surface **freshwater** (0.02 per cent of all water): A small, but important, store. This includes bodies of water on the surface of the land, including rivers, lakes and freshwater wetlands. In their natural state they have very low salinity, but they can be easily become polluted with farming, industrial and household wastes. Although this is a much smaller store than those listed above, surface water is often our most accessible source of freshwater.

Relatively minor stores include:

- The atmosphere (variable, but averaging approximately 0.001 per cent of global water): Water evaporates from surface waters and the oceans, and enters the atmosphere as water vapour. The amount of water vapour in the air varies widely from time to time, and between different locations. Water vapour is an important greenhouse gas, making a vital contribution to keeping the Earth's atmosphere warm enough to support life.

- Living organisms (0.0001 per cent of all water): Water is fundamental to life. All living organisms contain water in their cells and tissues.

You do not need to memorize these exact percentages, but you should understand the relative sizes and importance of each store compared to others.

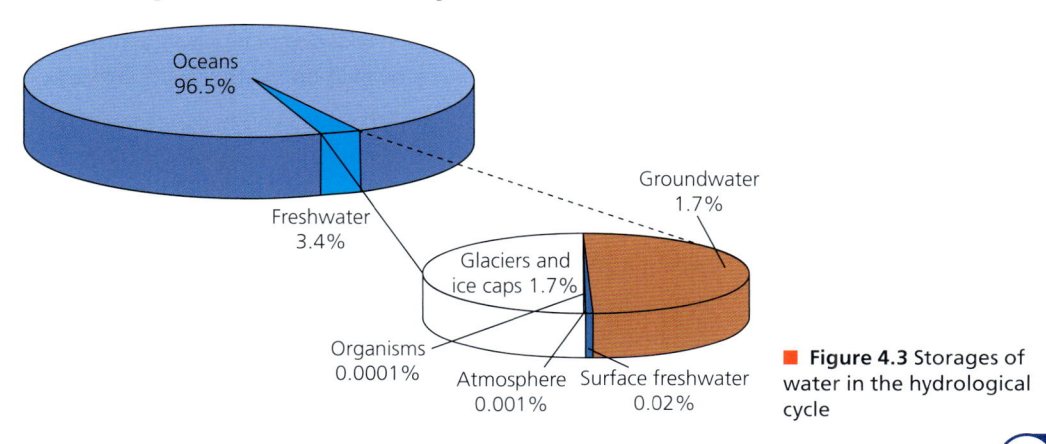

■ **Figure 4.3** Storages of water in the hydrological cycle

Oceans 96.5%
Freshwater 3.4%
Groundwater 1.7%
Glaciers and ice caps 1.7%
Organisms 0.0001%
Atmosphere 0.001%
Surface freshwater 0.02%

4.1 Water systems

4.1.4 Major flows in the hydrological cycle

The flows in the hydrological cycle include both transfers and transformations. Note that all the transfers and transformations are driven by either heating, cooling, or gravitational pull. Evaporation, transpiration, sublimation and melting are all driven directly by heat energy from the sun. Advection is driven by the wind, but wind is produced due to heating of the atmosphere by the sun, so this is also a result of solar radiation. Freezing and condensation happen when temperatures fall low enough and water molecules lose energy to their environments. Surface runoff, streamflow, infiltration, percolation, groundwater flow and precipitation are all downward movements of water, driven by the pull of gravity (see Figure 4.4).

■ Water cycle transfers

- Surface runoff: Water on the land, for instance due to rainfall, flows downhill over the surface, eventually entering surface waterways such as lakes and rivers. This usually happens in small quantities, but can involve large volumes of water if there has been flooding.
- Streamflow: Water in rivers and streams flows, following the river or stream channel. It eventually enters lakes, wetlands or the sea.
- Infiltration: If water stays long enough on the surface of the land, some of it may soak into the soil. This is more likely to happen on flat or gently sloping land, or land where there is vegetation that slows down the surface runoff. The extent of infiltration may also depend on the nature of the land surface. Paving the land and tarmac coverage often prevents infiltration.
- Percolation: Once water has infiltrated the soil, it will then slowly flow between the soil and rock particles, until it reaches the water table – the level at which the soil or rocks are saturated with groundwater.
- Groundwater flow: Groundwater can flow, usually very slowly, through porous rocks, in a downhill direction. The flow stops when it reaches impermeable rocks, which prevent further movement. Where the water table meets the surface of the land, groundwater may emerge on to the surface. Here it will form a spring, which may develop into a stream that flows into a river.
- Advection: Water vapour and clouds in the atmosphere can be moved by the wind. Water vapour and clouds may be transported/travel great distances from where they entered the atmosphere.

■ Water cycle transformations

- Evaporation: Waters, both on land and at sea, can undergo evaporation. Evaporation occurs when heat energy forces the bonds that hold water molecules together to break, causing the water to change from its liquid state to its gaseous state (water vapour). Water evaporates easily at its boiling point but much more slowly at its freezing point. Heat and wind tend to increase the rate of evaporation, whereas high humidity usually slows it down.
- Transpiration: This occurs when plants take up liquid water from the soil and release water vapour into the air from their leaves. As water molecules diffuse away from the leaves, water is drawn up from the soil, through the plant's tissues, and into the leaves to replace the lost water. Transpiration contributes significantly to the flow of water from the soil into the atmosphere, and can have a significant effect on local soil and air conditions. The term evapotranspiration is sometimes used to describe the combination of transpiration and evaporation from the land surface.

> **Link**
> The distinction between transfers and transformations is covered in Chapter 1.2 (page 24).

- Condensation: When conditions are right, water vapour can turn back to liquid water. This may happen if the air temperature drops, but it is also affected by the humidity of the air, and the presence of certain particles, which may encourage water molecules to come together to form water droplets. Condensation often happens at high altitudes, where the air temperature is low, and contributes to the formation of clouds.
- Sublimation: Under the right conditions, ice can turn directly into water vapour, or water vapour can turn directly to ice. Both conversions are called sublimation.
- Precipitation: When water vapour condenses or sublimes to ice, it is often brought back to the surface of the land as rain, snow, hail or sleet. Precipitation also includes the formation of dew or frost directly on the land, as well as fog in ground-level air.
- Melting and freezing: Ice on land or at sea can melt if the temperature rises sufficiently, and water can freeze if the temperature falls. When glaciers melt, the meltwater flows into rivers, and eventually to the sea.

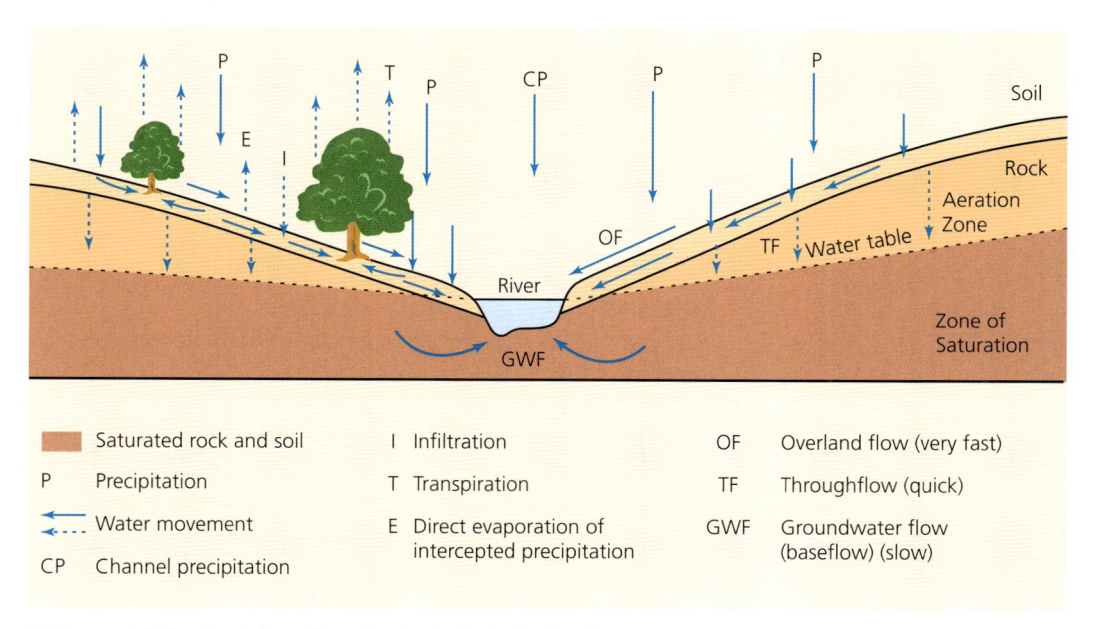

■ Figure 4.4 Gravity-driven flows in the hydrological cycle

Tool 4: Systems and models

System diagrams

In a system diagram, boxes show storage. Arrows show flows, which represent either transfers or transformations. The inputs are shown with arrows into the storage, and the outputs are shown with arrows out of the storage. When drawing a system diagram, processes must be labelled on the arrows.

ATL ACTIVITY

Assign and accept specific roles during group activities

Investigate the stores and flows of one major river from its source to its mouth, and the different locations, with examples of stores, transfers and transformation. You should focus on one specific river as a case study.

4.1 Water systems

Get into groups of three. Each member investigates one of the following components or flows:
- Stores
- Transfers
- Transformations

Once you have decided on the river you will investigate, use online tools such as Google Earth (**https://earth.google.com/web**) or River Runner Global (**https://river-runner-global.samlearner.com**) to find locations along the river to illustrate your component or flow.

Make a copy of the table below:

	Stores	Transfers	Transformations
1 Name of location(s)			
2 Explanation			
3 Role of store(s)/transfer(s) within the hydrological system			

1. In row 1, give details about the locational context of your component and/or flow process. This should include:
   - The Google Earth placemark link or coordinates of your example(s).
   - A screenshot of the location(s). This can come from Google Earth or Google Images.
   - Any useful sources of information about the location(s) and component/flow processes involved.

   The same location along the river can be used to illustrate different stores and flows.

2. In row 2, add an explanation of the store and/or flow processes identified in the context of the location(s). Include visuals in your explanation.

   Example: For the transformation process of transpiration, you should explain how the type of vegetation affects the rate of evaporation.

3. In row 3, explain the role of the identified store(s) or transfer(s) within the hydrological system of your chosen river. Make sure you focus your explanation on the interactions as part of the system.

Share your findings with the rest of the class as an informal group presentation. Make sure you include visuals from row 2 of your table.

4.1.5 Human activities can alter these flows and stores

Link

The environmental impacts of agriculture are explored further in Chapter 5.2 (page 456).

Humans have always interacted with water in various ways. These interactions increased in scale when we started farming, settling in villages, towns and cities, and as our industrial activities became more intense and with population increase in general. These activities often involve changing the flow and possibly the course of water. There are also indirect impacts, caused by changes in the way we use the land. We will examine some of these impacts here.

◆ **Irrigation** – the application of water to soil in order to help plants grow. This is usually done where there is not enough natural rainfall to supply plants.

◆ **Abstraction** – the process of taking water out of surface or groundwater sources for human use.

Agriculture

Farms often use a great deal of water. This might be taken from rivers or lakes, or from the groundwater via wells. The water may then be used to **irrigate** crops, by spraying or pumping the water on to the fields. In some forms of **agriculture**, such as some types of rice cultivation, entire fields are flooded. Livestock also require large volumes of water, often more than is needed for an equivalent quantity of crops. **Aquaculture** – the production of aquatic food products such as fish – can use even more water. Sometimes dams are built across rivers in order to ensure a continuous supply of water for farms.

Abstraction is the process of taking water from surface or groundwater for human use. Abstraction of water from surface waters is likely to reduce the volume of water in the surface water store, and the rate of streamflow. A famous example of this is the shrinking of the Aral Sea in Central Asia.

REAL-WORLD EXAMPLE

The shrinking of the Aral Sea

The Aral Sea was a large inland body of saltwater in the former Soviet Union, at the border of the modern nations of Kazakhstan and Uzbekistan. A number of rivers flowed into the sea. It later experienced a high rate of evaporation, due to the desert-like climate of the region. Continuous evaporation concentrated the salts in the water, resulting in the unusually high salinity.

■ **Figure 4.5** The shrinking of the Aral Sea

In the past, evaporation and inflow from the rivers balanced each other, so that the volume of the sea remained approximately constant. However, in the 1960s, significant volumes of water were diverted from the two main inflowing rivers, in order to supply cotton crops in the surrounding agricultural land. This reduced the rate of inflow, while evaporation rates continued. As a result, the volume of the sea steadily decreased. Today, despite attempts to reduce agricultural water use, all that remains are two much smaller bodies of water, and a large expanse of salt-contaminated soil that is unsuitable for agriculture.

Figure 4.6 Fishing boats stranded by the shrinking of the Aral Sea

Abstraction of water from the ground using wells can significantly reduce the volume of water stored as groundwater, and also the rate of groundwater flow. This can in turn reduce the availability of groundwater for use in other activities and areas.

Agriculture also affects the water cycle by changing the use and character of the land and soil. Creation of new farmland often involves removal of natural vegetation, which can affect the rates of transpiration, surface flow and infiltration. We will examine these effects next.

Deforestation

Effect on surface flow and infiltration

Trees have a significant effect on the flow of water in the hydrological cycle. Trees and other vegetation slow down the overland flow of water, and so encourage it to infiltrate the soil. As a result, vegetation tends to reduce surface runoff, and to increase infiltration and percolation. This means that more water enters the groundwater stores, and less enters surface waters. This helps to replenish depleted groundwater, and makes surface flooding less likely and less severe when it does happen. This is helpful as water is retained inland, rather than running off into the seas and oceans.

The removal of trees has the opposite effect: it means surface water is more likely to flow over the land and cause floods, and is less likely to percolate into the soil. This is particularly true on sloping land such as hills and mountains. The increased surface flow may cause increased soil erosion.

Effect on evapotranspiration

Trees increase the rate of evapotranspiration. Water is drawn up from the soil through the trees, and flows into the air during transpiration. Water that may have fallen as rain on the trees also evaporates, returning to the air. Together, this results in increased humidity in the local area. Water vapour commonly contributes to clouds forming and to more rain being produced. Effectively, trees help to keep the water cycling within the local area, rather than flowing away. In this way, they help to maintain the local microclimate.

Figure 4.7 Clouds forming over a rainforest

When trees are removed, the extra surface flow away from the area means that less water vapour is returned to the atmosphere. Over time, deforestation causes the local area to dry out, sometimes leading to the formation of desert where there was previously forest.

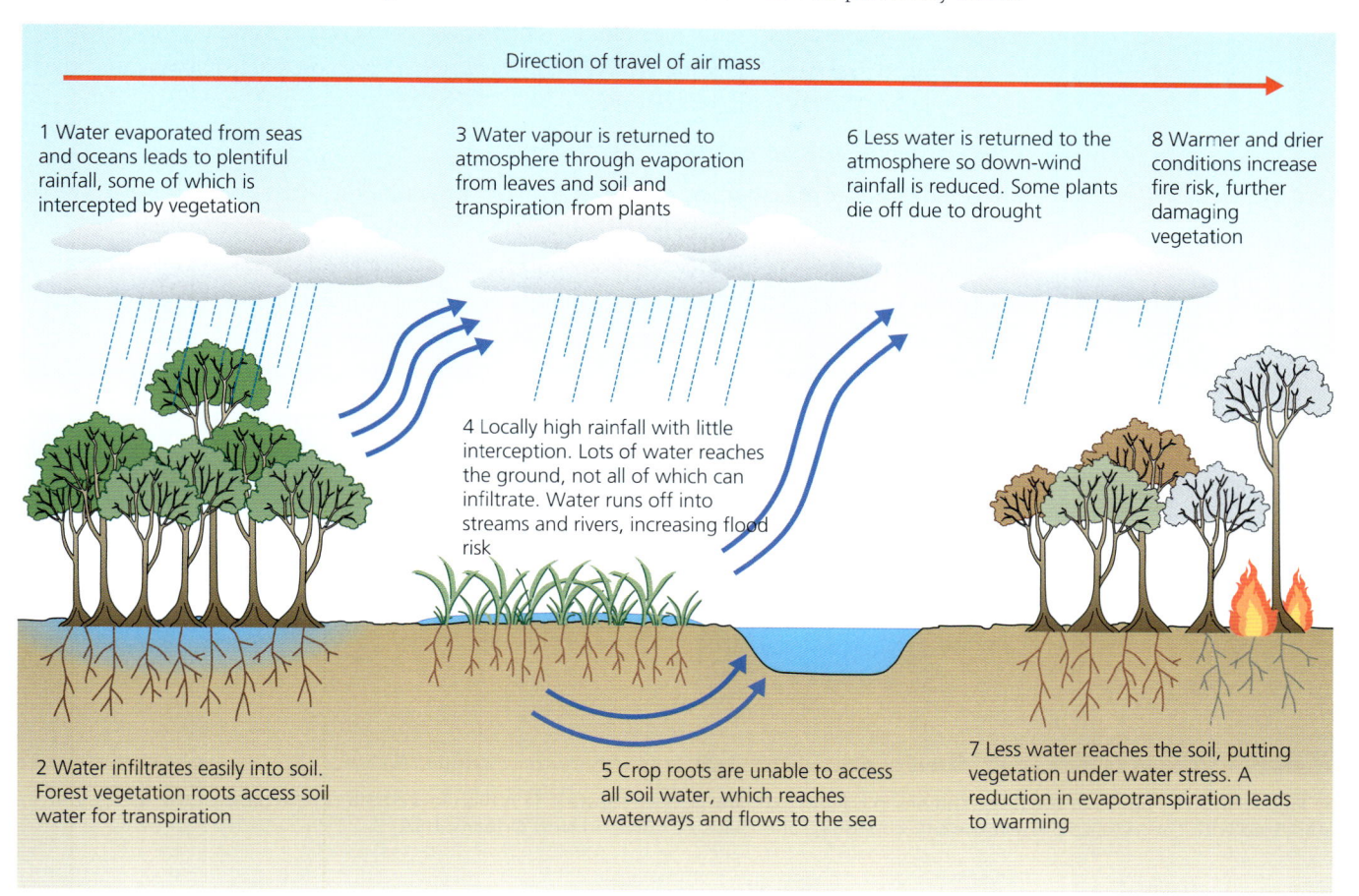

Direction of travel of air mass

1 Water evaporated from seas and oceans leads to plentiful rainfall, some of which is intercepted by vegetation

2 Water infiltrates easily into soil. Forest vegetation roots access soil water for transpiration

3 Water vapour is returned to atmosphere through evaporation from leaves and soil and transpiration from plants

4 Locally high rainfall with little interception. Lots of water reaches the ground, not all of which can infiltrate. Water runs off into streams and rivers, increasing flood risk

5 Crop roots are unable to access all soil water, which reaches waterways and flows to the sea

6 Less water is returned to the atmosphere so down-wind rainfall is reduced. Some plants die off due to drought

7 Less water reaches the soil, putting vegetation under water stress. A reduction in evapotranspiration leads to warming

8 Warmer and drier conditions increase fire risk, further damaging vegetation

■ **Figure 4.8** The effects of deforestation on the water cycle

■ **Figure 4.9** Flooding in New Orleans, USA during Hurricane Katrina in 2011

Urbanization

As human settlements expand, the surface of the land is typically changed. Urban land is often paved if not built on, and areas of natural vegetation and soil tend to be small and fragmented. Rivers and streams running through towns are often modified for the benefit of the human settlement. Towns and cities also use large volumes of freshwater, and produce large volumes of wastewater. Each of these processes affects the local hydrological cycle.

4.1 Water systems

Effect of paving and building

Paved surfaces are impermeable to water. When land is paved, such as to make roads, footpaths and courtyards, this prevents water from infiltrating into the soil. Instead, the water is forced to flow over the paved surface, until it can find a route away – perhaps into a drain. When large areas are paved, the risk of flooding is increased. The water is also more likely to pass into a river or stream rather than into the groundwater.

Effect of river modification

Towns and cities are often built on the banks of rivers. Rivers are useful as sources of water, and also as a route for transporting goods and people. However, rivers can also overflow causing flooding, and riverbanks can erode. To protect the nearby buildings and roads, engineers often stabilize the beds and banks of rivers with concrete or other hard materials, and raise high levees over the natural banks. This encourages water to flow faster and straighter through the city, meaning it reaches the sea faster, and natural flows of water in and out of the river may be blocked.

■ **Figure 4.10** The banks of the River Thames in London, UK are stabilized with concrete, to protect the city's buildings and roads

■ **Figure 4.11** A high levee built on the bank of the Mississippi River in New Orleans, USA

Effect of water use and disposal

The people living in towns and cities use a lot of water. The majority of this water must be brought from either surface or groundwater stores (some could come from rainwater collection or desalination). After the water has been used, it is then released into the environment as wastewater or sewage.

> **Link**
>
> Wastewater is generally contaminated with various pollutants, and may need to be treated before it is released. Water pollution and wastewater treatment are covered in Chapter 4.4 (page 373).

Link

The concept of steady-state equilibria is covered in Topic 1.2.9 (page 37).

4.1.6 The steady state of any water body can be demonstrated through flow diagrams of inputs and outputs

As we saw when we examined the water cycle in Topic 4.1.2 (page 270), each store in the system has inputs and outputs. The relative rates of these flows will determine whether the store increases or decreases, or remains constant, over time. This can be applied to any of the stores in the hydrological cycle.

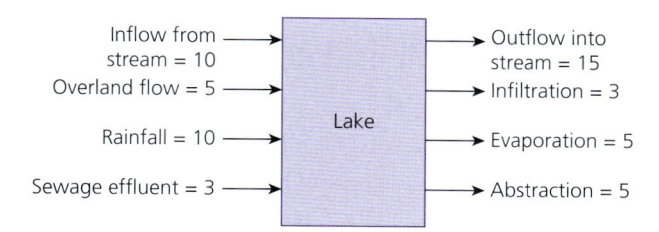

```
Inflow from stream = 10  →  ┌──────┐  →  Outflow into stream = 15
Overland flow = 5        →  │      │  →  Infiltration = 3
Rainfall = 10            →  │ Lake │  →  Evaporation = 5
Sewage effluent = 3      →  │      │  →  Abstraction = 5
                            └──────┘
```

■ **Figure 4.12** Balanced in- and outflows produce a steady-state equilibrium in a lake

If the total rate of inflow is greater than the total rate of outflow, the water body will increase in size. For the water body to remain constant, the inflows must be balanced by the outflows. Imagine a bathtub with the tap adding water at exactly the same rate as the water is flowing down the drainpipe: the level of water in the bath should remain constant, even though water is continuously flowing through it. Similarly, the volume of water in a lake will only remain constant if the combination of all the inflows is equal to the combination of all the outflows, as shown by Figure 4.12.

Tool 4: Systems and models

Modelling a real-world system

The diagram in Figure 4.12 is a model of a real-world system. By using a model like this, we can better understand how the system works and make predictions about how it might change under different circumstances.

If one of the inflows or outflows changes, this will change the state of the lake. For instance, if the surface flow into the lake increases because of heavy rainfall upstream, the volume of water in the lake may increase and the level of water may rise. Alternatively, the rate of the outflow from the stream might increase and lower the level of water.

To get a more complete understanding of the system, we could change the boundary of our model and take a wider view, perhaps to include the upstream and downstream elements, or the entire watershed.

Inquiry process

Inquiry 2: Collecting and processing data

Use the FAO's online database AQUASTAT, at **https://data.apps.fao.org/aquastat**, to collect data on the water budget of a selected country for selected years, and draw a system flow diagram of that country's water resources.

1. Using the Area and Year functions on the website, refine your search to include only the country and years you are interested in.
2. Use the Variable function to choose the variables you need for your investigation. To select the variables, click on Select Variables to reveal the Variables dropdown menu.
   - Within the Water Resources section, select the relevant items from the Internal Renewable Water Resources to obtain values for water stocks within the country.
   - Select from the Precipitation section for values of inputs from precipitation.
   - Select from the External Renewable Water Resources section for flows of water in to and out of the country.
   - Use the Water Use section for data on human withdrawals of water from the system.
3. Collect all the relevant data and collate them in a suitable table. Be sure to include units in your column or row headings.
4. Draw a system flow diagram to show the flows of water in and out of the country's water system. Label all stores and flows with the numerical values obtained from the database.
5. Examine your diagram. Do the numbers add up? Are there any flows that have not been quantified or identified through the database? Can you estimate the values of any missing flows using the numbers you have?

Concept

Sustainability

If we understand the rates of inflow and outflow from a lake or aquifer, we can work out how much water we can safely take without using up all the water. This helps us achieve sustainable water use.

4.1 Water systems

(HL) 4.1.7 Water has unique physical and chemical properties that support and sustain life

Water is essential for life: when scientists look for signs of life on other planets, they often look for signs of liquid water on the planets' surfaces. Water possesses a number of physical and chemical properties that make it very different from other molecules. These properties are the result of the structure of the water molecule, and the ways in which water molecules interact with each other and with other substances.

TOK

Exobiologists are scientists who are interested in whether there is life in other parts of the universe. Since water is so important for life, one approach they use is to work out whether liquid water is present on other planets. Some scientists have suggested that there may be other molecules with similar properties, which might perform the roles of water in other places.

Water molecules and hydrogen bonding

◆ **Covalent bond** – a chemical bond between two atoms, where the atoms share a pair of electrons.

◆ **Electronegativity** – the tendency of an atom to draw the electrons in its covalent bonds towards itself. In a water molecule, oxygen is more electronegative than hydrogen.

◆ **Polar molecule** – a molecule, like water, that has one end with a slight positive charge, and one end with a slight negative charge.

◆ **Hydrogen bond** – a relatively strong force of attraction between two molecules, which happens when there is a hydrogen atom in one molecule, and a strongly electronegative atom, like oxygen, in the other molecule.

A water molecule is made up of one oxygen atom and two hydrogen atoms, **covalently bonded** together. The arrangement of electrons around the oxygen atom forces the hydrogen atoms to make a V shape, with the oxygen atom at the point of the V (see Figure 4.13).

The electrons that make up the covalent bonds are not evenly shared between the oxygen and hydrogen atoms. Oxygen is more **electronegative** than hydrogen, which means that the electrons between the atoms spend most of the time closer to the oxygen atom, and further from the hydrogen atoms. Since the hydrogen atoms are both on the same side of the molecule (at the open ends of the V), this means that the electrons spend more time at one end of the molecule (the oxygen end, at the point of the V) and less time at the hydrogen end (at the open ends of the V). This makes water a **polar molecule**. This means it has a slight negative charge at the oxygen end, and a slight positive charge at the hydrogen end. The symbols δ+ and δ– in Figure 4.14 show the slight charges.

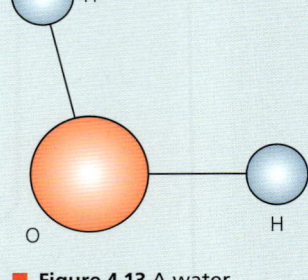

■ **Figure 4.13** A water molecule

Opposite charges attract. The slightly negative oxygen end of one water molecule will be attracted to the slightly positive hydrogen end of other water molecules nearby. The particular structure of the oxygen and hydrogen atoms means that a special type of intermolecular force called a **hydrogen bond** can form between adjacent water molecules. There are a few other molecules that form hydrogen bonds, but water is particularly good at it.

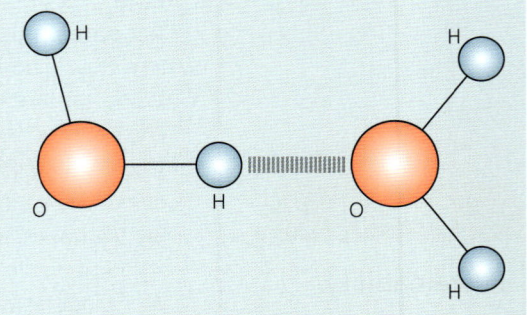

■ **Figure 4.14** A hydrogen bond between two water molecules

- ◆ **Cohesion** – the tendency of molecules to 'stick to each other' due to being attracted to each other.
- ◆ **Surface tension** – a force of attraction between molecules on the surface of a liquid as a result of cohesion, which holds the molecules together. Surface tension can resist weaker forces that push or pull against the surface.
- ◆ **Ion** – a charged particle, which is formed when an atom or molecule gains or loses an electron.
- ◆ **Solvent** – the substance that dissolves the **solute** in a solution.
- ◆ **Solute** – the substance that is dissolved in a solution.
- ◆ **Solution** – a mixture formed when particles of one substance are dissolved in another substance.
- ◆ **Soluble** – able to dissolve in a particular solvent, such as water.
- ◆ **Insoluble** – unable to dissolve in a particular solvent, such as water.
- ◆ **Hydrophobic** – repelled by, or not attracted to, water due to polarity. Hydrophobic substances are usually insoluble in water.
- ◆ **Hydrophilic** – attracted to water due to polarity. Hydrophilic substances are usually soluble in water.

Cohesion and surface tension

Since water molecules are attracted to one another and form hydrogen bonds, the molecules tend to stick together, and it can be relatively difficult to separate them. This is why water forms relatively large drops, and why a drop of water lying on a flat surface will sometimes remain more or less round, instead of spreading out flat on the surface (although this depends on the nature of the surface).

This 'stickiness' of water molecules is called **cohesion**. Water molecules tend to stick together, or cohere, unless there is something else nearby that they are also attracted to. Cohesion also produces a phenomenon called **surface tension**. Water molecules at the surface of a body of water are more attracted to other water molecules than to the nearby air. They 'pull' against each other, creating a force that makes the surface difficult to disrupt. As a result, surprisingly dense objects, including insects like water striders, can sometimes float on water supported by the surface tension.

■ **Figure 4.15** Drops of water tend to form a spherical shape

■ **Figure 4.16** Surface tension allows water striders to float on the surface

Adhesion and solvent properties

Water molecules can also be attracted to other substances. Other polar molecules, as well as charged particles (**ions**), attract water and will tend to 'stick' to water molecules. This is particularly true of substances with hydrogen-bonding potential. When water is mixed with small polar or charged particles, such as sodium chloride, these tend to mix in with the water and become dissolved. The water acts as a **solvent**, and the mixture is described as a **solution**. When water comes into contact with molecules that are too large to dissolve, the water may spread out and stick to the large molecule. This is called adhesion.

Many different substances are **soluble** and can dissolve in water, so water is sometimes described as a universal solvent. However, most non-polar substances, such as oils and waxes, do not dissolve in water. Non-polar **insoluble** substances can be described as **hydrophobic**. Polar and ionic soluble substances can be called **hydrophilic**.

Ionic substances

Certain compounds are made up of charged particles. Sodium chloride (NaCl), also known as table salt or sea salt, is made up of positive sodium ions (Na$^+$) and negative chloride ions (Cl$^-$). When salt is mixed with water, those ions separate from each other and become surrounded by water molecules. It may look like the salt has disappeared, but it has really been separated into tiny particles, which are far too small to be seen, and dissolved in the water. If we taste the water, we will know that it contains salt.

Most (but not all) ionic substances will dissolve easily in water. Other examples include ammonium nitrate (NH$_4$NO$_3$), a popular fertilizer, and amino acids, which are important molecules in living organisms.

Calcium carbonate (CaCO$_3$), found in limestone, chalk and marble, is an ionic compound that is *not* soluble in water.

4.1 Water systems

Polar substances

Some molecules with covalent bonds are polar, just like water, and some even have hydrogen-bonding potential. Molecules like these are often soluble in water.

Ammonia (NH_3), ethanol (C_2H_5OH), and simple sugars, for example glucose ($C_6H_{12}O_6$) and sucrose ($C_{12}H_{22}O_{11}$), all have hydrogen-bonding potential and dissolve easily in water. These are all naturally occurring substances that have industrial and domestic uses.

Some polar molecules are so strongly attracted to water that they actually split apart and become ions when they dissolve. If these molecules contain a hydrogen atom that splits apart from the rest, they will form an acidic solution. For example, hydrogen chloride is a gas that dissolves in water to form hydrochloric acid.

Solubility of gases

Non-polar molecules are usually insoluble in water, but some small non-polar molecules dissolve to some extent. Some of the gases of the atmosphere, including oxygen, carbon dioxide and nitrogen, can do this. Their molecules are not particularly attracted to water, but they are small enough to fit between the molecules of water and dissolve.

How much gas dissolves depends on environmental conditions. If the gases are placed under high pressure, more molecules will be forced into the water and dissolve. The gases also become more soluble at low temperatures. Beverage manufacturers use these facts to help them produce carbonated drinks. The liquid is cooled to low temperatures, and carbon dioxide is pumped into the beverage at high pressures. When someone opens a bottle or can the pressure is relieved, and much of the gas comes out as bubbles or 'fizz'.

Deep sea divers need to be aware of the effect of high pressure on nitrogen solubility. When they are deep underwater, the nitrogen from their air tanks tends to dissolve in their blood. As they rise to the surface, the pressure is reduced and the nitrogen comes out of solution. If they rise too quickly, the nitrogen forms bubbles in their blood vessels, causing a painful condition called decompression sickness or 'the bends'. This can be fatal. Divers avoid this by rising slowly and carefully, which gives enough time for the nitrogen to emerge into the lungs, so it can be breathed out.

The solubility of oxygen in water is very important for aquatic animals. They usually obtain oxygen from the water through their gills, to use in respiration. Since oxygen is only slightly soluble in water, the gills must be very efficient to extract as much oxygen as possible. If the water has less oxygen than normal, this can cause fish and other aquatic animals to die. Oxygen depletion can happen because of bacteria using up the oxygen, or because the water is too warm.

Adhesion: when things are too big to dissolve

Some materials and objects are made of large molecules, which may be attracted to water but are too large to mix with it and dissolve. Plant cell walls, for instance, are made of cellulose. This is a large carbohydrate molecule (a polysaccharide), which is arranged in bundles in the cell walls. Smaller carbohydrates, such as the simple sugars glucose and sucrose, dissolve easily in water. Cellulose, however, is too large, and is usually part of an even larger structure – the cell. When water makes contact with cellulose it is attracted to the molecule, forming hydrogen bonds with it, but the cellulose is too big to break apart and dissolve. The water therefore sticks, or adheres, to it. This is called **adhesion**.

In fact, water is so strongly attracted to cellulose that it can 'climb' up a vertical structure made of cellulose, such as a sheet of paper. If we hang a piece of paper into water, the water will gradually rise up through the paper, until it is all wet. In living plants, this helps to draw water up through the xylem vessels, from the soil to the leaves. Water also adheres to many types of cloth, which is why normal clothes don't keep you dry in the rain.

◆ **Common mistake**

Make sure you know that for most solid substances, solubility increases at higher temperatures (for example, more salt dissolves in warm water than in cold water), but the opposite is usually true for gases: warmer water holds less dissolved oxygen and carbon dioxide than cold water does. This in turn has consequences for warming of seas and oceans.

Link

The causes and effects of oxygen depletion are explored in Chapter 4.4 (page 373).

◆ **Adhesion** – the tendency of molecules to stick to other substances. Water molecules adhere to many substances, such as cellulose.

♦ **Specific heat capacity** – the amount of energy needed to raise the temperature of 1 g of a substance by 1°C. Water has an unusually high specific heat capacity.

♦ **Specific latent heat** – the amount of energy needed to change the state of 1 g of a substance.

♦ **Transparent** – allows light to pass through.

♦ **Euphotic zone** – the water near the surface of a lake or the ocean and down to about 200 m deep, where light can easily penetrate. Sometimes called the sunlight zone.

♦ **Dysphotic zone** – the water from about 200 m to 1,000 m deep, where light is rapidly reduced. The deeper the water, the less light is present. Sometimes called the twilight zone.

♦ **Aphotic zone** – the water below 1,000 m, where no light is present. Sometimes called the midnight zone.

Heat capacity and latent heat

Water also shows some important behaviours when it is heated or cooled. Since water molecules are so strongly attracted to each other, it takes a surprisingly large amount of energy to cause water to heat up or to evaporate.

The energy needed to warm 1 g of a substance by 1°C is called its **specific heat capacity**. Water's specific heat capacity is approximately 4.2 J/g°C (joules per gram per degree Celsius), or 4,200 J/kg°C, or 4.2 kJ/kg°C. This is higher than that of most molecules of similar size. The high specific heat capacity means that it takes a lot of energy to change water's temperature. This makes its temperature very stable – it can absorb a lot of energy before it becomes warmer. This also means it has to lose a lot of energy before it cools down. This is useful for aquatic organisms as it means the temperature of the aquatic environments of rivers and oceans do not change much. Even terrestrial organisms benefit from this because their cells are full of water, so body temperatures also don't change much when the air around them does.

The energy needed to change one gram of a substance from one physical state to another is called its **specific latent heat**. Water's specific latent heat of vaporization (the energy needed to turn liquid water to water vapour) is approximately 2,260 kJ kg. This means that it takes a lot of energy to evaporate water, so ponds and lakes don't easily dry up in the sun. Animals also use this property of water to help keep them cool, by sweating or panting. The evaporating water takes away a lot of heat from the animal. Similarly, when plants lose water by transpiration from their leaves, it helps to cool the leaves – otherwise they would get very hot from being out in the sun all day!

Transparency

When light is shone on water, almost all of it passes through – water is **transparent**. This allows light to pass through water to reach photosynthetic organisms like algae, and also allows fish to see. In plants, the transparency of water allows light to pass through layers of cells to get to the photosynthetic structures inside leaves.

However, the transparency of water is not perfect. Light cannot penetrate to the bottom of the ocean, because small amounts of light are absorbed or scattered as it passes through the water. Below about 200 metres the ocean is usually quite dark, although some light can reach to about 1,000 metres deep under certain conditions. The ocean can be divided into different zones, depending on how much light is present: the **euphotic** (sunlight), **dysphotic** (twilight) and **aphotic** (midnight) zones, as shown in Figure 4.17.

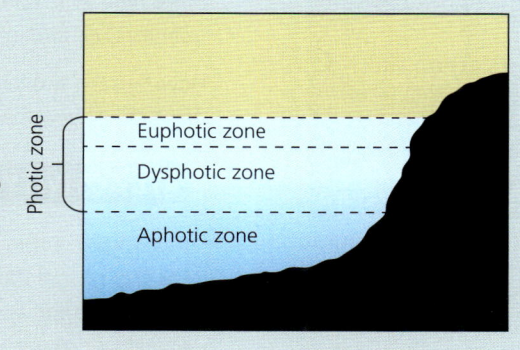

■ **Figure 4.17** The euphotic, dysphotic and aphotic zones of the ocean

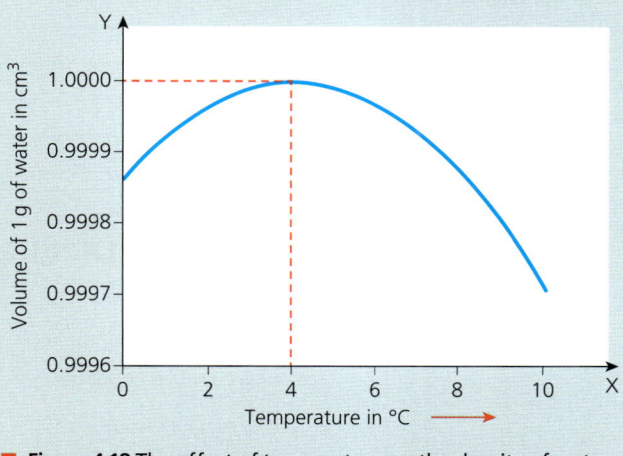

■ **Figure 4.18** The effect of temperature on the density of water

4.1 Water systems

Link

The carbon cycle is explored further in Chapter 2.3 (page 139). The role of carbon sinks and carbon sequestration as mitigation for climate change is explored in Chapter 6.3 (page 562).

◆ **Carbon sink** – a process that takes carbon dioxide out of the atmosphere.

◆ **Carbon sequestration** – the removal and storage of carbon dioxide from the atmosphere by natural or artificial means.

◆ **Saturated** – having the maximum amount of a solute in a solution. When a solution is saturated, no more of the solute can dissolve.

◆ **pH** – a measure of how acidic or alkaline a solution is. A pH of 7 is neutral; less than 7 is acidic, and more than 7 is alkaline.

The buoyancy of ice

Most substances become more dense as they get colder. This is because their particles have less energy, move around less, and come closer together. Objects that are denser than water will sink.

Water's behaviour when it cools is quite unusual. If water cooled down, starting at room temperature or warmer, it will become gradually denser until its temperature reaches 4°C. Below that temperature, it starts to become *less* dense. This happens because of the hydrogen bonds between the water molecules. As the temperature falls, the molecules have less energy. They are held in fixed positions by the hydrogen bonds. These bonds actually push the molecules slightly further apart in its liquid, so that water below 4°C is actually less dense than water at 4°C.

When water freezes, ice forms at the surface and floats on top of the liquid water. In cold climates, lakes may freeze over, leaving a layer of ice on top of the water. At the North Pole, and in the ocean around Antarctica, ice floats on the surface of the ocean. Fish and other organisms can continue to live under the ice.

4.1.8 The oceans as a carbon sink

As we have seen, carbon dioxide can dissolve in water. Increased use of fossil fuels and deforestation are leading to an increase in atmospheric CO_2. This means that as the amount of carbon dioxide in the atmosphere increases, more of it dissolves in the ocean. The ocean is acting as a **carbon sink**, meaning that atmospheric carbon dioxide levels have not risen as quickly as we might expect.

The carbon that enters the ocean does not contribute to climate change, although it can cause other problems. Carbon that is stored away from the atmosphere is said to be sequestered. Scientists see **carbon sequestration** as one way of helping to reduce or prevent climate change.

However, the ocean cannot absorb all the carbon dioxide we put into the atmosphere. At some point, it will become **saturated**, so that no more carbon dioxide can dissolve.

ATL ACTIVITY

Investigate other carbon sinks around the world to find out how important they are for the global carbon budget, and in what ways humans have affected their carbon storage capacity. Fill in the table below with your findings. Add more rows if you find any other examples.

Carbon sink	Location example	Processes involved in storing carbon	Estimated total carbon storage	Human impacts on sink
Oceans				
Forests				
Soils				
Wetlands				
Permafrost				

4.1.9 Carbon sequestration in oceans over the short and long term

When carbon dioxide dissolves in water, some of it reacts with water to form carbonic acid, H_2CO_3, as shown by the equation:

$$CO_2 + H_2O \rightarrow H_2CO_3$$

Theme 4: Water

◆ **Ocean acidification** – a decrease in the ocean's pH that happens when carbon dioxide dissolves in ocean water.

◆ **Calcium carbonate** – $CaCO_3$. A compound found in the shells of molluscs and crustaceans, coral reefs and in rocks such as limestone.

◆ **Biomass** – the matter that makes up living organisms.

◆ **Stratification** – the formation of layers in a liquid or gas.

◆ **Persistent** – remaining for a long time.

◆ **Thermocline** – a layer of water, beginning a few metres below the surface, where the temperature drops rapidly with depth. The thermocline separates the warm surface water from the deeper, colder water.

Link

The carbon cycle is explained fully in Chapter 2.3 (page 139).

● Common mistake

People are often confused by the meaning of a pH value. Remember that acids have low pH, and alkalis have high pH.

As more carbon dioxide is added to the ocean, the concentration of carbonic acid increases, and the **pH** of the water is reduced. Natural seawater is slightly alkaline (with a pH of around 8.1), but as carbon dioxide is added, the water becomes more acidic (with a pH closer to 7). This is called **ocean acidification**.

Ocean acidification is harmful to many marine organisms. Many shelled animals, such as crustaceans (for example, crabs, shrimp and lobster) and molluscs (for example, snails and oysters) have shells made of **calcium carbonate**, $CaCO_3$. Calcium carbonate reacts with acids, breaking down into weaker, more soluble calcium oxide, CaO, and carbon dioxide:

$$CaCO_3 + H_2CO_3 \rightarrow CaO + H_2O + CO_2$$

On the other hand, the same sorts of animals also use carbon dioxide to build their shells, and aquatic plants and algae use the carbon dioxide in photosynthesis. So over time, some of the carbon dioxide dissolved in the oceans is taken in by living organisms and stored in their bodies and shells, as **biomass**.

When these plants and animals die, their bodies sink to the bottom of the ocean, where they may begin to decompose. Decomposition releases carbon dioxide back into the water, but some of the carbon remains in the muddy sediment on the seafloor, especially the harder parts of the organisms such as shells and skeletons. Over millions of years, the sediments may be compressed and heated, and gradually turned into rocks such as limestone, or into the fossil fuels, coal, oil and natural gas.

Under natural conditions, these fuels and rocks act as long-term stores of carbon. When humans extract fossil fuels from the ground and burn them, we release this sequestered carbon back into the atmosphere, millions of years after it was first dissolved in the ocean.

4.1.10 Temperature and density differences produce stratification in water

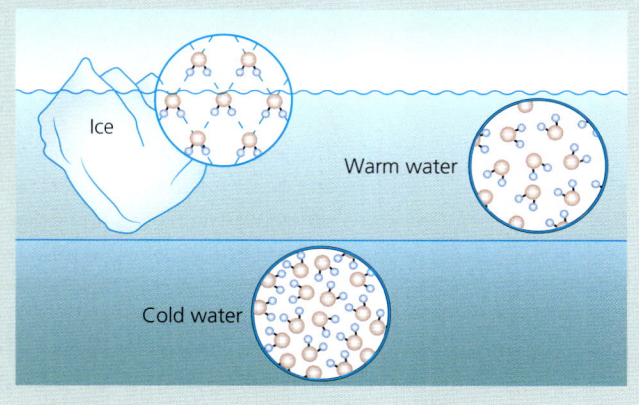

■ **Figure 4.19** Stratification of deep water

Oceans and lakes tend to be warmest near the surface, and to get colder deeper down. Warmer, less dense water tends to float on top of colder, denser water. Since warm water tends to rise, and cold water tends to sink, bodies of water tend to form distinct layers, which do not mix. This layered structure is called **stratification**.

Since the deeper water is colder and denser, it remains below the shallow, warmer, less dense water. As such, it is difficult for the different layers to mix. This means that anything that falls into the deep water will generally stay there for a very long time. We say that the stratification is **persistent**.

Salinity is another factor that affects the density of water. Saltier water is more dense than less salty water. Therefore, the deeper layers of water tend to be both colder and saltier than the shallower layers. Salinity also affects the temperature at which water freezes, lowering it slightly.

4.1.11 Stratification and thermoclines

Water near the surface of a lake or sea receives heat energy from the sun. This keeps it warm and less dense, so it floats on the colder water below. In some situations, there is a sudden drop in temperature a few metres below the surface, called the **thermocline**.

4.1 Water systems

Above the thermocline is the warmest water, which receives the most sunlight. Algae grows well in this warm, clear water, and this forms the base of aquatic food chains. Below the thermocline, the water is cooler and darker. Less heat and light reach these depths, and less algae grows. Animals in the deeper water depend mostly on food that falls from above.

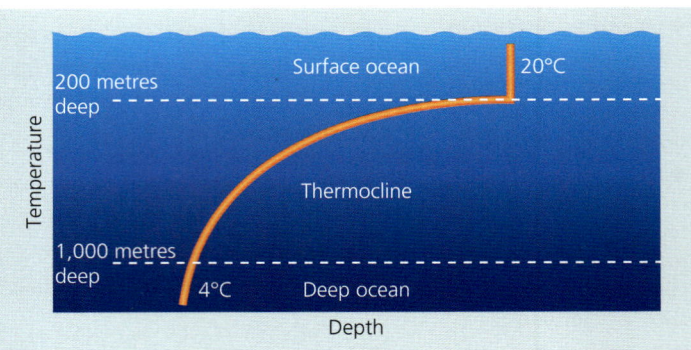

■ **Figure 4.20** A temperature profile of the ocean, showing the thermocline

The sinking particles of dead organisms gradually decompose, releasing minerals into the water. Decomposition is a slow process and most of the minerals are released into deeper water, below the thermocline. Because the water below the thermocline cannot mix with the upper layer of warm water, the minerals tend to remain in the deeper water. Therefore the upper layer of water usually has a low concentration of minerals.

At the surface, water is exposed to the air and to wind. The wind moves the water, producing waves and currents. Air becomes mixed into the water, and oxygen and carbon dioxide dissolve in the water. The winds and currents move the dissolved gases about within the upper layer of water, above the thermocline. As a result, the upper, warm layer of water often contains a high concentration of dissolved oxygen. However, stratification prevents this oxygen from reaching the lower layers. At the same time, most decomposition occurs at the bottom of the ocean, using up the little oxygen that is there. This means that the deep ocean water usually has very low oxygen content.

The exact patterns in the water layers depend on the specific geography of each region. Deep lakes and oceans tend to be very stratified. Coastal areas and areas of sea surrounded by land may show different patterns of stratification, depending on currents, surface temperatures and freshwater inflows. Estuaries, where fresh river water flows into the sea, often have a significant freshwater layer, floating on top of salty deep water.

4.1.12 Global warming and salinity changes have increased the intensity of ocean stratification

As the Earth's atmosphere gets warmer, more heat is added to the surface layers of the oceans. This means that the surface water becomes warmer and less dense. This makes ocean stratification more pronounced: the difference in temperature and density between the surface waters and deep waters is greater than it was before global warming. As a result, there is even less mixing between the warm, oxygen-rich surface water, and the cold, oxygen-poor deep water.

Global warming also affects the salinity of ocean water. In the polar regions, such as around Antarctica, the ice caps are melting. This adds extra freshwater to the sea, reducing the salinity and making it less dense. In warm areas, such as around the equator, evaporation has increased. When water evaporates, the salt is left behind, so salinity and density increase.

> **ATL ACTIVITY**
>
> **Sea surface temperature and salinity**
>
> Use an online database to find data on sea surface temperature and salinity in a selected geographic area – perhaps the sea closest to your home. How have temperature and salinity changed over recent years? What processes may have contributed to these changes?

Theme 4: Water

4.1.13 Upwellings in oceans and freshwater bodies can bring cold, nutrient-rich waters to the surface

Stratification usually produces a warm, nutrient-poor surface layer, and a separate, cold, nutrient-rich deep layer. In some situations, the cold deep water can be drawn up to the surface via upwelling.

> **REAL-WORLD EXAMPLE**
>
> ### Peru's anchovy fishery
> Along the Pacific coast of South America, the pattern of winds causes the surface waters of the Pacific to be blown away from the land, to the west. As the surface water moves away, colder water is pulled to the surface from deep below the ocean. This cold upwelling brings nutrients from the deep sea up to the surface.
>
> ■ **Figure 4.21** Upwelling off the coast of South America

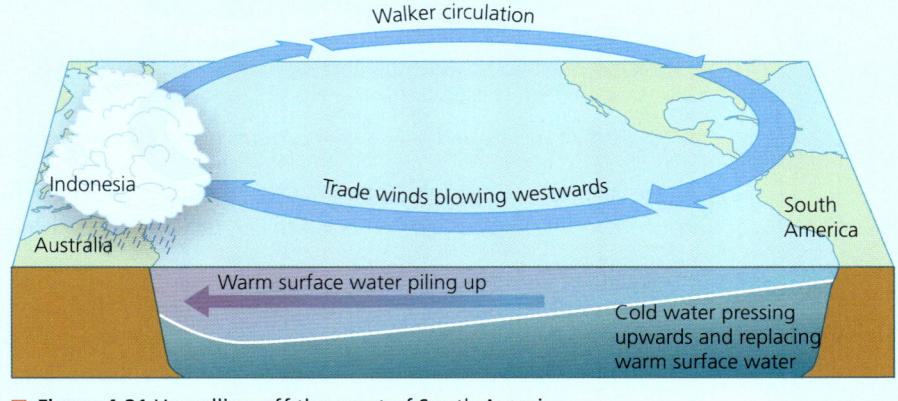

The nutrients are taken in by algae, which grow and reproduce. Without the upwelling, the algae would not have enough nutrients to grow.

The algae provide food for large numbers of anchovies – small fish that are used as food, and as an ingredient in many food products. The anchovies support an important commercial fishery in Peru, providing income to many people.

Occasionally, the upwelling off Peru's coast stops. No nutrients are brought to the surface, so the algae cannot grow, and there is no food for the anchovies. When this happens, Peru's anchovy fishery fails, and many people lose their main source of income. This happens because of a change in wind direction known as El Niño. El Niño weather patterns (see Topic 2.4.10, page 174) have become more common in recent decades, possibly due to global climate change.

4.1.14 Thermohaline circulation and the global ocean conveyor belt

Currents occur in all oceans, on the surface as well as in the deep. Wind and the rotation of the Earth both affect the direction and speed of the surface currents.

However, differences in temperature and salinity, and therefore differences in density, produce a global system of surface and deep-water currents called the thermohaline circulation. These currents form a continuous, interconnected flow, which is sometimes called the global ocean conveyor belt.

In cold regions, near the poles, water cools and becomes denser. At times, it is cold enough that some of the water freezes. When seawater freezes, the salt remains in the liquid sea water. This cold, salty water is even denser, and sinks to the bottom of the ocean.

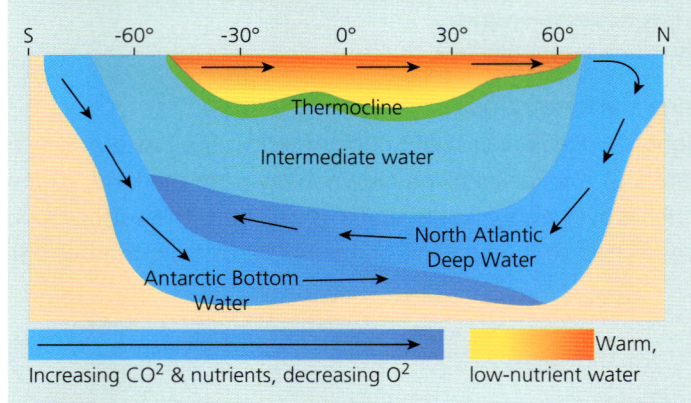

■ **Figure 4.22** Surface and deep-water currents in the Atlantic Ocean

As the cold, salty water sinks, it pulls surface water from warmer regions to replace it. Some of this warm water evaporates, leaving behind colder, saltier water, which again sinks when it reaches the polar regions.

The warm surface water moving towards the poles pulls more water behind it like a conveyor belt – wherever water goes, other water follows behind to fill the space it has left. Eventually, cold, deep water must be pulled back up to the surface, to replace the surface water that sank at the poles. The entire Earth is connected through this slow, steady flow, driven by differences in temperature and salt content.

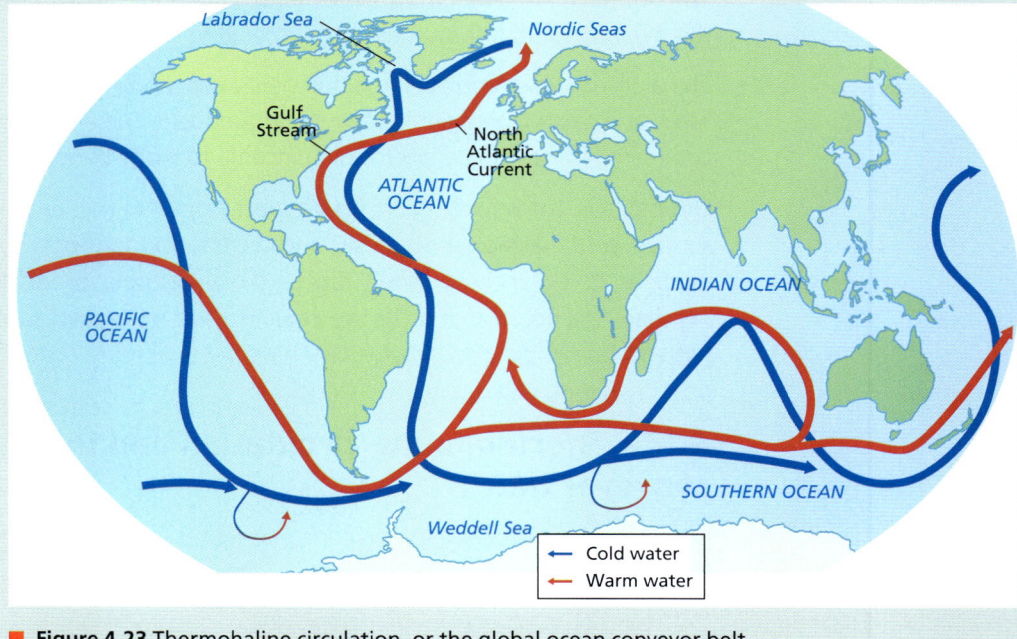

■ **Figure 4.23** Thermohaline circulation, or the global ocean conveyor belt

> **REAL-WORLD EXAMPLE**
>
> ### The Gulf Stream
>
> One section of the global thermohaline circulation is the Atlantic Meridional Overturning Circulation, or AMOC, which includes a surface current known as the Gulf Stream. The Gulf Stream flows from the region of the Gulf of Mexico, north and east across the Atlantic towards Europe, carrying warm water from near the equator.
>
> This warm water transfers heat energy from the tropics to the western parts of Europe, including Spain, France, Ireland and Great Britain. This heat transfer gives Western Europe a much warmer climate than would be expected from its latitude. Both Madrid and New York City are at about 40° North, but Madrid is much warmer than New York. London, at 51° North, is far warmer than Toronto, at 44° North.
>
> Beyond mainland Europe, the Gulf Stream continues towards Greenland and Iceland. Here it has lost much of its heat, and some of it has also evaporated. In the winter the water may become cold enough to freeze. The water in the Arctic is much colder, saltier and denser. It sinks and starts to flow back south, along the global conveyor belt.
>
> ■ **Figure 4.24** The Gulf Stream and the Atlantic Meridional Overturning Circulation
>
> There is some evidence that the AMOC, and perhaps the entire global thermohaline circulation, is slowing down, possibly due to global warming. Rising surface temperature means sea ice at the poles is melting and adding freshwater to the AMOC, just where it typically sinks. Since freshwater is less dense, the water sinks more slowly. It is possible that the sinking will stop entirely – although scientists are uncertain about this. If it does stop, it may have a major impact on global climate, as well as on the flow of nutrients and oxygen to and from the deep ocean.

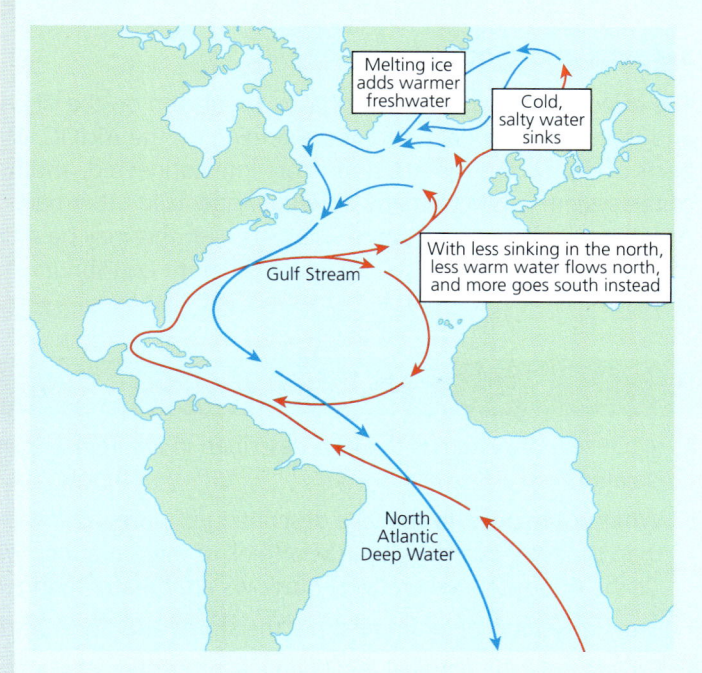

4.1 Water systems

HL.a.9: Environmental law

The international treaty of the Convention for the Establishment of the Intergovernmental Oceanographic Commission, which was adopted in 1960 at a UNESCO general conference, resulted in the creation of the Intergovernmental Oceanographic Commission (IOC). The commission is a major player in supporting global research in oceanography and ocean science. This includes research into El Niño–Southern Oscillation (ENSO) and the Atlantic Meridional Overturning Circulation (AMOC). The IOC facilitates the exchange of data, information and expertise from member nations and organizations. Several academic institutions, government agencies and research centres from different countries are actively engaged in the study of AMOC and ENSO.

> **ATL ACTIVITY**
>
> Visit the IOC website and explore the range of projects the agency is promoting by reading the Our Updates section of its website. Make notes on two projects of your choice in order to create a case study about the IOC.

Concepts

Systems and models

The effect of the global thermohaline circulation on the climate of Western Europe, and the possible impact of global warming on the AMOC, illustrate how different parts of the Earth's climate system are interconnected, and how a change in one part of the system can have many different impacts in other parts of the system. If the AMOC stops flowing, as some scientists fear, this may be a tipping point that will move the global climate system into a different equilibrium state.

REVIEW QUESTIONS

1. What sort of system is the lake described in Figure 4.2 (page 270): open, closed or isolated?
2. What will happen to the lake described in Figure 4.2 if the total inflows are greater than the total outflows? Can you think of any situations where this might happen? Or the opposite – where total outflows are greater than total inflows?
3. Draw a systems diagram to show the stores and flows of the hydrological cycle.
4. Explain how the physical and chemical properties of water make it essential to life.
5. Describe how the conditions in deep ocean water are different from those in water near the surface.
6. Explain why there is often a low concentration of oxygen and a high concentration of nutrients in the water at the bottom of the ocean.

4.2 Water access, use and scarcity

> **Guiding questions**
>
> - What issues of water equity exist and how can they be addressed?
> - How do human populations affect the water cycle, and how does this impact water security?

SYLLABUS CONTENT

This chapter covers the following syllabus content:
- ▶ 4.2.1 Water security is having access to sufficient amounts of safe drinking water.
- ▶ 4.2.2 Social, cultural, economic and political factors all have an impact on the availability of, and equitable access to, the freshwater required for human well-being.
- ▶ 4.2.3 Human societies undergoing population growth or economic development must increase the supply of water or the efficiency of its utilization.
- ▶ 4.2.4 Water supplies can be increased by constructing dams, reservoirs, rainwater-catchment systems and desalination plants, and through enhancing of natural wetlands.
- ▶ 4.2.5 Water scarcity refers to the limited availability of water to human societies.
- ▶ 4.2.6 Water-conservation techniques can be applied at a domestic level.
- ▶ 4.2.7 Water-conservation strategies can be applied at an industrial level in food-production systems.
- ▶ 4.2.8 Mitigation strategies exist to address water scarcity.

HL ONLY
- ▶ 4.2.9 Freshwater use is a planetary boundary, with increasing demand for limited freshwater resources causing increased water stress and the risk of abrupt and irreversible changes to the hydrological system.
- ▶ 4.2.10 Local and global governance is needed to maintain freshwater use at sustainable levels.
- ▶ 4.2.11 Water footprints can serve as a measure of sustainable use by societies and can inform decision making about water security.
- ▶ 4.2.12 Citizen science is playing an increasing role in monitoring and managing water resources.
- ▶ 4.2.13 Water stress, like water scarcity, is another measure of the limitation of water supply; it not only takes into account the scarcity of water availability but also the water quality, environmental flows and accessibility.
- ▶ 4.2.14 Water stress is defined as a supply of less than 1,700 cubic metres of clean, accessible water per year per capita.
- ▶ 4.2.15 The causes of increasing water stress may depend on the socioeconomic context.
- ▶ 4.2.16 Water stress can arise from transboundary disputes when water sources cross regional boundaries.
- ▶ 4.2.17 Water stress can be addressed at an industrial level.
- ▶ 4.2.18 Industrial freshwater production has negative environmental impacts, which can be minimized but not usually eliminated.
- ▶ 4.2.19 Inequitable access to drinkable water and sanitation negatively impacts human health and sustainable development.

4.2.1 Water security

Water is an essential part of all living organisms. Large organisms like humans need water for processes like transporting substances (in the blood), excreting wastes (in urine), and regulating body temperature (with sweat). Exactly how much water a human needs depends on factors such as age and activity level, and environmental factors like temperature and humidity.

Human societies often obtain water from rivers, lakes or aquifers, but not every society has easy access to these sources. Even if there is abundant water, it may not be safe to drink. Access to the sea, for instance, does not necessarily provide **water security**, since we cannot drink saltwater. Other sources may be polluted, making them unsafe as well.

Water society can be considered at a personal, society, or country level. If some people in a society do not have enough clean water, we could say that the society lacks water security. Different factors can contribute to water insecurity, including geographic, climatic, economic and political factors.

◆ **Water security** – having access to enough safe drinking water to meet people's needs.

> **Concept**
>
> **Sustainability**
>
> Water security is a state of having enough safe drinking water for a healthy life. In many places, while some people have enough safe drinking water, others do not. Water security can therefore be an issue of equity and justice, and is an essential component of sustainability.

> **ATL ACTIVITY**
>
> **Critically examine and evaluate evidence**
>
> Explore the definition of water security according to different multi-governmental organizations (such as the UN, WHO, World Bank) and non-governmental organizations (such as Water.org, Water Aid, World Water Council, WWF). In groups, compare and contrast the definitions you discovered and suggest reasons for the similarities and differences in different stakeholders' perspectives.

4.2.2 Factors affecting availability of and access to freshwater

In some countries, certain groups of people have greater water security than others.

Certain factors may combine to determine whether a particular group of people has access to sufficient and safe freshwater. These factors can be categorized as social, cultural, economic or political, but these categories may overlap. For instance, social differences may reflect economic differences, and these may be produced by political decisions. In many cases, multiple factors contribute to a particular water-security problem, as shown by the Real-world examples below.

> **REAL-WORLD EXAMPLE**

Flint, Michigan, USA: Political, economic and social factors

In 2014, in the midst of a financial crisis, authorities in the city of Flint in the state of Michigan decided to switch the water supply, to reduce the cost of supplying water. They switched from the Detroit water system, which used water from Lake Huron, to the Flint Water Treatment Plant, which obtained water from the Flint River.

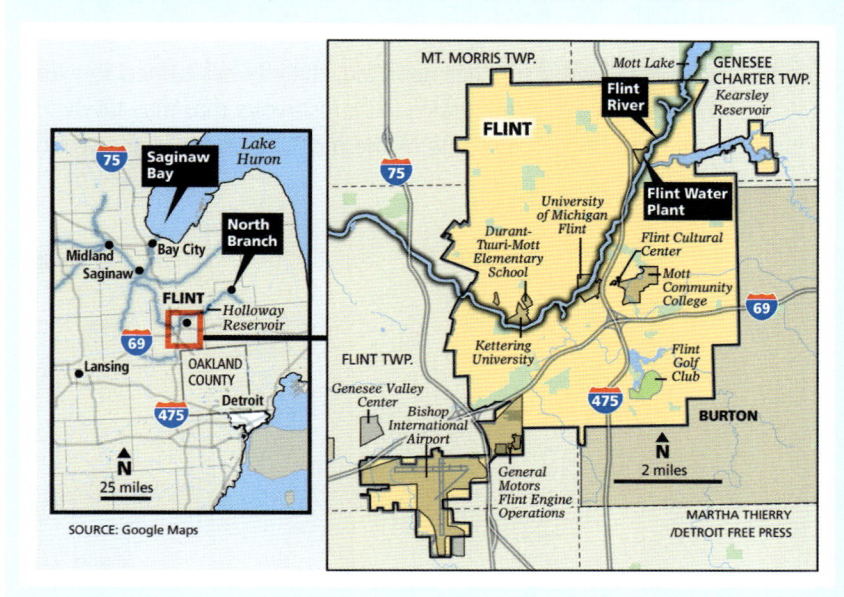

■ **Figure 4.25** The city of Flint, Michigan showing the two alternative water sources

Soon after the switch, a number of health concerns were raised. Residents complained about the water's unpleasant taste and appearance, and reported experiencing rashes and skin irritation. An outbreak of Legionnaires' disease occurred from 2014–15, killing at least 12 people. High concentrations of lead, which is toxic to humans, were also found in the tap water.

These effects were explained by differences in the quality of the source water, and in the treatment processes used. In particular, the new treatment plant did not use corrosion inhibitors, so lead from the city's pipes dissolved and entered the water supply.

Flint is a city with an industrial history, but it has suffered economic decline. Over 40 per cent of its population lives in poverty. The residents are predominantly African-American, in a country where African-Americans are a minority ethnic group and often socially and economically marginalized. A taskforce commissioned by the Governor of Michigan concluded that in this instance, the residents of Flint did not receive the level of service, or the protection from environmental and health problems which other, less disadvantaged, communities receive.

Equitable distribution of water can be particularly challenging in large, growing cities. With expanding populations, and increasing demand for food production and industry, it can be difficult to ensure that everyone has access to sufficient water. Poorer communities, living on the edges of cities in unplanned areas, tend to experience the fastest population growth, and also the poorest water supply.

> **REAL-WORLD EXAMPLE**
>
> ### Mexico City, Mexico: Economic and social factors
>
> Mexico City is a large city, located at 2,300 m altitude and surrounded by tall mountains. It was built in an area with several lakes and a high water table. Over time, as the city has grown, both the surface and groundwater have been severely depleted. Over-abstraction of groundwater has led to severe subsidence: the city is currently sinking at a rate of 7 cm per year.
>
> With local water supplies depleted, the city has turned to external sources. Water is now pumped over the mountains from reservoirs that may be over 100 km away. This makes water in Mexico City some of the most expensive in the world.
>
> ■ **Figure 4.26** Mexico City and its water sources
>
> Some areas of Mexico City have a better water supply than others. The wealthier, western neighbourhood known as Miguel Hidalgo receives a steady, high-pressure water supply. In contrast, the poorer, eastern neighbourhood of Iztapalapa, with its large and growing population, has an intermittent and unreliable water supply. High costs and physical challenges of providing a pipe-borne supply to a crowded neighbourhood with limited infrastructure mean that residents of this underprivileged community cannot easily meet their needs for water.

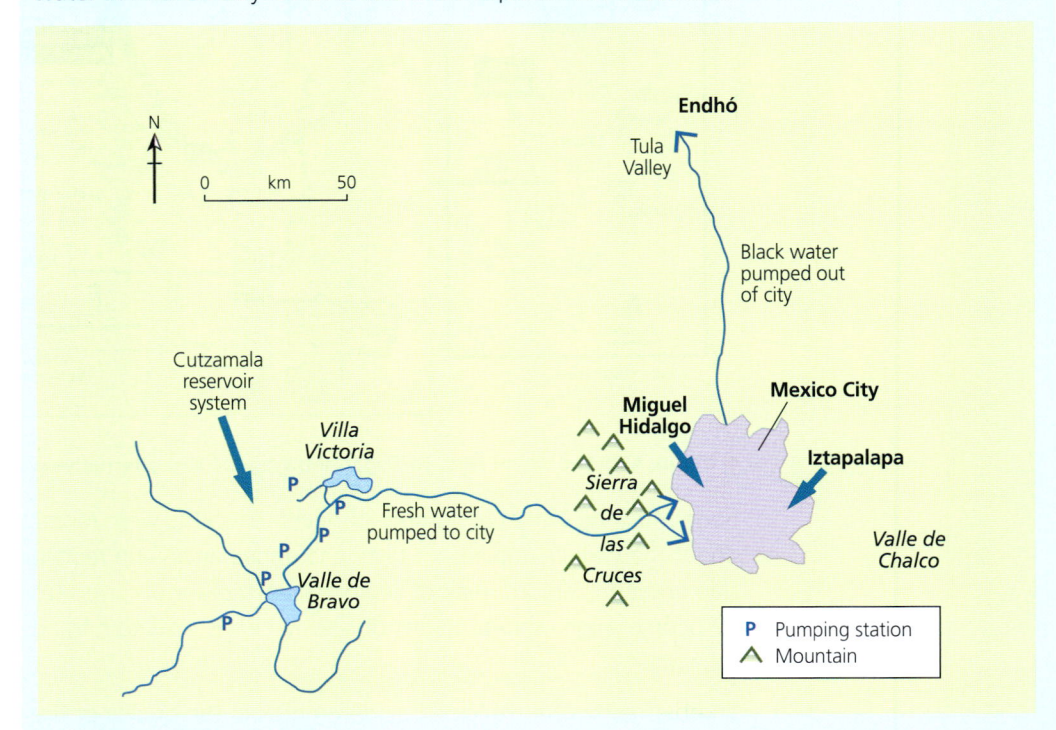

Cultural factors relate to the way people live. Culture includes the ideas, customs, beliefs and common behaviours of a society. Many of these relate to the way people use and treat water. Some cultures, for instance, tend to be more wasteful with water than others. Culture may also be influenced by social and economic factors, for example wasteful habits sometimes arise because of easy access to resources, which may be due to affluence.

> **REAL-WORLD EXAMPLE**
>
> ### Ganges River, India: Cultural, economic and political factors
>
> The Ganges River in northern India is one of the largest and most heavily populated river basins in the world. Hundreds of millions of people rely on the river for drinking water and to irrigate their crops. The river is also highly polluted, due to untreated sewage, industrial and agricultural runoff, and the use of the river for traditional Hindu cremations.
>
> ■ **Figure 4.27** The Ganges River basin
>
> Hindu believers consider the river to be sacred. They refer to the river as Mother Ganga, and ritually bathe in it to wash away their sins, and drink its water for good fortune. Many cremations are carried out on the banks and the ashes are scattered in the river.
>
> Water security in the Ganges River basin is threatened by the high level of pollution and the increasing extraction of water for the irrigation of crops. Climate change may also have an effect, as the Himalayan glaciers that feed the Ganges are getting smaller. These factors make it harder to ensure a safe, regular supply of water to the people of northern India.
>
> Cultural practices and industrial activities have contributed to the pollution of the Ganges. Industry and farming reduce the availability of water. Population growth – a social factor – has also increased the demand for drinking water and crop production.
>
> Poor water quality has an impact on human health, as people often use untreated or poorly treated water. This exposes them to water-borne diseases such as cholera and typhoid fever. Poorer people, including farmers and fisherfolk, are more vulnerable to disease as they are more likely to consume untreated water.
>
> Cultural and political factors, on the other hand, might help improve the situation. Partly because of the high cultural and religious value of the river, the government has established policies and committed significant funds to cleaning up the river.

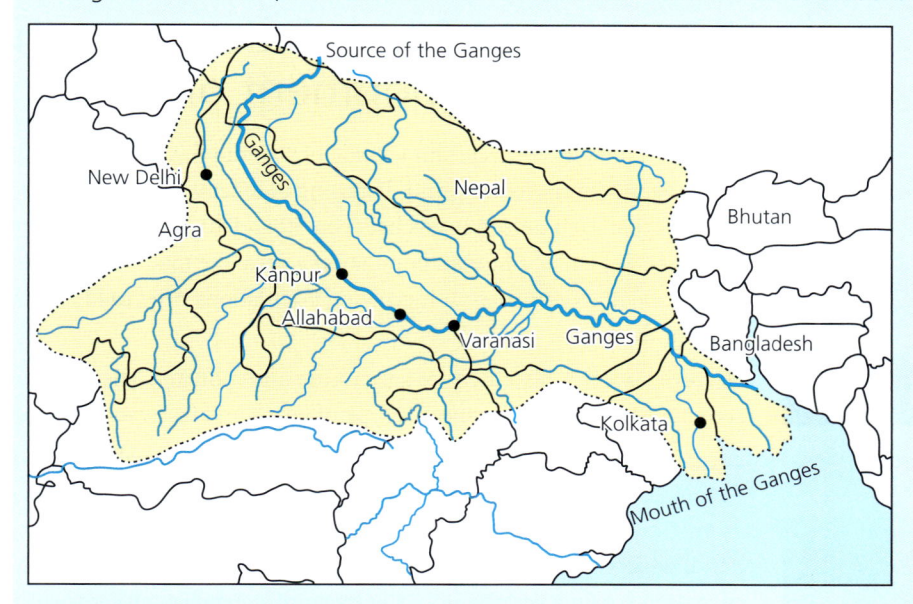

4.2 Water access, use and scarcity

> **ATL ACTIVITY**
>
> ### The availability of water
> Choose an area or country with which you are familiar and investigate the availability of water to different parts of the society. Are there some groups of people who receive less reliable water supplies than others? What sorts of factors contribute to these inequalities?

4.2.3 Effect of population growth and economic development on water use

A single human needs a certain volume of water to meet their individual needs: such as for drinking, washing, cooking and disposing of toilet waste. The World Health Organization suggests that each person needs between 50 and 100 litres of water per day to meet their domestic needs. Therefore, as a human population grows, the total water needs of the society will increase proportionally. Countries with growing populations need to take this into consideration as they plan for the future.

However, water is also used for agriculture and industry. The volume of water needed for these activities depends on the specific type of farms and factories involved.

■ **Figure 4.28** Irrigation on commercial farms

Large-scale commercial farming uses more water than smaller-scale or subsistence farming. Commercial farms are more likely to use irrigation systems, which pipe or spray water over the crops. Small-scale or subsistence farms are more likely to depend on rainfall and watering by hand, which is more sustainable but is labour-intensive. Livestock farming also uses more water than crop farming, since animals need more water than crops do.

Similarly, industries vary in their water use, depending on the scale and nature of the industrial processes. Some processes require more water than others.

■ **Figure 4.29** Irrigation on subsistence farms

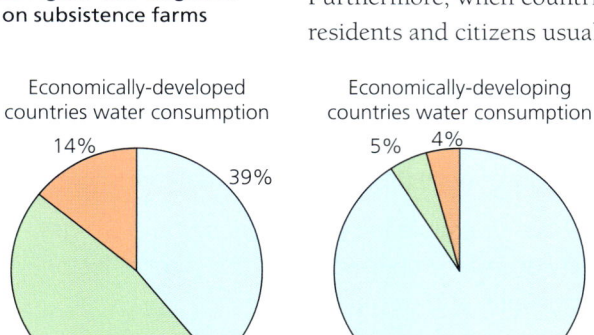

Furthermore, when countries become more economically developed, the average income of residents and citizens usually rises. With more economic activity, there is more money circulating in the economy, and people's incomes usually increase. As people become wealthier, their use of water changes. There is a tendency to be more wasteful, compared to people who have no pipe-borne water supply, or who struggle to pay their water rates. They may be more inclined to use water for washing cars or watering lawns. They may wash clothes more frequently. They may engage in more water-intensive leisure activities like visiting swimming pools or playing golf.

Economically-developed countries water consumption
- 39%
- 47%
- 14%

Economically-developing countries water consumption
- 91%
- 5%
- 4%

☐ Agriculture
☐ Industry
☐ Domestic

■ **Figure 4.30** Domestic, agricultural and industrial use of water in economically-developed and developing countries

Therefore, as countries become more economically developed, they implement more large-scale and high-tech factories and farms, which increases the demand for water. At the same time, individuals and households tend to be more wasteful with water and to engage in water-intensive activities. Overall, demand for water increases.

Theme 4: Water

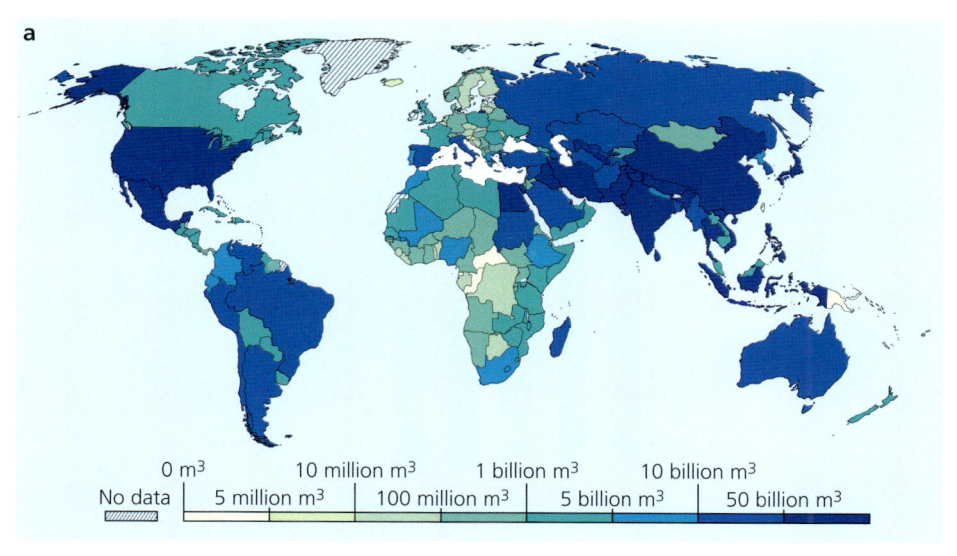

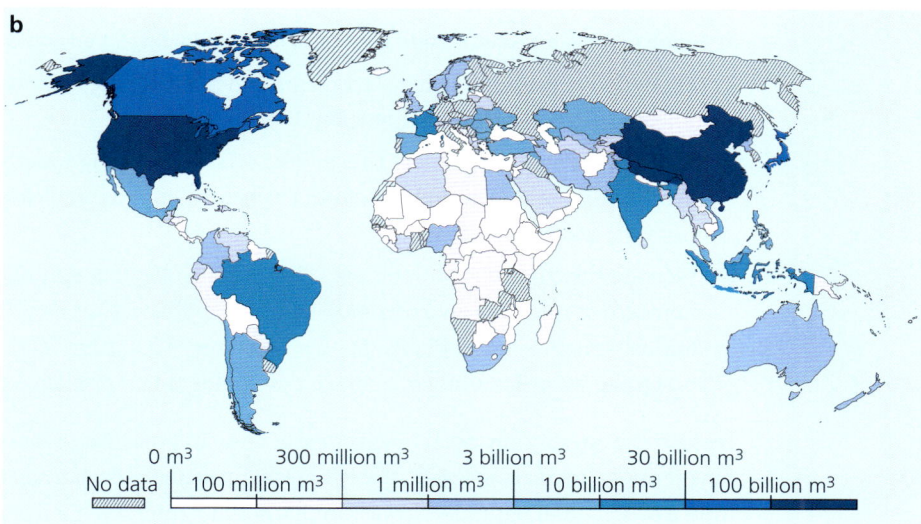

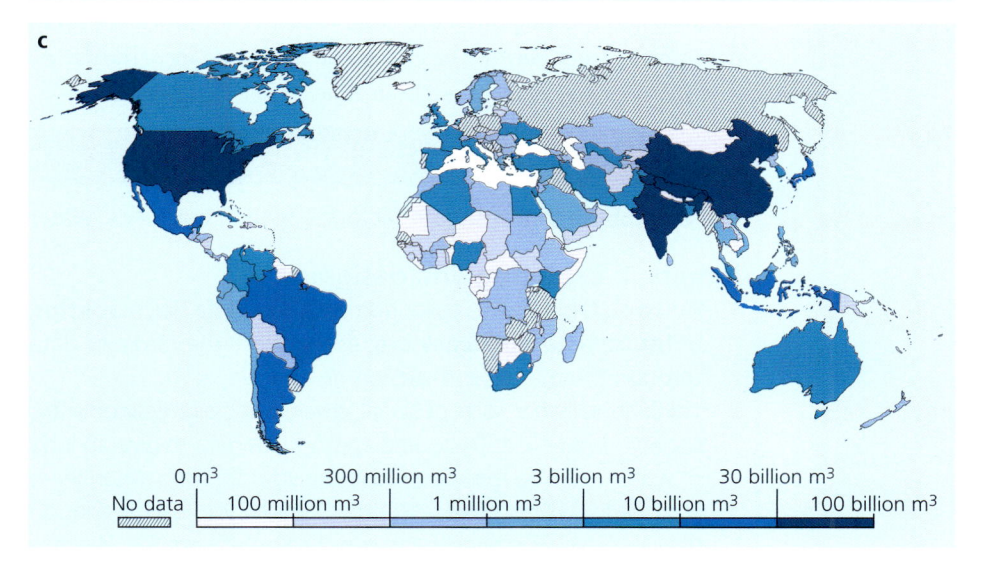

■ **Figure 4.31** Water use by sector around the world: (a) agriculture, (b) industry, (c) domestic (source: Our World in Data/Food and Agriculture Organization of the United Nations – AQUASTAT)

4.2 Water access, use and scarcity

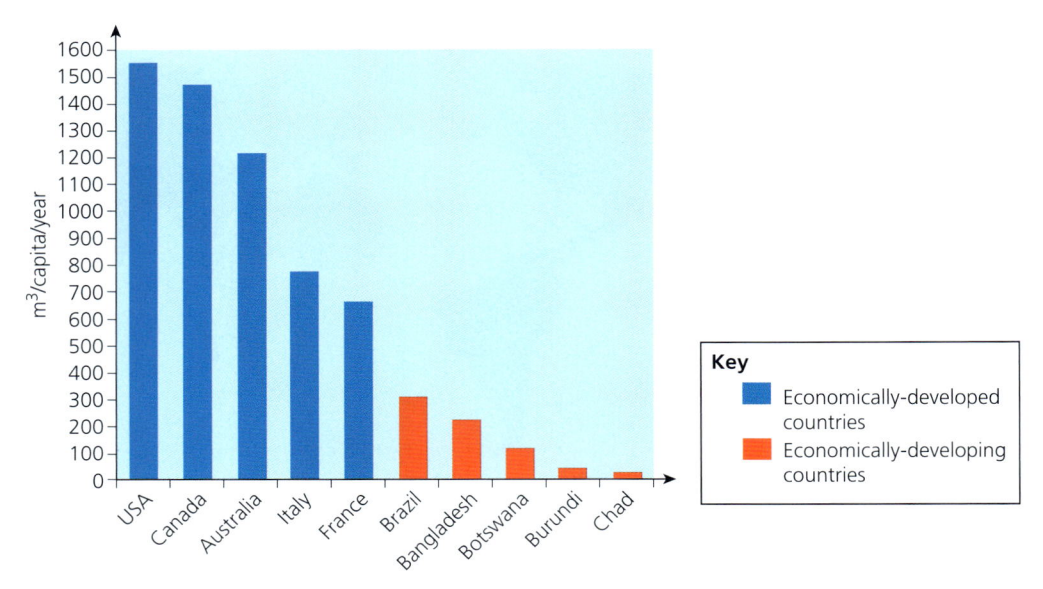

■ **Figure 4.32** Water consumption per capita in selected developed and developing countries

There are two approaches that governments and planners may take as their populations increase and their societies develop economically:

1. Increase the supply of water: This might involve extracting water from new sources, such as deeper or more remote aquifers, from rivers further away from where the water is needed, or from the sea, by desalination.

2. Increase the efficiency of water use: This may involve redesigning systems that use water. In homes, this might mean using more efficient appliances, or recycling used, slightly dirty water to flush toilets or water plants (see Grey water in 4.2.6, page 304). In factories, it could involve designing water-recycling systems or developing different manufacturing processes.

Increasing water supply often involves extra cost, and may cause new environmental impacts. Furthermore, increasing water supply may be unsustainable, if it involves removing more water from a source than naturally flows into that source. Improving efficiency may cost money initially, but in the long run it can often reduce costs. Efficiency is also generally better for the environment, but it may not be enough on its own to meet the society's growing needs. In many cases, societies may need to use both strategies at the same time.

Inquiry process

Use a database to investigate factors affecting countries' water use.

Inquiry 1: Exploring and designing
1. Find a suitable online collection of global data, such as **https://ourworldindata.org** or **https://data.worldbank.org**, and explore the range of data available by scanning through the menus and section headings.
2. Think about what sorts of social, demographic or economic factors might affect a society's water use. Try to find appropriate data to give an indication of these factors, and of water use. You may need to refine your thinking when you see what data is available.
3. Formulate a hypothesis relating to one factor you have identified, and one measure of water use. The hypothesis may be something like, 'As factor A increases, water-use statistic B will decrease.' Identify which of these factors will be your independent variable and which will be your dependent variable.

4 Refine your hypothesis by limiting it to a certain context, such as a geographic region, a timeframe, or societies with particular characteristics. If you restrict your search to certain conditions, you are introducing controlled variables into your investigation.
5 Formulate a clear, focused research question, which relates to the variables you want to investigate, and states the context or limitations of your investigation.

Inquiry 2: Collecting and processing data
1 Download your selected data from the database and open the file.
2 Extract the relevant data points and place them into a suitable table. Ensure that your table columns have clear headings and correct units.
3 Give your table a heading, including the source of your data.
4 Create a graph, using a spreadsheet programme. If you have investigated the relationship between two variables, you might use a scatter plot (XY) graph and insert an appropriate trendline.
5 Consider how best to analyse your data. For example, if you have looked at two factors that you think might be correlated, you might carry out a Spearman's rank correlation test.
6 State null and alternative hypotheses for your statistical test.
7 Analyse your data. For example, if you are using Spearman's, calculate the value of r_s. Use an online calculator, a spreadsheet, or a table and calculator to do this.
8 Use the appropriate table of critical values to determine whether there is a significant correlation or difference between the two factors you investigated. State whether you can reject your null hypothesis.

Inquiry 3: Concluding and evaluating
1 Draw a conclusion based on your research question and your statistical test result.

4.2.4 Increasing water supplies

Dams and reservoirs

A reservoir is an artificial store of water. River water that would normally flow downstream into the sea is pumped or diverted into an artificial tank or basin, where it is held until it is needed. This is a popular approach in places where rainfall and river flow are variable – such as in areas with long dry seasons. The reservoir allows water to be stored for times when it is needed.

One way of creating a reservoir is to build a dam across a river. A dam is a strong wall that is built across a river channel. Water accumulates behind the dam, creating an artificial lake, which works as a reservoir. In addition to storing water, dams can be used to produce hydroelectric power.

■ Figure 4.33 A dam and reservoir in Wales, UK

REAL-WORLD EXAMPLE

The Three Gorges Dam, China

China completed construction of the Three Gorges Dam across the Yangtze River in 2006. The main purpose of the dam was to produce hydroelectric power and control flooding, but it has also helped provide freshwater for the growing agricultural and industrial activities in the area. The dam stores water during the rainy season, preventing flooding in downstream areas. During the dry season, much of the stored water is released into the river, providing a steady, predictable supply of freshwater.

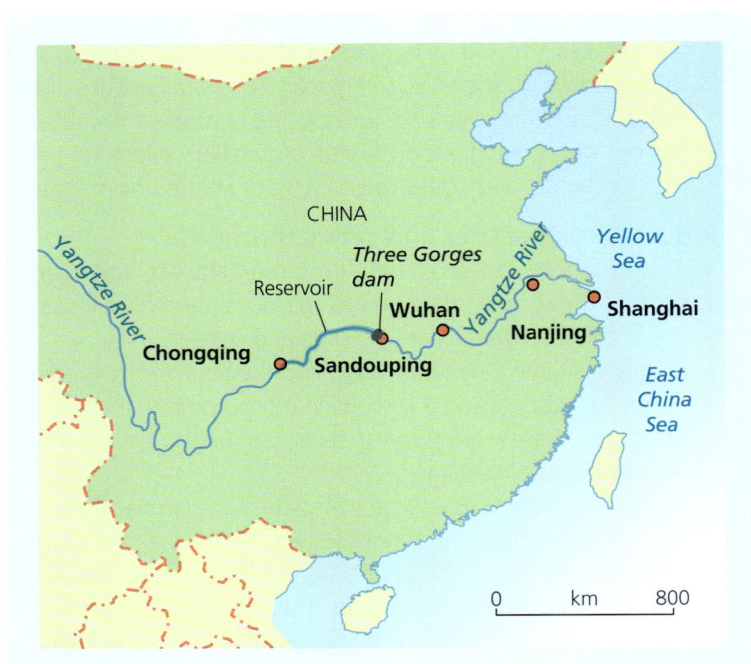

Construction of the dam produced a number of social and environmental impacts. Changes to the river ecosystem may have caused serious ecological damage. Some environmentalists have claimed that several important species have been affected, with some thought to be extinct (see Table 4.1).

■ **Figure 4.34** Location of the Three Gorges Dam

■ **Table 4.1** Species suspected of being affected by the Three Gorges Dam

Common (English) name	Latin binomial	IUCN conservation status	Year assessed
Chinese paddlefish	*Psephurus gladius*	Extinct	2019
Chinese sturgeon	*Acipenser sinensis*	Critically Endangered	2019
Yangtze sturgeon	*Acipenser dabryanus*	Extinct in the Wild	2019
Chinese river dolphin	*Lipotes vexillifer*	Critically Endangered (Possibly Extinct)	2017

Well over a thousand human settlements, including cities, towns and villages, were submerged in the reservoir behind the dam, and over 1 million people were relocated to new homes. Movements such as these can prove challenging for people, who have to adjust to new social and cultural environments, and find new jobs in an unfamiliar environment.

Link

Hydroelectric power is covered in Chapter 7.2 (page 634).

● HL.c.10, HL.c.13: Environmental ethics

The relocation of people to make way for a development project such as a dam raises difficult ethical questions. A consequentialist approach may be taken to determine the most ethical outcome.

In the case of the Three Gorges Dam, many people living downstream of the dam have benefited from its construction. They are at a lower risk of flooding and have a steadier supply of water. This may also provide economic benefits in terms of greater agricultural and industrial production, which might benefit people living in those areas. More widely, the production of hydroelectric power can have economic and environmental benefits for the entire country. The environmental benefits may even be global if it allows China to reduce its greenhouse gas emissions.

Using a consequentialist approach, the benefits outlined above may be compared to the possible harm caused to the people who had to be relocated, and the ecological damage to the Yangtze River ecosystem. If the overall benefits are thought to be greater than the overall harm, then it can be argued that the dam was an ethically good development.

A social justice approach, however, may come to a different conclusion. In many cases around the world, the people who are forced to leave their homes come from marginalized or disenfranchised groups, such as ethnic or cultural minorities. Moving people who have no power to argue against their removal, and who are likely to face extra challenges when they are relocated, might be considered to be ethically wrong, and to contribute to inequity and injustice in the society.

■ Rainwater-catchment systems

Rainwater-catchment systems catch rainwater that falls on houses or on to the ground, and channel it into a storage tank. From the tank, it can be pumped to where it can be used.

Rainwater can be used for a variety of purposes in homes, gardens and farms. It is not recommended to drink rainwater unless it has been treated to remove contaminants and disease-causing organisms, but it is usually safe for flushing toilets, watering plants or general cleaning. By using rainwater for these activities, we can save the treated water in the mains supply for drinking and cooking. Effectively, we are reducing our consumption of treated water, without necessarily reducing our overall use of water.

■ Desalination

Desalination is the process of producing freshwater from saltwater. It can be a very expensive process, requiring a lot of energy, so it is usually only used in very dry places where there is a cheap supply of energy.

There are two general approaches to desalination:

- Distillation: In this process, seawater is heated by burning fuel, and the steam is collected and condensed. The salt does not vaporize, meaning that pure freshwater is collected. Distillation of water is often used to produce pure water for use in car batteries and science laboratories. It is also the process used to produce alcoholic spirits, such as whiskey and rum. However, its use for desalination is relatively uncommon.

- Reverse osmosis: This, more popular, process involves forcing seawater through a semipermeable 'membrane', which effectively filters out the salt. The 'membrane' is actually a porous substance with tiny pores that selectively allows water molecules, but not salt, to pass through. A powerful electrical pump is used to force the water through the tiny pores. Reverse osmosis can also be used to treat wastewater.

● **Common mistake**

The membrane used in reverse osmosis is not like the thin, flexible membranes that surround living cells, or the tubing used in dialysis machines. Instead it is a tough material that is strong enough to withstand very high pressures.

■ Figure 4.35 The membrane used in reverse osmosis

Both methods of desalination produce a highly concentrated salt solution (brine) as a waste product. The salinity of the brine is often too high for living organisms, so releasing it into the natural waterways can have a negative environmental impact.

> **REAL-WORLD EXAMPLE**
>
> ### Ras Al-Khair Power and Desalination Plant, Saudi Arabia
>
> Saudi Arabia is a very dry, but very wealthy, country, with abundant fossil fuel reserves. In order to meet its growing demand for water, it uses its relatively cheap energy resources to carry out desalination of seawater.
>
> The Ras Al-Khair desalination plant uses both distillation and reverse osmosis methods to extract freshwater from seawater. It also generates electrical energy during the distillation process by using the heated gases to turn electrical turbines. This improves the efficiency of the entire plant.

■ Enhancing wetlands

Wetlands are areas of land that are at least partially covered by water, and that have various types of vegetation – trees, shrubs or herbs – growing out of the water. They include swamps, marshes, bogs and fens. They can be extremely productive and biodiverse ecosystems. Wetlands also play several important roles in the natural hydrological cycle. These roles can be described as ecosystem services.

- Water storage: Wetlands slow down the flow of water over the land, and hold it in one location, just as a lake or reservoir does. The wetland vegetation plays a role in slowing down the flow of water.
- Groundwater recharge: Since the water is held in place over time, more of the water will infiltrate the soil and percolate into the groundwater.
- Sediment removal: As the water slows down, much of the sediment it carries will settle to the bottom. The water that flows out of the wetland will therefore be clearer.
- Pollution reduction: Wetland vegetation will often absorb nutrients from the water, reducing problems of eutrophication. In some cases, wetland vegetation also absorbs other pollutants, including toxic chemicals. The water that flows out of the wetland may be cleaner than the water that flowed into it.

> **Link**
>
> Ecosystem services are explored more fully in Topic 7.1.5 (page 613).

Wetlands therefore help improve water security in several ways – both increasing the volume of water available and improving the water quality. They can also play roles in protecting biodiversity, and in removing carbon dioxide from the air.

Many wetlands have been degraded over time. People often see wetlands as wasted space that could be used for housing, farming or other activities. People may fill in wetlands, creating dry ground that can be used for other purposes.

Increasingly, people are more aware of the value of wetlands, and there is a growing trend to rehabilitate or restore damaged wetlands, to enhance or improve existing wetlands, or even to create new wetlands.

> **ATL ACTIVITY**
>
> ### Approaches to meet societies' water needs
>
> Select two developing nations from different parts of the world and investigate how they each meet their society's water needs. Discuss why they use those approaches and whether any other options might be available to them.

4.2.5 Water scarcity refers to the limited availability of water to human societies

◆ **Physical water scarcity** – limited availability of water because there is insufficient water in an area.

In many parts of the world, water scarcity means that people do not receive enough water to meet their needs. Water scarcity may be caused by either physical scarcity or economic scarcity. Water may be scarce continuously or for specific periods, such as during the dry season.

Physical scarcity occurs because there is not enough water in the area where people live. This typically occurs because of geography or climate – some places do not receive enough rainfall, and do not have enough surface and groundwater storage to meet the society's needs.

REAL-WORLD EXAMPLE

Jordan: Physical water scarcity

The Hashemite Kingdom of Jordan, in the Middle East, has a hot, dry climate and very limited water resources. It has the capacity to provide less than 100 cubic metres of water per person per year, making it one of the most water-scarce countries in the world. Almost all households receive pipe-borne, treated water, but about half only receive water once per week. Many homes have rooftop tanks to store water and ensure access for the rest of the week. A rapidly growing population means that the water resources will become even more stretched in the future.

Jordan obtains water from the Jordan and Yarmouk Rivers, but these are shared with neighbouring countries, such as Syria, so Jordan receives only a limited supply. Groundwater resources are also used, but these are being depleted faster than they can recover. To meet the population's needs, water is also purchased from neighbouring countries and wastewater is treated for reuse.

Jordan has limited access to the sea, so even desalination is not an easy option. Various proposals have been made to pipe water from the Red Sea coast to inland desalination plants, or perhaps to the Dead Sea.

◆ **Economic water scarcity** – limited availability of water because a society lacks the ability to store and transport water.

Economic scarcity occurs when a society is unable to store or transport enough water to meet everyone's needs. There may be enough water, but not enough tanks, reservoirs and pipes. A society may be able to overcome economic scarcity by investing more money into building the infrastructure needed to store and distribute more water.

REAL-WORLD EXAMPLE

Haiti: Economic water scarcity

The Republic of Haiti occupies the western side of the island of Hispaniola in the Caribbean Sea. It has a hot and humid tropical climate, with abundant rainfall, but almost 70 per cent of its population does not have direct access to safe drinking water. More than 70 per cent lack basic sanitation services. Haiti has experienced several outbreaks of waterborne diseases, such as cholera (source: https://rcc.cimh.edu.bb/files/2018/06/Country-Profile-Haiti.pdf, https://chrgj.org/wp-content/uploads/2009/06/wochnansoley.pdf and www.globalwaters.org/sites/default/files/external_haiti_gws_hpc_plan_2023.pdf).

Haiti has a long history of economic and political challenges (as well as certain natural hazards such as earthquakes) which have resulted in its being among the poorest, most unequal and least developed nations in the world. These challenges have made it difficult for its government to provide water services for its entire population.

Figure 4.36 The location of Haiti

4.2.6 Water conservation at a domestic level

It may be possible to save enough water for the most important things, like drinking and cooking, by reducing how much water is used for less important or non-essential things, like cleaning. Homeowners can use several strategies for water conservation.

Metering

Where water is delivered to homes in pipes, a water meter can be installed to measure the volume of water being used in the home.

Water companies usually use meters to work out how much each homeowner should pay. Typically, they set the price at a certain amount per litre, so that the more water used, the more the homeowner pays.

If homeowners have to pay for water by the litre, this may encourage them to reduce their water use in order to save money. Where metering is not used, water prices may be set at a constant value each month or year. In this case, there is no incentive for the homeowner to conserve water.

Figure 4.37 A domestic water meter

Even if homeowners do not pay by the litre, having a water meter may allow them to track their water use and identify ways to reduce water use. A meter reading might be the first clue that there is a leaking pipe in the house, for instance. If the meter keeps going up, even when no one is using water, it may be that a pipe is leaking water into the ground.

Rationing

When water is scarce, homeowners might decide to ration or limit the amount of water they use for various activities. There are several ways to ration water in the home:

- Setting a time limit on showers or a depth limit on baths.
- Washing dishes in a bowl or plugged sink, rather than with a running tap.
- Setting appliances like dishwashers and washing machines to use the least water possible.
- Limiting how often the dishwasher or washing machine is used.
- Restricting the use of water for non-essential things like washing cars and watering plants.
- Restricting toilet flushing (for example, no flushing for liquid-only waste).

Rationing like this can require self-control and usually involves changing habits. Technological measures can also help. For instance, taps which automatically close after a set time are a simple and effective measure.

Water-saving devices

Many of the systems in our homes which use water can be redesigned to reduce their water use. A popular example is the low-flush toilet. These use less water than a conventional toilet, as they are designed to flush with less water. Dual-flush toilets have low-flush and standard-flush options. One button activates the low flush, for liquid waste, and another button produces a standard flush, for solid waste.

Many electrical appliances, such as dishwashers and washing machines, can be designed to use less water. Modern appliances often use less water than older ones, and have 'eco' options that allow for even lower water consumption.

Greywater recycling

Greywater is water that has been used and slightly dirtied, but is still clean enough for certain uses. It includes water that has been used for bathing, handwashing, dishwashing or laundry. This water normally runs into the drains as wastewater. Greywater is no longer clean enough to drink, but it can still be used in various ways, such as for watering plants and flushing toilets.

For greywater recycling, pipes can be run from sinks and showers to a storage tank, and then to toilets or other places where it can be used. In some homes a small water-treatment system is included, which cleans the greywater. This may allow it to be used for a wider range of purposes.

Even without pipes and storage for greywater, some recycling may be possible. A person might collect their used bathwater, for instance, and carry it in buckets to other parts of the home for reuse.

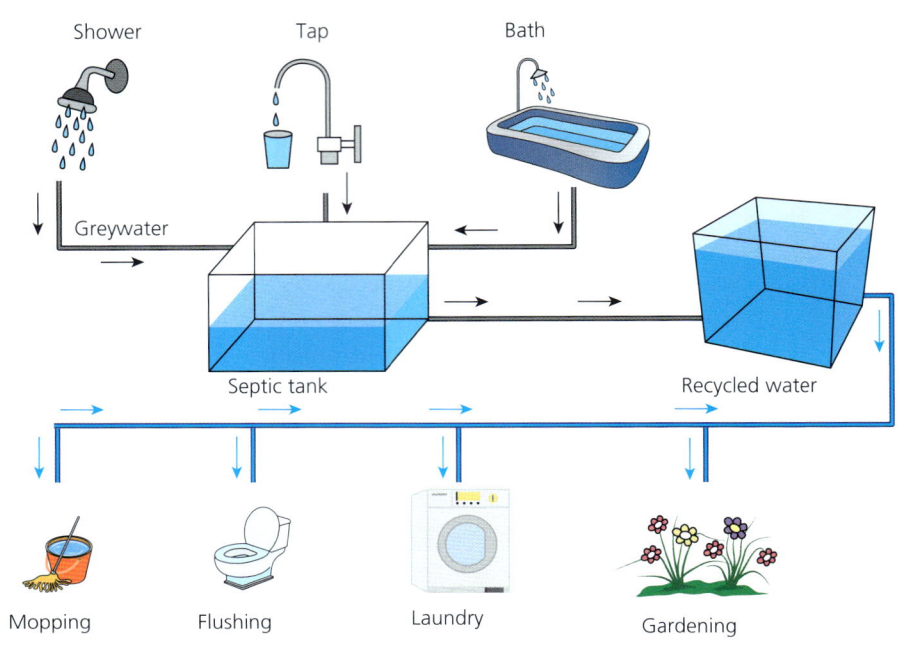

■ Figure 4.38 An example of a greywater-recycling system

Rainwater harvesting

In places where water is scarce, homes may be equipped with rainwater catchment systems (see 4.2.4, page 299). This usually involves drains running from the roof and other surfaces, into storage tanks called cisterns that hold the water until it is needed. Rainwater may not be safe enough to drink, but it can be used in the same ways as greywater – such as for flushing toilets and watering plants, as well as for washing and bathing.

■ Figure 4.39 Rainwater harvesting in Uganda

Link

Different agricultural systems and practices are explored in Chapter 5.2 (page 456).

4.2.7 Water-conservation strategies in food production

Farms use a lot of water. Crops and livestock need water to grow. In areas where water is scarce, farms may need to implement water-conservation strategies to reduce their consumption of water while still producing food. This is more important for large-scale commercial or industrial farms. Several strategies are available.

■ Harvesting and recycling rainwater

Many farms rely directly on rain to provide water for crops. However, during dry seasons or in drier locations, farmers may need to collect and store rainwater. In smaller farms, they may build ponds or ditches. On bigger farms, larger tanks may store water collected from the roofs of farm buildings.

Link

Aquaculture is explored in Chapter 4.3 (page 330).

■ Aquaponics

Aquaponics is a farming system which combines aquaculture with hydroponics.

In hydroponics, fruit and vegetable crops are grown in water, rather than soil. The farmer can carefully control the quantity and quality of the water, including the concentration of minerals, to precisely match the needs of the crops. A carefully designed and operated hydroponics system can use less water than conventional soil-based farming.

Aquaculture is the rearing of aquatic animals like fish or shrimp in tanks or ponds. It is often considered a more productive, and perhaps more sustainable, method of producing fish for human consumption. However, aquaculture typically requires large volumes of water. If fish are kept in a small tank or pond, their waste will quickly accumulate, which may cause the fish to become ill. In conventional aquaculture, the water in the ponds must be regularly flushed out and replaced. This means that large volumes of clean water are used, and large volumes of polluted water are discharged into rivers and streams.

In aquaponics, fish are grown in the same water that is used for crops. The crops take nutrients out of the water, cleaning it in the process. The fish therefore provide the fertilizer for the crops, and the crops help to clean the water for the fish. In this way, aquaponics can reduce the total amount of water needed, while efficiently producing both crops and animal products.

■ **Figure 4.40** Fish and crops grown together in an aquaponics project

Link

Soil salinisation is described in Chapter 5.1.4, page 423.

Drip irrigation

Crops require a constant supply of water, which farmers provide by irrigating the land. There are many approaches to irrigation, including small-scale watering from a bucket, watering can or hose; surface irrigation, where water is allowed to flow over the land on to the crops; or sprinkler irrigation, where water is sprayed over the crops. Each of these methods has advantages and disadvantages.

- Small-scale watering is time consuming and requires manual labour, so it can only work on small farms.
- Surface irrigation can be quick and easy but it may waterlog the soil, and can even cause soil salinisation as the water evaporates, leaving small amounts of salt behind in the soil, which build up over time. It can also be quite wasteful of water.
- Sprinkler irrigation is a popular option. It is quicker and less labour intensive than watering by hand, and less wasteful and less likely to cause waterlogging or salinization than flooding the fields. However, a large percentage of the sprayed water does not get to the plants, as it is blown away by wind, lands in the wrong place or evaporates before it reaches the crops.

Drip irrigation solves many of these problems. Pipes or hoses are laid on the ground or buried in the soil, running alongside the rows of crops. Small holes in the pipes allow water to slowly seep out into the soil, close to the base of the crops. The farmer can even control the rate of flow of the water, to precisely match the needs of the plants. Almost all of the water seeping out of the drip-irrigation system is taken up by the plants, so it is far less wasteful than spraying or flooding, less harmful than flooding, and less labour intensive than watering by hand.

Drought-resistant crops

Each species and variety of crop has its own specific requirements and tolerances. Some crops are better able to withstand dry conditions than others. These drought-resistant crops can help reduce water use on farms, since they need less irrigation.

Some species of crop plants are naturally adapted to living in drier conditions. Many common crop species also have varieties that are drought resistant. Scientists and farmers around the world are working on developing new drought-resistant varieties through both conventional breeding methods and genetic engineering.

As the human population grows, there is a need for more food. This will mean growing crops in new areas, which may be less suitable for farming, such as dry environments. As the global climate changes, we may also find that areas that currently receive plenty of rainfall become drier. For these reasons, finding new drought-resistant crops and varieties is an important area of research.

Opting for crops over livestock

Livestock – animals raised on farms – usually require more water than crops do. While crops take small quantities of water slowly from the soil into their roots, animals drink large volumes of water at intervals throughout each day. Livestock also eat grass or other plants, which need to be watered. Overall, to produce 1 kg of meat may take up to ten times as much water as it takes to produce 1 kg of crops. The specific numbers depend on the particular crops and type of meat, but it is generally true that farms producing crops use much less water than farms producing meat.

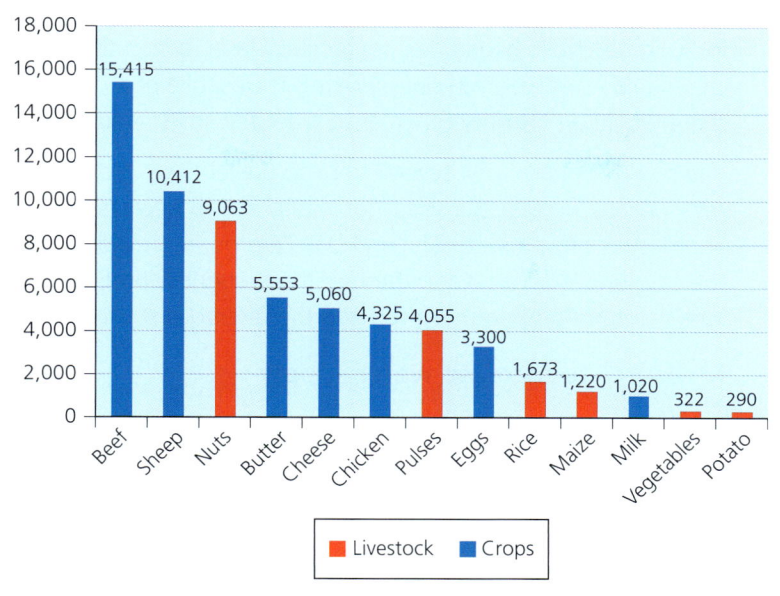

■ **Figure 4.41** The water footprint of a variety of agricultural products

This is one reason why many people choose to be vegetarian. Farmers, and officials who help develop farming policies, may choose to focus on producing crops rather than livestock for the same reason.

4.2.8 Mitigation strategies exist to address water scarcity

Water scarcity is a problem faced by many societies. It can be a temporary problem, or one that continues for a long time. Governments may need to take strong measures to manage periods of water scarcity, to ensure that everyone gets the water they need.

In managing water scarcity, authorities need to balance the demand for water with the supply. They must consider where the water is coming from – rainfall, surface waters or groundwaters, for instance – and how people are using it – for farming, domestic or industrial use. Supply of water is difficult to control, as it often depends on natural processes such as rainfall. Managing people's use of water can also be challenging, as this involves changing people's behaviours.

REAL-WORLD EXAMPLE

Cape Town, South Africa: Day Zero

Cape Town's water supply comes mostly from surface reservoirs in the mountains around the city. The water is used for domestic purposes in the city, and for irrigation of farms in the countryside. The water supply depends on rainfall, so when a period of drought struck in 2015 the water levels in the reservoirs started to fall.

■ **Figure 4.42** One of Cape Town's reservoirs during the 2018 water crisis

4.2 Water access, use and scarcity

In 2017, reservoirs began nearing a critical level, and authorities started to predict that the city would effectively run out of water. They spoke of a possible 'Day Zero' – the day when there would be insufficient water to meet the city's needs.

The city authorities took several steps to prevent Day Zero from happening. These included measures to increase the supply of water, and to decrease its use. Together, these approaches proved successful and Day Zero never arrived. The measures managed to keep the city supplied with water through the drought period, which ended in June 2018, when rainfall returned to normal levels and the reservoirs started to fill up again.

Measures to increase water supply

The following measures were considered to increase water supply:
- Purchasing and transporting water from neighbouring regions.
- Introducing a number of desalination plants.
- Implementing wastewater-recycling projects.
- Releasing water that had been stored for agricultural use into the city water supply.

These methods allow water to be moved quickly from one place to another, but can be expensive. They may also require sacrificing some activities, such as farming, in order to meet immediate needs. This may not be sustainable.

On the other hand, building new facilities like desalination and water-recycling plants takes time, so these measures ideally should be put in place before droughts happen. If authorities can anticipate changes in demand for water, due to population growth or economic development, they should build their capacity to supply water accordingly. They should also consider the risks of drought, and other changes due to weather patterns, in deciding how much capacity they need.

Measures to reduce water use

Attempts to reduce how much water people use are called demand management. Several approaches were attempted in Cape Town, including:
- increasing water tariffs (the price that people pay for water)
- restricting water usage
- public education and outreach.

Increasing tariffs can be effective, especially if the price of water is set based on the volume used. The more water a person or business uses, the higher the price per litre. This encourages people to use less water, so that they pay the lower price.

Authorities in Cape Town put in place rules that limited how much water each person could use. The strictest rules applied allowed only 50 litres of water per person per day. Specific uses like watering lawns and filling pools were banned. Enforcing these rules was difficult, but the authorities took the controversial approach of publicly 'naming and shaming' areas and households that were using more than their allowed quotas. In some cases a device was installed on water supplies to lock off the flow of water when a household's limit was reached.

In promoting the new restrictions, the authorities encouraged several water-conservation strategies, such as reducing showering, reusing greywater for flushing toilets, and using hand sanitizer instead of handwashing. These ideas were promoted through public announcements and an educational campaign in schools.

(HL) 4.2.9 Freshwater use is a planetary boundary

When the planetary boundaries framework was first proposed in 2009, freshwater use was identified as one of the nine planetary boundaries. Water was considered a key component of the global ecosystem.

In the first assessments of the freshwater planetary boundary, the focus was placed on human use of freshwater. The concern was that increasing human use of water could threaten the integrity of the water cycle. If the water cycle was significantly changed, this would have effects on various ecosystems, as well as on human well-being.

The major concern was that the demand for freshwater was increasing, due to population growth and economic development. This was expected to create water stress, which could result in sudden, permanent change to the water cycle.

Up until the 2015 assessment of the planetary boundaries, freshwater use was considered to be in the 'safe operating space'. This meant that the boundary had not been exceeded. Although there was concern about this issue for the future, it was not an immediate cause for concern. However, in 2022 a new assessment of the freshwater boundary came to a different conclusion, and suggested that freshwater was now beyond the boundary, in the 'zone of increasing risk'.

TOK

The change in the status of freshwater in the planetary boundaries framework reflects the nature of science as a growing body of knowledge. As we learn new things, and reinterpret evidence, we can draw new conclusions.

The 2022 assessment of freshwater recognized that direct human use of freshwater is not the only reason for concern. Humans have also caused changes in the amount of water available to plants. The amounts of water stored in soil, transferred in rainfall and transformed in evaporation have all changed due to human use of water, changes in land use such as deforestation, and weather changes caused by global warming.

The new assessment distinguishes between 'blue water', which humans can use directly, and 'green water', which is available to plants. Blue water includes water stores such as rivers, lakes and groundwater, from which humans can extract water. Green water includes water in the soil and atmosphere, which are not easily accessed by humans but are vital for plant health. According to the 2022 report, some ecosystems, such as the Amazon rainforest, have experienced a decrease in green water, whereas in others it has increased. These changes put terrestrial ecosystems such as forests at risk, and also affect the flows of matter between soils and other parts of the biosphere. One major concern is that soils may not be able to store as much carbon as they previously did.

The authors also recognized that blue water and green water affect each other. If humans use more blue water, there may be less green water available for plants. The flows of water through soils, plants and the atmosphere also affect the availability of water in surface and groundwater.

As a result of this new analysis, the words 'freshwater use' in the planetary boundaries framework have been changed to 'freshwater change', and blue and green waters have been addressed separately. More importantly, the new analysis places freshwater change in the zone of increased risk. This suggests that freshwater is now an urgent concern which should be addressed immediately.

Link

The planetary boundaries framework is covered in Topic 1.3.19 (page 72).

Link

The properties of soils, including water and carbon content, are explored in Chapter 5.1, page 416. Water movements through soils are also considered as part of the hydrological cycle in Chapter 4.1, page 268.

4.2 Water access, use and scarcity

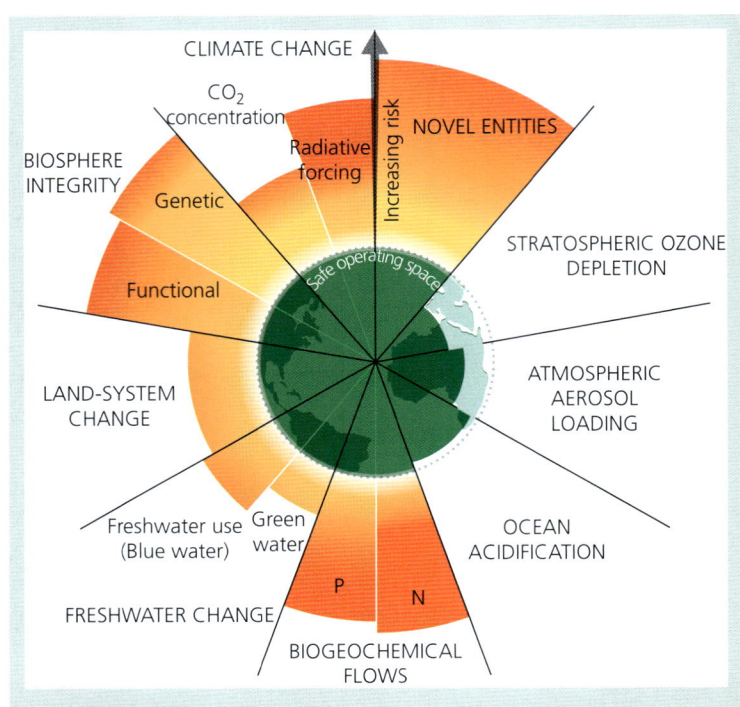

■ **Figure 4.43** The status of the planetary boundaries in 2023

Measuring freshwater change

The best way to assess the status of the freshwater planetary boundary is to measure the stores and processes of the water cycle. To measure blue water we must measure the volumes of water in surface and groundwater stores, and the rates of flow between them. This includes measuring human use. On a local scale this is fairly straightforward, but it requires a great deal of effort to do this on a global scale.

Measuring green water can be more difficult, and scientists have only recently started to do this. At present, the approach is to measure the amount of water in part of the soil where roots grow.

Measuring the stores and flows of the water cycle is the work of a hydrologist. They use data from a variety of sources, including records of human water use, and satellite imagery to determine how much water is in each store and how fast it flows between stores. Computers are used to process the large volume of data, and to make predictions about changes in the water cycle, using models.

> **Concepts**
>
> #### Systems and models
>
> Complex systems such as the water cycle are often modelled to understand how they function and how they might change in different scenarios. The most complex models require computers to process the necessary data.

Managing freshwater change beyond the boundary

It is urgent that we address freshwater change, as it has exceeded the planetary boundary. We must take immediate steps to reduce use of blue water, and to restore green water to sustainable levels. At the same time, societies may need to adapt to and manage the effects of exceeding this boundary.

Perhaps the most important step we should take is to reduce our water consumption. This will be quite difficult as the population continues to grow, and economies continue to develop. It will become increasingly important to use appropriate water-conservation strategies in domestic, industrial and agricultural settings. It will also be necessary to minimize our impact on natural processes of the water cycle, to ensure that enough green water is available for natural ecosystems to continue to function.

4.2.10 Local and global governance is needed to maintain freshwater use at sustainable levels

Local water governance

Governments have a vital role in protecting the natural systems which maintain life on Earth. They must establish policies and laws, and implement plans and programmes to help maintain the freshwater resources so that we can sustainably meet the needs of all people, while protecting the natural flows of the water cycle. Since the water cycle is a global system, governments need to co-operate to ensure that globally we can get back beneath the freshwater planetary boundary.

REAL-WORLD EXAMPLE

Water regulations in the UK

The United Kingdom has a reputation for experiencing a lot of rainfall, but in fact it often experiences water scarcity during the dry summer months. After periods of warm, dry weather, reservoirs are often depleted. Non-essential activities like using hoses to water plants or wash cars are often banned during these periods.

Regulations have also been introduced to promote greater water-use efficiency in homes. New homes built since 2010 are required to incorporate efficient water fixtures and appliances, so that water consumption should fall below 125 litres per person per day. A more recent proposal suggested this should be reduced to 110 litres per person per day.

Other regulations that are still under consideration include mandatory water efficiency labels on appliances like washing machines and dishwashers, requiring water suppliers to take more responsibility for fixing leaks in customers' pipes, and promoting rainwater harvesting and greywater reuse.

REAL-WORLD EXAMPLE

New Zealand's national freshwater policy

In 2020, the government of New Zealand published a new National Policy Statement for Freshwater Management, which incorporated some key insights from the indigenous Māori people. A central concept recognizes the importance of healthy freshwater for the well-being of all. The policy sets out a hierarchy of obligations:
1   First, to protect the health and well-being of water bodies and freshwater ecosystems.
2   Second, to protect the health and meet the basic needs of people.
3   Third, to meet the social, economic and cultural needs of people.

◆ **Moral standing** – deserving of, or entitled to, ethical consideration. If something has moral standing, we should consider the effect of our actions on it.

◆ **Deontological ethics** – an approach to ethics that considers the rights of different people or things. In this system, an action may be ethically wrong if it infringes on the rights of others.

People using and managing freshwater resources in New Zealand are required to follow this hierarchy.

> ### Concepts
>
> #### Perspectives
>
> The New Zealand freshwater policy incorporates the viewpoint of the indigenous Māori people. Many environmental campaigners argue that indigenous peoples have valuable knowledge and opinions on natural systems. They argue that indigenous peoples have often lived sustainably for very long times, and that their perspectives on nature can help to achieve sustainability in the modern era.

HL.c.5, HL.c.7: Environmental ethics

New Zealand's new freshwater policy implies that freshwater ecosystems have intrinsic value, and therefore **moral standing**. The concept of intrinsic value is most often used in reference to individual living organisms, such as an animal or plant, but some people believe that ecosystems or landscapes also have intrinsic value. For instance, American naturalist and philosopher Aldo Leopold proposed his 'land ethic' in the mid-twentieth century, which argued that all components of nature have the right to exist.

According to one line of reasoning, intrinsic value comes from having systems, structures and processes that are organized for survival and reproduction. Living organisms clearly meet this condition, but not everyone would agree that it is true of ecosystems or landscapes. However, the Māori people believe that even non-living things can have a 'life force' or 'vital essence', which they call *mauri*. If true, this would imply that these non-living things also have intrinsic value.

If an entity has intrinsic value, it can be argued that it has moral standing. A rights-based or **deontological ethical** view would suggest that if an entity has moral standing, we should consider how our actions could affect that entity. The New Zealand freshwater policy takes this position when it prioritizes the well-being of freshwater ecosystems above the interests of people.

International water governance

Many freshwater stores, such as rivers, lakes and aquifers, cross international boundaries. The River Nile and its tributaries, for instance, pass through 11 countries. When more than one country has access to a water source, there may be conflict. Each country may want to use the water, but taking water out of a river, changing its rate of flow, or putting wastes or other pollutants into it may have impacts on the other countries.

In 1992, the United Nations adopted the Convention on the Protection and Use of Transboundary Watercourses and International Lakes, also known as the Water Convention. The convention encourages countries to negotiate and collaborate on transboundary water issues, and to use international waterways in a sustainable and equitable way, recognizing the rights of other countries. It encourages countries to form specific agreements to deal with the management of particular international waterways. A later convention, the Convention on the Law of the Non-navigational Uses of International Watercourses (also known as the Watercourses Convention), was adopted in 1997.

REAL-WORLD EXAMPLE

The Paraná River, South America

The Paraná River begins in Brazil, and then runs along the international border between Brazil and Paraguay. It follows the border between Paraguay and Argentina, before flowing through Argentina and into the Atlantic Ocean. Its estuary forms part of the border between Argentina and Uruguay.

In the 1970s, Brazil and Paraguay began construction of the Itaipu Dam on their shared stretch of the river. The dam is used to generate hydroelectric power. Argentina, however, objected to the construction, as they thought it might reduce its water supply. In 1979, the three countries signed an agreement that ensures the flow of water into Argentina is continued, and that the three countries will cooperate on issues of water management in the Paraná River.

■ **Figure 4.44** The Paraná River in South America

■ **Figure 4.45** The Itaipu Dam on the border of Brazil and Paraguay

REAL-WORLD EXAMPLE

The Grand Ethiopian Renaissance Dam on the River Nile

In 2011, Ethiopia began construction of the Grand Ethiopian Renaissance Dam (GERD) on the Blue Nile, upstream of both Sudan and Egypt. The dam is intended to produce hydroelectric power for Ethiopia, but Egypt is concerned about its water supply, as the Nile supplies almost all of Egypt's water.

While the Renaissance Reservoir fills up, the flow of water to Sudan and Egypt will be reduced. Any reduction in flow into Egypt could also affect its own electricity production at the Aswan High Dam, close to the Egypt–Sudan border.

■ **Figure 4.46** The River Nile, showing the locations of the Grand Ethiopian Renaissance Dam and Egypt's Aswan High Dam

4.2 Water access, use and scarcity

The flow of water in the Nile has been a source of disagreement between Egypt, Sudan and Ethiopia for many years. In 1959, Egypt and Sudan signed an agreement allocating shares of the water to those two countries. However, the agreement did not include consideration for Ethiopia or any other countries.

As of 2023, no agreement has been reached between Egypt and Ethiopia, and tensions between the two countries continue. Ethiopia sees the dam as vital to its economic development. Egypt sees it as a threat to its own water and energy security. These are such important issues that some observers fear the conflict could lead to military action between the two countries.

■ **Figure 4.47** The Grand Ethiopian Renaissance Dam (GERD)

Tool 4: Systems and models

The rate of flow of water to Egypt

While the Renaissance Reservoir is filling up, the flow of water to Egypt will be reduced. The faster the reservoir is filled, the slower the flow of water downstream into Egypt will be. When the reservoir is filled to capacity, however, the rate of flow to Egypt will return to normal – unless other inflows or outflows have changed in the meantime. Use three simple systems diagrams, based on Figure 4.48, to explain how the rate of flow to Egypt changes as the Renaissance Reservoir is filled. Your diagrams should show the situation before, during and after filling. Choose arbitrary numbers to represent different possible values for the rates of flow in to and out of the reservoir, and suggest how the rates you have chosen will affect the amount of water stored in the reservoir.

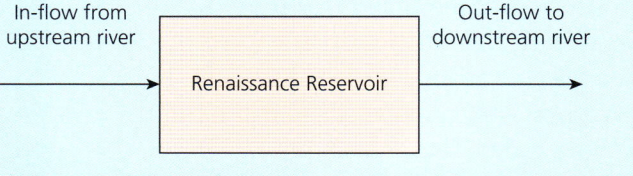

■ **Figure 4.48** Flows of water into and out of the Renaissance Reservoir

4.2.11 Water footprints

A water footprint is a measure of how much water is used to carry out a particular process. It aims to identify all the direct and indirect water use during the entire process.

Water footprints can be calculated for the production of an item, the operation of a business, or for the life of an individual, a city or a country. They can be used to compare the water consumption of different products, processes or societies. A water footprint can be compared to the size of the stores and flows of water in the water cycle, to determine whether a process or community is sustainable. If the combined water footprint of a country or community is bigger than the supply of available water, then the water use is unsustainable.

Information from a water footprint analysis can be used to decide how to become more sustainable. For instance, the water footprint of beef production is much greater than the water footprint of soybean production. Information about water footprints might influence governments to shift the focus of their agricultural policy from livestock to crop production. Individuals might be influenced to modify their diets.

> **ATL ACTIVITY**
>
> ### Water footprints
>
> 1. Search online for a water-footprint calculator and use it to estimate your own, or your family's or school's, water footprint. Think about what you could do to reduce your footprint. How difficult would it be to make a significant difference?
> 2. Select two similar countries, cities or products that you can compare. Use the internet to research their water footprints. How can this information be used to make decisions about water security and sustainability?

4.2.12 Citizen science and water resources

Citizen science, also known as community science or crowdsourced science, is the use volunteers from the public to collect scientific data. Various countries have established citizen science programmes to collect data on different environmental parameters, including water quality. Citizen science can collect large amounts of data, at relatively low cost, since hundreds or thousands of citizen scientists can be involved. Even a small number of citizen scientists can also help to collect data from one location, giving continuous information about that place. Citizen scientists are often highly motivated to help collect data about their local environment.

Professional scientists might provide volunteers with simple equipment for collecting and testing water, such as test strips for measuring nitrate or phosphate pollution. Mobile devices like smartphones have made it possible to submit data to a database which professional scientists have access to. A volunteer might install an app on their device, and use it to collect the right kind of data, in the right format, for the scientists to use.

A common challenge is standardizing the quality of the data, to ensure that it is useful. If citizen scientists do not follow the data-collection protocols correctly, the professional scientists may not be able to use it.

4.2.13 Water stress is another measure of the limitation of water supply

Water scarcity occurs when there is not enough water to supply the needs of the people in an area. The word 'scarcity' means *lack* – if something is scarce it means there is not much of it.

'Stress' has a broader meaning and can refer to any sort of pressure, strain or tension. **Water stress** refers to difficulty meeting the needs of humans or ecosystems for water. Water stress might be due to water quantity, **quality** or **accessibility**.

◆ **Water scarcity** – limited availability of water. Scarcity may be physical or economic.

◆ **Water stress** – difficulty meeting the water needs of people or ecosystems, due to lack of availability, poor water quality, or inability to access water.

◆ **Water quality** – a measure of the properties of water relative to what the water is needed for. For instance, drinking water should be clean, clear, colourless and odourless, and have no toxic chemicals or disease-causing organisms in it.

◆ **Water accessibility** – a measure of how easy it is for people to obtain water. Accessibility may be affected by the location of the water or the cost of obtaining it.

> **TOK**
>
> Sometimes technical terms are used differently in different contexts. This is the case for water stress, which is used differently by different organizations. It is important when using such terms to make it clear what you mean by them. For the purposes of this course, you should use the definition given here.

Link

Positive feedback is covered in Topic 1.2.10 (page 38).

The most water-stressed countries on Earth tend to be among the least developed nations. Low economic output makes it difficult for governments to provide access to clean drinking water, as well as to wastewater treatment, which helps protect drinking water supplies. This may lead to high levels of water-borne diseases, which further hinders development in a vicious, positive feedback cycle.

> **REAL-WORLD EXAMPLE**
>
> ### Water stress in Niger
>
> Niger in West Africa is one of the most water-stressed countries on Earth. It receives little rainfall and has suffered a series of severe droughts in recent decades. Surface water stores include the River Niger, River Komadougou and Lake Chad. In its dry, desert-like environment, the rate of evaporation is very high, and the smaller rivers sometimes dry up completely. The rivers also contain high levels of sediment, which means the water requires treatment before it can be used for drinking.
>
> ■ **Figure 4.49** Location of Niger
>
> High levels of poverty also mean that most of Niger's residents do not have toilets, so bodily wastes are often released untreated into surface waters. This leads to the spread of water-borne diseases.
>
> Several factors contribute to Niger's water stress, including the climate, poor water treatment and waste disposal facilities, and low level of economic development. Surface waters are limited, variable, and of poor quality. With a rapidly growing population, there will be an increasing need for water.
>
> Overcoming water stress in Niger will involve improving sanitation facilities, curbing population growth and promoting economic development. Fortunately, recent satellite and aerial searches have identified much larger groundwater resources than had previously been known in the region. If these resources can be accessed, it may be possible to meet the people's water needs.

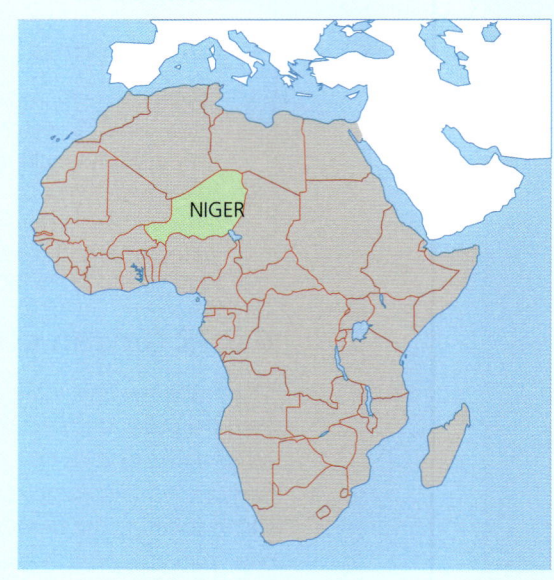

4.2.14 Measuring water stress

To help us understand how widespread water stress and water scarcity are, it is useful to define these concepts in numerical terms. This allows us to identify where these problems occur, and which societies need the most support. Water stress can be defined as occurring where people have access to less than 1,700 cubic metres of clean water, per person per year. Water scarcity is defined as having less than 1,000 cubic metres of clean water, per person per year.

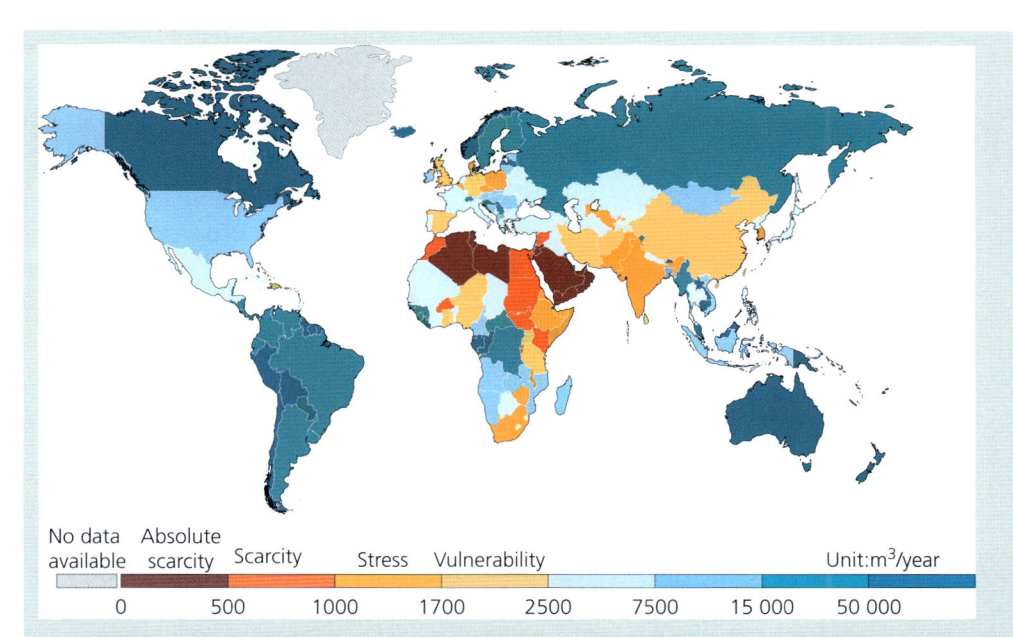

■ **Figure 4.50** Global freshwater availability; countries with less than 1,700 m³ per person per year are experiencing water stress

4.2.15 The causes of increasing water stress may depend on the socioeconomic context

Many different factors may contribute to growing water stress. The specific causes of water stress in any one country may depend on the socio-economic situation within that country, and the political decisions made about water usage.

Water is a finite resource. For most countries, there is a limit to the amount of water available for human use. The amount of available water may decrease as a result of overexploitation, pollution or climate change. Increases in water availability are rare, occurring only if new groundwater resources are discovered, or if rainfall increases significantly due to climate change. In most situations, any increase in water use is likely to increase water stress.

In some countries, water stress may increase due to population growth. With increasing numbers of people, the share of water available per person decreases. High population growth rates are most often found in poorer, less economically developed countries. Such countries may also struggle to provide the infrastructure to deliver safe water to all people.

In nations experiencing rapid industrial and economic development, increased use of water for industrial activities may cause increased water stress, even where population growth is relatively slow. The political decisions of the government may prioritize industrial or economic development, leading to an increased share of water resources being used in factories or on farms. Consequently, a smaller fraction of the country's water is available for human, domestic use, and water stress increases.

Link

The complex interaction of factors associated with population growth is covered in Chapter 8.1 (page 694).

Concepts

Perspectives

Decisions on how to use water resources are influenced by different perspectives regarding the value of economic development. Some people argue that economic development ultimately benefits everyone. As societies develop and become wealthier, they are better able to meet their population's needs. For instance, they may justify using scarce water resources for industrialization because of the potential

4.2 Water access, use and scarcity

long-term benefits. Others might argue that the immediate priority is to meet people's needs for water, and that only after that is done should excess water be used for industrial purposes.

Both perspectives have merit and are valid arguments. There is also a range of viewpoints between these two extremes. A person's perspective on this may be influenced by their environmental value system. Ultimately, decisions on the allocation of water resources tend to be made by political and business leaders, though they can also be influenced by voters and consumers.

4.2.16 Water stress and transboundary disputes

Since water resources are often shared among neighbouring countries, decisions made by one country, such as to increase its use of water from the shared water source, can add to water stress in its neighbours. The Real-world examples of Jordan (Topic 4.2.5, page 303), and the Nile and Paraná Rivers (Topic 4.2.10, page 313), illustrate how important it is for neighbouring countries to manage their resources cooperatively. International agreements and the United Nations Water Conventions (Topic 4.2.10, page 313) can help to protect countries from harmful decisions made by their neighbours.

REAL-WORLD EXAMPLE

Turkey, Syria and Iraq: The Euphrates–Tigris River Basin

The Tigris and Euphrates Rivers both begin in Turkey, and merge in Iraq, forming the Shatt al-Arab River, before flowing into the sea at the Persian Gulf. The Euphrates also flows through Syria before entering Iraq.

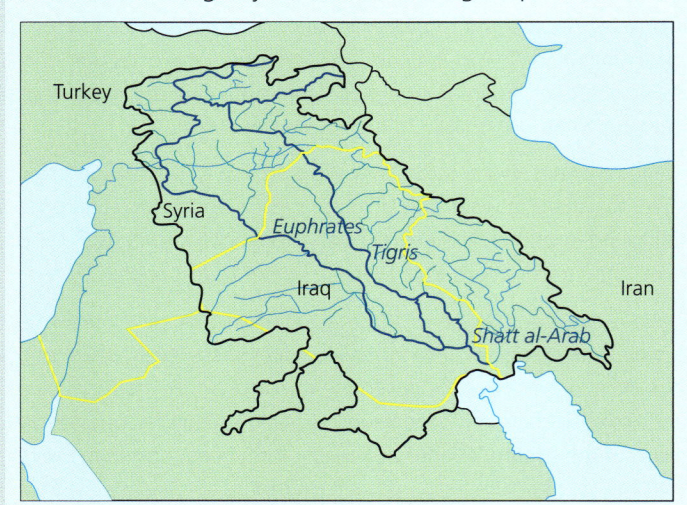

■ Figure 4.51 The Euphrates–Tigris River Basin

In the early 1960s, with populations growing rapidly, each of the three countries increased its use of water from the rivers, without consulting one another. Large-scale irrigation projects, a hydroelectric project in Turkey, and attempts to control flooding all changed the flow of water in the two rivers. Since the rivers begin in Turkey, it was Syria and Iraq that were most affected. The conflict between the countries may have become violent if not for mediation provided by Saudi Arabia.

Climate change and political differences made the situation more complex. Periods of drought in the 1970s increased the water stress in all three countries, and the conflict became more intense. Local territorial and political conflicts, and conflicting positions during the Cold War – with Turkey joining NATO, and Syria and Iraq being allied with the Soviet Union – made it more difficult to resolve disputes over water.

International agreements, such as the UN Water Convention and Watercourse Convention, can help to avoid these sorts of conflicts and settle disputes when they arise. However, not all countries agree to these conventions. Turkey, for instance, has not signed the 1997 Watercourses Convention.

HL.a.8: Environmental law

The UN Water Convention, which came into effect in 1996, was adopted in Helsinki in 1992 and deals with the protection and use of transboundary watercourses and international lakes.

The convention signatory countries must guarantee the sustainable management of transboundary waters, as well as prevent, control and minimize transboundary impact. Countries sharing transboundary waters are required to work together by creating joint bodies and signing special agreements.

ATL ACTIVITY

UN Water Convention projects

Create a case study of two implementation projects of the UN Water Convention. Search online for 'UN water convention projects examples' to find examples of projects.

4.2.17 Water stress can be addressed at an industrial level

Countries dealing with water stress have a number of options available to them. These industrial-scale strategies can help to meet the society's water needs, but each of them has its own economic, social and environmental costs. These must be balanced against the potential benefits.

Dams

A free-flowing river may carry large volumes of water over the course of a year, but at any one time there is only a limited amount of water in the river. Damming a river slows down the flow, and holds a larger volume of water in one place. This adds a new store to the local water cycle, which can be used to help meet the society's water needs.

When a dam is first constructed, and while it is filling with water, the rate of flow in the river downstream is reduced. Once it is filled up, the river may return to its normal flow rate. Constructing a dam can be expensive, and may flood land where people live, or that was used for other activities, such as farming. People may need to be relocated and ecosystems may be destroyed. Water stored behind a dam may also evaporate more quickly than in a river, and the flow of sediments and nutrients may be changed. These impacts are illustrated by the Real-world example of the Three Gorges Dam in Topic 4.2.4 (page 299).

Water storage in estuary barrages

Estuary or tidal barrages are movable structures that block the flow of water at the mouth of a river. They are mostly used to hold the sea back during high tides, preventing coastal flooding, or to produce electricity, using turbines built into the barrage. In some cases a barrage may be used to store water in the estuary for human use. A barrage used in this way might be linked to a desalination plant, since the water in the estuary is likely to have some salt content.

Estuaries can be very productive, biodiverse ecosystems. A barrage built in an estuary can cause significant ecological disruption. In considering a barrage, the potential benefits should be weighed against any social, environmental and economic costs.

REAL-WORLD EXAMPLE

Nagaragawa Estuary Barrage, Japan

The Nagara River basin in Japan does not include many sites that are suitable for conventional dams. This led to the decision to build a barrage in the estuary of the river, which was completed in 1995. It provides many benefits and stores water for domestic and industrial use in the nearby areas, including Nagoya City. It also helps to protect against tsunamis and storm surges during typhoons, which are significant risks in the area, and to reduce saltwater intrusion into the river and the surrounding land.

■ Figure 4.52 The Nagaragawa Estuary Barrage in Japan

Inter-basin water transfers

If one part of a country experiences water stress while another part has lots of water, it might consider moving some of the water from one area to another. This often means moving water from one river basin to another, so it may be called inter-basin water transfer. This is usually done with pipes, canals and aqueducts. It can be a huge, complex engineering feat, and may involve moving water uphill, using pumps.

REAL-WORLD EXAMPLE

South–North Water Transfer Project, China

In 2003, the Chinese Government began construction of a massive project to transfer water from the large, southern Yangtze River, to the more arid northern regions including the cities of Beijing and Tianjin. Three separate routes have been considered, and construction is expected to take many years.

■ Figure 4.53 The proposed routes for China's South–North Water Transfer Project

Figure 4.54 A new water diversion channel for the South–North Water Transfer Project

The economic cost of large-scale, long-distance water transfer can be very high. It may also have significant social and environmental costs, as water is moved away from one part of society, and from one ecosystem to another. There are also concerns about spreading pollutants and various species of organism between basins.

In some situations, building canals, pipes and aqueducts is not feasible. Instead, water may be transported in tankers – either trucks or ships. This tends to be a last resort, often used during drought, when a country or region has a temporary shortage of water.

Water treatment plants

Water taken from surface or groundwater stores may be clean enough for human use, but it often contains substances or organisms which may cause illness in people who drink it. If this is the case, the water should be treated before it is distributed to people's homes. By treating water appropriately, we can increase the amount of water available for human use.

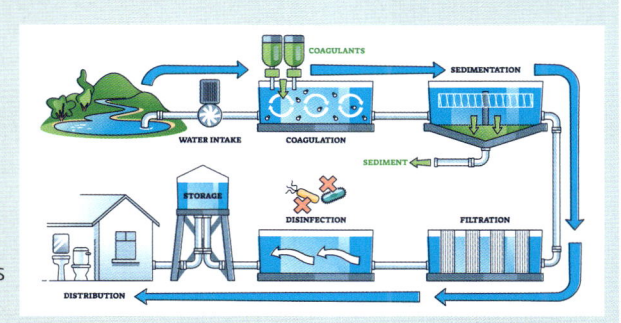

Figure 4.55 The main processes involved in drinking water treatment

Water treatment for human use may involve a variety of processes. A water treatment plant might use various combinations of these processes, depending on the quality of the water which it is intended to treat, and the purpose which the water is intended to be used for.

The most commonly used treatment processes include:
- Coagulation: A chemical is added to the water, which makes small suspended and dissolved particles clump together, so that they can be removed by sedimentation or filtration.
- Flocculation: After a coagulant is added, the water is slowly stirred. This helps the coagulant to mix in thoroughly, and speeds up the clumping (coagulating) process.
- Sedimentation: The water is allowed to stand in a tank for some time (often after coagulation and/or flocculation), so that solid particles settle at the bottom and can be removed.
- Filtration: The water is passed through a filter containing substances like gravel, sand and activated carbon. Solid particles become trapped in the filter. If activated carbon is used, many dissolved substances can also be removed.
- Disinfection: The water is treated to kill any microorganisms that might cause disease. The most common method of disinfection is adding chlorine. Alternatives include bubbling ozone through the water or shining ultraviolet light through it.

If water is going to be used by humans, disinfection may be the most important part of the treatment process, as this can prevent the spread of water-borne diseases. The other processes are only likely to be used if there is a specific need for them. For instance, water collected from a river is likely to have some sediment in it, so sedimentation might be used to remove that. If the river is polluted, perhaps because it flows through industrial or agricultural land, then filtration and coagulation might be necessary.

The methods of water treatment can even be used with wastewater. Normally wastewater treatment plants clean the water so that it can be released safely into natural waterways. But if freshwater is relatively scarce, wastewater treatment can be taken further, to produce drinking-quality water.

> **REAL-WORLD EXAMPLE**
>
> ### Recycling wastewater in Singapore
>
> Singapore is a small, tropical island nation, with a high population density. Although it receives abundant rainfall, there are no natural lakes and limited surface water. The country is not able to capture and store enough water to meet the demands of its population.
>
> Singapore uses several methods to provide water for its population, including reservoirs, rainwater harvest and estuary barrage storage. In the past, it also imported water from neighbouring Malaysia. However, since 2002, a significant portion of the water used in homes and industries has come from wastewater recycling.
>
> The wastewater-recycling plants in Singapore use repeated filtration, through different media, and disinfection with ultraviolet light. The plants are so efficient that Singapore releases hardly any wastewater into the sea.

Aquifer recharge, and aquifer storage and recovery

If water is continuously extracted from an underground source (an aquifer), the aquifer may run out of water. This happens if the rate of removal of water is faster than the rate of water entering the aquifer by infiltration or groundwater flow. This would be an unsustainable use of the resource, and would cause water scarcity or stress.

To replenish the reduced water resource, water may be pumped into the depleted aquifer to refill it. This is called aquifer recharge, and it is usually done using an injection well. This is where a hole is dug into the aquifer and water is pumped into the ground.

A similar strategy, aquifer storage and recovery, is used when a society wants to store excess water for use later. Usually dams and reservoirs are used for storage, but if there is an empty aquifer available, this may be a better option. An injection well can also be used in this case.

Desalination and solar distillation

Desalination is the removal of salt from seawater, to produce freshwater for human use. It is most often used in very dry places, where water stress is high. Conventional desalination is described in Topic 4.2.4 (page 299). It can be achieved by distillation, where seawater is boiled and the freshwater condensed, or by reverse osmosis, where the water is forced through a semipermeable membrane that excludes the salt.

Distillation is an alternative approach to producing freshwater, which can be used to treat either salty or contaminated water. Solar distillation uses solar energy to boil the water, making it more sustainable than using conventional energy sources.

> **ATL ACTIVITY**
> 1. Find out where the water for your home or school comes from and investigate how the water is treated. You may be able to find this out from online sources, or you may need to contact the relevant water company.
> 2. Evaluate the source or sources you used to answer Question 1.

Small-scale distillation can be carried out using a device called a still. A solar still works by exposing salty or contaminated water to the sun. The water evaporates, and the water vapour condenses on a surface placed above the water and is collected. This sort of still can be built cheaply and produces a small volume of water.

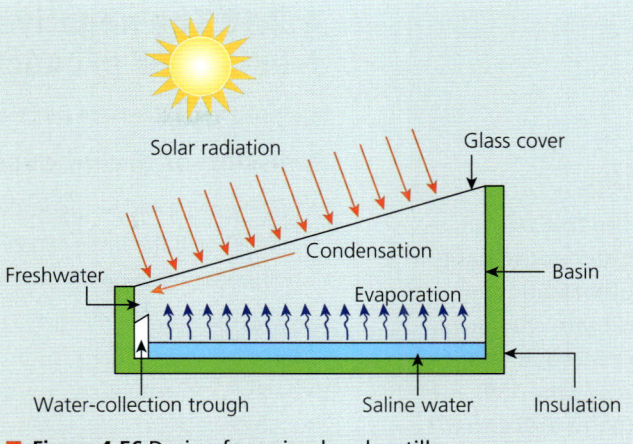

To produce freshwater on a larger scale, solar energy can be used to power a more conventional distillation or reverse osmosis plant.

■ **Figure 4.56** Design for a simple solar still

For distillation, a modified solar furnace can be used to boil the water. For reverse osmosis, solar photovoltaics can be used to generate electricity, which is used to drive a pump that forces the water through the membrane. Large-scale solar desalination plants are currently planned for various locations around the world, notably in Saudi Arabia the United Arab Emirates and Chile.

Link

Solar thermal and photovoltaic energy systems are covered in Chapter 7.2 (page 634).

Artificial weather modification

Cloud seeding is an experimental technique that is used to bring rain to dry regions. Tiny aerosol particles containing substances such as silver iodide, potassium iodide or solid carbon dioxide ('dry ice') are sprayed into clouds. The aerosol particles attract water molecules, causing them to come together and condense into rain or snow, which then falls to the ground. Experiments on this process, sometimes called rainmaking, have been carried out since the mid-twentieth century, but there is insufficient scientific evidence to conclude that it works.

Rainmaking can be expensive, as the 'seeds' are either sprayed out of aircraft or propelled by rocket launchers into the sky. There are concerns about the pollution, as the substances used would not normally be found in nature and it is not known if they cause harm.

Atmospheric water harvesting

The atmosphere contains variable amounts of water vapour. If a relatively cool surface is introduced into relatively warmer air, water vapour may condense into liquid on that surface. This approach is known as dew harvesting or fog harvesting.

Dew or fog harvesting is a simple, cheap method, but it produces relatively small volumes of water, depending on the humidity of the air. In arid regions where water is most scarce, humidity tends to be low so the method may have limited success.

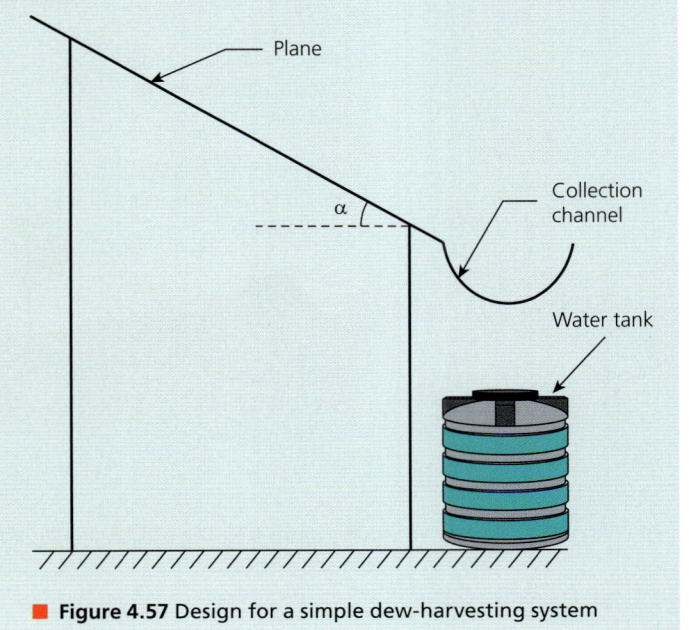

■ **Figure 4.57** Design for a simple dew-harvesting system

4.2 Water access, use and scarcity

4.2.18 The negative environmental impacts of industrial freshwater production

Each of the methods for producing freshwater has some negative impact on the environment.

Emission of greenhouse gases

All methods for producing freshwater involve using energy. In most cases, this will involve burning fossil fuels and therefore releasing carbon dioxide into the air. This can be reduced, or even avoided, if renewable alternative energy sources are used.

Desalination requires the most energy, either to boil the water, or to force it through the membranes. Where distillation is done, such as in Saudi Arabia and the United Arab Emirates, the energy source is usually oil or natural gas. For reverse osmosis, the pumps run on electricity, which can be produced from renewable sources. However, around the world, most electricity is still produced using fossil fuels.

Reservoirs can also be a source of greenhouse gas emissions. When land is flooded, any plant matter which is left on the land will decompose underwater. If decomposition is anaerobic, methane – a powerful greenhouse gas – will be emitted.

Effects of depleting aquifers

When groundwater is pumped out of rocks, it leaves behind pores within the rocks which fill with air. These air-filled spaces can create environmental problems, such as subsidence and saline or saltwater intrusion.

Subsidence happens when the rocks of the empty aquifer can no longer support the weight of the rocks above them, as well as any man-made structures built on the land. The land starts to sink, usually gradually, leaving structures like roads and homes twisted or leaning to one side.

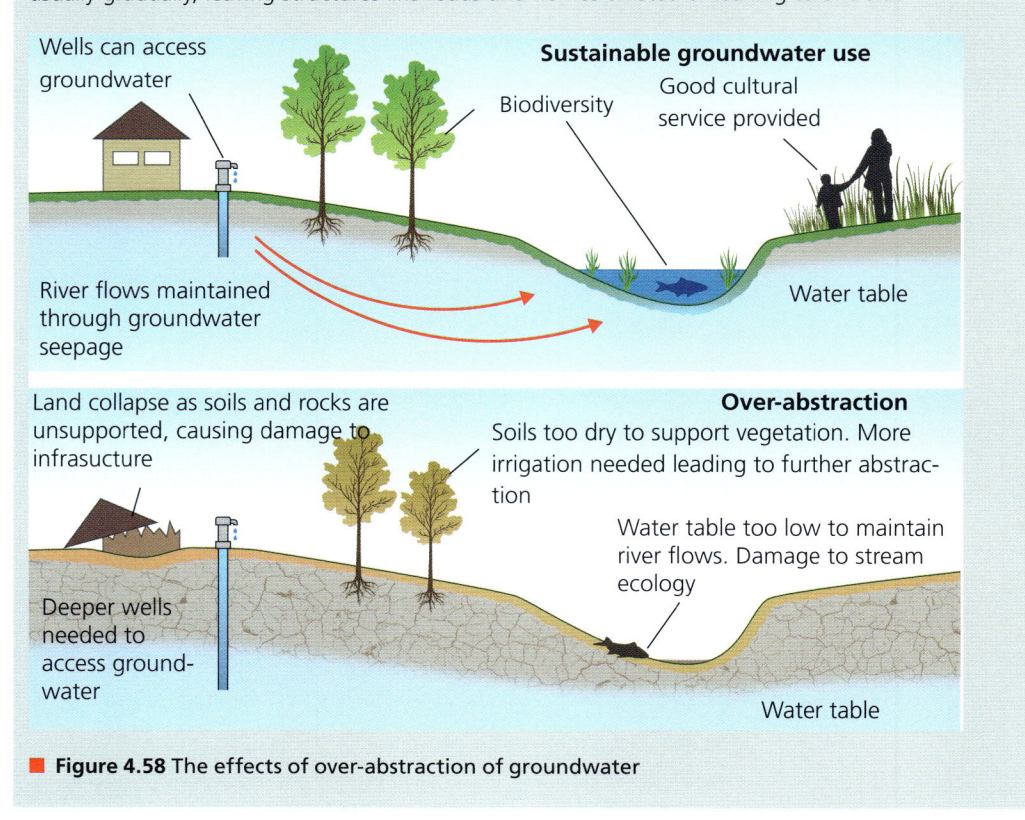

> **Link**
>
> The role of greenhouse gases in global warming and climate change is covered in Chapters 2.3 (page 139), 6.1 (page 514) and 6.2 (page 528).

> **Link**
>
> The effects of decomposition in water are covered in Chapter 4.4 (page 373).

■ Figure 4.58 The effects of over-abstraction of groundwater

Theme 4: Water

Saltwater intrusion most often happens on small islands or near the coast. As fresh groundwater is drawn up and out of the aquifer, saltwater is pulled in to replace it. Continuing to remove water from the aquifer will lead to the water becoming salty.

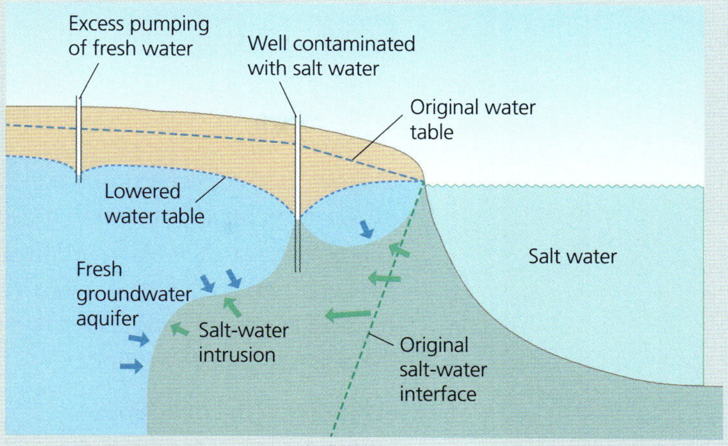

Production of wastes

Some methods of water treatment produce potentially harmful wastes, which may have negative impacts on the environment.

■ **Figure 4.59** Saltwater intrusion

Desalination, for instance, produces clean and usable water and leaves behind a concentrated solution of salt in water, called brine. This is normally too salty for living things to survive in. If released into the environment, it could cause serious harm to aquatic ecosystems. To avoid this, desalination plants may dilute the brine, by combining it with waste freshwater from other processes, or allow the remaining water in the brine to evaporate completely, producing salt for seasoning food.

Conventional water treatment plants also produce waste products, including the semi-solid sludge collected from various settling tanks. This sludge is usually dried and disinfected before disposal. In some cases it may be used to produce organic fertilizer or energy from biomass.

Air and noise pollution

Any industrial process that uses machinery will inevitably produce noise pollution, which can be a nuisance for nearby communities, as well as for wildlife. Careful design of the water treatment plant may help to reduce the noise, or prevent it from affecting others.

The most significant air pollution problems associated with water treatment plants come from the burning of fossil fuels. In addition to greenhouse gases, particulate matter, and oxides of sulphur and nitrogen may be released. Most water treatment plants use electricity as their main source of energy, so this is impact usually occurs at the power generation plant, rather than the water treatment plant. Air pollution may occur at a water treatment plant if there is an accidental release of chlorine gas or other chemicals. This can usually be avoided by using proper safety practices and carrying out risk assessments.

Unpleasant smells may arise from treatment plants, especially if smelly sediment or sludge is removed from the water during treatment. This is unlikely to be harmful, but it can be a nuisance to people living nearby. It can be managed by drying and disinfecting the sludge, and by appropriate disposal of any other solid waste, such as taking it to an incinerator or landfill. The sludge can also be used as fertilizer on farmland.

4.2.19 Water, equity and sustainable development

Safe, clean drinking water and sanitary wastewater disposal are essential for a healthy life. However, not all people have equal, or equitable, access to water or sanitation. People who are marginalized in various ways often have poor or limited access. This may include people with low incomes, those living in rural or isolated communities, or those of particular racial or ethnic identities, such as various groups of indigenous people around the world.

> **REAL-WORLD EXAMPLE**
>
> ### Lilongwe, Malawi
>
> Lilongwe is the capital city of Malawi, in southeastern Africa. It has experienced rapid population growth in recent decades, making it difficult for the authorities to meet all its citizens' needs. Malawi is also one of the poorest nations in Africa.
>
> Beginning in the mid-1960s, Lilongwe underwent rapid, planned development, which created three distinct zones in the northern, central and southern parts of the city. City administration, commerce, and high-income housing were concentrated in the northern and central areas, while the southern zone was designated for low-income, high-density housing.
>
> About 70 per cent of Lilongwe's population lives in the southern zone, but the supply of water to that area is poor compared to the north and central areas. The capacity of the reservoirs in the south is significantly lower than those for other areas, even though the population of the south is much higher.
>
> In 2018, 22 per cent of Lilongwe's population, mostly in the south, did not receive pipe-borne water supply. Those people bought water from public water kiosks, and had to carry the water home in buckets. Carrying water is far more difficult and time consuming than receiving water via pipes, so people in these areas consumed far less water than those in the north and central. Women are normally responsible for collecting the 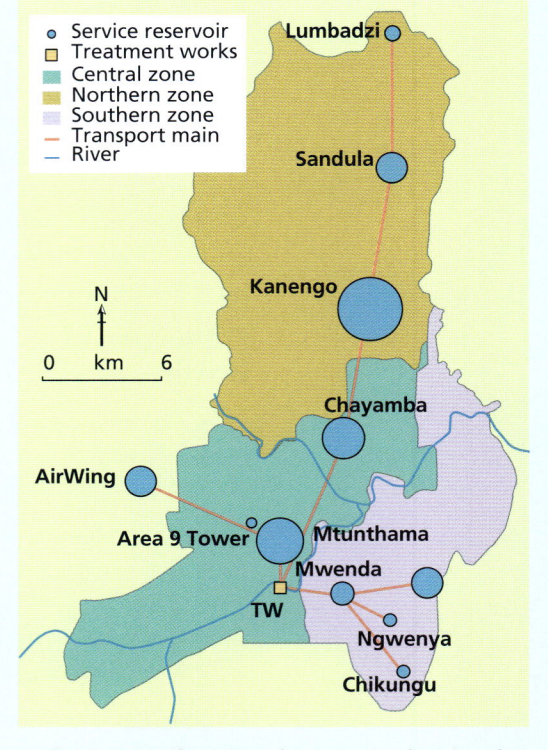
>
> ■ **Figure 4.60** Lilongwe's three zones, showing the water-distribution system 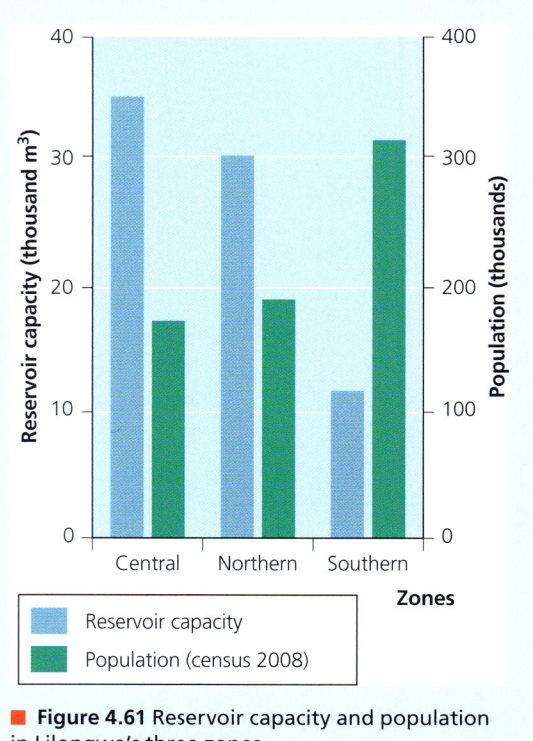
>
> ■ **Figure 4.61** Reservoir capacity and population in Lilongwe's three zones

Theme 4: Water

family's water, limiting their prospects for economic independence, and perpetuating gender inequity in society. Studies have also found that using open buckets and transferring water between containers can lead to water contamination and the spread of water-borne diseases.

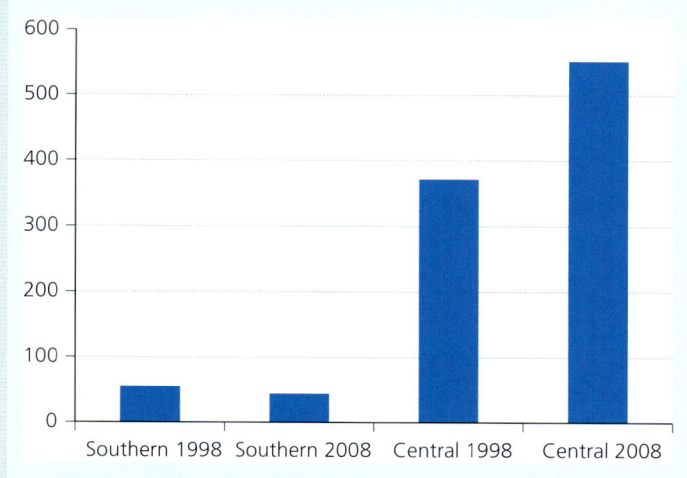

■ Figure 4.62 Average daily per capita water consumption in the central and southern zones of Lilongwe, in 1998 and 2008

During the dry season, when water supplies are limited, the kiosks in the south are the first to have their supply cut off. This leaves the largest, most vulnerable part of the population with no water supply, forcing them to store water for long periods in unsanitary containers and to reduce their overall water use. Meanwhile, the wealthier residents of the north and central areas are less likely to have their supplies cut.

Sustainable development is difficult, perhaps impossible, to achieve in societies where some people receive insufficient safe drinking water. Using the Brundtland Commission definition, sustainable development cannot be achieved if people's needs for water are not met. Rapid population growth makes meeting future generations' needs will be even harder. Inequalities in water supply, or in other essential resources, can lead to stunted economic growth and social unrest.

The United Nations Sustainable Development Goal (SDG) number 6 is Clean Water and Sanitation, to ensure the availability and sustainable management of water and sanitation for all. While inequitable supplies of water exist amongst the population, this goal cannot be met. It also might prevent other SDGs from being met, such as number 3 Good Health, number 11 Sustainable Cities and Communities, and number 16 Peace and Justice. Others such as number 5 Gender Equality might be affected if women are restricted to the work of carrying water. Number 14 Life Below Water might be affected if water resources are depleted or polluted due to inequalities.

Link

Sustainable development and the Sustainable Development Goals are explored in Topic 1.3.18 (page 70).

HL.c.10, HL.c.11, HL.c.13: Environmental ethics

Failure to provide sufficient water for a population can be interpreted through the lenses of consequentialist and rights-based (deontological) ethics, as well as social justice.

From a consequentialist perspective, if many people do not receive sufficient water, they are likely to suffer ill health. The wider society suffers due to the increased burden on the healthcare system and reduced economic growth.

The people who receive sufficient water may benefit in the short term, but their benefits may be outweighed by a larger number of people who do not receive water. In the long run, the benefit to them may not last, as the society as a whole suffers. Either way, the greatest common good is served by ensuring that everyone receives sufficient water for their needs.

A deontological or rights-based ethical perspective might argue that people have a right to sufficient water to meet their basic needs. If we assume that all people have a right to water, failure to provide water may be seen as ethically wrong.

The ethical effects of water inequity can also be seen as a social justice issue. In many places, the people who receive insufficient water are also marginalized or disenfranchised in other ways. They may be poor or from disadvantaged ethnic groups. Water poverty may add to their challenges, and make it harder for them to overcome the disadvantages that society has placed on them.

REVIEW QUESTIONS

1. Describe two domestic and two agricultural water-conservation strategies.
2. Compare and contrast the use of rainwater harvesting in domestic and agricultural settings.
3. What environmental value systems are illustrated by New Zealand's freshwater policy and by the UK's water-conservation regulations?
4. Using named examples, distinguish between water security, water scarcity and water stress.
5. Describe four different technological approaches to managing water stress, and discuss how appropriate each would be for a named country.

4.3 Aquatic food-production systems

> **Guiding questions**
>
> - How are our diets impacted by our values and perspectives?
> - To what extent are aquatic food systems sustainable?

SYLLABUS CONTENT

This chapter covers the following syllabus content:
- ▶ 4.3.1 Phytoplankton and macrophytes provide energy for freshwater and marine food webs.
- ▶ 4.3.2 Humans consume organisms from freshwater and marine environments.
- ▶ 4.3.3 Demand for foods from freshwater and marine environments is increasing due to the growth in human population and changes in dietary preferences.
- ▶ 4.3.4 The increasing global demand for seafood has encouraged the use of unsustainable harvesting practices and overexploitation.
- ▶ 4.3.5 Overexploitation has led to the collapse of fisheries.
- ▶ 4.3.6 The maximum sustainable yield (MSY) is the highest possible annual catch that can be sustained over time, so it should be used to set caps on fishing quotas.
- ▶ 4.3.7 Climate change and ocean acidification are having impacts on ecosystems and may cause collapse of some populations in freshwater or marine ecosystems.
- ▶ 4.3.8 Unsustainable exploitation of freshwater and marine ecosystems can be mitigated through policy legislation addressing the fishing industry and changes in consumer behaviour.
- ▶ 4.3.9 Marine protected areas (MPAs) can be used to support aquatic food chains and maintain sustainable yields.
- ▶ 4.3.10 Aquaculture is the farming of aquatic organisms, including fish, molluscs, crustaceans and aquatic plants. The industry is expanding to increase food supplies and support economic development, but there are associated environmental impacts.

HL ONLY
- ▶ 4.3.11 Productivity, thermal stratification, nutrient mixing and nutrient loading are interconnected in water systems.
- ▶ 4.3.12 Accurate assessment of fish stocks and monitoring of harvest rates are required for their conservation and sustainable use.
- ▶ 4.3.13 There are risks in harvesting fish at the maximum sustainable yield (MSY) rate, and these risks need to be carefully managed.
- ▶ 4.3.14 Species that have been overexploited may recover with cooperation between governments, the fishing industry, consumers and other interest groups, including NGOs, wholesale fishery markets and local supermarkets.
- ▶ 4.3.15 According to the UN Convention on the Law of the Sea (UNCLOS), coastal states have an Exclusive Economic Zone stretching 370 km out to sea, within which the state's government can regulate fishing. Almost 60 per cent of the ocean is the high seas outside these coastal zones, with limited intergovernmental regulation.
- ▶ 4.3.16 Harvesting of seals, whales and dolphins raises ethical issues relating to the rights of animals and of indigenous groups of humans.

Link
Producers, consumers, and energy flow through ecosystems are covered in Chapter 2.2 (page 119).

4.3.1 Phytoplankton and macrophytes provide energy for freshwater and marine food webs

Primary production in aquatic ecosystems is carried out mostly by aquatic plants and algae, using the process of photosynthesis. The plants and algae form the base of the aquatic food chains. The food they produce can be eaten by consumers such as animals.

■ Phytoplankton

Aquatic producers can be found floating in the water or growing from the sediments at the bottom of the water body. Floating producers are called phytoplankton. Most producers growing at the bottom are called macrophytes.

- Plankton is defined as organisms which float and drift in water. They are generally incapable of active swimming, so they drift with the currents, although some can swim over short distances. Most plankton are tiny, microscopic organisms, although some larger organisms like jellyfish and some seaweeds are also considered plankton.
- Plankton are usually classified into phytoplankton and zooplankton:
  - Phytoplankton are plant-like plankton, which carry out photosynthesis. They include photosynthetic bacteria such as cyanobacteria, single-celled algae and some free-floating seaweeds.
  - Zooplankton are animals and animal-like organisms. They are consumers and eat either other zooplankton or phytoplankton. They include single-celled protozoa, as well as tiny crustaceans and the larvae of many aquatic animals.

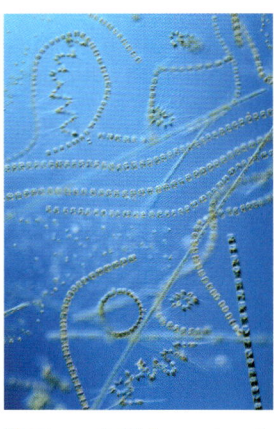

■ Figure 4.63 Examples of phytoplankton: diatoms

■ Figure 4.64 An example of zooplankton: a copepod

■ Macrophytes

Macrophytes are aquatic plants and algae that are large enough to be seen with the naked eye. Most macrophytes live attached or rooted to the bottom of the water body. They include plants such as seagrasses, reeds and water lilies, and algae such as kelp and sea lettuce. Some, such as seagrass, grow completely submerged under water. Others, such as water lilies, grow up through the water, reaching the air. Macrophytes that grow up out of the water are described as emergent. A few macrophytes, such as duckweed and sargassum, float on the water.

> **TOK**
>
> Although phytoplankton and macrophytes are usually considered to be different, if we use their technical definitions there are a few species that fit into both categories, such as the various free-floating seaweeds. This raises the question of the value and accuracy of scientific definitions. Can we always rely on them? Do they always serve useful purposes? Do defined categories of living organisms accurately reflect nature?

■ **Figure 4.65** An example of macrophytic algae: kelp

■ **Figure 4.66** Mangrove trees are an example of emergent vegetation

4.3.2 Humans consume organisms from freshwater and marine environments

The human diet is extremely variable. Although every society has its own typical dietary practices, aquatic organisms commonly form part of the human diet – sometimes a very important part.

Globally, many types of fish are eaten. Most of these are caught at sea, and most are relatively large, predatory fish. A smaller percentage comes from freshwater sources, such as salmon, trout, catfish and perch. Aquaculture, or fish farming, is an increasingly important source of fish and other aquatic foods.

> **REAL-WORLD EXAMPLE**
>
> ### Tilapia: popular freshwater fish
>
> Tilapia *(oreochromis niloticus)* are native to various tropical areas of Africa and the Middle East, including the Nile basin, central and west Africa and Israel. They have been fished for centuries throughout their native range. In more recent years, they have become one of the most popular farmed fish, and an important part of the diet in many countries.
>
> Tilapia are hardy and fast-growing fish which can be produced relatively cheaply by fish farms. These characteristics make it a popular choice around the world.
>
> ■ **Figure 4.67** Red tilapia

Invertebrates are also an important, though smaller, part of the human diet. These include crustaceans such as shrimp, lobster and crab from the sea, and crayfish from freshwater. Molluscs such as snails, squid and octopus are delicacies in particular regions.

4.3 Aquatic food-production systems

REAL-WORLD EXAMPLE

Queen conch: local marine delicacy

The queen conch, *Aliger gigas* (also known as *Strombus gigas*) is a large marine snail found in the Caribbean Sea and nearby parts of the Atlantic Ocean. They are herbivores, which graze on seagrass. The large, thick, usually pinkish shell is a popular souvenir, and often used as an ornament.

The queen conch is a popular delicacy in many Caribbean islands, where it is known as lambí or lambie, and can be prepared in a variety of ways. As a large, slow-moving animal found in shallow water, it is very easily caught. The popularity of the shell makes it extra attractive to fishers, as both the meat and shell can be sold for profit. Their slow growth and low reproductive rate make them vulnerable to overfishing, and they are protected national laws and international agreements, including the Convention on the International Trade in Endangered Species (CITES).

■ **Figure 4.68** Queen conch grazing in a seagrass bed

Link

The role of CITES and the IUCN in the conservation of species is covered in Chapter 3.3, page 245.

● HL.a.2, HL.a.6: Environmental law

Natural resources can be protected by national and international laws. For example, aquatic species, like the queen conch, are protected by specific legislation, making it illegal to capture or kill them, or may limit how many can be caught.

CITES, the Convention on the International Trade in Endangered Species, prohibits the import and export of certain species that are listed in its appendices. These lists are influenced by the Red List produced by the International Union for the Conservation of Nature (IUCN).

ATL ACTIVITY

IUCN Red List

Use the IUCN Red List website to find one example each of aquatic species that are listed as Vulnerable, Endangered or Critically Endangered. For each species, do further research to find out if it is protected by any national or international laws.

Mammals such as whales and seals, and reptiles such as turtles, were once popular local delicacies. They are less popular today, as people have become more concerned about their numbers and have raised ethical questions about hunting and eating them.

Seaweeds and aquatic plants are generally less widely eaten than fish and other animals. However, some species form an important part of the human diet (as shown in the Real-world examples below).

> **REAL-WORLD EXAMPLE**
>
> ### Watercress: globally popular salad
>
> Watercress, *Nasturtium officinale*, is an aquatic or semi-aquatic member of the cabbage family. It is a popular vegetable, often eaten raw. It grows in shallow freshwater, where its hollow stems help its leaves and flowers to float. It can also be cultivated in very wet soils or hydroponically. It is native to Europe and Asia, but is cultivated in temperate regions all around the world.
>
> ■ **Figure 4.69** Watercress growing in a river
>
> **REAL-WORLD EXAMPLE**
>
> ### Nori: popular Japanese alga
>
> Nori is a marine red alga or seaweed, belonging to the genus *Pyropia*. It is popular in Japanese cuisine, where it can be eaten on its own, or used as a wrap in sushi. It can be harvested from the wild, but is mostly cultivated today.
>
> ■ **Figure 4.70** Sushi rolls wrapped in nori

4.3.3 Changing demand for aquatic foods

Globally, consumption of both marine and freshwater organisms is increasing steadily. The main factors driving this increase are population growth and increased affluence (wealth).

Between 1998 and 2018, the world's population increased from about 6 billion, to about 7.7 billion. During that time, consumption of fish increased from about 90 million tonnes, to about 150 million tonnes. The increase in consumption of fish has been faster than the increase in population. This means that the per capita consumption of fish has increased from about 15.6 kg per person per year, to over 20 kg per person per year. In other words, on average, each person on Earth in 2018 was consuming almost 5 kg of fish more each year than they had been 20 years earlier.

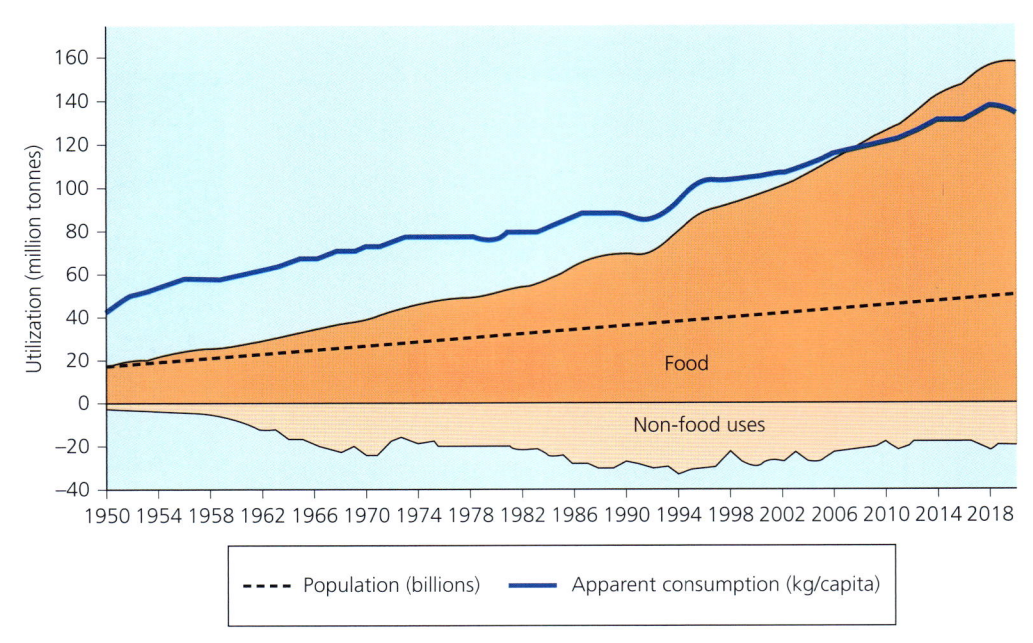

■ **Figure 4.71** Population growth and seafood production, 1950–2020

> **ATL ACTIVITY**
>
> 1 Explain how the graph in Figure 4.71 illustrates an increase in per capita seafood consumption.
> 2 What sorts of 'non-food uses' do aquatic organisms have? Do some research to find out.

Most of the growth in seafood consumption over this period has occurred in China and other parts of East Asia. In these areas, population growth has generally been slow, but economic growth and average personal wealth have increased significantly during this time. In many African countries, where population growth is relatively high, the increase in seafood consumption more closely matches the increase in population growth.

Overall, consumption of all types of seafood is expected to increase rapidly in the coming decades. However, the pattern is complicated. Seafood includes a wide range of organisms, including thousands of species of fish, shellfish and algae. The specific seafoods eaten by people in different societies depends on a range of factors, including cultural preferences, availability and price. The price of seafood is highly variable. In some places, locally produced varieties of seafood are cheaper and more affordable than meat, so they are a major source of dietary protein. Some globally traded species, such as bluefin tuna, can be very expensive, making them luxury items only available to wealthier consumers.

There are several reasons why wealthier people might choose to eat more seafood. Some seafoods are thought to provide health benefits. Oily fish such as salmon are often rich in omega-3 fatty acids, which are known to reduce the risk of heart disease. Some people choose seafoods for environmental reasons. Wild-caught fish, for instance, can have a much lower carbon footprint than farm-bred meat, although some consumers are more concerned about the impact of overfishing. In other cases, the choice of luxury seafood is often driven by taste, or cultural phenomena. Sushi, for instance, is a traditional Japanese delicacy that has become increasingly popular in Europe and North America.

Link

The concept of carbon footprints is covered in Chapter 1.3 (page 47).

Inquiry process

Inquiry 1: Exploring and designing

Use an online data source such as **https://ourworldindata.org** to investigate the relationship between consumption of aquatic foods and socioeconomic factors.

1. Choose a socioeconomic factor which you think might affect demand for seafood, for example an aspect of population growth or economic development.
2. Choose a specific aspect of demand for aquatic food. This might be consumption, capture or production of one particular type of food.
3. Consider the context in which you would like to focus your study. You might choose to look at countries within a particular geographic region of the world, or countries with particular characteristics, such as population size, climate or economic development. You should be able to make a reasonable comparison between the countries you select. Ensure that you have a reasonable number of countries within your selection.
4. Check your data source to see if you can obtain data on your chosen factors. If not, you might consider exploring other data sources, or changing your focus to other factors.
5. Formulate a clear, focused research question, identifying the variables you want to investigate and the context of the countries you want to study.
6. Identify the independent and dependent variables in your research question.
7. Consider whether there are any variables that you think you should control in your study. How do you plan to control them?
8. Formulate a hypothesis for your study. What do you expect to find and why?
9. Consider how you plan to analyse the data. How will you present it? Can you illustrate it graphically? What sort of graph should you use? How will you analyse it, to test your hypothesis? What statistical test can you use?

Inquiry 2: Collecting and processing data

1. Use the database to retrieve the data you need for your investigation.
2. Extract the relevant data and present it in a suitable table. Be sure to include appropriate column headings and units, and to present numerical data in a consistent format.
3. Use a spreadsheet programme to construct an appropriate graph of your data. If you are investigating the effect of one variable on another, you may want to carry out a correlation. What sort of graph is best for correlations? Be sure to include a title and axes labels for your graph.
4. Carry out your chosen statistical test, and use a table of critical values to interpret the result.
5. Draw a conclusion related to your hypothesis and research question.

◆ **Artisanal fishing** – small-scale, non-commercial fishing, typically using traditional fishing methods. Relatively small numbers of fish are usually caught, to be used in the home or to sell for household income.

◆ **Commercial fishing** – fishing for profit, typically using large-scale, modern vessels and methods, and catching large numbers of fish.

Figure 4.72 Artisanal fishing in Lake Victoria, East Africa

4.3.4 Unsustainable harvesting and overexploitation of seafood

A range of methods are available for catching fish, at a variety of scales. Small-scale, **artisanal fishing** may involve an individual with a single fishing line, using bait and hooks, or with a small net that is cast into the water. Large-scale, **commercial fishing** may involve fleets of large ships, equipped with elaborate technology, dragging huge nets, or trailing kilometres-long lines with thousands of baited hooks.

Unsustainable fishing methods

The increasing demand for fish has led to the use of new, more efficient methods to catch fish. Some of the methods currently in use are harmful to fish stocks and the wider environment. These practices may catch so many fish that they cannot replenish their populations, or they may damage the environment the fish depend on. Some of the methods cause wider damage to other, non-target species, or to the wider ecosystems, which could cause ecosystem collapse.

■ Table 4.2 Unsustainable fishing practices

Method	Description	Issues
Bottom trawling	A large net is dragged behind a boat, along or close to the bottom of the water.	Bycatch Disturbance of sediment Habitat destruction
Ghost fishing	Abandoned or lost fishing nets and lines are left in the water, and continue to catch fish.	Ghost gear continues to catch and kill fish and other animals for a very long time.
Cyanide fishing	A low dose of sodium cyanide is squirted at fish to stun them so they can be caught alive, often for the aquarium trade, or for restaurants that offer live fish for customers to select.	Can kill other, non-target species, and cause significant habitat and ecosystem damage. Higher prices for live fish makes the practice very attractive to fishers.
Blast fishing	Explosives such as dynamite are detonated in the water, stunning fish that can then be caught.	Highly dangerous Habitat destruction Kills many non-target species Many fish killed this way sink and are not collected.

■ Figure 4.73 A trawler pulling its net

Bottom trawling

Trawling involves dragging a large net behind a boat. Wing-like trawl doors attached to the net hold the net open. It is an industrial method that can capture large numbers of fish. Trawl nets can be made with different sizes of mesh, to capture different sizes of fish. Many countries have regulations about what size mesh can be used.

However, even with regulated mesh size, trawling is a non-selective method of fishing, which captures many non-target species. Turtles, dolphins and whales are often caught in trawl nets, along with many unwanted fish. These unwanted animals are called **bycatch**. They often die in the nets and are dumped back into the sea.

Trawl nets can be dragged at different depths. One approach, called bottom trawling, uses a weighted net that is dragged along the bottom of the sea in order to catch bottom-dwelling or **benthic** animals, such as shrimp and flounder.

◆ **Bycatch** – unwanted fish or other animals which are unintentionally caught, and usually thrown back in the water. These may be members of endangered or protected species, or animals which have no commercial value.

◆ **Benthic** – related to the bottom of the sea or other body of water. The term can be used to describe the habitat or the species that live there.

Bottom trawl nets can cause significant damage to the seafloor. On sandy or muddy bottoms, they drag through the sediment, kicking it up into the water, and catching or killing everything that lives there. The disturbed sediment also causes problems for fish and other animals swimming in the sea. Habitats like rocky bottoms, coral reefs, seagrass beds and kelp forests can be destroyed by bottom trawling.

Ghost fishing

Fishing nets and other fishing gear are commonly lost or abandoned at sea. These nets, traps or fishing lines may remain in place, continuing to trap and kill fish and other animals such as turtles and dolphins. Modern fishing gear is often made from synthetic materials, which do not decay, so they continue to kill wildlife for many years.

The abandoned fishing gear is known as ghost gear, and the continued killing of wildlife is called ghost fishing. The continuous death of many generations of fish and other wildlife harms populations and ecosystems.

■ **Figure 4.74** Ghost fishing gear

Cyanide fishing

Sodium cyanide is a well-known poison. At high enough doses it can kill rapidly. At low doses it can stun a person or animal temporarily.

In some parts of the world, notably the Philippines and other parts of Southeast Asia, sodium cyanide is used to stun and catch live fish. This is often done to supply aquariums with tropical reef fish and also to supply restaurants. Some restaurants keep live fish in tanks and allow their customers to choose the fish they would like to eat.

■ **Figure 4.75** Cyanide fishing

Cyanide fishing is an effective way to catch large numbers of fish alive. Fishers swim underwater, using masks and snorkels or similar equipment. They carry a squeezable bottle containing a mix of sodium cyanide dissolved in water. They squirt this mixture on to the fish they would like to capture. If the method works, the fish is stunned and can easily be collected.

However, the cyanide squirted at the fish does wider damage to the environment. Cyanide is toxic to corals and algae, as well as other animals. Even in small amounts, it causes significant damage to the wider ecosystem, making the practice unsustainable.

Cyanide fishing is illegal in most countries where it is practised, but it is very difficult to enforce the laws, and the earnings from the practice are attractive. Live fish can sometimes be sold at up to five times the price of dead fish caught with more conventional methods.

Blast fishing

Another unusual and unsustainable fishing practice is blast fishing. In this technique, explosives such as dynamite or ammonium nitrate are detonated in the habitat being fished. The shockwaves produced stun or kill the fish, which can then easily be collected. The practice is quite dangerous, and people have been injured or killed while fishing in this way.

Blast fishing is highly inefficient and non-selective. All the fish near the blast are stunned or killed, but many of the dead fish sink so they cannot be collected. Many non-target species, and juvenile fish that are too small to use, are also killed. The practice also does significant damage to habitats where it is used, such as coral reefs.

■ **Figure 4.76** Blast fishing

4.3 Aquatic food-production systems

The indiscriminate killing and habitat damage make both blast fishing and cyanide fishing highly unsustainable practices. Damaged habitats, with depleted biodiversity, cannot sustain viable populations of fish. Like cyanide fishing, blast fishing is generally illegal, but it can be very difficult to enforce the laws.

Overfishing

Fish are, in principle, a renewable resource. As living organisms, they can reproduce and replenish their numbers. However, if fish are caught faster than they can reproduce, then the numbers of fish will decline. Exploitation of a resource at a rate that prevents it from replenishing itself is overexploitation. Overexploitation of fisheries resources is called overfishing.

If a fish stock is continuously exploited faster than it can grow, the stock will decline – that is, the numbers of fish in the population will fall. With smaller numbers of fish in the sea, it becomes harder to catch fish. There will therefore be fewer fish available for people to buy. Smaller numbers of fish also means that the rate of population growth will decrease. With fewer fish remaining in the population, there is less chance for the population to recover.

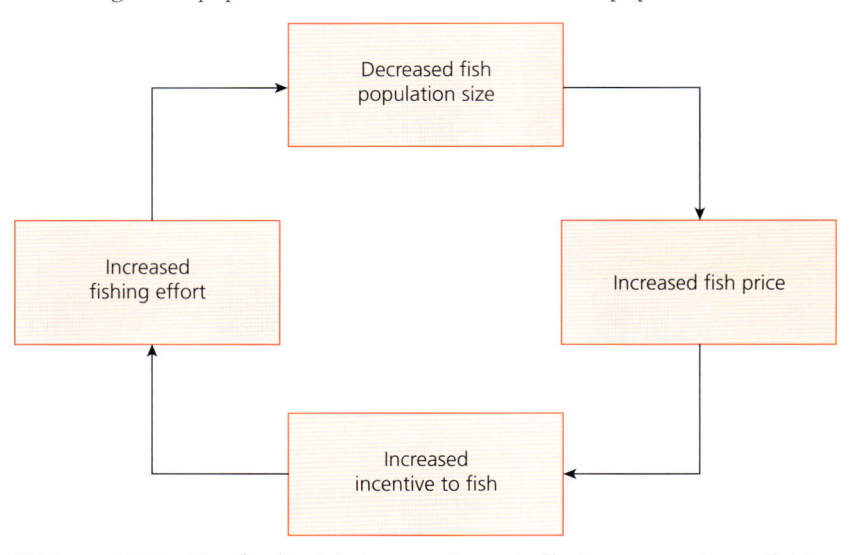

■ **Figure 4.77** Positive feedback between price and effort can promote overfishing

Following the principles of economics, as fish become scarcer while demand for fish rises, the price of fish will rise. This can make it even more attractive for fishing communities and businesses to try to catch more fish. They may increase their fishing effort by sending out more boats, spending more time fishing, and using improved fishing techniques and technologies. This can produce a vicious cycle, or positive feedback loop, that produces greater pressure on fish stocks, and makes it more likely that fisheries will collapse.

In practice, this scenario does not necessarily occur, as many other factors affect the fishing effort, including the cost of fishing and government subsidies paid to the fishing industry. The higher price may also mean that consumers can no longer afford to buy the fish, so demand falls. This represents negative feedback and can stabilize fish production. Each of these factors may increase or decrease the rate of exploitation.

> **ATL ACTIVITY**
>
> Draw a diagram similar to Figure 4.77 to show how rising price and falling demand can produce a stabilizing, negative feedback loop.

Link

Tragedy of the commons is introduced in Topic 3.2.7, page 231.

> ● **HL.b.3, HL.b.6: Environmental economics**
>
> Overfishing can be interpreted as an example of market failure. The dynamics of the free market make rare fish expensive. This encourages people to catch those rare fish, making a significant profit. However, as the fish stock is depleted, the overall fishing community suffers and the ecosystem is also damaged. These may be seen as external costs or externalities. The people who catch the last fish benefit, and do not pay the price of the damage they cause to the stock.
>
> Overfishing is also often seen as a case of the tragedy of the commons. Since no one owns the fish in the sea, they are accessible to everyone. If there is no regulation of fishing, individuals or companies may take the decision to catch as many fish as they can. This leads to overexploitation, and depletion of the resource. In the long term, everyone suffers.

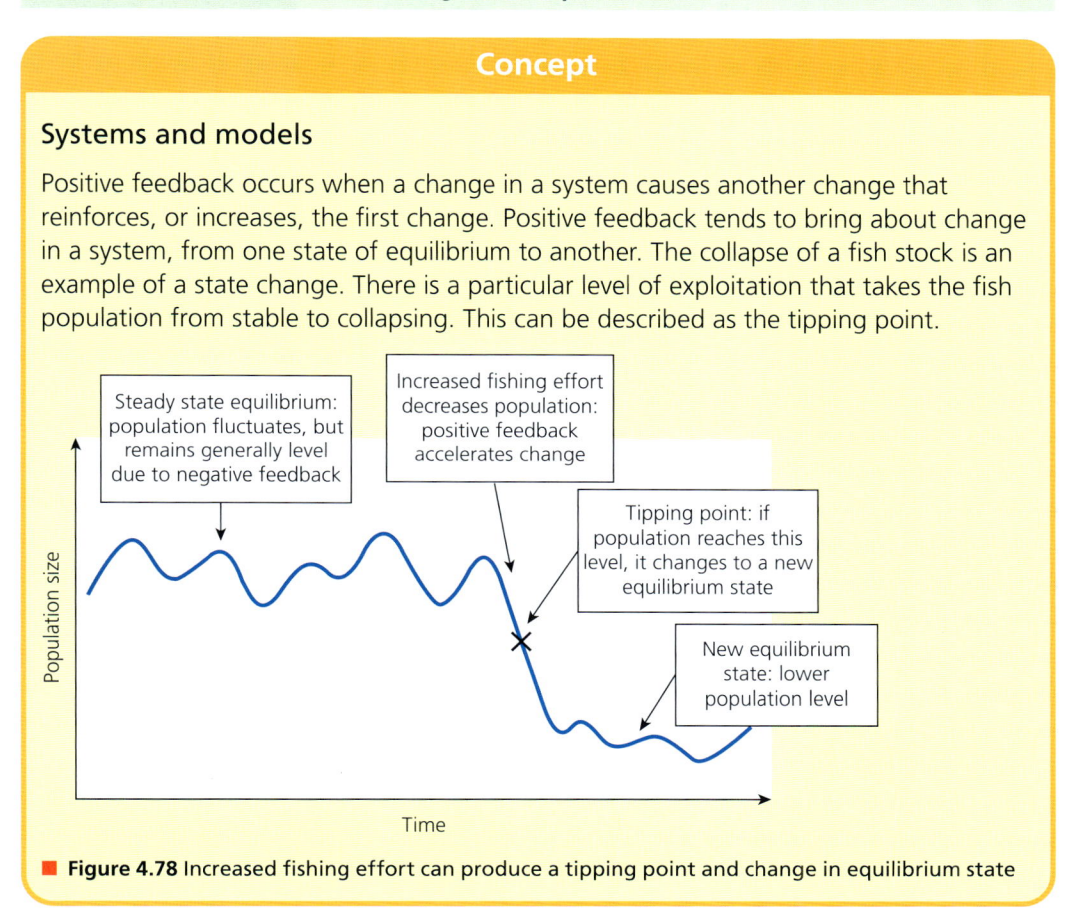

Concept

Systems and models

Positive feedback occurs when a change in a system causes another change that reinforces, or increases, the first change. Positive feedback tends to bring about change in a system, from one state of equilibrium to another. The collapse of a fish stock is an example of a state change. There is a particular level of exploitation that takes the fish population from stable to collapsing. This can be described as the tipping point.

Link

Feedback loops, tipping points and their effects on system equilibria are covered in Chapter 1.2 (page 24).

■ **Figure 4.78** Increased fishing effort can produce a tipping point and change in equilibrium state

4.3.5 Overexploitation has led to the collapse of fisheries

The term fishery may refer to an area where fish are caught, also known as a fishing ground. More broadly, fishery may refer to all the fishing practices of a particular fishing ground. The location, fish stock, people, equipment and methods involved may be considered part of a particular fishery. A fishery may collapse if the business of fishing stops being productive, perhaps because of rapid depletion of the fish stock.

There are many examples of fisheries that have collapsed under the pressure of overfishing. The most famous of these was the collapse of the northwest Atlantic cod fishery, centred in Newfoundland, Canada.

4.3 Aquatic food-production systems

> **REAL-WORLD EXAMPLE**

Cod fishery in Newfoundland, Canada

The Atlantic cod, *Gadus morhua*, is a popular food fish that is found in the temperate regions of the North Atlantic. It swims in large shoals near to the bottom of the sea. Trawling is the most common method used in fishing for cod. In the eastern part of its range, between northern Europe and Iceland, its population is generally stable and healthy. In the western Atlantic, however, it has been heavily overfished.

■ Figure 4.79 Catching cod off Newfoundland, Canada

Throughout most of its history, the northwest Atlantic cod fishery produced a stable but gradually increasing catch. Beginning around the 1960s, however, new technologies allowed a sudden increase in fish catch. New fishing vessels were introduced, equipped with navigation and tracking technologies such as radar and sonar, and capable of travelling further and fishing deeper. Collectively, they produced a rapid increase in the size of the cod catch.

The increased catch meant that fewer cod were left in the sea, and the rate of reproduction dropped. At the same time, the new ships caught increasing numbers of unwanted fish, or bycatch. Bycatch included various species that were part of the cod's ecosystem, including important prey species. The wider ecological impacts also contributed to slower recovery of the cod population.

In 1992, the cod fishery in the northwest Atlantic suddenly collapsed. Fish catch fell dramatically, and fishing communities and businesses lost their main source of income.

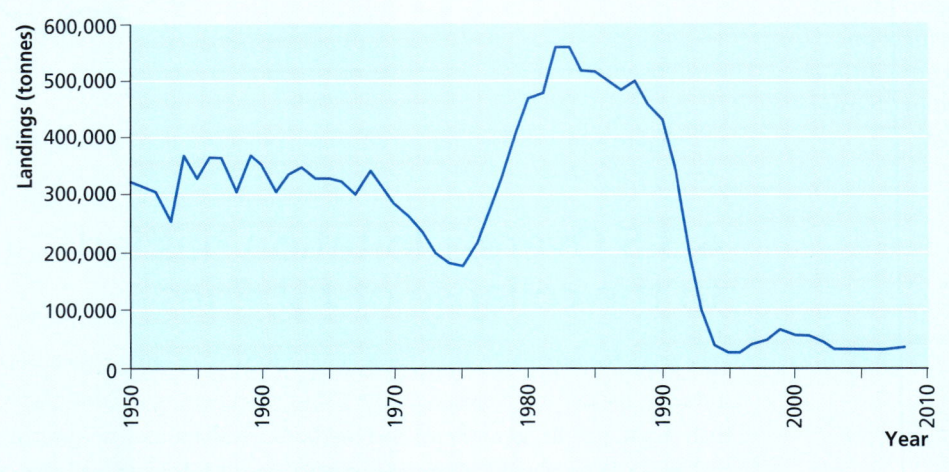

■ Figure 4.80 Annual harvest of northwest Atlantic cod, 1950–2008

Theme 4: Water

There are many other examples of overexploitation of marine and freshwater resources. Table 4.3 lists some well-known examples.

■ Table 4.3 Examples of overexploited marine and freshwater fish

Common name	Latin name	Distribution	IUCN status
Atlantic bluefin tuna	Thunnus thynnus	Atlantic Ocean and Mediterranean Sea	Least Concern (raised from Endangered in 2021)
Nassau grouper	Epinephelus striatus	Western Atlantic Ocean and Caribbean Sea	Critically Endangered
Beluga sturgeon	Huso huso	Rivers, estuaries and inland seas of Eastern Europe and western Asia	Critically Endangered
Scalloped hammerhead shark	Sphyrna lewini	Tropical and subtropical coastal seas around the world	Critically Endangered

The United Nations Food and Agriculture Organization (FAO) reported that, as of 2019, 35.4 per cent of global marine fisheries were overfished, and 17.5 per cent of global seafood came from unsustainable stocks.

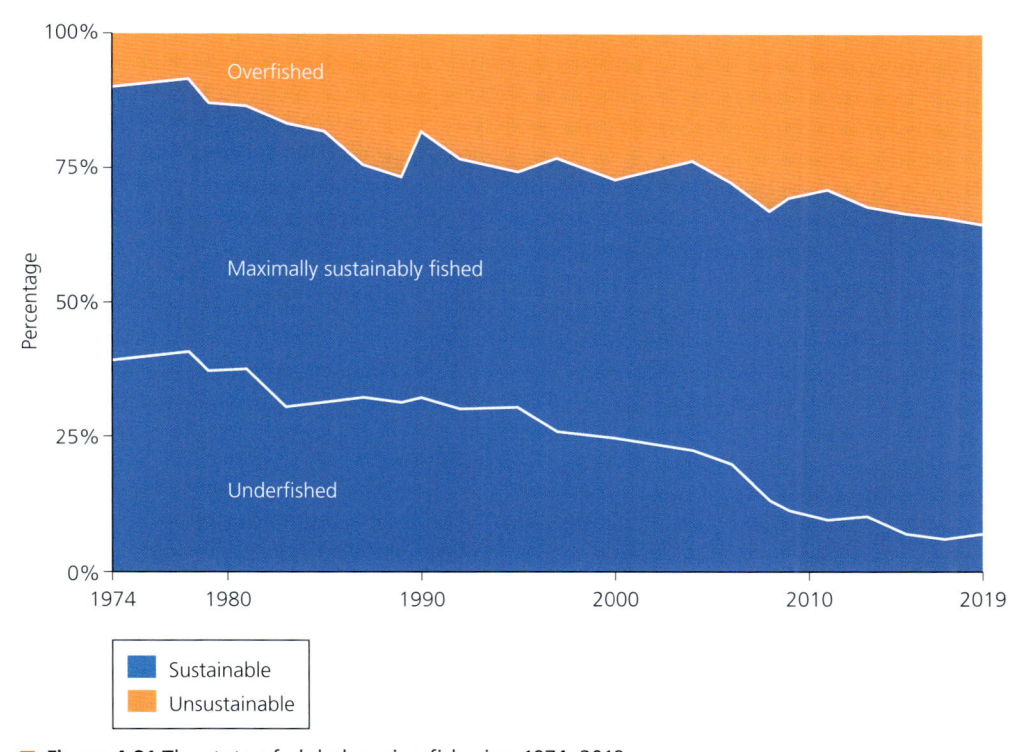

■ Figure 4.81 The state of global marine fisheries, 1974–2019

ATL ACTIVITY

1. Use Figure 4.81 to determine what percentage of global fishery stocks were being fished at the 'maximally sustainable' levels. What do you think 'maximally sustainable' means?
2. What does Figure 4.81 suggest about the prospect for increasing global fisheries production?
3. Explore the latest State of World Fisheries report on the FAO website, **www.fao.org/home/en**, to find out about fisheries activities in your part of the world.

Link

The concept of a steady-state equilibrium is covered in Chapter 1.2 (page 24).

4.3.6 The maximum sustainable yield is the highest possible annual catch that can be sustained over time, so it should be used to set caps on fishing quotas

To avoid overfishing, we should aim to match the number of fish we catch with the number of fish added to the population. This would allow the population to remain in a steady-state equilibrium.

In many cases, this would mean reducing how many fish we catch. However, this would reduce the income of the many people who work in the fishing industry, as well as the amount of fish available for people to eat. Fisheries scientists have developed an approach that aims to find a balance between the need to maximize fish production while maintaining healthy fish populations. This approach is called **maximum sustainable yield** (MSY).

Link

Logistic population growth is covered in Chapter 2.1 (page 84).

MSY involves setting a maximum number of fish that can be caught, based on the highest possible growth rate of the fish population. The calculation is based on the S-shaped **logistic population growth** curve.

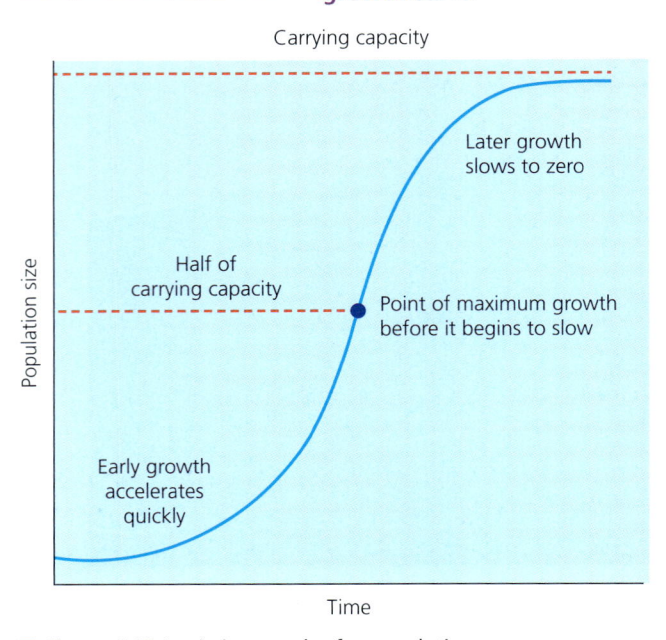

■ Figure 4.82 Logistic growth of a population

The rate of population growth is shown by the gradient, or steepness, of the curve. As the graph in Figure 4.82 shows, the steepest point on the graph occurs when the population size is half of the **carrying capacity**. When the population is less than half the carrying capacity, growth is slow because there are too few fish, so that the number of new fish produced is small. At above half the carrying capacity, the rate of growth is slowed by **density-dependent factors** such as disease and competition for resources.

In theory, if the population of fish remains at half the carrying capacity, then it will grow at the fastest rate possible. If we catch the same number of fish as are added to the population, we can keep the population growth at its highest. This level is the maximum sustainable yield, or MSY. In theory, if we match these numbers precisely, we can catch the highest possible number of fish without depleting the fish stock.

In practical terms, a more useful way of looking at this is by looking at how **fishing effort** affects **yield**. This produces the graph in Figure 4.83. Fishing effort is a measure of how much effort is put into catching fish. A greater effort may be achieved by more larger boats, spending more time at sea, and using more elaborate technology. Yield refers to the amount of fish caught. It is usually measured as biomass of fish.

◆ **Maximum sustainable yield** – the highest yield that it is possible to achieve without depleting the fish stock.

◆ **Logistic growth** – growth of a population that shows a typical S-shaped or sigmoid curve. Growth starts slow, then speeds up, becoming exponential, then slows down, approaching zero.

◆ **Carrying capacity** – the maximum population size which can be supported by a particular environment. If the population rises above the carrying capacity, resources will become depleted and the population will fall.

◆ **Density-dependent factors** – factors which influence population growth and that change as the population increases in size, such as stress, disease, predation and competition.

◆ **Fishing effort** – a measure of the level of fishing activity. It takes into account how many vessels are active, how much time they spend fishing and what sort of fishing gear they use.

◆ **Yield** – the total production of a product. In fisheries and agriculture, it is usually measured in mass of product (e.g. tonnes of fish or tomatoes).

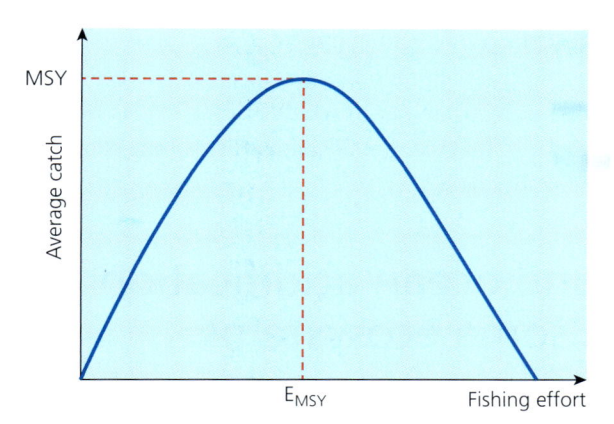

Figure 4.83 Determining the maximum sustainable yield

Figure 4.83 shows the yield as average catch, measured over a period of time, such as a year. At low fishing effort, the catch is low because there are few boats trying to catch fish. There will be a lot of fish in the sea, and they will be reproducing slowly, to replace the fish that are caught. This results in a stable population of fish and a sustainable yield, but the yield is low so income from fishing is low. This is not attractive to the fishing community.

At high levels of fishing effort, the average catch is low because the resource has been overexploited. Too many fish are caught at any one time, so the fish population falls to less than half of the carrying capacity and the rate of population growth is low. Even if we catch a lot of fish at first, we cannot sustain that yield, because the population growth is too slow.

The theoretical solution is to maintain fishing effort somewhere in the middle. In theory, we set this fishing effort at a point that maintains the population at half of the carrying capacity. The rate of population growth is the highest it can be, and we are able to produce the greatest possible yield without depleting the resource. If we can achieve this, we might say we are using the natural income without depleting the natural capital.

Link

The ideas of natural capital and natural income are explored in Topic 7.1.4 (page 611).

Concept

Sustainability

Maximum sustainable yield is an attempt to balance people's needs with the needs of the environment, while maintaining a stable, long-term supply of a resource. It aims to ensure that we produce the largest possible amount of resource, to meet people's current needs, while maintaining a large enough stock of resources to fulfill people's future needs.

Concept

Perspectives

Maximum sustainable yield can be seen as an attempt to find compromise between those who believe natural resources are there to be used for economic purposes, and those who believe natural resources have intrinsic value, in and of themselves. A person's environmental value system influences their beliefs on matters such as this.

ATL ACTIVITY

Evaluate claims about sustainability and identify the barriers to attaining them

Work in groups. Take turns to make a claim about the maximum sustainable yield (MSY) approach (an explanation or interpretation of some aspect of MSY), then identify support for your claim (ideas, concepts or evidence you know that support your claim), and finally ask a question related to your claim (What isn't explained? What new reasons or issues does your claim raise?).

Now reflect on the activity. What new thoughts do you have about the value of MSY in regulating global fish production?

◆ **Quota** – a limit placed on the amount of a resource that can be extracted (for example, the amount of fish that can be caught). It might be applied to individuals, companies or countries.

Maximum sustainable yield is usually applied in practice by setting **quotas**, that is limits on how much fish can be caught in a given area or time. Fishing communities must ensure that they do not exceed their assigned quota by stopping fishing when they reach their limit. In this way, setting quotas indirectly reduces the fishing effort, so that the overall community fishing effort stays at the level that produces maximum sustainable yield.

4.3.7 Climate change and ocean acidification are having impacts on some ecosystems

Overfishing is not the only factor that causes fish populations to decline. Population size can also be affected by environmental factors, such as climate change and ocean acidification.

Climate change

Climate change is affecting rainfall patterns and ocean currents, which in turn affect surface and deep-water temperatures, which then impacts fish populations.

Changes in temperature can affect fish directly or indirectly. Many fish have a limited tolerance for changing water temperature. If the temperature is too high or too low, they may die or migrate to more suitable areas. Changing temperature can also affect reproduction. When we are calculating maximum sustainable yield, we need to consider the effect of warmer seas on reproduction.

Link
The effects of increased algal growth are covered in Topic 4.4.10 (page 397).

◆ **Coral polyp** – tiny, individual animals that coral colonies are made up of. Each polyp looks a little like a small sea anemone, with tentacles around its mouth.

◆ **Zooxanthellae** – single-celled algae that live inside the tissues of coral polyps, giving the coral their colour. Zooxanthellae and corals have a mutualistic relationship, where the algae use the corals' waste products and in turn provide food to the corals.

Warmer water can also affect other organisms within the fish's ecosystem. It may cause more rapid algal growth, for instance, meaning there is more food for herbivorous fish. Changing numbers of herbivores can have wider effects on other fish, producing higher or lower numbers of the fish we are interested in. Some algae are also harmful and release toxins into the water that can kill fish.

On a coral reef, excessive growth of algae can also harm the corals by growing over them and preventing light from reaching the corals. Corals form the base of the coral reef food web, and also provide an amazing array of microhabitats for fish to live and reproduce in. If the corals die, it is likely that the entire ecosystem will collapse. The fish we are interested in catching will probably disappear.

Furthermore, on coral reefs warmer water may be harmful to the entire ecosystem by causing coral bleaching. Bleaching occurs when the **coral polyps** are stressed by some environmental factor, such as unusually high temperatures. Healthy coral polyps contain single-celled algae called **zooxanthellae** that live symbiotically inside their tissues. The algae give the coral their different colours. They use sunlight to photosynthesise and provide the polyp with some of the sugars they produce. During bleaching, coral polyps eject the zooxanthellae from their bodies. With no algae, the corals appear white and their growth is severely reduced. Bleached corals can recover if the environmental stress is removed, but they are weakened by the bleaching and may die.

■ **Figure 4.84** Bleached corals

Changing patterns in water temperature and ocean currents can change the distribution of food organisms such as plankton. The fish that eat the plankton will therefore also move. This can have positive effects on some fishing communities, where the availability of fish may increase, but elsewhere communities may lose their source of income.

Changing ocean currents can also change the availability of key factors like nutrients and oxygen. If upwellings stop, nutrients may not be brought to the surface and fish populations will die, as happened to the Peruvian anchovy fishery in El Niño years (see Topic 4.1.13, page 403). If downward flow of water is slowed or stopped, oxygen may no longer be brought to the deep water and deep-water fish may die.

Ocean acidification

Ocean acidification occurs because of excess carbon dioxide dissolving in the water. As carbon dioxide levels rise due to human activities, more and more carbon dioxide dissolves in the ocean, making it more acidic. This can have devastating effects on marine ecosystems. Prior to the Industrial Revolution, the pH of ocean water was approximately 8.2, which is slightly alkaline. Today, excess carbon dioxide dissolved in the water has reduced it to 8.1. This may seem a small change, but pH is a measured on a logarithmic scale, and a 0.1 change in pH represents a 26 per cent increase in acidity.

A lower pH in the ocean can affect many different species, which are adapted to living in the slightly alkaline ocean. The biggest effect may be felt by organisms that grow shells using calcium carbonate. Calcium carbonate is dissolved by acids, so the shells of these animals will wear away, and the growth process of the shells will slow down (see Topic 2.3.10, page 150).

Crucially, many members of the zooplankton are crustaceans, including important groups such as copepods and krill. These provide food for a wide range of fish and other larger animals.

Coral skeletons are also made of calcium carbonate, so ocean acidification slows coral growth, indirectly affecting coral reef ecosystems. The increased concentration of carbon dioxide in the atmosphere harms coral reefs through both increased water temperature and ocean acidification.

■ **Figure 4.85** Antarctic krill

REAL-WORLD EXAMPLE

Great Barrier Reef, Australia

The Great Barrier Reef, off the coast of Queensland, Australia, is the largest area of coral reef on Earth, and a designated World Heritage Area. It is made up of over 3,000 individual reefs and is home to an impressive array of biodiversity. It supports a number of commercial fisheries, and is an important site for recreation and tourism.

■ **Figure 4.86** The location of the Great Barrier Reef

The fisheries of the Great Barrier Reef suffer from multiple threats, including overfishing, bycatch, global warming and ocean acidification.

Commercial, recreational and indigenous fishing communities all use the Great Barrier Reef fisheries, leading to conflicts and difficulty in regulating the industry. Quotas exist, but they are difficult to enforce. The large gillnets that are sometimes used on the reef have caused bycatch of animals, including endangered sea turtles, dugongs and dolphins.

4.3 Aquatic food-production systems

> The Great Barrier Reef has experienced seven significant mass coral bleaching events since 1998. The scale and frequency of these events seems to have increased: there were four events in the seven years from 2016 to 2022. It was reported that 91 per cent of the 3,000 reefs were affected by the 2022 event.
>
> Ocean acidification has also been observed on the Great Barrier Reef, where lowered pH has caused a decrease in the rate of growth of corals. The rate of growth of corals measured between 1990 and 2005 was the slowest it has been for the previous 400 years.

4.3.8 Unsustainable exploitation of freshwater and marine ecosystems can be mitigated through policy legislation addressing the fishing industry and changes in consumer behaviour

As we have seen, marine and freshwater fisheries have been depleted by overfishing and by the use of unsustainable fishing practices. Maintaining these important sources of food and income is essential for a sustainable future. Many countries have established national policies and legislation to help manage fisheries more sustainably. Governments, non-governmental organizations and international agencies have also made efforts to educate the public, so as to change consumer behaviour.

Fisheries legislation

Many governments have used legal tools to reduce the impacts of different threats to sustainable fisheries. Some regulations aim to control the types of fishing gear and fishing methods that are allowed. Others regulate the types and amounts of fish that can be caught.

Legislation can be used for several different purposes, as shown in Table 4.4.

■ Table 4.4 Examples of objectives of fisheries legislation

Legislation type	Description	Issues and challenges
Quotas	Limits are placed on catch size. These may be applied to individuals or to entire fisheries. Ideally, they should be based on sound data, and an understanding of population and ecosystem dynamics. MSY is usually enforced through quotas.	Difficult to obtain accurate data on populations and ecosystems. Conditions in populations and ecosystems are dynamic. Difficult to enforce quotas.
Fishing gear regulations	Rules limiting the types of equipment that are permitted to be used. Can include limits to net type or size, or mesh size, to reduce bycatch and protect younger fish. Can also require use of equipment to limit bycatch or other environmental impacts (see Figure 4.88).	Can increase the cost of gear, so may meet resistance from fishers. Subsidies and education may help. Enforcement may be difficult.
Fishing permits	Individuals or companies are required to purchase a permit allowing them to fish in a particular location or time. A limited number of permits are sold, in order to limit fishing pressure. Permits may include a requirement to obey quotas and other regulations. Can be used for both recreational and commercial fishing.	People who see fish as a common resource may oppose the need for permits. Others may be unaware of the need for one, especially for recreational fisheries. Enforcement may be difficult.
Fisheries management zones	Geographic areas are allocated for different uses, such as fishing for different species, or using different techniques. Some areas may be only for recreational use; others may be 'no-take zones', where no fishing is allowed. This can protect particularly vulnerable areas, or areas where fish reproduce, while permitting fishing elsewhere.	Can be difficult to enforce. May cause conflict between different users.
Fishing seasons	Fishing is restricted at certain times of year. Can be used to protect fish while they are breeding or migrating.	Enforcement may be difficult.

Figure 4.87 A diver removes a crown-of-thorns starfish from a coral reef

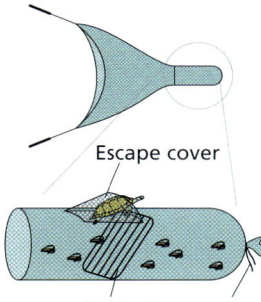

Figure 4.88 A turtle excluder device (TED) in a trawl net. The grid allows fish through, while excluding larger bycatch such as turtles. In some places, their use is required by law

Figure 4.89 The Marine Stewardship Council's Blue Fish Label

HL.a.2: Environmental law

Laws regulating fisheries provide a good example of environmental law. Their purpose is to manage a natural resource, in the interest of economic, social and environmental sustainability. Failure to manage fisheries can result in loss of income for individuals, companies and nations. Societies that depend on fisheries for their livelihoods may collapse, with many people losing their jobs, falling into poverty and possibly moving away. Modern fisheries laws often recognize the importance of protecting aquatic environments for the benefit of societies and economies.

International cooperation

National legislation can only operate within a country's borders, but fish populations regularly extend across borders and into international water. Various international initiatives have been used to ensure that different countries cooperate in the management of fish stocks.

Many of these are bilateral agreements and collaborations between neighbouring states, such as between the European Union and the United Kingdom, or between the United States and Canada. Others are formal multilateral agreements, such as the United Nations Fish Stocks Agreement, which requires that the best available scientific evidence, and a precautionary approach, are used in managing international fish stocks. The Convention on Biological Diversity, the International Whaling Commission, the Convention on International Trade in Endangered Species, and the United Nations Convention on the Law of the Sea all contribute to international cooperation on fisheries management.

HL.a.7: Environmental law

International law, and international organizations, play an important role in managing fisheries. In 1995 the United Nations established the Fish Stocks Agreement (UNFSA) in order to manage fish stocks that cross international borders. A follow-up agreement set out specific strategies to ensure that international fish stocks are effectively managed. This agreement recognized that enforcing fisheries laws is essential to the sustainable management of fish stocks. Illegal, unreported and unregulated (IUU) fishing undermines fisheries management and can lead to overexploitation.

Education for consumer change

Well-informed consumers can make educated choices about which seafoods are most sustainable. Education and product labelling help them to make these decisions.

The Marine Stewardship Council (MSC) is an international non-governmental organization that aims to prevent overfishing, and helps to ensure sustainability in seafood production. It reviews fishing practices and scientific data on fisheries all around the world. Seafood products can receive an MSC Blue Fish Label if they meet certain criteria. The label is easily recognized, and explained on food packaging, to help consumers make informed purchasing decisions.

To receive an MSC Blue Fish Label, a product must come from a fishery that is sustainably exploited, has minimal negative environmental impact and is managed effectively. The label is internationally recognized.

ATL ACTIVITY

Consider the various approaches to reducing overexploitation presented in this section. Evaluate the effectiveness of each approach for the sustainable management of fisheries. For each approach, you should aim to find both strengths and weaknesses. Consider also that some measures may be effective in some situations and ineffective in others.

Conduct a class debate on the effectiveness of one or more of the measures described. Randomly assign members of the class to argue the case in favour of or against each measure.

4.3 Aquatic food-production systems

4.3.9 Marine protected areas can be used to support aquatic food chains and maintain sustainable yields

Protected areas have been used to protect biodiversity on land for many years. Marine protected areas (MPAs) are a more recent development. They are areas of the sea that are designated and given special legal protections.

Certain types of activities can be restricted or prohibited in an MPA, depending on the purpose of the protection. In the most strictly controlled MPAs, humans are not allowed to enter the area. In others, specific activities such as fishing may be highly regulated or banned. In some larger MPAs, different activities may be allowed in different areas or zones within the MPA.

Many MPAs are designed specifically to protect fisheries resources. Some ban any sort of fishing, to allow fish populations to recover. 'No-take' MPAs such as this are often established to protect spawning grounds – areas where fish gather to reproduce. By protecting fish reproduction, they encourage population recovery.

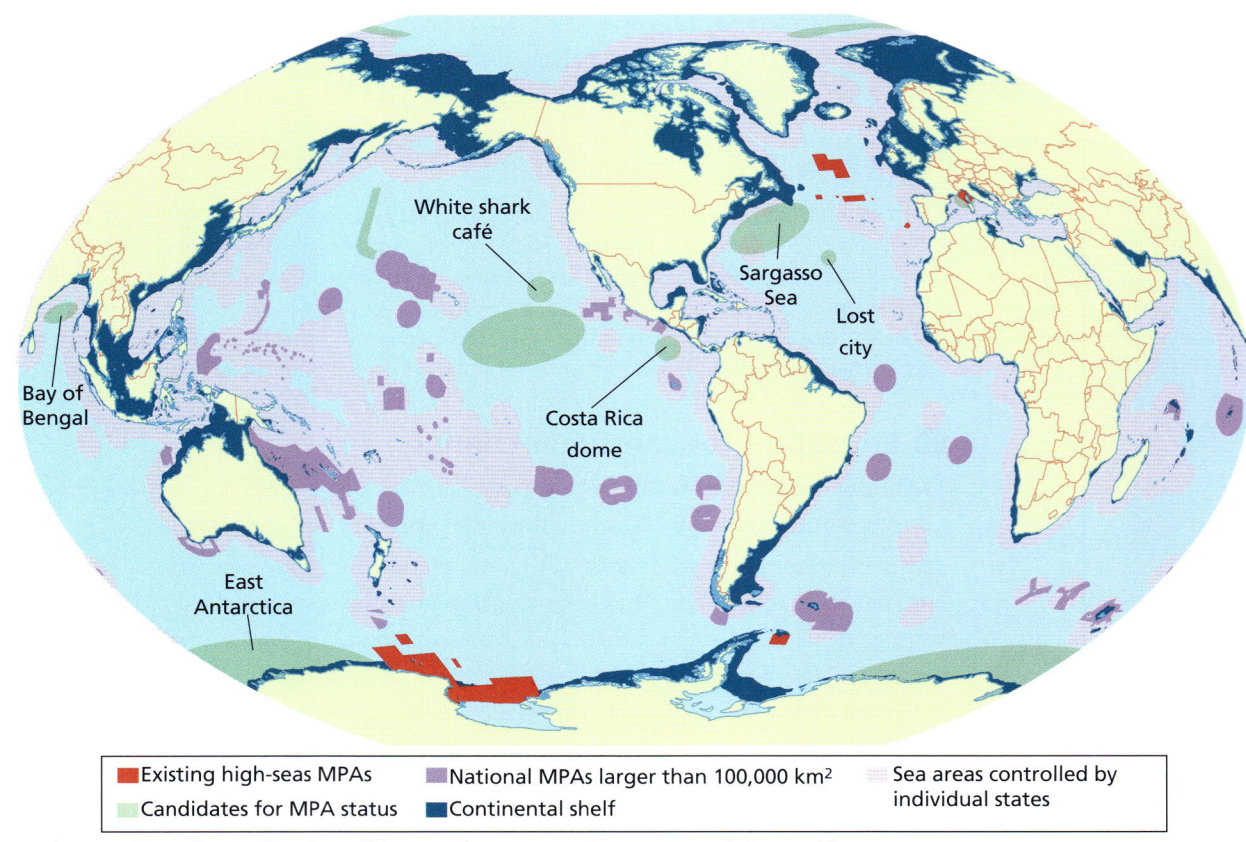

■ **Figure 4.90** Designated and candidate marine protected areas around the world

> **REAL-WORLD EXAMPLE**
>
> ### Great Barrier Reef Marine Park
>
> The Great Barrier Reef Marine Park is one of the largest, and perhaps most complex, of the world's MPAs. It encompasses 14 different, interconnected coastal ecosystems, including coral reefs, lagoons, mangroves and islands. It claims to protect 10 per cent of the world's coral reefs. The reef is vulnerable to several threats, including climate change and overfishing.
>
> One approach used in the park is zoning. Some zones are designated as 'no-take' or 'green' zones, meaning that no fishing or other extractive activity is allowed, but boating, snorkelling, diving and other recreational activities are permitted. Green zones make up more than 33 per cent of the MPA. Other zones allow multiple uses, including fishing, but these are highly regulated to ensure sustainability.
>
> ■ **Figure 4.91** Zoning in the Great Barrier Reef Marine Park (supplied by Spatial Data Centre, Great Barrier Reef Marine Park Authority, © Commonwealth of Australia (GBRMPA))

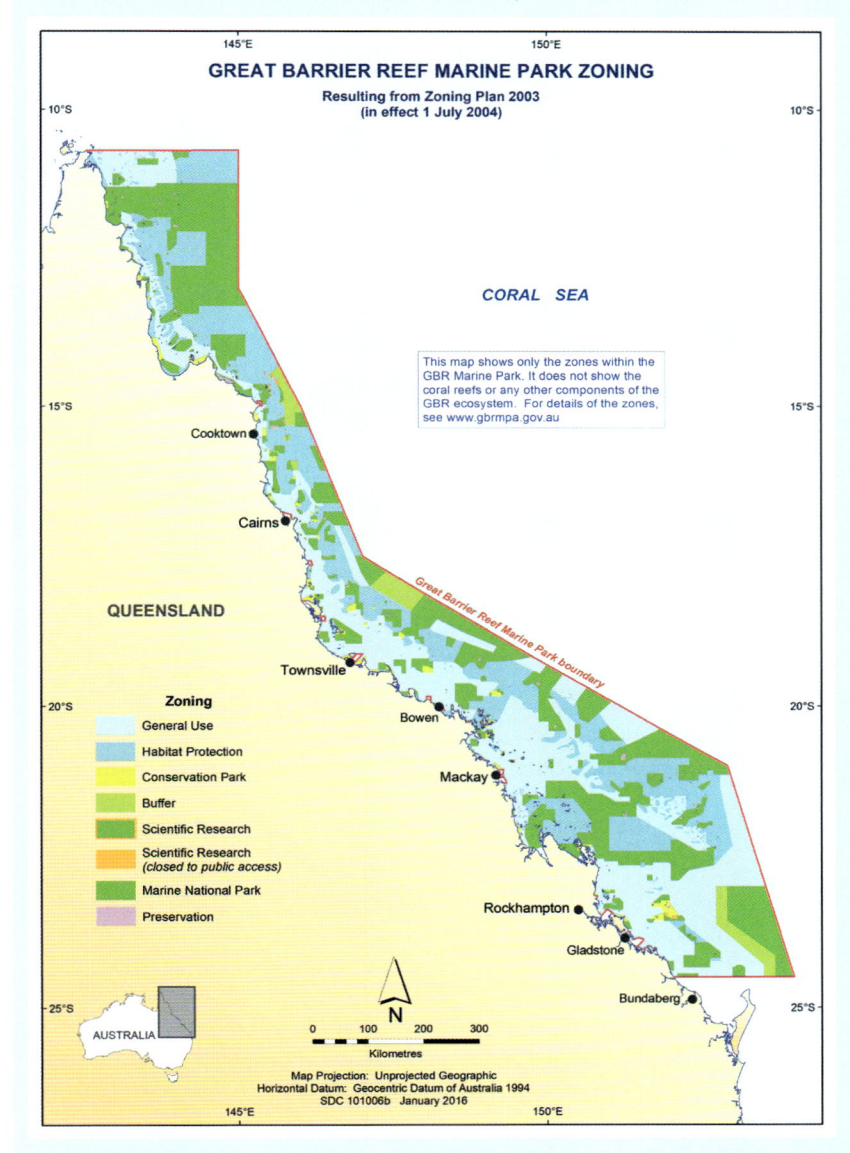

4.3 Aquatic food-production systems

Other MPAs may allow fishing, but with limitations, such as size of fishing vessels, or types of fishing methods or gear permitted. Fishing restrictions can cause conflicts with the fishing industry. However, MPAs have generally proven beneficial for fisheries.

No-take MPAs, and no-take zones within MPAs, have been shown to increase the individual size, total biomass and number of fish compared to nearby unprotected areas. These extra fish can then spread into surrounding areas, where they improve the fisheries yield. Protected areas also show improvements in overall ecosystem characteristics, such as biodiversity and improved ecosystem services.

REAL-WORLD EXAMPLE

Goat Island Marine Reserve, New Zealand

Goat Island Marine Reserve, also known as Leigh Marine Reserve or the Cape Rodney-Okakari Point Marine Reserve, was New Zealand's first marine reserve. It was established in 1975 on the coast of North Island, New Zealand as a no-take reserve. It has been well studied by a nearby marine science laboratory.

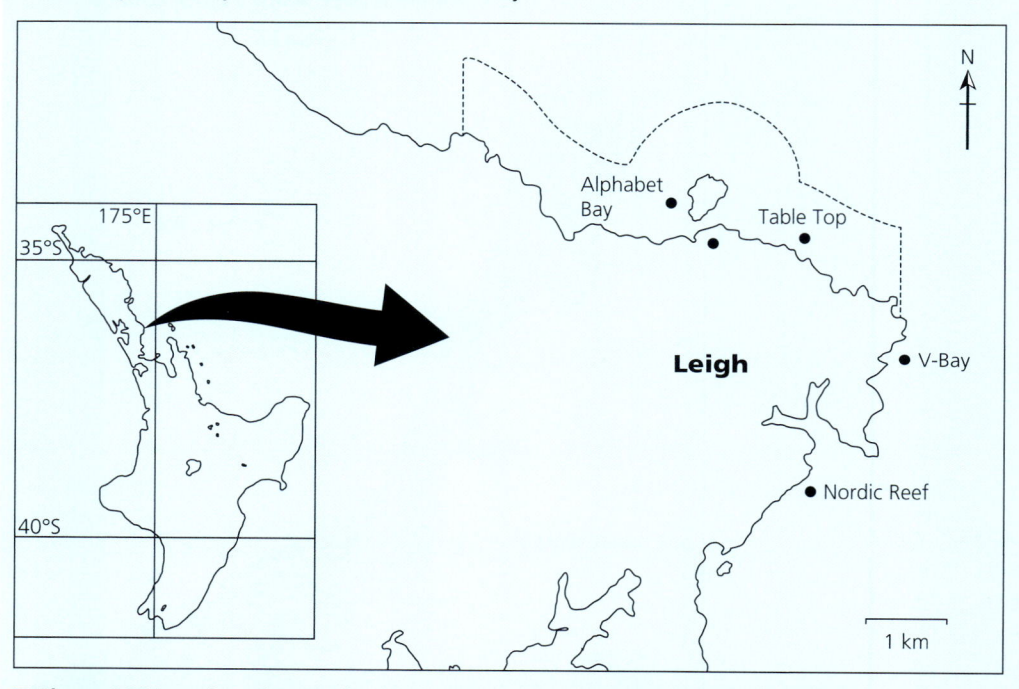

■ Figure 4.92 Location of Goat Island Marine Reserve, on North Island, New Zealand

Before the reserve was established, the area had been severely overfished, with species like rock lobster and Australasian snapper showing major declines. The ecosystem at the time was dominated by a herbivorous sea urchin, which had significantly reduced the amount of algae.

Studies have shown that since fishing was banned in the area, several larger predatory species have returned and their populations have recovered. The predators have reduced the sea urchin population, which has allowed algae and kelp to recover. With greater algal growth, productivity, habitat diversity and overall biodiversity have all increased. These widespread effects of returning predators, which pass down the trophic levels, are known as trophic cascades.

4.3.10 Aquaculture is the farming of aquatic organisms. The industry is expanding but there are associated environmental impacts

Aquaculture is increasingly being used to meet the rapidly rising demand for fish and other aquatic foods whilst protecting marine and freshwater ecosystems that are threatened by overfishing, habitat destruction and climate change.

Many different species of fish, crustaceans, molluscs, algae and other aquatic plants are farmed around the world, using a variety of different approaches and scales.

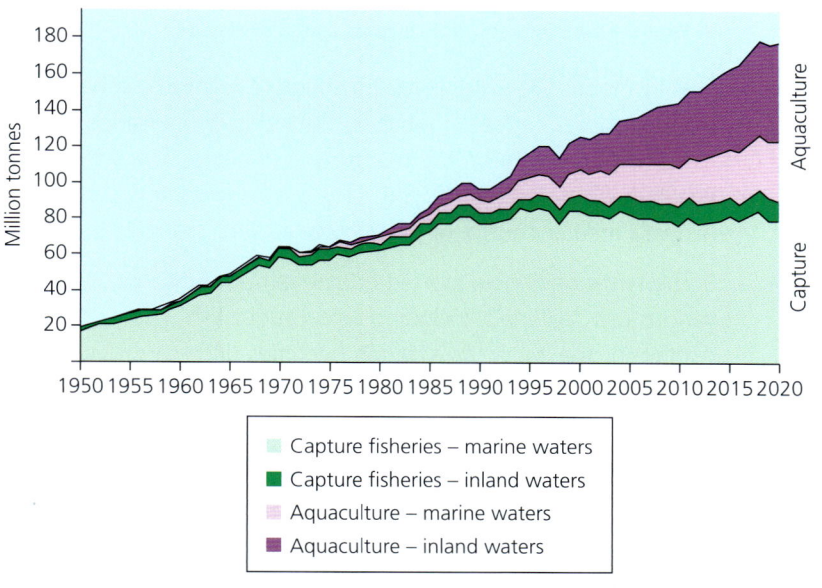

■ **Figure 4.93** Global fisheries production, 1950–2020; since about 1990, the increase in production has been mostly due to aquaculture

Link

Aquaculture faces many of the same issues as farming on land, which is covered in Chapter 5.2 (page 456).

■ Intensive aquaculture

Intensive aquaculture involves building tanks or ponds, stocking fish at high densities, and providing them with large amounts of high-quality food and water. Intensive aquaculture can be highly productive as it uses a small area of land, but it also produces severe environmental impacts, such as habitat destruction, water pollution, depletion of water resources, and the release of farmed species into the wild.

REAL-WORLD EXAMPLE

Shrimp farming in Ecuador

Ecuador is a global leader in the production of shrimp. Beginning in the 1970s with small-scale, extensive farming, production has increased steadily, partly through expansion and partly through intensification.

Around the world, shrimp farms are often built by clearing areas of mangrove wetlands and converting the natural waterways into shallow saltwater ponds for shrimp production. Globally, shrimp farming is one of the main causes of mangrove habitat loss, mostly in Ecuador and Southeast Asia.

■ Figure 4.94 Shrimp farms in a coastal area of Ecuador

Initially, ponds were stocked with shrimp caught in the wild, and no extra food was added to the ponds. Gradually, however, the industry intensified. Agricultural fertilizer was added to some ponds to promote the growth of algae, which the shrimp could eat. Hatcheries based in laboratories started to breed juvenile shrimp, which were supplied to farms. This allowed greater control of production and for selective breeding to improve the stock.

Some farms have now started adding extra food to ponds. The food is produced from various raw materials, including wild-caught fish. The extra food reduces the oxygen content of the water, so aeration becomes necessary. With these changes, more shrimp can now be grown in the same size ponds. The costs of operation have increased significantly, but sales have increased as well. Overall, farmers now make greater profits.

■ Figure 4.95 An aerator working on a shrimp farm in Ecuador

Greater intensity usually results in lower water quality, both in the ponds and in wastewater flowing out of the farm. Higher stock density means more animal waste is produced. Additional food which is not eaten becomes extra organic matter. The extra food and animal waste undergo decomposition, which leads to oxygen depletion. In some farms, chemicals such as antibiotics and fungicides might be added to the water to help control disease. As a result, water flowing out of the ponds carries all these pollutants into the environment. Since shrimp farms are usually in coastal environments, the wastewater may be released into the sea, often without any treatment.

■ Extensive aquaculture

Extensive aquaculture uses more natural ponds or wetlands with natural water flows. Fish are added to the ponds at lower densities, and feed on the naturally occurring algae, plankton or organic matter. Extensive fish farms are typically less productive and use more space, and have much smaller environmental impacts.

> **Link**
>
> Water quality and water pollution are covered in Chapter 4.4 (page 373).

> **REAL-WORLD EXAMPLE**
>
> **Veta la Palma, Spain**
>
> Veta la Palma Estate is a large aquaculture project, based in the intertidal estuary of the Guadalquivir River in southern Spain. It occupies a large area of river and wetland habitat, and produces a variety of fish, including sea bass, bream and grey mullet, as well as shrimp.
>
> ■ Figure 4.96 Veta la Palma Estate, Spain
>
> The fish farm was established in a restored wetland area. Water enters and flows through the site as the tides rise and fall. Juvenile fish are added to the water and they feed on the natural algae in the water. Fish densities are much lower than on a conventional fish farm. This semi-natural ecosystem is highly productive, and also supports a wide variety of birds and other wildlife. The water quality is high and there are no significant pollution problems.
>
> The owners of Veta la Palma aim to produce food in an entirely sustainable way, with minimal inputs, in a naturally self-regulating system. To achieve this, they have opted to maintain a low level of production, which means their potential sales are limited. To make their business economically viable, they market their products as high quality, environmentally friendly, sustainable delicacies, and charge relatively high prices for them.

Tool 4: Systems and models

Intensive and extensive aquaculture systems

Draw systems diagrams to represent the flows in to and out of intensive and extensive aquaculture systems.

● Common mistake

People often confuse the terms intensive and extensive farming. These terms refer to the quantity of food produced per unit area of land. Extensive farming uses a larger (more extensive) area of land to produce a given amount of product. It tends to be more natural, use fewer inputs and produce less pollution. Intensive farming aims to produce the *same amount* of product on a smaller area of land. It requires more inputs and usually produces more pollution.

■ Mariculture

Most aquaculture projects occur in ponds or tanks constructed on land, or in natural or semi-natural surface waters such as wetlands. The water is mostly fresh or brackish. However, aquaculture can also occur in the sea, where it is called marine aquaculture or mariculture. Many mariculture operations use pens or cages to keep the fish from escaping, while letting seawater wash freely through.

■ **Figure 4.97** Marine aquaculture, or mariculture, using pens

REAL-WORLD EXAMPLE

Salmon farming in British Columbia, Canada

Salmon is a popular and nutritious fish, which has suffered from overfishing in various parts of the world. In North America, salmon are an important food source for wild bears, and the nutrients from the dead bodies provide important minerals for the forest soils and trees. Salmon also play an important part in the cultural traditions of the indigenous peoples of the west coast of North America.

Wild salmon stocks around the world are generally either overfished or managed for maximum sustainable yield. It is not possible to sustainably increase salmon production from wild fisheries. Instead, the salmon-farming industry is growing in several parts of the world.

Salmon farming began in British Columbia (BC), on the west coast of Canada, in the late 1960s. Early farms used cage-like pens made of wood and fishing nets to hold the fish, positioned in sheltered areas. Farmers fed the fish with pellets made of ground-up wild-caught fish and used antibiotics to control the spread of disease. Waste from the pens, including excess food, chemicals and the wastes from fish, easily escaped from the pens, often settling at the bottom of the sea, and affecting local ecosystems.

Farm operations have changed due to advances in technology, greater investment, and greater concern for the environment. Pens are larger, made of more durable, modern materials and generally located further from shore. Feeding is automated and closely monitored to reduce waste. Farms are switching to more plant-based feed. Antibiotic use has been reduced, partly because vaccines have been introduced against a number of common fish diseases. These measures have reduced the environmental impact of farms.

However, concerns remain about the impact of farmed salmon on wild salmon populations. There are concerns that sea lice, viruses and other pathogens, might be passed to wild salmon from farmed salmon. This is particularly worrying where non-native Atlantic salmon are farmed, as these might introduce non-native parasites and pathogens.

Many indigenous First Nations peoples of coastal BC are concerned about the health of the wild salmon, as salmon are so important to their cultural traditions and way of life. In the past, First Nations peoples were not always consulted, and their concerns were not addressed in decisions about salmon farming. More recently, authorities are making a greater effort to include them in the discussion. As a result, a number of fish farms operating in particularly sensitive locations will be closed.

Concept

Perspectives

The example of salmon farming in British Columbia illustrates the different perspectives of salmon farmers, government, environmentalists and indigenous peoples. Salmon farming is a lucrative business that brings in a great deal of income and employment to coastal communities. These benefits must be balanced against the risks to wild salmon, given their ecological and cultural importance. Opinions differ between those who value the economic benefits over the ecological and cultural concerns, and those who feel otherwise. In this case, there have been some attempts to reconcile those differences and make compromises for the benefit of all, and for overall sustainability.

Link

Water pollution and wastewater treatment are covered in Chapter 4.4 (page 373).

Human impacts on biodiversity, including the effect of non-native invasive species, are explored in Chapter 3.2 (page 219).

■ Environmental impacts of aquaculture

Table 4.5 describes some environmental impacts of aquaculture.

■ Table 4.5 Environmental impacts of aquaculture

Impact	Description	Mitigation
Habitat destruction	Wetlands, rivers and coastal habitats may be converted to aquaculture ponds.	Policy and legislation may regulate the conversion of habitats, and promote habitat protection or restoration. Some extensive farms may actively rehabilitate natural habitats.
Depletion of water resources	Large volumes of water may be used to house fish, especially in intensive farming.	Closed containment systems can be used. These recycle all water, reducing the need to take water from natural sources.
Water pollution	Nutrients, organic matter, antibiotics, pesticides and antifouling chemicals can be released from farms into natural waterways.	Wastewater should be treated before being released. Closed containment systems aim to achieve this. Alternatives to pesticides and other chemicals (e.g. vaccines, biological control) can be used. Feeding should be carefully regulated, so that excess food does not pollute the water.
Impacts on wildlife	Diseases and pests can spread from farms to wild populations. Non-native farmed species may escape into the wild and become invasive.	Farming only native species reduces the risks of disease and invasive species. Reducing stock densities reduces the risk of disease. Securing the farm reduces the risk of farmed animals escaping.

◆ **Productivity** – the rate of accumulation of energy in living organisms. Usually expressed in kilojoules per square metre per year (kJ m^{-2} y^{-1}).

◆ **Limiting factor** – an environmental factor that determines the rate of a process, often because it is lacking (for example, in dim light, the rate of photosynthesis is slow; when the light gets brighter, the rate increases).

◆ **Production** – the process of accumulation of energy in living organisms. The term can also refer to the amount of energy stored in living organisms. Usually expressed in kilojoules (kJ).

◆ **Thermal stratification** – the formation of layers with different temperatures in water. Above 4°C, warmer water is less dense, and rises above cooler water.

◆ **Turnover** – the reversal of thermal stratification which occurs when the temperature falls below 4°C, resulting in the coldest water and ice being at the surface, with slightly warmer water below.

Link
Productivity is explored fully in Chapter 2.2 (page 119).

Link
The release of nutrients into water by humans is discussed in Chapter 4.4 (page 373).

(HL) 4.3.11 Productivity, thermal stratification, nutrient mixing, and nutrient loading are interconnected in water systems

Productivity is the rate of accumulation of energy in an ecosystem. Almost all primary production in aquatic ecosystems is carried out by photosynthetic organisms such as plants and algae. Factors such as temperature, nutrient availability and light affect the rate of photosynthesis, and therefore affect the productivity of ecosystems. If there is not enough heat, light or nutrients, productivity may be limited.

Limiting factors affecting aquatic productivity

Photosynthesis in aquatic ecosystems is affected by a number of **limiting factors**, including light, temperature and nutrients.

Light

Plants and algae use light for photosynthesis, but light can only penetrate through the first few metres of water. Therefore most primary **production** happens near the surface of water. The amount of light entering water also depends on the season. During summer, the days are longer, so more light enters the water. This allows for more primary production.

Temperature

Photosynthesis is generally faster at warm temperatures, as the molecules involved have more energy to react. Primary productivity is therefore usually higher in tropical regions than in temperature or polar ones, and in summer rather than winter.

Common mistake

At very high temperatures, enzymes become denatured and biological processes like photosynthesis stop. This is true, but the temperatures needed to denature enzymes are usually well over 40°C. Since such high temperatures do not occur very often in natural aquatic ecosystems, it is usually true that higher temperatures produce faster rates of photosynthesis.

Above 4°C, the density of water decreases as the temperature rises. Within this temperature range, warmer water tends to float above cooler water, and the warmest water is usually found at the surface. This warm surface water is also heated by the sun, making it less dense and less likely to sink. This produces **thermal stratification**, where the water forms layers. The uppermost layer nearest the surface has the warmest temperatures and the most sunlight. Therefore, this is where most primary production occurs.

If the temperature falls below 4°C, however, water's density decreases as the temperature falls. During winter, when the air is cold, and there is less heat from the sun, water temperature falls. As it approaches its freezing point of 0°C, the coldest water rises to the surface and slightly warmer water sinks. If ice forms it covers the surface of the water. This reversal of positions of warm and cold water is described as **turnover**, and results in the mixing of the water.

Nutrients

Minerals like nitrates and phosphates are critical for the growth of plants and algae, and for the process of photosynthesis. In waterways that have low nutrient levels, primary productivity is low.

Water flowing over land and through rocks slowly dissolves minerals, and washes them into surface waters and eventually to the sea. The addition of minerals to water is called nutrient loading. It is generally a slow process, although humans can speed it up by releasing nutrients in wastewater and in farm runoff.

Theme 4: Water

If there is enough light and warmth, nutrients in the water will be taken in by plants and algae, and primary production will increase. The nutrients will pass up the food chain. Eventually, nutrients will be returned to the water when organisms die, or release wastes like faeces, as dead organic matter.

In aquatic ecosystems, dead organisms and faeces usually sink to the bottom of the water body, and then decompose. The muddy sediment at the bottom of a sea or lake is usually rich in decaying organic matter. As the organic matter decays, the nutrients, including nitrates and phosphates, are released into the water.

The released nutrients tend to accumulate in the deep water, or in the sediment itself. If the water is stratified, the nutrients will generally remain there. However, if the water is mixed or if there are flows of water from the deep to the surface, then nutrients may be returned to the surface.

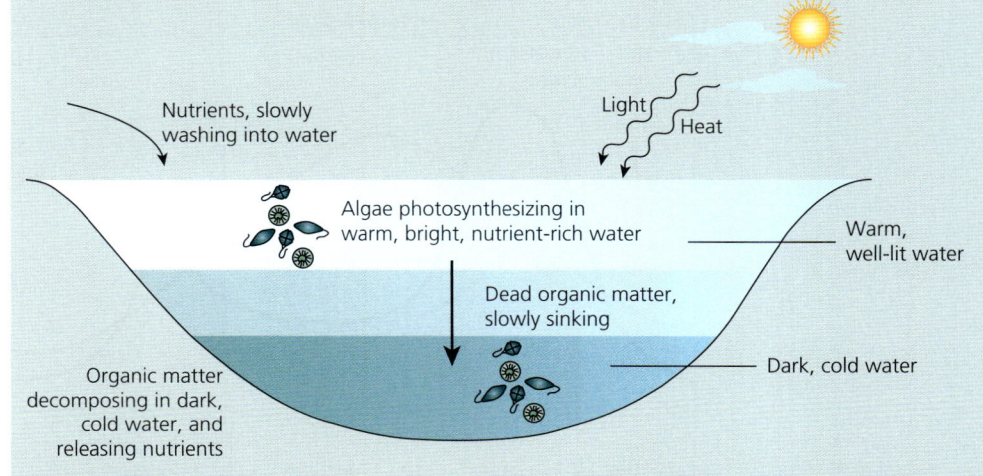

■ **Figure 4.98** The effects of light, nutrients and thermal stratification on productivity in deep water

Stratification, mixing and upwellings

Since warm water tends to rise and cold water tends to sink, the layers formed in water do not mix easily. Anything that is present in one layer tends to stay in that layer. Therefore, nutrients in the deeper, colder water cannot easily get to the surface.

However, sometimes the layers can mix. If the water is shallow enough, the wind might be strong enough to mix the layers. In deeper water in temperate regions, mixing might happen when the seasons change. As winter approaches, the air temperature falls and the surface water cools off. If it gets cold enough, the surface water may start to sink, forcing the deep water upwards. This is called turnover, and it may happen in deep lakes if the surface water temperature falls below 4°C.

Turnover allows substances that were trapped in the surface layers to move to the bottom of the lake, and substances that were trapped at the bottom to rise to the surface. Therefore, as winter approaches, turnover brings nutrients from the deep water to the surface and moves oxygen from the surface water into the deep water.

Upward flow of water can also happen as part of the global thermohaline circulation, or due to upwellings (see Topics 4.1.13, page 287, and 4.1.14, page 287). Upwellings bring cold, nutrient-rich water to the surface, where algae use the nutrients and primary production increases.

Patterns in productivity in aquatic environments

The three limiting factors, sunlight, temperature and nutrients, combine to determine the level of productivity in any given place, at any given time (see Figure 4.98). All three need to be high in order to have high productivity. If any one of them is too low, productivity will be low.

Seasonal changes

Deep water in temperate and polar regions tends to show predictable changes in productivity over the course of a year.
- In the warm months of summer, temperature and light conditions are favourable, but nutrient levels are low, so productivity tends to be low.
- In the winter, if turnover happens, nutrient levels may be high, but light and temperature are low. This again leads to low productivity.
- In the spring, however, the nutrients that were brought up during the winter turnover are still there, and the temperature and sunlight are increasing. Productivity will increase, as the water gets warmer and receives more sunlight. At some point, though, the nutrients will all have been used up. The algae will die and settle to the bottom of the lake again.

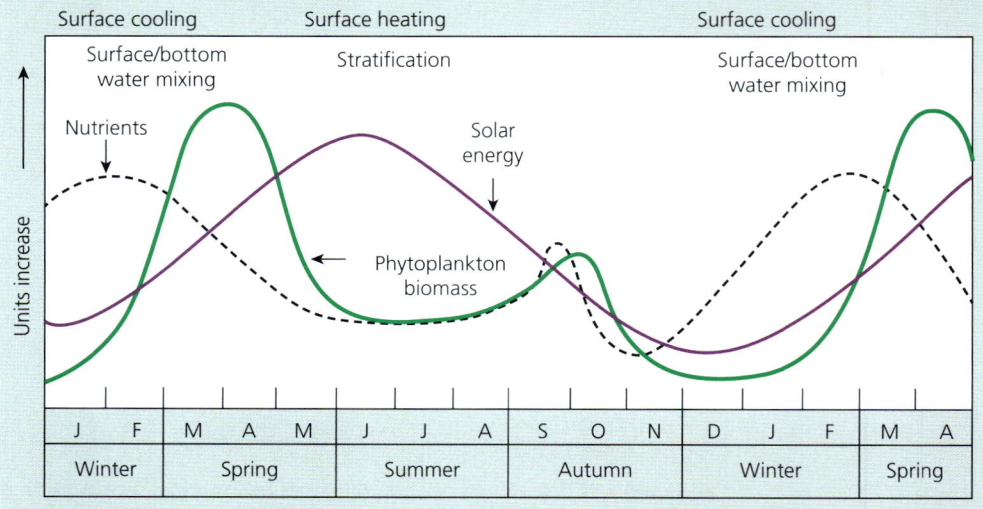

■ **Figure 4.99** Seasonal changes in light, temperature, nutrients, stratification and productivity in temperate waters in the Northern Hemisphere

ATL ACTIVITY

Examine the graph in Figure 4.99 carefully and then answer the following questions.
1. What is the relationship between phytoplankton biomass and production?
2. Explain why the nutrients level rises during the winter.
3. Explain why phytoplankton biomass rises in the spring and then falls in the summer.

Geographic patterns

Productivity tends to be highest in areas with a combination of warm water, abundant sunlight and high nutrient concentration. Some of these locations include:
- Shallow coastal areas, especially in the tropics: Shallow water tends to be fairly well mixed, and there is usually some nutrient loading from the land in coastal areas. The temperature and sunlight in the tropics also help.
- Estuaries, where nutrient loading is high due to runoff from the land: Estuaries also often have tidal flows, so that water comes from the sea as well, possibly bringing more nutrients.
- Areas where upwelling occurs: Upwelling tends to happen off the coasts of continents, such as the Pacific Coast of South America, where the wind causes the surface current to flow away from the land, and deep water is brought to the surface, bringing nutrients with it. The surface current might then take the nutrients out into the open ocean.
- Areas where vertical mixing happens, such as colder parts of the ocean: In these areas most production happens in the spring, when nutrients have been brought to the surface by mixing, and the temperature and sunlight are favourable.

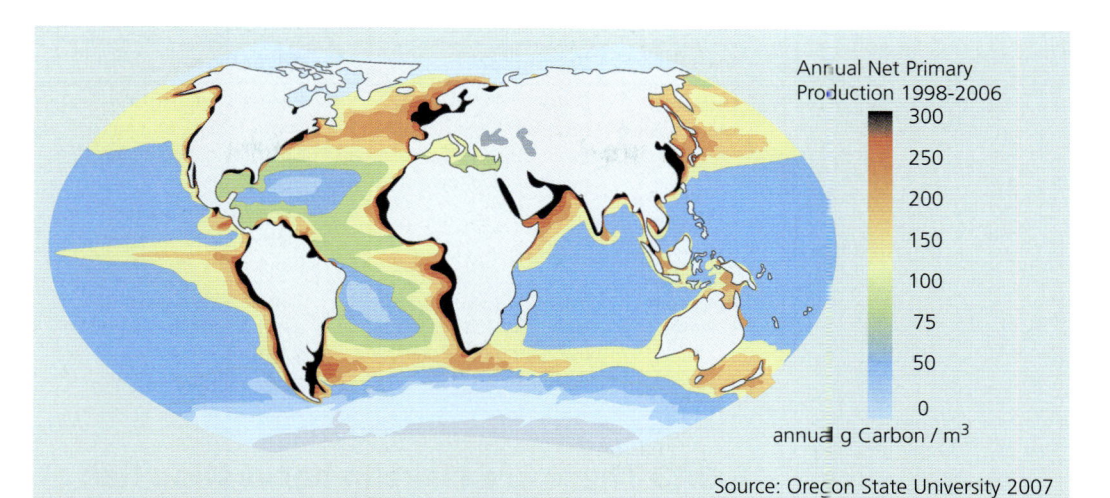

Figure 4.100 Average net primary production in the ocean

4.3.12 Accurate assessment of fish stocks and monitoring of harvest rates are required for their conservation and sustainable use

In order to ensure that we do not overexploit fish stocks, we need three types of data:
1. The size of the fish population.
2. The population's age and sex structure, so we can predict its growth rate.
3. The rate of harvest or the size of the catch.

With these three types of data, scientists and fisheries managers can assess the state of the resource. They can work out whether the stock is increasing or decreasing, and what is a sustainable rate of harvest.

Stock assessment and monitoring

The most easily accessible data on fish stocks comes from the fishing industry itself. Fisheries managers can obtain data from commercial and recreational fishers, and they can use this to work out the status of the fish stocks. At the same time, this data allows them to monitor the harvest. The most commonly used approaches are:
- landing records
- portside sampling
- onboard observation.

Data about fish stocks can also be collected directly by the fisheries managers, without relying on fishers. They do this through research surveys and using appropriate sampling methods.

Table 4.6 Methods used in assessing fish stocks

Method	Description	Advantages	Limitations
Landing records	Data is collected from official port records. When a vessel brings in its catch, it is weighed and recorded.	Easy to obtain in well-regulated fisheries.	Do not include data on bycatch, illegal or unreported catches, or age and sex structure. Often reported as mass, rather than number of fish.
Portside sampling	Catches are examined on the port and samples are taken.	Direct measurement provides more accurate and detailed data. Can be used to estimate age and sex structure.	No data on bycatch, illegal or unreported catches. Requires personnel to visit ports.

4.3 Aquatic food-production systems

Method	Description	Advantages	Limitations
Onboard observation	Entire fishing process is observed from onboard fishing vessels. Samples can be taken and measured.	Provides detailed and accurate data on catch. Can provide estimates of bycatch. Can provide data on fishing methods and whether regulations are followed.	Requires substantial time and effort by qualified personnel. This limits how often this method can be used. No data on illegal or unreported catches.
Research surveys	Scientific surveys conducted from research vessels, using a range of sampling methods.	Can provide a great deal of data, directly measured by scientists. Does not depend on activities of fishing vessels.	The most expensive method, requiring the most specialized personnel and equipment.

4.3.13 There are risks in harvesting fish at the maximum sustainable yield rate, and these risks need to be carefully managed

The concept of maximum sustainable yield (MSY) was covered in Topic 4.3.6 (page 344). MSY can potentially be very useful, but there are risks involved. These risks come from a combination of incomplete or inaccurate data, and difficulty in monitoring what actually happens on fishing vessels.

Data availability and accuracy

In order to achieve MSY, we need to ensure that the correct number of fish are caught in order to keep the population growing at its maximum growth rate. To work out what the correct number of fish is, we need to answer the following questions:
- What is the carrying capacity for this species of fish, in this environment, at this time?
- What is the current population size of this species?

Carrying capacity depends on the condition of the environment. How much food is available for the fish? How many predators are present? Are there other species competing for the same resources? Are there any diseases or parasites present? Does the number of predators, competitors or parasites change as the size of the fish population grows?

None of these questions is easy to answer. Our knowledge of these conditions, and these interrelationships, is usually limited. Where we do understand them, the interrelationships can be very complex, involving many different positive and negative feedback loops. We also need to consider the possibility that unexpected change may occur. For instance, we may have a year of unusual weather. If the sea is colder than normal, how will that affect the carrying capacity?

We also need to know about the population itself. What is the population size? How fast is the population growing? We might be able to project future population growth if we know the age structure of the population, but this is difficult data to collect.

Enforcement of quotas

Based on our knowledge of the population size and the carrying capacity of the environment, we can set quotas to ensure that we catch the right number of fish to produce MSY. We must then ensure that the fishing industry sticks to these quotas.

Enforcing quotas is difficult. Most fishing happens far away from land, where it is difficult to monitor. We must rely on fishing vessels to accurately report what they catch. While most vessels will do so, it is possible that some might exceed their quota and fabricate their reports, or that unregistered vessels may be fishing a particular stock but not reporting it.

> **TOK**
>
> Do commercial fishing businesses have different ethical obligations and responsibilities with regard to respecting quotas, compared to small-scale or artisanal fishing boats?

In the long run, fishing crews should benefit from co-operating with quotas. If everyone catches only what they are entitled to, then the fish stock can be managed sustainably, and everyone's income will be protected for the future. However, this may require crews to sacrifice their income in the short term, by catching less than they would like to. If the quotas mean that their incomes are too low to live comfortably on, they may be tempted to exceed the quota.

The quota system also relies on mutual trust between stakeholders, including different fishing crews. If one crew believes that another crew is 'cheating' and catching more than their allowable quota, they may be tempted to do the same. If crews have competed in the past, it can be difficult to develop trust between them.

The risk of getting it wrong

If fisheries scientists' knowledge and understanding of a fishery is wrong, they may set inappropriate quotas. If the authorities cannot enforce the quotas then fishing crews may harvest more than they are allowed. These are real risks, which cannot be avoided, though we can try to minimize them.

If these sorts of mistakes are made and too many fish are harvested, the population will fall to a level below half the carrying capacity. The longer we fishing continues at that rate, the faster the population will decline. In only a few years, the population may collapse to the point where the stock is overfished. This is an example of positive feedback (see Topic 3.3.11, page 261), bringing about a change in the equilibrium of a system.

Reducing the risk

Most of the data we need is obtained by sampling procedures and therefore is unlikely to be entirely accurate. We must always be aware of the uncertainties in our data and take them into account when we make decisions.

To reduce the risk of incomplete and inaccurate data, we can make all reasonable efforts to ensure that our data is accurate and complete by engaging in more thorough research. However, we will never achieve perfect knowledge and understanding.

To reduce the risk of fishing vessels exceeding their quotas, we could improve our enforcement and monitoring, while also facilitating collaboration and communication between fishing crews and fisheries managers, as well as among the different fishing crews. Again, achieving this would be very difficult.

Given the difficulties and uncertainties, it is often recommended that we err on the side of caution, and set our quotas slightly lower than the level needed to achieve MSY. The result is that we would harvest fewer fish, but would protect the stock from overexploitation. Even if we make mistakes in our data collection or calculations, or some crews exceed their quotas, there will be a margin of error. The population is less likely to fall below half the carrying capacity, so it is unlikely to collapse.

● HL.b.2: Environmental economics

Reducing the fishing quota might seem like an unpopular decision, but considering the economics of fishing, it might actually increase a crew's profits. With a lower quota, crews will put less effort into fishing. Their expenses will therefore be lower, as they would, for instance, need less fuel for their boats. If we add a line showing the cost of fishing on to the MSY graph shown in Figure 4.101, we see that a little to the left of the peak of the yield curve, the difference between cost and yield is actually highest. In theory, this is the point of maximum profit, sometimes called the maximum economic yield (MEY).

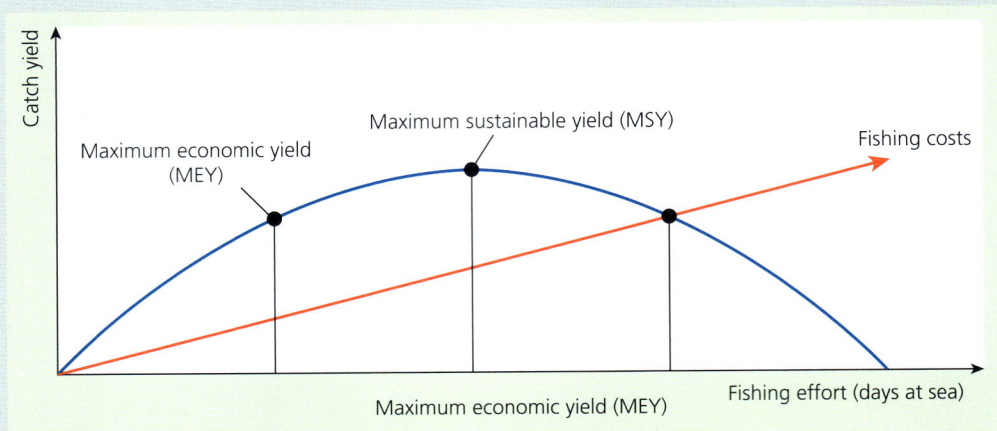

■ **Figure 4.101** Maximum economic yield (source: SPICe, adapted from http://sustainablefisheries-uw.org)

The graph is based on the idea that as fishing effort increases, the cost of fishing increases, as shown by the fishing costs straight line. The income from sales is proportional to the yield. The more fish they catch, the more they sell and the more they earn. Profit is the difference between income and costs, so the point on the graph where these two lines are furthest apart is the point where the fishing crew earns the most profit.

This point is always a little to the left of the MSY curve's peak. If the quota is set at this point, then the population should remain at about half the carrying capacity, and the harvest should be sustainable. At the same time, the fishing crews should earn more profit than if the quota was set at the MSY.

4.3.14 Species that have been overexploited may recover with cooperation between governments, the fishing industry, consumers and other interest groups

When a species is overexploited, there is a significant risk that it will die out locally, or even go extinct. However, if appropriate conservation measures are taken, the species might be able to recover to a level where the population is stable and it can be sustainably harvested again.

Stakeholder perspectives

Any decisions and changes that occur in the fisheries will affect different people in different ways. In some cases, there might be disagreement and conflict between these groups. However, to achieve sustainable recovery of an overexploited resource, it may be necessary to work through these disagreements, and find ways to collaborate and co-operate.

■ **Table 4.7** Stakeholders in fisheries management

Stakeholder	Role and interests	Potential conflicts
Government	Responsible for the economy and the well-being of citizens. Capable of passing and enforcing legislation, setting policy and establishing institutions. Concerned about voter opinions.	Interest groups such as fishing industry and environmental NGOs may disagree with decisions. Relations with other countries may be harmed by decisions.
Fishing industry	Interested in profits, economy and sustainable income. Concerned about livelihood and community dynamics. Possesses direct knowledge of fisheries and fish stocks.	Competition within the industry may cause conflict between crews or communities. Often distrusts government and may dispute government decisions.

Stakeholder	Role and interests	Potential conflicts
Foreign nations and businesses	May seek to exploit resources in foreign waters. Interested in profits, so may be more inclined to overexploit one location and move to another, rather than sustain a local fishery.	Pay local government for permission to fish. Competition with local businesses and crews. Relations with governments may be complicated.
Traders (e.g. wholesalers)	Buy and sell products from fishing crews. Interested in profits and sustainability of their businesses. Dependency on fisheries is variable, as they may be able to change their businesses.	Competition among traders. Disputes over pricing with fishing industry. Disputes with government over regulations and taxation.
Consumers	Buy products from traders or direct from fishing crews. Interested in price and quality of product. May have concerns about sustainability, which may affect purchasing decisions.	Disputes over pricing. May also be members of other stakeholder groups.
Non-governmental organizations (NGOs) and other interest groups	Campaign for various causes, including environmental causes, social or economic justice, consumer rights and sustainability. Often engage in public awareness campaigns and protests. May collaborate with other stakeholders, such as fishing industry members.	Disputes with fishing industry about methods and quotas. Disputes with government about regulations and enforcement.

> **REAL-WORLD EXAMPLE**
>
> ### Cod moratorium, Canada
>
> The first step to saving an overexploited species is often to stop harvesting the species. This can happen if the government decides to establish a moratorium on the fishery. A moratorium is a period of time during which a particular activity is stopped. If this negatively affects people's livelihoods, a moratorium is likely to be unpopular.
>
> When the cod fishery off the coast of Newfoundland, Canada dramatically collapsed in 1992, the government declared a moratorium on the fishing of cod, which caused a great deal of hardship to local fishing communities. When the announcement was made, members of the communities angrily confronted the politicians.
>
> Several members of the communities later admitted that they knew the fishery was collapsing, as they had seen their catches declining for some time. Even so, the moratorium created great difficulty for them. Some suggested that the government should have intervened sooner. However, the disagreements and lack of trust between the government and the fishing communities had existed before the cod collapse, as there had been many years of conflict about how the fishery was managed.

For the fishing communities, the moratorium was a disaster. Over 30,000 people lost their jobs in fishing and related industries. The government provided some degree of compensation, financial support, and opportunities to retrain in other industries or to return to education. Still, this was widely found to be too little, and not everyone who was affected received support, leading to ongoing anger at the government. Eventually, most of the people moved away from the coastal areas. They experienced a significant change to their way of life and cultural identity. The communities were effectively destroyed.

The cod moratorium illustrates a failure of communication, collaboration and scientific foresight. Earlier intervention might have avoided both the fishery collapse and the need for the moratorium. Furthermore, the moratorium did not work. The cod population in that part of the Atlantic Ocean is still far too small to support a fishery, even after more than 30 years. This suggests that either the stock had been depleted to such a low level that recovery was impossible, or that other factors, such as climate change, habitat destruction or pollution, might be hindering the recovery of the cod population.

Conservation measures

Several different measures can be used to help conserve overexploited aquatic species. Many of these require co-operation and collaboration between the different stakeholders. Measures can include:
- restrictions on catch size, implemented through quotas or total allowable catches
- restrictions on the use of particular types of fishing gear
- restrictions on the size of fish that may be caught
- restrictions on where fishing can take place, possibly through establishing marine protected areas
- temporary bans or moratoria on fishing in particular areas, or for particular species.

These measures are often best implemented by governments or intergovernmental agencies. However, they are generally most successful if they are supported by other stakeholders, especially the fishing industry.

REAL-WORLD EXAMPLE

Eastern Atlantic bluefin tuna

The Atlantic bluefin tuna, *Thunnus thynnus*, is a popular food fish. Increasing demand in the late twentieth century, however, along with changing fishing methods, led to severe overexploitation. The IUCN declared the species Endangered in 2011.

Atlantic tuna fisheries are managed by an intergovernmental organization called the International Commission for the Conservation of Atlantic Tuna (ICCAT). In 1998, with the tuna population declining and fish catches remaining high, ICCAT introduced quotas in the form of a total allowable catch (TAC), which was shared between several countries.

For the first few years, the TACs agreed were significantly higher than the levels advised by ICCAT's scientists and enforcement of the limits was ineffective. The scientists found evidence that fishing vessels were under-reporting their catches. Collectively, they may have been catching up to 20,000 tonnes more tuna than were allowed under the TAC. ICCAT received a great deal of criticism for this.

In 2006, following continued warnings from the scientific committee, and growing concern from the public, ICCAT introduced a new recovery plan for the Eastern Atlantic tuna. The plan made changes to the length of the fishing season and the minimum size of fish allowed, as well as improving control and enforcement measures. Over a series of years the TAC was reduced, until it was aligned with the scientists' recommendations.

By 2014, the scientists found that the stock was increasing and suggested that the TAC could be increased again. By 2021, the IUCN reassessed the Atlantic bluefin tuna from Endangered to Least Concern.

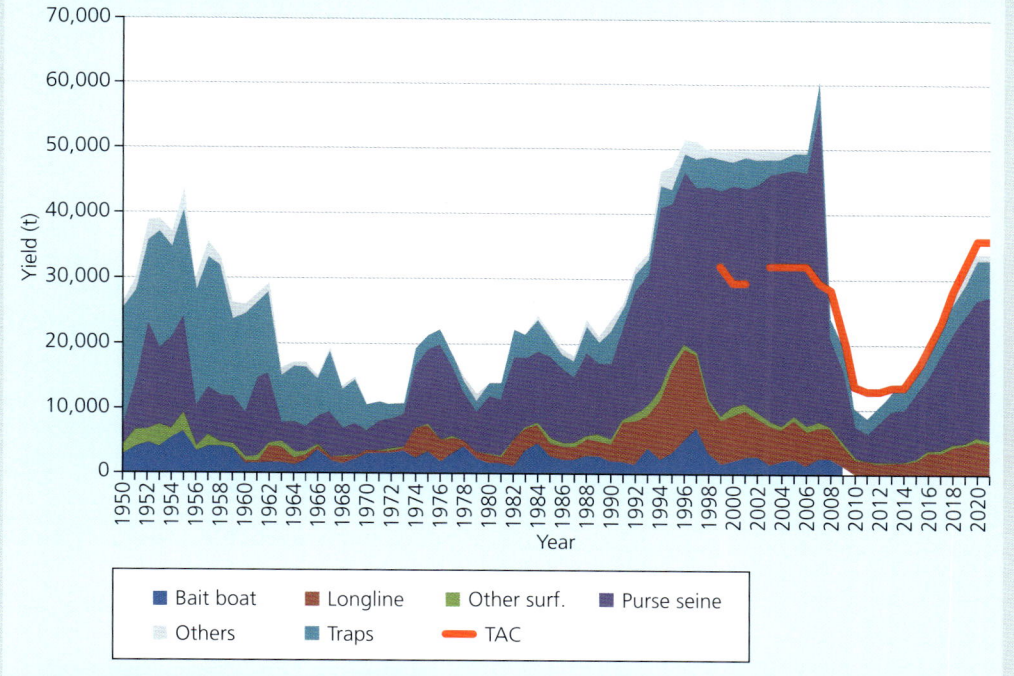

■ **Figure 4.102** Eastern Atlantic bluefin tuna yields, 1950–2021, showing the total allowable catch (red line) established by ICCAT

The successful recovery of the Eastern Atlantic tuna was only possible through the collaboration and cooperation of various stakeholders, including the governments of several countries, the scientific community, and the fishing and related industries. Pressure from the public and environmental NGOs also helped push ICCAT to introduce more stringent controls and to produce the successful recovery plan.

HL.a.3: Environmental law

ICCAT's initial attempts to introduce quotas on the Atlantic bluefin tuna met resistance from the fishing industry in many of its member countries. Industry representatives successfully lobbied their governments and ICCAT to set the quotas higher than the scientists' recommendations. This allowed each country to catch more fish and earn more income in the short term. However, it weakened the management of the species and allowed overexploitation to continue. The fish population continued to decline.

The initial quotas set were unsustainable. In the long term, the fishing industries themselves would have suffered. Over time, with fish catches declining and public pressure increasing, ICCAT was able to agree quotas that allowed the fishery to recover.

4.3 Aquatic food-production systems

4.3.15 Coastal states have an Exclusive Economic Zone, within which the state's government can regulate fishing. Almost 60 per cent of the ocean is outside these coastal zones

Most fisheries are based in the sea. This makes them difficult to manage. Traditionally, human societies, and the nation states we live in, have operated mainly on land. Countries' laws have not always held any power at sea. The sea has often been an unregulated resource, which belongs to no one country, and where anyone can do whatever they like. It is sometimes described as a common resource, or simply as part of 'the commons', and it is often used as an example of the tragedy of the commons.

Starting in 1973, the United Nations held a series of conferences on the protection of ocean resources, and the regulation of human activities at sea. After lengthy and complex negotiations, the United National Convention on the Law of the Sea (UNCLOS) was agreed at a meeting in Montego Bay, Jamaica, in 1982.

Link

The tragedy of the commons is covered in Topic 3.2.7 (page 231).

● HL.c.3: Environmental ethics

UNCLOS is a complex agreement, covering many aspects of human activities at sea. Parts of it have been controversial, including one section dealing with mining of the deep-sea floor. It took nine years of negotiations before the final document was agreed and many countries have hesitated before signing and ratifying it, and some, including the USA, have refused to sign it.

The USA's resistance to joining UNCLOS stemmed in part from opposition to the regulation of sea-floor mining and in part from ideological differences. It was opposed to the regulation of a common, freely accessible resource. It was also concerned about reducing its economic and political power over the ocean, and expressed concern that smaller, less powerful nations could use UNCLOS to collaborate against US interests.

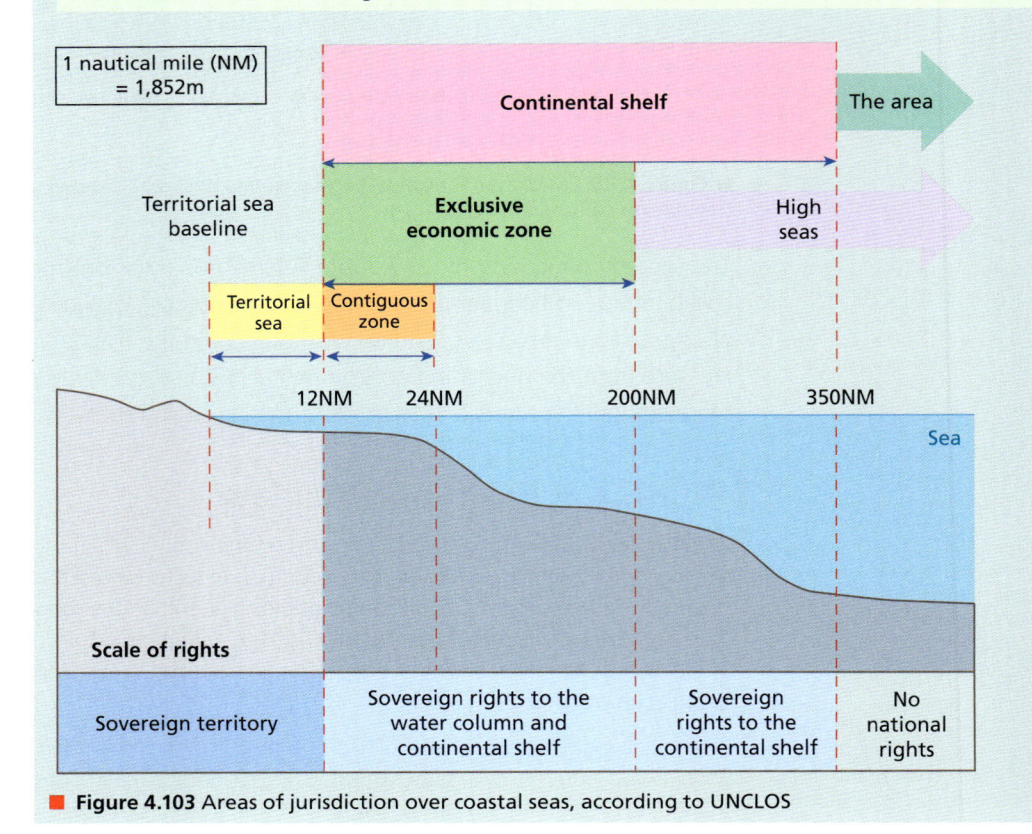

■ Figure 4.103 Areas of jurisdiction over coastal seas, according to UNCLOS

One of the most important aspects of UNCLOS is the definition of the Exclusive Economic Zone (EEZ). This is an area of sea up to 200 nautical miles (370 km) from the coastline. According to UNCLOS, countries with coastlines have the right to exploit the resources with their EEZ. This includes fisheries as well as minerals, such as oil and natural gas. UNCLOS and EEZs have played an important part in resolving disputes about fishing rights at sea.

Ships, airplanes and pipelines belonging to other countries may pass through a country's EEZ, but other countries must obtain permission to take resources from the EEZ. Typically, governments or citizens of foreign states would be required to pay for the right to extract resources from another country's EEZ. However, the EEZ can be a very large area of water, which makes enforcing a country's rights quite difficult.

EEZs cover almost 40 per cent of the world's oceans, including almost all of the accessible marine fisheries. The ocean beyond 370 km from the coast is defined by UNCLOS as the high seas. No country has exclusive rights to resources in the high seas.

Conflicts can arise when foreign fishing vessels are given permission to fish in a country's EEZ. The foreign vessels may be larger, with more capacity than local vessels, especially where the local vessels are part of a small-scale or artisanal fishing industry. Local artisanal fishing crews may find it difficult to compete with larger foreign vessels. If the foreign vessels deplete the fish stocks, the local fishers may not be able to earn enough income to support themselves.

4.3.16 Harvesting of seals, whales and dolphins raises ethical issues relating to the rights of animals and of indigenous groups of humans

History of hunting marine animals

People have hunted whales, dolphins, seals and other marine animals for thousands of years. Traditional hunting methods involved using large spears called harpoons to capture and kill the animal. Over time, technological developments changed the nature of marine hunting in two significant ways. First, non-food uses of whale products, such as for whale oil, increased the demand for whales. Secondly, whaling became highly mechanized, meaning that more whales could be harvested more easily.

As a result of these developments, whaling became a large-scale industry and overexploitation became a problem. Populations of all species of hunted whales fell dramatically, and it was widely recognized that the whaling industry was unsustainable.

HL.a.9: Environmental law

International discussions on whaling led eventually to an agreement called the International Convention for the Regulation of Whaling (ICRW), and the formation of the International Whaling Commission (IWC). The IWC initially set quotas for whale hunting. In 1982, it went further and declared a moratorium on hunting of the 'great whales'. This prohibited hunting of the larger, more commercially valuable species of whales.

This moratorium is still in place, but there are two situations in which limited whaling is allowed:
- 'Scientific whaling', which involves capturing and killing a limited number of whales for scientific research. The purpose of the hunt should be scientific, but the meat of the whales caught can still be sold as food.
- Small-scale whaling by specific indigenous peoples who have established cultural traditions of whaling.

> **REAL-WORLD EXAMPLE**
>
> ### Indigenous whaling in northern Alaska
>
> The Iñupiat are a group of indigenous people who live in the icy north of Alaska, USA. Agriculture is impossible in their Arctic tundra environment and plant growth is limited. Fish and hunted animals, including seals, walrus, polar bears and whales, form a large part of their diet.
>
> Whaling forms an important part of Iñupiat culture and way of life. The main target species is the bowhead whale, *Balaena mysticetus*. The IWC carefully regulates the hunt and allocates an annual quota. Hunted whales are shared freely by the entire community, according to a traditional system of allocation. Cooperation, sharing and community spirit play a major part in the harvest. Replacing the whale products with food imported from elsewhere could cost several million dollars each year.
>
> The whale harvest is celebrated at the time of harvest and at an annual whaling festival called *Nalukataq*. There is a spiritual element to the hunt. Hunters give thanks to the spirit of the whale for its sacrifice, and believe that wasting any part of the whale is disrespectful.
>
> The IUCN assessed the bowhead whale in 2018 and found its global population is increasing. It listed the bowhead whale as Least Concern.

Ethical questions

The original purposes of the IWC's regulation of whaling, and of the 1982 moratorium, were to protect whale populations from overexploitation and extinction. Over time, however, people have become increasingly concerned about ethical issues surrounding whaling. The ethics of scientific and indigenous whaling can be particularly controversial. Several ethical questions can be asked, but in most cases people disagree strongly about the answers to those questions.

> **Concept**
>
> ### Perspectives
>
> Ethical questions are often answered differently depending on people's perspectives. The issues discussed here illustrate significant differences of opinion among those who take anthropocentric, technocentric and ecocentric positions. The positions of people in industrialized western societies are also often different from those in traditional indigenous societies. This reflects the different ethical frameworks that exist in these societies. The different perspectives make these ethical questions very difficult to resolve.

Ethics of sustainable resource use

The original reason for the IWC limits on whaling was to protect the whales as a useful resource. It was widely recognized that commercial whaling was unsustainable, and would lead to the depletion and potential extinction of whales. This is both a practical, economic argument and an ethical one. The loss of a resource can potentially cause harm to future generations, who will not have access to that resource. This raises the question of whether we have a duty or responsibility to look after the interests of others, including people who have not yet been born.

> **TOK**
>
> Is it unfair to judge indigenous people's values and practices by the standards of industrialized nations?

Theme 4: Water

> ### ● HL.c.4: Environmental ethics
>
> The idea that we should protect the whales because of their usefulness recognizes the instrumental value of whales. If we accept that we do have this responsibility, then it is ethically correct to restrict the use of the resource and prevent its depletion. By this argument, scientific and indigenous whaling are ethically acceptable, since they can be done without overexploitation.

Intrinsic value

Intrinsic value is the value of a thing in and of itself, regardless of whether it can be bought or sold, or whether people value it. Many people would agree that whales have intrinsic value. If we accept that animals have intrinsic value, then we must provide a sound reason to justify killing them.

> ### ● HL.c.5: Environmental ethics
>
> Whales having intrinsic value raises questions related to the reasons we hunt whales.
> - Can we hunt them for economic gain? If the money is earned by poor people, who need it to survive, is that acceptable? What if it is earned by a business that employs many people?
> - Can we kill them for scientific research purposes? Does the purpose of the research matter? If it helps to produce medicines or useful products, is it justifiable? How many whales may we kill for these purposes?
> - Can we kill them for food? Is it acceptable if it is to feed people who have no other reasonable or affordable source of food? If it is a traditional delicacy, is that justifiable?
> - Can we kill them as part of our cultural practices or heritage? If killing whales helps to keep communities together and to maintain their cultural identities, can it be justified?
>
> These questions do not have easy, clear or even objective answers. What one person or society might justify, others might find unjustifiable. But starting from the position that whales have intrinsic value means we have to ask and answer these questions.
>
> Interestingly, many indigenous peoples who hunt whales and other animals believe strongly in the intrinsic value of their prey. This is why they take the hunt very seriously, and may pray or give thanks to the animal they have killed. But this also means that they are cautious and conservative in their hunting. They hunt enough for their needs and are careful not to waste.

Questions of animal rights

There is a growing agreement, at least in western, industrialized societies, that animals have certain rights. These rights might include a right to life, and a right to freedom from suffering and pain.

The right of animals to be free of pain and suffering raises questions about how we kill animals, and how we treat them before we kill them. In hunting animals, we might aim to ensure that the animal is killed as quickly as possible, and does not suffer for long. This cannot usually be guaranteed, but modern tools and knowledge may help to ensure, for instance, that we kill the animal with the first shot or strike. It may be more difficult to achieve using traditional tools and methods, so this may be a concern raised about indigenous hunting rights.

HL.c.11: Environmental ethics

Ethical questions raised by hunting commonly involve the question of whose rights are more important. Rights are not absolute, and there are many situations where one person's rights interfere with those of another. In discussing the ethics of hunting, we might consider which is more important: the human rights to eat, investigate, earn money or practise cultural traditions, or the animal's right to live and be free of suffering. A person's perspective on this is likely to be influenced by their environmental value system.

HL.c.13: Environmental ethics

The debate over indigenous whaling is a complex one, with strongly held opinions on both sides. One important perspective on it takes a social justice viewpoint, related to the position of indigenous peoples within the wider society.

Many indigenous peoples have suffered social, economic and political marginalization in their native countries. Their populations have dwindled and their cultures have been diluted by the influence of more 'modern' societies. These cultures are at risk of dying out.

Social justice campaigners may argue that banning indigenous whaling would continue the historic marginalization of indigenous peoples. Their right to practise their culture is essential to maintain their cultural identity and protect them from cultural dilution. Many environmental campaigners agree with this perspective, and argue that respect for indigenous cultures is consistent with respect for the environment.

Some environmental campaigners go further and argue that many indigenous peoples practise a way of life that is far more sustainable than that of modern industrialized societies. They see indigenous cultures as a model for others to imitate.

REVIEW QUESTIONS

1. What does the term maximum sustainable yield mean?
2. Describe how yield is affected by increasing fishing effort.
3. What is a quota? How are quotas used to achieve maximum sustainable yield?

4.4 Water pollution

Guiding questions

- How does pollution affect the sustainability of environmental systems?
- How do different perspectives affect how pollution is managed?

SYLLABUS CONTENT

This chapter covers the following syllabus content:
- ▶ 4.4.1 Water pollution has multiple sources and has major impacts on marine and freshwater systems.
- ▶ 4.4.2 Plastic debris is accumulating in marine environments. Management is needed to remove plastics from the supply chain and to clear up existing pollution.
- ▶ 4.4.3 Water quality is the measurement of chemical, physical and biological characteristics of water. Water quality is variable and is often measured using a water-quality index. Monitoring water quality can inform management strategies for reducing water pollution.
- ▶ 4.4.4 Biochemical oxygen demand (BOD) is a measure of the amount of dissolved oxygen required by microorganisms to decompose organic material in water.
- ▶ 4.4.5 Eutrophication occurs when lakes, estuaries and coastal waters receive inputs of mineral nutrients, especially nitrates and phosphates. It often causes excessive growth of phytoplankton.
- ▶ 4.4.6 Eutrophication leads to a sequence of impacts and changes to the aquatic system.
- ▶ 4.4.7 Eutrophication can substantially impact ecosystem services.
- ▶ 4.4.8 Eutrophication can be addressed at three different levels of management.

HL ONLY
- ▶ 4.4.9 There is a wide range of pollutants that can be found in water.
- ▶ 4.4.10 Algal blooms may produce toxins that threaten the health of humans and other animals.
- ▶ 4.4.11 The frequency of anoxic/hypoxic waters is likely to increase due to the combined effects of global warming, freshwater stratification, sewage disposal and eutrophication.
- ▶ 4.4.12 Sewage is treated to allow safe release of effluent, by primary, secondary and tertiary water-treatment stages.
- ▶ 4.4.13 Some species are sensitive to pollutants, or are adapted to polluted waters, so these can be used as indicator species.
- ▶ 4.4.14 A biotic index can provide an indirect measure of water quality based on the tolerance to pollution, and the relative abundance and diversity of species in the community.
- ▶ 4.4.15 Overall water quality can be assessed by calculating a water-quality index (WQI).
- ▶ 4.4.16 Drinking water-quality guidelines have been set by the World Health Organization (WHO), and local governments can set statutory standards.
- ▶ 4.4.17 Action by individuals or groups of citizens can help to reduce water pollution.

4.4.1 Water pollution has multiple sources and has major impacts on marine and freshwater systems

Pollution refers to the release of harmful substances or energy into the environment. The substances or energy released are called pollutants. Many human activities cause pollution, which in turn affect aquatic systems, both freshwater and marine.

Pollutants may be released from a distinct location, such as a factory or one particular drain. This type of source is called a point source. In contrast, if the pollutants come from a large area, such as a farm or an entire town, or from a moving source like a vehicle. These are non-point sources. Point sources are usually easier to manage, since the source can be clearly identified and mananged.

■ Agricultural runoff

Farms are often non-point sources for water pollution. Rainfall or irrigation water can flow over the farmland, or perhaps **infiltrate** and **percolate** through the soil, and carry various pollutants with it, into either surface or groundwaters. The runoff can include any of the **agrochemicals** used on the farm, as well as natural substances such as animal wastes and eroded soil.

Agrochemicals include any natural or man-made chemical substances which are used on farms, such as **pesticides** and **fertilizers**. Some agrochemicals can cause pollution if they are released into waterways, such as when rainfall and surface runoff carry them into streams and wetlands, or when they are **leached** from the soil by percolating water.

Most agrochemicals are synthetic or human-made substances, which do not occur in nature, but are produced in factories. Others are natural substances, which are collected, processed and packaged in various ways. Synthetic substances may be more harmful in the environment, since living organisms may be unable to break them down into simpler, less harmful substances.

- ◆ **Infiltration** – the entry of water into the soil.
- ◆ **Percolate** – the downward movement of water through the soil.
- ◆ **Agrochemical** – any chemical substance that is used in farming, including fertilizers, pesticides, hormones and antibiotics.
- ◆ **Pesticide** – a substance that is used on farms (for example, sprayed on to crops or into the soil), to kill (or sometimes repel) pests such as insects or fungi that would otherwise harm the crops.
- ◆ **Fertilizer** – a substance that is added to soil or irrigation water to promote plant growth. Typically, fertilizers include minerals that plants need for growth. Farmers use them when the soil does not contain enough of the minerals to meet the plants' needs.
- ◆ **Leaching** – the washing away of substances, for example from the soil, as water flows over or through it.
- ◆ **Non-target organism** – an organism which is not intended to be harmed by a pesticide, because they are not pests.
- ◆ **Bioaccumulation** – the build-up of pollutants inside a living organism over time.
- ◆ **Biomagnification** – the increase in concentration of pollutants in living organisms at higher and higher trophic levels in a food chain.

Pesticides

Pesticides are designed to kill pests, but they may also kill or harm other, **non-target** organisms, and they can be very damaging to the environment. Synthetic pesticides are particularly dangerous, as living organisms are usually incapable of breaking them down or excreting them. Non-target plants and animals may absorb synthetic pesticides into their bodies continuously through their lives, building up higher and higher concentrations of pesticide in their tissues. This process is called **bioaccumulation**. Eventually, the pesticide concentration inside the living organisms may be high enough to cause illness or death.

> **TOK**
>
> The word 'organic' has multiple different meanings, depending on the context in which it is used. Investigate the use of this word in the following phrases:
> - Dead organic matter
> - Organic food
> - Organic chemistry
> - Organic growth.
>
> What does this tell us about the importance of language and context in knowledge?

Link

Agriculture and its environmental impacts are covered in Chapter 5.2 (page 456).

While pesticides are bioaccumulating in plants and animals, those organisms might be eaten by other animals. If contaminated algae are eaten by a herbivore, for instance, the herbivore will take in all the pesticide from the algae. Since one herbivore may eat many algae, the herbivore may end up with an even higher concentration of pesticide in its body. When herbivores eaten by carnivores, the process is repeated and even higher concentrations are produced. This process is called biomagnification. Biomagnification results in increasing concentrations of pollutants such as pesticides in higher and higher trophic levels. Top predators may have such high concentrations that they die or are severely harmed.

REAL-WORLD EXAMPLE

Biomagnification of DDT

DDT, or dichlorodiphenyltrichloroethane, is a synthetic pesticide that was first developed in the 1940s. It was used on farms to control insect pests, and in natural areas for controlling diseases spread by insects, such as malaria. As it is a synthetic substance, living organisms are not able to break it down or excrete it, so it accumulates in their tissues, in particularly in fatty tissues. DDT undergoes both bioaccumulation and biomagnification.

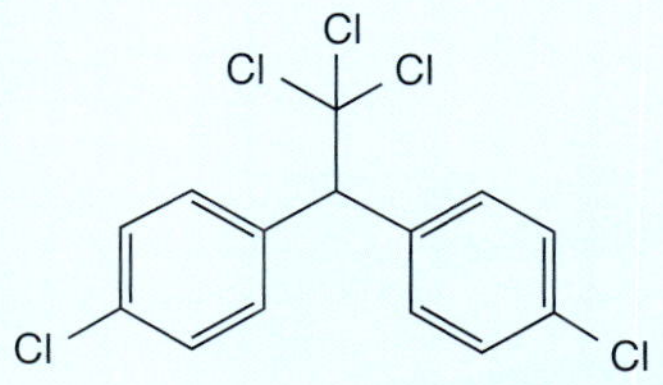

■ **Figure 4.104** The structure of DDT

When first sprayed on farms or in swamps, its concentration is high enough to kill many of the target insects. Initially, scientists, farmers and health officials who used it assumed that it would then disperse and dilute in nature, and not cause any further harm. However, whenever DDT molecules came into contact with a living organism, they were absorbed and stored in the organism in surprisingly high concentrations. Over time, DDT also passed up the food chain – from small animals on the farms and in the swamps, on to larger animals such as fish.

As it went up the food chain, larger fish took it away, down rivers and into the sea. Eventually, it passed into birds, and ultimately to the larger birds of prey, such as eagles and hawks, at the highest trophic levels. The concentration of DDT in these predators had become high enough that the birds' health was affected. One impact was that birds with high DDT content in their tissues tended to lay eggs with thin shells. Many of these eggs cracked before they could hatch, so fewer chicks were born and the bird populations started to fall.

DDT sprayed on farms and in swamps has contributed to declining numbers of birds in a range of habitats. The role of DDT in the disappearance of birds is the subject of Rachel Carson's classic environmental book *Silent Spring*. As a result of these concerns, since the late 1960s various countries around the world banned the use of DDT. In 2001, the Stockholm Convention on Persistent Organic Pollutants banned the use of DDT except for controlling the spread of diseases such as malaria.

HL.a.2, HL.a.6–HL.a.8: Environmental law

Bans on DDT have been enacted in national law in many countries around the world, and through an international convention, the Stockholm Convention on Persistent Organic Pollutants. The bans were specifically enacted to prevent harm to birds and the ecological impact of bird decline.

The Stockholm Convention has also been an important tool in the control of DDT and other pollutants, because they can spread over long distances and cross international borders. The convention came out of a conference held in Stockholm, Sweden in 2001, following several years of investigation and discussion between countries. The convention bans or restricts the production and use of certain specified pollutants, and requires existing pollutants to be disposed of in an environmentally safe way.

After signing the convention, individual countries had to ratify it by creating their own laws and regulations to implement the rules of the convention. The convention came into effect and became legally binding 90 days after the first 50 countries had ratified it. As of 2023, 185 countries plus the European Union have ratified the convention.

Fertilizers

The minerals in fertilizers, such as nitrates and phosphates, can help plants to grow, but they are often washed from the soil surface of farms when it rains or when the farms are irrigated. If nitrates and phosphates are washed into surface waters, they can cause algae to grow faster, producing an algal bloom. The algae eventually die and decompose. The decaying algae are extra organic matter in the water, and as they decompose, the bacteria feeding on them use up the oxygen in the water, causing oxygen depletion. The process of adding extra nutrients to a body of water is called eutrophication (see Topic 2.2.20, page 130 and 4.4.5, page 388.).

■ **Figure 4.105** Algal bloom in a eutrophic lake in Devon, UK

Synthetic fertilizers tend to cause more eutrophication than organic ones, such as manure or compost. Organic fertilizers tend to hold on to their minerals more tightly, so the nitrates and phosphates in them are released slowly, and there is less runoff of nutrients when it rains or when the crops are watered. Synthetic fertilizers tend to contain nitrates and phosphates in more soluble forms, so that as soon as water is added, they dissolve and flow away.

Nutrients from fertilizers can also be washed downwards into the groundwater when water infiltrates and percolates through the soil. Nitrates can then get into drinking water. Very high levels of this may cause a condition called methaemoglobinaemia (MetHb), also known as blue baby syndrome. It most often occurs in babies who are fed formula made with nitrate-contaminated water. MetHb interferes with the red blood cells' ability to carry oxygen around the body. It can cause serious illness or death.

■ **Figure 4.106** Synthetic fertilizers are often produced and sold as solid granules

■ Sewage

◆ **Sewage** – any form of wastewater from human activities.

Sewage is any water that has been used by humans and is then released into the environment. This includes water from kitchen and bathroom sinks, baths and showers, appliances like washing machines and dishwashers, as well as toilets. Wastewater from homes is called domestic sewage. Wastewater from industrial processes can also be called sewage, although it is usually considered separately.

Common mistake

Many people think sewage refers to toilet waste only. However, all wastewater is considered sewage.

People sometimes confuse sewage and sewerage. Sewage is wastewater. Sewerage refers to the systems of pipes, drains, pumps and treatment facilities that transport, treat and dispose of sewage.

♦ **Pathogen** – a disease-causing organism, such as viruses, some bacteria, and some other single-celled organisms.

♦ **Water-borne disease** – a disease that is contracted by drinking or sometimes bathing in contaminated water. Most water-borne diseases are caused by pathogens in the water.

♦ **Oxygen depletion** – a reduction in the concentration of oxygen that is dissolved in water. It is often caused by nutrient or organic matter pollution, and may cause the death of aquatic animals.

♦ **Sediment** – small pieces of insoluble material that can be carried by water. Sediments may come from natural erosion of rocks and soils, but can also be released in sewage and in agricultural runoff.

♦ **Organic matter** – substances from living organisms, which can decompose. It can come from natural sources, such as falling leaves, but can also include waste from toilets, farms and kitchens.

Domestic sewage may contain many different pollutants. Toilet waste may contain disease-causing organisms or **pathogens**. If toilet waste is released without any treatment, people may be exposed to these pathogens, and get sick with **water-borne diseases** including cholera, typhoid and various forms of dysentery.

When organic matter (including toilet waste) is released into water, it decomposes. Bacteria break it down into simpler substances. In the process, the bacteria use up some of the oxygen dissolved in the water. This can cause **oxygen depletion**, killing many of the animals that live in water.

Flushing other waste items down the toilet, including tissues, sanitary products and condoms, can block the pipes. These items should be placed in bins. In some places, people are advised not to put toilet paper into the toilet for the same reason.

Wastewater from other parts of the house, such as from sinks, baths and washing machines, does not usually contain many pathogens. However, it often contains detergents, **sediments** and **organic matter**.

- Detergents include soap, dishwashing liquid and laundry detergent. These may be harmful to living organisms, and also may contain organic matter and nutrients such as phosphates, which can contribute to eutrophication. Many modern detergents are marketed as being free of phosphates and non-polluting.
- When we wash dirt or mud off our hands or clothes, we are washing sediments into the drains. Sediments that accumulate in a home's drainpipes can cause blockages. The quantities of sediments coming from one home are not likely to cause major problems, but the combination of sewage from all the homes in a village, town or city can add a lot of sediment to waterways.
- When we wash dirty dishes or throw away excess foods like sauces, we may add organic matter to the sewage. If possible, it is usually best to use food waste to make compost. Fat and oily foods in particular shouldn't go down the drain because they tend to solidify and block drains, especially during cold weather, when the oil may freeze.

■ Industrial effluent

Liquid wastes released from factories are called effluent. Industrial effluents may be treated at the point where they are released, so that the environment is not harmed significantly. In some cases, though, not all the contaminants can be fully treated. If treatment is not possible, the liquid waste may be stored on the site of the factory. There have been many examples of untreated and unsafe liquid wastes being released into the environment. Many of these cases involve accidental releases, but some have been intentional.

◆ **Carcinogen** – a substance that causes cancer. Although many pollutants are suspected of being carcinogenic, it is not always easy to prove that they are.

> **REAL-WORLD EXAMPLE**
>
> ### Erin Brockovich and chromium VI
>
> During the 1950s and 1960s, in the small town of Hinkley, California, USA, an energy company operated a compressor station on a natural gas pipeline. Cooling towers were used to keep the compressor station from overheating. The water used in the cooling tower contained a chemical called hexavalent chromium, or chromium VI. Hexavalent chromium is thought to be a **carcinogen**. After it was used, the water was stored in ponds on the site.
>
> Several years later, it was reported that an unusual number of illnesses were occurring among people living in the Hinkley area. It was found that hexavalent chromium had leached out of the storage pond and into the groundwater. Residents of the area obtained their drinking water from the local aquifers. Fearing that the illnesses were caused by the contamination, residents filed a lawsuit against the energy company. The case was eventually settled, and the company was required to pay damages to the residents and to clean up the contaminated area. The case was dramatized in the 2000 film *Erin Brockovich*.

Industrial effluent may include many different substances and cause different types of environmental problems. Each type of industry has its own processes and produces its own types of wastes. To some extent, each industry must develop its own waste treatment and disposal processes. Some examples of substances found in industrial effluents are:

- organic compounds, including solvents, lubricants and byproducts of factory processes
- acids, alkalis and salts
- heavy metal ions and compounds
- other toxic compounds, such as cyanide, pesticides, pharmaceuticals and others.

> **REAL-WORLD EXAMPLE**
>
> ### Minamata Disease
>
> During the 1950s and 1960s, a chemical factory in the city of Minamata, Japan, released a small but steady amount of methylmercury in its effluent. Methylmercury was a byproduct of the production of acetaldehyde, which was an important industrial chemical. Initially, the raw effluent was released into the sea in Minamata Bay without any treatment. The effluent also contained other harmful substances.
>
> A few years later, people in fishing villages around Minamata Bay started suffering from a strange disease, which affected their brains, nerves and muscles. It took many years of research to determine that methylmercury was the cause, and that the people were suffering from a form of mercury poisoning.
>
> Methylmercury undergoes both bioaccumulation into living organisms and biomagnification up the food chain. Eventually, the methylmercury was transferred to people in the nearby fishing villages as fish caught in Minamata Bay formed a large part of their diet.
>
> When the problem was recognized, the factory was forced to pay compensation to victims of the disease and to workers in the fishing industry. It was also required to install a treatment system for its effluent, although it is uncertain whether this reduced the release of methylmercury. In 1968, the factory changed its production process, so that methylmercury was no longer released in its effluent.

◆ **Market failure** – the concept in economics by which the free market fails to allocate resources efficiently. In environmental economics, many cases of environmental harm can be described as market failures.

◆ **Externality** – a cost or benefit for a business that is not included in the normal accounts of the business. External costs and benefits are felt by people or environments outside of the business.

◆ **Polluter-pays principle** – the principle that the person or business that causes pollution (or other forms of environmental harm) should bear the cost of the damage it causes.

This example illustrates the difficulty in identifying the cause of an environmental problem, as well as in managing it. The difficulty stemmed in part from the length of time which passed between the release of the pollutant and the first cases of disease. Eventually there were consequences for the company that caused the problem, but the damage was already done to the people of Minamata.

HL.b.3, HL.b.4: Environmental economics

Environmental economists might see the release of pollutants into the environment by industrial processes, such as the release of methylmercury in Minamata, as a **market failure**. The free market failed to prevent the release of the pollutant.

The release caused harm to the environment and to the people living there. The cost of this harm was not, at first, paid by the company that caused the release. The cost could therefore be described as an external cost or **externality**, because it did not increase the cost of running the factory. Instead, it caused cost to the wider community, including the economic cost of providing healthcare to sick people and loss of productivity due to their illnesses.

Eventually, the **polluter-pays principle** was applied when the company was forced to pay compensation to the victims. The company was required to take economic responsibility for its action. Economists might describe this as an attempt to internalize the external cost, or to correct the market failure.

Thermal pollution

Industrial effluent may cause thermal or heat pollution. Many factories use processes that involve high temperatures. To avoid overheating, these factories usually use water from a nearby river or lake to take away excess heat. The cooling water used in this way becomes hot, and is usually released back into the natural waterway.

Hot water released like this can have multiple effects on the natural ecosystem, including:

- Direct harm to animals and plants, which cannot tolerate the higher temperature.
- Algal bloom, as the higher temperatures may cause algae to grow rapidly.
- Oxygen depletion, since hot water holds less dissolved oxygen than cold water does.

■ Urban runoff

Rainfall or snowmelt in urban environments like towns and cities typically flows into drains. Larger cities often have large storm drains running below the streets. Water enters the storm drains through inlets, typically located at the sides of roads. These inlets usually have a grill over them, which stops larger pieces of debris from entering, but smaller pieces of rubbish and anything dissolved or suspended in the water flows easily into the drain.

Many different substances are washed by stormwater into drains. These may include:

- sediments, like dust, soil, and worn-off pieces of roads, buildings, cars, tyres and other objects
- liquids like oil, petrol, coolants and other fluids that have leaked from vehicles
- any small pieces of litter, including food, which have been dropped on the ground
- salt that may have been spread on streets to reduce winter ice
- fertilizers and pesticides from lawns and gardens
- detergents or other cleaning agents.

◆ **Open dump** – a site where garbage is deposited, but not covered.

◆ **Sanitary landfill** – a carefully constructed site designed for the disposal of garbage. The ground is usually excavated to form a large pit, and the bottom of the pit is lined with an impermeable material. Garbage is evenly spread within the pit and covered with soil.

◆ **Leachate** – a liquid that washes away as water flows over something, usually into the soil.

◆ **Hazardous waste** – garbage that contains specific, dangerous substances or items, such as toxic, carcinogenic or explosive materials.

◆ **Crude oil** – petroleum as it is found underground. It is a mixture of many different substances called **hydrocarbons**. Crude oil must be refined, or separated into its components, before it is useful.

◆ **Hydrocarbons** – substances made up mostly of carbon and hydrogen. These are the main constituents of all fossil fuels.

Storm drains usually run underneath the entire city, carrying water downhill to the nearest stream or river. The water typically runs straight into this watercourse without any treatment. When sediments and litter accumulate in storm drains they may block the flow of water, causing street flooding

Leachate from dumps and landfills

Solid waste, including the garbage produced by households, is often disposed of in **open dumps** or **sanitary landfills**. Over time, the waste starts to decay and any liquids produced trickle down into the ground below. This is called leaching and the liquid produced is **leachate**. Rain falling on uncovered waste may speed up the process.

Leachate may contain all sorts of substances, including organic matter, nutrients or toxic chemicals. In an open dump, where the ground below the garbage is usually ordinary soil, the leachate may infiltrate into the soil and percolate to the groundwater. Groundwater polluted by leachate may not be safe to drink, especially if the garbage contained **hazardous wastes**.

The soil beneath a sanitary landfill is usually lined with some sort of impermeable material, such as compressed clay, concrete or strong plastic. These are often used together for extra security. This reduces the risk of leachate reaching the groundwater. Instead, the leachate accumulates at the base and can then be collected and taken to a special waste-treatment facility to make it safer.

Oil spills

Occasionally, the ships or pipes that transport petroleum leak. This may be due to an accident, such as a ship running aground. Some of the largest ships in the world are giant oil tankers, which carry millions of barrels of oil. If one of these spills its contents into the sea or a lake, the ecosystem can be badly damaged.

Oil tankers and pipelines are used to transport **crude oil** from the oil wells where it is brought up from underground, to refineries where it will be turned into useful products. Crude oil is a complex mixture of substances, some of which are highly toxic. Most components of crude oil are insoluble in water, so they usually float on the surface. Light and oxygen may not easily pass through this layer of oil, so algae and animals living in the water may die.

If seabirds such as pelicans or seagulls land on the oily water, their feathers may get covered by oil, leaving them unable to fly or to control their body temperatures. Marine mammals like seals and whales, and marine reptiles such as turtles, may be exposed to oil when they surface to breathe. Unless they are rescued and cleaned, these animals may die. Other organisms are likely to be poisoned by the chemicals, which may stay in the environment for a long time.

■ Figure 4.107 Cleaning a seagull covered in oil

REAL-WORLD EXAMPLE

Deepwater Horizon oil spill, Gulf of Mexico

In 2010, an explosion occurred on the Deepwater Horizon oil rig, in the Gulf of Mexico, killing 11 workers. The explosion left a leaking oil well, 2 km below the surface. During the following weeks, the well released enough oil to cover over 6,500 km² of water. This incident caused the largest oil spill in history.

■ Figure 4.108 Fire on the Deepwater Horizon oil rig

Theme 4: Water

The oil spill had a major impact on wildlife in the Gulf of Mexico, with turtles, dolphins, birds and various shellfish all at risk. Coastal and deep-water marine habitats were threatened with contamination. The fishing, tourism and oil industries were all impacted, which had a significant effect on the local economy. Oil is likely to remain in these habitats for many years.

Stopping the oil leak proved particularly difficult, partly because of the depth of the leaking well. It was eventually closed off more than three months later, during which time 4.9 million barrels, or 780,000 m³ of oil are estimated to have leaked into the sea. Attempts to control the spread of oil included using skimmers to collect oil from the water's surface, portable barriers to prevent the oil from washing on to the coastlines, and pipes lowered into the leaking well, to draw the flowing oil into tankers on the surface. About 1 million tonnes of dispersant was also used to break up the oil, diluting it so that its impact was less concentrated. However, the dispersants themselves are toxic and may have caused even more harm to wildlife.

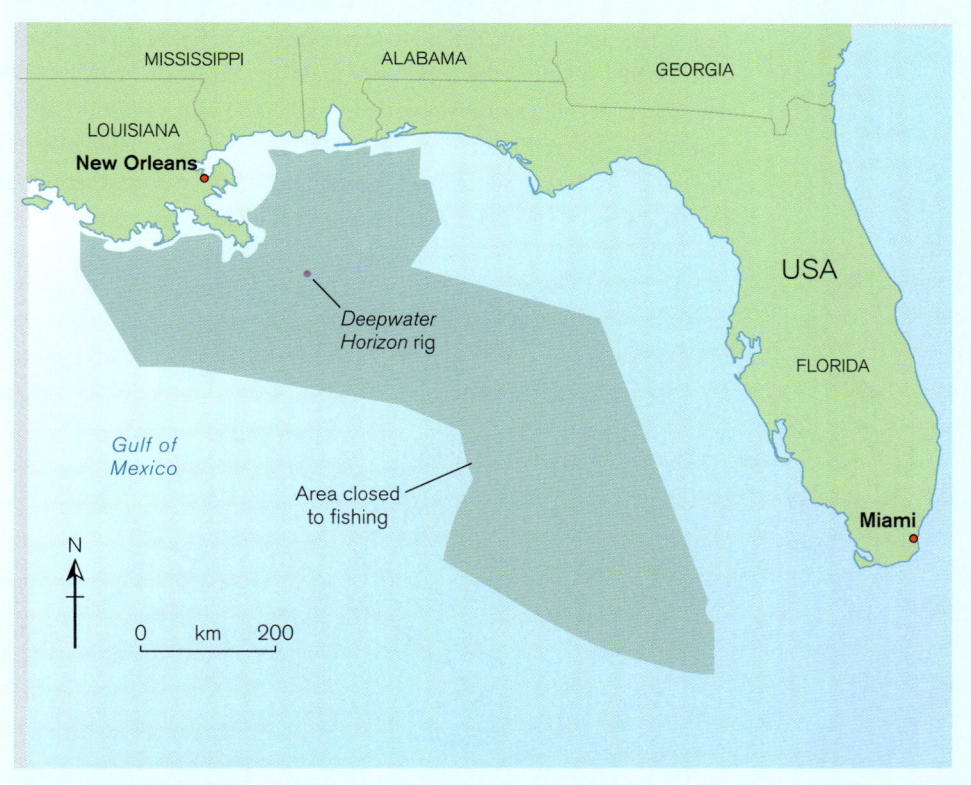

■ **Figure 4.109** The site of the Deepwater Horizon oil rig, showing the area that was closed to fishing due to the oil spill

Large oil spills such as Deepwater Horizon are usually heavily reported by the news media, so they are often well known to the public. However, smaller oil spills occur much more regularly, from leaking wells and pipes. All ships and boats use oil in their engines, so small spills can occur from any vessel.

One of the better solutions for oil spills is **bioremediation**. This method uses microorganisms such as bacteria to break the oil down into harmless substances. Some bacteria, which live in areas where oil naturally seeps into the sea, have evolved the ability to break down some of the chemicals in oil, using it as an energy source. Bioremediation is a relatively new method and scientists are working on improving it.

◆ **Bioremediation** – the use of living organisms, particularly bacteria, to break down or remove pollutants from the environment.

4.4 Water pollution

> **ATL ACTIVITY**
>
> **Bioremediation**
>
> Find out more about bioremediation and how it can be used to clean up a range of different pollution problems. You may also investigate phytoremediation (remediation using plants) and mycoremediation (remediation using fungi). Investigate one specific case study and present it to your class.

4.4.2 Plastic debris is accumulating in marine environments; management is needed to remove plastics from the supply chain and to clear up existing pollution

Plastic has become an extremely important part of modern life. Plastic items are useful, cheap and convenient. We often use them once and then throw them away. Disposing of plastic items can be difficult, especially since we produce and use so much plastic. Far too often, plastic waste is not disposed of properly, and ends up in drains, rivers and eventually the sea. Since plastic cannot be broken down by microorganisms, it remains in the environment for a long time. It can entangle or trap animals, or enter their bodies, physically harming or killing them.

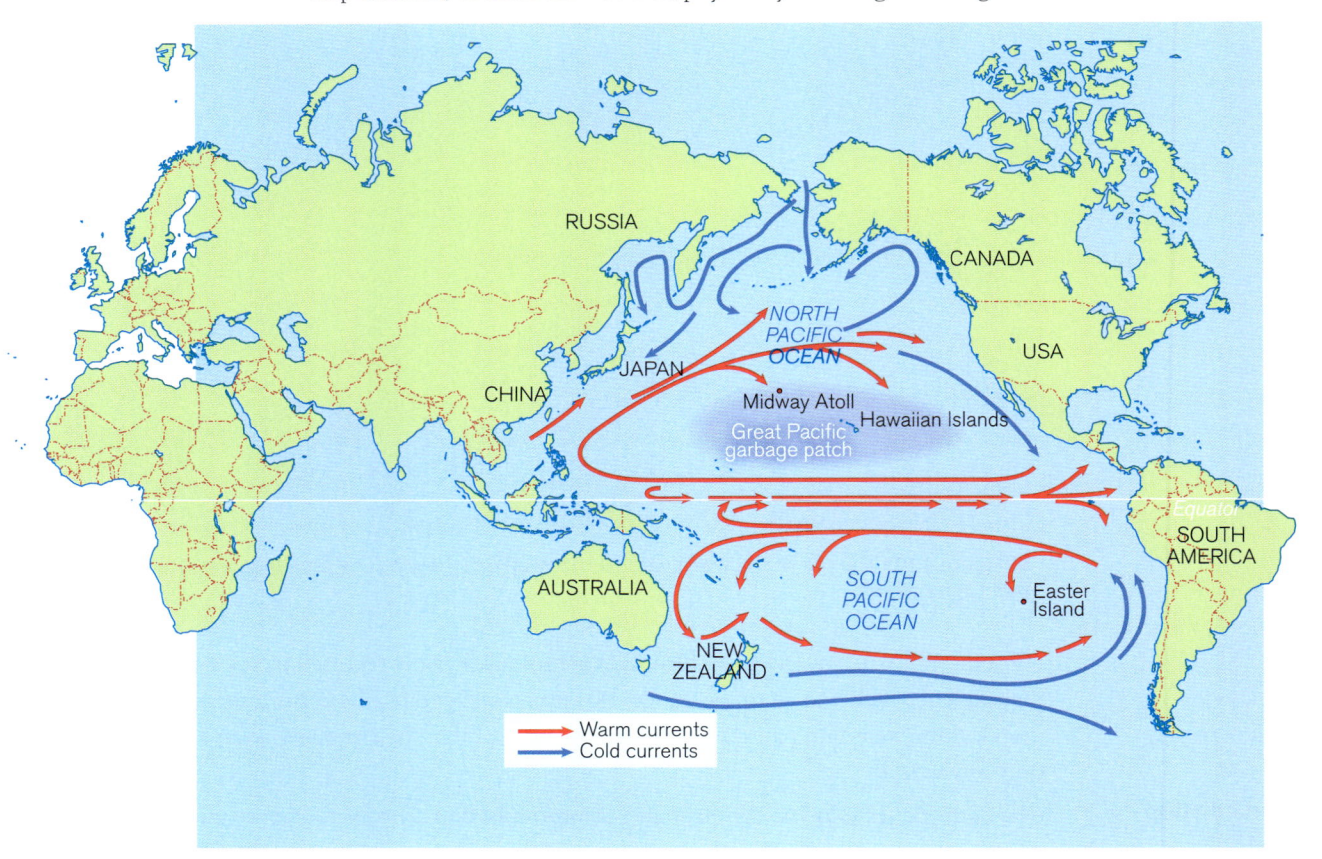

■ Figure 4.110 The Great Pacific Garbage Patch

Figure 4.111 Microplastics

Plastic items often float in water, so they may be transported over long distances in rivers and the sea. Thousands of pieces of plastic have begun to accumulate in central areas of each of the major oceans. Ocean currents typically form natural spirals or gyres, which trap debris in a circular flow around the calm, central part of the ocean. The most famous of these is the Great Pacific Garbage Patch.

Although most plastic cannot be broken down by microorganisms, it can be physically break down into smaller and smaller pieces, eventually forming microplastics. Microplastics are so tiny and so abundant that it would be very difficult to remove them from the environment. Over time, microplastics in water gradually sink to the bottom of the water body and accumulate in the sediments there.

Collectively, the tiny particles of microplastics have a very high surface area. As they move about in the environment, they may come into contact with various toxic pollutants, which may become adsorbed or attached to the surface of the microplastics. If these microplastic particles are ingested or absorbed by living organisms, they may cause significant harm to the organisms. In some cases, they may undergo biomagnification up the food chain.

Cleaning up microplastics from the ocean is practically impossible. Even collecting and removing larger pieces of plastic debris is incredibly difficult, especially since we keep producing and throwing away more and more plastic. Ultimately, the solution to our plastic waste problem is to reduce the production and use of plastics – especially single-use, disposable ones. One approach to achieve this is the circular economy.

Link
The circular economy is explored in Topic 1.3.21 (page 77).

4.4.3 Measuring and monitoring water quality to inform management strategies for reducing water pollution

Since water is such an important resource for humans and other lifeforms, it is important to be able to measure how suitable water is for human use and for natural ecosystems. Scientists have developed various methods to measure water quality. We can use these standard methods to monitor how water quality changes over time.

Water-quality parameters

Important water-quality factors that scientists can measure and monitor include:

- temperature
- turbidity
- suspended sediments
- dissolved oxygen
- pH
- nitrates
- phosphates
- dissolved metals
- faecal coliform.

Dissolved oxygen

Oxygen is essential for most organisms that live in water. If levels of dissolved oxygen become, many organisms will die. Water movements, temperature and decomposition of organic matter can each affect the quantity of dissolved oxygen in water, as shown in Table 4.8.

■ **Table 4.8** Impact of various factors on dissolved oxygen

Factor	Effect on dissolved oxygen
Water movement	Rapidly moving or splashing (turbulent) water tends to absorb a lot of oxygen from the air. Water that has recently flowed over a waterfall or weir, for instance, usually has high dissolved oxygen. Standing or stagnant water is more likely to have low dissolved oxygen.
Temperature	Oxygen dissolves more easily in cooler water than warmer water. As the temperature of water rises, oxygen molecules come out of solution, and diffuse into the air.
Decomposition	Bacteria carrying out decomposition use up the dissolved oxygen as they respire. The more decomposition is taking place, the less oxygen is likely to remain dissolved in the water.
Photosynthesis	Oxygen is produced during photosynthesis. If there are plants and algae in water, and there is a lot of light available, then there is likely to be a lot of oxygen in the area where the plants are. However, this relationship is complicated. If there is too much algae or plant biomass, such as during an algal bloom, there is likely to be a lot of decomposition happening and high oxygen concentration near the algae and at the surface of the water, but deeper down, near the sediments, the oxygen concentration might be very low. (See Topic 4.1.11 on stratification, page 285.)

Dissolved oxygen (often abbreviated as DO) can be measured using a DO probe. This is a small electrical device which is lowered into the water and which measures the dissolved oxygen directly. It may give readings in milligrams per litre (mg/l) or in percentage saturation – 100 per cent saturation means that the water holds the maximum possible amount of dissolved oxygen at that temperature.

DO can also be measured using the Winkler method of titration. This involves mixing known quantities of specific reagents with the water and observing colour changes. The volume of one of the reagents that is needed to produce a particular colour change can be used to determine the concentration of oxygen. The Winkler titration can be carried out in a laboratory, but there are also portable, easy-to-use kits which can be used in the field.

pH

pH is a measure of how acidic or alkaline the water is. Natural, unpolluted freshwater typically has a pH in the range of 6 to 8. It is likely to be closer to 6 if the water has flowed through acidic soils or rocks, or closer to 8 if it has flowed through basic soils or rocks. If the pH lies outside this range, or does not match the natural soil conditions, this may be an indication of pollution.

pH is best measured using a pH probe, which is similar to a DO probe. An electrical device is lowered into the water and the pH reading is taken. pH probes need to be regularly calibrated against solutions with a known pH, to ensure that they produce accurate readings.

Other methods, such as pH paper and universal indicator, can also be used, but these give less precise readings and interpreting the colour changes can be subjective. A pH probe is generally the better option if one is available.

Temperature

The temperature of water is typically reasonably stable. We can usually predict the temperature of a body of water if we know where the water has come from and what the weather has been like. If we measure the temperature and it is unusually warm or cool, this may suggest that an effluent has been released into the water. Cooling water from a factory, for instance, could cause the temperature to be unnaturally warm.

Temperature can be measured with a thermometer. Many types of thermometers are available. DO probes often have a built-in electronic thermometer.

Turbidity

Turbidity refers to the murkiness of water. If there are suspended sediments or algae in the water, light may not be able to pass very far through it. Turbid water is not clear. High turbidity may be an indication of too much sediment in the water or of eutrophication.

The most common method for measuring turbidity in deep water is to use a Secchi Disk. This is a weighted circle of 20 cm diameter, painted in black and white quarters, and attached to a rope. The disk is lowered slowly into the water until it can no longer be seen from the surface. The depth of the disk can then be determined by measuring the length of rope which was lowered into the water. If the water is very turbid, the disk will disappear at a shallow depth.

The Secchi disk method can be quite subjective. It can also be affected by how bright the lighting in the area is. Brighter sunlight may mean the water appears clearer, but it may also cause reflections on the surface that make it harder to see the disk.

To measure turbidity in shallow water, a turbidity tube or transparency tube can be used. This is a clear plastic tube with a small version of the Secchi disk at the bottom. The tube is filled with water from the river or stream, and then, the water is allowed to drain out through a hole near the bottom of the tube, until the disk becomes visible. The depth of the water remaining in the tube is then measured using a scale on the side of the tube.

Nitrates and phosphates

◆ **Nitrates** – compounds containing the nitrate ion, NO_3. They are an essential nutrient for plants and algae and often included in synthetic fertilizers.

◆ **Phosphates** – compounds containing the phosphate ion, PO_4^{3-}. They are an essential nutrient for plants and algae and often included in synthetic fertilizers.

Nitrates and **phosphates** are mineral nutrients that are often found in agricultural fertilizers. Plants need these substances to grow, and farmers may add them to the soil to help their crops grow faster. Nitrates and phosphates are also released when organic matter decomposes. When added to water, nitrates and phosphates cause algal bloom, which eventually leads to oxygen depletion. Unusually high levels of nitrates or phosphates may indicate pollution from agricultural runoff, or from sewage containing organic matter.

The most precise way to measure nitrate and phosphate concentration in water is to carry out a titration, either in a laboratory or using a portable field kit. A quicker, often more convenient option is to use the appropriate test strip. This is a small piece of paper which reacts with particular chemicals in water and change colour. The colour is compared to a chart, which indicates the concentration of the pollutant. Different reagents and strips are needed to test for nitrates and phosphates.

Dissolved metal ions

Dissolved metals in water may be an indication of pollution, perhaps from an industrial source. Some metals are quite dangerous and can cause serious illness.

Testing for dissolved metals can be done using test strips, similar to the strips used for nitrates and phosphates, or with more complicated titrations in a laboratory. A different test would be needed for each of the metals. It is unlikely that non-scientists would test water for a specific metal, and this is rarely done unless there is a special reason to think that that metal might be present in the body of water.

Suspended sediments

Suspended sediment refers to insoluble particles that are carried by the water. Eroded soil is the most common form of sediment in water bodies. A high suspended-sediment measurement might be due to soil erosion, or the release of sediments from activities such as construction or quarrying. Suspended-sediment measurements may be higher after heavy rain, when more soil is washed into rivers.

A rough idea of the amount of suspended sediment can be quickly obtained using a Secchi disk or turbidity tube. However, the best method for measuring suspended sediment is to filter the water in the laboratory.

> ## Tool 1: Experimental techniques
>
> ### Determine total suspended solids (TSS)
> 1. Collect a large sample of water, without disturbing the bottom of the stream, and take this back to the lab.
> 2. Weigh a clean, dry piece of filter paper, using an accurate laboratory balance (scale).
> 3. Shake the collected sample to re-suspend any particles that might have settled, and pour a known volume (e.g. one litre or five litres) of the sample through the filter paper.
> 4. Collect the filter paper and dry it by warming it gently in an oven for a few hours.
> 5. Allow the filter paper to cool in a desiccator, to ensure that it is completely dry.
> 6. Weigh the dried filter paper, to determine the mass of residue that was filtered out of the sample.
> 7. Calculate the mass of sediment in one litre of the collected sample. To do this, divide the mass of the residue, in grams or milligrams, by the volume of the filtered water, in litres.

Faecal coliforms

Faecal coliforms are bacteria that are often found in human and animal faeces. Some of them may be pathogenic, which means they cause disease. If faecal coliforms are present in water, this suggests that human or animal faeces, such as from farm animals or untreated sewage, may be present in the water.

The presence of faecal coliforms in water can be assessed by growing the bacteria on a suitable medium in a petri dish. A known volume of water is spread on the petri dish and incubated at a standard temperature. Each bacterium in the sample grows into a visible colony. The number of colonies indicates how many bacteria were present in the initial sample of water.

Faecal coliforms are not necessarily pathogenic, but they can be. More importantly, if they are present in water, it is likely that other microorganisms, including pathogens, might be present. Water with high faecal coliform counts is not suitable for drinking, and may be unsafe for washing or swimming.

■ Water-quality indices

Various water-quality can be measured and combined into one number, called a water-quality index (WQI). This requires some mathematical manipulation of different test results to come up with a single number. Different water-quality indices have been developed. One commonly used index, the Vernier WQI, gives an overall value between 0 and 100, with 100 being the highest possible quality.

> ## Inquiry process
>
> ### Inquiry 1: Exploring and designing
>
> Investigate the effects of human activity on water quality. Select a river or stream near your home or school, and plan an investigation to determine how human activities have affected its water quality.
> 1. Identify a watercourse that you can access safely in at least two, preferably more, locations.
> 2. Find out what sorts of human activities occur near the watercourse. You may do this using maps or satellite imagery, and/or by exploring the area in person. You may find, for example, that there is farming or industry in the area. Identify the activities as specifically as possible.

3. Do some literature-based research to determine how the activities you have found are likely to impact the quality of the water. As far as possible, use only peer-reviewed journals for your research, to ensure that your sources are reliable. You may need to ask your school librarian for help accessing journals. From the literature, identify any specific water-quality parameters that are likely to be affected, and record any values obtained within published studies you find.
4. Choose at least three of the water-quality parameters discussed in this chapter that the literature suggests might be affected by the activities you have found, and that you are able to test. Consider the equipment and materials that are available to you. Ask your teacher or lab technician if you can test the parameters you are interested in.
5. Formulate a hypothesis on the effect of the activities you found at your study site on the water-quality parameters you have chosen.
6. Design an investigation to test your hypothesis. Identify your independent, controlled and dependent variables, and describe clearly how you will carry out your study.
   a. For your independent variable, consider testing the water quality at different locations along the watercourse.
   b. For your controlled variables, consider how you can ensure that your independent variable is the only factor that is different at the different sites. Bear in mind that it is often impossible to control variables perfectly during field investigations. For variables that you cannot control, but that could affect your results, you may need to record measurements and observations, so that you can determine whether they affect your test results.
   c. For your dependent variables, consider how you can measure their values, using the equipment and within the timeframe available to you.
   d. If time and equipment allow, aim to take enough repeat measurements to allow you to test your results statistically. Consider what sort of statistical test would be appropriate, and how many measurements are needed for that test.
7. Write out the method for your investigation in full, clear detail.
8. Before beginning your investigation, carry out a detailed risk assessment to ensure that you stay safe during your fieldwork. Consider the risks of working outdoors, near water, and of working with the equipment and materials you have selected. Consider also any ethical or environmental concerns with your study, including any possible impacts on the well-being of animals and other people.

Inquiry 2: Collecting and processing data

1. Carry out your procedure as you have planned it, being sure to follow the guidance of your risk assessment and the detailed procedure you described. If you have to adjust your method on site, be sure to record any changes you make. Try to ensure that you follow the same procedure for each measurement you take.
2. Record your data in appropriate tables and graphs, and carry out any statistical tests that you can.
3. Draw a conclusion based on your results, incorporating the results of any statistical tests you used.
4. Compare your results to any findings you found during your literature review.

4.4.4 Biochemical oxygen demand is a measure of the amount of dissolved oxygen required by microorganisms to decompose organic material in water

One of the most important, and widely used, tests of water quality is the biochemical oxygen demand (BOD) test. This test gives an accurate indication of how much biodegradable organic matter is in a sample of water.

Organic matter may be present in farm runoff, domestic sewage and some sorts of industrial effluent (for example, from the food industry). In the water, the organic matter decomposes or breaks down into simpler substances.

Decomposition is carried out by organisms called decomposers. In aquatic ecosystems these are usually bacteria, which use oxygen from the water for respiration. As decomposition occurs, the dissolved oxygen content of the water decreases.

> **Link**
> Decomposition of organic matter in the carbon cycle is covered in Chapter 2.3 (page 139).

Tool 1: Experimental techniques

Determine the BOD of a water sample
1. Collect a sample of water in a sealed, dark glass bottle.
2. Test the sample immediately for dissolved oxygen, using either a DO probe or the Winkler titration.
3. Store the sample in the dark, at 20°C, for five days.
4. Test the sample again, using the same method.
5. Calculate the five-day BOD for the sample by subtracting the final DO measurement from the initial DO measurement.

The BOD test result is measured in milligrams per litre (mg/l). In stating the result, it is important to state how many days the sample was incubated for. If it is incubated for five days, the result is given as the five-day BOD.

The BOD test result tells us how much oxygen was used up during the five-day incubation period. We would normally expect the dissolved oxygen concentration to decrease over five days as any living organisms in the water will use up oxygen during respiration. Since the sample is kept in the dark, there should be no photosynthesis taking place, so no oxygen should be added. The higher the result of the BOD test, the more organic matter is present in the water.

4.4.5 Eutrophication often causes excessive growth of phytoplankton

Eutrophication is widely considered to be the single biggest water-pollution problem around the world. Many rivers, lakes and estuaries, as well as some coastal seas, experience eutrophication at some point in time, and its effects can be severe.

Eutrophication refers to the presence of excess nutrients in water. Most bodies of water have very low nutrient concentrations, and are described as **oligotrophic**. When water contains significantly more nutrients than normal, it is described as eutrophic.

Nitrates and phosphates are the most common cause of eutrophication in freshwater ecosystems. These are both found in most agricultural fertilizers. When fertilizers are used on farms, nitrates

◆ **Oligotrophic** – having low concentrations of nutrients. Most naturally occurring waterways are oligotrophic.

Figure 4.112 A eutrophic watercourse

Link
Exponential or logarithmic growth is covered in Topic 2.1.12 (page 94).

◆ **Algal bloom** – the rapid growth of algae, including phytoplankton and macroalgae. The algae often form a dense layer covering the surface of the water, or make the water appear green.

◆ **Anoxia** – the complete absence of dissolved oxygen in water, due to extreme oxygen depletion.

◆ **Hypoxia** – the condition of having a very low oxygen concentration in water due to oxygen depletion.

◆ **decomposition** – the breakdown of dead organic matter by decomposers such as bacteria and fungi, using oxygen. The process usually produces substances containing oxygen, such as carbon dioxide, nitrates and phosphates.

◆ **Anaerobic decomposition** – the breakdown of dead organic matter by special anaerobic decomposers, including some bacteria and fungi. The process usually produces substances containing no oxygen, such as methane, ammonia and hydrogen sulphide.

and phosphates may wash off into nearby bodies of water when it rains. They are also released when organic matter decomposes, so they may be present in water polluted with domestic sewage or food waste. Phosphates may also be present in detergents used for washing clothes or dishes.

The immediate impact of eutrophication is usually the rapid growth of algae. This is called an **algal bloom**. In most natural, oligotrophic water, algal growth is limited by a lack of nitrates or phosphates. If these minerals are added – and if other factors like temperature and sunlight are suitable – the algae will start to grow and reproduce rapidly.

In eutrophic water, plant and algal growth is usually concentrated near the surface of the water. As more and more plants and algae grow, less and less light can penetrate to the deeper water, so plants and algae living below the surface will die, while those living at the surface will thrive. Eutrophic water often has a thick layer of green algae or pondweed floating on its surface.

4.4.6 Eutrophication leads to a sequence of impacts and changes to the aquatic system

Eutrophication often follows a predictable pattern, characterized by algal bloom and dieback, oxygen depletion and **anoxia**.

■ Algal bloom and dieback

The growth of algae during an algal bloom can follow the typical J-shaped curve of logarithmic or exponential growth. Exponential growth cannot, however, continue indefinitely. When the population becomes too large for its environment, resources such as water, oxygen or nutrients may be depleted. At this point, many of the algae die.

The dead algae sink to the bottom of the water and begin to decompose. Bacteria break down the dead algae into simpler substances, releasing smaller molecules and ions, such as carbon dioxide into the air, and nitrates and phosphates into the water. The released nutrients may then be used by more algae to promote a second wave of algal bloom, and the process is repeated.

■ Oxygen depletion

The decomposing bacteria also need oxygen for respiration. The more organic matter that is present, the more oxygen is used up by the bacteria. This rapid use of oxygen leads to oxygen depletion. Water that has lost most of its dissolved oxygen can be described as **hypoxic**. Eventually, all the oxygen might be depleted. Water with no oxygen is described as anoxic.

Oxygen is essential to most animals that live in the water, including fish and many invertebrates. In hypoxic or anoxic water, most or all of these animals will die and cause the ecosystem to collapse. The dead bodies of these animals also add to the decaying organic matter.

■ Anoxia

If all the oxygen in a body of water is used up, but there is still organic matter present, **anaerobic bacteria** may take over the process of decomposition (**anaerobic decomposition**). Anaerobic bacteria do not use oxygen for respiration, and release different products into the environment, such as methane (a greenhouse gas) and hydrogen sulphide (a toxic gas with an unpleasant smell).

Link

Anaerobic decomposition is explored further in Chapter 2.3, (page 139) and Chapter 5.1, (page 416). Methane's role as a greenhouse gas is covered in Chapter 6.1, (page 514).

> **Tool 4: Systems and models**
>
> ### Construct a flowchart of eutrophication
>
> Use the information in this section to construct a flowchart illustrating the process of eutrophication, and showing how positive feedback can cause the problem to get worse over time.
>
> You might start with a box stating 'Nitrates and phosphates are added to water'. Include the impacts on algae and animals, and the activity of decomposers. Try to include the action of anaerobic decomposers.
>
> For extra insight, see if you can include the effect of humans adding organic matter, such as food or farm waste, into the water.

REAL-WORLD EXAMPLE

The Gulf of Mexico Dead Zone

The Gulf of Mexico is a large body of ocean water, which is almost entirely surrounded by large land masses, as shown by Figure 4.113.

■ **Figure 4.113** The Gulf of Mexico

As a result of this, the water of the Gulf of Mexico is partially trapped there, and only small amounts of water are exchanged with the wider ocean. Some of the water in the deeper parts of the Gulf is thought to have been trapped there for well over 200 years. This long **residence time** means that any pollutants that enter the Gulf will be trapped there for a long time. The effects of pollutants will also last a very long time.

The Mississippi River is the main freshwater input into the Gulf of Mexico. The Mississippi drains a very large area of the continental United States, including densely populated urban areas, large areas of intensively farmed land and centres of industrial activity. Domestic sewage, agricultural runoff and industrial effluent all flow into the river and drain eventually into the Gulf of Mexico.

◆ **Residence time** – the average length of time that water remains in one location, such as a lake, aquifer or enclosed area of the sea. A high residence time means that pollutants in that area tend to accumulate more than they would if the water flowed freely away.

♦ **Dead zone** – an area of water where oxygen levels have been severely depleted and few animals can survive. These often occur in enclosed parts of the ocean where pollutants such as nitrates and phosphates accumulate.

Due to the long residence time in the Gulf of Mexico, nutrients from urban, industrial and especially agricultural areas accumulate and cause severely hypoxic conditions, where few animals can live. Each year a large area near the mouth of the Mississippi experiences such significant oxygen depletion that it is described as a **dead zone**. Dead zones have been reported in the Gulf since the mid-twentieth century, ranging in size from about 5,000 to over 20,000 square kilometres. They reach their maximum size each year in the late summer, when warm temperatures, river output and nutrient-rich runoff from farms combine to provide optimum conditions for eutrophication.

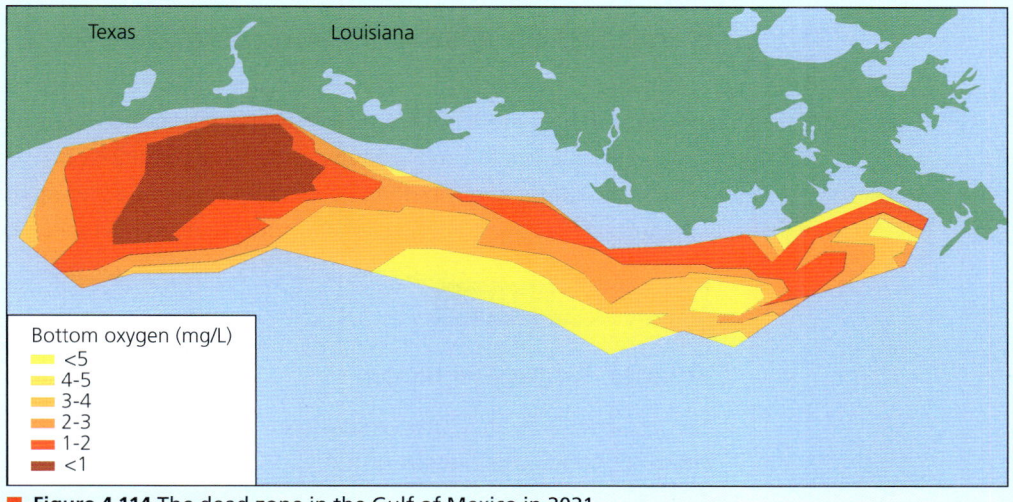

■ **Figure 4.114** The dead zone in the Gulf of Mexico in 2021

4.4.7 Eutrophication can substantially impact ecosystem services

Link

Ecosystem services are covered in Topic 7.1.5 (page 613).

Functioning ecosystems provide many benefits to living organisms, including humans. These benefits are sometimes described as ecosystem services. For aquatic ecosystems, the services that humans benefit from may include:

- abundant, clean drinking water
- fish for food
- aesthetically pleasant environments
- opportunities for recreation and leisure activities
- opportunities for cultural and spiritual activities
- control of floods, erosion and sequestration.

Eutrophication, algal blooms and oxygen depletion can affect an ecosystem's ability to provide these services.

■ Impact on fisheries

When eutrophication leads to oxygen depletion, most fish die or migrate away from the eutrophic water. If there are no fish to catch, the fisheries industry will suffer.

4.4 Water pollution

> **REAL-WORLD EXAMPLE**
>
> ### Gulf of Mexico fisheries
>
> As we saw in Topic 4.4.6, eutrophication has produced a dead zone in the Gulf of Mexico. With low-oxygen (hypoxic) conditions in this area, most fish will die or swim elsewhere. Fishing vessels therefore have to travel further from port to catch enough fish to make a profit. Travelling further means using more fuel and takes more time, so costs are also increased.

Impacts on aesthetics and recreation

Algal blooms can be unsightly, spoiling the appearance of an area. Water surfaces can become covered by a thick green or brown layer. People may be less likely to use eutrophic rivers and lakes for activities like boating and swimming. Hypoxic conditions may impact on recreational fishing. On beaches, excessive seaweed might wash up on shore, covering the sand. The rotting seaweed looks and smells unpleasant, which may affect economic activities at the beach, such as tourism.

> **REAL-WORLD EXAMPLE**
>
> ### Atlantic *Sargassum* bloom
>
> *Sargassum* is a floating brown seaweed often found in coastal areas of the tropical Atlantic Ocean. Beginning around 2011, the quantity of *Sargassum* in the Atlantic increased significantly, producing large rafts of seaweed every year. Although the cause of this bloom is not entirely clear, some studies have shown a correlation with nutrient levels in the ocean, which may be due to activities such as farming and deforestation on land.
>
> ■ **Figure 4.115** *Sargassum* blooms in the Atlantic Ocean in March 2023
>
> From time to time, ocean currents have washed large quantities of *Sargassum* on to the Atlantic beaches of the Caribbean islands and Mexico. The unsightly, spiky and smelly seaweed has affected the use of these beaches for recreation and tourism, and impacted the economies of some of these societies, which depend heavily on tourism for income.
>
> ■ **Figure 4.116** *Sargassum* washed up on a beach in Mexico

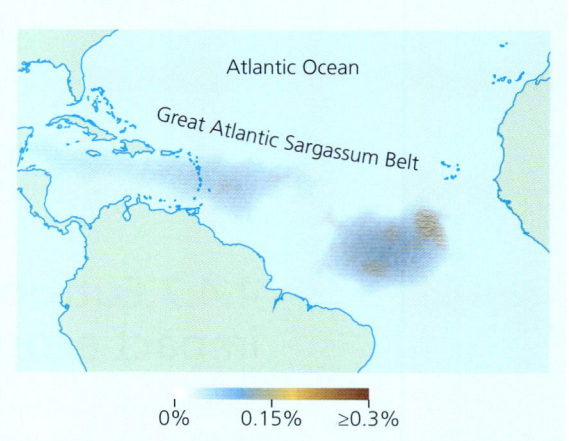

Impacts on human health

There are many different types of algae. Some algae produce chemicals that are harmful to humans, as well as to fish and other animals. If one of these species of algae experiences rapid growth, these harmful substances may be released into the water, making it unsafe to drink. In some cases, the harmful algae may be eaten by other organisms, such as mussels or oysters, which then become toxic. If humans eat these shellfish, they may become seriously ill.

If these algal blooms occur in freshwater, humans drinking the water may become sick. Some of the toxins produced by freshwater algae are difficult to remove during water treatment.

4.4.8 Eutrophication can be addressed at three different levels of management

Eutrophication is a global problem, which produces negative effects in a variety of societies. The following general mitigation strategies can be used to address this problem:

- Changes in human behaviour, so that fewer pollutants are produced or used.
- Actions designed to prevent the release of the pollutants into the environment.
- Actions designed to remove the pollutants from the environment, and repair the damage done.

Reducing the use of pollutants

If farmers change the way they use fertilizers, less nutrients will be washed into surface waters. There are a number of options available, including:

- Applying smaller amounts of fertilizer, so that most is used by the plants and less is washed away.
- Using slow-release forms of fertilizer, so that only a little is released at a time, allowing plants to take up as much as possible.
- Using organic fertilizers, such as compost and manure, which release nutrients slowly and are less likely to be washed away by the rain.
- Using methods like crop rotation and intercropping with nitrogen-fixing plants, so that less fertilizer is needed.

Other sources of nutrients can also be addressed this way. For instance, we can stop using detergents containing phosphates. Many detergents are already marketed as being 'phosphate free'.

Reducing the release of pollutants

Farmers can also reduce the loss of fertilizer from their farms by considering how water flows off the farmland. They can build drains to direct runoff to a holding pond, wetland or treatment facility. Wetlands are particularly useful, as they can provide a habitat for wildlife, reduce the risk of floods and help to treat runoff. The plants growing in the wetland use up some of the nutrients, so that less is released into rivers and streams.

Farmers can also reduce nutrient loss by avoiding growing crops or keeping livestock near to rivers and streams. They can establish a buffer zone between their fields and the natural waterways. They can also protect the banks of rivers by planting suitable vegetation along the edges of rivers. Soil erosion from the banks of rivers may add nutrients to the river if fertilizers have been used on that soil.

Link

Different approaches to farming, and to using fertilizers, are covered in Chapter 5.2 (page 456).

Nutrient pollution can also come from other sources, such as domestic and industrial wastes. In these cases, it may be relatively easy to direct all wastewater into a treatment facility, where the nutrients can be removed. Many wastewater-treatment facilities use bacteria to remove organic matter and nutrients from sewage.

■ Removing the pollutants and repairing the damage

Eutrophic waterways can sometimes be restored to their natural state. Restoration involves interrupting the cyclical process of algal growth, death and decomposition. Table 4.9 shows some options for restoring eutrophic waterways.

■ Table 4.9 Options for restoring eutrophic waterways

Method	Description
Dredging	The muddy sediment at the bottom of the waterway is dredged out, removing much of the organic matter that would otherwise decompose, deplete the oxygen and release more nutrients.
Aeration	Air or oxygen is added to the water to help speed up the process of decomposition without causing oxygen depletion. A common method of aeration is to use a paddle wheel that stirs up the surface, allowing oxygen to dissolve in the water.
Bioremediation	Living organisms such as aquatic plants or shellfish can be grown in the water. These absorb the excess nutrients. They can then be harvested to remove the nutrients from the waterway.
Chemical treatment	Non-toxic chemicals such as aluminium sulphate (also known as alum) can be added to the water, to encourage the nutrients and organic matter to clump together (coagulate) and settle at the bottom. Coagulation makes the nutrients less easily accessible to living organisms.

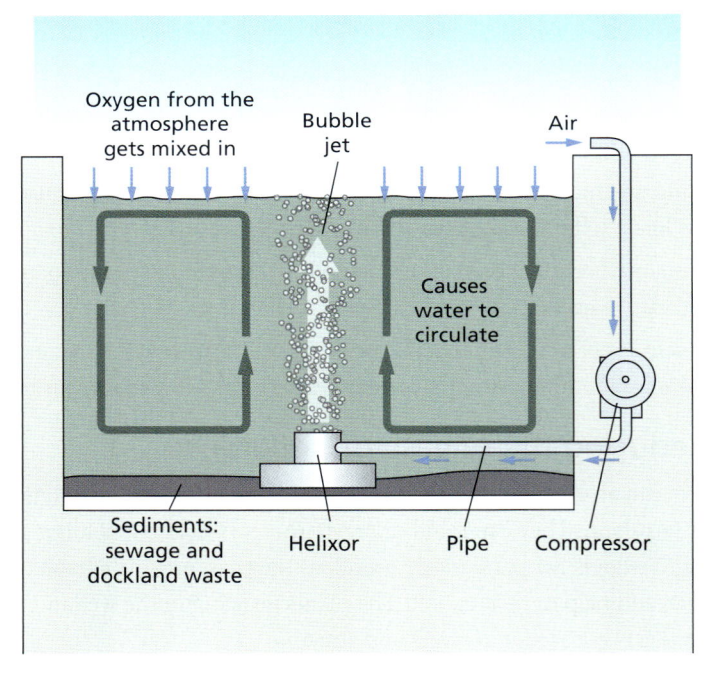

■ Figure 4.117 An aeration system used to address eutrophication at Salford Docks in Manchester, UK

The removal of pollutants from an ecosystem is called remediation. If we use living organisms in the process, it is known as bioremediation. Once remediation has been achieved, we may then try to restore the ecosystem by reintroducing some of the native plants and animals that may have died out when eutrophication began. Bringing a degraded ecosystem back to its natural state is known as ecosystem restoration.

> ### Concepts
>
> #### Systems and models
>
> Aquatic ecosystems, like all ecosystems, have a degree of stability or resilience. They can, to some extent, remain in the state they are in despite being disturbed by external factors, such as pollution. However, if the external factors are strong enough, they can cause the ecosystem to cross a tipping point and switch to a new stable state.
>
> When unpolluted water has nutrients added to it, initially the nutrients are used up by the algae in the water, and the ecosystem remains in the same, healthy state. However, when enough nutrients are added, this can take the system over the tipping point, and switch the system to a new, eutrophic, oxygen-depleted state.
>
> The eutrophic state is also stable, however. It can take a great deal of effort at remediation before the system crosses the tipping point and returns to the original healthy state.
>
> | Healthy state | | Polluted, eutrophic state |
> |---|---|---|
> | Clear water; little algae; high oxygen content; abundant, diverse animals | Very hard to change → Lots of pollution → | Murky, green water; algal bloom; low oxygen content; few animals |
> | **Help clean the water and provide habitats** | ← Remediation efforts | **Produce toxins and cause the water to stratify** |
>
> ■ **Figure 4.118** Aquatic ecosystems can switch between two stable states

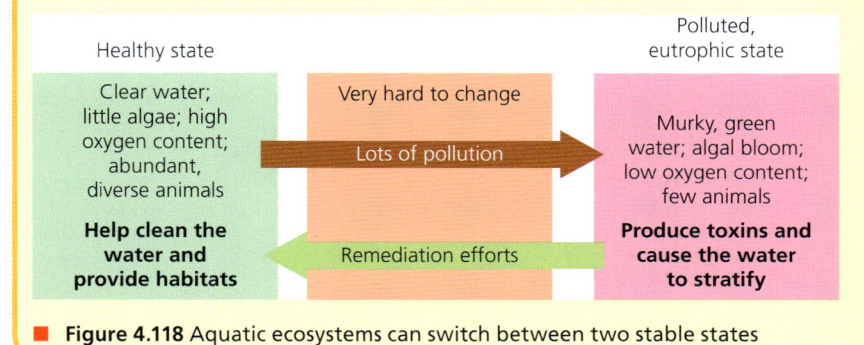

● (HL) 4.4.9 There is a wide range of pollutants that can be found in water

Pollutants such as nutrients, organic matter and oil have already been discussed. Some further, more specific, examples are provided in Table 4.10.

■ **Table 4.10** A selection of important water pollutants

Pollutant	Sources	Effects
Nutrients: • Nitrates • Phosphates	Farm runoff Domestic sewage Decomposition of organic matter	Eutrophication: • Algal bloom, oxygen depletion, growth of microorganisms • Ecosystem collapse • Toxins produced by algae affect human and animal health.
Biodegradable organic matter	Farm runoff Domestic sewage Industrial waste (e.g. food industry)	Oxygen depletion Growth of microorganisms Release of nutrients – eutrophication Ecosystem collapse
Pathogens (disease-causing organisms): • *Vibrio* spp. • *Giardia* spp. • Norovirus	Untreated domestic sewage Open defecation	Various diseases, mostly gastrointestinal, with symptoms such as diarrhoea and vomiting. Many can be fatal: • Cholera • Giardiasis.
Synthetic organic chemicals: • Polychlorinated biphenyls (PCBs) • Dichlorodiphenyl-trichloroethane (DDT) • Dioxin	Various industrial processes Some are found in common household products, so may be present in domestic sewage.	May be toxic or carcinogenic. Often undergo bioaccumulation and biomagnification, and become toxic at the concentrations found in higher trophic levels. Often cannot be broken down by living organisms, so they remain in the environment for a long time: these are described as persistent organic pollutants.
Heavy metals and their compounds: • Arsenic • Lead • Methylmercury • Tributyltin	Various industrial processes Some found in products with domestic uses. Also often found in natural sources, such as specific rock types.	May be toxic or carcinogenic. Some cause specific illnesses. Often undergo bioaccumulation and biomagnification, and become toxic at the concentrations found in higher trophic levels.
Plastics	Manufactured from petroleum. Commonly used in a wide range of industrial and domestic activities.	Non-biodegradable, so they persist in the environment for a long time. May cause direct harm to wildlife, e.g. if swallowed. Eventually break up into microplastics, which may collect other, more toxic substances on their surfaces. Microplastics may be eaten by animals.
Heat energy	Cooling water from factories.	Direct harm to living organisms Oxygen depletion Algal bloom (thermal eutrophication)

◆ **Heavy metal** – a metal element with relatively high density. They are mostly found in the lower central part of the Periodic Table.

◆ **Toxic** – causing harmful effects on health. How significant the effects are may depend on the chemical nature of a substance, the chemical form it is in, its concentration and the duration of exposure.

◆ **Biofouling** – the growth of living organisms such as algae, oysters and barnacles on surfaces underwater, such as the hulls or propellers of ships.

◆ **Antifouling** – a substance used to prevent biofouling, such as certain types of paint.

Tributyltin

Tin is a metallic element, symbol Sn, which is classified as a **heavy metal** because of its high density. Most heavy metals are **toxic** in high enough concentrations.

Tributyltin, or TBT, is a group of organic compounds containing tin and three carbon-containing butyl groups. In the past, TBT was widely used as a component of paints used on the outside of ships and boats to prevent the growth of marine organisms such as algae and barnacles.

The growth of organisms on man-made surfaces is called **biofouling**. Biofouling can increase the drag on ships and boats, slowing them down and increasing their fuel consumption. To prevent this, **antifouling** agents such as TBT can be painted on the hull of a vessel. TBT is only slightly soluble in water, so it mostly remains in the paint, where it prevents organisms from growing. When it does dissolve in the water, it mostly settles at the bottom, becoming attached to sediments, or bioaccumulates in living organisms. It can also undergo biomagnification up the food chain.

TBT is known to affect the function of animals' endocrine (hormonal) systems. This can result in changes in the growth, development and reproduction of the animals. It may also affect the immune system, making animals more vulnerable to infections. As a result of its toxicity, the use of TBT in antifouling paints has been banned in several countries and by international agreements.

Polychlorinated biphenyls

Polychlorinated biphenyls (PCBs) are a group of synthetic liquids with a range of uses in various industries. They are known carcinogens and their manufacture has been banned in many countries.

When PCBs are used, there may be some leakage into the environment, causing pollution. Pollution also occurs when they are disposed of inappropriately. PCBs are very stable, unreactive substances, which means that they persist in the environment for a long time. They are also not very soluble in water, but bioaccumulate easily in living organisms. Once in living organisms, they can undergo biomagnification up the food chain.

Although the manufacture of PCBs has mostly stopped, products and machinery containing PCBs are still in use. Since the substances are so persistent, there are many environments that are still contaminated with PCBs.

Since they are so stable and persistent, special methods must be used to dispose of them. For instance, treatment with certain chemicals can convert them to harmless substances. Incineration at very high temperatures also destroys them, but if the temperature is too low, even more harmful substances can be produced. Bioremediation, using microorganisms, may also be possible.

4.4.10 Algal blooms may produce toxins that threaten the health of humans and other animals

Algal blooms due to eutrophication typically cause oxygen depletion, which can lead to the death of aquatic animals. Some species of algae also produce toxic substances that are released into the environment, and which can cause illness or death in animals, including humans. These are called harmful algal blooms.

Cyanobacteria

Most harmful algal blooms in freshwater ecosystems are caused by cyanobacteria, such as the species *Microcystis aeruginosa*. Cyanobacteria are photosynthetic bacteria, containing a blue–green pigment. They are sometimes called blue–green algae, but they are not really algae. In eutrophic water, they grow and reproduce rapidly, and may produce a bright green or blue–green scum on the surface.

There are several species of cyanobacteria. Some of them live in colonies of several individual cells stuck together in a gel-like substance. Many float close to the surface of the water, blocking out the light. Because of this, they can dominate the water, preventing other algae from growing below.

Cyanobacteria produce a variety of toxins, which may affect different organ systems such as the liver, nerves or skin, and may cause gastrointestinal illness. Algal blooms due to cyanobacteria can also be a problem in fish farming. If the water becomes eutrophic, a cyanobacteria bloom can occur, which can cause illness and death in the farmed fish.

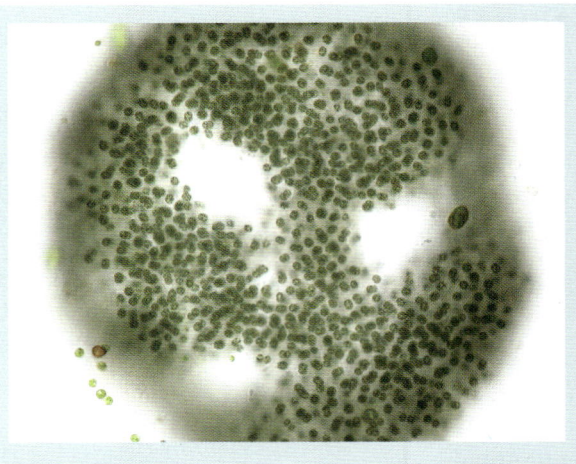

■ **Figure 4.119** *Microcystis*, an example of cyanobacteria

Dinoflagellates, diatoms and red tides

In marine environments and estuaries, harmful algal blooms are most often caused by single-celled algae such as dinoflagellates or diatoms. Diatoms are extremely common marine algae, which produce a glass-like shell made of silica around their cells, sometimes forming amazingly complex structures. Dinoflagellates are also common. They possess a variable number of tail-like flagella, which allow them to move in the water. Both dinoflagellates and diatoms are types of phytoplankton.

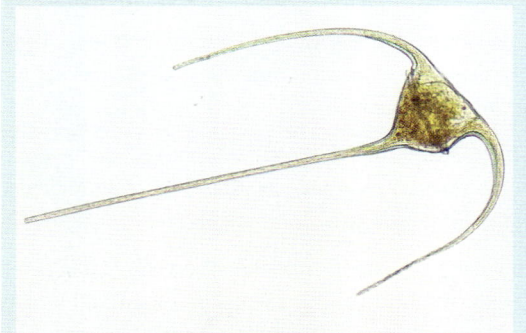

■ **Figure 4.120** A dinoflagellate and a diatom

Some dinoflagellate and diatom blooms give the water a reddish or brown colour. People in coastal and fishing communities sometimes call this effect a 'red tide' – although it is not actually associated with tidal movements. They know that red tides are associated with poor fish catches. Some species produce bioluminescence – a ghostly glow in the dark. Bioluminescence can be very attractive, and many people enjoy going to see it.

■ **Figure 4.121** A red tide is an algal bloom that gives the water a reddish appearance

If humans eat animals that have been exposed to harmful dinoflagellate or diatom blooms, they may become ill. During a harmful algal bloom, some filter-feeding shellfish such as mussels or oysters may ingest large numbers of the algae. The toxins remain in the tissues of the shellfish, due to bioaccumulation. If humans eat these shellfish, they may develop one of various diseases, such as paralytic (PSP), diarrhetic (DSP), neurotoxic (NSP) or amnesiac shellfish poisoning (ASP).

Another disease, known as ciguatera poisoning, occurs due to biomagnification. The dinoflagellate *Gambierdiscus toxicus* produces a group of toxic chemicals called ciguatoxins. These toxins bioaccumulate in animals that feed on the algae. When those animals are eaten, biomagnification occurs, until the concentration in the largest carnivorous fish is high enough to affect people who eat those fish. Ciguatera poisoning is one of the most common foodborne diseases, especially in tropical areas where *Gambierdiscus toxicus* lives. Large, predatory fish such as grouper, snapper, mackerel and barracuda are possible sources of poisoning.

4.4.11 The frequency of anoxic/hypoxic waters is likely to increase due to the combined effects of global warming, freshwater stratification, sewage disposal and eutrophication

We have seen how these three different types of pollution can lead to oxygen depletion in water:
- Eutrophication: This is most often a consequence of the use of fertilizers on farms, but domestic sewage and urban runoff can also contribute to it.
- Biodegradable organic matter: This pollution typically comes from domestic sewage, farm waste, and certain types of industrial waste, such as from the food industry.
- Thermal pollution: This is usually caused by industrial activity, where cooling water is used to control the temperature of factories.

The concentration of oxygen dissolved in water can also be affected by natural processes. Some processes add oxygen to water, while others remove it.

Photosynthesis produces oxygen, but only occurs near the surface of water, where there is plenty of light. More oxygen is added by water movements. When the surface of water is disturbed, such as by the wind causing ripples and waves, or when water splashes over rocks or down a waterfall, a greater surface area of water is exposed to the air, and oxygen diffuses more rapidly into the water, and dissolves.

The main process that removes oxygen from water is decomposition. Bacteria use oxygen to break down organic matter into simpler substances. Most decomposition in water occurs in sediments at the bottom of rivers, lakes and the ocean. Dead organisms, or parts of organisms, sink to the bottom of the water and settle there, decomposing.

Freshwater stratification

We have seen that deep bodies of water, such as lakes and the ocean, tend to be stratified. Differences in temperature and density produce distinct layers with different characteristics, and these layers do not usually mix. One common difference relates to the concentration of dissolved oxygen. The surface layer typically has a high oxygen content, due to photosynthesis and diffusion of oxygen from the air. Lower layers, especially the bottom layer, tend to have a low oxygen content, since this is where decomposition mostly happens.

Oxygen depletion due to eutrophication or organic matter pollution is most likely to happen at the bottom of deep water. Replenishing the oxygen levels in the deepest layers involves water movements, such as the global thermohaline circulation, or ocean conveyor belt. This system of interconnected ocean currents transports heat, nutrients and oxygen all around the globe. In places where surface water is forced down into the deep, oxygen can be transported with it. This, however, is a very slow process.

Link
Decomposition is covered in Chapter 2.3 (page 139).

Link
Stratification of water bodies is covered in Topic 4.1.11 (page 285).

Link
The global thermohaline circulation, or great ocean conveyor belt, is covered in Topic 4.1.14 (page 287).

Climate change

As global warming, economic development and population growth continue, local incidents of oxygen depletion are likely to occur more often.

With growing population and economic development, use of fertilizers on farms is likely to increase, meaning there will be more frequent incidents of eutrophication. The growing population and increased urbanization will also probably increase domestic sewage production, as well as organic matter pollution from farms and the food industry. These will contribute directly to oxygen depletion.

At the same time, with higher global surface temperatures, there is likely to be increased stratification of oceans and lakes. With less mixing between layers, there will be less transport of oxygen into the deep water.

Furthermore, some climate change models suggest that global thermohaline circulation may change. It is possible that the downward flow of surface waters in places like the North Atlantic Ocean will slow down and possibly even stop. If this happens, there will be less transport of oxygen into the deeper parts of the ocean. It is possible that larger areas of the deep ocean will experience significant hypoxia or even anoxia for longer periods of time, creating long-term or even permanent dead zones.

4.4.12 Sewage is treated to allow safe release of effluent, by primary, secondary and tertiary water-treatment stages

Sewage may contain several different pollutants, including pathogens, organic matter, nutrients and sediments. If wastewater is allowed to flow directly into surface waters or the sea, all of these pollutants will be released into the environment. Where possible, sewage should always be treated to remove any harmful pollutants, before the cleaned wastewater is released.

Sewage treatment facilities are built to remove pollutants such as pathogens, nutrients and organic matter from wastewater. Water is usually brought into a sewage treatment plant via underground pipes or drains called sewers. The water in sewers comes from homes and other buildings in the area, and may also include rainwater which washes off the streets into drains – although this 'stormwater' is often kept apart and released without treatment. Effluent from a sewage plant should be clean enough so that it causes no harm to humans, wildlife and ecosystems.

Although individual facilities may vary, there are certain common steps to the sewage treatment process. The process usually begins with screening out large pieces of litter and debris, before moving on to more technical stages. The rest of the process is often divided into three stages or levels: primary, secondary and tertiary. However, not all plants carry out all three stages.

Primary treatment

Primary treatment involves using physical processes to separate solid matter from water. Several steps may be involved, as shown in Table 4.11.

■ Table 4.11 Possible stages in primary treatment of wastewater

Stage name	Description
Screening	A screen, often made up of metal bars, is used to trap larger solid objects, such as litter and branches, as the water enters the treatment plant.
Grit removal	The water is held in a tank called a grit chamber for a short period, to allow larger sediments such as sand and gravel to settle.
Sedimentation	The water is held in a tank called a primary clarifier for several hours, allowing smaller particles of suspended sediment to settle out. Coagulation and flocculation may be part of this process.
Coagulation	Coagulants, such as aluminium sulphate or alum, are added to the water. These encourage the smaller particles to clump together, or coagulate. The coagulated particles, called flocs, settle out of the water faster than individual particles would.
Flocculation	The water is slowly stirred, encouraging the small particles and coagulants to combine. This speeds up the process of coagulation.

Most of the suspended sediments are removed during sedimentation in the primary clarifier. They settle at the bottom of the primary clarifier, forming a soft sludge, which can be scraped off by the machinery at the treatment facility.

While the water is held in the primary clarifier, any buoyant substances, like grease and oil, will float to the top, forming a scum on the surface of the water. This can also be scraped off.

Primary treatment removes most of the sediments, and some of the organic matter and nutrients, from the water, using mostly physical processes. This may be sufficient treatment, and the water may be released into the environment after primary treatment. In However, in some facilities, the process may continue with secondary treatment.

Secondary treatment

Secondary treatment usually uses bacteria to remove the remaining organic matter and nutrients, as well as some of the pathogens in the water. This can be done in a few different ways, as shown in Table 4.12. The bacteria used are usually aerobic, so secondary treatment often involves exposing the water to air or oxygen.

■ Table 4.12 Processes commonly used in secondary treatment

Treatment process	Description
Activated sludge process	Wastewater is held in an aeration tank and oxygen and 'activated sludge' are added. Activated sludge contains aerobic bacteria that decompose any organic matter that is suspended in the water. After a short time, the water is transferred to a secondary clarifier, where it is held for some time, and the bacteria and any solid matter settles to the bottom, forming sludge. A sample of the sludge is reused as the activated sludge in the aeration tank. The rest of the sludge is transferred to a solid-waste management facility for treatment and disposal.
Trickling filter beds	Wastewater is slowly sprinkled over rocks, gravel or similar material with a high surface area, allowing the water to be exposed to the air. A thin layer of bacteria called a biofilm grows on the material, using and breaking down the organic matter in the water. The water then flows into a clarifier tank, where the sludge settles and can be removed.
Sewage lagoons	Wastewater is held in a series of natural or semi-natural ponds or lagoons, where naturally occurring bacteria break down the organic matter, and support a natural ecosystem with plants and animals. Lagoons are usually shallow enough that oxygen is able to dissolve from the air and mix throughout the water.

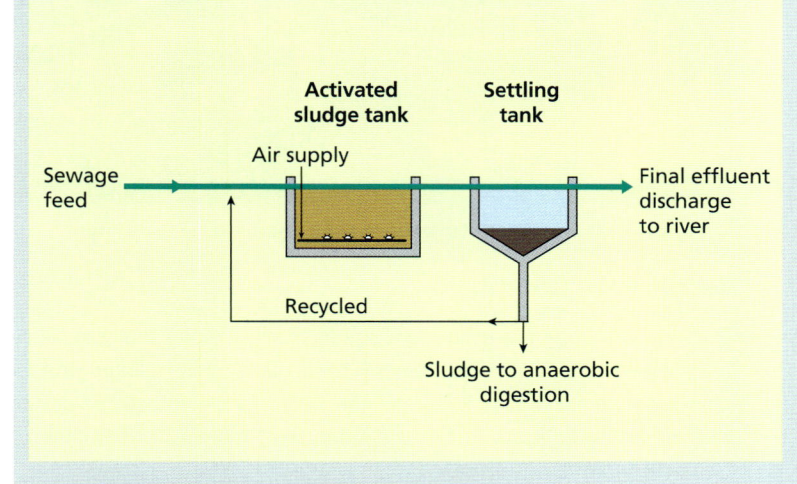

■ Figure 4.122 The activated sludge process

■ Figure 4.123 An aeration tank in a sewage-treatment facility

■ Figure 4.124 Trickling filter beds in a sewage-treatment facility

Tertiary treatment

Secondary treatment is usually enough to produce water that can safely be released into the environment. However, if the water is particularly badly polluted, the receiving environment is particularly sensitive to pollution, or the water is likely to be used as human drinking water, tertiary treatment may be used. Tertiary treatment uses more advanced chemical and physical processes to clean the water even further.

Some of the methods used in tertiary treatment are presented in Table 4.13.

Link

The treatment of water for human use is covered in Chapter 4.2 (page 291).

■ **Table 4.13** Methods used in tertiary treatment

Treatment process	Description
Filtration	Wastewater is passed through one or more tanks filled with a fine-grained material that filters out the smallest particles. The material may include activated carbon (also called activated charcoal or porous carbon). This has millions of tiny pores, which create an extremely large surface area. Even dissolved solids can become trapped by the activated carbon.
Chlorination	Chlorine is added to wastewater to kill any living organisms that might remain in the water. This is cheap and effective, but may create harmful chemicals that must then be removed.
Ultraviolet treatment	Ultraviolet light is shone through the water to kill microorganisms. This is only effective if the water has been thoroughly cleaned of any solid particles, otherwise these would block the light.
Ozonation	Ozone gas (O_3) is bubbled through the water, killing microorganisms. This is more expensive than chlorination or ultraviolet treatment, but can be more effective and it produces no harmful byproducts.

◆ **Disinfection** – a process intended to kill microorganisms and therefore prevent the spread of diseases.

Link

The ozone used in water treatment is the same gas that forms a protective layer in the atmosphere, as covered in Chapter 6.4 (page 594), and which is part of photochemical smog, as covered in Chapter 8.3 (page 740).

Chlorination, ultraviolet treatment and ozonation are different approaches to **disinfection**. This is an essential part of water treatment where the water is going to be used by humans, since it is the only sure way to remove microscopic pathogens.

Cost and availability of sewage treatment

Regardless of which approaches are used, sewage treatment always involves some costs. Sewers must be built to collect wastewater and transport it to the treatment facility. Sewers may be designed to allow water to flow downhill, by gravity, to the plant, but in some situations, pumping stations may be needed to raise the water. Treatment facilities themselves must be built and maintained. Operating a treatment facility requires energy and materials such as coagulants, activated carbon and chlorine. The more polluted the water is, and the more treatment is required, the greater the cost.

For this reason, sewage treatment is not readily available in all societies. Where the money is not available, or where populations are scattered or remote, it may be too expensive to provide sewage treatment to all people. As a result, the United Nations Environment Programme (UNEP) estimates that over 80 per cent of global wastewater is released into the environment without treatment.

Where sewage treatment is not available, natural environments can become more polluted, putting humans at greater risk of water-borne disease.

4.4.13 Some species are sensitive to pollutants, or are adapted to polluted waters, so these can be used as indicator species

The number of different pollutants found in water makes it difficult to test water quality effectively. Different tests are needed to investigate the presence of pathogens, nutrients, organic matter and each of the different toxic chemicals. The fact that water moves, carrying pollution with it, makes this even harder, since a pollutant that is present today may have moved to another location by tomorrow.

However, there is a quick and easy way to determine if a body of water is polluted. Instead of testing for individual pollutants, we can look at what organisms are present in the water. Species that give an indication of whether the environment is polluted are called indicator species.
- Some organisms are highly sensitive to pollution. They require very high quality water, and will die or move if they are exposed to even low concentrations of pollution. If we find these types of organisms in a body of water, we can infer that the water is relatively clean and unpolluted.
- Other organisms are highly tolerant of pollution and can live in very polluted waters. If we find only pollution-tolerant species in water, we can infer that the water is relatively polluted.

4.4 Water pollution

Stonefly nymphs are found all around the world, and are widely used as an indicator species of good water quality. Stoneflies are flying insects belonging to the order Plecoptera. Their young, called nymphs or naiads, are wingless and live in water. They are easily recognized by their two long 'tails' or cerci. The nymphs typically have many feathery gills along both sides of their bodies.

■ Figure 4.125 A stonefly nymph

Stoneflies require high concentrations of dissolved oxygen, so they are vulnerable to even slight levels of oxygen depletion. If they are present in a water body, this means that the water is continuously well oxygenated. It is unlikely that the water would have experienced any of the sorts of pollution that cause oxygen depletion, such as eutrophication, organic matter, or thermal pollution. If stoneflies are absent, it may be an indication that the water is polluted, but this cannot be said with certainty.

A commonly used indicator of poor water quality is dronefly larvae, or rat-tailed maggots. These can live in highly polluted, hypoxic or anoxic water. Their 'tail' is actually a breathing tube, which allows them to obtain oxygen from the air. They feed on organic matter and the microorganisms that live in organic matter. If dronefly larvae are present, it suggests that the water is rich in organic matter, and lacks oxygen.

■ Figure 4.126 A dronefly larva, or rat-tailed maggot

4.4.14 A biotic index can provide an indirect measure of water quality based on the tolerance to pollution, and the relative abundance and diversity of species in the community

The presence of a single indicator species can provide some evidence of water quality. However, many different factors can affect the presence or absence of a single species. Predators, food availability and weather conditions, for instance, may affect the rat-tailed maggot or the stonefly nymph. Stoneflies might be absent even though the water is good quality, perhaps because the habitat conditions are unsuitable.

Instead of looking at the presence or absence of individual species, it might be more useful to consider the variety of species in a water body. A variety of different biotic indices have been developed for this purpose. A biotic index is a number, usually between 1 and 10, which gives an indication of environmental conditions, based on the presence, absence and sometimes relative abundance of several different species.

Tool 1: Experimental techniques

The Trent biotic index

The Trent biotic index was one of the first biotic indices developed for freshwater. It was first developed for use in the Trent river basin in the United Kingdom. Ideally, for use in other locations, it should be modified to reflect local river communities. This index relies on identifying seven key groups of animals, in six categories:

- Stonefly nymphs (order Plecoptera)
- Mayfly nymphs (order Ephemeroptera)
- Caddisfly larvae (order Trichoptera)
- Freshwater shrimp (genus *Gammarus*)
- Water hoglouse (genus *Asellus*)
- Tubifex worms (genus *Tubifex*) and non-biting midge larvae (order Chironimidae).

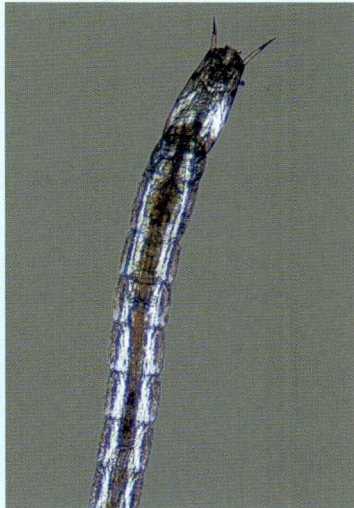

■ Figure 4.127 Two of the seven key animals of the Trent biotic index

To use the Trent index, a sample of animals is collected from the water body by lick sampling, as described below:
1. Hold a kick net (also called a D-net) in the water, pressed as close to the stream bed as possible and facing upstream.
2. While wearing waterproof boots, use the feet to agitate the stream bed in the area immediately upstream of the net, to disturb any animals. Be sure to disturb any stones or sediment on the stream bed. any dislodged animals should be carried by the stream flow into the net.

3   Transfer the animals into a collection tray filled with water from the stream.
4   Sort the animals by appearance. Transfer each type of animal into a separate container and identify them using an identification key.
5   Once the data is recorded, carefully return the animals to the area of the stream from which they were collected.

■ Figure 4.128 Kick sampling for aquatic macroinvertebrates

For the Trent biotic index, only the seven key groups of animals need to be identified, but it is also necessary to count the total number of different animal groups. For some of the groups, it is also necessary to count how many different types or species are present (for example, how many different types of caddisflies).

To work out the value of the Trent index, Table 4.14 is used.

■ Table 4.14 Determining the Trent biotic index (data from www.researchgate.net)

Key indicator group	Diversity of fauna	Total number of groups of animals present				
		0–1	2–5	6–10	11–15	16+
Stonefly nymphs	More than one species of stonefly	-	7	8	9	10
	Only one species of stonefly	-	6	7	8	9
Mayfly nymphs	More than one species of mayfly	-	6	7	8	9
	Only one species of mayfly	-	5	6	7	8
Caddisfly larvae	More than one species of caddisfly	-	5	6	7	8
	Only one species of caddisfly	4	4	5	6	7
Freshwater shrimp (*Gammarus*)	All above groups absent	3	4	5	6	7
Water hoglouse (*Asellus*)	All above groups absent	2	3	4	5	6
Tubifex worm and/or Chironimid larvae	All above groups absent	1	2	3	4	-
All above groups absent	Other organisms may be present	0	1	2	-	-

Theme 4: Water

To use the table, we need to answer the following questions:
- Which indicator species are present?
- How many different types of each indicator species are present?
- How many different types of animals are present?

■ Table 4.15 Three worked examples of the Trent biotic index showing the presence of species (number of species where relevant) in each sample

	Sample A	Sample B	Sample C
Stonefly	3	0	0
Mayfly		0	0
Caddisfly		1	0
Gammarus			Yes
Asellus			
Tubifex and/or Chironimid			
All species	12	7	6
Trent biotic index	9	5	5

Note that we only need to know the number of species from the first group we find on the list, starting from the top, and the total number of species found. The greyed-out cells in Table 4.15 could have any numbers in them, but they are irrelevant to the analysis.

As this Tool box shows, the Trent biotic index works on two principles:
1. That as pollution levels increase (or water quality decreases), each of the key indicator animals disappears, in the sequence stoneflies > mayflies > caddisflies > *Gammarus* > *Asellus* > Tubifex and Chironimid.
2. That higher water quality (or less pollution) will support a higher diversity of species.

Other biotic indices are in use in different parts of the world. Some are simpler, making them easier for volunteers or citizen scientists to use, whereas others require more work to identify a wider range of species, making them more suitable for professional scientists to use.

Biotic indices are a useful tool for assessing water quality because they are relatively quick and easy to use, and they give an overall impression of water quality over an extended period of time. Animals living in the water are exposed to all the changes in water quality that happen during their lives. If an animal is present, the water must have been clean enough for that species during its lifetime.

Inquiry process

Inquiry 1: Exploring and designing
1. Identify a shallow stream near to your school or home that you can safely and easily access.
2. Write out a detailed plan to use kick sampling at your selected stream.
3. Carry out a risk assessment on your plan and modify it to minimize any risks.

Inquiry 2: Collecting and processing data
1. Collect a series of kick samples at your selected stream.
2. Examine your samples carefully, and identify the animals present, using an identification key for aquatic macroinvertebrates.

> **TOK**
>
> Both conversion curves and tables and weighting factors are developed based on the judgement of experts. They reflect their interpretation of what makes good quality water, and how important each of the different parameters is. As a result, different water-quality indices may produce different results for the same water. This makes comparison between different WQIs impossible, but it allows for different WQIs to be developed for different situations and locations.
>
> Do you think the judgement of experts is an appropriate way to assess environmental quality? Is it appropriate that WQIs are context-specific?

> 3 For each group of animals, distinguish between animals that appear to be different species. Distinguishing species can be very difficult and common identification keys may not help you, so list animals as belonging to different species if there are significant observable differences between them.
> 4 Tabulate your data.
> 5 Use Table 4.14 (page 406) to determine the Trent biotic index for each of your samples.
>
> **Inquiry 3: Concluding and evaluating**
> 1 Draw a conclusion about the quality of the water at your site, based on your calculation of the Trent biotic index.
> 2 Write an evaluation on the usefulness of the Trent biotic index method, considering any challenges you had in carrying out the sampling procedure and analysis.

4.4.15 Overall water quality can be assessed by calculating a water-quality index

There are many different components of water quality. A biotic index is one way to gain an overall impression of water quality, by considering what species are present and absent in the water. Another approach is to use a water-quality index (WQI), which combines the values of several different water-quality tests, weighing each test based on its importance and coming up with an overall numerical value.

For example, the Vernier water-quality index tests nine different elements of water quality:
- temperature
- pH
- turbidity
- total solids
- dissolved oxygen
- biochemical oxygen demand
- phosphates
- nitrates
- faecal coliform.

Testing each of these parameters was covered in Topic 4.4.3 (page 383). Each parameter is assigned a weighting factor that reflects its importance for overall water quality. Each weighting factor is a number between 0 and 1, where the sum of all the weighting factors equals 1.

Once each of the nine measurements is taken, a conversion graph or table is used to convert each measurement to a Q-value between 0 and 100. Each Q-value is then multiplied by the weighting factor for that parameter. The results are added together, to produce an overall WQI value between 0 and 100, where 100 is the highest quality water.

4.4.16 Drinking water-quality guidelines have been set by the WHO and local governments can set statutory standards

WHO guidelines and state legal requirements

Access to safe drinking water is considered essential for a healthy life, and is an important component of sustainable development. With this in mind, the World Health Organization (WHO) publishes guidelines on water quality, with recommendations for countries to follow.

The guidelines include recommended maximum values for many water-quality parameters. They also include guidance on how to achieve high quality drinking water, considering all the stages of production, including sourcing, treating, distributing, testing and disposing of water.

The WHO guidelines have no legal power in individual countries, but most governments have produced their own standards and regulations for drinking water. The agencies responsible for providing drinking water must follow these guidelines, and can be legally liable if these standards are not met or the regulations are not followed.

Statutory regulations on drinking water usually include rules about where water can be obtained from, how it should be treated and distributed, and when, how often and how it should be tested. Statutory standards identify legal thresholds that should not be crossed. These might include the maximum concentration of particular solutes, contaminants or microorganisms.

If a developer proposes to build a new water-treatment plant to produce drinking water for the public, it must follow the statutory standards and regulations. These standards often require water to be tested regularly for particular parameters, such as concentration of various contaminants.

Regulation of pipe-borne drinking water

In more economically developed countries, most people receive drinking water through pipes. Consumers typically pay a relatively small fee to receive this tap water. The companies producing and supplying the water may be state agencies or private companies.

When a private company or state agency proposes to construct a new water-treatment plant and to supply water through pipes to people's homes, it must follow the state's drinking water regulations and prove that it can meet the drinking water standards. In most situations it will be required to carry out an Environmental Impact Assessment (EIA) to show that its proposal will have generally positive impacts on the environment. Part of the EIA would normally relate to the quality of water provided to consumers. Once the plant is built and operating, it will normally be required to carry out regular testing, including at the point of end use – the taps in people's homes – to ensure that it is conforming to the legal standards.

Regulation of bottled water

The production, sale and use of bottled drinking water is highly controversial. Bottled water can be produced in the same way as conventional pipe-borne water, but is distributed differently. Bottled water can be shipped very long distances, and is often sold at a much higher price than tap water. Bottled water is most commonly packaged in plastic bottles or larger containers, although glass bottles are occasionally used.

Bottled water is promoted as a safe alternative to tap water, especially where people do not receive a regular piped supply, or where people doubt the safety of tap water. Bottled water is useful in water-scarce regions and emergency situations, such as times of warfare or natural disasters. In wealthier and more stable regions, bottled water is more likely to be a luxury or convenience item.

Link

Environmental Impact Assessments (EIA) are covered in Chapter 7.1 (page 608).

> **Concept**
>
> ### Perspectives
>
> Attitudes towards piped and bottled water vary widely. Bottled water may sometimes be seen as a status symbol, especially when it is branded and marketed. However, in many parts of the world, people do not trust the quality of the pipe-borne water they receive. There may be stories about people getting sick from drinking tap water, or about water companies failing to follow regulations or to meet legal standards. In cases like these, people may prefer to drink bottled water.

One major environmental criticism of bottled water concerns the use and disposal of plastic packaging. Plastic is a non-renewable, non-biodegradable product, which is not easily recycled and causes significant water pollution. If safe water can be provided through pipes, these problems are eliminated.

The distance bottled water travels is also a concern, especially where bottled water is treated as a luxury. Some brands of bottled water are marketed on the basis that the water has come from an exotic, apparently pristine location, possibly thousands of kilometres away. Critics are concerned about the amount of energy used and carbon dioxide emitted in transporting the water. Water produced more locally should have a much lower carbon footprint.

Other criticisms relate to the economic, social and political questions raised by bottled water. Almost all bottled water is produced by private companies, rather than state agencies. The cost of bottled water to consumers can be many times higher than the cost of tap water. Critics argue that drinking water is a human right, and should be provided with no concern for profit. State agencies are usually best positioned to do this.

Another concern relates to the regulation of water quality in the bottled water industry. The private, competitive and international nature of the bottled water market makes it difficult to regulate production or to monitor standards. Many scientific studies have shown that the quality of bottled water is highly variable. Differences in various water-quality parameters have been found between bottled waters produced by different companies, marketed under different brands, produced from different locations and produced at different times. Some studies have even found differences between individual bottles from the same batch. Despite these findings, though, evidence from actual disease outbreaks suggests that water-borne diseases are far more likely to be caused by contaminated or poorly treated tap water rather than bottled water.

TOK

These studies raise questions about people's perceptions and beliefs, or 'common knowledge', compared to scientific evidence.

The nature of the scientific evidence is also worth considering. Evidence from testing water suggests that bottled water is less well regulated, but epidemiological evidence of actual outbreaks suggests that tap water may still be less safe. Which evidence is stronger? Do we need to consider other factors as well?

The regulation of the bottled water industry can be further complicated when powerful international companies are operating in less economically developed countries. These countries may have weaker regulations, or less capacity to enforce standards and regulations, than the home countries of the companies. The companies concerned may also have influence over political decisions, as they are able to offer a source of income for the host country.

HL.a.6: Environmental law

Investigate the laws and regulations that apply in your country regarding both tap and bottled water.
- What rules and standards are applied to each?
- Are there different local and national rules, perhaps established by the national or federal government, as opposed to by local councils?

4.4.17 Action by individuals or groups of citizens can help to reduce water pollution

Regional and national governments, as well as privately owned businesses and international agencies, have a great deal of responsibility for reducing and preventing water pollution. However, citizens, acting either individually or in organized groups, can also be effective in helping to solve the problem.

Individual choices

Individual people can make several choices to reduce their personal contribution to water pollution. These can include:
- Choosing products that have less impact on water quality, such as phosphate-free detergents.
- Reducing the use of fertilizers and pesticides in gardens.
- Making informed choices about waste disposal, such as composting food waste instead of washing it down the drain.
- Avoiding using sink-installed garbage disposal units.
- Ensuring that household wastes, especially hazardous wastes, are disposed of correctly.
- Picking up litter from streets and natural areas, and disposing of it correctly.
- Installing small-scale wastewater treatment facilities at home.

Citizen science and clean-up campaigns

Individuals and groups can engage in voluntary citizen science projects to monitor water quality in rivers, streams, lakes and the sea. The data collected can be shared with state environmental agencies, non-governmental organizations or research institutions working on water-quality issues.

Various clean-up campaigns, such as the annual International Coastal Cleanup®, and similar initiatives in cities and along rivers and streams, can have a significant impact on the flow of litter, especially plastics, into waterways. These are also an excellent source of data for research purposes. The data can help to inform policy decisions about waste collections, siting of bins, and product and packaging use and disposal.

Lobbying and legal action

In democratic societies it may be possible for citizen groups to communicate with lawmakers, such as members of parliament, to argue for action on particular issues such as water pollution. Lobby groups can also go to the press and seek publicity through social media, in order to gain support for their cause. Lawmakers in a democracy are usually voted in to office, and can be voted out again at the next election. If there is enough public interest in water pollution, this may influence lawmakers to take decisions that promote improvements in water quality, such as new standards or regulations, or better enforcement of existing ones.

REAL-WORLD EXAMPLE

Clean Water Action
In the United States, an advocacy group called Clean Water Action successfully campaigned for the passing of the 1972 Clean Water Act. Fifty years later, the group continues to operate, campaigning around the country, on issues such as water access, safety of drinking water and pollution from energy projects. It is active on social media and claims a membership of over 1 million people.

Individuals and lobby groups can also take legal action, hiring lawyers to argue a case against a company or a state that they believe is causing water pollution.

> **REAL-WORLD EXAMPLE**
>
> ### Carolyn Roberts vs UK Water Companies
>
> In August of 2023, a British university professor and environmental activist named Carolyn Roberts began legal action against six of the biggest water companies in the United Kingdom. Her lawsuit alleged that releases of untreated sewage by these companies into rivers and beaches had not been accurately reported to the Environment Agency. The case is currently still ongoing, but if successful the companies may be required to pay millions of pounds in compensation to over 20 million water customers.

Protest action

Protest action can be used to support or supplement lobbying or legal action, or it might be a last resort if those methods fail. Protests aim to draw attention to an issue, such as water pollution, and to encourage decision-makers to take action.

> **REAL-WORLD EXAMPLE**
>
> ### Anti-sewage protests in the UK
>
> In 2023, regional, national and international campaign groups held protests across the United Kingdom, against the reported release of untreated sewage into rivers and seas. For instance, Surfers Against Sewage held protests on beaches all around the UK, called the Paddle Out Protest. Protesters held placards on the beaches and on their surfboards on the water.
>
> ■ **Figure 4.129** Members of Surfers Against Sewage protesting at a beach
>
> In a related protest, members of the international campaign group Extinction Rebellion held protests at the headquarters of two water companies. They wore protective clothing and poured fake 'sewage' on to the steps at the entrance to each office.

> ### Concept
>
> **Perspectives**
>
> Occasionally protests may become violent, and people may be hurt or property may be damaged. In other cases, protests cause a significant nuisance to members of the public, by blocking access to facilities or causing traffic congestion. People often disagree on what makes an effective protest, and on whether violence and inconvenience to others can be justified.
>
> Violence or a willingness to cause inconvenience might be an indication that protesters are especially angry or desperate, and believe strongly in their cause. The strength of that conviction might be related to their environmental value system. Consider the effect of anthropocentric and ecocentric perspectives on people's opinions about protest. Design and carry out a survey to determine respondents' environmental value systems, and their attitudes towards disruptive or violent protests. Analyze the responses to determine whether attitudes to protests are affected by EVS.

REVIEW QUESTIONS

1. Outline the use of biochemical oxygen demand to evaluate water quality.
2. Describe four pollutants which might be found in domestic sewage.
3. Explain what is meant by the term biotic index.

Link

Values and related concepts are covered in Topic 1.1.4 (page 7).

EXAM-STYLE QUESTIONS

1. Outline how deforestation can affect local flows in the hydrological cycle. [4 marks]
2. Evaluate the use of desalination as a means of achieving water security for a named country. [7 marks]
3. Outline three water-conservation strategies that can be used in agriculture. [3 marks]
4. Explain how economic development can lead to an increase in demand for freshwater. [4 marks]
5. Discuss the use of quotas as a tool for managing overexploited fisheries. [7 marks]
6. To what extent can aquaculture provide a sustainable source of food to meet global demand? [9 marks]
7. Explain how human activities can cause oxygen depletion in aquatic ecosystems. [6 marks]
8. Describe a programme for monitoring the effects of sewage pollution in a river. [4 marks]
9. Describe the different ways in which eutrophication can be managed. [6 marks]

HL only

10. Outline the role of the ocean as a carbon sink. [4 marks]
11. Discuss the ethical arguments surrounding the hunting of marine mammals by indigenous peoples. [9 marks]
12. Compare and contrast the use of biotic indices and water-quality indices for assessing water quality. [5 marks]

Theme 5
Land

5.1 Soil

> **Guiding questions**
> - How do soils play a role in sustaining natural systems?
> - How are human activities affecting the stability of soil systems?

SYLLABUS CONTENT

This chapter covers the following syllabus content:
- 5.1.1 Soil is a dynamic system within the larger ecosystem that has its own inputs, outputs, storages and flows.
- 5.1.2 Soil is made up of inorganic and organic components, water and air.
- 5.1.3 Soils develop a stable, layered structure known as a profile made up of several horizons, produced by interactions within the system over long periods of time.
- 5.1.4 Soil system inputs include those from dead organic matter and inorganic minerals.
- 5.1.5 Soil system outputs include losses of dead organic matter due to decomposition, losses of mineral components and loss of energy due to heat loss.
- 5.1.6 Transfers occur across soil horizons, in to and out of soils.
- 5.1.7 Transformations within soils can change the components or the whole soil system.
- 5.1.8 Systems flow diagrams show flows in to, out of and within the soil ecosystem.
- 5.1.9 Soils provide the foundation of terrestrial ecosystems as a medium for plant growth (a seed bank, a store of water and almost all essential plant nutrients). Carbon is an exception; it is obtained by plants from the atmosphere.
- 5.1.10 Soils contribute to biodiversity by providing a habitat and a niche for many species.
- 5.1.11 Soils have an important role in the recycling of elements as a part of biogeochemical cycles.
- 5.1.12 Soil texture defines the physical make-up of the mineral soil. It depends on the relative proportions of sand, silt, clay and humus.
- 5.1.13 Soil texture affects primary productivity through the differing influences of sand, silt, clay and dead organic matter, including humus.
- 5.1.14 Soils can act as carbon sinks, stores or sources, depending on the relative rates of input of dead organic matter and decomposition.

HL ONLY
- 5.1.15 Soils are classified and mapped by appearance of the whole soil profile.
- 5.1.16 Horizons are horizontal strata that are distinctive to the soil type. The key horizons are organic layer, mixed layer, mineral soil and parent rock (O, A, B and C horizons).
- 5.1.17 The A horizon is the layer of soil just beneath the uppermost organic humus layer, where present. It is rich in organic matter and is also known as the mixed layer or topsoil. This is the most valuable for plant growth but, along with the O horizon, is also the most vulnerable to erosion and degradation, with implications for sustainable management of soil.
- 5.1.18 Factors that influence soil formation include climate, organisms, geomorphology (landscape), geology (parent material) and time.
- 5.1.19 Differences between soils rich in sand, silt or clay include particle size and chemical properties.

▶ 5.1.20 Soil properties can be determined from analysing the sand, silt and clay percentages, percentage organic matter, percentage water, infiltration, bulk density, colour and pH.
▶ 5.1.21 Carbon is released from soils as methane or carbon dioxide.

5.1.1 Soil is a dynamic system

Soil is a complex mixture of organic and inorganic matter, water, air and living organisms, which forms on land and rests on top of the rocky lithosphere.

The various components of soil interact with each other and with other parts of the environment in complex ways. In this way, thereby acting as a system within the larger ecosystem and biosphere. Within the soil, transfers such as percolation and absorption, and transformations such as decomposition and respiration, occur. Stores within the soil include living organisms, inorganic minerals and organic matter.

The soil also experiences inputs and outputs of matter and energy, such as heat, infiltration, leaching and evaporation. Figure 5.1 summarizes some of the flows between different components of the soil, and between the soil and other parts of the environment. Each of these stores and processes will be explored further in this chapter.

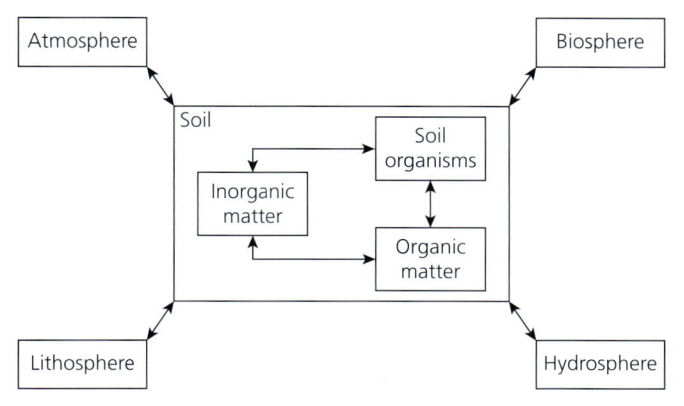

■ **Figure 5.1** Interactions among soil components and other parts of the environment

Soils form an important part of all terrestrial ecosystems. They provide a habitat and a source of resources such as mineral ions and water for living organisms. Human societies rely heavily on soil to support food production in agriculture. Soil quality varies widely, from highly fertile **loams**, to waterlogged clays, to dry sandy soils. Soil quality affects the ways in which humans can use the soil.

◆ **Loam** – soil made up of particles of various sizes, and containing a significant percentage of organic matter.

Link

The concepts of systems, stores, transfers and transformations are covered in Chapter 1.2 (page 24).

◆ **Weathering** – the process of rocks breaking up into smaller and smaller fragments, through the action of water, substances dissolved in it, and living organisms, as well as temperature changes.

5.1.2 Soil composition

Soils in different locations may contain different combinations of the various soil components. Some may be rich in sand, others contain mainly clay, and others have a combination of both. The amounts of organic matter, air and water in soil also vary widely.

■ Inorganic components

Soil forms when rocks are broken up into smaller pieces of inorganic material in a process called **weathering**. Weathering can be caused by:

- mechanical process (for example ice wedging, where water collected in cracks in rocks freezes and expands, increasing the size of the crack and eventually breaking the rock apart),
- chemical process (for example weathering by acid, where rainwater dissolves carbon dioxide from the air, forming carbonic acid, which reacts with certain rocks such as limestone, breaking it down chemically),
- biological process (for example weathering by animals digging or boring into rocks, which creates holes and produces soil particles).

5.1 Soil

◆ **Parent rock** – the rock which is broken down to provide the mineral components of a particular soil.

◆ **Clay** – soil made up mostly of tiny particles, less than 0.002 mm in diameter.

◆ **Silt** – soil made up mostly of intermediate-sized particles, from 0.002 to 0.05 mm in diameter.

◆ **Sand** – soil made up mostly of relatively large particles, from 0.05 to 2.0 mm in diameter.

◆ **Gravel** – stony or rocky particles in soil, which are bigger than 2.0 mm in diameter.

Concept

Systems and models. Perspectives

Connect, Extend, Challenge in soil components learning

You may have been to a beach or desert and know sand pretty well. Sand is one of the key components of soil. Sands are broken down from crystals of quartz, feldspars, mica, etc. An experiment such as grinding quartz into sand can explain the weathering of rocks by abrasion.

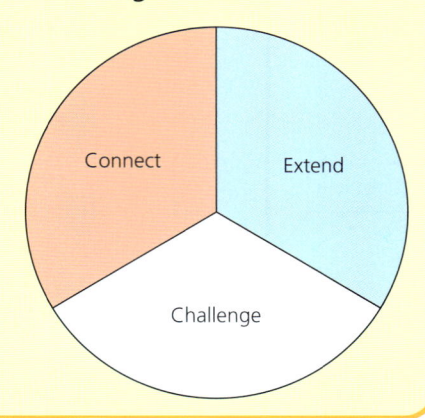

You can also look for sand and other fractions from local soil using sieves. Other fractions (excluding stones) are smaller than sand. A soil texture triangle diagram is useful for soil classification. (Organic materials are not included.)

The nature of the inorganic particles depends on the **parent rock** that they come from, and they may include many different chemical compounds.

Soil particles also vary by size, as shown in Table 5.1 and Figure 5.2.

■ **Table 5.1** Classification of soil particles by size

Particle class	Minimum size (mm)	Maximum size (mm)
Clay	-	0.002
Silt	0.002	0.05
Sand	0.05	2
Gravel	2	-

Each of the particle classes looks and feels quite different:

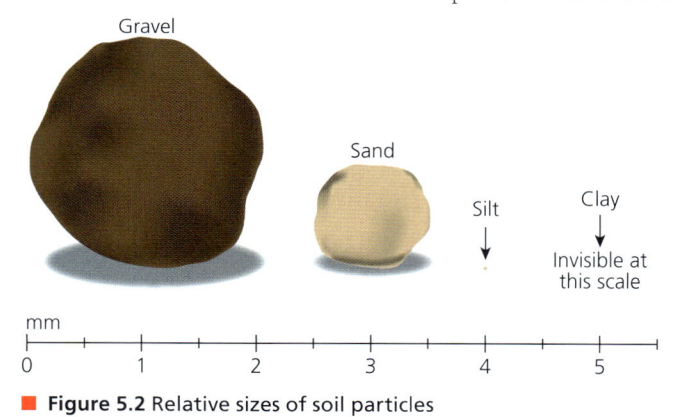

- Sand is grainy and individual particles are visible to the naked eye.
- Silt particles are too small to see without a microscope. Silt feels smooth and the particles slide freely between the fingers.
- The smallest clay particles are so small that they can only be seen using an electron microscope. They are so small that they have an extremely high surface-area-to-volume ratio, which makes them tend to stick together, giving clay a sticky feel and a clumpy appearance. Clay particles also tend to stick and hold on to various soluble substances in the soil.

■ **Figure 5.2** Relative sizes of soil particles

Because of the differences in texture of the mineral components, it is possible to distinguish soil types by feel, although some of the differences are very subtle, and it takes an experienced soil scientist to use this method consistently.

Follow these steps and the flow diagram to determine a soil's texture:

- Take about a dessert spoonful of soil.
- If dry, wet up gradually, kneading thoroughly between finger and thumb until soil crumbs are broken down.

Common mistake

People often confuse the terms *weathering* and *erosion*. Weathering is the breakup of rocks into smaller pieces, while **erosion** is the movement of weathered particles from one place to another. Erosion can only happen after weathering has occurred.

- Enough moisture is needed to hold the soil together and to show its maximum stickiness.
- Follow the paths in the diagram below to determine the texture.

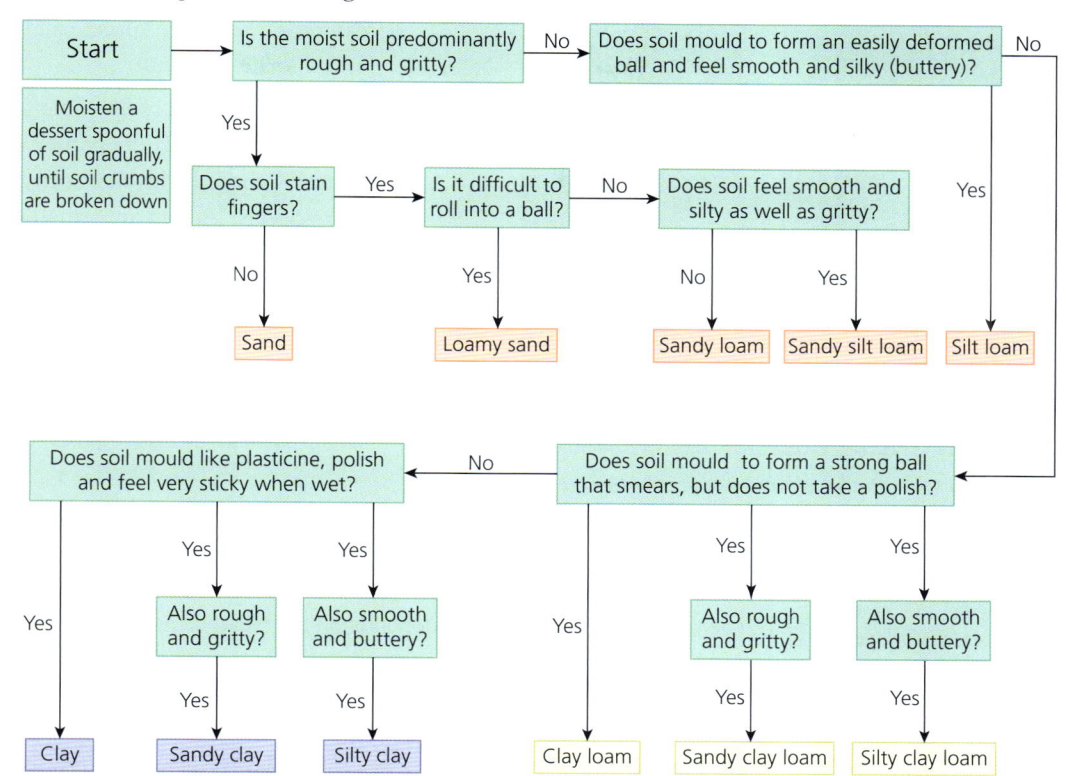

■ **Figure 5.3** A key for identifying soil type by feel

◆ **Erosion** – the movement of broken-down particles of rock from one location to another, due to the movement of agents such as wind, water or gravity.

◆ **Exudate** – a fluid secreted by a living organism.

◆ **Decomposer** – a microorganism that breaks down organic matter by secreting enzymes on to it, and absorbing the products of decomposition.

◆ **Detritus** – partially decomposed organic matter.

◆ **Detritivore** – an animal that eats detritus.

◆ **Humus** – fully decomposed organic matter, made up of a sticky, dark substance, which often coats the mineral soil particles.

■ Organic components

As mineral particles accumulate due to the weathering of rock, living organisms often come to live in and on the soil. These organisms add organic matter to the soil in various ways. Animals and plants shed body parts, such as skin cells and leaves. Plant roots may secrete fluids called **exudate**, containing molecules such as sugars and amino acids, into the surrounding soil. Animals also release waste products such as urine and faeces.

At the same time, microorganisms such as bacteria and fungi come to live in the soil. Many of these are **decomposers**, which break down the organic matter, releasing smaller molecules and ions, which can be absorbed by plants. Partially broken-down organic matter, or **detritus**, provides food for a variety of animals, called **detritivores**, such as earthworms and beetle larvae. Fully decomposed organic matter, called **humus**, contributes to the structure and properties of soils. Humus-rich soil tends to have a rich, dark colour.

Common mistake

Many people confuse detritivores and decomposers. Decomposers are microorganisms, such as many fungi and bacteria, which break organic matter down into simpler substances. Detritivores are animals that eat the decomposing organic matter, breaking it down into smaller pieces. Detritivores then return the organic matter to the soil as faeces, where decomposers can break it down further. Together, decomposers and detritivores help to recycle minerals into the soil, where plants can use them.

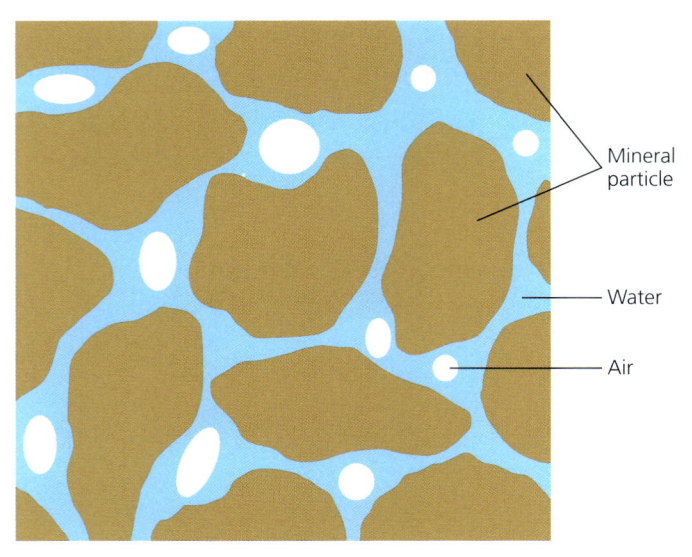

Figure 5.4 Air and water in the spaces between soil particles

Water and air

The spaces, or **pores**, between the mineral particles in soil can be found. These spaces can be filled with air or water, or a combination of both.

The size and shape of the mineral particles in soil determine how closely together the particles can fit. Smaller, more regularly shaped particles, such as clays, may fit together more closely, leaving only small pores between them, while larger, more irregularly shaped particles, such as sand, generally have larger pores. Soils containing particles with a mixture of sizes and shapes may have pores of variable size.

Organic matter can also influence the size of soil pores, as it may cause the mineral particles to stick together, forming larger particles or **aggregates**, thereby creating larger pores.

◆ **Pore** – a space between the solid mineral particles of soil, where air or water may be found.

◆ **Aggregate** – a clump of soil formed when mineral particles stick together.

5.1.3 Soil profiles and soil horizons

The process of soil formation depends on the weathering of rocks and the accumulation of organic matter, and may take several hundreds or even thousands of years. Over time, if soil is left undisturbed and allowed to form naturally, the components of the soil tend to settle into a regular arrangement made up of a series of layers called horizons. We can see these layers if we dig a deep, vertical hole into the soil. The exposed layers form a soil profile.

> ### Common mistake
>
> It is easy to get confused between soil profile and soil horizon because both terms are closely related and integral to understanding the composition and characteristics of the soil. However, they refer to different aspects of the soil structure. To distinguish between the two terms:
> - Soil profile: Think of it as the entire vertical section of the soil, like a layered cake representing the different horizons.
> - Soil horizon: Consider each horizon as an individual layer within that cake, each with its own set of unique characteristics.

Soil profiles usually show at least four distinct layers or horizons. Starting from the surface, these are named O (or 'organic'), A, B, and C The nature of these horizons is explored further in Topic 5.1.16 (page 441).

These horizons may vary in properties such as texture, mineral composition, pore size and organic matter content. For instance, the organic matter content is highest in the uppermost, O horizon, since most biological activity occurs near the surface. Organic matter is added by plants shedding leaves, branches, flowers and fruits onto the soil, and by animals depositing faeces and other waste products, mostly at or near the surface. The organic matter may then move gradually down through the soil profile, carried by percolating water, or the growth and movements of plants and animals. Typically, organic matter content decreases with increasing depth, as there are fewer plants and animals in deeper soil.

Figure 5.5 A soil profile is made up of different horizons or layers

Inquiry process

Inquiry 1: Exploring and designing

Carry out a thorough risk assessment for the experimental technique described below, to measure the moisture and organic matter content of soil.

1. Read carefully through the instructions below, taking note of any hazards or potentially dangerous aspects of the procedure.
2. For each hazard, identify any possible risks. These are events that could occur, which might be dangerous or cause harm to yourself or others.
3. Think about any risk-reduction strategies, or steps you could take to avoid or minimize each risk you have identified.
4. Consider any emergency responses you might need to take in the event that each risk arises. These might include measures to deal with harm or damage caused, such as first aid measures or seeking medical attention.
5. Copy the table below and record all of the above in it, giving as much detail as possible.

Hazard	Risk	Risk-reduction strategies	Emergency responses

Tool 1: Experimental techniques

Measure the moisture and organic matter content of soil

Before beginning this procedure, check with your teacher whether you have access to crucibles, an oven and a furnace, and carry out a risk assessment, as described above.

The moisture content of soil can be determined by heating the soil until all the moisture evaporates, leaving dry soil.

1. Weigh a clean, dry crucible, using a precise laboratory balance.
2. Place a sample of soil in the crucible and weigh it again. Calculate the mass of the soil by subtracting the mass of the crucible.
3. Place the crucible and soil sample in a drying oven, at about 105°C for about one hour.
4. Remove the soil sample from the oven and allow it to cool in a dry place, such as a desiccator containing a water-absorbing substance.
5. When safely cooled, weigh the dried crucible and calculate the mass of the dried soil by subtracting the mass of the crucible.
6. If time permits, return the crucible to the oven for a further period of drying, and then cool and weigh again. If the mass is the same after drying twice, then all the moisture has been removed.
7. Determine the percentage moisture content, as follows:

$$\text{Percentage moisture} = \frac{(\text{Initial mass of soil}) - (\text{Final mass of soil})}{(\text{Final mass of soil})} \times 100$$

The organic matter content of soil can then be determined by heating dry soil to a high temperature. The organic matter undergoes thermal decomposition, leaving only mineral particles.

1. Place the dried soil and crucible into a furnace, set at 550°C for about 30 minutes. If a furnace is not available, the crucible can be heated on a Bunsen burner for about one to two hours.
2. Allow it to cool to a safe temperature in the furnace, before moving it to a desiccator to cool to room temperature.
3. When safely cool, weigh the soil and crucible again, and determine the percentage organic matter content as follows:

$$\text{Percentage moisture} = \frac{(\text{Initial mass}) - (\text{Final mass})}{(\text{Final mass})} \times 100$$

Inquiry process

Inquiry 1: Exploring and designing

Measure soil water retention

Water retention is the ability of a soil to hold on to water in its pore spaces, and to resist drainage. Soils with high water retention tend to drain slowly and may become waterlogged easily. Soils with low water retention tend to drain rapidly and may dry out quickly. Water retention can be measured by passing a known quantity of water through a known quantity of soil, and collecting the water that drains through it. Design an laboratory procedure using this principle, to compare the water retention of soils collected from different sites. In your design, consider the following issues:

1. What apparatus would you use and how would you set it up?
2. How would you ensure that you make a fair and valid comparison between the soils? What variables would you need to control and how would you control them?
3. What measurements should you take, using what apparatus?
4. How would you ensure that your findings are reliable?
5. How would you present, analyse and interpret your data?

ATL ACTIVITY

Compare soil from different sites or different horizons

Various simple tests are available to investigate different properties of soil. Some of these can be done with simple, readily available equipment, but others require special tools. Consider carrying out each of the following tests of different soils, if the equipment is available to you. Before carrying out each test, make a prediction about how the different soil samples will compare.

Drainage

Dig a square hole in the soil, 30 cm deep and 30 cm on each side. Fill the hole with water and allow it to drain completely overnight. Then re-fill it with water and allow it to drain again, but this time measure the depth of the water at one-hour intervals and plot a graph of depth over time.

Nitrogen, phosphorus and potassium (NPK)

Use a standard test kit, which can be obtained from a scientific or agricultural supplier, to determine the approximate concentrations of these macronutrients in the soil.

Aeration

If available, use an electrode to measure the oxygen diffusion rate (ODR) of the soil.

> ◆ **Inorganic matter** – substances that do not come from living organisms. In the soil, inorganic matter mostly comes from the underlying bedrock, or from the air or water. It can include a wide range of substances, containing many different elements and compounds.
>
> ◆ **Anthropogenic** – produced by, or originating from, human action.
>
> ◆ **Organic matter** – in general, refers to substances that come from living organisms. This might include dead parts of plants and animals, or substances released from their bodies, such as faeces or other waste products. Organic matter is usually made up of carbon-rich compounds such as carbohydrates, proteins or fats.

5.1.4 Soil system inputs include those from dead organic matter and inorganic minerals

Soil functions as an open system, receiving inputs of both matter and energy. Inputs include **inorganic** and organic matter and energy, and can come from natural or **anthropogenic** sources.

■ Inputs of dead organic matter

Natural sources

Organic matter is added to soil by all of the organisms that live in or on the soil. Dead organisms or parts of organisms may fall to the ground and be added to the O horizon, or leaf litter. This includes leaves, twigs, seeds and fruits from plants, and feathers, fur and skin of animals. Animals also add faeces to the soil. Some of these inputs may come from some distance away, as animals can move, and plant seeds and leaves may be blown by the wind, or carried by animals, over long distances. Birds and bats, for instance, provide organic matter when they deposit their faeces, called guano, on to soil.

Organic matter is also added to the soil by plants, in the form of exudates. Plant roots often secrete fluid containing organic molecules such as sugars and amino acids that were produced in the leaves of the plant using carbon dioxide from the air.

Organic matter in the soil provides food for detritivores and decomposers, and forms the base of many soil food chains. Decomposing organic matter releases nutrients into the soil, which plants can use for growth. Organic matter also changes the characteristics of the soil, including colour, texture, and properties such as water retention and drainage properties.

Anthropogenic sources

Humans also add organic matter to soils, especially on farms, in the form of manure or compost. Manure is faecal matter collected from livestock, such as cattle, horses or chickens. Compost is decaying plant matter, including cuttings from crops or weeds that have been allowed to partially decay. Both manure and compost are used as natural or organic fertilizers, to help promote crop growth.

■ Inorganic inputs

Natural inorganic inputs

■ Table 5.2 Natural inorganic inputs into soil

Process	Input	Significance
Weathering	Mineral particles	Bedrock weathers slowly, adding smaller mineral particles to the soil, contributing to the composition and characteristics of the soil, and gradually deepening the soil.
Deposition	Mineral particles	Mineral particles produced by weathering, and transported by erosion, are deposited from water or the air. These may add substances that can affect soil fertility positively or negatively.
Decomposition	Mineral ions and molecules	Organic matter is broken down by bacteria and fungi, releasing mineral ions and small molecules into the soil. These are often important for plant growth.
Precipitation	Dissolved minerals	Rainfall and other forms of precipitation may bring dissolved substances from the air into the soil. These substances may be helpful to plants and soil organisms (e.g. nitrites produced when nitric oxide dissolves and reacts with rainwater), or harmful (e.g. acids produced when nitrogen dioxide and the oxides of sulphur dissolve to form acid precipitation – see Chapter 8.3, page 740).
Diffusion	Gases, including oxygen, nitrogen and water vapour	Air can diffuse into the spaces between soil particles. Air comprises several gases, including nitrogen, oxygen and water vapour. Oxygen is needed for aerobic respiration by soil organisms. Nitrogen is used by specific bacteria to carry out nitrogen fixation (see Chapter 2.3, page 139).
Solar radiation	Heat energy	Solar radiation warms the soil. In cold climates, this may make it more suitable for living organisms. It may also increase the rates of processes such as evaporation and decomposition of organic matter.

Anthropogenic inorganic inputs

Anthropogenic inputs of inorganic matter are also mostly associated with farming. Deliberate inputs include:

- Irrigation: Water is added to soils, especially in dry areas, to promote plant growth.
- Agrochemicals: Fertilizers and pesticides are applied to soils or crops. Inorganic fertilizers include a variety of mineral ions, such as nitrates, phosphates and potassium, which promote plant growth. Pesticides are substances used to kill or repel pests, to reduce the damage they do to crops.

Unintentional anthropogenic inputs can also occur, in farming and elsewhere. For instance, excessive irrigation in dry environments can lead to the addition of salt to the soil, in a process called **salinization**. When the irrigation water evaporates, it leaves behind small amounts of salt. Over time, more and more salt can accumulate, until the soil is incapable of supporting plant growth.

In industrial or urban areas, various industrial and domestic waste products might be unintentionally (or sometimes intentionally) released into soils, causing **soil contamination** or soil pollution. Contaminated soil may be unsuitable for farming and other uses. Pollutants can also leach from the soil into ground and surface waters.

◆ **Salinization** – the process of addition of salt to soil, usually as a result of excessive irrigation.

◆ **Soil contamination** – the addition of harmful substances or pollutants to soil, either through deliberate disposal or by accidental or unintentional leakage of wastes.

5.1.5 Soil system outputs include losses of dead organic matter due to decomposition, loss of mineral components and loss of energy due to heat loss

■ Loss of organic matter

As organic matter in soil decomposes, it is broken down into simpler substances, including minerals such as nitrates and phosphates, and gases such as carbon dioxide or, under anaerobic conditions, methane. These gases diffuse out of the soil into the air.

Carbon dioxide and methane both contain atoms of carbon, and they are both greenhouse gases. Soil has the capacity to store a great deal of carbon and is a major sink in the carbon cycle. However, when soil organic matter decomposes fully, it can increase the emission of these greenhouse gases into the atmosphere and contribute to global warming.

■ Loss of mineral components

Inorganic parts of soil can be lost through erosion, leaching, absorption by plants and evaporation.

Minerals in the soil can be absorbed by plant roots. The plants draw the minerals in via diffusion or active transport, and use the minerals in their biological processes. For instance, plants use magnesium to make chlorophyll, nitrates to make proteins and potassium to perform various cellular functions.

Various substances also leave the soil by simple physical processes such as diffusion and evaporation. Water, for instance, evaporates, and carbon dioxide diffuses out of the soil.

Link

The carbon cycle is covered in Chapter 2.3 (page 139). The greenhouse effect and global warming are discussed in Chapters 6.1 and 6.2 (pages 514 and 528).

Figure 5.6 Soil erosion by water

Erosion is the movement of matter from one place to another. Wind and water can both cause erosion of soil, carrying both organic matter and mineral particles away. Soil erosion is particularly rapid on exposed soils, such as farmland where the crops have been harvested. Ploughing also encourages soil erosion, by breaking up the topsoil, making it easier for the wind or surface flow of water to carry the particles away. Loss of topsoil by erosion is a major threat to global food production.

Leaching happens when water percolates through the soil. Soluble minerals, such as nitrate ions, can dissolve in the water, and be washed away into surface waters such as streams and rivers, or downwards into the groundwater. Leaching can reduce soil fertility and contribute to water pollution.

> **Link**
>
> The importance of soil for food production is covered in Chapter 5.2 (page 456). Water pollution due to excessive nutrients is explored in Chapter 4.4 (page 373).

Loss of heat energy

Heat energy enters or leaves the soil depending on the amount of solar radiation falling on the soil, and the temperature of the soil relative to the atmosphere.

Where the soil is exposed to direct sunlight, it will absorb the sun's energy and become warmer. Likewise, if the air is warmer than the soil, heat energy will flow from the air into the soil.

However, if the air is cooler than the soil, heat energy will flow by radiation from the soil into the air.

5.1.6 Transfers occur across soil horizons, in to and out of soils

Transfers are movements of matter or energy from one location to another, without changing state or form. Many transfers occur within soils, and between soils and other parts of the ecosystem. Transfers include:

- Movements of water: Water enters the soil from the surface by infiltration, and then drains through the soil by percolation into the groundwater. Groundwater can also flow through the soil.
- Leaching: As water percolates through the soil, some minerals dissolve and are washed downwards in a process called leaching. Leached minerals may accumulate in lower soil horizons, or be washed away into the groundwater or nearby surface waters.
- Soil erosion: Soil particles are moved from one location to another, often by wind or water.
- Aeration: Gases such as oxygen and carbon dioxide can diffuse into or out of soils, depending on their relative concentrations. Oxygen usually diffuses into the soil, since it is being used by soil organisms for respiration. Carbon dioxide usually diffuses out of the soil. Ploughing and the action of burrowing animals such as earthworms can encourage soil aeration.
- Biological mixing: Various living organisms can cause parts of the soil to move vertically or horizontally. Burrowing animals such as earthworms, for instance, move soil and create spaces for substances like water and air to flow through the soil. Dung beetles move organic matter from the surface down into the soil. These movements can be very important for cycling materials, and for making substances such as oxygen and minerals available for living organisms.

◆ **Respiration** – the breakdown of organic molecules by living cells, to produce energy. Respiration may be **aerobic** or anaerobic.

◆ **Aerobic** – involving or using oxygen. When aerobic respiration occurs, carbon from the organic molecules is combined with oxygen, and carbon dioxide is produced.

◆ **Anaerobic** – occurring without oxygen. When anaerobic respiration occurs, carbon from the organic molecules may be released as methane.

Link

Weathering, especially biological weathering by lichens, is an important step in primary succession, which is discussed in Chapter 2.5 (page 178).

5.1.7 Transformations within soils can change the components or the whole soil system

Transformations are system processes that involve a change of form or state. Many transformations occur in soils.

Biological decomposition

When bacteria and fungi break down organic matter, they convert it into simpler, smaller molecules. The process can be a very complex one, involving many different microorganisms, and many different chemical reactions. In general, it takes places in two stages: the formation of humus, followed by mineralization.

- Humus (see Topic 5.1.2, page 417) coats the soil's mineral particles. Humus is relatively stable and can take a long time to break down further.
- Mineralization is the slow conversion of humus to mineral particles such as nitrates and phosphates, which can be taken in and used by plants. During mineralization, the remaining carbon in the humus is combined with oxygen, to produce carbon dioxide.

The microorganisms carrying out decomposition obtain food from the organic matter. They then break the food down in the process of **respiration**, releasing carbon dioxide if oxygen is present. If there is no oxygen present, such as in waterlogged soils, the decomposers respire **anaerobically** and produce methane.

Weathering

Rocks and larger mineral particles in soil can be broken into smaller fragments through the process of weathering. Weathering may be:

- mechanical: physical processes such as heating and cooling, or freezing and thawing of ice, cause rocks and soil fragments to break apart.
- chemical: chemical reactions occur between mineral particles and water, oxygen or acids in rain. The breakdown of limestone by carbonic acid in rain is an example of chemical weathering.
- biological: living organisms, including microorganisms, plants and lichens can cause rocks and mineral particles to break apart. Weathering by lichens is often one of the first steps in the formation of soil from bare rock.

■ **Figure 5.7** A combination of mechanical, chemical and biological processes can break rocks into smaller and smaller pieces, in a process called weathering

In the agricultural heartland of the Midwest, USA, a noteworthy shift towards sustainable practices is transforming the traditional landscape dominated by monoculture and conventional farming. One real-life scenario involves the adoption of cover crops, including legumes and grasses, during the non-growing seasons by local farmers. This strategic use of cover crops serves as a proactive measure against the erosive forces of weathering on soil health. They act as protective shields, mitigating the impact of heavy rainfall and wind erosion. As these cover crops decompose, they enrich the soil with organic matter, fostering improved fertility and structure. Importantly, this approach also cultivates diverse microbial communities in the soil, contributing to nutrient cycling and disease suppression.

Theme 5: Land

> **TOK**
>
> In what ways does technological advancement contribute to our understanding and manipulation of soil and weathering?
> - Explore the role of technology in soil analysis, precision agriculture and geotechnical engineering.
> - Discuss the ethical considerations associated with technological interventions in managing soil and weathering.

Beyond environmental benefits, the economic resilience of farmers is heightened, as improved long-term productivity and decreased reliance on external inputs become evident. This underscores the intricate relationship between soil and weathering, showcasing how deliberate agricultural interventions can positively influence the dynamic interplay between these vital processes, promoting sustainability and resilience in the face of environmental challenges.

■ Nutrient cycling

Inorganic substances released in the soil through processes such as weathering and decomposition can take part in a range of chemical reactions. These form part of the global biogeochemical cycles, such as the carbon and nitrogen cycles.

Many nutrient cycling processes in the soil are caused by microorganisms. For instance, in the nitrogen cycle, bacteria cause nitrogen fixation, nitrification, denitrification and mineralization.

■ Salinization

Salinization is the accumulation of salt in soil. It often occurs when farms in dry environments are irrigated. If too much water is added to the soil, some of it may evaporate instead of infiltrating and being absorbed by plants. Although farmers generally use freshwater for irrigation, there is always a little salt dissolved in water. When the water evaporates, the salt is left behind as salt crystals in the soil.

Salinization can also occur if the water table is close to the surface. Water may rise through the soil from the water table to the surface, where it evaporates, leaving salt behind.

Salinization is a transformation since the salt changes state from dissolved to solid crystals. Salinized soil is unsuitable for agriculture, since most plants cannot tolerate salty environments.

5.1.8 Systems flow diagrams show flows in to, out of and within the soil ecosystem

The transfers and transformations described in the preceding sections can be important for the movements and changes of various substances around the biosphere. Many biogeochemical cycles, including the carbon, nitrogen and hydrological cycles, include important processes that occur in the soil, such as:

- Carbon cycle: Accumulation and decomposition of organic matter.
- Nitrogen cycle: Nitrogen fixation, nitrification and denitrification.
- Hydrological cycle: Infiltration, percolation and evaporation.

Many of these processes also overlap and interact in the soil. For instance:

- Percolation of water may cause the leaching of nitrogenous compounds.
- Waterlogging of soils may encourage anaerobic decomposition and denitrification.
- Decomposition of organic matter releases both carbon dioxide and nitrogenous compounds such as ammonium ions.

All of the processes that occur in soils can be illustrated by a systems diagram.

◆ **Macronutrient** – a substance that a plant needs in large amounts. Nitrogen, phosphorus and potassium (N, P and K) are macronutrients for all plants, and these are the main constituents of plant fertilizer.

◆ **Micronutrient** – a substance that a plant needs in relatively small quantities.

5.1.9 Soils provide the foundation of terrestrial ecosystems as a medium for plant growth

Soils are an important component of terrestrial ecosystems. Plants are the main primary producers in terrestrial ecosystems, and most plants have an intimate relationship with soil. As primary producers, plants must obtain all the raw materials they need to produce food from the environment. Almost all of these raw materials come from the soil.

Plant roots grow through the soil, absorbing water and mineral ions from the soil. The **macronutrients** nitrogen, phosphorus and potassium, as well as many other essential and beneficial **micronutrients**, are obtained from the soil. These nutrients exist as minerals in the soil. They may come from the inorganic components of the soil, through weathering or from the organic matter, through decomposition.

■ **Table 5.3** Macronutrients and micronutrients that plants obtain from the soil

Element	Needed for plants?	Role in plant nutrition
N	Essential	Producing proteins, nucleic acids and membranes
P	Essential	Producing proteins, nucleic acids and membranes
K	Essential	Regulation of water potential between cells
S	Essential	Producing proteins, nucleic acids and membranes
Ca	Essential	Water potential regulator, cell wall formation, cell signalling
Mg	Essential	Major component of chlorophyll molecules
B	Micronutrient	Cell wall component, translocation, reproduction
Cl	Micronutrient	Water regulation, controls photosynthesis
Mn	Micronutrient	Component of chlorophyll synthesis enzymes
Fe	Micronutrient	Component of chlorophyll synthesis enzymes
Ni	Micronutrient	Converting urea to useable N (ammonia)
Cu	Micronutrient	Component of chlorophyll synthesis enzymes, seed production
Zn	Micronutrient	Component of chlorophyll synthesis enzymes, growth regulation
Mo	Micronutrient	Converting nitrite/nitrate to useable N (ammonia)
Na	Beneficial element	Can help chlorophyll synthesis
Al	Beneficial element	May stimulate root and shoot growth
Si	Beneficial element	Boosts resistance to both drought and fungal infection
V	Beneficial element	Can help chlorophyll synthesis, shoot growth
Co	Beneficial element	May help stem or bud formation, nitrogen fixation
Se	Beneficial element	Component of antioxidants, anti-senescence agent
I	Beneficial element	May influence antioxidant enzyme synthesis

Soil also contains seeds of plants. Seeds may germinate, growing into new seedlings, or they may remain **dormant** in the soil, sometimes for a very long time. Soil is sometimes described as a natural **seed bank**. Seeds in soil can facilitate the process of secondary succession.

◆ **Dormant** – in a state similar to sleep. Dormant seeds have very little biological activity going on inside them.

◆ **Seed bank** – a store of living, but usually dormant, seeds, which can grow into new plants. Artificial seed banks are widely used in conservation of plant biodiversity. Soil can act as a natural seed bank.

Common mistake

Plants also need to take in carbon in large quantities from their environment. However, carbon is not absorbed from the soil through the roots. Instead, carbon, in the form of carbon dioxide, is taken in by leaves from the air.

Link

Secondary succession is covered in Chapter 2.5 (page 178).

Theme 5: Land

5.1.10 Soils contribute to biodiversity by providing a habitat and a niche for many species

■ Soil biodiversity

Many living organisms can be found in soil. The soil habitat is dark, cool and moist, making it suitable for many organisms that could not tolerate a more exposed environment. The habitat can vary in the types and sizes of mineral particles, the amount of organic matter and the amount of water, meaning that different soils can be homes for different organisms.

ATL ACTIVITY

Animals that live in soil can be easily explored by looking under leaf litter and fallen logs, and digging into the soil with a trowel or small shovel. In healthy soil, you will encounter many interesting animals such as earthworms, woodlice (also called pillbugs) and millipedes.

For a more detailed look at soil organisms, you might use a pitfall trap. Dig a small hole in the soil and place a container, such as a plastic cup or beaker, into the hole. Fill in the space around the cup with soil, and leave the trap for a few hours or overnight. If the soil is reasonably healthy, you should find a variety of interesting animals in your trap.

When handling animals, always take care to avoid injuries such as stings or bites, and wash your hands thoroughly afterwards. Also, be careful not to cause any harm or stress to the animals!

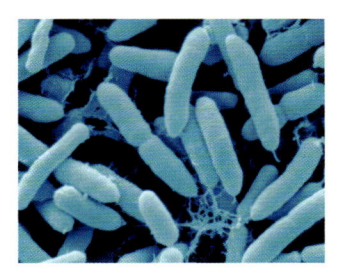

■ **Figure 5.8** A selection of common soil organisms

● HL.c.7: Environmental ethics

Search online for, or ask your teacher to show you, the IB Animal Experimentation Policy, and read what it says about working with animals. The IB takes the perspective that animals have moral standing. This means that their interests should be considered when deciding how to behave towards them. Specifically, the IB position is that we should always try to avoid causing harm or stress to animals.

This position is not universally accepted in science or education. For instance, some methods for studying soil organisms use stressful stimuli to drive animals out of the soil and into a preservative such as alcohol, which kills and preserves them for later study. Consider the ethical issues surrounding the use of methods like this. Are there any potential benefits to be gained from killing animals to study them? Can this approach be justified, perhaps using virtue, consequentialist or rights-based ethical arguments?

Many soil organisms are microorganisms, such as bacteria, fungi and protozoa. Some of these are very important for soil ecology and terrestrial ecosystems, but many have not been identified by scientists and their roles in the soil ecosystem are not well understood.

■ Ecological niches in soil

The organisms in soil perform many different functional roles and occupy many different niches in the soil ecosystem. Table 5.4 outlines some of these roles, but it is important to appreciate that each species has its own niche and plays its own part in the ecosystem.

■ Table 5.4 Examples of functional roles of organisms in the soil

Function role	Examples	Description
Producers (photosynthetic)	Grasses, herbs, shrubs, trees Mosses, liverworts, algae Lichens Cyanobacteria	Most plants have parts that grow on or in the soil. Mosses, liverworts, lichens, some algae and cyanobacteria can live on the surface of the soil. Plants produce food via photosynthesis, using water and nutrients from the soil, and carbon dioxide from the air. Underground parts of plants, such as roots, respire, using up oxygen in soil air spaces and producing carbon dioxide.
Producers (chemosynthetic)	Nitrifying bacteria, such as Nitrobacter and Nitrosomonas	These bacteria live in the soil, using ammonium and nitrite ions as a source of energy, and carbon dioxide as a source of carbon, to produce organic molecules. The process is similar to photosynthesis, but uses chemical energy instead of light energy.
Herbivores	Root-eating nematodes Insects and insect larvae, e.g. some beetle larvae	Most soil herbivores feed on the roots of plants. On farms, they may be seen as pests.
Decomposers	Bacteria and fungi	Break organic matter (detritus) down into smaller, simpler compounds.
Detritivores	Animals, including earthworms, millipedes, woodlice and springtails	Eat detritus to obtain nutrients directly from the detritus and from the decomposers living on it. Much of the detritus is indigestible, so it is released as faeces. In this way, detritivores break detritus up into smaller pieces, with larger surface areas, enabling decomposers to break the organic matter down faster.
Carnivores	Protozoans, nematodes, centipedes, scorpions, mites, moles	May feed on different prey, depending on size, including bacteria, nematodes and insects. Above-ground carnivores such as birds also feed on soil animals.
Parasites	Fungi, bacteria	Live and feed on living organisms, such as plant roots or animals. Some cause diseases in crops and are considered pests on farms.
Mutualists	Fungi, bacteria	Live in or on other organisms, providing benefits to the host and receiving benefits from the host. These include the mycorrhizal fungi, which absorb water and minerals from the soil and pass them to plant roots in exchange for sugars, and nitrogen-fixing bacteria, which live in the root nodules of legumes.

◆ **Parasite** – a close, **symbiotic** relationship between two species in which one species benefits and the other suffers.

◆ **Mutualism** – a relationship between two species in which both species benefit.

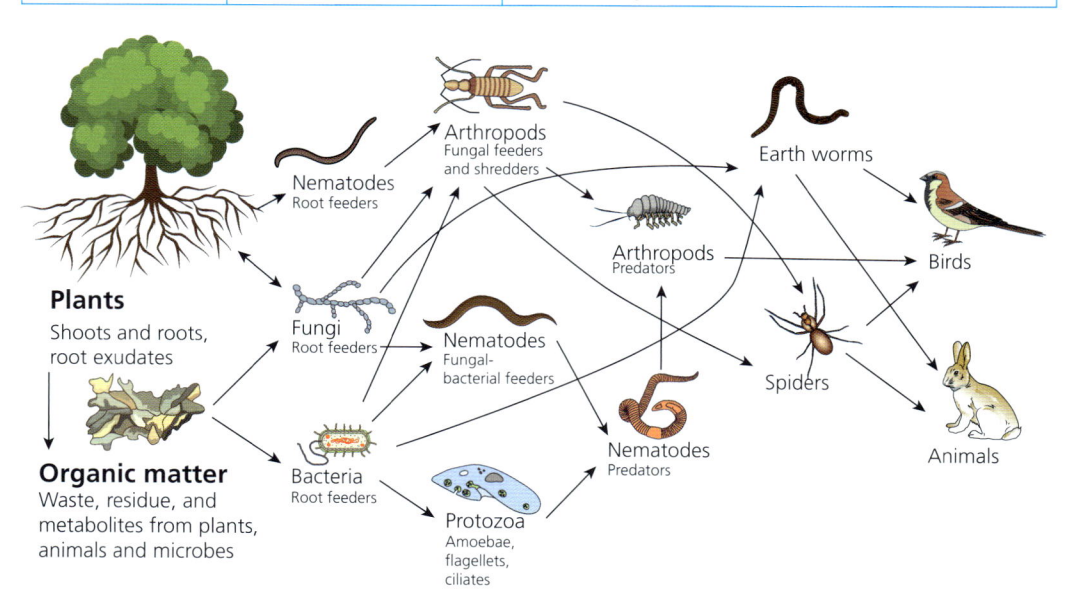

■ Figure 5.9 A soil food web

◆ **Mycorrhizae** – a large group of fungi that have a strong mutualistic relationship with many plants, exchanging water and minerals for organic molecules, and possibly forming network-like connections between plants. The name mycorrhizae comes from words meaning fungus (*myco*) and root (*rhiza*).

◆ **Hyphae** – thin, branching, thread-like structures of fungi.

REAL-WORLD EXAMPLE

Mycorrhizae

Mycorrhizae are an extremely important group of fungi which live in soils all around the world. Their thread-like **hyphae** wrap around or even penetrate into the smallest roots of trees and other plants, and grow outwards into the soil. Hyphae are only one cell thick and branch out many times, producing an enormous total surface area. In this way, they reach further, and absorb more minerals and water, than the roots would be able to on their own. They pass some of the water and nutrients directly to the plants, and receive organic molecules such as sugars and amino acids in return. This allows the plants to absorb more than they would be able to without the mycorrhizae.

The relationship between the plants and mycorrhizae is a mutualistic one, where each partner benefits from the other one. Studies have shown that plants grown with mycorrhizae in the soil grow far faster than those grown in soil without mycorrhizae.

Further studies have also shown that mycorrhizal hyphae can connect different plants, either of the same or different species, and can share resources and possibly send signals between different plants. Canadian researcher Suzanne Simard, a pioneer in mycorrhizal research, has suggested that they form a network of cooperative relationships between plants and fungi, which she called the 'Wood Wide Web'.

■ **Figure 5.10** White threads of mycorrhizal fungi extend away from yellow and brown roots of trees

■ **Figure 5.11** Plants grown with mycorrhizae grow faster than those without

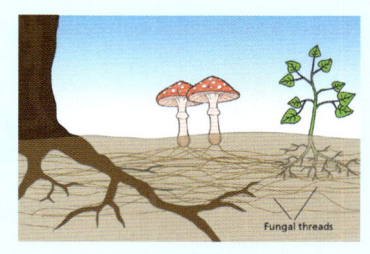

■ **Figure 5.12** Mycorrhizae can make connections between different plants

5.1.11 Soils have an important role in the recycling of elements as a part of biogeochemical cycles

Link

The biogeochemical cycles are explored in Chapter 2.3 (page 139).

The biogeochemical cycles, such as the carbon, nitrogen and hydrological cycles, describe the transfers and transformations of substances around the Earth's biosphere. Decomposition of organic matter in soil plays a major part in many of these cycles. Living organisms contain many elements in their cells and tissues, and decomposition allows these elements to be released into the environment so that they can be used again.

In the carbon cycle, the breakdown of dead organic matter releases carbon dioxide into the atmosphere, where it can be used again by plants carrying out photosynthesis. Soils can also store a great deal of carbon, in the form of humus. This makes them an important factor in the Earth's carbon budget, and in understanding and combating global warming.

For most other cycles, such as the nitrogen and phosphorus cycles, the breakdown of dead organic matter releases soluble forms of the elements into the soil, where they can be absorbed by plant roots. For instance, the breakdown of proteins results in the release of ammonium salts into the soil, while the breakdown of structures such as bones, cell membranes and DNA release phosphates into the soil.

A number of other interesting steps in the nitrogen cycle also occur in the soil. When decomposers release ammonium from dead organic matter, another group of bacteria called nitrifying bacteria convert the ammonium to nitrites and then nitrates. Nitrates and nitrites are then turned into nitrogen gas by denitrifying bacteria. Finally, nitrogen can be returned to the soil by nitrogen-fixing bacteria, which combine gaseous nitrogen with other elements, making them available for plants to use.

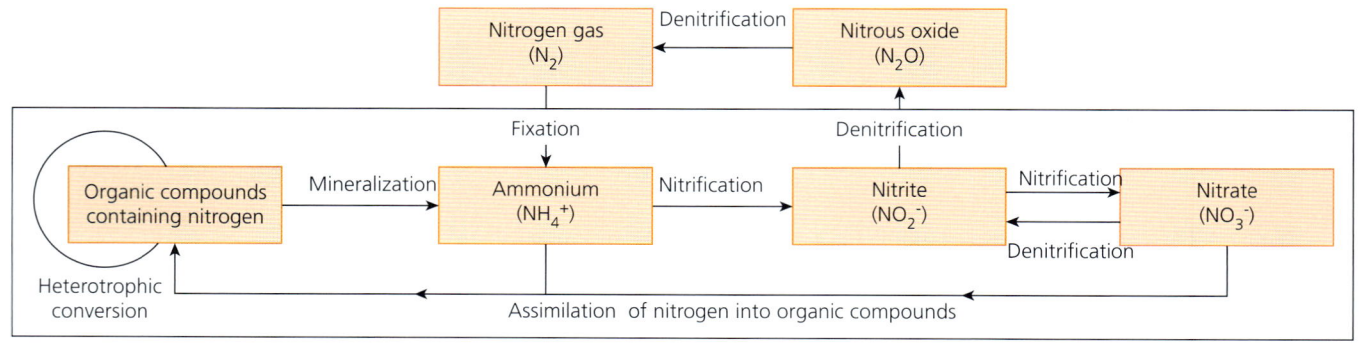

■ **Figure 5.13** The nitrogen cycle. The box shows the portions of the cycle that occur in the soil

REAL-WORLD EXAMPLE

Nitrogen-fixing bacteria

Nitrogen-fixing bacteria are some of the most important organisms on Earth, as they convert very unreactive nitrogen into more reactive forms that can be used by plants. Nitrogen is an essential element for life, since proteins and DNA contain nitrogen.

Many nitrogen-fixing bacteria live free in the soil or water. Some species, however, live in close mutualistic relationships with plants. *Rhizobium* is a genus of nitrogen-fixing bacteria that lives in swellings called nodules in the roots of plants in the legume family, which includes peas, beans, clover and many other species. The bacteria take nitrogen from the air and use their own enzymes to convert it into a form that the plant can use. The plant provides sugars produced in photosynthesis, which the bacteria can use to obtain the energy needed for the conversion.

Legumes were the first group of plants that were found to have mutualistic nitrogen-fixing bacteria in their roots. However, more recent discoveries have shown that several different types of plants have similar relationships, with different species of nitrogen-fixing bacteria.

■ Figure 5.14 Nodules in the roots of a soybean plant

5.1.12 Soil texture defines the physical make-up of the mineral soil. It depends on the relative proportions of sand, silt, clay and humus

The relative abundance of clay, silt, sand and humus in soil determines the soil texture. Soil texture has a significant impact on the overall properties of soil, including whether it is suitable for agriculture or construction, for instance.

Experienced soil scientists can often recognize the different types of soil by observing the appearance and feel of it. Keys, such as the one shown in Figure 5.3 (page 419), can be used to identify which class of soil a sample represents (see Topic 5.1.2, page 417). A more reliable means of classifying soil is to separate the mineral particles and allow them to settle in a column of water.

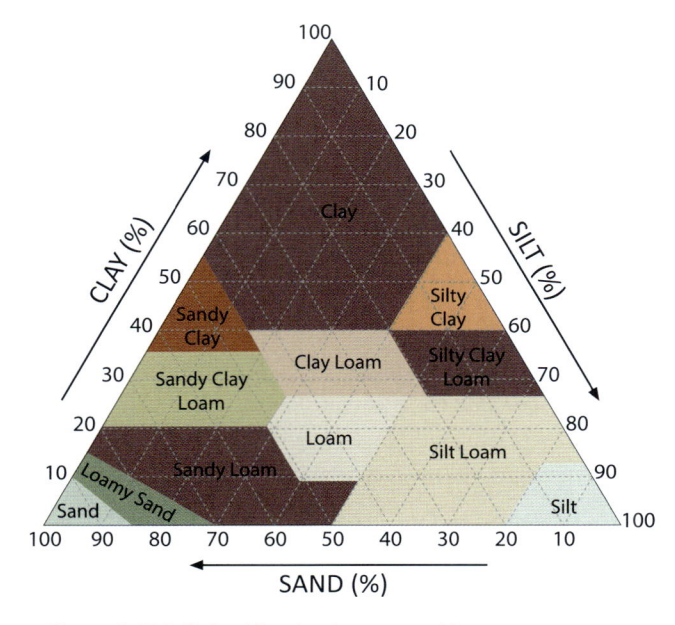

■ Figure 5.15 Soil classification by composition

> **Tool 1: Experimental techniques**
>
> **Classify a soil sample**
>
> Use the soil classification chart in Figure 5.15 to determine the classification of a soil sample.
> 1. Collect a soil sample from a site near to you. You may also consider collecting more than one sample, from different sites, for comparison.
> 2. Sift the soil through a sieve with a mesh of at least 2 mm, to remove any larger particles like gravel or organic debris.
> 3. Place the soil in a large measuring cylinder, to about one-third of cylinder's capacity.
> 4. Add clean water to the cylinder, until it is almost full, leaving just a little space at the top.
> 5. Cover the measuring cylinder as securely as possible, with a rubber bung or similar stopper.
> 6. Shake the measuring cylinder vigorously, to mix the soil thoroughly with the water.
> 7. Place the cylinder on a flat surface and leave it for 24 to 48 hours. Make observations for the first few minutes, then again after about two hours, and then again after 24 to 48 hours.
> 8. After about one to two minutes, there should be a layer of large-grained sand at the bottom. Measure the depth of this layer, using the scale of the measuring cylinder or a ruler.
> 9. After about two hours, there should be a new layer, containing silt. Measure the depth of this layer.
> 10. After 24–48 hours, a third layer, containing clay, should have settled. Measure the depth of this layer and of the entire column of soil.
> 11. At this stage, you should be able to see at least three distinct layers in the soil. Measure the depths of these layers. They should match with the three measurements you made previously, but there may be other bands if there are particles of different size classes within each of the categories (for example, there may be fine sand and coarse sand, which form separate layers).
> 12. Use the depths of the three main layers to determine the percentage of sand, silt and clay in your soil sample:
>
> $$\text{Percentage sand} = \frac{(\text{Depth of sand})}{(\text{Total depth})} \times 100$$
>
> 13. Use the chart in Figure 5.15 to determine the classification of your soil sample.

TOK

Test another portion of your soil sample by feeling it, and use the key in Figure 5.3 (page 419) to determine the soil classification. Did you get the same result with the two methods? Compare the methods in terms of ease of use, objectivity and reliability. Is the knowledge obtained from the two methods equally useful?

5.1.13 Soil texture affects primary productivity through the differing influences of sand, silt, clay and dead organic matter, including humus

The relative abundance of sand, silt, clay and humus in soil affects the ability of the soil to support plants. In the most fertile soils, plants have all their needs met and they can photosynthesize rapidly, giving the ecosystem a high primary productivity. If soil is unable to provide all the plant's needs, then primary productivity will be lower.

Plants need to obtain water, minerals and oxygen from the soil. Water is needed for photosynthesis, but it also helps transport materials through the plants and provides support to plant tissues. Minerals are needed for various functions within the plant, such as carrying out photosynthesis, or synthesizing the proteins and other substances needed. Oxygen is needed to allow plants' roots to respire, so that they have the energy to absorb minerals and carry out other life processes.

Each of the components of soil has different impacts on these important factors.

1 The large pores between sand particles mean that water and air can freely flow through the soil. Sandy soils are well aerated, meaning that oxygen is freely available, but also drain rapidly, meaning that water does not remain in the soil for very long. It may be necessary to water sandy soils frequently. The rapid flow of water through sandy soil also means that any minerals present are easily leached away. Due to lack of water and minerals, sandy soils are not very fertile.

2 Clay particles are so small that they pack together tightly, meaning the spaces between the particles are very small. Water does not flow freely through clay. Some clay soils have such low permeability that water sits on top of the soil and only infiltrates extremely slowly. Other clays may allow water to enter and fill up the tiny spaces, but it does not drain away easily. In either case, the soil become waterlogged, so there is no space for air to enter the soil. This means roots may not get enough oxygen to carry out respiration. In fact, clay soils can be so compact that roots may not even grow into the soil.

Clay soil may contain many useful minerals present, but they are often tightly bound to the soil particles, and with the lack of oxygen for respiration, these minerals are often unavailable to plants. Overall, due to the lack of available oxygen and the tightly compacted mineral particles, clay soils are often unsuitable for agriculture.

3 Humus provides many benefits to overcome some of the problems with sandy or clay soils. Humus coats the mineral particles, allowing them to clump together, forming relatively large aggregates. These aggregates may contain particles of various sizes, including sand, silt and clay. The large aggregates create larger pore spaces, which allow the free flow of water and air into the soil, but the nature of the humus means that water is held within the soil, and does not drain away too rapidly. Humus also gradually decays, releasing useful minerals that plants can absorb.

■ **Figure 5.16** Sandy, clay and loam soils. The different colours may be due to differences in the parent rock, or in the organic matter content, of the soils.

Humus-rich soil therefore provides many benefits, which make it more fertile and productive than either sandy or clay soils. It allows water to enter it, but doesn't dry out easily. It has large air spaces, allowing oxygen to diffuse in, so that plant roots can respire. The gaps between particles are large enough that waterlogging is unlikely. The humus retains sufficient water for plants and restricts leaching. Finally, the humus provides a slow, steady supply of minerals, allowing plants to absorb what they need.

The best, most fertile soils, which allow for the highest possible primary productivity, tend to be mixtures of sand, silt and clay, with a relatively high humus content. Soils such as this are described as loams, and they are found in the centre of the soil classification chart in Figure 5.15.

5.1 Soil

5.1.14 Soils can act as carbon sinks, stores or sources, depending on the relative rates of input of dead organic matter and decomposition

Soil contains carbon in the form of living organisms and dead organic matter. Taken together, all the world's soils form a significant store of carbon, and an important part of the global carbon budget. Carbon enters the soil when photosynthesizing plants accumulate carbon in their biomass, and transfer some of it into the soil as new root growth and exudates from roots. Plant biomass can also be passed into the soil when plants or parts of plants die and fall to the soil, or when soil organisms feed on plants. Carbon leaves the soil as humus undergoes mineralization and as soil organisms respire.

If soil takes in more carbon than it releases, it can be considered a carbon sink. If it releases more carbon than it takes in, it acts as a carbon source. Factors such as temperature and soil water can contribute to switching soil from being a source to a sink.

Global carbon storage in soils

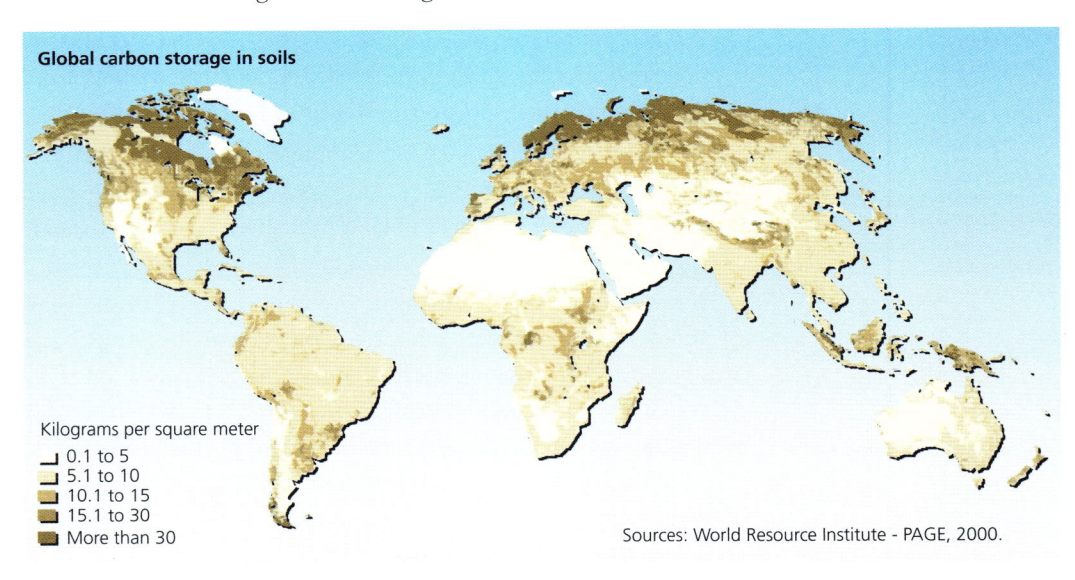

Kilograms per square meter
- 0.1 to 5
- 5.1 to 10
- 10.1 to 15
- 15.1 to 30
- More than 30

Sources: World Resource Institute - PAGE, 2000.

■ **Figure 5.17** Global carbon stores in soil

- At low temperatures, the rate of decomposition of soil organic matter is low, and soil acts as a carbon sink or store. As temperature rises, the rate of decomposition increases and carbon is released into the atmosphere, and the soil becomes a carbon source.
- Waterlogged soils are also more likely to act as carbon sinks. When soils are filled with water, there is insufficient oxygen for aerobic decomposition. Anaerobic decomposition tends to be a slower process, so waterlogged soils may accumulate organic matter, acting as a sink. This is particularly true where the temperature is low, which reduces the rate of decomposition even more.

■ Wetlands and peatlands

Naturally waterlogged soils are most often found in wetland ecosystems, including marshes, bogs and mangrove forests. The specially adapted plants living in these ecosystems can be very productive, and accumulate a great deal of carbon in their biomass and in the soil. The organic matter in the waterlogged soil often decomposes very slowly, so the soil acts as a sink and long-term store of carbon. In many cases, a layer of undecomposed organic matter called peat accumulates in the soil. Peat can form layers, from a few centimetres to several metres deep. These areas are called peatlands.

■ **Figure 5.18** A tropical peatland with rich, productive vegetation and waterlogged soil

> **REAL-WORLD EXAMPLE**
>
> ### Peatlands in the UK
>
> Peatlands, such as bogs and fens, occupy over 10 per cent of the land surface of the UK, but contain about half of all soil carbon stored in the country. At over 5 billion tonnes, this is far more than the carbon stored in plant biomass.
>
> Peatlands are threatened by climate change, drainage for agriculture, fires and excavation of the peat for use as a growing medium by gardeners and horticulturalists. Historically, dried peat was also used as a household fuel. Degradation of peatlands results in the release of carbon dioxide into the air. Degraded peatlands therefore become carbon sources.
>
> Currently the UK has approximately 3 million hectares of peatland, but only 22 per cent of that is in a 'near natural' condition. Given the importance of these as carbon stores, the UK government has committed to protecting and restoring peatlands as part of its commitment to the Paris Agreement on climate change, signed in 2015. It aims to restore and conserve peatlands to achieve 2 million hectares of healthy peatland by 2040.
>
> ■ **Figure 5.19** Peatland in the UK

The Arctic tundra

In the cold tundra of the Northern Hemisphere, peat has accumulated in the soil and been frozen. Where the soil remains continuously frozen for thousands of years, it is called permafrost.

The permafrost has accumulated organic matter for thousands of years, adding more each summer when plants photosynthesize rapidly, and freezing it in the winter. Frozen organic matter cannot decompose. It is estimated that the top three metres of the permafrost holds at least 1.4 trillion tonnes of carbon. This represents a substantial, long-term store.

Global warming, however, has changed the balance between accumulation and release of carbon. Arctic temperatures have risen significantly faster than the global average. This is causing larger areas and deeper layers of the permafrost to thaw for the first time in millennia. The thawed organic matter can now undergo decomposition, releasing carbon into the atmosphere.

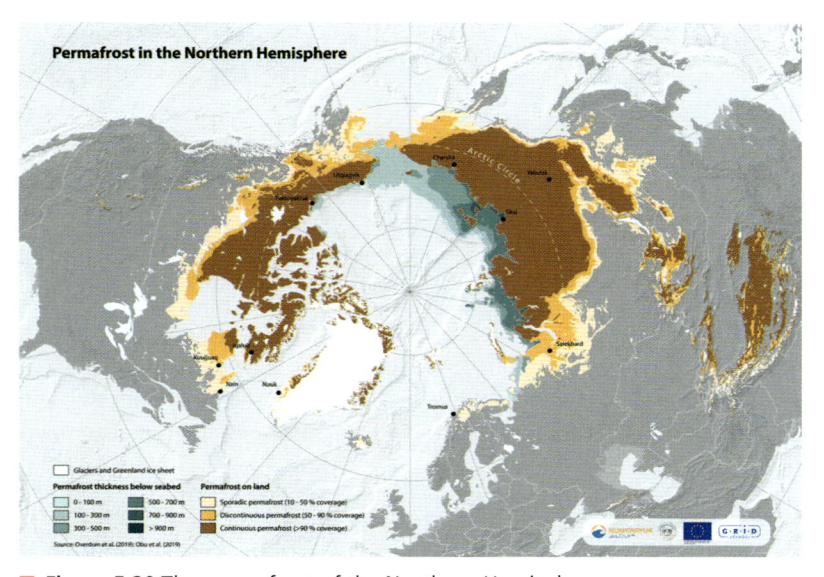

■ **Figure 5.20** The permafrost of the Northern Hemisphere

5.1 Soil

> **Link**
>
> Global warming is covered in Chapters 6.1 (page 514) and 6.2 (page 528). Systems and feedback loops are explored in Chapter 1.2 (page 24).

Furthermore, as the ice of the permafrost thaws, the liquid water floods and waterlogs the soil, so that there is limited oxygen present in the soil. This results in anaerobic decomposition, producing the powerful greenhouse gas methane. The impact of the thawing permafrost, therefore, is to release a gas that traps even more heat in the atmosphere, causing more warming. This is a clear example of positive feedback.

> **Tool 4: Systems and models**
>
> **Illustrate the accumulation and release of carbon in the permafrost**
> 1. Draw a pair of systems diagrams to illustrate the accumulation and release of carbon in the permafrost under historical (freezing) and current (warming) conditions. Use arrows of different sizes to illustrate the relative size of the carbon inputs and outputs.
> 2. Draw a flow diagram to illustrate the positive feedback between Arctic thawing and global warming. What sort of change in temperature is expected if positive feedback operates on it?

Temperate grasslands

Grasslands, especially in temperate climates, are another valuable carbon store. Grasses grow rapidly, accumulating significant amounts of carbon during photosynthesis. They tend to transfer most of this carbon – around 85 per cent – in their roots. In temperate grasslands, this allows much of the energy accumulated by plants during the summer to be stored through the winter, and used again for new growth the next year. Root storage also protects the grasses from disturbances such as grazing and fires – the roots remain largely untouched by these, and the stored energy allows the grasses to grow back after the disturbance.

The storage of carbon in roots makes temperate grasslands an important carbon store, especially considering how widespread they are globally. The major threat to grasslands is the conversion to agriculture or urban development. When the native grasses are removed, the accumulation of carbon in the soil will stop, but decomposition will continue, turning the soil from a carbon sink to a carbon source. Eventually all the stored carbon will be returned to the atmosphere.

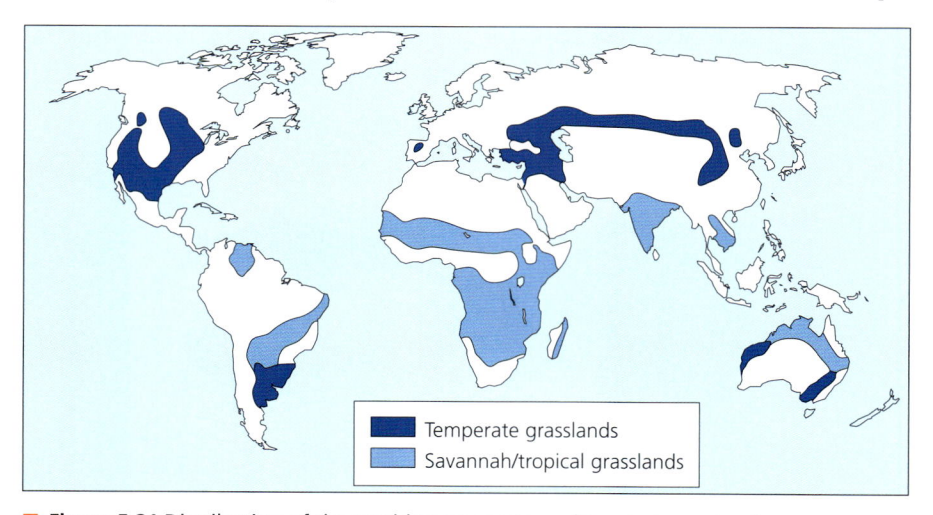

■ **Figure 5.21** Distribution of the world's temperate and tropical grasslands

Tropical forests

The tendency of grasses to store carbon in the roots distinguishes them from trees. Most carbon storage in trees occurs in above-ground biomass, such as the trunks and larger branches.

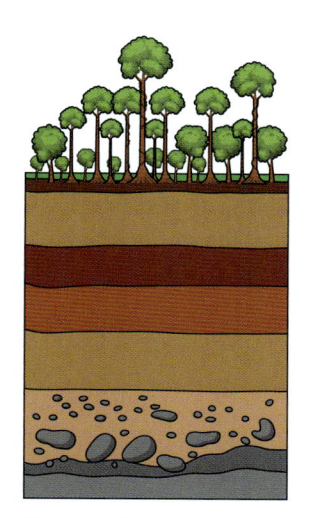

Figure 5.22 The thin topsoil and poor subsoil of a tropical rainforest

When trees are cut down, these parts gradually decompose, returning the carbon to the air. If trees are burned, the carbon returns to the air almost immediately. This makes carbon storage in trees highly vulnerable to disturbance.

While forest trees can have extensive root systems, they tend not to be used for storage. In fact, many tropical forest soils have quite low organic-matter content. They may have a thin layer of humus in the topsoil (A horizon), but the subsoil is often nutrient-poor and carbon-poor. This is because decomposition and nutrient cycling are so rapid in the warm, moist, tropical environment. Leaf litter decomposes rapidly and nutrients are returned almost immediately to the trees. Tree roots tend to spread horizontally, rather than penetrating deep into the soil.

Tropical forests can play an important part in moderating the greenhouse effect, by absorbing carbon dioxide and storing it in biomass. However, that biomass can be easily destroyed, returning the carbon to the atmosphere. In order to ensure the long-term storage of carbon in a tropical forest, the living trees must be protected.

(HL) 5.1.15 Soils are classified and mapped by appearance of the whole soil profile

Soils around the world vary widely in their characteristics and properties. To compare soils found in different places, soil scientists look at the complete soil profile, drawing diagrams showing the depths and character of each horizon or layer. This sort of analysis has shown that soil characteristics differ in predictable ways. For instance, different biomes often grow on soils with different profiles.

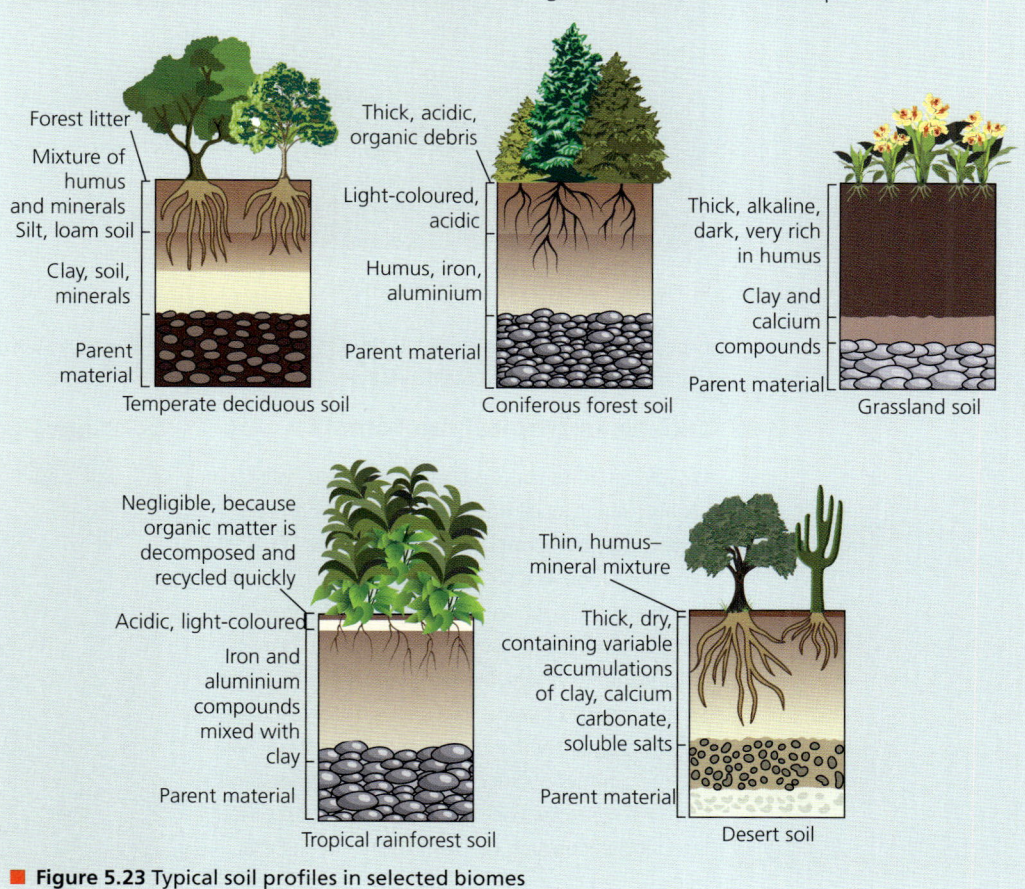

Figure 5.23 Typical soil profiles in selected biomes

Link

Biomes are explored in Chapter 2.4 (page 163).

Brown earths in temperate deciduous forests

The typical soil in temperate deciduous forests of Europe, Asia and North America are called brown earths, brown soils or brown forest soils. Brown earths typically have:

1. A relatively thin organic layer of decomposing leaf litter and mildly acidic humus.
2. A relatively thick layer of dark brown topsoil, with a mixture of humus and mineral particles, and many animals and plant roots.
3. A lighter brown layer of subsoil, mostly made up of mineral particles weathered from the parent rock, along with some organic matter that has been mixed in by animals and plant roots. Due to the mixing, the boundary between the darker topsoil and the lighter subsoil is not always clear.
4. An underlying layer of weathered parent material.

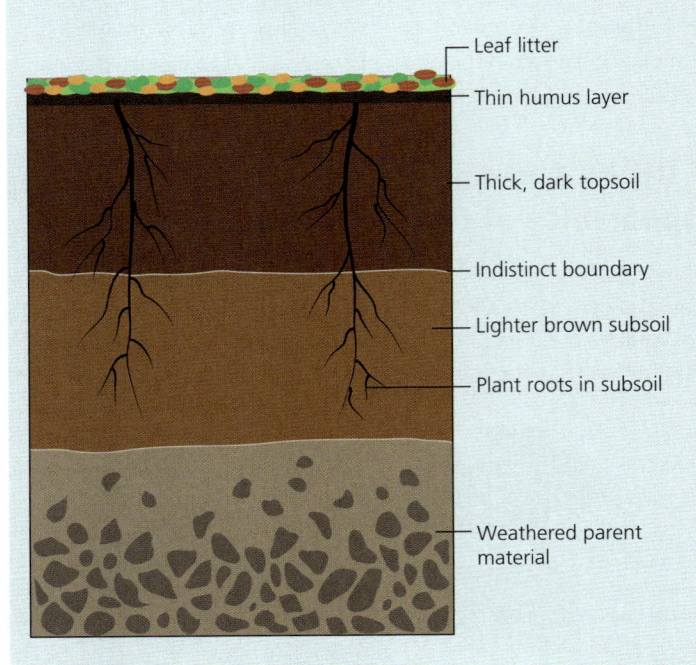

■ **Figure 5.24** Soil profile of brown earth

Brown earths are widespread in temperate regions, where they support forests under natural conditions. They are typically relatively fertile, and have often been cleared for cultivation.

Oxisols in tropical rainforests

Oxisols are reddish or yellowish soils with very thin organic and topsoil horizons. Below the topsoil, further layers are not easily distinguished. They are typically found in equatorial regions of Africa and South America, where they support the growth of tropical rainforests. Oxisols are old soils, formed in environments where the climate has not changed for thousands of years. They are heavily leached due to regular rainfall during their formation, so they retain few useful nutrients and have generally low fertility. However, rapid decomposition of leaf litter and absorption by plant roots, means that undisturbed oxisols can support rich vegetation. When used for farming they usually require large inputs of fertilizer.

■ **Figure 5.25** Oxisols have reddish or yellowish colour and indistinct layers

Soil maps

Soil maps are useful tools for identifying where different soils are found. Producing soil maps requires looking at entire soil profiles. Soil maps have been produced for many parts of the world, in part to help plan appropriate approaches to farming in different areas.

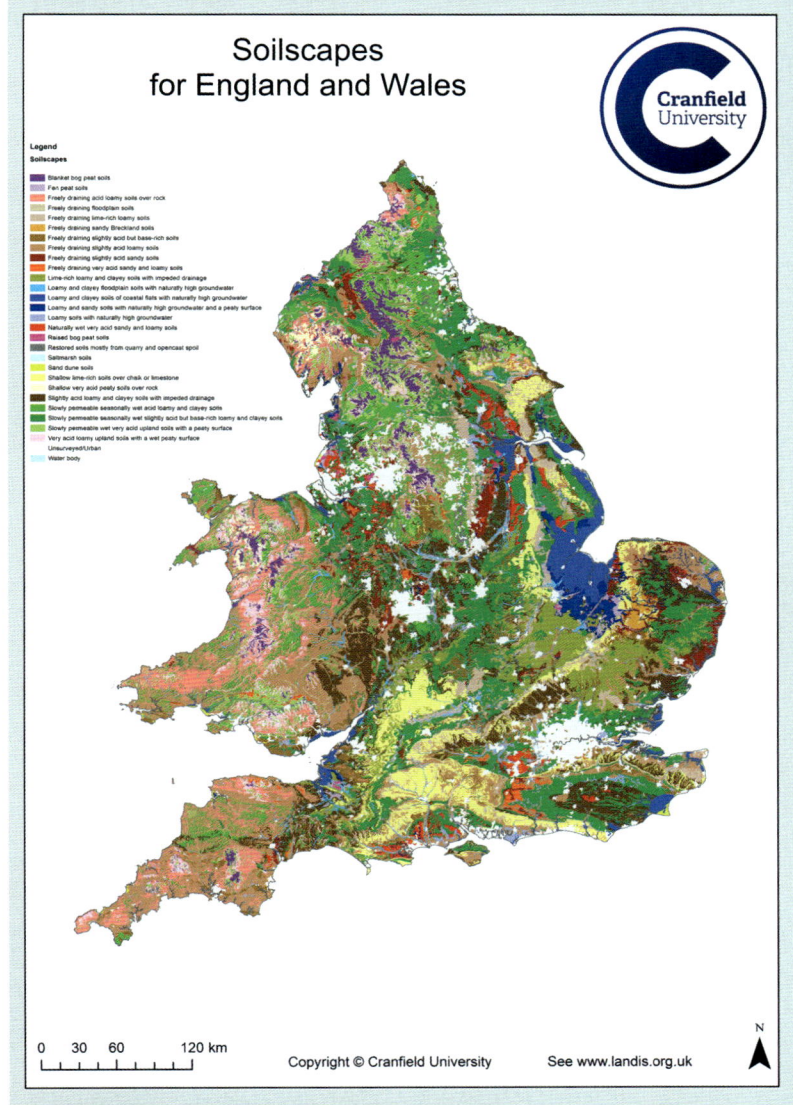

■ **Figure 5.26** Soil map of England and Wales

In Figure 5.26, different colours represent different soil types. The legend shows information about the different layers found in each soil type.

5.1.16 Horizons are horizontal strata that are distinctive to the soil type

Natural soil profiles

Although soil profiles can differ widely, as illustrated by the brown earths and oxisols, there are some common patterns that have allowed soil scientists to identify a number of general layers or horizons that are found in most soils. These horizons are named O (for organic), A, B and C. Beneath the C horizon lies the solid bedrock or parent material, which is sometimes known as the R horizon (R for rock).

5.1 Soil

Table 5.5 shows the nature of the five generalized soil horizons, O, A, B, C and R.

■ **Table 5.5** Typical soil horizons

Horizon	Name	Description
O	Organic layer	A layer of decomposing organic material, such as leaf litter. Sometimes called the humus layer.
A	Topsoil or mixed layer	Weathered mineral particles mixed with a significant amount of organic matter. This horizon contains the most decomposers, is the most fertile and important for plant growth.
B	Subsoil or mineral layer	Mostly mineral particles that have washed downwards from the upper layers. Contains little organic matter.
C	Parent material	Broken up, or weathered, fragments of the underlying rock.
R	Bedrock	Underlying rock that cannot easily be broken up with a spade or shovel.

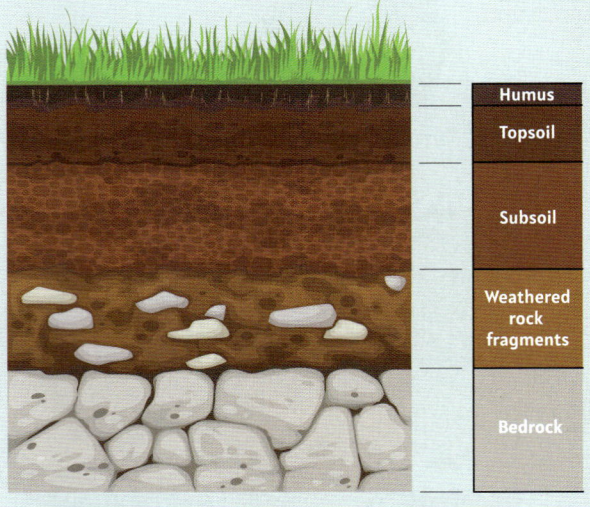

■ **Figure 5.27** A soil profile showing the five main horizons

Common mistake

Eluviation and **illuviation** are very similar words, but they have roughly opposite meanings. The first letter makes the difference. The E in eluviation is a prefix that means 'away' or 'out', as in the E from *exit*. The I in illuviation is a prefix meaning 'into'. Therefore, eluviation is washing *away* and illuviation is washing *into*.

◆ **Eluviation** – the process of washing substances downwards, *away* from a soil layer, into a layer below.
◆ **Illuviation** – the process of accumulation of substances *in* a soil layer, which have been washed away from a layer above.

In some soils, an additional horizon forms between the A and B horizons, called the E or eluviated horizon. This layer is left behind when certain substances, including the smaller, clay particles, are washed downwards by water, into the subsoil, or B horizon. The remaining E horizon tends to be rich in silts and sands.

■ **Figure 5.28** A soil profile showing a pale E (eluviated) horizon

Soil profiles in agricultural systems

Intensive, continuous cultivation can have significant impacts on soil profiles. In particular, ploughing tends to remove the uppermost O and A profiles, leaving only the subsoil (B) and parent material (C) horizons.

Farming typically involves clearing the natural vegetation from the land and ploughing the top layers of the soil. The removal of the vegetation exposes the organic (A) horizon to erosion by wind and water. Ploughing breaks up the soil surface and mixes the top few centimetres of the soil. The broken up surface is more easily eroded, and the mixing results in the A and B layers (and E if present) becoming combined into a man-made 'plough layer'.

Figure 5.29 Ploughing breaks up and mixes the top layers of soil

5.1.17 The A horizon (mixed layer or topsoil) is the most valuable for plant growth but, along with the O horizon, is also the most vulnerable to erosion and degradation

The A horizon, also known as topsoil or the mixed layer, is made up of a mixture of mineral particles and organic matter, and lies immediately below the organic layer of leaf litter and humus, known as the O horizon. The topsoil is typically rich in organic matter, which moves downwards from the O horizon, and is the site of most activity by decomposers. As a result, the A horizon contains a high concentration of useful nutrients, and is typically full of plant roots, fungi, bacteria and various other soil organisms.

Since the topsoil is so near the surface, and contains so many burrowing animals and plant roots, it is also rich in oxygen. Air flows freely into the spaces within the topsoil, providing oxygen for the animals, plants and decomposers.

For these reasons, topsoil is the most important component of soil for plant growth, as well as for carbon storage, nutrient cycling and soil biodiversity. The quality of topsoil determines how fertile the soil is.

However, topsoil, and the O horizon, are also highly vulnerable to damage and loss, especially when soils are intensively farmed. Ploughing encourages soil erosion and blending of the topsoil with the lower, subsoil layers. Over time, this leaves the soil less fertile, meaning that farmers need to add more and more fertilizers to the soil each year.

> **Concept**
>
> **Sustainability**
>
> The erosion and degradation of topsoil, leading to decreasing soil fertility and increasing use of fertilizers, is an indication of unsustainability in soils and agriculture. Sustainable farming maintains soil quantity and quality, and even achieves improvements over time, and does not require ever-increasing input of fertilizers.
>
> Given the growth of the human population, sustainability in agriculture also means increasing the quantity of food produced globally. But this would have to be done without degrading soils or using more fertilizers – which are expensive, cause pollution, and require the use of energy and natural resources for their production.

5.1.18 Factors that influence soil formation include climate, organisms, geomorphology (landscape), geology (parent material) and time

Many factors can influence the way a particular soil forms and develops. A widely used model for understanding soil formation lists five key factors and uses the mnemonic CLORPT, which stands for:

- **Climate**
- Organisms
- **Relief**
- Parent material
- Time.

■ Table 5.6 Five factors that influence soil formation: CLORPT

Factor	Definition	Influence on soil development
Climate	The long-term patterns of weather, including temperature and precipitation patterns	Temperature (and temperature changes) and precipitation have a direct effect on weathering, speeding up or slowing down processes in mechanical, chemical and biological weathering.
		Climate also affects the types of ecosystems that develop in an area, which can themselves affect soil formation.
Organisms	Animals, plants and microorganisms such as fungi and bacteria	Animals and plant roots burrow through soil, aerating it and mixing its components in a process called **bioturbation**.
		Microorganisms secrete enzymes that chemically change soil components, including decomposing organic matter, and leaching ions from mineral particles.
Relief	Variation in heights or elevation of the land, and direction (or **aspect**) and gradient of slopes. Relief is one element of the **geomorphology** or form of the land.	Land structure affects exposure to sunlight, temperature changes, precipitation and drainage. These can affect both weathering and erosion, and **deposition** or accumulation of eroded materials in particular places.
Parent material	The raw materials that undergo weathering to produce soil. Parent material of a soil may include the underlying bedrock or eroded sediments that have arrived from another site.	The chemical composition of the parent material determines the composition of the mineral particles in the soil, and the physical and chemical processes that they can undergo. Some types of parent materials, such as **calcareous** rocks, may produce nutrient-poor, infertile soils, while others, such as volcanic rocks, may produce nutrient-rich, fertile soils.
Time	How long a soil has existed.	The age of a soil affects the degree of weathering it has experienced. Older soils are likely to have undergone more weathering than younger ones.
		The length of time during which a soil is undisturbed by processes such as **glaciation** also affect the soil's nature.

- ◆ **Climate** – long-term weather patterns, including temperatures, winds and precipitation. Climate varies around the globe and can change over time. Some soils may have experienced different climates over the course of their formation.
- ◆ **Relief** – variation in height or elevation of the land. A large change in elevation over a short distance produces a steep slope or gradient which might be prone to erosion, due to gravity or fast-flowing water.
- ◆ **Bioturbation** – the disturbance or mixing of soils by living organisms, such as burrowing animals.
- ◆ **Aspect** – the direction an area of land, such as a slope or mountainside, faces. The aspect may be towards or away from the sun, wind or rain, for instance. This may affect the processes of weathering and erosion.
- ◆ **Geomorphology** – the shape of the surface of the Earth, including mountains, valleys, plains, etc.
- ◆ **Deposition** – the accumulation of sediments in a particular place, after they have been transported from elsewhere. Deposition may occur in low-lying areas, at the bottoms of slopes or in the lower course of a river.
- ◆ **Calcareous** – describing a rock or soil that contains a high percentage of calcium carbonate.
- ◆ **Glaciation** – the effects of glaciers on the rocks and soils. Glaciers flow slowly over the land, scouring the soil away from where it first forms and depositing it elsewhere.

REAL-WORLD EXAMPLE

Ferralsol in the Amazon Basin

Various different soils are found under the Amazon rainforest of South America, but the most common type is an oxisol called ferralsol. Ferralsols are made up of unreactive clay particles, which are rich in iron and aluminium, but have very low levels of other useful minerals. They are reddish or yellowish as a result of iron particles, and can be very deep – up to 20 m in some places, with no visible horizons between the topsoil and the parent material.

The formation of the Amazon's ferralsols is related to time and climate, as well as to the underlying parent material. This area of the world has remained within the humid topical climate zone for many millions of years, allowing continuous, uninterrupted weathering of the parent rock. The parent material is described as basic rock, with relatively high iron and aluminium content, but low levels of silica (quartz). The process of formation is very slow, so ferralsols are only found in areas that are geologically stable. Continuous rainfall has leached the nutrients from the soil, leaving it generally infertile.

■ **Figure 5.30** Iron-rich ferralsols are found in humid tropical zones, in particular in the Amazon basin

REAL-WORLD EXAMPLE

Chalk soils of the South Downs, England

The bedrock underlying much of the South Downs National Park in southern England is chalk. Chalk, and the related rock limestone, are sedimentary rocks formed from the accumulation of the shells of dead marine organisms, and made up primarily of the compound calcium carbonate, $CaCO_3$. Rocks and soils containing a high percentage of calcium carbonate are described as calcareous.

When chalk weathers, it produces white, chalky soils, which drain freely and are nutrient poor. Chalk soils are therefore generally infertile and unsuitable for agriculture, but can produce very biodiverse meadows. Chalk meadows in the South Downs often support a wide variety of small, slow-growing plants, including many attractive wildflowers. Since the soil's fertility is low, plant growth is slow and many different plants can coexist without outcompeting each other.

■ **Figure 5.31** Wildflowers growing in a chalk meadow in the South Downs, England

REAL-WORLD EXAMPLE

Volcanic soils of Naples, Italy

Near the city of Naples in southern Italy stands the volcano Mount Vesuvius, which is well known for having destroyed the ancient city of Pompeii in the year 79 CE. Eruptions of Vesuvius, including particularly large ones approximately 12,000 and 35,000 years ago, have provided the parent material for much of the soil in the area around Naples.

These eruptions blanketed the area with fragments of volcanic rock called tephra. Over time, the tephra has weathered, producing a rich volcanic soil, which has supported a successful farming industry for thousands of years. Similar rich soils are found in the vicinity of volcanoes all around the world. Farming communities may be attracted to such areas, despite the risk of eruptions.

■ **Figure 5.32** Grape vines growing on volcanic soil produced by eruptions of Mt Vesuvius, near Naples, Italy

In contrast, the surrounding areas outside of the volcano's influence are dominated by calcareous soil formed from limestone bedrock. These areas are much less productive in terms of agriculture.

5.1.19 Differences between soils rich in sand, silt or clay include particle size and chemical properties

The mineral particles in soil occur in different sizes, classified into sand, silt and clay (see Topic 5.1.2, page 417). These particles may be made up of different chemicals, which behave in different ways.

Sand and silt, for instance, are typically composed of the mineral quartz. Quartz is a very common mineral, found in a variety of rocks. It forms large molecules of the elements silicon and oxygen, with the formula SiO2. The chemical name for this compound is silica.

Clay particles, however, are made of various complex compounds called silicates. Silicates also contain silicon and oxygen, along with various other elements. Clay silicates might include aluminium, magnesium or iron, for instance. The precise chemical makeup of a particular clay depends on the nature of the parent material it weathered from. Table 5.7 gives the names and chemical formulae of a few common clay minerals. NB You do not need to memorize these examples.

■ Figure 5.33 Granite rocks such as this are rich in quartz, and may undergo weathering to form sand

■ **Table 5.7** Selected minerals found in clays

Mineral name	Chemical formula
Kaolinite	$Al_2Si_2O_5(OH)_4$
Pyrophyllite	$Al_2Si_4O_{10}(OH)_2$
Talc	$Mg_3Si_4O_{10}(OH)_2$
Chlorite	$(Mg,Fe)_3(Si,Al)_4O_{10}(OH)_2(Mg,Fe)_3(OH)_6$
Vermiculite	$(Mg,Fe,Al)_3(Al,Si)_4O_{10}(OH)_2 \cdot 4H_2O$

The more complex chemical composition of clay particles means that they take part in a wider range of chemical reactions in the soil. One process that is particularly important for the fertility of soil is called cation exchange.

Cation exchange in soils

Many nutrients that plants need to obtain from soils are positively charged ions, or **cations**, such as calcium (Ca^{2+}), magnesium (Mg^{2+}) and ammonium (NH^{4+}). These cations are attracted to negative charges that are found on the surface of the complex silicate compounds in clay particles. Since clay particles are very small, they have a very high total **surface area**, so they have a very high capacity to attract these cations. The cations become loosely attached, or **adsorbed**, to the surface of the clay particles.

◆ **Cation** – a positively charged ion, most often formed when an atom loses one or more electrons.

◆ **Surface area** – the space available on the surface of an object. Many small objects of a given mass would have a higher total surface area than a few larger objects of the same total mass.

◆ **Adsorption** – the process of becoming attached to the surface of an object.

● Common mistake

It is easy to confuse cations and anions, and to forget which is positive and which is negative. There are a few different memory tricks to help remember which is which. One is to replace the t in cation with a +, which looks similar: *cation* becomes *ca+ion*. Another is to note the first and last letters in **a n**egative **ion** spell out *anion*.

5.1 Soil

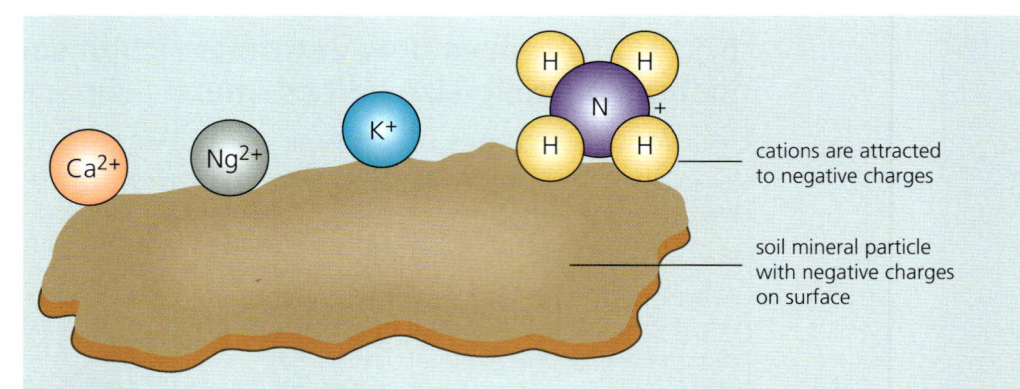

Figure 5.34 Cations adsorbed to a soil particle

Since the cations are attracted to the clay particles, they are unlikely to be leached, or washed away, when water flows through the soil. This means that the soil holds on to the useful substances that plants need.

However, this also means that the cations are not easily released for plants to take in. Plants can overcome this by supplying alternative cations, usually hydrogen ions, H⁺, which can replace the cations on the soil particles. This exchange of useful cations for H⁺ ions is called **cation exchange**. Soils such as clays that have the ability to hold lots of cations have a high **cation exchange capacity (CEC)**. Soils with a high cation exchange capacity are generally more fertile than others.

◆ **Cation exchange** – the transfer of positive ions from soil particles into plants, in exchange for positive ions such as hydrogen ions, which are released by the plants.

◆ **Cation exchange capacity (CEC)** – the ability of a soil to hold cations. CEC is a measure of how many negative charges are present in the soil. A high cation exchange capacity indicates a fertile soil.

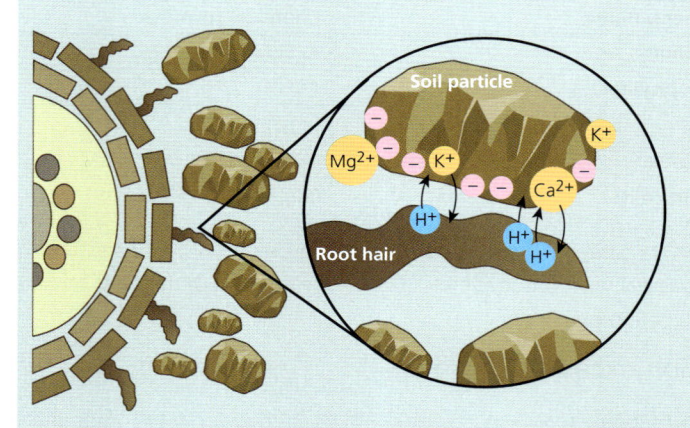

Figure 5.35 Exchange of cations between a soil particle and the root of a plant

● Common mistake

Clay soil typically has a high cation exchange capacity, meaning that many cations are present in the soil, and the soil can be described as fertile. However, plants may not be able to absorb the cations easily. Since the cations are strongly attracted to the soil particles, plants must expend energy to pump hydrogen ions into the soil, in order to obtain the useful cations. This energy is released from food by the process of respiration, but since clay soils are often poorly aerated due to the small pore size and tendency to waterlogging, aerobic respiration is often limited, so plants may not have sufficient energy for this process. As a result, although clay soils are often fertile, they may also be difficult to cultivate, and unsuitable for agriculture.

Theme 5: Land

Sand and silt form much larger particles than clay, so there are fewer particles in a given volume of soil. As a result, there is typically a smaller total surface area available for cations to become attached to (see Figure 5.36). The quartz or silica that they are made of is also less likely to have negative charges on its surface. As a result, sand and silt particles have much lower cation exchange capacity than clay does. Cations in sandy or silty soils are only weakly held, and will easily wash away when water passes through the soil.

> ● **Common mistake**
>
> The relationship between surface area and the size of an object can be confusing. It is true that one large object will have a larger surface area than one small object. However, if we break one large object up into many smaller objects, we increase the total surface area of all the small objects, taken together.
>
> Surface area = 6 m²   Surface area = 12 m²   Surface area = 24 m²
>
> ■ **Figure 5.36** Total surface area increases when an object is broken into smaller pieces
>
> In soils, if we compare one cubic centimetre of sand particles to one cubic centimetre of clay particles, the total surface area of the clay particles may be thousands of times higher than that of the sand particles.

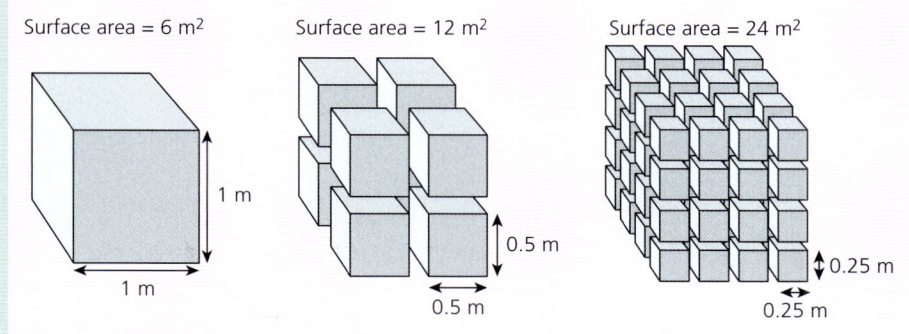

5.1.20 Determining soil properties

Soil composition

Soil composition refers to the proportions of sand, silt and clay in soil. It can be investigated using the method described in Topic 5.1.12 (page 433). Depending on the proportions, soil can be classified as sandy, silty, clay, or some combination of these, using the soil triangle in Topic 5.1.3 (page 420).

The type of soil in a particular location may influence the types of crops grown there. Different plants tend to have different needs, in terms of drainage, water retention and nutrient availability, for instance.

> **ATL ACTIVITY**
>
> Choose five food crops that form part of your own normal diet. Consider fruits and vegetables, as well as crops such as wheat, corn or barley, which are used to produce foods you eat.
> - Use the internet to find out what sorts of soils each of your chosen crops grow best in.
> - Try to find out why each crop prefers that soil type.

Organic matter content

Organic matter improves many soil characteristics, including:
- water retention in sandy soils
- drainage in clay soils
- nutrient availability
- number and diversity of microorganisms.

◆ **Buffer** – a substance that resists changes in pH. When acid or alkali is added to a buffer, the pH remains relatively constant.

Link

The hydrological cycle is explored in Chapter 4.1 (page 268).

Organic matter influences soil's physical structure by sticking mineral particles together to form aggregates. This tends to improve drainage and aeration of soils, which benefits plants and other soil organisms. It improves the biological function of the soil by providing food for microorganisms, in particular those involved in nutrient cycling, and nutrients for plants. It influences the chemical conditions of the soil by increasing the soil's ability to hold on to useful nutrients, preventing them from leaching away. The organic matter also acts as a **buffer**, which helps to maintain a stable pH in the soil.

Moisture content

Water is an important component of soil. It is necessary for plant growth and provides the medium that plant nutrients dissolve in. However, excessive water, or waterlogging, can harm plants by preventing them from obtaining oxygen for respiration. Soil water is also an important factor in the hydrological cycle, as it relates to the infiltration, percolation and underground flow of water.

Soil moisture content is highly variable and is affected by the amount of rainfall or other forms of precipitation, the rate of evaporation and the rate of percolation through the soil. Percolation is affected by the nature and structure of the soil and its organic matter content. Clay soils tend to retain water in their tiny pores and stay wet, while sandy soils drain readily and become dry quickly. The presence of impermeable soil or rock in a lower horizon may prevent the topsoil or subsoil from drying out. Organic matter in soil can help to retain water in sandy soils or to drain clay soils.

If the soil is covered by a thick layer of organic matter such as leaf litter, this may reduce the rate of evaporation and keep the soil moist. Vegetation cover can also provide shelter from the sun and reduce evaporation, but plants also draw water out of the soil and can increase evapotranspiration.

Measuring water quality can help farmers to determine whether and how much they need to irrigate. It is also important for understanding the flows of water in the hydrological cycle.

Infiltration rate

Infiltration rate refers to the rate at which water enters the soil. It can be affected by the nature of the soil, the initial moisture content and the degree of compaction. A low infiltration rate may be an indication of a soil that is waterlogged, compacted or poorly aerated, making it unsuitable for many crops.

Water added to soil with a low infiltration rate is likely to stand on, or run over, the surface, leading to flooding or erosion. Farmers might use the infiltration rate to decide how much water to apply when irrigating the land.

Common mistake

Infiltration rate can be easily confused with drainage or percolation rate. Infiltration refers to the entry of water into the soil, whereas percolation or drainage refer to the passage of water through the soil into deeper layers. A method to measure drainage (or percolation) rate is described in Topic 5.1.3 (page 420).

Tool 1: Experimental techniques

Measure the basic infiltration rate of soil

1. Hammer a short length (approximately 25–30 cm) of pipe with a relatively large diameter (approximately 20 cm) into the soil, until it is about halfway into the soil.
2. Pour water into the pipe, to a depth of about 10 cm, and mark this level on the pipe.
3. Use a metre rule to measure the depth of water in the pipe at 1- or 2-minute intervals.
4. As the water level falls, and in between readings, add more water to the pipe to return the level to approximately the original level.
5. Each time water is added, record the new level and continue to measure the level at 2-minute intervals.

■ **Figure 5.37** Measuring the infiltration rate of soil

Unless the soil is waterlogged, the initial infiltration rate will probably be fairly high as water first enters the soil. Over time, however, the rate should slow down until it becomes constant. As the rate slows down, you may take measurements less often. When the fall in water level remains the same after each interval, record this rate as the basic infiltration rate, expressed in millimetres per hour (mm/h).

Bulk density

Bulk density refers to the density of a set volume of soil. The mineral particles in soil generally have similar densities, so bulk density is mainly influenced by the soil's porosity and its organic matter content. Porosity is a measure of how much space there is between the solid particles in the soil. These spaces, or pores, may be filled with air or water. Soil with high porosity tends to have a low bulk density. Organic matter is also less dense than the mineral particles, so soil with high organic matter content will also have low bulk density.

To measure bulk density, a known volume of soil is collected, usually using a metal ring of known volume. The collected soil is dried completely in an oven and weighed. The bulk density is calculated by dividing the dry mass by the volume. The units of bulk density are g/cm^3.

■ **Figure 5.38** A soil sample in a bulk density ring

In general, low bulk density is advantageous for plant growth, and therefore for farming. Low bulk density usually indicates high organic matter content and/or high porosity, both of which help plants grow. High bulk density due to low porosity can restrict the growth of plant roots. High bulk density may result from compaction by people walking or driving heavy machinery on soil.

ATL ACTIVITY

Bulk density and organic matter content of soils

Compare the bulk density and organic matter content of a variety of soils, using the method outlined here to measure bulk density and the procedure described in Topic 5.1.3 (page 420) to measure soil organic matter content. You may choose to compare soils from different areas, such as forest, grassland or farmland.
- Do your results support the hypothesis that high organic matter content produces low bulk density?

Inquiry process

Inquiry 1: Exploring and designing

Suggest how to extend the above investigation to determine whether there is a correlation between bulk density and organic matter content in soil.
1. Consider what statistical test you would use and how many samples you would need to take.
2. How would you ensure that your samples:
    a. provide you with a wide enough range of values of organic matter
    b. can provide a valid comparison, which is unaffected by other variables?

Colour

Examining the colour of soil can provide a quick and easy way to evaluate soil. Soil colour can be influenced by several factors, as shown in Table 5.8.

■ **Table 5.8** Factors affecting soil colour

Factor	Effect on soil colour
Parent material	Soil colour may reflect the colour of the parent material.
Organic matter content	Soil with high organic matter content is likely to be black, due to the presence of humus.
Moisture content	Moist soil tends to be darker. Specific patterns of colours might indicate changes in water content, such as past waterlogging.
Iron and oxygen content	Iron-rich soil may occur in different colours, which are influenced by how well-aerated the soil is. Yellow is the most common colour. Red occurs in soil which is highly aerated. Greyish, bluish or greenish colours occur in soils which are poorly aerated, perhaps due to regular waterlogging.

Colour can be evaluated by comparing the soil to a standard soil colour chart. The most commonly used chart is the Munsell colour chart. This allows a degree of objectivity in the assessment of colour.

■ **Figure 5.39** A page from the Munsell colour chart

Colour provides a number of clues about the soil. A scientist or farmer might use an evaluation of colour as a first step, which might be followed by more precise testing.

pH

Soil pH has significant effects on plant growth and nutrient availability. Several factors can contribute to the pH of soil, including:
- Rainfall adds carbonic acid to soil; acid rain adds stronger acids, such as sulphurous and nitric acids.
- Respiration by plants releases carbon dioxide, which forms carbonic acid in soil water.
- The release of hydrogen ions (H^+) by plants in exchange for cation uptake (see Topic 5.1.19, page 447) makes the soil more acidic.
- Weathering of some minerals produces either acidic or alkaline products.
- Use of fertilizers containing ammonium can cause acidity.
- Addition of limestone to soil can reduce acidity or cause alkalinity (this is sometimes done to counteract the acidifying effect of fertilizers).

The pH of the soil can affect the availability of various nutrients. Different nutrients are released into the soil water more readily at different pHs. For instance, iron remains attached to soil mineral particles at neutral pH, but dissolves in the water under acidic conditions.

5.1 Soil

Extremely high or low pH can also be toxic for plants or cause the soil to become toxic. In strongly acidic soils, aluminium ions from mineral particles may dissolve in the soil water, where they inhibit the growth of plants' roots.

5.1.21 Carbon is released from soils as methane or carbon dioxide

Scientists are very concerned about the release of carbon from soils, and how this affects the global carbon cycle and contributes to global warming. Soils can contain a significant amount of carbon, stored as organic matter (see Topic 5.1.14, page 436). Depending on conditions, carbon may be stored in the soil, or released as either carbon dioxide (CO_2) or methane (CH_4). Both gases are major greenhouse gases. Carbon dioxide is of greatest concern as more of it is released. Methane is released in smaller quantities, but it has much higher global warming potential.

Carbon accumulates in soils as humus, a stable form of carbon-rich organic matter. Humus can remain in soil for a long time, but eventually it may decompose to release either methane or carbon dioxide. Decomposition occurs when microorganisms such as bacteria and fungi use the organic matter as a source of energy. Ultimately, they break down the carbon-rich compounds during respiration, producing either carbon dioxide or methane.

Carbon dioxide is produced by decomposers carrying out aerobic respiration. This process requires oxygen, so it can only occur if the soil is well aerated. Methane is produced by anaerobic respiration, which can occur in poorly aerated soils, such as waterlogged fields (for example, rice paddies) or wetlands.

Both aerobic and anaerobic respiration are affected by conditions in the environment, such as temperature and soil chemistry. The rate of both processes increases at higher temperatures, so it is likely that the rate of decomposition in soils will increase as the Earth's temperature rises due to global warming.

This can lead to a positive feedback loop, causing global warming to accelerate. Higher temperatures causes faster decomposition, which releases more greenhouse gases, leading to further temperature increases.

Link

The global carbon cycle is discussed in Chapter 2.3 (page 139). Global warming is explored in detail in Chapter 6.1 (page 514).

> **Tool 4: Systems and models**
>
> Positive feedback loop
>
> Draw a flowchart to illustrate the possible positive feedback loop between the Earth's temperature and rates of decomposition.

Certain human activities can also accelerate decomposition in soil. **Ploughing** of farmland breaks up the soil surface, and exposes organic matter to more air and warmer temperatures. This allows the aerobic bacteria to break down the humus more rapidly, and more carbon dioxide is released.

Drainage of **wetlands** can also expose organic matter to the air. Historically, wetlands have often been seen as useless areas of land with little value. Many wetlands have been drained by building ditches to take water away from the area. This allows air to enter the previously waterlogged soil. Aerobic decomposition can then increase and more carbon dioxide is emitted.

Methane and methane clathrates

Methane is produced during anaerobic respiration in soils. Methane release from poorly aerated soils is significant for global warming. Under some circumstances, however, the methane may remain in the soil.

◆ **Ploughing** – using heavy cutting or digging tools to break up the surface of the soil, removing any vegetation, and mixing (or turning over) the top few centimetres of the soil. Also spelled plowing.

◆ **Wetland** – an area of land that is regularly or permanently flooded by water, such as a swamp, marsh or bog.

Theme 5: Land

◆ **Methanotroph** – a microorganism that uses methane as a source of energy and carbon to carry out primary production. Methanotrophs convert methane into more complex, energy-rich organic molecules, which generally remain in the soil.

In some poorly aerated soils, a group of microorganisms called **methanotrophs** use the methane as their source of energy and convert it to more complex organic molecules. While these microorganisms live in the soil, methane release into the air is reduced, and much of the carbon remains in the soil.

In other circumstances, methane may combine with water to form a complex solid called a **clathrate**. A clathrate is a complex structure somewhat like a cage. In a methane clathrate, a molecule of methane is enclosed within a cage formed by water molecules. Since the clathrate is made up of methane surrounded by water, the substance is known as **methane hydrate**.

Methane clathrates form at relatively low temperatures and moderately high pressures. They are mostly found in sediments deep below the seafloor and, to a lesser extent, frozen in the soil of the Arctic tundra, in what is known as the **permafrost**.

As global warming causes the permafrost to thaw and the ocean to warm, climate scientists are concerned that the methane clathrates may break down, releasing their methane suddenly into the atmosphere. If this were to happen, this could produce a climate tipping point, setting off rapid and irreversible global warming.

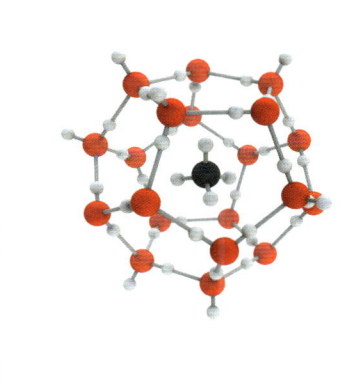

■ **Figure 5.40** Methane hydrate in the form of a methane clathrate

◆ **Methane clathrate** – a solid, ice-like structure formed by methane hydrate, where a molecule of methane is enclosed within a cage-like structure formed of water molecules.
◆ **Methane hydrate** – a compound formed by methane associated with water molecules.
◆ **Permafrost** – an area of permanently frozen ground, where all the water in the soil is frozen throughout the year. Permafrost is mostly found in the Arctic tundra.

Link

Tipping points in systems are covered in Chapter 1.2 (page 24). The tundra biome is described in Chapter 2.4 (page 163).

● TOK

Research is ongoing on the question of whether global warming can cause methane clathrates to set off a climate tipping point. Scientists are investigating the properties, locations and quantities of methane clathrates, and using various computer models to simulate changes in temperature. At the time of writing, the evidence suggests that release of methane from clathrates will not cause a climate tipping point under any of the predicted scenarios of global warming.

REVIEW QUESTIONS

1 Describe the organic and inorganic components of soil.
2 Outline a method for measuring the organic matter content of soil.
3 Describe the properties of sandy and clay soils, as they relate to plant productivity.
4 Outline how humus forms in soils.
5 Explain how a high percentage of humus in the soil can increase the productivity of a terrestrial ecosystem.

5.2 Agriculture and food

> **Guiding question**
>
> - To what extent can the production of food be considered sustainable?

SYLLABUS CONTENT

This chapter covers the following syllabus content:
- ▶ 5.2.1 Land is a finite resource, and the human population continues to increase and require feeding.
- ▶ 5.2.2 Marginalized groups are more vulnerable if their needs are not taken into account in land-use decisions.
- ▶ 5.2.3 World agriculture produces enough food to feed 8 billion people, but the food is not equitably distributed and much is wasted or lost in distribution.
- ▶ 5.2.4 Agriculture systems across the world vary considerably due to the different nature of the soils and climates.
- ▶ 5.2.5 Agricultural systems are varied, with different factors influencing the farmers' choices. These differences and factors have implications for economic, social and environmental sustainability.
- ▶ 5.2.6 Nomadic pastoralism and slash-and-burn agriculture are traditional techniques that have sustained low-density populations in some regions of the world.
- ▶ 5.2.7 The Green Revolution (also known as the Third Agricultural Revolution in the 1950s and 1960s) used breeding of high-yielding crop plants – combined with increased and improved irrigation systems, synthetic fertilizer and application of pesticides – to increase food security. It has been criticized for its sociocultural, economic and environmental consequences.
- ▶ 5.2.8 Synthetic fertilizers are needed in many intensive systems to maintain high commercial productivity at the expense of sustainability. In sustainable agriculture, there are other methods for improving soil fertility.
- ▶ 5.2.9 A variety of techniques can be used to conserve soil, with widespread environmental, economic and sociocultural benefits.
- ▶ 5.2.10 Humans are omnivorous, and diets include fungi, plants, meat and fish. Diets lower in trophic levels are more sustainable.
- ▶ 5.2.11 Current global strategies to achieve sustainable food supply include reducing demand and food waste, reducing greenhouse gas emissions from food production, and increasing productivity without increasing the area of land used for agriculture.
- ▶ 5.2.12 Food security is the physical and economic availability of food, allowing all individuals to get the balanced diet they need for an active and healthy life.

HL ONLY
- ▶ 5.2.13 Contrasting agricultural choices will often be the result of differences in the local soils and climate.
- ▶ 5.2.14 Numerous alternative farming approaches have been developed in relation to the current ecological crisis. These include approaches that promote soil regeneration, rewilding, permaculture, non-commercial cropping and zero tillage.
- ▶ 5.2.15 Regenerative farming systems and permaculture use mixed farming techniques to improve and diversify productivity. Techniques include the use of animals like pigs or chickens to clear vegetation and plough the land, or mob grazing to improve soil.
- ▶ 5.2.16 Technological improvements can lead to very high levels of productivity, as seen in the modern high-tech greenhouse and vertical farming techniques that are increasingly important for supplying food to urban areas.

- 5.2.17 The sustainability of different diets varies. Supply chain efficiency, the distance food travels, the type of farming and farming techniques, and societal diet changes can all impact sustainability.
- 5.2.18 Harvesting wild species from ecosystems by traditional methods may be more sustainable than land conversion and cultivation.
- 5.2.19 Claims that low-productivity, indigenous, traditional or alternative food systems are sustainable should be evaluated against the need to produce enough food to feed the wider global population.
- 5.2.20 Food distribution patterns and food quality variations reflect functioning of the global food supply industry and can lead to all forms of malnutrition (diseases of undernourishment and overnourishment).

5.2.1 Land is a finite resource, and the human population continues to increase and require feeding

There is a total of approximately 149 million km² of land on Earth, accounting for about 29 per cent of the Earth's surface. Of this, only 106 million km² is considered to be habitable – the rest is either barren or covered by glaciers. 46 per cent of the habitable land, about 48 million km², is used for agriculture. The remainder is occupied by natural and managed forests and other ecosystems, with small amounts – approximately 1 per cent each – covered by surface water and urban development.

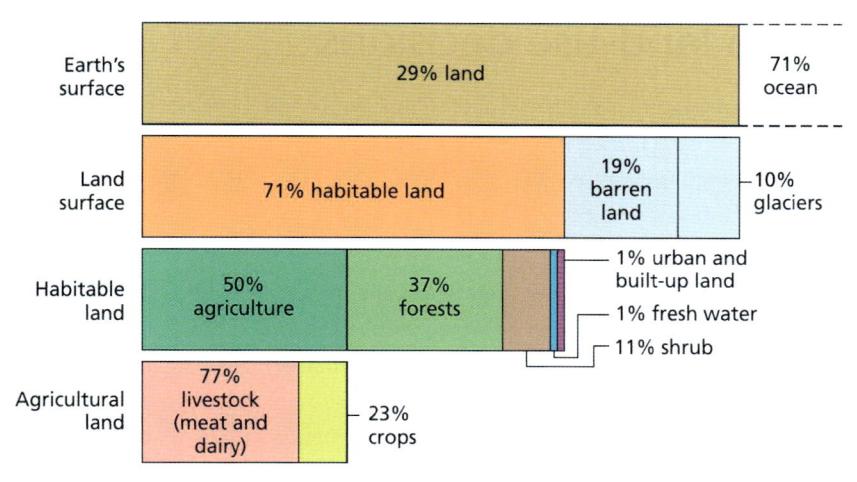

■ Figure 5.41 Global availability of land

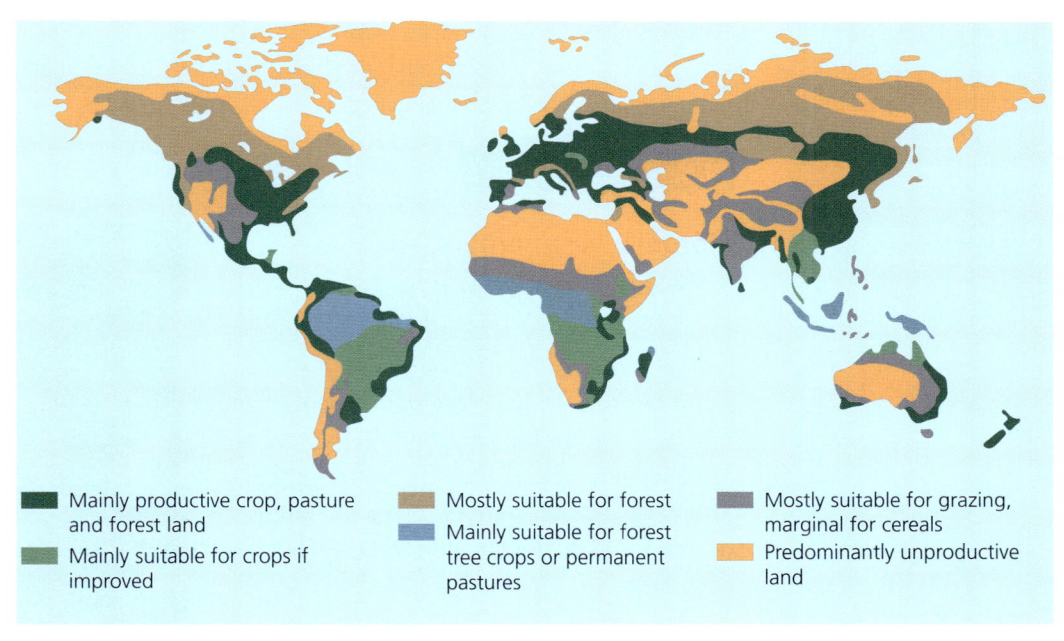

■ Figure 5.42 Potential land use around the world

5.2 Agriculture and food

- ◆ **Arable** – related to the growth of crop plants.
- ◆ **Pastoral farming** – related to farming animals or livestock.
- ◆ **Livestock** – animals raised on farms, for food or other products such as wool.
- ◆ **Commercial farming** – generally large-scale farming as a business, aimed at earning a profit.
- ◆ **Subsistence farming** – small-scale farming primarily to provide food for the farmer's family, sometimes with a little excess sold for income.
- ◆ **Cash crop** – a crop that is grown, typically in large quantities, primarily for sale, often on the international market.
- ◆ **Biofuel** – a fuel produced from living organisms, such as ethanol or biodiesel produced from corn, sugar cane or vegetable oils.
- ◆ **Biodiesel** – a fuel produced by chemical modification of vegetable oil, which can be used as a substitute for diesel fuel.
- ◆ **Marginal land** – land that is considered to be unsuitable for agriculture. The term is not well-defined, however, and can be applied to land used for grazing or small-scale farming, in order to justify converting them to other uses.

Barren land refers to land that lacks suitable soil for farming, such as deserts, sandy beaches and areas of exposed rock. Even among the areas of habitable land, not all areas are appropriate for **arable**, or crop, farming. Some mountainous areas are too steep or their soils are too thin. Some areas have soil, but it lacks the necessary nutrients. Some areas with poor soils may be used for **pastoral farming** or grazing **livestock**.

However, the need for food continuously increases, as the human population continues to grow. Increasing food production, given a finite amount of land available, would involve either spreading agriculture into new land, or increasing the efficiency of production on the land we currently use. Both of these options are problematic. If we start farming in new areas, we risk losing the ecosystems and biodiversity that currently exist there. If we intensify our production on existing farmland, we risk increasing the negative effects of agriculture.

5.2.2 Marginalized groups are more vulnerable if their needs are not taken into account in land-use decisions

With continuously increasing demand for food, and a finite amount of suitable land, conflict may arise about how land should be used. Decision-makers may want to increase food production by creating farms on land that is used in other ways, or by converting small-scale farming to larger-scale, more **commercial** agriculture. Conflicts also arise when agricultural land is needed for other purposes such as industry or housing.

Decisions about land use are typically made by local or national governments, driven by political or economic goals. Government decisions tend to favour the interests of businesses and industries which produce large amounts of income for the society, or employ many people. Small-scale or **subsistence** farmers, and sectors of society which contribute less to the economy, may be negatively affected by these decisions. Marginalized communities, such as groups of indigenous people, may be the worst affected.

The term land grab can be used to describe the transfer of ownership or control of large areas of land from local populations to wealthy, powerful investors. Many modern examples involve converting small-scale farms, often producing food for the local market, to large-scale plantations, producing either **cash crops** for export or for **biofuels**.

REAL-WORLD EXAMPLE

Biofuel investment in Tanzania

In 2008, over 80 km² of rural land in Kisarawe, Tanzania was leased by a British biofuel firm, aiming to produce **biodiesel** from jatropha, a small tree that produces an oily seed. At the time, jatropha was seen as a sustainable option for biofuel production, as it is easy to cultivate on **marginal land**, which is unsuitable for other types of agriculture.

■ Figure 5.43 The fruit of the jatropha plant

The villagers agreed to the transfer of mostly forested land on the basis of increased investment in local facilities such as schools, roads, wells and hospitals, and of employment on the plantation. Through the mediation of the government, control of the land was transferred to the Tanzanian government, which then leased it to the British company on a 99-year lease. In transferring the land the villagers lost access to the forest, which was a source of resources such as water, timber, clay, medicine and honey.

In the end, the investment was a commercial failure, as the yield of oil from the jatropha plant was disappointingly low. The British firm collapsed and transferred control of the land to a Tanzanian-owned cattle-farming company. The villagers continue to be excluded from the land and the promised local developments were never delivered.

The overall result was the conversion of locally useful forested land to commercial farming, and the further marginalization and impoverishment of the local community.

(Source: https://theworld.org/stories/2014-06-27/picking-pieces-failed-land-grab-project-tanzania and https://news.mongabay.com/2023/04/jatropha-the-biofuel-that-bombed-seeks-a-path-to-redemption)

Link
Biofuels are discussed in Topic 7.1.17 (page 628).

HL.a.3, HL.a.4: Environmental law

A number of British and European investments in biofuel production in Africa, Asia and Latin America were driven by a 2003 European Union directive that promoted the use of biofuels in transport. To meet this new demand, several companies set out to increase biofuel production, through investment in less economically developed countries. This had the unintended consequence of marginalizing rural communities around the world, as illustrated by the above Real-world example from Tanzania.

The EU's directive on biofuels was updated in 2009 and 2023 to ensure that land that supports high biodiversity or significant carbon storage should not be converted to biofuel production. However, these more recent directives make no provisions to protect the rights of people currently occupying land.

5.2.3 World agriculture produces enough food to feed 8 billion people, but the food is not equitably distributed and much is wasted or lost in distribution

Global food production and hunger

Globally, enough food is produced to feed all 8 billion people on Earth. Some estimates suggest that there is enough to feed up to 10 billion people. However, many people around the world still suffer from **hunger**. The United Nations estimates that between 691 and 783 million people, or about 9 per cent of the world's population, were **undernourished** in 2022.

Hunger exists because the food which is produced does not reach all people. Most of the world's food production occurs in a relatively small number of countries, such as China, India, the United States and Brazil. To reach all people around the world, food may need to be transported over long distances, which can be costly and raises the price of food for the consumer. People in areas which

◆ **Hunger** – discomfort or pain caused by insufficient food consumption. World hunger refers to the percentage of people who are undernourished.

◆ **Undernourished** – when habitual food consumption provides insufficient dietary energy to lead a normal active and healthy life.

do not produce enough food may need to buy expensive, imported food to meet their needs. High food prices make it difficult for poorer people to obtain enough food.

Furthermore, approximately one-third of all food produced on Earth is lost or wasted and never consumed. Losses happen all along the supply chain, from the farm to the final consumer. They may occur at several steps along the way. Products may:

1. be left unharvested and remain in the field, where they rot
2. be lost or damaged during packaging and transportation, perhaps being overlooked, misplaced or damaged when being handling
3. be deliberately discarded because they are not of high enough quality. Products of undesirable shape, size or colour, for instance, might be discarded
4. spoil during transport, storage or display in retail stores.

Food loss may also occur because of adverse weather (for example,. storms or unseasonal frost) or inefficiencies in the transport network (for example, delays between shipments). Losses are likely to be greater where food is transported in large quantities or over long distances.

At the end of the supply chain, food waste may occur in the hands and homes of the consumer. Food that is stored inappropriately or for too long can spoil on the shelves, in the cupboards or in the fridge. Preparation of food may involve removing parts that are edible, such as potato or carrot peel. When too much food is prepared and served, food is also discarded from the plate.

TOK

Waste from the kitchen may include things that are not usually eaten, such as bones or certain types of seeds or peels. This inedible food waste might be unavoidable, but it still represents a percentage of agricultural production that is not eaten, so it can be seen as an inefficiency in the food supply chain. Inedible waste may or may not be included in estimates of food waste, so it is important to look closely at how terms like 'waste' and 'loss' are defined for any particular data set.

Food and the Sustainable Development Goals

Hunger and food waste are mentioned explicitly in two of the 17 Sustainable Development Goals:

- SDG 2: Zero hunger
- SDG 12: Responsible consumption and production.

SDG 12, in particular, makes reference to food waste. Target 12.3 states: 'By 2030, halve per capita global food waste at the retail and consumer levels and reduce food losses along production and supply chains, including post-harvest losses' (source: **https://sdgs.un.org/goals/goal12#targets_and_indicators**).

The target distinguishes between food loss and food waste:

- Food loss is defined as losses between production on the farm and delivery to retailers. It does not include loss that occurs at the retailers or in the hands of the consumers.
- Food waste refers to losses occurring in households, retailers, restaurants and similar food services.

The aim is to reduce per capita food waste to 50 per cent of the 2007 level, and to reduce food loss across the supply chain, by 2030.

Link

The United Nations' Sustainable Development Goals (SDGs) are discussed in Topic 1.3.18 (page 70).

5.2.4 Agriculture systems across the world vary considerably due to the different nature of the soils and climates

Soil and climatic conditions have a significant influence on farming. Different crops and livestock have different needs, and tolerate different conditions. Farmers take these into consideration when planning their agricultural operations. They must decide what crops to grow, when to plant and harvest, and how to fertilize, irrigate and control pests. In some soil and climatic conditions, livestock farming may be more suitable than crop farming.

REAL-WORLD EXAMPLE

Bananas in Latin America and the Caribbean

Bananas are one of the most popular fruits around the world, and one of the most widely cultivated crops in tropical areas. Several large-scale banana plantations operate in parts of Latin America, including in Ecuador, Costa Rica and Colombia. In the Caribbean, bananas are more often produced on small, family-owned farms on islands such as Jamaica, Dominica and St Vincent.

Bananas grow in large bunches on tall, tree-like herbs. They require continuously warm temperatures and abundant rainfall, making them suitable for cultivation in moist tropical climates. They prefer dark, fertile soils, so they benefit from the volcanic soils on the Caribbean islands and the mountains of South and Central America. For larger plantations, heavy use of fertilizer is common. Typically, each plant bears one bunch of fruit and is then cut down. New plants then grow from the base of the old plant. They do not grow from seeds: their reproduction is strictly asexual.

■ **Figure 5.44** Bunches of bananas on a plantation in Cuba

Banana growth is not particularly affected by seasons, which allows production to continue through the year in the tropics. However, pests are also active throughout the year, which means pest control can be challenging, and pesticide use is common. The fact that banana plants only reproduce asexually means that there is very little genetic diversity on banana plantations which allows pests to spread rapidly through an entire crop, leading to crop failure. On small-scale farms, other crops may be grown between the banana plants in order to reduce the spread of pests. Worldwide, bananas are currently threatened by a fungal disease called Black Sigatoka, which spreads easily through plantations.

> **REAL-WORLD EXAMPLE**
>
> ### Wheat in North America and Europe
>
> Wheat is an annual crop, which means it is grown and harvested in one year, and must be replanted each year. It is a type of grass, with a starch-rich seed that humans use to make flour. It is one of the most important crops globally, as wheat flour is used in producing bread and pasta. It is the most widely grown and traded agricultural product in the world, but it is almost exclusively grown in temperate climates.
>
> There are two main varieties of wheat in cultivation: winter and spring wheat. Winter wheat can tolerate reasonably low temperatures, but requires abundant rainfall. It is planted in the autumn, grows through the winter and is harvested the following summer. It requires the temperature to fall below about 4°C before it produces seeds, but it cannot tolerate the harshest winters, so is more often grown in the southern parts of Europe and North America. It is unaffected by dry summers.
>
> ■ **Figure 5.45** Wheat
>
> Spring wheat cannot tolerate the colder winters in the north, so it is planted in the spring and harvested in the autumn. It requires relatively continuous rainfall through the summer, so this limits its range.
>
> Wheat is mostly grown in large monocultures, on relatively flat land, within the temperate grassland biome. The North American prairies provide the appropriate climate and topography, along with the dark grassland soil, which is naturally rich in organic matter and nutrients. Most wheat production involves the use of machinery to plough the soil, plant the seeds, and spread fertilizer and pesticides. Irrigation is used in drier environments.

> **ATL ACTIVITY**
>
> Choose an agricultural product (crop or livestock) that you are interested in, and research the conditions it requires and the methods used to farm it. Create an information-rich poster or write a detailed report on your chosen crop. You may choose from the following examples, or select one of your own:
> - crops: rice, grapes, strawberries, cacao (also known as cocoa), maize (corn), millet, sorghum
> - livestock: cattle (you may choose a particular breed), chickens, sheep.

5.2.5 Agricultural systems are varied, with different factors influencing the farmers' choices. These differences and factors have implications for economic, social and environmental sustainability

Agricultural systems can be classified in terms of what the system produces, what the purpose of the system is, or how the system operates.

Classification by farm output

Historically, farms tended to produce a variety of different outputs, but in modern times the trend has been towards single products.

Monoculture describes the practice of producing a single product (usually a crop) on a farm. Monocultures are most common in large-scale, commercial farms, where a single crop, such as wheat, might be grown over a large expanse of land.

■ Figure 5.46 A palm oil monoculture in Indonesia

Polyculture describes the production of many crops on one farm. Small-scale, subsistence farms are typically polycultures. Polycultures is also common on farms practising organic farming and permaculture.

Many farms around the world focus on arable farming, producing crops. Arable farming usually requires relatively flat areas with mild climates and fertile soils, while livestock or pastoral farming is more flexible, and may occur on steeply sloping land, with harsher climates and poorer soils. Some farms may produce both crops and livestock: this practice is called **mixed farming**.

◆ **Mixed farming** – farming for the production of both crops and livestock.

Classification by purpose

Farming is always about producing food or other products from plants and animals. However, farmers may practise agriculture for two broad reasons: either to feed themselves and their families, or to earn income.

Farms that are designed to earn income by selling their product are described as commercial farms. A commercial farm operates as a business, where the farmer must ensure that the income from sales is higher than the costs of running the farm. To do this, commercial farmers tend to operate on a relatively large scale, which allows for greater profits. To manage a large farm, they often need to hire workers, and use machinery such as tractors and combine harvesters.

TOK

Although we make a clear distinction between commercial and subsistence farming, these are actually two ends of a continuum or spectrum of farming practices. Many farms operate somewhere between the two extremes.

What are the benefits and drawbacks of classifying items or processes into two discrete categories, when many examples actually fall between the two extremes?

■ **Figure 5.47** A combine harvester being used to harvest the cereal grain, barley

Subsistence farms are operated mainly to provide food for the farmer's family. They are usually small, and produce a variety of crops and livestock. Most work is done by family members, using only simple tools such as hoes. Although the primary motivation is to produce food for the family, subsistence farmers will often sell excess products in a local market, where they will also buy products from other farmers who grow different produce.

■ **Table 5.9** Typical characteristics of commercial and subsistence farming

Commercial	Subsistence
Large scale	Small scale
Highly mechanised (uses machinery such as tractors and harvesters)	Labour intensive (most work done by hand, with simple tools)
Heavy use of fertilisers, pesticides and irrigation	Limited use of fertilisers, pesticides and irrigation
Focused on cash crops, such as cereals	Diverse produce, such as fruit and vegetables

The purpose and practice of farming also relate to the way of life of the farmer, which may in turn relate to the nature of the farmer's environment. Nomadic farming is practised by people who regularly move from one location to another. This is a relatively rare way of life in today's world, but was much more common in the past. Nomadic farmers are mostly pastoralists, or herders, who take their livestock with them. They typically move in a seasonal pattern, taking their animals to graze or obtain water in different locations as the seasons change.

> **REAL-WORLD EXAMPLE**
>
> ### Nomadic herders of Mongolia
>
> Mongolia is a large, sparsely populated country in Asia, with harsh and variable landscapes and climatic conditions, which makes arable farming difficult. About 50 per cent of the population live partially or fully nomadic lives, mostly involving herding livestock such as sheep and goats. They build lightweight, portable homes called *gers* (sometimes called *yurts* by foreigners), which can be erected and dismantled easily, and carried from one place to another. They move seasonally, taking advantage of conditions in different locations to find food and water for the livestock.
>
> ■ Figure 5.48 A flock of sheep at a seasonal settlement in Mongolia

The vast majority of farmers are settled in one place and farm the same land continuously. Farms such as these are described as sedentary, which means they do not move.

■ Figure 5.49 Crops being irrigated via a system of mobile pipes that are rolled over the land

■ Figure 5.50 Hydroponic production of lettuce crops

Classification by inputs

Farming can involve a wide variety of methods, each of which requires different inputs. Inputs can include energy, water, fertilizer, pesticides, as well as labour and finance (money). One approach to classifying farming systems is by considering what inputs are provided by the farmer, and how important the inputs are for the farm operation.

Farms that depend on many significant inputs from the farmer may be described as intensive farms. These farms could not function without the farmer's inputs, such as artificial fertilizers and pesticides, irrigation water, and energy to run the machinery that the farm depends on. Extensive farming, on the other hand, depends on fewer inputs from the farmer, and can to some extent continue to produce even if the farmer stops working.

Intensive farms, for instance, may be irrigated, while extensive farms are more likely to be rain-fed. In irrigated agriculture, extra water is provided to the crops, either from wells or from nearby surface waters. The water may be brought via pipes and delivered by spraying over the crops.

Traditionally, farms are soil-based, requiring land with appropriate, fertile soil. An alternative to soil-based farming is hydroponics. In a hydroponics farm, crops are grown in water, which has mineral nutrients added to it. No soil is required, which means farming is not limited to places with rich soil, and it does not contribute to soil erosion or nutrient depletion. Hydroponics, however, requires the farmer to construct pipes or troughs to hold the water and crops, and to provide either a steady input of water and nutrients, or a mechanism for treating and recycling the water.

5.2 Agriculture and food

Figure 5.51 The bollworm larva is a pest of the cotton plant

Most modern farms, especially commercial ones, require regular input of agrochemicals, such as artificial fertilizer and pesticides. Modern commercial farming practices, in particular the practice of monoculture, may lead to depletion of soil nutrients and the spread of pests. As crops grow, they take nutrients from the soil. Growing the same crop repeatedly, and across a large area of land, tends to remove the same nutrients from the soil, while not allowing time for them to be replaced by natural processes. Farmers usually overcome the depletion of nutrients by applying artificial fertilizers to the soil.

Monoculture also promotes pests. Different pest species prefer to feed on different crops, but if only one crop is grown, then a single pest species can spread rapidly through the farm and cause extensive damage. The modern, commercial solution to this is to spray pesticides to kill the pests.

Common mistake

Both fertilizers and pesticides are described as agrochemicals, because they are chemical substances that are used in agriculture. They are often spoken of together and people sometimes confuse them. However, they have very different uses and raise different environmental issues.
- Fertilizers contain nutrients that plants need to grow, such as nitrogen, phosphorus and potassium. They are used to promote crop growth on farms and they are the main cause of eutrophication, which is a form of water pollution (see Topic 4.4.5, page 388).
- Pesticides contain chemicals that are designed to kill pests such as insects, nematodes and fungi. They are used on farms to prevent or reduce the loss of crops to disease or consumption. They are toxic, and may harm other, non-target species, such as bees or even humans.

Agrochemicals such as pesticides and fertilizers cause a number of environmental problems, so some farmers prefer not to use them. Farms that avoid all agrochemical use may be described as organic farms. Organic farms have fewer artificial inputs, but they may still use natural fertilizers, such as manure or compost, and biological pesticides, such as extracts from particular plants. Organic farming may also require more manual labour than inorganic farming, where agrochemicals are used. For instance, to avoid using pesticides, organic farmers may opt to plant polycultures, which are less vulnerable to pests. With multiple different crops growing on the farm, machinery such as combine harvesters cannot usually be used. More generally, organic farmers may need to pay closer attention to individual plants, physically removing pests or diseased plants. On a conventional farm, the use of pesticides eliminates much of this labour.

REAL-WORLD EXAMPLE

The neem tree (*Azadirachta indica*) is a member of the mahogany family, which is native to India and parts of Southeast Asia but has been introduced into other parts of the world. The tree produces various chemical substances which can be used in agriculture as a natural pesticide or pest repellent. Extracts from the leaves, seeds and fruits can be purchased and applied on farms. As a natural substance, it is considered to have fewer environmental impacts than artificial pesticides and may be used in organic farming.

TOK

The term *organic* is used differently in different contexts, which may be confusing.
- In chemistry, organic means roughly 'made up mostly of carbon atoms'.
- In biology, it usually means 'coming from or made by a living organism'.
- In agriculture, it means 'avoiding all use of artificial agrochemicals'.

There are other meanings in other contexts as well. Use a dictionary to find the different ways in which this word is used.

Sustainability implications

Each agricultural system has its own implications for economic, social and environmental sustainability. While it is possible to generalize about different systems, it is important to appreciate that the sustainability of agricultural systems may depend on the specific social, economic and environmental context of the farm.

Economic sustainability

Agricultural systems can be economically sustainable if they produce sufficient profits to allow for reinvestment of funds into the farm, while providing suitable income for farmers and farm workers. Profit is the difference between income and expenditure, so economic sustainability is affected by the cost of farm operation and by the income from selling farm produce.

Larger, more commercial farms are designed to maximize profits by reducing the unit cost of production. They operate on the economic principle of economies of scale, which suggests that by producing many identical items, the cost of production of each individual item is reduced. However, in practice, large commercial farms often require many expensive inputs such as fertilizers, pesticides and fuel for farm machinery. Larger farms also typically sell their produce at wholesale prices to food processing, distributing and retail companies. This reduces the income per individual item sold. In some cases, commercial farms can only make a profit because some or all of their running costs are subsidized by the country's government. This is economically unsustainable, as it means that agricultural production is dependent on funding from other, unrelated sources.

Smaller-scale farms, including subsistence and organic farms, may have lower operating costs, since they use fewer agrochemicals and machinery, but they also typically have lower income, since they produce smaller quantities of products. They can improve their profitability by selling their produce directly to consumers, via farmers' markets. Some consumers are also willing to pay higher prices for food produced in more socially or environmentally sustainable ways. Organic produce, for instance, is often sold at a higher retail price than conventional produce from commercial farms.

The economic sustainability of agriculture is also vulnerable to external factors such as the price of fuel, international trade agreements, pests, diseases, natural disasters and climate change. Smaller-scale farms may be less vulnerable to some of these threats since, for instance, they may use less fuel or sell their produce locally, and they tend to practise more resilient forms of agriculture such as polyculture.

Social sustainability

Social sustainability in agriculture relates to the effect of farming on people and communities. Issues of social and economic sustainability may be closely linked. Employment in agriculture, and the wellbeing of farm workers and farming communities can be affected by farming systems. Larger-scale, commercial farms may provide satisfactory employment for workers, but the number of workers may be reduced as farms incorporate the use of more machinery.

Where machinery is inappropriate or unavailable, commercial farms may employ large numbers of manual labourers to do mundane, sometimes backbreaking work such as harvesting fruits. This sort of work is typically seasonal and poorly paid, so provides little social or economic stability for the workers. The workers may come from marginalized communities and have limited options for better employment. In some more economically developed countries, seasonal workers are hired from less economically developed countries, and given temporary work permits to do jobs that local workers are unwilling to do. Where jobs in farming are unpopular or unattractive, it is unlikely that agriculture can be socially or economically sustainable.

Alternative farming models which seek to promote social sustainability include community, cooperative farms, including those established in urban environments. These farms aim to engage people from local communities in a collective effort to produce food for themselves, while selling surplus production for income. These systems are closer to the subsistence end of the subsistence-commercial spectrum of farming systems. The objective is to enable people to become self-sufficient, while providing them with meaningful employment and a stable, healthy source of affordable food. In this way, healthy, sustainable communities may develop.

Environmental sustainability

Agriculture can be described as environmentally sustainable if its overall impact on the quality of the environment and the availability of natural resources, is positive. Maintenance of biodiversity and soil fertility and release of pollutants are major environmental sustainability concerns.

Farms which rely heavily on synthetic fertilizers, typically including larger-scale commercial farms, contribute to both soil degradation and environmental pollution. Smaller-scale farms which promote soil fertility through organic fertilizers and methods such as cover crops and crop rotation produce less pollution, and may increase soil fertility over time, making them more environmentally sustainable.

Pesticide use on farms contributes to pollution, soil contamination and loss of biodiversity. Organic farms which operate without the use of synthetic pesticides may be more environmentally sustainable than commercial farms.

Clearance of land for agriculture and the adoption of monoculture on farms both reduce biodiversity. The use of a small number of selectively bred crop or livestock varieties also reduces biodiversity in agricultural systems, making farms more vulnerable to disease and environmental changes, and therefore less sustainable. Systems which use polyculture, and generally promote diversity on farms, are more environmentally sustainable.

> **ATL ACTIVITY**
>
> Select two farms operating with different agricultural systems, preferably in your local area, or alternatively, ones which you can find information about online. Carry out a comparison of the economic, social and environmental sustainability of the two farms, and write a report discussing which is more sustainable overall.

> **Inquiry process**
>
> ### Inquiry 2: Collecting and processing data
>
> Investigate the environmental sustainability of different agricultural practices.
> - Discuss issues related to soil degradation, water usage, biodiversity, and the use of pesticides and fertilizers.
> - Include a discussion of the challenges faced by farmers in adopting sustainable agricultural practices.
> - Highlight potential opportunities and innovations that can contribute to sustainable farming.
> - Propose policy recommendations at local, national or international levels to promote sustainable agricultural practices. Also consider incentives, regulations and support mechanisms for farmers.

5.2.6 Nomadic pastoralism and slash-and-burn agriculture are traditional techniques that have sustained low-density populations in some regions of the world

When humans first started farming, about 12,000 years ago in the Middle East, the total human population on Earth was probably no more than about 10 million. Before agriculture, people lived in small, generally nomadic groups, hunting animals and gathering edible parts of plants for food. Global population did not reach 1 billion until about 1700 CE.

Today, needing to feed a population of over 8 billion, the scale of global agriculture is huge, and its impact on the environment is substantial. Increasingly, farmers, environmentalists and academics examine their agricultural practices, seeking to find methods that produce enough food for all people, while maintaining the integrity and health of the environment.

Some approaches to agriculture that were sustainable in the past may no longer be so today. Two such systems are nomadic pastoralism and slash-and-burn agriculture. These approaches can work well in situations where population density is low and land is abundant. They both involve moving around large areas of land.

Nomadic pastoralism

Pastoralism is the practice of raising livestock herds in open fields or pastures. It is often done on marginal land that is unsuitable for arable farming because of poor fertility or difficult terrain. Livestock such as sheep, cows or reindeer are allowed to graze on the natural vegetation, and the farmers use the animals for meat, milk or other animal products, such as wool.

In some situations, the livestock may use up too much of the vegetation and may need to be moved. On sedentary farms, they may be moved from one field to another, allowing the first field to recover before it can be grazed again. In **nomadic** pastoralism, the farmers – sometimes called **herders** – move, along with the livestock, to entirely new areas, where they establish new, temporary settlements.

In situations where population density is low and land is freely available, nomadic pastoralism can be sustainable. Moving herds from one location to another ensures that one area of land is never **overgrazed**. Instead, each area of land is lightly grazed and then given time to re-grow.

This approach cannot work, however, if the herds are too large, or if the surrounding land is occupied by farms, settlements or even other herders. In many modern contexts there are too many people, and the land is largely occupied and frequently privately owned. Nevertheless, a small number of successful societies still practise nomadic farming in particular situations where conditions are appropriate.

◆ **Pastoralism** – a form of **animal husbandry** in which animals are grazed in open fields.

◆ **Animal husbandry** – the practice of farming and caring for animals.

◆ **Nomadic** – related to people who move from place to place as a way of life. Nomadic people do not have one settled, permanent home.

◆ **Herder** – a pastoralist farmer who raises herds of livestock.

◆ **Overgrazed** – the state of a field or other habitat that has had too much of its vegetation eaten by herbivorous livestock. The vegetation on overgrazed land is depleted to the extent that it cannot grow back.

REAL-WORLD EXAMPLE

Maasai herders in East Africa

The nomadic Maasai herders live in the Great Rift Valley of Kenya and Tanzania. They raise *zebu*, a locally-adapted breed of cattle, which provides them with meat, milk and blood, and which is an important part of their economy, culture and identity.

The area is arid, making it unsuitable for arable farming, and making finding water a challenge. The Maasai's nomadic way of life allows them to move their herds between different parts of their range, leaving each area of land to recover for several months at a time. The practice protects the fertility and carbon content of the soil, and the biodiversity of the entire region.

The Maasai have lived in this part of Africa for hundreds of years. The success of their lifestyle depends on the availability of land, and it is threatened by the spread of people from surrounding areas into their rangeland. In some areas they have been forcibly excluded from parts of their traditional territory and grazing lands. Climate change has also caused increasingly long dry seasons, making water harder to find. Their continued survival may be helped by the fact that parts of their range have been declared protected areas, reducing some of the pressures of modern development, although in some places protected area status has meant that grazing has been further restricted.

■ Figure 5.52 A Maasai herder in Tanzania

■ Slash-and-burn agriculture

Slash-and-burn agriculture involves clearing and burning areas of natural vegetation in order to plant crops. Typically, each area of cleared land is farmed for two or three years, and then abandoned to recover naturally. The farmers then move to a new area and repeat the process. It is also known as shifting cultivation, swidden or fire–fallow cultivation.

Slash-and-burn works because the ash produced by burning the vegetation fertilizes the soil, and because vegetation naturally regenerates by secondary succession if a cleared area is left alone for long enough. It has been practised for thousands of years, notably in the tropical forests of South America, Africa and Asia, and is still common today in some regions. It is generally done on a small scale, using simple tools, local crops and few inputs.

Link
Secondary succession is covered in Topic 2.5.6 (page 184).

Given abundant land, to allow farmers to leave each burned patch of land for long enough to recover fully, and enough time for each farmed patch of land to recover, slash-and-burn can be sustainable. However, with increased population density in many parts of the world, it has become difficult to leave land **fallow** long enough to recover. Increasing numbers of farmers using the practice also means that more and more areas of forest are burned, leading to increased soil erosion, deforestation and carbon emissions.

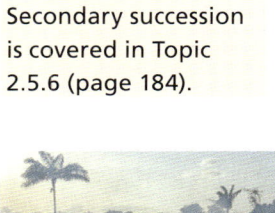

■ Figure 5.53 Slash-and-burn agriculture in the Amazon rainforest

◆ **Fallow** – soil left uncultivated in order to recover fertility. In conventional farming, land may be left fallow for a time after it has been ploughed. In slash-and-burn farming, land must be left fallow for long enough for the ecosystem to recover. This may take decades.

Theme 5: Land

> **REAL-WORLD EXAMPLE**
>
> ### Milpas in Mexico
>
> Milpa is a traditional, sacred form of slash-and-burn polyculture that is widely practised by the indigenous Maya people of Mesoamerica, particularly in Mexico's Yucatan Peninsula. The Maya have lived in this region for thousands of years, using milpa to produce their food.
>
> Milpa involves a ten-year cycle, involving two years of cultivation and eight years of fallow. Cultivation is carefully planned and timed, with different crops planted at different intervals. An essential component of milpa is **intercropping**, where different crops are planted between each other. Carefully designed intercropping can make efficient use of soil resources, as different crops use different nutrients, perhaps from different depths of soil. In some cases, certain crops may even add nutrients to the soil, which other crops can use. Maize (corn), squash and different types of beans are common crops in many milpas.
>
> ■ Figure 5.54 A traditional milpa in the Yucatan Peninsula, Mexico

5.2.7 The Green Revolution used breeding of high-yielding crop plants to increase food security, and has been criticized for its sociocultural, economic and environmental consequences

During the 1950s and 1960s, global population was rising faster than it ever has before or since. There was a great deal of concern that humans were heading for a major disaster, when food production would not keep pace with population. Commentators such as the controversial US biologists Paul and Anne Ehrlich, in their 1968 book *The Population Bomb*, predicted the coming death of millions by starvation.

In response to these concerns, a great deal of effort was put into agricultural research, in the hope of increasing food production. Some of the earliest successes came from a research team working in Mexico in the 1940s and 1950s, led by another US biologist, Norman Borlaug.

Borlaug's team used **selective breeding** to improve the Mexican wheat crop. This involved carefully controlling the breeding of plants, by physically removing the male parts of thousands of wheat flowers to prevent undesirable crosses. In this way they were able to cross-breed plants with different desirable traits, in the hope of producing offspring that combined all the desirable traits, and eliminated any undesirable ones. The result was the production of high-yield, fast-growing, pest-resistant strains of wheat, and a dramatic increase in Mexico's wheat production.

◆ **Intercropping** – an agricultural practice in which different crops are planted in close proximity to one another, in order to make efficient use of the many resources in the soil, and to capitalize on mutually beneficial relationships between crops.

◆ **Selective breeding** – a process of cross-breeding individual plants or animals with desirable characteristics or traits, to produce offspring with those favourable traits. The method has been used since the very beginnings of agriculture, but was improved by developments in the modern field of genetics.

● Common mistake

Selective breeding may be confused with genetic engineering, or genetic modification. Selective breeding is an ancient practice which involves selecting animals or plants with favourable characteristics, and allowing them to reproduce sexually. Genetic engineering is a modern technology, which involves modifying the genes inside an organism, typically by transferring genes from one species of organism into another. (See 5.2.11)

Borlaug's successes in Mexico led to the adoption of the new Mexican strains of wheat in India in the 1960s, and the implementation of selective breeding of rice in the Philippines, with similar results. Building on these results, other researchers in various parts of the world started developing other technological solutions to some of the challenges of agriculture, including synthetic fertilizers, pesticides and high-tech irrigation systems. Over the 1960s, 1970s and 1980s, these technologies were widely adopted in many parts of the world, especially in developing nations where population growth was rapid. The rapid developments at this time came to be known as the Green Revolution, or the Third Agricultural Revolution.

Common mistake

The use of the word *green* in Green Revolution often causes confusion. Today, 'green' usually refers to things that are beneficial to the environment. However, this use of 'green' had not yet arisen in the 1960s, so the Green Revolution was called green because of the colour of crop plants. In fact, the Green Revolution caused or worsened many environmental problems, and many environmentalists see it as a generally negative change.

TOK

The word *revolution*, derived from the word *revolve*, is often used to describe a significant, perhaps fundamental, change – as if something has spun around, or revolved, to face a different direction. It often refers to political events, such as the French and American Revolutions, which may or may not be violent. However, it can also refer to sociocultural or economic events, such as the Scientific and Industrial Revolutions.

The Green Revolution is a revolution because it caused widespread, significant changes to the way agriculture was practised. It is also known as the Third Agricultural Revolution. The First Agricultural Revolution, also called the Neolithic Revolution, refers to the origins of agriculture, about 12,000 years ago. The Second Agricultural Revolution refers to a period starting in the seventeenth century and overlapping with the Industrial Revolution, when agriculture moved from being mostly for subsistence to being a commercial enterprise, characterized by larger farms, run as businesses for profit, employing many workers and using newly developed machinery.

Benefits of the Green Revolution

Since the 1960s there has been a steady and significant increase in the global production of certain crops, especially cereals such as wheat. Remarkably, this has happened without a significant increase in the area of land used for farming these crops. Effectively, the production of cereals has increased as a result of the adoption of more intensive farming, meaning that a greater yield has been produced from a similar area, as shown by Figure 5.55.

As the graph shows, the increase in cereal yield has been consistently greater than the increase in human population. Therefore, in the time since the Green Revolution, not only has total food production increased, but food production per capita, or per person, has also increased. As a result, predictions of widespread famine and population collapse have not been fulfilled.

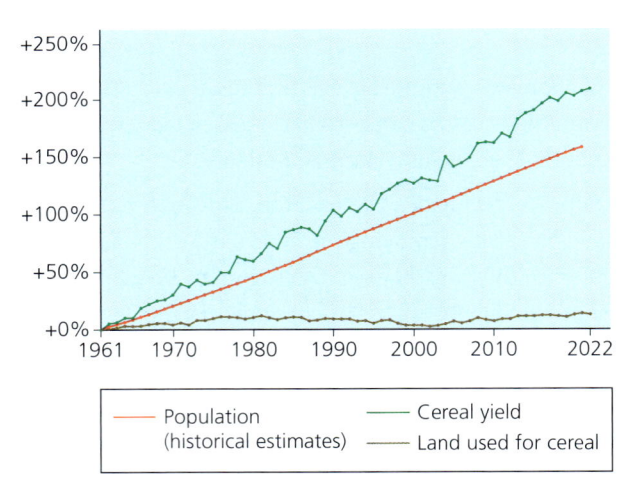

■ Figure 5.55 Increases in global cereal yield, land used for farming cereal, and human population, relative to 1961

■ Criticisms of the Green Revolution

The Green Revolution achieved its goal of increased food production by increasing the inputs required in farms. This has, arguably, made farming less sustainable, for several reasons:

- Farms became larger and more likely to be monocultures, reducing the biological diversity on farms. Crops became more susceptible to pests and diseases, and farmers became more dependent on global trade and susceptible to price fluctuations in the global market.
- Many farms around the world started adopting the same selectively bred varieties of crops such as wheat, corn and rice. Other crops, and other varieties of those crops, were often abandoned, leading to reduced genetic diversity in crops (making them even more vulnerable to pests) and reduced economic resilience in farms – if one crop fails or the price falls dramatically, the farmer's income may disappear.
- Farms became more dependent on fertilizers, especially nitrogenous ones. Producing nitrogenous fertilizers requires the use of fossil fuels as an energy source, to combine atmospheric nitrogen with hydrogen (also derived from fossil fuels) in an industrial process. Excess nitrogen in the soil also leads to increased emission of nitrous oxide, a powerful greenhouse gas. At the same time, farms now depend more on pesticides, also produced in industrial processes, and on machinery. The contribution of farming to global warming has therefore increased significantly.
- The indirect dependence on fossil fuels means that the price of food is now tightly tied to the price of energy. This has given more power to those countries that produce fossil fuels, and arguably reduced the food security of countries that must import fossil fuels, or fertilizers, to support their own farming.
- The environmental impacts of farming, such as eutrophication, soil degradation and disruption of food webs, have increased significantly due to the increased use of fertilizers, pesticides, irrigation and machinery.
- The increased scale of farms using Green Revolution technology has made it difficult for smaller farms to compete. Many smaller farms have been unable to continue to operate, leading to loss of income and livelihood for farmers and farm workers, and the decline of many agricultural communities. The global trade in food has also meant that smaller countries cannot compete with larger ones in the food market, and many have become even more dependent on food imports, reducing their food security.
- The increased focus on production of certain crops such as cereals has arguably reduced the availability of many other crops, making it harder for people to obtain some crops which may have been important in their traditional diets. For some people this may have made it harder to eat a healthy, balanced diet.
- Increased food production has not occurred in all parts of the world or benefited all people. Most notably, the Green Revolution has had relatively little impact on Africa. Moreover, increased global production has had no impact on problems of inequitable distribution of food. We produce more food than we need, but many people still go hungry.

◆ **Synthetic fertilizer** – a fertilizer that does not occur in nature, but is manufactured by humans from natural raw materials. They usually contain specified amounts of nutrients, particularly nitrogen (N), phosphorus (P) and potassium (K).

◆ **Nitrogenous fertilizer** – a fertilizer that is rich in nitrogen compounds, such as ammonium salts and nitrates.

◆ **Nitrogen-fixing bacteria** – bacteria found in soils, and sometimes living in plants, which take nitrogen from the air and convert it into soluble compounds that add fertility to the soil. A lack of nitrogen-fixing bacteria means that soils lack a natural source of nitrogen.

◆ **Nitrifying bacteria** – bacteria found in soils that convert ammonium (NH_4^+) ions to nitrite (NO_2^-) and nitrate (NO_3^-) ions.

◆ **Denitrifying bacteria** – bacteria found in soils that convert nitrate and nitrite ions to nitrous oxide and nitrogen gas. An excess of denitrifying bacteria reduces the nitrogen content of the soil, and increases the emission of nitrous oxide.

◆ **Nitrous oxide** – N_2O, a greenhouse gas emitted from soils. Use of nitrogenous fertilizers increases the emission of nitrous oxide from soil.

Link
The nitrogen cycle is discussed in Chapter 2.3 (page 139). Global warming and the greenhouse effect are covered in Chapter 6.2 (page 528).

5.2.8 Synthetic fertilizers are needed in many intensive systems to maintain high commercial productivity at the expense of sustainability. In sustainable agriculture, there are other methods for improving soil fertility

Much of the increased food production that was achieved by the Green Revolution can be credited to the increased use of fertilizers. This is particularly true where fast-growing varieties of crops are grown in large monocultures. As the crops grow, they rapidly deplete the nutrients available in the soil. To overcome this loss, farmers apply **synthetic fertilizers** (also called artificial or inorganic fertilizers) to their crops.

Synthetic fertilizers are manufactured industrially from raw materials obtained in nature. The manufacturing process requires energy, which typically comes from fossil fuels such as coal or natural gas. This makes fertilizers more expensive and increases agriculture's contribution to climate change.

Link
Eutrophication is covered in Chapter 4.4 (page 373).

The use of inorganic fertilizers is also highly wasteful and polluting. Studies consistently show that most fertilizer applied to farms dissolves in rainwater and washes away, either into the groundwater or into nearby surface waters. In surface waters, fertilizers contribute to eutrophication, which is the most widespread global water pollution problem.

Furthermore, adding fertilizers to soil changes the biological and chemical conditions in the soil, and negatively affects the natural fertility of soils. **Nitrogenous fertilizers**, for instance, cause changes in the balance of bacteria taking part in the nitrogen cycle. **Nitrogen-fixing bacteria** are reduced, while **nitrifying** and **denitrifying bacteria** are increased. The loss of nitrogen-fixing bacteria means that the natural fertility of soil is reduced, and farmers become even more dependent on synthetic fertilizers. The increase in nitrifying and denitrifying bacteria means that there is increased emission of **nitrous oxide** from the soil. Nitrous oxide is a powerful greenhouse gas and contributes significantly to global warming.

Concept

Systems

The changes in the nitrogen cycle which occur when nitrogenous fertilizers are added to the soil may be explained by the systems concept of negative feedback. When extra nitrogen is added to the soil, populations of nitrifying and denitrifying bacteria may increase and convert the nitrogen to other forms. In particular, denitrifying bacteria remove the nitrogen from the soil, returning it to the air. Since the action of denitrifying bacteria counteracts the addition of nitrogen to the soil, it can be described as a form of negative feedback.

The use of nitrogenous fertilizers by farmers can be seen as an example of positive feedback. When fertilizers are added to soil, the action of nitrogen-fixing bacteria in the soil is reduced, making the soil less fertile. As a result, farmers may need to add even more fertilizer to the soil. The decline in soil fertility increases in a vicious cycle of positive feedback.

Theme 5: Land

Alternatives to synthetic fertilizers

Farms that depend on synthetic fertilizers are unsustainable, as they will always need to input fertilizers, typically in increasing amounts over time, and they cause significant environmental damage. Fortunately, there are alternative ways to improve soil fertility that are more sustainable. A selection of these are described in Table 5.10.

■ Table 5.10 Selected alternatives to synthetic fertilizers

Method	Description	Benefits
Organic fertilizer	**Manure** collected from livestock or human waste (humanure) **Compost**, including farm, garden and kitchen waste	Contains a wide range of nutrients, which are slowly released into the soil. Promotes healthy soil microorganism community. Improves soil texture, drainage, water retention. Less leaching than synthetic fertilizers.
Mixed herbal leys	A mixture of grasses, herbs and **legumes** grown on the land.	Increases soil organic matter, carbon storage and healthy microorganism community. Can provide forage for livestock. Legumes add nitrogen to the soil. May include other species that benefit the soil, and help control pests.
Mycorrhizae	Soil fungi, which form a mutually beneficial relationship with plants. Mycorrhizae draw nutrients from the soil and provide them to plants. They may naturally exist in soils, but can be added as spores if they are lacking.	Increases ability of plants to draw nutrients from the soil. Promotes healthy soil microorganism community. Improves soil structure and organic matter.
Agroforestry	The inclusion of trees on farms. Trees may provide useful products, such as fruits or **forage** for livestock, or may be included as shelter or windbreaks.	Trees' roots draw nutrients up from deep in the soil: the nutrients are then released into the topsoil when trees' leaves, fruits and branches fall. Trees protect soil from wind and water erosion.
Continuous cover forestry	An approach to **forestry** that avoids clear-cutting any areas. Trees are harvested individually or in small groups, ensuring that the land is constantly covered either by trees or other vegetation.	Ensures that there is constant cycling of nutrients between trees and soil, including maintaining soil microorganisms such as mycorrhizae.

◆ **Manure** – faecal matter of animals, usually livestock or human, which can be used as an organic fertilizer.

◆ **Compost** – partially decomposed plant matter, which can be produced by collecting farm, garden or kitchen waste in a suitable container, and allowing it to decompose. Compost can be used as organic fertilizer.

◆ **Ley** – an area of arable land which is temporarily converted to grassland.

◆ **Legume** – a member of the plant family *Fabaceae*, which possesses root nodules (swellings) containing symbiotic nitrogen-fixing bacteria. Legumes include peas and beans, as well as a variety of herbs and trees that are not typically used as food crops. Well-known, non-food crop examples include clover, acacia and poinciana trees, and the sensitive plant *Mimosa pudica*.

◆ **Forage** – living plant matter, such as grasses or leaves, which livestock can feed on.

◆ **Forestry** – the management of forests for human use. This can include the cultivation of forests for timber or paper production, and the management of forests for conservation.

■ Figure 5.56 Agroforestry

5.2 Agriculture and food

Several of the soil conservation methods described in Topic 5.2.9 (page 476) can also improve soil fertility.

> **Inquiry process**
>
> ### Inquiry 2: Collecting and processing data
>
> Select two alternatives to synthetic fertilizers listed in the above table to analyse factors influencing farmers' choices. Examine the economic, social and environmental implications of these differences. Present your findings with a focus on sustainability, highlighting the varied factors shaping agricultural practices and their broader impacts.

5.2.9 A variety of techniques can be used to conserve soil, with widespread environmental, economic and sociocultural benefits

Certain common agricultural practices can be particularly harmful to the soil. A selection of these is described in Table 5.11.

■ **Table 5.11** Harmful cultivation methods

Method	Description and effects
Cultivating marginal land	Marginal land has poor quality soil, which cannot support intensive cultivation. If it is cultivated, it is likely to be further degraded rapidly. **Desertification** may occur if marginal lands in dry environments are cultivated.
Overgrazing	If too many animals are kept on one patch of land, they will use up too much of the vegetation, so that it cannot regrow. Animals' hooves can also damage the vegetation and the physical structure of the soil. This can cause compaction and erosion.
Overcropping	Continuously growing the same crops on land for many years with no breaks can deplete the nutrients in the soil. The impacts are worsened if the cultivation is intensive.
Ploughing	Breaking up the surface of the soil makes erosion more likely, and may lead to the loss of the topsoil.
Using heavy machinery	Continuous use of heavy machinery such as tractors, harvesters, etc., presses down on the soil causing **soil compaction**.

◆ **Desertification** – a form of soil degradation that may occur in dry climates. The loss of vegetation cover and organic matter, and the accumulation of salt, may result in the soil becoming desert-like and being unable to support vegetation.

◆ **Soil compaction** – a result of heavy loads pressing down on soil. The spaces between mineral particles and organic matter are lost, meaning that air and water cannot move through the soil easily, and plants' roots cannot grow easily.

■ **Figure 5.57** Desertification in the Sahel region of Africa, due to overgrazing of marginal land

To combat soil degradation, a number of alternative cultivation methods have been developed, which seek to protect and even enhance soil. Some of these are described in Table 5.12.

■ **Table 5.12** Alternative cultivation methods

Method	Description	Benefits
Mixed cropping, intercropping or polyculture	Different crops are planted in the same field at the same time. Crops have different root systems and different nutrient needs. Legumes are often included in order to add nitrogen to the soil.	Each crop uses different nutrients from the soil, and adds different substances to the soil. Makes more efficient use of the soil. Mixing crops also reduces the spread of pests.
Crop rotation	Different crops are planted in the same land, in successive seasons. The sequence of crops is designed so that each crop provides benefits for the following one. A fallow period may be included.	Soil fertility is maintained over time, or even increased. Pest life cycles are interrupted by the changing crops.
Strip cultivation	The land is divided into narrow strips, perpendicular to either the wind or the slope of the land. Different crops are planted in each strip. Crops that use or add different nutrients, and that are affected by different pests, are used. Crops may be rotated on to different strips from one season to the next.	Similar benefits to mixed cropping and crop rotation. Also reduces soil erosion by wind and/or water, and reduces spread of pests across the field.
Conservation tillage	Farming involving less ploughing than conventional farming – may be no-till or reduced-till. Seeds are placed straight into the soil without breaking up or clearing the surface with a plough.	Soil erosion and compaction are reduced, and soil structure is maintained.
Agroforestry	Including trees on farms, usually mixed in with other crops.	Trees use their roots to draw nutrients up from deep in the soil, incorporating them into their branches, leaves and fruits. When these eventually fall off the tree and decompose, the nutrients are added to the soil surface, where crops can use them. Trees can also provide shade from the sun and shelter from wind.

◆ **Tillage** – the practice of ploughing the land.

■ **Figure 5.58** Strip cultivation of corn and other crops, following the contour of the land

5.2 Agriculture and food

■ Table 5.13 Methods for reducing soil erosion

Method	Description and benefits	Benefits
Contour ploughing	Sloping land is ploughed along the contours (perpendicular to the slope).	Removal of topsoil water is reduced, as the water is slowed down by the ploughed lines.
Terracing	Steeply sloping land is shaped into step-like platforms or terraces.	Creates flat land that can be cultivated. Water flows along terraces, rather than downhill, so less topsoil is carried away.
Drainage systems	The system of drains and ditches should include settling ponds or similar structures at a point before the water leaves the farm.	Water runoff is slowed or stopped, and sediments settle out before the water flows onwards.
Bunding	A retaining wall is built around the boundaries of a field, particularly on the downhill side.	Rainwater and any eroded soil is trapped by the bund. The water can also be used for irrigation.
Cover crops	Plants are grown on the soil in between crops, instead of leaving the ground bare.	Cover crops cover the soil and hold soil particles together, reducing erosion by both wind and water. Carefully chosen cover crops may add nutrients and organic matter to the soil, especially if they are ploughed in before the crop is planted.
Windbreaks	Tall and/or thick trees and hedges are planted around the borders of the farm, and perhaps at strategic locations within the farm.	Trees and hedges act as a barrier to the wind, slowing down air movements over the farm and reducing removal of topsoil by the wind. Windbreaks can also protect crops from damage by strong winds.

■ Figure 5.59 Terracing on a rice plantation in Vietnam

■ Table 5.14 Soil-conditioning methods for improving fertility

Method	Description	Benefits
Lime	Ground-up limestone or chalk (calcium carbonate rock), or similar substances, are added to the soil. This is often done in soil that has been fertilized.	Lime neutralizes acid in the soil, increasing the pH. This improves the ability of plants to absorb minerals including nitrogen, phosphorus and potassium in acidic soils. May also provide a source of calcium for plants.
Organic matter	Manure and/or compost is added to the soil.	Organic matter slowly decomposes, releasing nutrients into the soil. It also improves the soil texture, water retention and drainage, and growth of microorganisms.
Green manure	Cover crops are grown on the land between crop seasons, then ploughed into the soil before the main crop is planted.	The organic matter and all nutrient contents of the cover crop are incorporated into the soil, providing all the benefits of both a cover crop and organic matter.

Theme 5: Land

5.2.10 Humans are omnivorous, and diets include fungi, plants, meat and fish. Diets lower in trophic levels are more sustainable

Biologically speaking, humans are omnivores, as we eat plants, animals and fungi. Our digestive systems are adapted to break down and absorb a range of foodstuffs.

The human diet is highly variable. Each of us typically eats a range of different foods, produced in a variety of ways. Cultural, personal and socioeconomic differences result in widely differing diets.

■ **Figure 5.60** Humans eat a range of foods

Figure 5.60 shows a selection of some common human foods – though what is 'common' for one group of people may not be common for others! The photo shows meat, fish, eggs, dairy products, oil, mushrooms, fruits and vegetables including beans, nuts, root vegetables and leafy greens. Most of these would probably have been produced on a farm, though the processes for growing them can be quite different.

A major difference occurs between the production of fruits and vegetables, which come from plants, and meat and dairy products, which come from animals. Animals must be fed, so the farmer must either grow, find or purchase food for them. Most farmed animals are herbivores, so this means either growing crops, finding natural vegetation for them to graze, or buying feed from other farmers or food producers.

■ Livestock production

Pastoralists graze their herds on open fields, where grasses and other plants grow wild. In some cases they may deliberately cultivate grasses or other plants for their livestock. Typically, they move the animals around between different fields, or even live a nomadic life, moving with their herds from one area to another. Overall, pastoral production of livestock is generally extensive, so there are relatively few inputs, but a large area of land is needed.

■ **Figure 5.61** Animals raised on pastures require large areas of land for grazing

Intensive livestock farming typically involves keeping large numbers of animals in relatively small areas, perhaps even indoors, and bringing food and water to them. The farmer might grow crops specifically for the livestock or purchase food grown on other farms. Pest and disease problems tend to increase if animals are kept in higher densities.

Livestock farming is associated with particular environmental impacts associated with it. Waste from animals (faeces and urine) must be disposed of – though it can often be used as manure. Many livestock farms produce unpleasant smells, which may be a nuisance to neighbours. Most significantly, some species of livestock, called **ruminants** (for example, cattle), emit methane, which is an important greenhouse gas and contributes to global warming.

◆ **Ruminant** – one of several species of herbivorous mammals, including cattle, sheep and goats, which have a special chamber in their stomachs called a rumen where some digestion of plant matter occurs. Ruminants pass food back and forth between their mouths and their rumens, repeatedly chewing and swallowing. Bacteria in the rumen break down plant matter and produce methane, which is released from the animal's mouth.

■ **Figure 5.62** An intensive dairy farm, with many cows kept indoors at high density

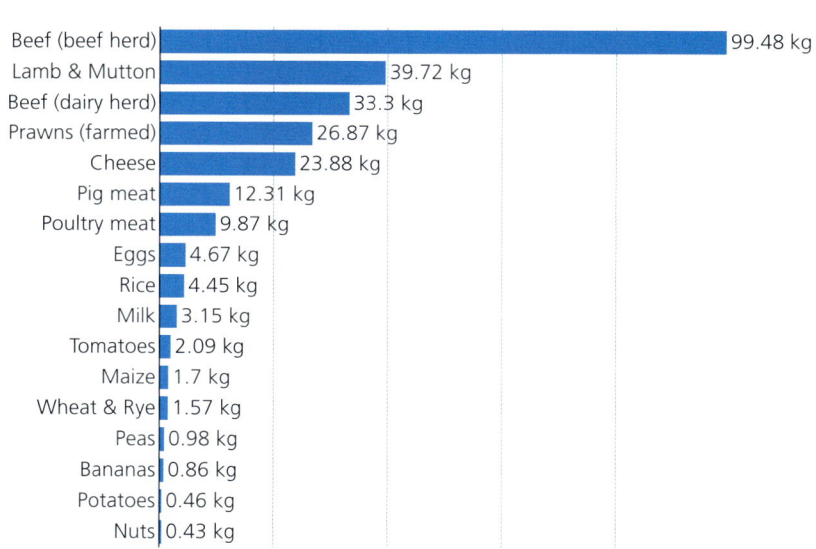

Food	kg
Beef (beef herd)	99.48 kg
Lamb & Mutton	39.72 kg
Beef (dairy herd)	33.3 kg
Prawns (farmed)	26.87 kg
Cheese	23.88 kg
Pig meat	12.31 kg
Poultry meat	9.87 kg
Eggs	4.67 kg
Rice	4.45 kg
Milk	3.15 kg
Tomatoes	2.09 kg
Maize	1.7 kg
Wheat & Rye	1.57 kg
Peas	0.98 kg
Bananas	0.86 kg
Potatoes	0.46 kg
Nuts	0.43 kg

■ **Figure 5.63** Greenhouse gas emissions from the production of various foods (carbon dioxide-equivalent per kilogram of food)

Livestock farming also requires the production of food for the animals, which can involve the use of large amounts of land, water and energy. On average, livestock production uses significantly more land and water than crop production.

> **ATL ACTIVITY**
>
> Use the information in Figure 5.64 to estimate the greenhouse gas emissions from the production of the food contained in your most recent meal.

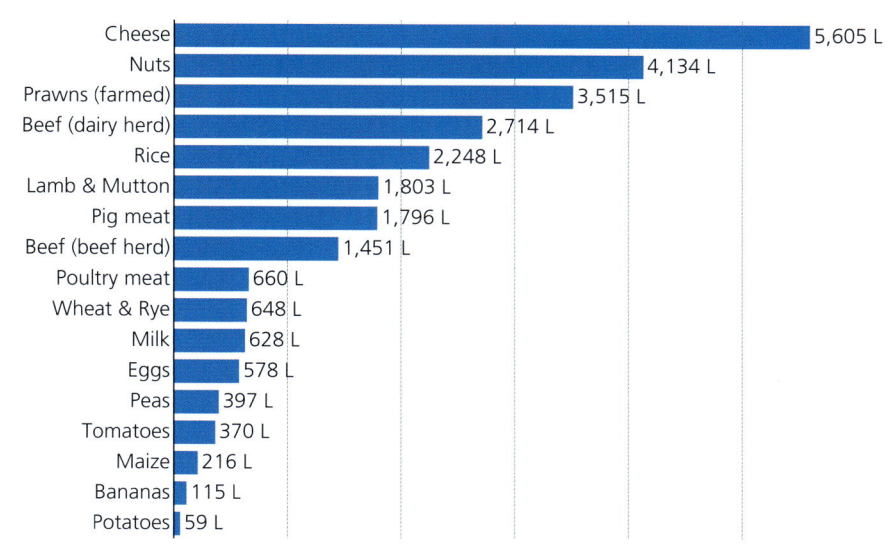

Food	Freshwater (L)
Cheese	5,605 L
Nuts	4,134 L
Prawns (farmed)	3,515 L
Beef (dairy herd)	2,714 L
Rice	2,248 L
Lamb & Mutton	1,803 L
Pig meat	1,796 L
Beef (beef herd)	1,451 L
Poultry meat	660 L
Wheat & Rye	648 L
Milk	628 L
Eggs	578 L
Peas	397 L
Tomatoes	370 L
Maize	216 L
Bananas	115 L
Potatoes	59 L

■ **Figure 5.64** Freshwater used in producing various foods (per kilogram of food)

> **Link**
>
> Energy flow through food chains and the laws of thermodynamics are covered in Chapter 2.2 (page 119).

Another major consideration is the efficiency of energy transfer through the trophic levels. The second law of thermodynamics shows that, on average, about 90 per cent of energy is lost between each trophic level, so to produce 1 kJ of beef, we need to feed the cattle 10 kJ of plant matter. Similar patterns are found when looking at land, water and other inputs: taking the animals' food into consideration, producing livestock is much less efficient than producing crops. As a result, consuming plants, from the first trophic level, can be seen as more sustainable than consuming animals, which come from higher trophic levels.

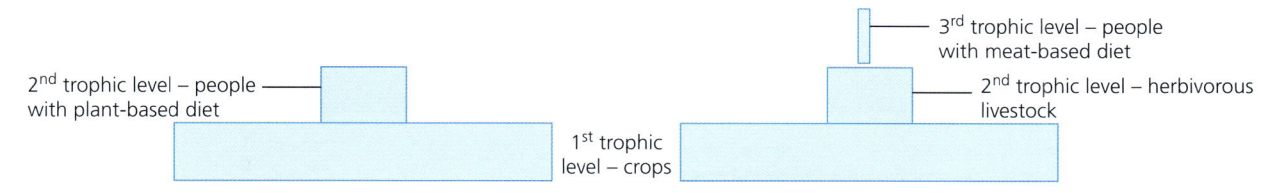

■ **Figure 5.65** Energy transfers between trophic levels on a farm

◆ **Vegetarian** – generally, a person who does not eat meat. The term is used inconsistently, however, as some vegetarians may eat some animal products, such as eggs, while others do not.

◆ **Vegan** – a person who does not eat any products made from animals. Some vegans also prefer not to use non-food animal products, such as leather. 'Vegan' is more consistently used than 'vegetarian', but there are still variations from person to person.

● Common mistake

It is not necessarily true to say that 1 kg of meat contains less energy than 1 kg of grain, fruit or vegetables. The important point is that to produce 1 kg of meat, we must first produce about 10 kg of plants. The overall impact of producing both the meat and the plants to feed the animals is much higher than the impact of producing just the plants.

Another way of thinking of it is that we are consuming a larger percentage of the Earth's total primary production when we eat from a higher trophic level. If we eat from a lower trophic level, we leave a larger percentage of the production available for other people, and other animals, to consume.

■ Converting to a plant-based diet

Many people already follow plant-based diets, whether **vegetarian** or **vegan**. Some people eat only certain types of meat, or limit how much meat they eat, or how often they eat it. Their reasons for this may be religious, cultural or personal preference, or they may have environmental or ethical concerns about eating meat.

5.2 Agriculture and food

HL.c.3: Environmental ethics

For many people, choosing to reduce their meat consumption is an ethical matter – though they may approach the matter in different ways.

For some, the choice is about animal rights and welfare. They argue that farming animals – especially in intensive 'factory farm' settings – is cruel and violates the rights of the animals. Others, with a more ecocentric viewpoint, might argue that animals are no different from humans, so farming them is fundamentally unethical.

For others, it is about the ethics of environmental impact and sustainability. They might argue that any action that harms the environment is ethically wrong, and any action that protects or improves it is right. This may come from either an anthropocentric or ecocentric position, depending on why they think it is wrong. It is anthropocentric if their concern relates to human well-being and how humans depend on the environment. It is ecocentric if they argue that the environment itself, or perhaps the biosphere, is worthy of protection, regardless of human needs.

For yet others, the ethical question is tied to their religious beliefs. Certain religions recommend abstaining from meat for various reasons, often for specified fasting periods. Some value animal life to the extent that any consumption of meat is prohibited. The ancient Indian religion of Jainism, for instance, strictly prohibits harming any humans or animals for any reason.

In general, meat production has a bigger environmental impact, and is less sustainable, than crop production. In theory, if fewer people ate meat, there would be less demand for meat, and less meat would be produced. This might reduce the impact of food production on the environment, and make production more sustainable. One study, published in 2018, estimated that if everyone converted to a vegan diet, the land used for agriculture would be reduced by 75 per cent, as shown in Figure 5.66.

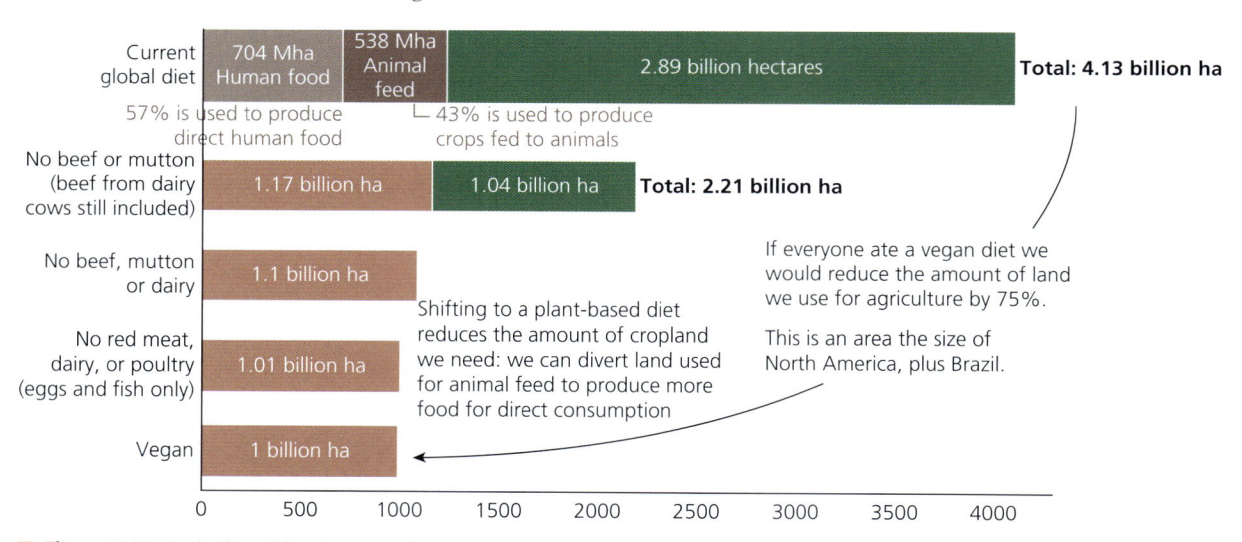

■ **Figure 5.66** Agricultural land use could potentially be reduced by over 3 billion hectares if all people adopted a plant-based diet.

The study showed, however, that most of the impact of livestock production comes from just two types of meat – beef and mutton – plus dairy production. Eliminating only those sources of animal products had almost the same impact as eliminating all animal products.

5.2.11 Current global strategies to achieve sustainable food supply include reducing demand, food waste, greenhouse gas emissions from food production, and increasing productivity without increasing the area of land used for agriculture

Achieving sustainability in agriculture is one of humankind's greatest challenges. The Green Revolution was an early attempt to match food production with the needs of the human population, but it introduced problems that mean agriculture remains unsustainable.

Modern agriculturalists and food producers continue to look for sustainable solutions aimed at increasing food production while reducing food waste and environmental impact.

Reducing demand

Meeting all of humanity's food needs is challenging when some people consume more than they need, while others receive less than they need. Reducing demand among people who have excess can, in principle, help to make more available for those who need more. Efforts might focus on specific foods where production is particularly unsustainable or environmentally problematic, such as meat, or on non-food agricultural products, such as tobacco. If the demand for these products is reduced, then, in principle, more of the global food production effort can go towards meeting the needs of those who do not have enough. However, reducing demand does not guarantee that more food will become available for those who need it. Any increases in production need to occur in locations where it is needed, and at an affordable price.

◆ **Subsidy** – a payment made by governments to producers (e.g. farmers) to lower the cost of production of an item. Effectively, the government pays part of the cost of production, so that the farmer's costs are lowered. Subsidies have often been used to encourage farmers to produce goods that the government sees as essential for the country.

◆ **Tax** – a payment made to the government by various parties within a country. Governments generally use taxes to raise funds to cover the cost of running the country, providing services to citizens, etc., but they can also use taxes to manipulate demand. High taxes on alcohol and tobacco, for instance, have both effects: they raise money for the government, and they raise the prices of those products so that fewer people might buy them.

HL.b.1: Environmental economics

The reality is more complex than this. Reducing the demand for luxury, unsustainable and non-food items might reduce the production of these things, but it does not guarantee more production of the things people need. Decisions about what to farm are generally made based on business and economic considerations.

The relative levels of supply and demand determine the price that people are willing to pay for an item. Farmers would normally choose to produce items that mean they can make a significant profit. This occurs where the cost of production is significantly lower than the selling price on the market.

Consumers have some power over this as they can increase or decrease the demand for items. Governments can arguably have more influence: they can affect the cost of production by **subsidizing** the production of desirable products, or reduce demand by **taxing** undesirable products.

As people have become more aware of sustainability issues associated with meat production, there has been growing interest in alternatives to meat. Meat provides important nutrients, including proteins and certain minerals, which may be challenging to obtain from a plant-based diet. Successful meat substitutes should be able to provide these nutrients.

The simplest substitute for meat is high-protein vegetables. Peas and beans – the seeds of legumes – have relatively high protein content, and have been used in this way for many years. Soybeans are particularly popular, as they are used to produce high-protein foods such as tofu and tempeh. Other meat substitutes are produced from fungi, milk or algae.

> **ATL ACTIVITY**
>
> Many large supermarkets have sections dedicated to providing meat alternatives for vegetarians, vegans, or those who prefer to reduce their meat consumption.
> - Visit a nearby supermarket to see if they have a section like this.
> - Explore the range of products available, reading the packaging to find out what they are made from.
> - Choose one interesting product and do further research online to find out about the production process.
> - You may also consider buying some of the products to try in your own diet.

Some food scientists are working on approaches that might seem radical to many people. One such approach is called single-cell protein or SCP. This involves growing single-celled organisms such as yeast and algae, which are then processed into edible protein. It can be shaped into any form, from powder supplements to steak-like chunks. Other ideas include culturing meat from cells, using cloning technology, and using 3D printers to produce meat-like products from cultured cell products.

A less technology-heavy approach is the farming of insects: some insects can be grown relatively easily with few inputs and few environmental impacts. They can be eaten whole, or crushed and reconfigured into more meat-like products.

■ **Figure 5.67** Insects may be a more sustainable source of protein than meat

Reducing food waste

Food waste is a serious concern for sustainability. Wasted food represents agricultural efforts that have no beneficial outcome. The most efficient use of agricultural production is to ensure that all agricultural products are used. When significant amounts of food are wasted, this means that the resources used in producing them have been wasted. At the same time, the production processes have continued to cause harmful environmental impacts.

Waste happens at all stages of the food supply chain, from the farm to the table. Producers, distributors, intermediaries on the supply chain, consumers, charities and government agencies can all make efforts to reduce losses. At the retail and consumer end of the supply chain, efforts to extend the shelf life of food, by preventing or delaying spoilage, can reduce waste.

Extended food shelf life can be achieved by careful management of the environment in which food is stored and transported. Conditions such as temperature, humidity and composition of the atmosphere should be controlled. Microorganisms such as fungi and bacteria can speed up the decay of food, so these should also be controlled.

> **ATL ACTIVITY**
>
> Research and conduct a debate on either of the following propositions:
> 1. Meat substitutes are a sustainable alternative to livestock farming.
> 2. Farming of insects as a source of protein is ethically preferable to farming animals.

Reducing greenhouse gas emissions

Greenhouse gas emissions are a major issue in agriculture and food production. They are produced in a number of different ways on farms. Specific concerns relate to the emission of nitrous oxide from soils, and of methane from cattle and waterlogged soils.

Reducing nitrous oxide emissions

Nitrous oxide is produced by bacteria in soil, mainly denitrifying bacteria. These bacteria use nitrate ions in the soil and convert it, first to nitrous oxide and then to nitrogen. Some of nitrous oxide is emitted into the air, before it can react further. Nitrates may be naturally present in soil, but the concentration is increased significantly by adding fertilizer to soil.

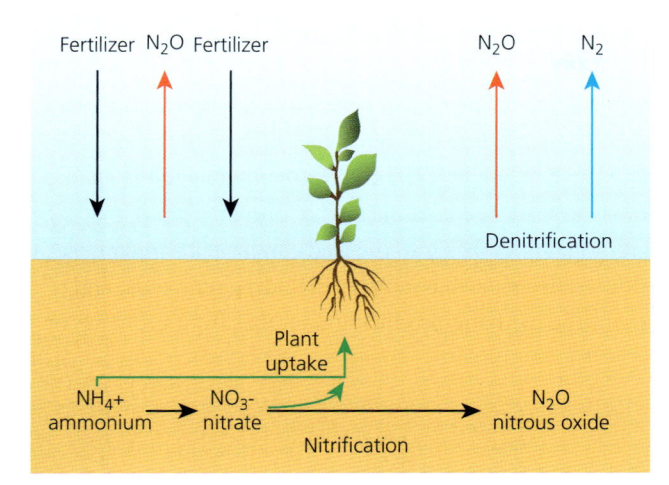

Figure 5.68 Nitrogen conversions in the soil

Figure 5.69 Rice cultivation in a flooded field

Figure 5.70 Furrow irrigation allows the raised parts of the land to be aerated

Nitrogen is one of the macronutrients that plants need in large quantities. Many years of evidence, especially since the Green Revolution, shows that adding nitrogen-rich fertilizer to soil can cause significant increases in crop growth rate and yield. Fertilizers usually include either ammonium (NH_4^+) or nitrate (NO_3^-) ions.

Both ammonium and nitrate can be absorbed by plants, but any remaining in the soil will provide the raw materials needed by bacteria. Ammonium is converted by nitrifying bacteria into nitrates. Nitrates are used by denitrifying bacteria and converted into nitrous oxide.

The key to reducing nitrous oxide emissions is to manage the use of nitrogenous fertilizers. Alternatives such as organic matter and nitrogen-fixing plants, and better use of methods like mixed cropping, crop rotation and cover crops can reduce the need for nitrogenous fertilizers. Where fertilizers are used, the amount should be carefully considered, to ensure that just enough is provided for the needs of the crops. This should reduce nitrous oxide emissions and leaching, which may lead to eutrophication.

Reducing methane emissions from crops

Rice is one of the world's most important crops. Conventional rice cultivation is carried out in flooded fields called paddies. The flooding suppresses weeds, while allowing rice, which is semi-aquatic, to thrive. However, the flooding creates anaerobic conditions in the soil, so organic matter, such as the remaining stalks from the previous crop, undergoes anaerobic decomposition, producing methane. Rice production occupies about 10 per cent of global arable land, and is responsible for about 10 per cent of global methane emissions.

Various strategies have been implemented to reduce methane emissions from rice cultivation. One approach, called alternate wetting and drying (AWD), reduces emissions by allowing the land to dry during particular phases of the crop growth. Alternatively, farmers might use **furrow** irrigation, where only narrow channels between rows of plants are flooded.

Reducing the amount of organic matter on the ground can also help. Most of the organic matter is **crop residue**: the remaining stalks from the previous harvest. If these can be removed, to be used as compost or animal fodder elsewhere, it would reduce the decomposition and methane production.

◆ **Furrow** – a trench or trough created by ploughing a field. Furrows are the lower part of the soil, which is typically used for irrigation. The raised soil between furrows, where the crops are normally planted, are called ridges.

◆ **Crop residue** – the leftover remains of crops after a harvest. Typically parts of the crop plant remain on the land after the useful parts (e.g. fruits or grain) are collected.

5.2 Agriculture and food

Reducing methane emissions from livestock

Ruminants such as cattle are one of the main sources of methane emissions into the atmosphere. Various suggestions have been proposed to reduce methane emissions by changing the environment in the animals' rumens.

Methane is produced when certain microorganisms in the rumen help with the digestion of plant matter. Various experiments have shown that by providing different types of food to the animals, the composition of the microbe community can be changed, and the emission of methane can be reduced. When cattle are fed on starch- or sugar-enriched feed, instead of ordinary grass, methane emissions are lower. Inserting specific, additional microbes into the feed can have a similar effect. There is ongoing research into specific additives from various sources including seaweeds, and the possibility of vaccinating cattle to prevent the methane-producing microbes from growing, or breeding cattle with lower emissions.

> **ATL ACTIVITY**
>
> Discuss and compare the potential benefits of reducing methane emissions from cattle, relative to the benefits of reducing the production of beef and dairy products. Consider also the difficulties or challenges of implementing each of these strategies.

■ Increasing efficiency of production

There is a finite amount of land available for farming. Large areas of land are unsuitable for farming, and other areas have been degraded by current farming practices, so that their productivity is significantly reduced. Therefore, one of the main strategies in the drive for sustainability in agriculture is to make more efficient use of the land remaining. This effectively means increasing the efficiency of production.

In general, improving efficiency means finding the best combination of inputs and conditions for each crop. In the past, this has often been achieved by supplying increasing amounts of fertilizer. However, this does not improve sustainability, as it increases the dependence on industrial inputs and reduces the quality of agricultural soils. Improving efficiency without damaging the environment that agriculture depends on will require innovative solutions, based on sound understanding of the entire agroecosystem.

Improved yield through genetic modification

One approach to improving agricultural productivity is through the use of **genetic engineering** technology to improve the yield.

Genetic engineering typically involves the transfer of individual **genes** from one organism to another. It is a powerful tool as genes can be transferred between any species. The technique is carried out in a laboratory, and produces a small number of individuals with new combinations of traits. These transgenic or genetically modified organisms (GMOs) can then be multiplied by tissue culture, cloning or conventional breeding, depending on the species. Another approach to genetic engineering is to modify the genes within the organism.

The technology has been used in agriculture for a wide range of purposes, many of which relate to improved yield or sustainability. A selection of examples is shown in Table 5.15.

◆ **Genetic engineering** – a laboratory technique for extracting individual genes from one organism and inserting them into another organism. Also called genetic modification or genetic manipulation. When successful, the organism that receives the transferred gene will have the selected characteristic from the organism that provided the gene. The organism that receives the genes is described as genetically modified, or transgenic.

◆ **Gene** – an individual stretch of DNA, usually found in the nucleus of an organism's cells. Genes influence the characteristics of the organism. Each gene typically influences one particular characteristic.

Table 5.15 Selected examples of genetically modified crops and livestock

Modified organism	Source of transferred gene	Objective of genetic engineering
Corn	Bacteria *Bacillus thuringiensis*	Resistance to certain insect pests. Pest infestations are reduced, so yield is increased. Pesticide use can be reduced.
Potato	Potato DNA has been modified to inactivate certain genes	Delayed or reduced browning when cut or bruised. Food waste is reduced, as consumers are less likely to throw out brown potatoes.
Papaya	Ringspot virus, PRSV	Resistance to ringspot virus, so yield is increased.
Banana	Rice	Resistance to black sigatoka fungus, so yield is increased.
Soybean	Bacteria *Agrobacterium* sp.	Resistance to specific herbicides, so that herbicides can be sprayed to kill weeds, without harming the crop. Yield is increased due to lack of weeds.
Tomato	Tomato	Delayed spoilage, extending shelf life. Food waste is reduced.
Rice	Daffodil and bacteria *Erwinia uredovora*	Production of carotene, which is converted to vitamin A when eaten. Improves the nutritional value of rice, to combat vitamin deficiency in less economically developed countries, reducing the risk of blindness in children, and various infections and illnesses in children and mothers.
Pig	Mouse and bacteria *E. coli*	Increased metabolism of phosphorus. Animal digests and absorbs more phosphorus, and less is released in faeces. Pollution due to animal faeces is reduced.

Figure 5.71 Unripe papayas showing signs of ringspot disease. Some papayas have been genetically modified for resistance to this virus

Figure 5.72 Conventional rice and 'golden rice', which has been modified to contain carotene as a source of vitamin A

5.2 Agriculture and food

Figure 5.73 Protestors in the UK campaigning against the use of genetic engineering in agriculture

Genetic engineering has great potential to improve agricultural productivity, but it is very controversial. Concerns range from ethical questions about the rights of humans to modify living organisms, to complex questions about the potential ecological and health effects of GMOs, and about access to and rights to use GMOs. Different countries have responded to these controversies in different ways, including regulating or restricting its use.

> **ATL ACTIVITY**
>
> Research the arguments for and against the use of genetic engineering. Compile a summary table outlining both perspectives.

> **Concept**
>
> **Perspectives**
>
> People's perspectives on genetic engineering vary widely, and can be influenced by their environmental value systems, as well as the political and cultural landscapes of their countries. For instance, general attitudes towards the technology in Europe and North America can be quite different. Explore the media in your own country, and carry out an informal survey of your friends and family, to find out how and why people's opinions on the topic vary.

HL.c.12: Environmental ethics

One of the arguments which may be made against the use of genetic modification is that it is unnatural and therefore ethically wrong. This is a questionable claim, and is considered by logicians and ethicists to be an example of the appeal to nature fallacy.

There are multiple problems with this argument. First, it is difficult to define what is 'natural'. It could be argued, for instance, that DNA is natural, and many organisms (example, viruses, bacteria and fungi) are capable to transferring DNA between individuals, even of different species. One could argue that most examples of genetic engineering simply harness these natural phenomena.

Furthermore, the claim that natural things are inherently good, and unnatural things are inherently bad, is questionable. Many examples can be found, for instance, of 'natural' phenomena that cause harm, such as poisons or disease, and 'unnatural' things that produce benefits, such as medicines, clothes and eyeglasses.

A more meaningful debate about the ethics of genetic modification might take a consequentialist approach, considering the likely outcomes of each example of genetic modification for humans, the organism and the environment. Applying a social justice lens to a consequentialist argument, the effect of GM on small farmers and consumers might be examined relative to the power of businesses and corporations that hold patents to genetically modified organisms, and control access to them.

Solar-powered, on-site fertilizer production

Providing fertilizers to crops has proven benefits, but when those fertilizers are produced using fossil fuels and are transported long distances from the factory to the farm, the environmental cost can be prohibitive. Furthermore, if too much fertilizer is applied, it can cause pollution and soil degradation.

In recent years, an experimental technology has been developed to produce small and steady amounts of nitrogenous fertilizer using solar energy, directly on the farm. The method involves solar photovoltaic panels to generate electricity, and a catalyst to cause a reaction between nitrogen gas and water.

Nitrogen gas is very stable and unreactive, and normally a reaction of this sort can only occur during thunderstorms (driven by the energy of lightning), in an industrial complex (typically driven by fossil fuels) or in nitrogen-fixing bacteria (with highly specialized enzyme complexes and a biological energy source). If this technology succeeds, farmers will be able to produce exactly as much nitrogenous fertilizer as they need, on site, without using fossil fuels or contributing to global warming. By controlling the rate of production and release into the field, they can also increase their crop yield, while reducing the risks of nitrous oxide emission and eutrophication.

5.2.12 Food security is the physical and economic availability of food, allowing all individuals to get the balanced diet they need for an active and healthy life

Food security is the ability of people to meet their needs for food. It is defined by the United Nations' Food and Agriculture Organization as 'when all people, at all times, have physical and economic access to sufficient safe and nutritious food that meets their dietary needs and food preferences for an active and healthy life'.

The FAO defines four dimensions of food security, as outlined in Table 5.16.

■ Table 5.16 The four dimensions of food security

Dimension	Explanation
Physical availability	The availability of sufficient, appropriate food, where people can obtain it. This depends on production of food and distribution to where people live, including international trade, where relevant.
Economic and physical access	The ability of individuals and households to obtain, and where necessary purchase, food. This is affected by people's incomes, food prices, inflation and the broader cost of living.
Utilization	The biological use of food by the body. It relates to the health and well-being of people, and can be affected by people's ability to prepare and consume a healthy variety of food.
Stability	The consistency of the other three dimensions. Changes in circumstances, due to factors such as changing weather, natural disasters, price fluctuations, and social and political upheavals can affect stability.

Food security can be evaluated at the global, national or household level. At the national level, food security is affected by factors such as poverty and inequality, and domestic agricultural capacity relative to dependence on food imports. Countries that are unable to produce sufficient food for themselves are vulnerable to external threats, such as price fluctuations and trade disruptions. Domestically, weather conditions have great influence on agriculture, so countries experiencing droughts or other extreme weather events may suffer.

> **REAL-WORLD EXAMPLE**
>
> Global effects of the conflict in Ukraine
>
> The armed conflict between Russia and Ukraine, which began in February 2022, has severely affected Ukraine's agricultural industry. Farms, farm equipment and agricultural products were damaged and destroyed. Transport networks were disrupted, making export of food produced in Ukraine more difficult and costly. Prior to the conflict, Ukraine produced and exported 10 per cent of the world's wheat. Its share of maize (corn), barley and sunflower exports were even higher. Russia also produced and exported large amounts of these and other foods.
>
> As a direct result of the conflict, exports of food from both countries have fallen drastically. This has reduced their supply on the global market, and food prices rose to record highs in mid-2022. Russia was also a major global exporter of fertilizers and natural gas, both of which have also increased in price. Natural gas is used to produce fertilizers, which support much of global agriculture, while high natural gas prices drive up the overall price of energy. When Russia's exports of these products declined, the cost of both energy and food production increased around the world.
>
> Much of Ukraine's wheat export – estimated at 92 per cent from 2016 to 2021 – was previously sold to poorer countries of Africa and Asia, where the number of people experiencing food insecurity is high. Reduced exports and increased prices have worsened the situation in such countries. In Egypt, for instance, domestic food prices rose by 30 per cent in 2022, affecting households' ability to afford food. As food became less affordable, people tended to switch from more expensive, more nutritious foods, to cheaper, less nutritious ones. As a result all four dimensions of food security declined.
>
> The conflict in Ukraine is only one of several factors that has contributed to current food insecurity in many parts of the world. The COVID-19 pandemic hurt economies and food production. Climate change has made food production more difficult in many areas, with droughts and unusual weather patterns hindering farmers' work. Together, these multiple factors provide significant barriers to food security around the globe.
>
> *(Sources: www.consilium.europa.eu/en/infographics/how-the-russian-invasion-of-ukraine-has-further-aggravated-the-global-food-crisis and www.csis.org/analysis/russia-ukraine-and-global-food-security-one-year-assessment)*

> **ATL ACTIVITY**
>
> Food security and hunger around the world
>
> The United Nations and various international charities and agencies publish regular reports on food security and hunger around the world. Many of these are available online, and provide data and background information that is worth exploring.
> - Start by searching for the most recent editions of the *Global Report on Food Crises* and the *Global Hunger Index*, to find out the situation in different countries and regions of the world.
> - Another useful source is the website **https://ourworldindata.org**, which can be searched using terms such as 'food security' and 'hunger'. There you can find maps and graphs such as those in Figures 5.74 and 5.75, showing global and regional patterns.

■ **Figure 5.74** Food insecurity around the world

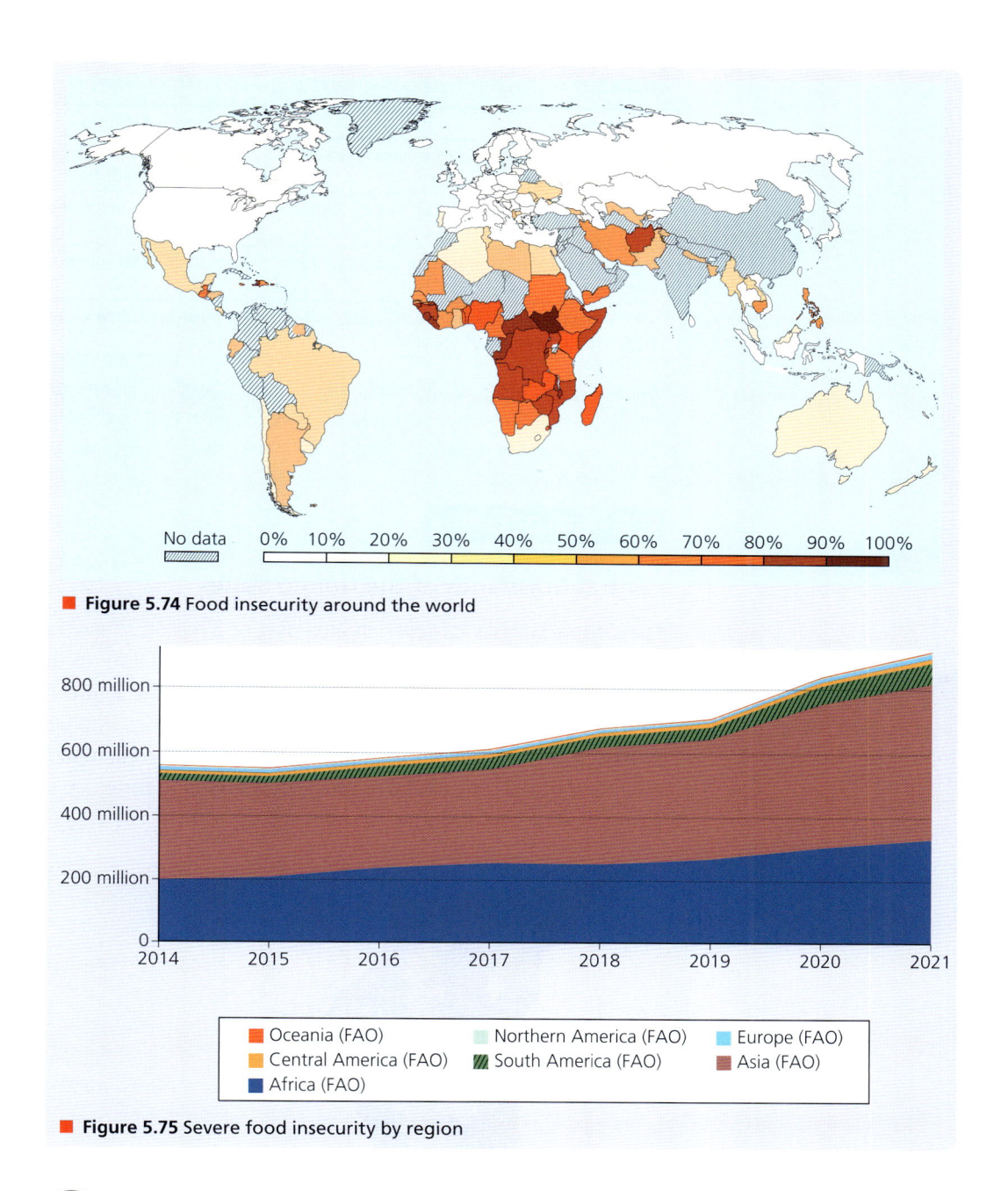

■ **Figure 5.75** Severe food insecurity by region

(HL) 5.2.13 Contrasting agricultural choices will often be the result of differences in the local soils and climate

Agricultural choices need to be made by farmers and depend on factors such as operating costs and potential earnings from sales, government policies (including subsidies), proximity to population centres, as well as climatic and soil conditions.

One fundamental choice is between growing crops, raising livestock, or doing both in a mixed farming system. Some considerations for making this choice are shown in Table 5.17.

5.2 Agriculture and food

Table 5.17 Selected differences between crop and livestock farming

Crop farming	Livestock farming
Crops are more vulnerable to changing weather.	Livestock are often more tolerant of different weather conditions.
Crops are directly dependent on healthy, fertile soils.	Livestock can be grazed on less fertile soil, but may require input of food grown elsewhere.
Crops can be left unattended for longer periods of time.	Livestock need continuous care.
Crop farming may be highly mechanized, using machinery to do much of the work.	Livestock farming usually involves more manual labour.

Crop production tends to be limited to areas with stable, moderate weather patterns and fertile soil, whereas livestock can often be raised on more marginal land, with poorer soils and harsher or more extreme weather conditions. In many parts of the world, livestock are primarily farmed in drier areas with poorer soils.

REAL-WORLD EXAMPLE

Central Grasslands of the United States

The Central Grasslands cover an area of over 2 million square kilometres, lying between the Rocky Mountains to the west and the Mississippi River to the east, and stretching from the Canadian border in the north, to Texas in the south. The native biome of the region is temperate grassland, known in North America as prairie. Prairie soils tend to be soft, rich **mollisols**, with high organic matter content in the topsoil. The area is a major site of agricultural production, accounting for much of the United States' production of wheat, corn and soybeans.

◆ **Mollisol** – a rich, soft, dark soil, typical of temperate grasslands, with high organic matter content in the A horizon, or topsoil. Mollisols often make fertile farmland.

◆ **Rangeland** – areas of land where wild or domesticated herbivores graze. Rangelands are often covered by grasses or shrubs, and herbivores are able to roam over wide areas, unless their movements are restricted, for instance by fencing.

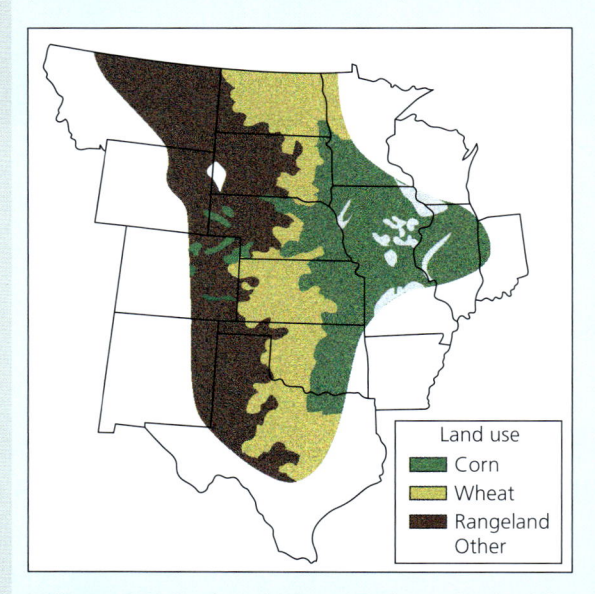

Figure 5.76 Land use in the Central Grasslands of the United State

Figure 5.76 shows a clear pattern in land use in the area. Travelling from east to west, land use changes from arable farming of corn, soybeans and wheat, to **rangelands** where cattle are raised. This pattern reflects changes in rainfall, as shown in Figure 5.77. Corn and soybeans are grown mostly where rainfall exceeds 800 mm per year, and wheat is grown where rainfall is between 500 and 800 mm per year. To the west, where rainfall is below 500 mm, most land is used for cattle and winter wheat. Small areas of the drier, western region are also cultivated with cotton and corn, but these require irrigation.

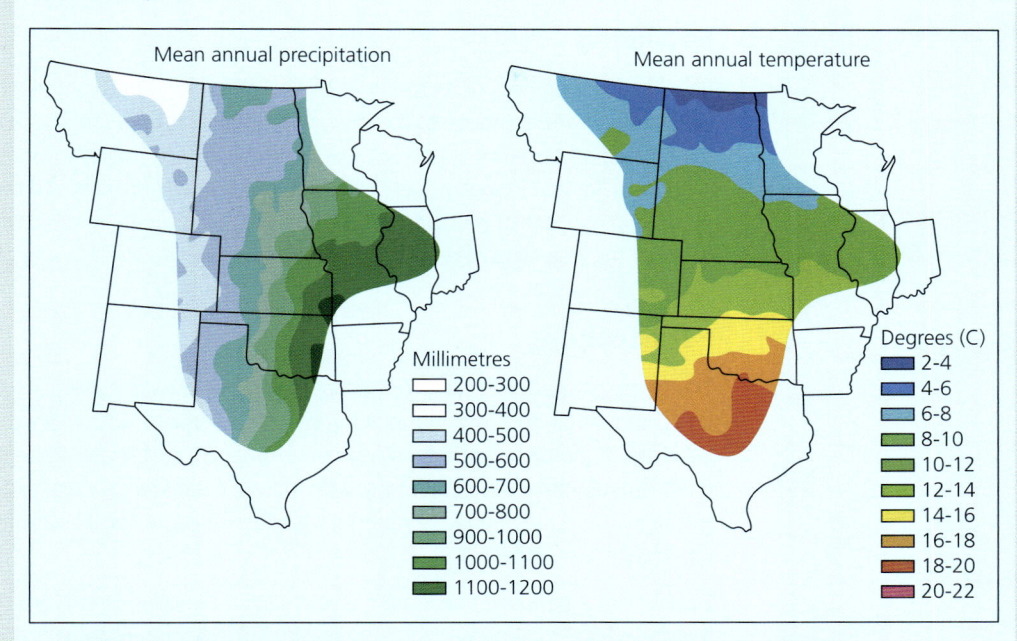

■ **Figure 5.77** Mean annual rainfall and temperature in the Central Grasslands region

(Source: www.researchgate.net/publication/233379722_Nitrogen_in_the_Central_Grasslands_Region_of_the_United_States)

5.2.14 Numerous alternative farming approaches have been developed in relation to the current ecological crisis

Some of the processes used in modern, conventional agriculture have resulted in negative environmental impacts, and there are concerns that such farming methods are unsustainable. This has led to the development of alternative approaches to overcome these problems. Farmers around the world have experimented with these new approaches, with varying degrees of success.

The methods described here can be used individually, but they are often used together on the same farms. Each of these methods provides benefits, including protecting and improving soils, increasing biodiversity and reducing water pollution. In some cases, they have potential economic benefits, by reducing farmers' costs while increasing their income. Any method that improves the soil has the potential to increase the long-term productivity of a farm.

Soil regeneration

Conventional agriculture can have several negative effects on farm soils, including depletion of organic matter and nutrients, compaction, waterlogging, and changes in soil chemistry and texture. Soil regeneration refers to a variety of practices that aim to restore soils to a healthy, natural state, repairing the damage done by farming and other human activities. Regenerative approaches may include adding organic matter as fertilizer, planting cover crops and green manure to stabilise the soil and add carbon via photosynthesis, and using plants such as legumes to add nutrients to the soil, and grasses to stabilize the soil and add carbon, are some approaches that can be used. Carefully managed use of livestock can help to add organic matter to soil, but overgrazing must be avoided. Regeneration can be a slow process, but if successful it can improve the soil, making it more productive and less dependent on inputs such as fertilizers, and therefore achieving sustainability.

Rewilding

Rewilding is a process of restoring disturbed or damaged ecosystems to a natural, wild state, by allowing natural processes to occur, with little or no human intervention. It may involve the reintroduction of native species of plants or animals, or the removal of human-made structures such as buildings, fences or walls. On farms, it typically involves avoiding activities such as ploughing and use of pesticides. It is often practised on abandoned or unproductive farmland. In general, the objective of rewilding is to improve the biodiversity and restore the natural ecological functions of a site.

Permaculture

The word 'permaculture' is a contraction of *permanent agriculture*. Its practitioners describe it as an ecological design system or philosophy, which aims to learn from and use natural interactions between organisms and their environment, to create sustainable systems for producing food and, more broadly, for living in harmony with nature. It aims to produce systems that are stable and self-sustaining, and to continuously produce a wide variety of foods, with minimal artificial inputs like fertilizers or pesticides.

Establishing a permaculture system involves carefully studying the local environment, selecting appropriate crops and livestock, and designing cultivation systems that make the most of mutually beneficial interactions between species. By their nature, permaculture systems are very diverse polycultures, which are designed to work in specific, local contexts. As a result, permaculture projects can be very diverse.

HL.c.3: Environmental ethics

Permaculture has grown in popularity, initially through the work of two Australian academics, Bill Mollison and David Holmgren, who published the book *Permaculture One* in 1978. In the years that followed, they and their students led courses in various countries that came to be known as Permaculture Design Courses (PDCs). Their teachings promote a series of ethical and design principles, which have been adopted by many permaculture practitioners around the world.

Their ethical principles are often summarized as:
1 care of the Earth
2 care of people
3 fair shares.

These principles could be seen as encompassing both ecocentric and anthropocentric viewpoints, as they promote the well-being of nature as well as of people. There is a clear recognition that people depend on, and are a part of, nature, or the community of Earth.

The third principle, fair shares, has evolved over time and been stated in different ways by different authors. In general, it refers to the need to support the well-being of all people and all of nature. It promotes a more just distribution of food, and a less competitive or commercial approach to food production.

Non-commercial cropping

Non-commercial cropping refers to the cultivation of crops that are not for sale, or not intended to make a profit. Farmers might practise non-commercial cropping for a number of reasons. For some, farming is a subsistence activity, where they produce a variety of foods for their families. Others may grow non-commercial crops as part of their commercial farming practices. In these cases, non-commercial crops may support the farm business in an indirect way.

Cover crops are an example of a non-commercial crop grown within the context of a commercial farm. Cover crops are used to protect and improve the soil, in between the cultivation of commercial crops. By covering the soil surface, they can prevent erosion, for instance, or add nutrients such as nitrogen to the soil. Some cover crops are used as green manure, and ploughed into the soil before the commercial crops are planted.

Non-commercial crops, including trees, can also be used to provide shade or fodder for livestock, or as windbreaks to prevent soil erosion. Some farmers plant a variety of non-commercial crops to increase the biodiversity of the farm. A wider range of plants on the farm can encourage beneficial insects such as pollinators and predators which eat pests, thereby reducing the need for pesticides. Some non-commercial crops release natural substances that repel pests, so they may be grown near the commercial crops to help reduce the need for pesticides.

Zero tillage

Zero-till or no-till farming is a form of conservation tillage, which aims to protect soils from the damaging effects of conventional ploughing. Ploughing breaks up the surface of the soil, leading to increased erosion, loss of organic matter, moisture and soil organisms, and compaction. Conservation tillage involves ploughing less frequently and less deeply. Zero tillage is the ultimate approach, where no ploughing occurs. These methods reduce soil erosion, and conserve organic matter, moisture and soil organisms, compared to ploughing. Since tractors are used less frequently, less energy is consumed, less pollution is produced, and the soil is less likely to become compacted.

In no-till farming, various methods are used to plant seeds directly into or on to the soil, with minimal disturbance to the topsoil. One approach, for instance, uses rotating discs pulled behind a light tractor to dig narrow trenches into which the seeds are deposited. No-till farming using methods such as this might be called direct drilling.

Since the land is not ploughed, the crop residues from the previous harvest, such as roots, stalks and leaf litter, remain in the soil, to help stabilize it and to provide organic matter. The crop residue can also function as a mulch, covering and sheltering the soil, and reducing weed growth and moisture loss by evaporation.

■ **Figure 5.78** Corn seedlings sprouting through crop residues in a no-till farming system

Ploughing is one of the most expensive processes in farming, as heavy machinery must be dragged across the land, sometimes multiple times. No-till and reduced-till farming can therefore reduce the cost, energy use and carbon dioxide emissions of farming. However, no-till farming may suffer from pests, including weeds, slugs and other animals that live in the soil. This is because crop residue left on the land can feed and protect these, whereas ploughing would normally remove them. Therefore, farmers practising no-till may need to take extra measures to protect their crops from pests, possibly including increased use of pesticides.

> **ATL ACTIVITY**
>
> ### Develop community and disseminate knowledge
> 1. In groups of three, choose a land and food production issue (options include food wastage, overgrazing, animal farming, beef production), and assign an indirect name or clue to represent your chosen issue. Create visual representations of the environmental problem on chart paper using at least five examples, and incorporate colours for a vibrant display. Set a time limit of 30 minutes for completion. Write the names of group members on the charts.
> 2. After completing the chart activity, exchange your chart with another group.
> 3. Write the names of group members on the charts.
> 4. Using the other group's chart, analyse and elaborate on the visuals, adding details and crafting a 400-word paragraph explaining the observed environmental issue. Conclude with an attempt to guess the identified problem.
> 5. Synthesize and reflect: Present your findings to the entire class, allowing for a guided discussion facilitated by the teacher or student-led questions from the audience.

5.2.15 Regenerative farming systems and permaculture use mixed farming techniques to improve and diversify productivity

Livestock farming can be quite damaging to the environment, because of the heavy dependence on resources such as water, and the release of organic wastes into water and methane into the air. However, livestock can also help achieve sustainability in some farming systems, if they are carefully managed. Many regenerative farming and permaculture systems, for instance, involve mixed farming, with crops and livestock, where the livestock play key roles on the farms.

Grazing herbivores such as cows, goats and sheep, as well as pigs and chickens, can be useful in clearing vegetation from land and controlling weeds. The animals' hooves can break up the soil surface, trampling crop residues or cover crops into the soil. In this way they act like a plough, but without the negative effects of ploughing such as soil compaction and erosion. At the same time, livestock add manure to the soil, increasing its organic matter content and promoting the growth of soil microorganisms. When livestock graze on live plants, such as cover crops or grasses, the plants may respond in different ways. Generally, light to moderate grazing may increase the rate of growth of plants. This might result in more root growth, and possibly more release of exudates, which ultimately add organic matter to the soil and promote microbe growth even more.

Effective use of livestock in regenerative farming requires careful management of the numbers of animals and the length of time they are allowed to graze. Excessive grazing can lead to some of the same problems that ploughing causes, such as soil compaction and erosion, and complete loss of vegetation. In dry climates, overgrazing can lead to desertification. However, if the number of livestock and the intensity of grazing are managed, livestock can encourage plant growth and improve the soil.

In general, livestock should be kept on one patch of land for only a short period of time – perhaps a few days. They should then be moved elsewhere, leaving the grazed area to recover for several weeks. One approach that may be used is mob grazing, also known as ultra-high stock density (UHSD). In this method a large number of animals are allowed to graze each area for a short period, sometimes just a few hours. The grazed area must then be allowed to rest without grazing for a long time, up to several weeks or months. This high-intensity, short-duration grazing, with long rest periods in between, cuts down much of the vegetation in that area, but encourages the plants to grow back rapidly, and to store more carbon in their roots and the soil.

Animals are often seen as an essential part of a permaculture farm, since they form part of the natural ecosystems that permaculture seeks to replicate. Permaculture farmers carefully select animals to perform specific roles on the farm. Chickens peck at the soil, breaking up the surface and removing various insect pests. Pigs can be used to eat garden or kitchen waste, and produce manure, while also digging out weeds from the soil.

Incorporating livestock into a farm can provide many benefits, but rearing them often requires extra work. Fencing and feeding and watering systems may be needed, and there is always the risk that animals can cause damage to the crops if they are stocked too densely, or if they are allowed to roam freely. Farmers must weigh the potential advantages and disadvantages, and consider the specific conditions of their farm, before deciding to include livestock in their regenerative or permaculture farm.

5.2.16 Technological improvements can lead to very high levels of productivity

The continuing increase in demand for food is an incentive to develop more efficient, productive farming systems. Many of these systems are designed to provide food to urban environments, where increasing numbers of people live. Cities often have limited land space available for farming, and depend on food transported a long distance from where it is produced.

Controlled-environment agriculture is one approach which uses technological innovations to increase agricultural productivity. Crops are cultivated inside enclosed spaces, where conditions such as temperature, light, humidity and atmospheric composition can be controlled. By farming in an enclosed space, conditions can be tailored to the specific needs of the crops, and higher efficiency can be achieved in the use of resources such as water and nutrients. Water and nutrients that are not used immediately by the crops can be collected and re-used, instead of flowing away and causing pollution problems. These systems often use hydroponics, where soil is not needed.

These sorts of systems can be very efficient and productive. They do not need large open spaces or fertile soils, so they can be established in relatively small spaces, often in urban environments, close to where the demand for food is. They are unaffected by the weather and pest control may be relatively easy in the enclosed space. The main disadvantage is the cost of building and operating the facilities. In particular, operating them may require a great deal of energy, which may contribute to greenhouse emissions unless alternative energy sources can be used.

Greenhouses

Greenhouses, also called glasshouses and hothouses, are a well-established method of controlled-environment agriculture. The simplest greenhouses are used to raise the temperature, providing more favourable conditions for crop growth. They are widely used in temperate regions for growing warm-weather crops during the colder seasons. Modern greenhouses can also include systems for controlling the light and carbon dioxide concentration, in order to promote crop growth. The brightness, wavelengths and duration of exposure to light can be manipulated by using artificial lights to suit the specific needs of the crops being grown. Increasing the carbon dioxide concentration within the greenhouse can also speed up crop growth, as they are able to photosynthesize faster.

■ **Figure 5.79** A large commercial greenhouse with artificial lighting

Vertical farms

Vertical farms are designed to make more efficient use of space. Crops are grown on multiple levels, directly above one another, within an enclosed space such as a greenhouse or building. Artificial lights are typically used, and water and nutrients can be recycled.

■ **Figure 5.80** Vertical farming using hydroponics and artificial lighting

Vertical farms can be expensive to establish and run. Real estate in urban areas can be expensive, and energy costs (to run artificial lighting and irrigation systems) can be high. As a result, vertical farms are most often established in abandoned structures such as warehouses, tunnels or shipping containers, in relatively affluent communities. They typically produce high value crops such as salad vegetables, which can be sold for a relatively high price.

5.2.17 The sustainability of different diets varies

Commercial farms produce food for people, and some agricultural systems and practices are more sustainable than others. Different systems have different environmental impacts, and some systems are more likely to deplete essential resources. The impact of agricultural systems may also extend beyond the farm. If many people choose to eat a particular type of food, this encourages farmers to produce more of that food. Therefore, people's dietary choices can have an effect on the sustainability of agriculture.

Food miles

In the modern, globalized economy, it is often possible to eat food that was produced a long way away. Tropical fruits such as mangos and bananas can be found in supermarkets in many temperate countries, while apples and pears grown in the temperate region can be bought in the tropics. Shrimp farmed in Ecuador and Bangladesh, and lamb reared in New Zealand are eaten all around the world.

These foods are transported over long distances, usually on ships. The further they travel, the more fossil fuels – usually diesel – are used, and the more carbon dioxide and other air pollutants are emitted. For some foods, such as dairy products and some fruits and vegetables, the ships must be refrigerated to keep the food fresh. This requires extra energy and therefore produces more emissions. A 2022 study estimated that the transport of food accounts for approximately 19 per cent of the greenhouse gas emissions associated with food-production systems. 'Food miles' refers to the distance food travels from the farm to the final consumer. One aspect of the environmental impact of a meal or a product made from many ingredients may be estimated by calculating the total food miles of all the ingredients in the meal or the product.

Food supply chains

However, transport is only one aspect of the food supply chain. A supply chain refers to all the processes that an item undergoes from production to consumption. For most products, the supply chain includes many steps, where the item is passed from one person or business to another several times before getting to the end user. For food, the supply chain typically includes production on a farm, processing in some sort of industrial facility, distribution via various forms of transport, repackaging and selling by wholesale and retail shops, and finally consumption.

At each step of the food supply chain, money is spent, and energy and raw materials are used to carry out processes such as transport, processing and repackaging. Money usually travels in the opposite direction to the food: from the consumer, through multiple intermediate steps, and eventually to the farmer. In general, the longer the supply chain, the more energy and raw materials are used, and the less money reaches the farmer.

Shortening the food supply chain by eliminating some of the intermediate parties, could have environmental, social and economic benefits. The shortest possible chain is one where the consumer buys directly from the farmer. **Farmers' markets** and **pick-your-own farms** are examples of this. **Vertical integration** of the supply chain, where the farmer takes over some of the roles of the intermediate parties, can also achieve this. Potential advantages and disadvantages of short supply chains are shown in Table 5.18.

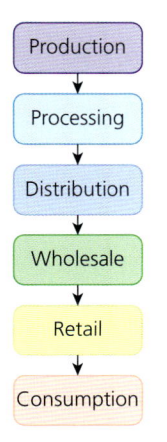

■ **Figure 5.81** A typical food supply chain

Production → Processing → Distribution → Wholesale → Retail → Consumption

◆ **Farmers' market** – a market where local farmers can sell their produce directly to consumers.

◆ **Pick-your-own farm** – a farm that allows paying customers to pick the crops they want directly from the fields.

◆ **Vertical integration** – a business model where one party carries out multiple functions in the supply chain. A farmer may package and process the food on site, before selling it to wholesalers, retailers or consumers.

■ **Table 5.18** Selected advantages and disadvantages of short supply chains

Advantages	Disadvantages
The farmer may earn more income, as they can sell at retail prices, rather than wholesale.	Farmers may sell less of their produce than if they had sold to intermediaries or wholesalers, potentially reducing their total income.
The consumer may spend less, as there are fewer intermediaries who each increase the price of food in order to make a profit.	Farmers may have to do more work or employ others to transport produce and run stalls or shops in order to sell directly to consumers.
Less energy and fewer resources are used in transporting and processing food from the farm to the consumer.	Consumers may have less variety of food available to choose from, and may find shopping in this way less convenient than using large supermarkets.
Consumers and local farmers may establish direct relationships, potentially improving social cohesion in the community.	Intermediaries such as food processors, shippers and wholesalers may lose business, leading to wider impacts on the economy.
May promote smaller farms, which tend to use more sustainable methods of farming.	Farmers who live far away from their consumers may lose income.

5.2 Agriculture and food

Figure 5.82 A farmers' market allows farmers to sell directly to consumers

Shortening the supply chain to the point where farmers and consumers interact directly can have broader benefits related to sustainability. Consumers who get to know local farmers may become more aware of and concerned about the social, economic, environmental and political issues relating to farming. This may lead to changes in consumer behaviour, such as purchasing more local produce and supporting initiatives for sustainability in farming.

Eating food out of season

Most crops' growth is affected by the seasons, so that crops are generally ready to harvest at particular times of year. In the past this usually meant that these foods were only available at certain times of year. Two features of modern food production have changed this: controlled-environment agriculture and international trade in food.

- Controlled-environment agriculture, such as farming in greenhouses, allows crops to be produced outside of their normal season, so that they can be available throughout the year.
- International trade means that products can be shipped from one part of the globe to another. Since different places experience seasons at different times of year, foods can often be obtained out of season if they are shipped from far enough away. For instance, apples, which are generally harvested in the autumn, may be available in the spring if they are shipped from the opposite hemisphere.

Both controlled-environment agriculture and international shipping of food can have significant impacts on the environment. People who are concerned about sustainability may therefore choose to eat only foods in their normal seasons.

Vegetarianism and veganism

Meat production is generally less efficient and has a larger environmental impact, than crop production (see Topic 5.2.10, page 479). In order to reduce the environmental impact of their diet, some people choose to reduce their meat consumption. Some eliminate meat from their diet to become vegetarian, while others eliminate all animal products, becoming vegan. Other dietary choices can be described as flexitarian (mostly vegetarian, but eating meat occasionally) or pescatarian (eating fish, but not meat).

Trends in meat consumption, and the adoption of vegetarian and vegan diets, are complex. For many years, meat consumption has increased worldwide, as societies and individuals become wealthier as a result of economic development. There are also cultural elements to meat consumption, as some societies, such as the United States and Brazil, have higher per capita meat consumption than others with similar levels of affluence. In some societies, such as India, meat consumption remains relatively low, partly because religious and cultural practices promote a vegetarian diet.

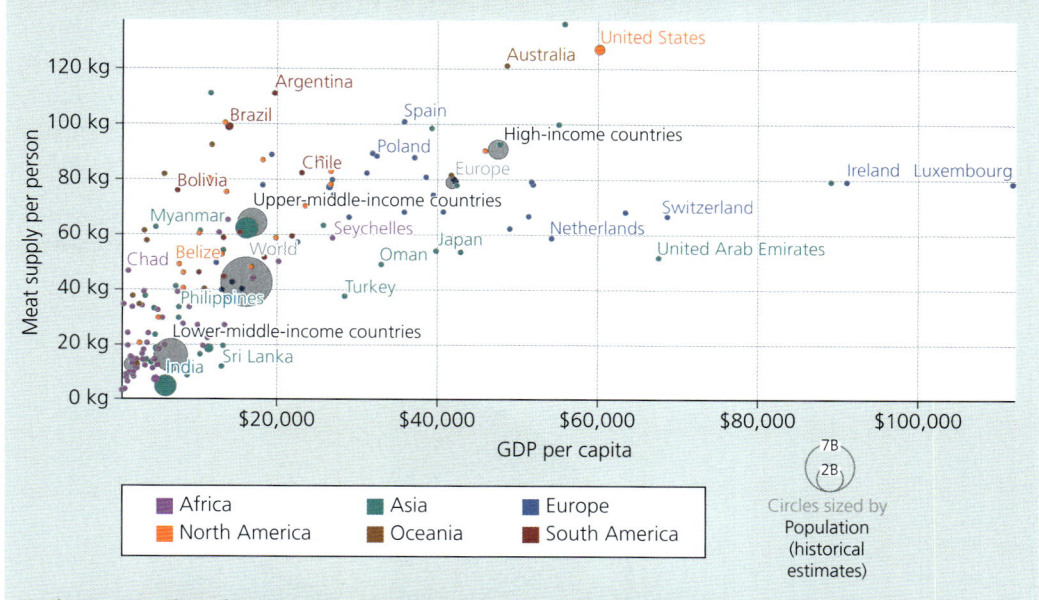

■ **Figure 5.83** The relationship between meat consumption and wealth, as measured by GDP per capita

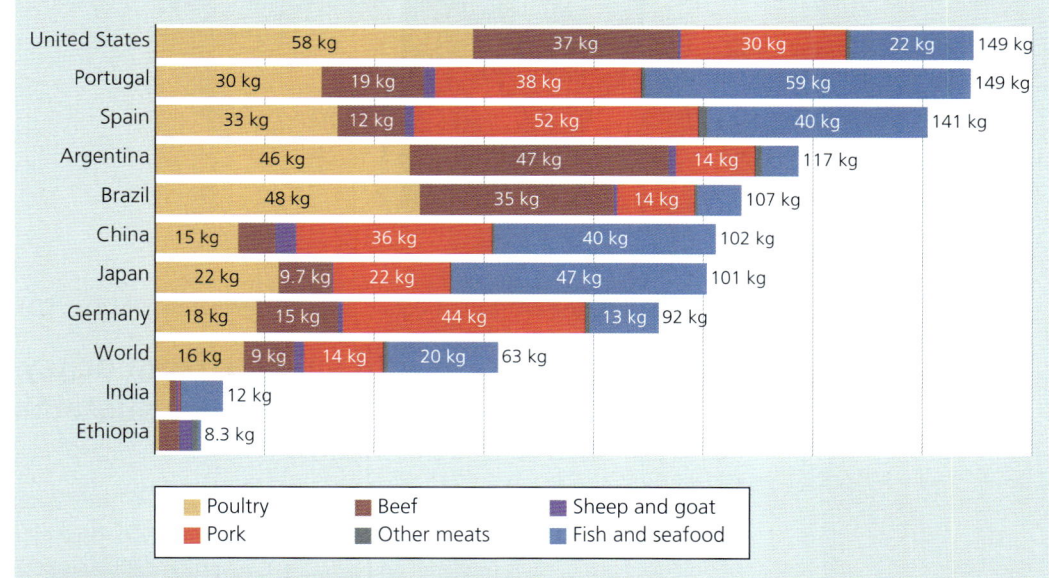

■ **Figure 5.84** Per capita meat consumption of selected countries

At the same time, however, a small but growing minority of people in many countries are choosing to reduce their meat consumption. A 2018 Ipsos survey conducted in 28 countries found that as many as 25 per cent of people described themselves as either flexitarian (14 per cent), vegetarian (5 per cent), vegan (3 per cent) or pescatarian (3 per cent). Environmental concern is only one of the reasons people make this choice. Concerns about animal welfare and the health impacts of eating meat are also common reasons.

5.2 Agriculture and food

The Planetary Health Diet

Diet is an important factor in human health. Changing one's diet for the sake of the environment, or for any other reason, can have positive or negative effects on health. For this reason, a non-profit organisation known as EAT collaborated with the renowned medical journal *The Lancet* to produce a guide to 'Healthy Diets from Sustainable Food Systems'. Their report, known as the EAT-Lancet Commission Report, or Food in the Anthropocene, takes into account the dietary needs of people and the sustainability of global food systems, as well as projected population growth. Its objectives are aligned with the United Nations Sustainable Development Goals.

One key outcome of the report is a flexible guide on the composition of a diet that is healthy for both people and the planet. The relative proportions of recommended foods are shown by the relative sizes of the portions on the plate in Figure 5.85.

■ **Figure 5.85** The EAT-*Lancet* Commission's recommendations for a Planetary Health Diet

5.2.18 Harvesting wild species from ecosystems by traditional methods may be more sustainable than land conversion and cultivation

Modern, commercial agriculture typically uses large areas of land for crop cultivation and rearing of livestock. In all cases, the land used in farming was once wild, covered by natural vegetation, and home to many species of plants and animals. Converting wild land to farmland can destroy many habitats, such as forests, grasslands and wetlands, displacing many species.

Before the Neolithic, or First Agricultural Revolution, when humans started farming, most humans lived as hunter–gatherers. They obtained food by harvesting wild species from their natural habitats. A few small groups of indigenous people still practise hunter–gatherer lifestyles, including the Hadza of Tanzania, the San of southwest Africa and several small groups in the Amazon rainforest.

In some places, hunting and gathering natural, wild food has developed into a commercial enterprise. This is the main way in which certain foods, such as Brazil nuts and truffles, are obtained. If these foods can be collected in sufficient quantities to meet the demand for them without causing damage to the environments in which they grow, then their harvest might be considered sustainable.

Theme 5: Land

> **REAL-WORLD EXAMPLE**
>
> ### Brazil nuts
>
> Brazil nuts grow on large trees of the species *Bertholletia excelsa*, which are native to South America. The trees are pollinated by insects that are only found in intact forests, so they cannot be cultivated easily. Almost all of the global production of Brazil nuts occurs in natural rainforest in Bolivia, Peru and Brazil, where local communities – often indigenous people – gather the fallen pods. The nuts are generally processed locally, and then exported, mostly to the United States and Europe. The global trade in Brazil nuts was estimated at US$350 million dollars in 2021 (source: **https://oec.world/en/profile/hs/brazil-nuts-fresh-or-dried**).
>
> ■ Figure 5.86 Brazil nuts and their pods
>
> Gathering of Brazil nuts provides direct employment for many thousands of people, and a popular source of food for people around the world. Furthermore, since Brazil nut trees are only found in healthy rainforests, the industry provides strong incentive to protect the rainforest and its biodiversity.

> **REAL-WORLD EXAMPLE**
>
> ### Truffles
>
> Truffles are the fruiting bodies, or reproductive structures, of a variety of underground mycorrhizal fungi, mostly belonging to the genus *Tuber*. They often have very strong flavours and scents and are highly prized, luxury food items. They can also be used to produce truffle oil, which is a lower-cost alternative for flavouring food.
>
> Most commercially valuable truffles live in a mutualistic relationship with a tree, often an oak. Early attempts
>
> ■ Figure 5.87 Truffles
>
> to cultivate them involved growing the appropriate species of tree. More recently, scientists have learned how to deliberately add truffle spores to the roots of tree

seedlings. This has allowed the spread of truffles from their native range in Spain, France and Italy, to many other regions of the world.

Since truffles grow underground, they can be difficult to locate. However, their strong scents can be easily detected by animals with a heightened sense of smell. Pigs were once the most commonly used animals for finding truffles, but nowadays specially trained dogs are considered a better choice.

Whether truffles are cultivated by adding fungal spores to tree seedlings or harvested from the wild, their harvest can be beneficial for soil and biodiversity. Truffle production depends on the presence of trees, so harvesters are encouraged to protect existing trees, and plant new ones. Meanwhile the mycorrhizal fungi provide benefits for both trees and soil organisms.

The main product from commercial forests is wood. This is a primary forest product. Other products that are gathered from forests, such as truffles, Brazil nuts, various berries, herbs, bamboo shoots and wild honey, are described as secondary forest products. In general, secondary forest products can be harvested more sustainably, since they do not require the destruction of trees.

ATL ACTIVITY

In pairs, research two other secondary forest products than Brazil nuts. Find out:
- Where they are found in the world (regions, climate types).
- How they are usually harvested.
- Whether there is a more sustainable method of harvesting them.

Share your findings with the rest of the class.

TOK

The way people name things may give an indication of what they think or believe about them. What can we infer about the perspectives, worldviews or values of people who consider wood to be a primary forest product, and fruits, nuts and honey to be secondary forest products?

Harvesting secondary forest products can be sustainable if the rate and methods of harvesting allow the resource to replenish itself, and do not degrade the resource. Some examples of hunting and gathering may be unsustainable if the resource being exploited is rare. This may be the case with animals whose populations are declining – sometimes because of overharvesting.

REAL-WORLD EXAMPLE

Pangolins

There are eight species of pangolin, found in different habitats across Africa and Asia. They have sharp, armour-like scales and resemble anteaters, though they are not closely related. They are widely hunted for their meat and scales, particularly in China and Vietnam, and are thought to be the world's most commonly trafficked animal.

Link

The IUCN classification of threatened species is covered in Topic 3.2.4 (page 226).

The IUCN lists three pangolin species as Critically Endangered, three as Endangered, and the remaining two as Vulnerable. All eight are experiencing population declines. Their main threats are habitat loss due to economic development and hunting.

■ **Figure 5.88** An Indian pangolin, *Manis crassicaudata*

Link

Sustainable use of resources is explored in Chapter 1.3 (page 47).

Hunting animals sustainably may be possible if wildlife populations and their habitats are stable, and the rate of hunting is relatively low. This may be possible for subsistence hunting, to meet the needs of small numbers of people. However, if hunting is a commercial enterprise, where hunters sell their catches in local markets or to international buyers, it is likely that demand for wildlife will exceed the capacity of the populations to replenish themselves. Larger animals, with slow growth and reproduction rates, are particularly vulnerable to unsustainable harvesting.

TOK

Meat from wild animals is sometimes referred to as bushmeat, but this term is almost exclusively used for animals hunted in tropical regions, in particular in the forests of West Africa. Many people in Europe and North America first encountered the term during the Ebola outbreak of 2014, when it was reported that the first human victim may have been infected by handling or eating wild-caught bats. For such people, the term may be associated with concerns about the spread of disease. Many species of tropical – especially African – animals, including pangolins, bats and monkeys, are described as bushmeat. The term is rarely if ever used for animals hunted in Europe and North America, such as deer, moose or pheasants, though these examples fit the definition of the word.

Some critics of the term have suggested that the term being commonly used for some types of animals and not for others, and the suggestion that there is a 'bushmeat problem' in Africa, have colonialist connotations. Can you think of any other examples of common environmental perspectives that may derive from or be influenced by (perhaps subconscious) biases against people from other places or cultures? Do some reading on the concepts of environmental racism and environmental fascism for a deeper understanding of these issues.

HL.c.3: Environmental ethics

Hunting animals may raise complex ethical questions, depending on the purpose of hunting. People with strong ecocentric or biocentric perspectives may see all hunting as ethically wrong, arguing that humans do not have the right to kill other animals. This opinion may be softened in the case of subsistence hunting, where indigenous people rely on hunting to provide food for themselves.

People with anthropocentric viewpoints may also oppose commercial hunting, if it is considered to be unsustainable. Where hunting risks depleting species' populations, and causing loss of biodiversity, an anthropocentrist might argue that these practices negatively impact the welfare of people, including future generations. Strong anthropocentrists, though, may take an individualist, libertarian position, and argue that the individual right to hunt supersedes the rights of others to use the wildlife resources.

Conversely, hunting to manage ecosystems – such as culling deer to reduce the impact of herbivores on a woodland – may be considered unethical by those primarily concerned with animal rights, but ethical by those who see the well-being of the ecosystem as more important than the well-being of individual animals.

> **ATL ACTIVITY**
>
> Write two to three paragraphs explaining your thoughts on the ethical questions of hunting animals, using the appropriate terms, for example, ecocentric.

5.2.19 Claims that low-productivity, indigenous, traditional or alternative food systems are sustainable should be evaluated against the need to produce enough food to feed the wider global population

Modern conventional agriculture, which follows the practices introduced by the Green Revolution, is widely criticized and generally considered unsustainable. Many of its critics have encouraged more widespread adoption of alternative models of agriculture, including the practices of subsistence farmers and indigenous peoples. However, most of these approaches have generally been used to produce relatively small quantities of food. While they are likely to have more positive outcomes for soil and the wider environment, questions remain over whether they can be scaled up to provide enough food for the global population of over 8 billion people.

The problem is that many models of sustainable agriculture are based on practices that are essentially extensive in nature: they rely on few inputs, but produce relatively low yields. They show tremendous promise for reducing the negative impacts of farming, such as soil degradation, pollution and biodiversity loss, but their ability to produce enough food for the world's growing population is uncertain. The potential solution is to find ways to intensify production in these systems, without relying on monoculture or heavy use of agrochemicals such as pesticides and fertilizers.

> **REAL-WORLD EXAMPLE**
>
> ### Companion planting: the Three Sisters
>
> Companion planting refers to a method of cultivating different species together, in a way which benefits each of the species. One well-known example of companion planting is the Three Sisters – a method used by many different indigenous peoples of North and Central America.
>
> The Three Sisters are maize (corn), beans and squash. Each of the 'sisters' provides specific benefits to the other two. The tall maize stalks provide support for the bean plants to climb up, for better access to sunlight. The beans' symbiotic root nodule bacteria add nitrogen to the soil. The squash vines grow along the ground, covering the spaces between the other 'sisters', suppressing weeds, protecting the soil from erosion and keeping it moist by reducing evaporation.

The Three Sisters system has clear environmental benefits, as it requires no artificial fertilizer, and protects the soil from degradation. It can also be used in crop rotation and polyculture systems, which further help manage the soil and control pests. There is potential for it to produce a higher total yield than a monoculture of any of the three individual species. Space is used more efficiently as three different products are produced in the same general space.

■ **Figure 5.89** The Three Sisters companion planting system widely used by North and Central American indigenous peoples

Indeed, some studies have confirmed this in specific cases. Yield can be affected by many different factors, however, so it is difficult to generalize, and individual farmers must always adapt their practices to their specific climatic and soil conditions. There is evidence that the Three Sisters can be as productive as modern commercial agriculture, at least on a relatively small scale. Mechanical methods of planting and harvest are not easily applied, so the system usually depends on manual labour, which restricts farm size.

ATL ACTIVITY

Organic farming

Organic farming, where no artificial agrochemicals are used, is often described as both sustainable and productive. Organic farms typically practise polyculture, making use of intercropping, companion crops, crop rotation, biological pest control and integrated pest management. The environmental benefits of organic farming are clear, but as with other alternative farming practices, it is uncertain whether it can produce enough food to replace conventional agriculture. Opinions on this question differ widely, often depending on a person's perspective.

Use your library's resources and the internet to conduct research on this issue, and present arguments for and against each position. Use your research to hold an in-class debate on the proposition: *Organic farming can feed the world*.

● TOK

People working towards sustainability in agriculture often refer to the importance of indigenous ecological knowledge. By definition, indigenous people have lived in their native area for a long time, often thousands of years. Their ability to survive for so long in their homelands is strong evidence that their ways of life are sustainable. It is often suggested that modern, industrialized societies should seek to understand and learn from indigenous ecological knowledge about how best to practise sustainable agriculture.

An important question remains, however, about whether indigenous knowledge can be applied appropriately to modern, large-scale agriculture. What works for small groups, in specific locations, at specific times, may not be applicable elsewhere.

5.2.20 Food distribution patterns and food quality variations reflect the global food supply industry and can lead to all forms of malnutrition

Globally, enough food is produced to feed all people. However, many people still eat poorly and suffer from **malnourishment**. Some have too little food or unreliable access to food, and suffer from undernourishment. Others find it difficult to access high quality food and instead eat unhealthy diets. One form of malnourishment is **overnourishment**, where a person eats too much of some nutrients, but not enough of others. A common form of overnourishment happens when people eat too much high-calorie food, such as starchy or sugary foods, often leading to **overweight** and **obesity**. **Nutrient deficiency** is another form of malnourishment, where people lack particular nutrients in their diets, such as specific vitamins or minerals.

- **Malnourished** – poorly fed; eating a diet that does not provide the right nutrients for a healthy life. Malnourishment can include both undernourishment and overnourishment, as well as nutrient deficiency.
- **Overnourished** – eating a diet that contains too much food overall, or too much of particular nutrients, leading to poor health.
- **Overweight** – the condition of having more than an average amount of body fat, generally considered a risk to health. Overweight is often defined as having a **Body Mass Index** of more than 25, but less than 30.
- **Body Mass Index (BMI)** – a widely used, simple measure, which relates mass to height. It is calculated by dividing a person's mass (in kilograms) by their height (in metres), squared. Depending on their BMI, a person may be classified as underweight, healthy weight, overweight or obese.
- **Obesity** – the condition of having significantly more body fat than average, often defined as having a Body Mass Index of over 30. Being obese significantly increases the risk of several health problems.
- **Nutrient deficiency** – illness caused by lack of particular nutrients, such as proteins, or specific vitamins or minerals. Deficiency of each nutrient leads to specific health conditions (for example, vitamin C deficiency causes scurvy).

● Common mistake

It is not correct to equate health and/or nourishment with weight. It is possible to be overweight or obese, and to be relatively healthy or to be undernourished. Likewise a person with a healthy BMI may be malnourished.

What a person eats is affected by multiple factors, including culture, personal taste and availability of foods. The price of different foods is also variable and affects what people can afford to buy. In some cases, low-quality, high-calorie food is cheaper than more nutrient-rich food. As a result, poorer people may eat lower-quality food than wealthier people in the same society, and be more likely to experience overweight and obesity.

Inquiry process

Inquiry 3: Concluding and evaluating

While widely used, BMI is increasingly considered a controversial measure and making links to health is not always accepted.

1. Research the history of the BMI and issues surrounding its use, including:
   - the statistical basis for the data used to calculate the BMI, particularly relating to ethnicity and gender differences
   - the relative mass of muscle and fat as represented by this index
   - the change in 1998 of US definitions of what BMI value is considered overweight.

   You may want to start by reading:
   - www.psychologytoday.com/us/blog/the-gravity-of-weight/201603/adolphe-quetelet-and-the-evolution-of-body-mass-index-bmi
   - www.abc.net.au/news/2022-01-02/the-problem-with-the-body-mass-index-bmi/100728416
2. Based on your research, outline the strengths and weaknesses of the use of the BMI at population or individual level.
3. Investigate if there are any measures that could be used as alternatives to BMI.

REAL-WORLD EXAMPLE

Obesity in the United States

The United States of America is one of the wealthiest nations on Earth. It has the highest total gross domestic product (GDP) and one of the highest per capita GDPs in the world. It is also the most unequal of the more economically developed countries, measured by the Gini index of inequality, and has one of the highest poverty rates of wealthy countries.

According to a survey conducted between 2017 and 2020 by the US Centers for Disease Control (CDC), the BMI of 41.9 per cent of US residents classified them as obese. This was an increase from 30.5 per cent in 1999–2000. World Health Organization (WHO) figures show that, globally in 2016, 13 per cent of adults were obese.

The relationship between poverty and obesity is complex, and various studies have found different patterns. The picture is complicated by geographic, ethnic, gender and educational differences, which each appear to influence obesity rates (see Inquiry process above). However, there is consensus that obesity is more significant among US residents in the lowest income brackets. One proposed explanation for this is the relatively high cost of purchasing healthy, fresh food, such as fruits and vegetables, compared to cheaper, highly processed, 'fast' or 'junk' food. Lower quality foods may also be more convenient – easier to access and quicker to prepare – making them more accessible to poorer people who may have less time for food shopping and preparation. Poorer people are therefore more likely to have unhealthier diets, and to consume more 'empty calories' – food with high energy content but little other nutritional value.

In some places, the reliance on imported food makes people vulnerable to fluctuating or high prices. This may occur because a country is unsuitable for farming because of its soil or climate, or because it has been suffering from poor weather. It can also happen in agriculturally productive countries, however, if local food production is focused on so-called cash crops for export (see 5.2.2, page 458). Cash crops are intended to provide income, but not necessarily food. Where a country or region focuses on producing cash crops for export, it may need to import most of its food from elsewhere.

◆ **Famine** – extreme, widespread scarcity of food. Multiple factors may contribute to famine, including adverse weather or natural disasters, which may cause crop failure, and poverty, economic downturns and war, which may limit a society's ability to import food.

When importing food is difficult and local food production is hindered by poor weather, such as droughts, or by crop disease or pests, people may experience severe undernourishment. If many people in a country are unable to obtain enough food, we may say that the country is experiencing **famine**.

REAL-WORLD EXAMPLE

Irish Potato Famine

Between 1845 and 1849, famine killed an estimated 1 million people in Ireland, and drove 1–2 million people to emigrate, mostly to Britain, North America and Australia. The famine is often described as the Irish Potato Famine because a major contributing factor was the destruction of much of the potato crop by a microscopic, fungus-like pest, *Phytophthora infestans*, also known as potato blight.

Other factors also contributed to the famine. At the time, Ireland was part of the United Kingdom of Great Britain and Ireland, and most agricultural land in Ireland was controlled by British landlords, and used for production of cereal crops such as wheat, primarily for export. Cereal production was unaffected by the blight and exports continued throughout the famine. Potatoes were primarily grown by the rural poor, who farmed small plots of land in less fertile regions. The poor were therefore the main victims of the famine.

REAL-WORLD EXAMPLE

Famine in East Africa

The East African nations of Somalia, Kenya and Ethiopia suffered an extended period of drought between October 2020 and March 2023. With little rainfall, crops and livestock failed, and food became scarce and needed to be transported in from elsewhere. An estimated 23 million people faced severe food insecurity. Many migrated in the hope of finding better conditions elsewhere, but the drought was widespread and migration led to conflicts between new arrivals and established populations. Displaced people such as these are some of the most vulnerable to food insecurity.

■ **Figure 5.90** The Horn of Africa, including Somalia, Ethiopia and Kenya, where severe drought caused famine, starting in 2020

510

Theme 5: Land

Climate modellers have related the extended dry period to warmer oceans and to recent La Niña events, which resulted in lower moisture levels in the air over eastern Africa. These conditions appear to be the result of global warming. Indeed, since 1999, this part of East Africa has experienced increasingly frequent periods of drought, leading to two previous incidents of famine, in 2011 and 2017–18.

Conflict also played a part in the famine, both locally and abroad. The 2020 drought coincided with war in Ukraine, which drove food prices up. A long history of local conflict had also already displaced many people, hindered the economic development of the region, and made living conditions difficult. These conditions left many people even more vulnerable to the food crisis.

REVIEW QUESTIONS

1. Discuss the extent to which global food production meets the needs of all people.
2. Describe four ways in which food loss occurs along the food supply chain.
3. State the objective of Sustainable Development Goal 12, with respect to food loss and food waste.
4. Draw a systems diagram showing the inputs and outputs from an intensive farm, practising commercial arable agriculture.
5. Distinguish between extensive and intensive agriculture.
6. Explain why fertilizers are needed on most commercial farms.
7. Describe four agricultural methods that were promoted as part of the Green Revolution.
8. Outline the negative effects of synthetic fertilizers on soil.
9. Describe three alternatives to synthetic fertilizers for improving soil fertility.
10. Explain how crop rotation can be used to reduce soil degradation.
11. Explain the benefits of conservation tillage for soil quality.
12. Outline three ways in which trees can be beneficial on farms.
13. Outline four ways in which livestock farming causes greater environmental impacts than crop farming.
14. Outline four ways in which agriculture contributes to greenhouse gas emissions.
15. Using a named example, explain how genetic modification can be used to make food production more sustainable.
16. Compare and contrast two strategies that are currently in use for achieving a sustainable food supply.
17. Compare and contrast two alternative approaches to farming that are intended to provide ecological benefits.
18. Evaluate the use of zero-till agriculture as a means of reducing the environmental impact of farming.
19. With reference to named examples, explain how harvesting wild plants or animals can be either sustainable or unsustainable.

EXAM-STYLE QUESTIONS

1. Evaluate the sustainability of a named technique for conserving soil in agriculture. [6 marks]
2. Describe the impact of commercial agriculture on soil quality. [4 marks]
3. To what extent has the Green Revolution contributed to global food security? [9 marks]
4. Describe how soil fertility can be improved without the use of synthetic fertilizers. [6 marks]
5. Outline three methods to reduce soil erosion that also promote soil fertility. [3 marks]
6. Describe two transfers and two transformations that occur in soils. [4 marks]
7. Explain how soil can act as both a source and a sink for carbon. [4 marks]
8. Describe the impact of organic matter on the properties of soil. [7 marks]

HL only

9. Discuss how technological advances can improve food production, while reducing the environmental impact of agriculture. [7 marks]
10. To what extent could the elimination of meat from the human diet contribute to sustainability of food production and global food security? [9 marks]
11. Discuss how indigenous agricultural practices have influenced the movement for sustainable agriculture. [7 marks]
12. Describe how livestock can be used in regenerative agriculture. [4 marks]

Theme 6
Atmosphere and climate change

6.1 Introduction to the atmosphere

Guiding question

- How do atmospheric systems contribute to the stability of life on Earth?

SYLLABUS CONTENT

This chapter covers the following syllabus content:
- 6.1.1 The atmosphere forms the boundary between Earth and space. It is the outer limit of the biosphere, and its composition and processes support life on Earth.
- 6.1.2 Differential heating of the atmosphere creates the tricellular model of atmospheric circulation, which redistributes the heat from the equator to the poles.
- 6.1.3 GHGs and aerosols in the atmosphere absorb and re-emit some of the infrared (long-wave) radiation emitted from the Earth's surface, preventing it from being radiated out into space. They include water vapour, carbon dioxide, methane and nitrous oxides (GHGs), and black carbon (aerosol).
- 6.1.4 The greenhouse effect keeps the Earth warmer than it otherwise would be. This is due to the broad spectrum of the sun's radiation reaching the Earth's surface, and infrared radiation emitted by the warmed surface then being trapped and re-radiated by GHGs.

HL ONLY
- 6.1.5 The atmosphere is a dynamic system, and the components and layers are the result of continuous physical and chemical processes.
- 6.1.6 Molecules in the atmosphere are pulled towards the Earth's surface by gravity. Because gravitational force is inversely proportional to distance, the atmosphere thins as altitude increases.
- 6.1.7 Milankovitch cycles affect how much solar radiation reaches the Earth. They lead to cycles in the Earth's climate over periods of tens to hundreds of thousands of years.
- 6.1.8 Global warming is moving the Earth away from the glacial–interglacial cycle that has characterized the Quaternary period, towards new, hotter climatic conditions.
- 6.1.9 The evolution of life on Earth changed the composition of the atmosphere. This in turn influences the evolution of life on Earth.

6.1.1 Composition and processes of the atmosphere

Life on Earth is supported by the mutual interactions between the lithosphere, hydrosphere and atmosphere. The atmosphere forms the boundary between the Earth and space, and it is a dynamic system consisting of inputs, outputs, flows, transformations, transfers and storages. It includes a mixture of gases such as nitrogen (78.1 per cent), oxygen (20.9 per cent), argon (0.9 per cent), and trace gases (0.1 per cent) such as carbon dioxide (0.04 per cent), liquids and some suspended solids (ash, soot, dust). The atmosphere is the layer surrounding the Earth consisting of all these gases, which are continuously redistributed by physical processes such as winds.

Plants in the biosphere maintain the balance in the atmosphere. They absorb carbon dioxide and, in the presence of sunlight, convert it into starch and oxygen. On Earth, carbon dioxide is found trapped in the permafrost, venting volcanoes, decomposition of organic matter and rocks such as limestones. Current anthropogenic activities have greatly affected this balance, resulting in an increase in the percentage of greenhouse gases (GHGs).

The atmosphere is divided into distinct layers, namely the troposphere, stratosphere, mesosphere and thermosphere. Figure 6.1 shows these different layers of the Earth's atmosphere.

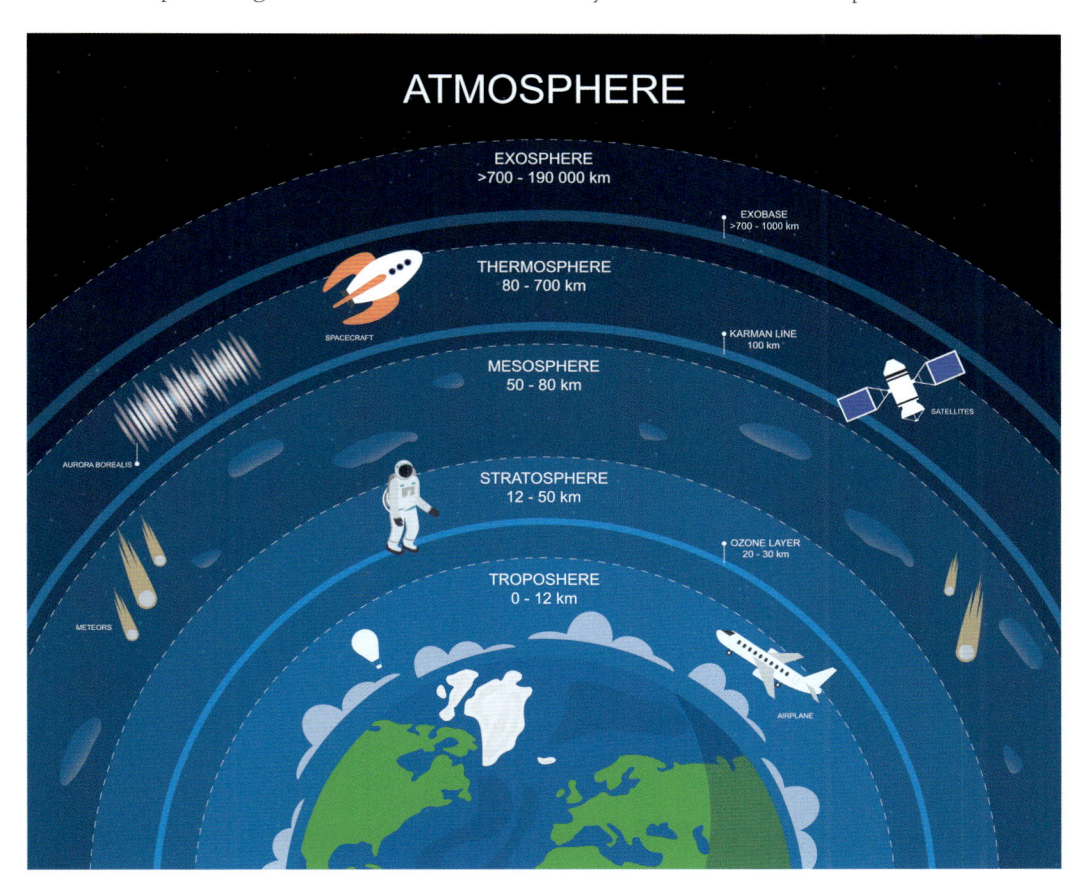

■ **Figure 6.1** The layers of Earth's atmosphere

Starting closest to the Earth's surface is the troposphere, where virtually all weather phenomena occur. The sunlight heats up the Earth's surface, which then radiates the heat to the air closest to the surface. This causes the air to rise, then cool and condense to form clouds and precipitation. The troposphere extends upwards from the ground to about 12 km, varying with location and climate.

Above the troposphere lies the stratosphere, extending from about 12 to around 50 km. The stratosphere is notable for its temperature inversion, wherein temperatures rise with increasing altitude due to the presence of the ozone layer. This thin band of ozone molecules at about 20-25 km absorbs and scatters harmful ultraviolet (UV) radiation from the sun, protecting life on Earth.

Beyond the stratosphere is the mesosphere, which goes up to approximately 80 km. In this layer, temperatures reach some of the coldest levels in the atmosphere. The mesosphere is also where meteoroids burn up upon entry, creating the luminous phenomena known as meteors or 'shooting stars'.

The thermosphere extends from around 80 km to the outer fringes of Earth's atmosphere, up to 700 km. In this region, there is an extremely low density of gas, which means that the few particles present can become highly energized by solar radiation. This is where the International Space Station orbits and it is a critical zone for satellite operations.

Everything beyond 700 km comprises exosphere. The exosphere marks the outermost layer of the Earth's atmosphere, where the thin air gradually merges with the vacuum of space. Gas molecules here are sparse and travel long distances before colliding with another particle. This region is crucial for the escape of some gases, such as hydrogen and helium, into space.

The distinct layers of Earth's atmosphere form a complex and interconnected system that regulates temperature, filters radiation and supports life as we know it. The interactions and processes occurring within these layers contribute to the dynamic and ever-changing nature of our planet's atmospheric environment.

> **Concept**
>
> **Systems and models**
>
> Systems are used to produce the predictive global circulation models, giving a better understanding of the nature of climate change.

6.1.2 Differential heating of the atmosphere creates the tricellular model of atmospheric circulation

■ Table 6.1 The albedo of different surfaces

Surface	Albedo
Clouds	0.15–0.8
Snow and ice	0.8–0.9
Forests and cities	0.1–0.2
Desert	0.35
Water	0.05–0.5

The Earth rotates on a tilted axis as it orbits the Sun. This causes unequal distribution of sunlight over the Earth's surface. Therefore, different regions on the Earth get heated to different degrees. Sunlight falls perpendicularly (at 90°) on the equator (see Figure 6.2). This causes air here to heat faster than the air at the poles, as these receive sunlight at an angle of less than 90°. The tropical regions receive more intense sunlight per unit area and a greater amount of heating than the polar regions. The albedo at the poles is high, due to the snow and cloud cover, causing most of the sunlight to be reflected (see Table 6.1). More solar radiation is absorbed near the equator than near the poles. As a result, the poles are very cold and the tropics are very hot. One might expect that the equator would become increasingly warmer and the poles would become increasingly colder. However, this is not the case.

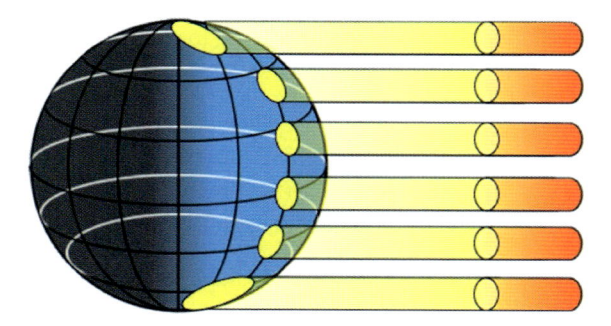

■ Figure 6.2 Differential heating of Earth's surface due to the tilt of the axis

Theme 6: Atmosphere and climate change

Differential heating of the atmosphere creates the tricellular atmospheric circulation (see Topic 2.4.5, page 170), which disperses energy across the planet, reducing the temperature at the equator and increasing the temperature in higher latitudes. The excess energy at the equator is transported towards the poles and the extreme cold of the poles is transported towards the equator, so moderating the climate at both extremes. These atmospheric convections create winds that blow over water to create waves. Ocean circulation also plays an important role in redistributing the heat across the planet.

> **ATL ACTIVITY**
>
> Create a systems diagram to represent the atmospheric system. Include descriptors of flows, storages, transfers, transformations and feedback loops.

6.1.3 GHGs and aerosols in the atmosphere absorb and re-emit some of the infrared radiation emitted from the Earth's surface, preventing it from being radiated out into space

The atmosphere envelops the Earth like a natural blanket and helps in maintaining an optimum temperature to provide a conducive living environment for all species. Greenhouse gases (GHGs), which consist of carbon dioxide, methane, ozone, sulfur dioxide, nitrous oxide, water vapour and black carbon aerosol, are naturally present in small amounts. Carbon dioxide and water vapour are the most abundant GHGs in the atmosphere. The current concentration of CO_2 is 0.04 per cent (400 ppm), which is far higher than it was in the Carboniferous age (~300 ppm). Methane has significant warming effects, even though it comprises only 0.00017 per cent of the atmosphere.

Black carbon aerosols are tiny particles of carbon (soot) suspended in the air, which are emitted by incomplete combustion processes. Greenhouse gases and aerosols absorb and re-emit some of the infrared (IR, or long-wave) radiation back towards the Earth's surface, thus increasing the overall temperature of the planet. Other gases also contribute to radiative forcing, which is the energy imbalance imposed on the climate system either externally or by human activities such as the emission of greenhouse gases, aerosols, and so on.

> **Inquiry process**
>
> **Inquiry 1: Exploring and designing**
>
> **Inquiry 2: Collecting and processing data**
>
> **Inquiry 3: Concluding and evaluating**
>
> Investigate the relationship between GHG emissions and the changing production/consumption levels of key resources, for example meat production. Look at data for five to ten countries. Formulate a focused research question and a testable hypothesis.
> - Identify the relevant independent, dependent and controlled variables.
> - Use the website **www.ourworldindata.org** to collect and record sufficient relevant secondary data in spreadsheets.
> - Identify any risks and ethical considerations associated with the investigation.

- Perform the appropriate statistical tests to process the data.
- Construct graphs and tables to represent the processed data.
- Provide a valid conclusion for the investigation, based on the results obtained.
- Evaluate the investigation and suggest realistic improvements.

6.1.4 The greenhouse effect keeps the Earth warmer than it otherwise would be

Greenhouse gases help in keeping the Earth warm by trapping the energy reflected by the Earth's surface. They act like the glass walls of a greenhouse and maintain the temperature and moisture levels. The greenhouse effect is a natural process that keeps the Earth warm enough for life to be possible.

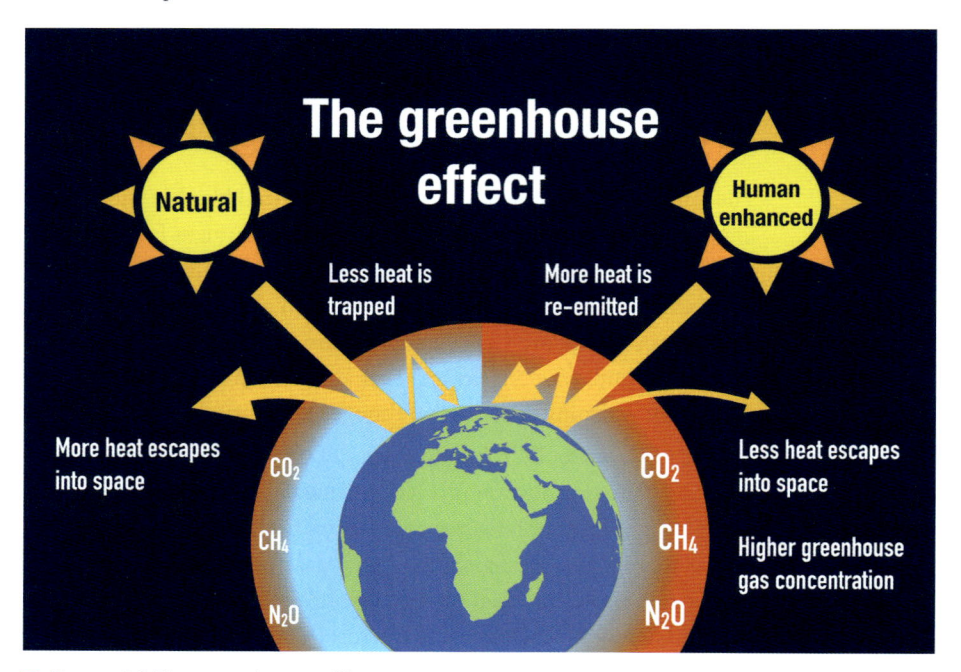

■ Figure 6.3 The greenhouse effect

The sun's radiation reaches the Earth's surface in the form of infrared (IR) waves and heats it. Part of the radiation is reflected back into the atmosphere by white clouds and never reaches the surface. Other light-coloured and shiny surfaces also reflect the radiation received from the sun. Snow-capped mountains, moist surfaces, tree tops, shiny metal structures and surfaces, glassed buildings with mirrors or reflectors have a high albedo. Darker surfaces like asphalt roads, concrete buildings have a low albedo. They absorb almost 95% of the radiation and then radiate it back into the atmosphere. The GHGs present in the atmosphere trap or re-radiate this reflected IR radiation.

All greenhouse gases cause global warming, causing the Earth's average temperature to increase year on year. Water vapour is a GHG and most of it is found in the troposphere (up to around 12 km from the Earth's surface). The distribution of water vapour in this region is controlled by the temperature of the Earth. Warmer temperatures lead to higher evaporation rates and increased humidity. Colder temperatures allow water vapour to condense and fall as rain, snow, hail and sleet. Beyond the troposphere, due to low temperatures, water in the air freezes so the stratosphere is unable to carry water vapour.

Although water vapour is a significant GHG, it is usually excluded from climate models for the following reasons:

- It has a very low tendency to cause an increase in the average global temperature, compared to other GHGs.
- Its mean lifetime is around two weeks, compared to years and centuries for the other GHGs.
- It is dynamic in nature (it changes).
- It is an essential requirement for living beings, and so cannot be mitigated.

The concentration of GHGs in the atmosphere determines the temperature of the Earth. The term "enhanced greenhouse effect" has been used in reference to the accumulation of GHGs from human activities. Clearing forests (deforestation) for building or farming, burning of fossil fuels, and an increasing number of vehicles and factories all contribute CO2 to the atmosphere (see Figure 6.4). Similarly, high demand for meat and meat products has resulted in an increase in animal farming. Since cattle and other ruminants release methane (CH4) as a by-product of the digestion of cellulose, the concentration of methane in the atmosphere has also increased.

The higher levels of GHGs has led to an increase in mean global temperature that is referred to as global warming. This change is having a negative effect on many ecosystems around the world. The glaciers are starting to melt faster, pouring more freshwater into the sea, and as a result causing a rise in the sea level. A collection of events, such as destruction of habitats, shifting of biomes and extinction of species, constitute what we know as climate change.

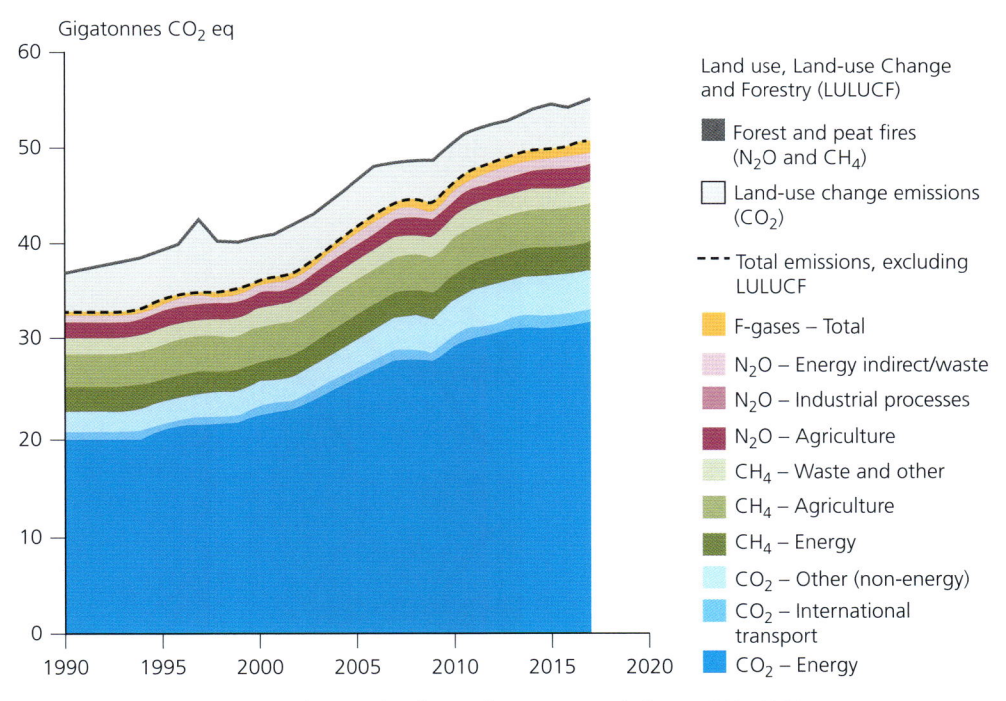

■ **Figure 6.4** The change in the trends of greenhouse gas emissions, 1990–2020

⊙ Common mistake

There is often confusion between the natural greenhouse effect and the enhanced greenhouse effect. The greenhouse effect the trapping of sun's heat at the Earth's surface. It is beneficial for life on Earth. However, the enhanced greenhouse effect caused due to emission of excess greenhouse gases is harmful for living organisms as it enhances the heat absorbed and retained in the atmosphere.

(HL) 6.1.5 The atmosphere is a dynamic system, and the components and layers are the result of continuous physical and chemical processes

The Earth's atmosphere is a dynamic system, continuously shaped by physical and chemical processes. Its composition and layers are not static but rather the outcome of ongoing transformations driven by various factors, including solar radiation, atmospheric pressure and the presence of gases.

The uneven heating of the Earth causes changes in the balance of warm and cool air, creating winds called the prevailing winds. These winds blow in regular patterns in the troposphere. The tricellular atmospheric circulation model explains how the atmosphere transforms at different latitudes on the Earth's surface.

- The air at the equator gets heated faster. It loses moisture, becomes lighter and moves upwards. There is a low pressure created due to this constant upwards movement of hot air. As a result, the air from the surrounding cooler, high-pressure areas rushes to fill this gap (vacuum). The hot air cools as it travels upwards, picks up moisture and becomes heavy. It then flows downwards at 30° latitude. This continuous movement creates the Hadley cells, which maintain the air circulation in the tropical regions. They are responsible for the trade winds in the tropics and low-latitude weather patterns.
- While coming downwards, the moisture-laden air can move either towards the equator or towards the poles. When this surface air moves towards the poles and meets the cool dry surface air coming from the poles, it creates the Ferrel cells. These maintain the distribution of temperature and moisture around the temperate region (between 30° and 60° latitudes). This happens because the temperate regions receive less heat than the tropical regions.
- The point at which these two surface air currents meet is called the polar front, where the ascending air and low pressure are created at 60° latitude. The main function of the Polar cells is to move cold air towards the equator.

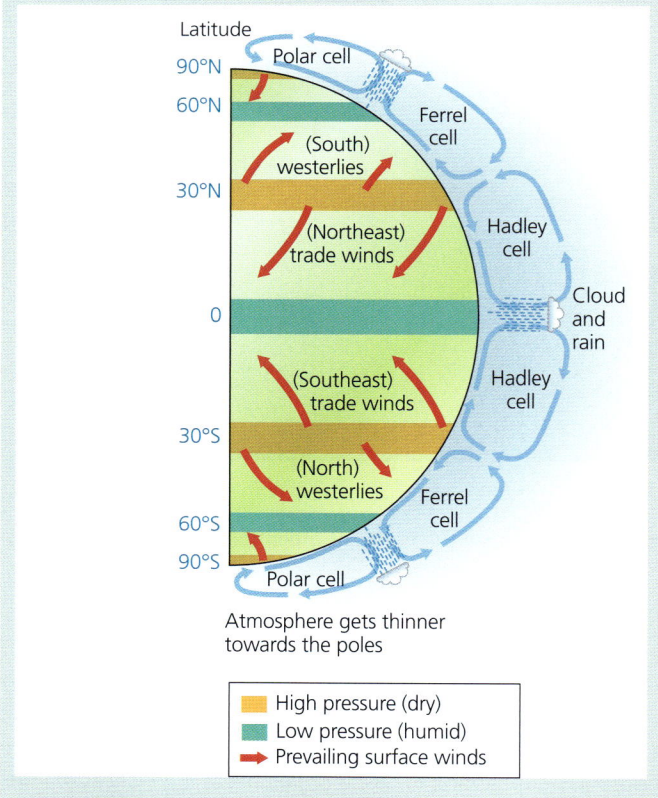

■ Figure 6.5 Atmospheric circulation of air

◆ **Advection** – the horizontal movement of air

The atmosphere contains several GHGs that are also carried by the wind currents and evenly distributed throughout the lower atmosphere. The distribution of GHGs is controlled by jet streams, atmospheric circulation and large weather systems (global winds, air masses, fronts, Coriolis effect). **Advection** and convection also contribute to the movement and distribution of GHGs in the atmosphere. According to NASA, the accumulation of CO_2, for example, is greater in the Northern Hemisphere than in the Southern Hemisphere. During the summer months in the Northern Hemisphere, the CO_2 is absorbed by plants for photosynthesis, keeping its concentration low. However, when deciduous trees shed their leaves in fall and winter, the rate of photosynthesis decreases, the CO_2 is re-released into the atmosphere as respiration continues and CO_2 concentration increases.

More CO_2 is released from soils as they begin to warm in the early spring. Anthropogenic CO_2 traps more of the outgoing radiation, causing the Earth's temperature to rise further. This is the main cause of the shifting snow and rainfall patterns, and the more extreme weather conditions such as heatwaves, flash floods, blizzards, cyclones and wildfires.

6.1.6 Molecules in the atmosphere are pulled towards the Earth's surface by gravity. Because gravitational force is inversely proportional to distance, the atmosphere thins as altitude increases

Air exerts a pressure on everything it surrounds because moving air molecules make a force when they collide with an object. Denser air has more molecules in a given volume and so exerts a greater pressure than less dense air.

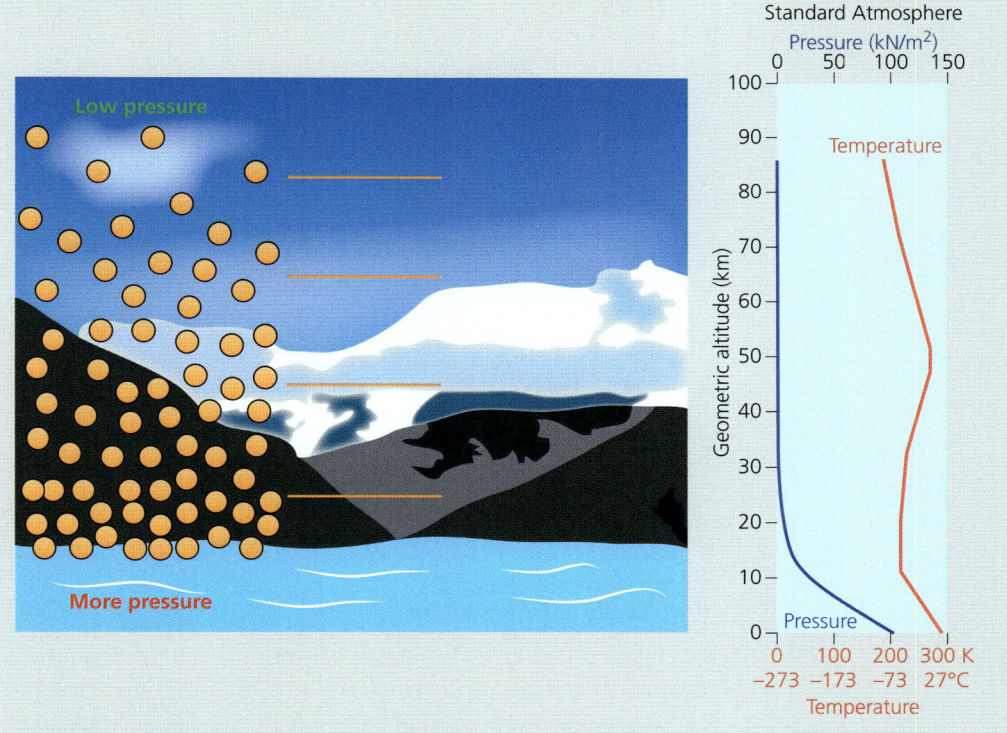

■ **Figure 6.6** The effect of altitude on atmospheric pressure International Standard Atmosphere (ISA)

The temperature and density of air decrease with increasing altitude. The standard lapse rate is approximately 1°C for every 100 m altitude. However, this rate is affected by convection, radiation and condensation. Two factors that govern the density of air are:

6.1 Introduction to the atmosphere

◆ **Inversely proportional** – if something is inversely proportional to something else, it increases as the other thing decreases, or vice versa.

- Gravitational force: Gas molecules in the atmosphere experience a constant downward vertical pull due to the gravitational force exerted by the Earth. As the gravitational force is **inversely proportional** to distance, the further we move away from the Earth's core, the lesser the density of air becomes. At higher altitudes, the gravitational force decreases and the molecules of air are only feebly attracted towards the Earth. This creates a thinning of the atmosphere as we move upwards. It is this gravitational force that keeps the atmosphere from escaping into open space.
- Air pressure from above: At higher altitudes, there is less air pushing down from above. As a result, the air is denser near the Earth's surface and less dense at higher altitudes. At lower altitudes, the air exerts pressure and pushes the molecules of air down, where they come closer to each other making the air dense.

6.1.7 Milankovitch cycles affect how much solar radiation reaches the Earth, and lead to cycles in the Earth's climate over tens to hundreds of thousands of years

Milankovitch cycles describe slow changes in the Earth's movement that lead to changes in the Earth's climate. The Earth revolves around the sun in an elliptical orbit. Over a period of tens of thousands of years, the eccentricity (roundness) of the Earth's orbit and the obliquity (angle of tilt) of its axis changes. The Earth also wobbles about its axis, a phenomenon known as axial precession. These changes significantly affect the distribution of solar radiation reaching Earth, thus causing cyclic fluctuations in the planet's climate causing noticeable patterns of glacial and interglacial epochs that have left their mark on Earth's climatic history.

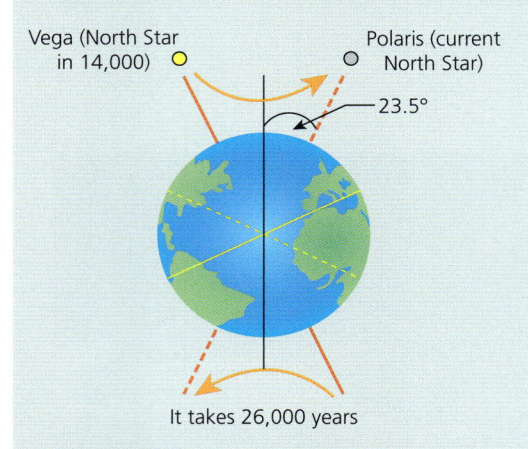

■ Figure 6.7 The precession of the Earth's axis

Eccentricity

The orbit changes from a near circle to an ellipse due to the gravitational force exerted the giant planets such as Jupiter and Saturn. Eccentricity is a measure of the roundness of an orbit. The larger the eccentricity, the more elliptical the orbit. The Earth's orbit undergoes a cyclical change from less eccentric to more eccentric and then back over a period that may last 100,000 or 400,000 years.

Due to this eccentricity, the distance between the Earth and sun changes over a year. The average distance, around 150 million km, is known as an astronomical unit (AU). The Earth is closest to the sun on 3 January each year and farthest from it on 4 July. As the distance between the Earth and sun changes, so does the amount of radiation reaching the Earth.

Changes in the eccentricity of the Earth's orbit over a period of several thousands of years lead to changes in climate.

The angle of tilt of the axis

The Earth spins about its axis, which is why we experience day and night. However, its axis is slightly tilted instead of being upright with respect to the orbital plane. Over the last million years, the angle of tilt varied between between 22.45° and 24.5°. Currently, the Earth is tilted at an angle of 23.5°, causing warm summers and cool winters.

The has changes over a period of 41,000 years. If the tilt is shifts to the lower end (towards 22.1°), we would warmer winters and cooler summers. This results in more accumulation of moisture and slows down the melting of ice sheets. In contrast, when the tilt is shifts the higher end (towards 24.5°), winters would become cooler and summers warmer, which results in increased melting of glaciers.

Precession

Earth rotates on its axis like a spinning top. If you observe a spinning top carefully you will notice that the top does not always remain upright, instead it keeps wobbling from side to side. In a similar way, the Earth wobbles around its axis, and takes around 26,000 years to complete one wobble cycle. This cycle governs the pattern of seasons that we experience on Earth. This cycle is also responsible for the celestial pole facing a different North Star – either Polaris or Vega. Currently, Polaris is our North Star.

> **● TOK**
>
> Empirical sciences are based on the logical theory of induction. If you collect enough data, you can legitimately induce, to a certain extent, a proposition that can be tested given the ability to collect more data in the future. This can support the theory or disprove it. The problem with the theory of Milankovitch cycles is that the cycles are very long, for instance 100,000 years. This means there has not been enough historical time to collect sufficient relevant data. Most of the data that could either support or disprove the theory can only be collected in the future.
>
> > **ATL ACTIVITY**
> >
> > With reference to two other areas of knowledge, discuss the relationship between a hypothesis and a theory that emerges from that hypothesis that is historically impossible to test because of the weaknesses of inductive reasoning.
> >
> > Hints:
> > - Consider intellectual structures of explanation that have similar strengths and weaknesses.
> > - What counts as a valid explanation in different areas of knowledge?

6.1.8 Global warming is moving the Earth away from the glacial–interglacial cycle that has characterized the Quaternary period, towards new, hotter climatic conditions

The Quaternary period began 2.58 million years ago. It is the third and the last of the three periods in the Cenozoic Era and consists of two epochs: the Holocene epoch and the Pleistocene epoch. We are living in the Holocene epoch. It encompasses numerous cycles of glacial growth and retreat, the extinction of many large mammals (for example, mammoths) and birds, and the introduction and spread of humans.

This period has shown a pattern of repeated changes between colder and warmer climates. During glaciations large ice sheets covered significant portions of the Earth's surface, and in interglacial periods the ice retreated and temperatures were milder. However, the influence of global warming is now leading the planet towards new and hotter climatic conditions that don't fit this familiar pattern. Global warming is causing a shift away from the well-known glacial–interglacial cycle that has been a defining characteristic of the Quaternary period.

Climate has changed over a geological timescale without human influence, however human actions such as the emission of greenhouse gases (for example carbon dioxide and methane) are altering the Earth's natural climate balance. The excess greenhouse gases trap more heat in the atmosphere, causing temperatures to rise. This warming effect is now shifting the Earth's climate away from the predictable cycles of the past. The consequences of these unprecedentedly rapid shifts are far-reaching and part of the Anthropocene epoch, which is a proposed geological epoch reflecting the profound and lasting impact of human activities on Earth's geology and ecosystems.

Unlike traditional geological epochs that are characterized by natural changes in the Earth's systems, the Anthropocene epoch acknowledges the dominant influence of human actions in shaping the planet's environment. It is associated with a range of observable markers in the geological record, such as the widespread distribution of plastic, the presence of radioactive isotopes from nuclear testing, and significant shifts in carbon dioxide and methane concentrations. These markers serve as evidence of the profound and lasting impact of human activities on the planet. Habitats and ecosystems that have evolved based on the traditional glacial–interglacial rhythm may face challenges as they try to adapt to these changing conditions. Sea levels, weather patterns, and the distribution of plant and animal species could all be affected.

Inquiry process

Inquiry 2: Collecting and processing data

Investigate the impact of albedo or different GHGs on the temperature of a closed system.

- Set up a closed system with a heat source (bulb), a temperature-measurement device (thermometer) and a container with a lid (bottle, plastic/cardboard box).
- Use black and white paper to study the albedo effect, and a source of carbon dioxide and water vapour to study the effect of GHGs.

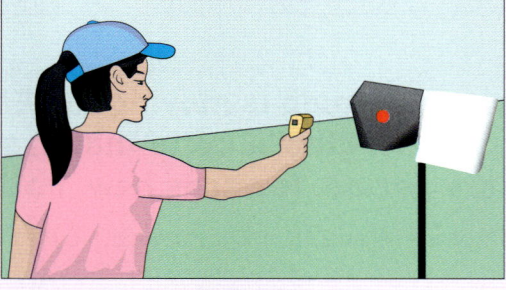

■ **Figure 6.8** Using black and white paper to study the albedo effect

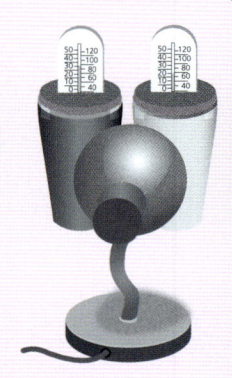

■ **Figure 6.9** Using a source of carbon dioxide and water vapour to study the effect of GHGs

Theme 6: Atmosphere and climate change

6.1.9 The evolution of life on Earth changed the composition of the atmosphere, which in turn influences the evolution of life on Earth

The study of rocks, ancient soils and fossils gives us knowledge about the formation of the atmosphere we see today. Earth was formed approximately 4.5 billion years ago and, at first, was molten and had no atmosphere. Extreme degassing of the Earth's mantle during the early years of its formation resulted in the creation of a primitive atmosphere as numerous volcanoes erupted all over the planet. Table 6.2 lists the gases that are emitted by some volcanoes today.

■ Table 6.2 Composition of volcanic gases

Name of volcano	Volume (mol%)			
	H_2O	CO_2	SO_2	H_2S
Masaya	94.2	3.4	1.4	-
Villarrica	95.0	2.0	2.1	<0.01
Erta Ale	79.4	3.9	6.8	0.6
Kīlauea	79.7	3.5	13.7	1.2

Figure 6.10 illustrates how the atmosphere developed into that which surrounds the Earth today. You may want to watch this video to gain more insights into this subtopic: **https://youtu.be/Gyn754vw8ZQ**

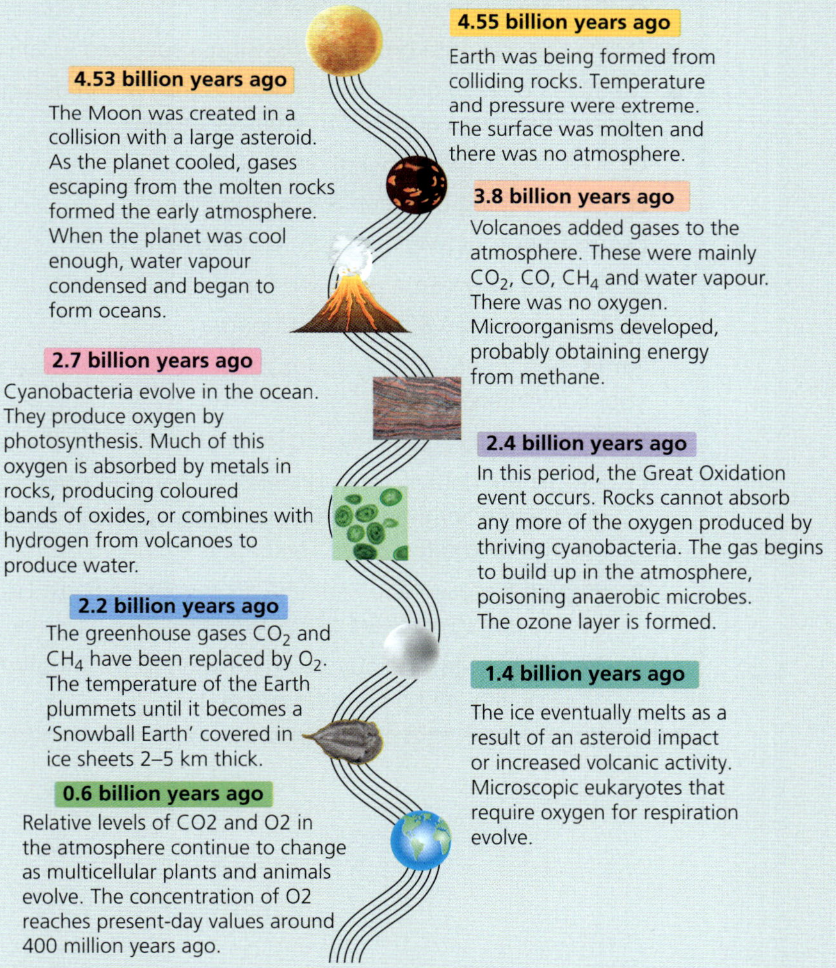

4.55 billion years ago
Earth was being formed from colliding rocks. Temperature and pressure were extreme. The surface was molten and there was no atmosphere.

4.53 billion years ago
The Moon was created in a collision with a large asteroid. As the planet cooled, gases escaping from the molten rocks formed the early atmosphere. When the planet was cool enough, water vapour condensed and began to form oceans.

3.8 billion years ago
Volcanoes added gases to the atmosphere. These were mainly CO_2, CO, CH_4 and water vapour. There was no oxygen. Microorganisms developed, probably obtaining energy from methane.

2.7 billion years ago
Cyanobacteria evolve in the ocean. They produce oxygen by photosynthesis. Much of this oxygen is absorbed by metals in rocks, producing coloured bands of oxides, or combines with hydrogen from volcanoes to produce water.

2.4 billion years ago
In this period, the Great Oxidation event occurs. Rocks cannot absorb any more of the oxygen produced by thriving cyanobacteria. The gas begins to build up in the atmosphere, poisoning anaerobic microbes. The ozone layer is formed.

2.2 billion years ago
The greenhouse gases CO_2 and CH_4 have been replaced by O_2. The temperature of the Earth plummets until it becomes a 'Snowball Earth' covered in ice sheets 2–5 km thick.

1.4 billion years ago
The ice eventually melts as a result of an asteroid impact or increased volcanic activity. Microscopic eukaryotes that require oxygen for respiration evolve.

0.6 billion years ago
Relative levels of CO2 and O2 in the atmosphere continue to change as multicellular plants and animals evolve. The concentration of O2 reaches present-day values around 400 million years ago.

■ Figure 6.10 The evolution of the atmosphere

6.1 Introduction to the atmosphere

Tool 1: Experimental techniques
Tool 2: Technology

Monitor air quality

The objective of this activity is to engage in a meaningful citizen science project to monitor air quality, focusing on the application of technology and data analysis, and investigating the relationship between air quality and time of the day.

Materials
- Carbon dioxide sensor
- Oxygen sensor
- Data analysis software or application (e.g. Excel, Google Sheets)
- Maps of the local area.

Instructions
- Choose three specific locations in the local area:
  - One residential area
  - One garden, park or forest area
  - One area with maximum traffic.
- Divide the class into groups and assign a specified location to each group.
- Members of each group must measure the carbon dioxide and oxygen content of the specified area at different times of the day:
  - Morning between 6 a.m. and 7 a.m.
  - Late morning between 8 a.m. and 10 a.m.
  - Afternoon between 12 p.m. and 2 p.m.
  - Evening between 4 p.m. and 6 p.m.
  - Night between 8 p.m. and 10 p.m.

  Times mentioned are only suggestions. You may choose to collect data as per your convenience.
- Record the data consistently at different times during the day for at least 2–3 weeks.
  - Record the data in spreadsheets and perform some statistical analysis, such as mean, median, mode, range, variance, standard deviation, Pearson correlation coefficient and hypothesis testing.
  - Prepare a presentation summarizing your findings. Use technology to create visually engaging slides or infographics.
  - Discuss how technology and data analysis contributed to your understanding of air quality and pollution sources.

REVIEW QUESTIONS

1. Explain the tricellular model of atmospheric circulation and how it contributes to redistributing heat from the equator to the poles.
2. What role do greenhouse gases and aerosols play in the atmosphere, and how do they influence the Earth's radiation balance?
3. Discuss the concept of global warming and its impact on Earth's climatic patterns.

6.2 Climate change – causes and impacts

Guiding questions

- To what extent has climate change occurred due to anthropogenic causes?
- How do differing perspectives play a role in responding to the challenges of climate change?

SYLLABUS CONTENT

This chapter covers the following syllabus content:
- ▶ 6.2.1 Climate describes the typical conditions that result from physical processes in the atmosphere.
- ▶ 6.2.2 Anthropogenic carbon dioxide emissions have caused atmospheric concentrations to rise significantly. The global rate of emission has accelerated, particularly since 1950.
- ▶ 6.2.3 Analysis of ice cores, tree rings and deposited sediments provides data that indicates a positive correlation between the concentration of carbon dioxide in the atmosphere and global temperatures.
- ▶ 6.2.4 The greenhouse effect has been enhanced by anthropogenic emissions of greenhouse gases. This has led to global warming and, therefore, climate change.
- ▶ 6.2.5 Climate change impacts ecosystems at a variety of scales, from local to global, affects the resilience of ecosystems and leads to biome shifts.
- ▶ 6.2.6 Climate change has an impact on (human) societies at a variety of scales and socioeconomic conditions. This impacts the resilience of societies.
- ▶ 6.2.7 Systems diagrams and models can be used to represent cause and effect of climate change with feedback loops, either positive or negative, and changes in the global energy balance.
- ▶ 6.2.8 Evidence suggests that the Earth has already passed the planetary boundary for climate change.
- ▶ 6.2.9 Perspectives on climate change for both individuals and societies are influenced by many factors.

HL ONLY

- ▶ 6.2.10 Data collected over time by weather stations, observatories, radar and satellites provides opportunity for the study of climate change and land-use change. Long-term data sets include the recording of temperature and greenhouse gas concentrations. Measurements can be both indirect (proxies) and direct. Indirect measurements include isotope measurements taken from ice cores, dendrochronology and pollen taken from peat cores.
- ▶ 6.2.11 Global climate models manipulate inputs to climate systems to predict possible outputs or outcomes. They do this by using equations to represent the processes and interactions that drive the Earth's climate. The validity of the models can be tested via a process known as hindcasting.
- ▶ 6.2.12 Climate models use different scenarios to predict possible impacts of climate change.
- ▶ 6.2.13 Climate models show the Earth may approach a critical threshold with changes to a new equilibrium. Local systems also have thresholds or tipping points.

▶ 6.2.14 Individual tipping points of the climate system may interact to create tipping cascades.
▶ 6.2.15 Countries vary in their responsibility for climate change and also in their vulnerability, with the least responsible often being the most vulnerable. There are political and economic implications, and issues of equity.

6.2.1 Climate describes the typical conditions that result from physical processes in the atmosphere

The atmosphere is a dynamic system. The conditions of the atmosphere at a particular place at a given time is called **weather**. **Climate** is the weather conditions prevailing for a long period of time in a given area. Seasonal variations in temperature and precipitation are the main factors that impact the climate in an area. Physical processes in the atmosphere are responsible for shaping the atmosphere. These processes include:

- Solar radiation: The difference in intensity of the solar radiation at the equator and at the poles sets many atmospheric processes in motion. As discussed in Topic 6.1.2 (page 516), the equator receives higher-intensity solar radiation than the poles due to the tilted axis of the Earth.
- Atmospheric circulation: Solar radiation heats up air at the equator faster and triggers a convection current of air that moves towards the poles. The Hadley, Ferrel and Polar cells (see Figure 6.5 page 521) redistribute heat to the poles and so influence global weather and climate. Different regions on the Earth therefore have different climates due to the transport of heat and moisture through the atmosphere.
- Convection currents: As the Earth's surface heats up unevenly, the hot air rises up and cool air descends, creating convection currents. This mostly results in the formation of thunderstorms in tropical regions.
- Condensation and cloud formation: At higher latitudes and altitudes, air carrying water vapour cools and condenses, resulting in the formation of clouds. These clouds play an important role in reflecting sunlight and trapping infrared radiation.
- Precipitation: Condensed water droplets in clouds coalesce to form bigger droplets that fall to the Earth as precipitation. There can be different forms of precipitation – such as rain, hail, snow, sleet – depending on the temperature and atmospheric conditions.
- Evaporation: Water from large water bodies on the Earth evaporates due to heat and wind. This enters the atmosphere in the form of water vapour. Humidity is the measure of water vapour in the atmosphere.
- Greenhouse effect: The presence of greenhouse gases such as CO_2, CH_4 and water vapour keeps the Earth warm enough to support life. However, several anthropogenic activities increase the concentration of these gases, enhance the effect and cause a rise in the annual average temperature of the Earth.

● Common mistake

It is possible to get confused between weather and climate because both deal with the same parameters. The basic difference between the two is the timeframe. Weather is the change in the atmospheric condition in a given region recorded over a short period of time, for example between day and night or over a few days, while climate is the change recorded over a longer period of time, for instance approximately 30 years.

Link

More information about the difference between weather and climate is available in Topic 2.4.1 (page 164). How different climatic conditions support different types of vegetation on the Earth is discussed in Topic 2.4.2 (page 264).

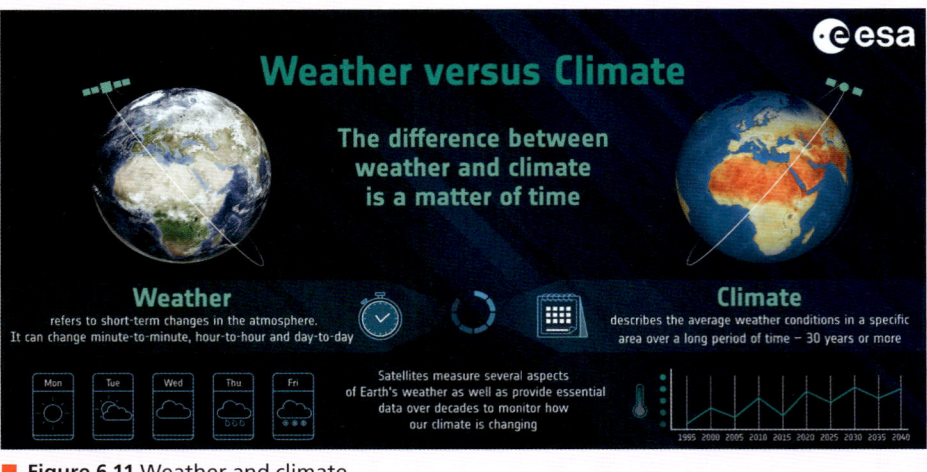

■ Figure 6.11 Weather and climate

6.2.2 Anthropogenic carbon dioxide emissions have caused atmospheric concentrations to rise significantly, particularly since 1950

Carbon dioxide accounts for 76 per cent of greenhouse gas emissions. At the advent of the **Industrial Revolution** in late eighteenth-century Europe, the atmospheric concentration of CO_2 was around 278 ppm. It increased to 292 ppm in 1880. The 20th century brought about the spread of industrialization, along with a dramatic increase in human population, and CO_2 levels crossed 300 ppm in 1915. In the 1950s, consumption of fossil fuel increased further and CO_2 concentrations reached 316 ppm. Figure 6.12 shows the Scripps **Keeling curve**, which depicts changes in CO_2 concentration in the atmosphere from 1958 to 2023 as recorded at the Mauna Loa Observatory in Hawaii, USA. Up until 2022 when measurements stopped due to the eruption of the Mauna Loa volcano, this observatory measured the elements in the atmosphere that contribute to climate change. The curve shows that CO_2 levels are now (2023) 422 ppm, 50% higher than in 1750. They will continue to increase unless we take necessary measures to reduce the anthropogenic (human-induced) contribution.

◆ **Industrial Revolution** – defined as the shift or change from an agrarian economy to one that is dominated by machine manufacturing and industries.

◆ **Keeling curve** – a graph that represents the concentration of carbon dioxide (CO_2) in the Earth's atmosphere since 1958, recorded at the Mauna Loa Observatory.

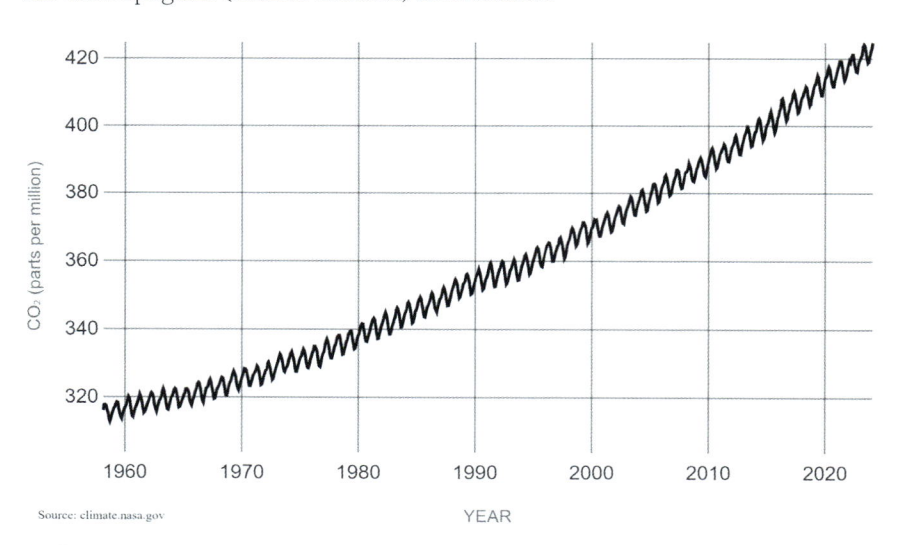

■ Figure 6.12 Annual carbon dioxide concentration in the atmosphere

Theme 6: Atmosphere and climate change

> **Concept**
>
> **Perspectives**
>
> Climate sceptic disagree with the mainstream scientific view on climate change and the impact of the Industrial Revolution. Some people believe that technology has the solution to all the carbon-related environmental issues that currently challenge the world. Many others believe that humans are too insignificant to have an appreciable impact on the global temperature. People can differ in their perspectives and degree of scepticism, which may range from mild, where they hold doubts about climate change, to extreme, where they demonstrate absolute denial of anthropogenic affects leading to climate change.

Link

More information is available in Topic 1.1.7 (page 12) about the effects of environmental value systems that lead to certain outputs. Inputs such as education, media exposure and culture have a significant impact on anthropogenic actions.

Technological and industrial advances go hand in hand with the gross domestic product (GDP) of a country. GDP is the measure of the monetary value of the goods and services produced by a country in a given period of time. There is a positive correlation between CO_2 emissions and GDP. This means that a growing per capita GDP leads to an increase in CO_2 emissions. The increase in CO_2 emissions is also related to the increase in human population as this creates greater demand for goods and services, thus increasing emissions of CO_2 and other GHGs to the atmosphere.

■ **Figure 6.13** The Mauna Loa Observatory – the atmospheric baseline station on the island of Hawaii

6.2.3 Analysis of ice cores, tree rings and deposited sediments provides data that indicates a positive correlation between the concentration of carbon dioxide in the atmosphere and global temperatures

Atmospheric CO_2 is trapped in ice cores, tree rings and sediments. Analysis and comparison of ice cores, tree rings and sedimentary rocks from the past to the present reveals much about the changing concentrations of CO_2. Ice compacts over time, and bubbles of atmospheric gases such as CO_2, O_2 and CH_4 get trapped in it, preserving a record. A close examination of air pockets formed millions of years ago can provide valuable data about the atmosphere that existed at that time. Trees are important sinks of CO_2 and tree rings grow wider in warm, wet years and thinner in cold dry years. They thus provide reliable information that can be used to reconstruct past climatic conditions, as well as to study the recent effects of climatic conditions on plant growth. There is a pattern in CO_2 measurements shown in Figure 6.14, which follows the seasonal cycles of the Northern Hemisphere. During the growing season, plants absorb more CO_2 for photosynthesis than they release through respiration, lowering the concentration in the atmosphere. CO_2 concentration increases during the non-growing season when photosynthesis stops or is reduced.

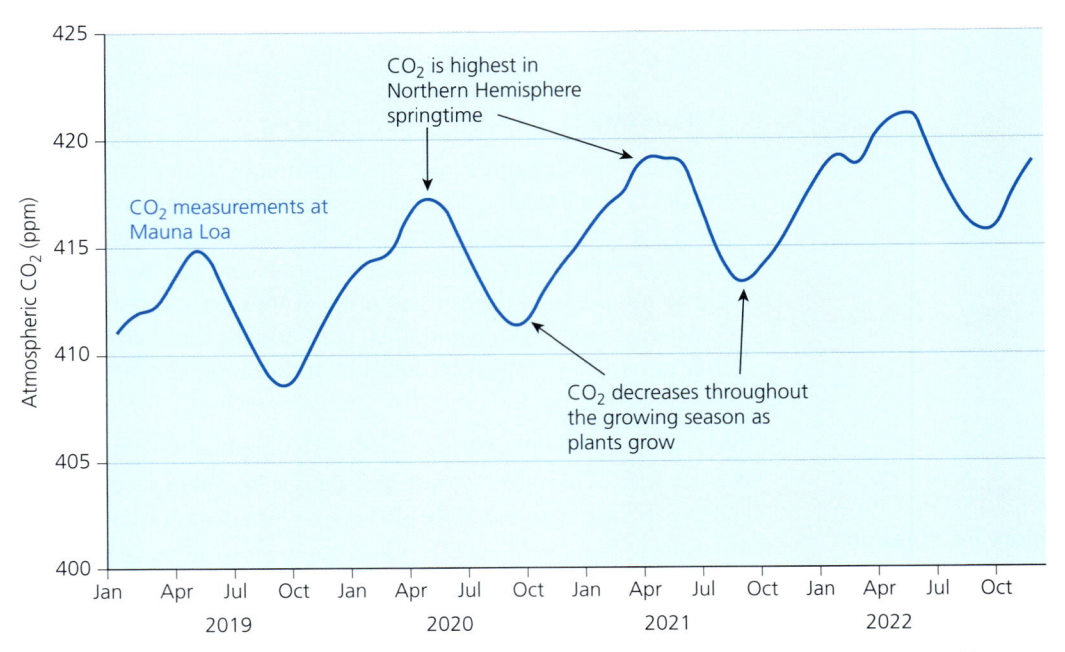

■ **Figure 6.14** The concentration of atmospheric CO_2 is highest in spring and lowest in autumn (data from NOAA Global Monitoring Laboratory)

There is a positive correlation between the concentrations of CO_2 and global temperatures. Figure 6.15 shows CO_2 levels in gas bubbles trapped in ice cores. It shows that, although the concentration of CO_2 has increased and decreased over the last 800 thousand years, it never exceeded 300 ppm until 1911.

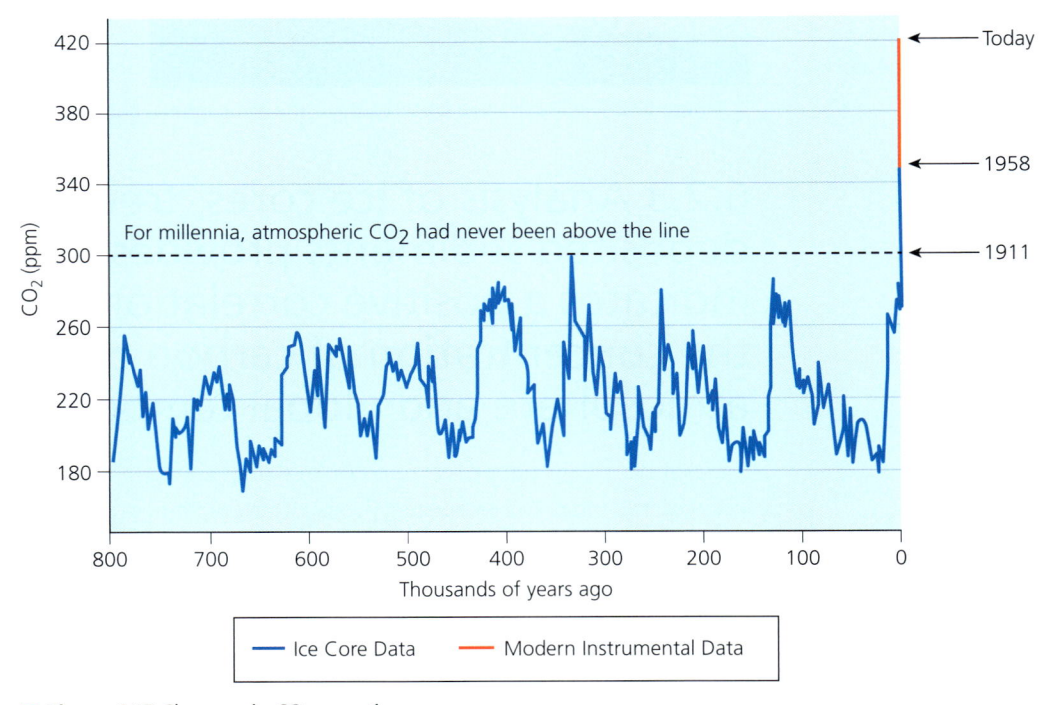

■ **Figure 6.15** Changes in CO_2 over time

ATL ACTIVITY

Explore glacial cycles through data graphs

1. Watch the video **https://youtu.be/rivf479bW8Q**. Search for and read the article 'The three-minute story of 800,000 years of climate change with a sting in the tail'.
2. Once you have done this, investigate graphs of data for the past 800,000 years and how the variables shown have changed during the glacial cycles.
3. Present your findings in the form of a presentation.

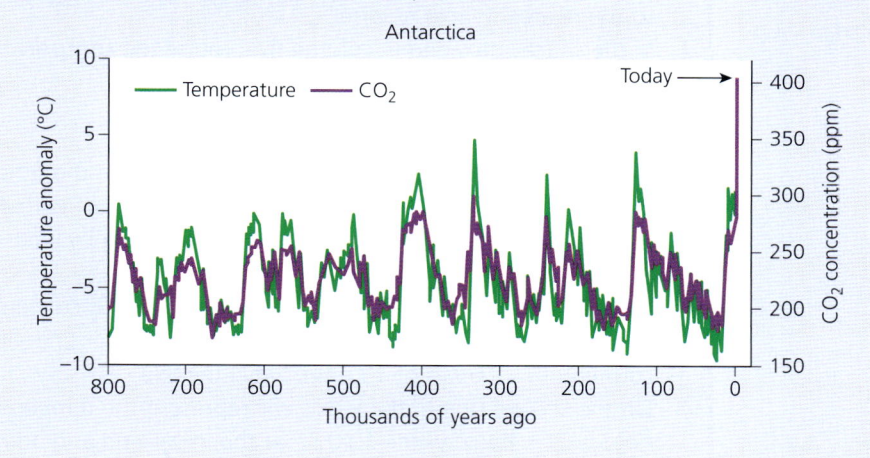

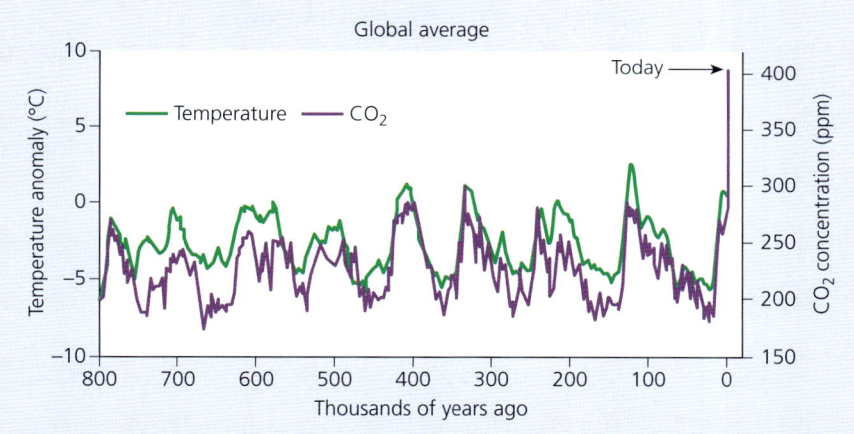

■ **Figure 6.16** Graphs showing the temperature and CO_2 concentration in Antarctica and the global average for the past 800,000 years

Inquiry process

Inquiry 2: Collecting and processing data

Inquiry 3: Concluding and evaluating

Investigate the relationship between carbon dioxide and temperature

Use the data given in pages 8 and 9 in the link below to plot appropriate graphs. The data has been taken from the Vostok, Antarctica, Ice Core Data. **https://archive.epa.gov/climatechange/kids/documents/temp-and-co2.pdf**

6.2 Climate change – causes and impacts

◆ **Temperature anomaly** – the difference between a measured temperature and a reference value or long-term average. It is positive if the measured temperature is above the reference value, and negative if the measured temperature is below the reference value.

Materials
- Spreadsheet
- Graph paper
- Coloured pencils

Instructions for filling out the Vostok, Antarctica, Ice Core Data data table

1. In the spaces provided in column three, round the carbon dioxide (CO_2) concentration to the nearest whole number.
2. In the spaces provided in column five, round the **temperature anomaly** to the nearest tenth of a degree.

Instructions for plotting the graphs

1. Create two graphs: one for CO_2 concentration and the other for temperature anomaly.
2. On both graphs, the x-axis will represent years. Start with 400,000 BCE on the left and number as far as the year 0 on the right, counting by intervals of 20,000 years. Label the axis.
3. On the first graph, the y-axis will represent the CO_2 concentration using units of parts per million (ppm). Begin with 100 ppm at the lower end, and number up to 400 ppm, counting by intervals of 10 ppm. Label the axis.
4. On the second graph, the y-axis will represent the temperature anomaly in degrees Celsius (°C). Begin with −10.0 °C at the lower end and number up to 2.0 °C, counting by intervals of 0.5 °C. Label the axis.
5. Using different coloured pencils, plot the points for CO_2 concentration and temperature anomaly.
6. Write a title on each graph.
7. Use your graphs or Figure 6.17 to add the correct information to Table 6.3.

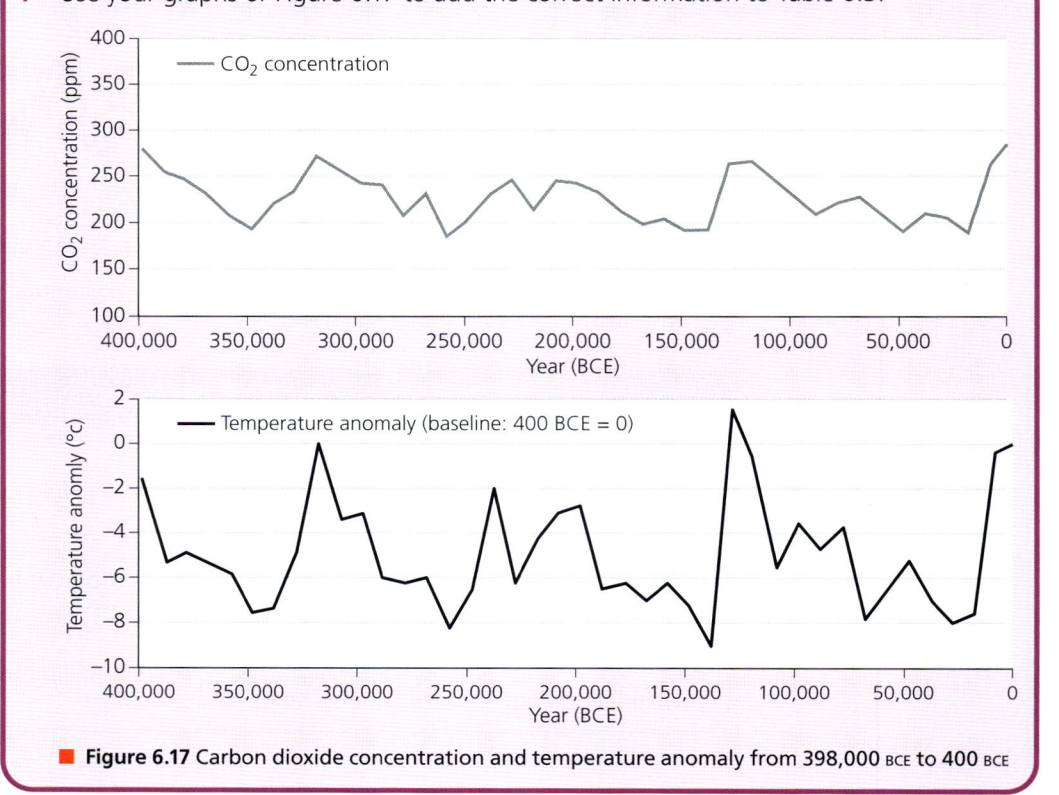

■ **Figure 6.17** Carbon dioxide concentration and temperature anomaly from 398,000 BCE to 400 BCE

Theme 6: Atmosphere and climate change

■ Table 6.3 118,000 to 400 BCE. Length of time: _____ years

Variable	Value in 118,000 BCE	Value in 400 BCE	Change	Rate of change per year
CO_2 concentration (ppm)				
Temperature anomaly (°C)				

8 Use Figure 6.18 below to fill in Table 6.4.

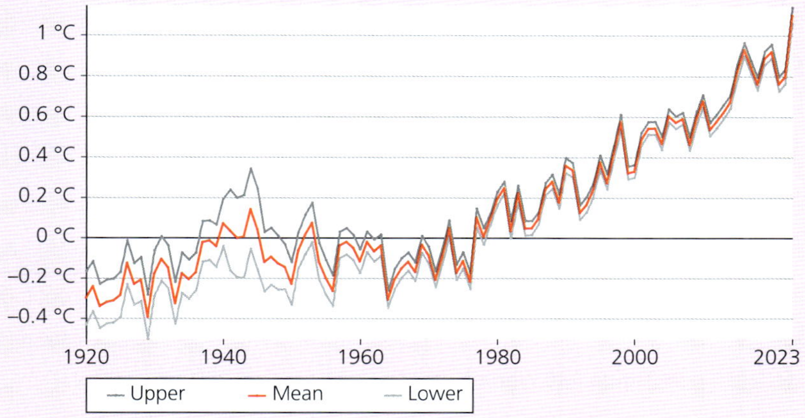

■ Figure 6.18 Average global temperature anomaly from 1920 to 2023

■ Table 6.4 1920 to 2023. Length of time: _____ years

Variable	Value in 1920	Value in 2023	Change	Rate of change per year
CO_2 concentration (ppm)				
Temperature anomaly (°C)				

9 Answer the following questions:
  a What is the relationship between CO_2 concentration and temperature in the two graphs that you have drawn?
  b Discuss the possible reasons for any changes observed in the rate of change per year in Tables 6.3 and 6.4.

6.2.4 The greenhouse effect has been enhanced by anthropogenic emissions of GHGs leading to global warming and, therefore, climate change

The greenhouse effect is a natural phenomenon, but it has been greatly enhanced by anthropogenic actions. The atmospheric concentration of GHGs, such as carbon dioxide (CO_2), methane (CH_4) and nitrous oxide (N_2O), has substantially increased over the past decades, causing the Earth's atmosphere to undergo rapid and unfamiliar changes.

The burning of fossil fuels for energy, deforestation, industrial processes, transport and agriculture is the primary sources of these emissions. CO_2, the most prevalent anthropogenic

6.2 Climate change – causes and impacts

GHG, is released predominantly from burning coal, oil and natural gas. Methane emissions arise from livestock and rice paddies as well as the use of fossil fuels. The number of livestock-rearing farms has increased over the past few decades to meet the demands of the growing milk and meat industries around the world. Globally, the animal-rearing industry contributes 40 per cent of methane emissions.

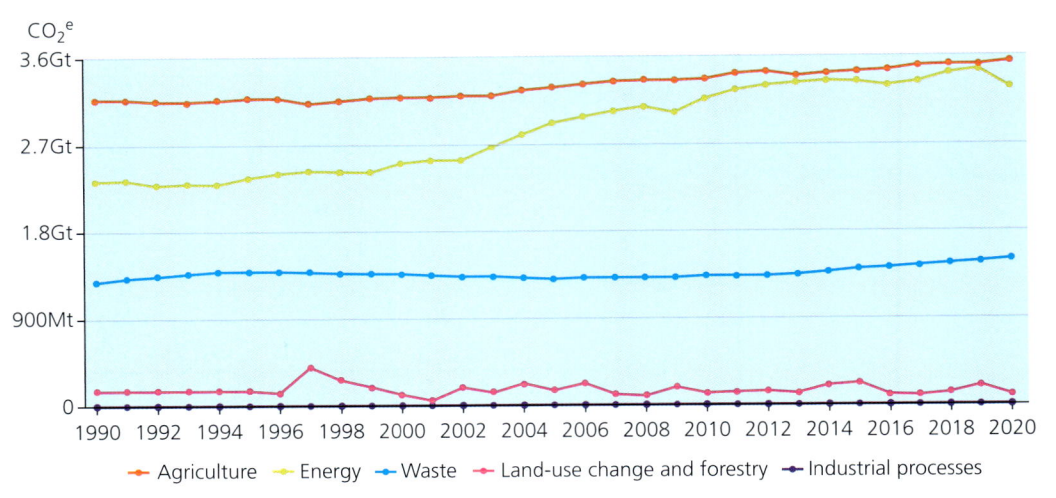

■ **Figure 6.19** World methane emissions from different sectors

N_2O is linked to agricultural and industrial practices, including the use of synthetic fertilizers. According to the Food and Agriculture Organization, N_2O from synthetic fertilizers contributes to approximately 13 per cent of total GHG emissions. Livestock manure left on pastures, and wildfires in forests and grasslands might also contribute to increased emissions of N_2O.

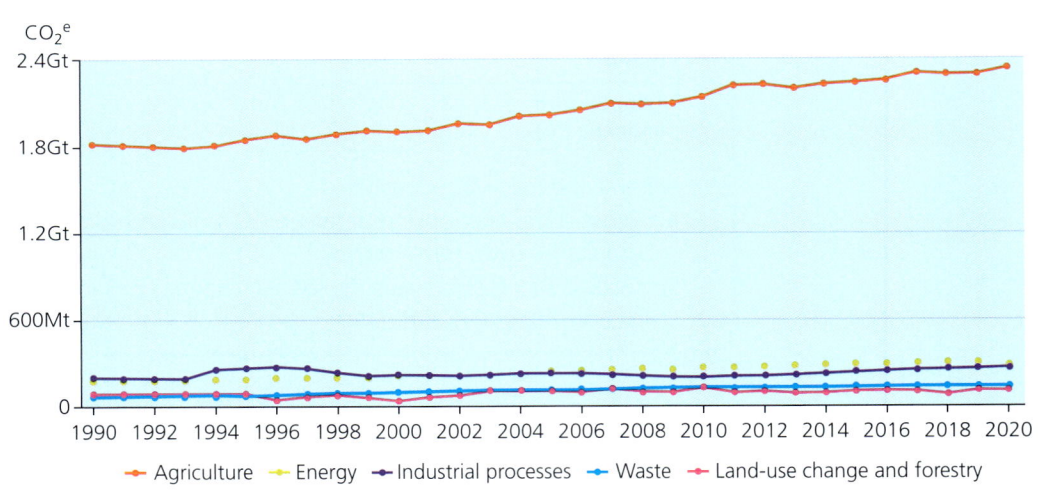

■ **Figure 6.20** World nitrous oxide emissions from different sectors

Increased concentrations of GHGs intensify the heat-trapping effect, raising average global temperatures. This warming contributes to numerous environmental shifts, including melting polar ice, rising sea levels, altered weather patterns, and more frequent and severe heatwaves, storms, flash floods and droughts.

Theme 6: Atmosphere and climate change

> **ATL ACTIVITY**
>
> **Explore the role of bias**
>
> Explore the role of bias in understanding the enhanced anthropogenic greenhouse effect.
> - Analyse various sources of information. Consider their potential biases, and reflect on the implications for knowledge production and dissemination.
> - Possible resources could include:
>   - excerpts from Intergovernmental Panel on Climate Change assessment reports
>   - public statements by major industrial stakeholders such as ExxonMobil
>   - belief statements from climate sceptics such as The Heartland Institute.
> - Critically examine the role of bias in the discourse surrounding climate change, particularly in relation to the enhanced anthropogenic greenhouse effect.
> - Discuss the role of bias in understanding the enhanced anthropogenic greenhouse effect.

6.2.5 Climate change impacts ecosystems at a variety of scales, from local to global, affects the resilience of ecosystems and leads to biome shift

Climate change has impact the way organisms interact with each other and their surroundings. Rising temperatures and reduced precipitation have affected several ecosystems around the world. These effects resonate through ecosystems, affecting their ability to adapt and recover, ultimately resulting in a significant change in biome distribution.

The consequences of climate change are often most noticeable at local scales. Ecosystems in specific regions face alterations in temperature and precipitation patterns, leading to changes in vegetation distribution, wildlife behaviour, and even the timing of flowering and migration. For instance, mountainous regions might experience a reduction in snowfall, which might affect water availability downstream, thus having an effect on the entire ecosystem structure. Global phenomena like rising average temperatures and sea levels come into play at a larger scale. These shifts have long-term effects on ecosystems, forcing species to move away from their geographical range and in extreme cases restructuring their habitats. Arctic ecosystems are warming twice as fast as the global average. Seals rely on sea ice to raise their pups. The early melting of sea ice forces the pups into water before they are able to survive independently. This may lead to a decrease in the population of seals and also that of polar bears for which seals are a key food source.

Resilience is a measure of the ability of an ecosystem to regain its basic structure, functions and processes after disturbances such as disaster events and the introduction of invasive species. Climate change can reduce this resilience by altering the factors that support ecosystem balance, such as temperature, water availability and nutrient cycles. As a result, ecosystems are increasingly vulnerable to challenges including invasive species, disease outbreaks and habitat loss, making recovery more difficult. If the stressful conditions and disturbances prevail, the ecosystems are forced to migrate. This is a **biome shift**.

◆ **Biome shift** – a change in the distribution of plant species causing a change in the biome.

> **REAL-WORLD EXAMPLE**

Severe coral bleaching endangers the Great Barrier Reef

The Great Barrier Reef is a coral reef in the Pacific Ocean off the northeast coast of Australia and is the largest structure in the world built by living creatures. It extends 2300 km along the coast and covers an area of approximately 350,000 sq km. The reef consists of around 2,100 individual reefs and some 800 fringing reefs. The large and intricate framework is formed by the calcareous remains of tiny polyps. The area was declared a Marine Park in 1975 and has been listed as a World Heritage Site by UNESCO since 1981.

The ecosystem of coral reefs is built by the symbiotic relationship between polyps (scleractinians) and single-celled dinoflagellate algae (zooxanthellae). The coral polyps provide a dwelling place for the algae and in turn the algae provide the coral polyps with energy in the form of food created by photosynthesis. Over the years, the Great Barrier Reef has suffered due to tourism and climate change. Coral bleaching is the main cause of destruction of the reef. The coral polyps are very sensitive to temperature and can survive only in a narrow range from 16–30 °C. Any change in the surrounding temperature causes the symbiotic relationship between the coral and the algae to break. As a result, the corals lose the colourful algae, become white and die of starvation. The mortality rate of corals due to bleaching can be as high as 80–100 per cent.

> **REAL-WORLD EXAMPLE**

Desertification of the Murray–Darling basin in Australia

The Murray–Darling basin is the largest interconnected system of rivers in the interior of southeastern Australia. It stretches across 1,061,469 sq km and is fed by the Murray River and its tributaries, the Darling and Murrumbidgee. It is the major freshwater supply for the southeastern part of the country. The basin attracts tourists from all over the world and contributes approximately $11 billion to the economy every year. Around 40 per cent of Australia's agricultural produce comes from this basin, including rice (100 per cent), grapes (74 per cent) and dairy (30 per cent).

Changes in rainfall patterns and a rise in temperatures have caused increased evaporation from the basin. Moisture from the oceans is unable to reach the basin as often as it should, leading to a reduction in rainfall. This has continued for almost 30 years and has led to the drying up of the basin. In 2020, the lowest ever rainfall was recorded for this area. Drought conditions in the Murray–Darling basin are a cumulative effect of the ocean and atmospheric circulation. Communities dependent on the Murray–Darling basin are greatly affected by the reduced water supply and dwindling biodiversity. A mass death of fish was reported due to lack of water in the basin in 2019. The current climatic conditions are reducing the size of the river basin and pushing it towards desert. The frequency of severe droughts and extreme weather events in Australia has led to the desertification of some of the most fertile lands in the southeastern parts of the continent.

■ Table 6.5 Summary of the impacts of climate change at a local scale

Type of impact	Examples
Phenological changes	Bird migration timing in a local wetland, e.g. warblers arriving earlier than usual in Central Park in New York City.
Changes in precipitation patterns	Droughts leading to water scarcity in a specific region, e.g. the California drought from 2012 to 2017.
Temperature extremes	Heatwaves impacting agriculture in a local area, e.g. the European heatwave in 2003 affecting crops and ecosystems.
Sea level rise	Coastal erosion affecting communities near sea coasts, e.g. erosion along the coast of Bangladesh.
Disruption of ecological succession and habitat	Alpine ecosystem changes in the Rocky Mountains.

■ Table 6.6 Summary of the impacts of climate change at an ecosystem level

Type of impact	Examples
Changes in ecosystem dynamics	Shifts in local plant and animal populations, e.g. changes in the distribution of plant species in a forest.
Ocean acidification	Impact on coral reefs and marine life, e.g. bleaching of the Great Barrier Reef due to increased water temperature.
Changes in biodiversity	Disruption of species interactions and ecosystems, e.g. decline in pollinator populations affecting ecosystems.
Altered migration patterns	Impact on migratory species in terrestrial ecosystems, e.g. changes in bird migration patterns.
Melting glaciers and ice caps	Loss of habitat for species dependent on icy environments, e.g. polar bears losing ice platforms for hunting.

■ Table 6.7 Summary of the impacts of climate change at a global scale

Type of impact	Examples
Changes to crop selection	Farmers choosing crops which are better suited to changed soil nutrient and moisture levels, and water availability e.g. drought-resistant crops such as sorghum and soy rather than wheat.
Melting polar ice	Accelerated melting of ice in the Arctic and Antarctic, e.g. melting of the Greenland Ice Sheet.
Changes in ocean circulation	Impact on global climate systems and weather patterns, e.g. El Niño events causing widespread climate anomalies.
Sea level rise	Threat to low-lying coastal regions and island nations, e.g. potential submersion of parts of the Maldives.
Disruption of global weather patterns	Changes in precipitation and storm patterns globally, e.g. altered monsoon patterns impacting Asia.

ATL ACTIVITY

Explore climate data graphs

Use the website **https://ourworldindata.org** to collect relevant data to investigate the effect of anthropogenic actions on the temperature and precipitation over a period of at least 50 years.

6.2.6 Climate change has an impact on (human) societies at a variety of scales and socioeconomic conditions, impacting the resilience of societies

The impact of climate change on human societies is far more widespread than it seems. It affects everything from local communities to global economies. Regions with limited resources or inadequate infrastructure are particularly susceptible to extreme weather events. Floods, droughts and storms disrupt livelihoods, damage homes and challenge essential services, worsening existing socioeconomic disparities. Low-income neighbourhoods in urban and sub-urban areas are often disproportionately affected by flooding, due to inadequate drainage systems.

■ **Figure 6.21** High temperatures such as those experienced in continental Europe in the summer of 2019 could become more extreme with climate warming

REAL-WORLD EXAMPLE

Rising sea levels and flooding in Bangladesh

Bangladesh is frequently affected by cyclones and flood. that cause widespread damage and displacement. The melting of glaciers due to global warming and increased ocean surface temperatures are leading to rising sea levels. According to the IPCC, Bangladesh may lose around 10% of its coastal region by 2050 due to rising sea levels.

Most of the urban population of the country is settled in the low-lying coastal regions in the Bay of Bengal which are most prone to coastal flooding. The intense monsoon contributes to the flooding. High temperatures, especially during summers, cause increased evaporation and the formation of thick moisture-laden clouds, which give rise to heavy precipitation.

Sea level rise increases the frequency and severity of these events, leading to the displacement of communities, loss of livelihoods and damage to infrastructure. In 2007, Cyclone Sidr caused significant devastation, resulting in the loss of thousands of lives and massive property damage. Children and low-income communities are at higher risk due to flooding. The entire infrastructure of a region including residential areas, schools, hospitals and transportation may be jeopardized.

Bangladesh is working to increase resilience to climate change by building seawalls and embankments, and by developing early warning systems. The country is also working to reduce its vulnerability to climate change by improving its agricultural practices and developing renewable energy sources.

Figure 6.22 Flooding in Bangladesh

Developed regions of a city experience a greater rise in temperature than undeveloped regions. This is called the urban heat island (UHI) effect. Among the most prominent factors are increased anthropogenic activities, reduced natural landscapes, urban infrastructure with dull and dark surfaces, and the weather and geography of the region. The UHI effect is more prominent at night-than during the day. At night, when all natural surfaces cool down, artificial surfaces such as concrete, asphalt and bricks release the heat that they absorbed throughout the day. In this way they create pockets of heat in the urban areas.

The differences in land cover in urban and suburban areas exposes the communities living there to different heat levels. People living in urban areas experience more heat-related illnesses and even death. The distribution of population in urban areas is mainly governed by demographic factors, such as income and race. Low-income people often live in neighbourhoods that have higher temperatures than other parts of the city. Excessive heat can be a financial burden on low-income communities. They may not be able to afford the high cost of cooling their homes. Often, their homes are not energy efficient. The inability to afford household energy needs, known as 'energy insecurity', makes it harder to stay cool, comfortable and healthy during periods of extreme heat.

REAL-WORLD EXAMPLE

California wildfires and urban heat islands

California, USA, faces a changing climate marked by prolonged droughts and increasing temperatures. These changes have contributed to the growing frequency and intensity of wildfires, as well as the emergence of urban heat islands in cities like Los Angeles. About 12 million people are considered to be socially vulnerable to wildfires. In California, wildfires have led to loss of life, destruction of homes and vast stretches of charred landscapes. The prolonged fire seasons strain emergency response systems and create health hazards due to smoke inhalation.

6.2 Climate change – causes and impacts

Meanwhile, urban heat islands result in higher temperatures in densely built-up areas, impacting public health and energy demand. Lack of proper urban planning is one of the major factors contributing to UHI effects. An increase in vehicle numbers leads to excessive use of fossil fuels. Los Angeles experiences worse UHI effects than any other city in California because of its wide expanse of skyscrapers, roads and other infrastructure. Adding to this is the poor distribution of trees in the city, which leaves more asphalt and concrete surfaces exposed to the heat of the sun. The burden of such extreme heat conditions is disparately borne by the populace. A major portion of the population comprises people of Latino ethnicity and they make up the high-risk communities exposed to and most affected by the increasing heat. It is predicted that by 2050 there will be more than 22 days with temperatures over 35°C in Los Angeles. Initiatives that help create a more equitable life for the most exposed population of Los Angeles include installation of cool pavements and planting of trees in the hottest residential areas.

■ **Figure 6.23** Wildfires in California

HL.b.6: Environment economics

Equity assumes a prominent role in safeguarding the most susceptible individuals from the detrimental impacts of global warming. It plays a secondary role in establishing fairness in resource allocation both for present and future generations. It guarantees an all-encompassing and open negotiation procedure. Native populations and individuals belonging to low-income communities are among those most susceptible to heatwaves. Other groups who are severely affected by climate change include homeless people, people who are climate exposed at work, people living in mobile homes, and people dependent on electrically powered medical devices.

The unequal distribution or availability of resources results in a situation of the tragedy of the commons (see Topic 3.2.7, page 231, and Topic 6.3.14, page 590), where privileged communities tend to exploit the resources more. However, the burden of their actions is borne by the low-income communities.

ATL ACTIVITY

Research and identify the communities around the world that are more prone to suffer the consequences of climate change. Present your findings in the form of a spider diagram.

HL.b.13, HL.b.14: Environment economics

Decoupling economic growth from environmental degradation is essential for sustainable development. Eco-economic decoupling can be achieved by investing in renewable energy, improving energy efficiency and reducing pollution. If decoupling is localized, then the burden of climate change may shift to other regions and communities. Therefore, steps must be taken at local, national and global level in order to minimize and bring equity in sharing the burdens of climate change.

ATL ACTIVITY

List at least five areas of economic growth that have led to environmental degradation.

■ Table 6.8 Summary of climate change impacts on human societies

Impact category	Specific impact	Description
Temperature extremes	Heatwaves and increased heat-related illnesses	Rising temperatures lead to more frequent and intense heatwaves, resulting in heat-related illnesses, such as heatstroke and dehydration.
	Cold snaps and increased vulnerability	Paradoxically, climate change can also contribute to cold snaps, affecting vulnerable populations, particularly those with limited access to heating resources.
Extreme weather events	Increased frequency and intensity of storms	More frequent and severe storms and hurricanes/cyclones can lead to devastating impacts, including flooding, infrastructure damage and loss of lives.
	Coastal erosion and threats to coastal communities	Rising sea levels and storm surges contribute to coastal erosion, posing threats to communities living near coastlines.
Water scarcity	Droughts and reduced water availability	Changes in precipitation patterns can lead to prolonged droughts, reducing water availability for agriculture, drinking and sanitation.
	Increased competition for water resources	As water becomes scarcer, competition for water resources escalates, leading to potential conflicts and challenges in managing water supplies.
Agricultural impacts	Crop failures and food insecurity	Changes in temperature and precipitation patterns can affect crop yields, leading to increased food insecurity and challenges in global food production.
	Disruption of livelihoods for farming communities	Farmers face challenges as traditional agricultural practices become less viable due to changing climate conditions, impacting livelihoods.
Health impacts	Spread of vector-borne diseases	Warmer temperatures and changing climatic conditions contribute to the expansion of vector habitats, increasing the spread of diseases like malaria and dengue fever.
	Heat-related illnesses and mortality	Increased temperatures can lead to a higher prevalence of heat-related illnesses, contributing to mortality, particularly in vulnerable populations.
Economic disruptions	Damage to infrastructure and increased reconstruction costs	Extreme weather events and rising sea levels can damage critical infrastructure, leading to increased costs for reconstruction and adaptation efforts.
	Disruption of supply chains and trade	Climate-related impacts on agriculture, transportation and energy can disrupt global supply chains and trade, affecting economies on a broader scale.
Migration and displacement	Climate-induced migration and increased refugee numbers	Rising sea levels, extreme weather events and resource scarcity can contribute to climate-induced migration and displacement, leading to increased refugee numbers.
	Social tensions and conflicts	Competition for resources, such as water and arable land, can exacerbate social tensions and contribute to conflicts within and between communities and nations.
Social inequality	Disproportionate impact on vulnerable communities	Vulnerable populations, including low-income communities and marginalized groups, often bear the brunt of climate change impacts, leading to increased social inequality.
	Differential access to adaptation and mitigation resources	Access to resources for adapting to and mitigating climate change impacts is often unequal, exacerbating existing social disparities.

6.2 Climate change – causes and impacts

> **ATL ACTIVITY**
>
> - Use the factsheets published by the Intergovernmental Panel on Climate Change to explore the relationships between two or more factors: **https://ipcc.ch/report/ar6/wg2/about/factsheets**
> - Present your findings in the form of a written report.
> - Include graphs and tables, and articulate your finds and insights using proper subject-specific terminologies.

6.2.7 Systems diagrams and models can be used to represent cause and effect of climate change with feedback loops, either positive or negative, and changes in the global energy balance

Systems diagrams and models (Chapter 1.2, page 24) can be used to represent the climate system, which is made up of the atmosphere, oceans, land, ice and biosphere. They enable us to visualize the impacts of various phenomena, including solar radiation variations, terrestrial albedo changes and methane gas release, while also accounting for the associated feedback loops. The feedback loops are a way of showing how the output of a system can affect its input. Positive feedback loops amplify the effects of a change, whereas negative feedback loops dampen the effects of a change.

Link
Feedback loops are discussed in Chapter 1.2 (page 36).

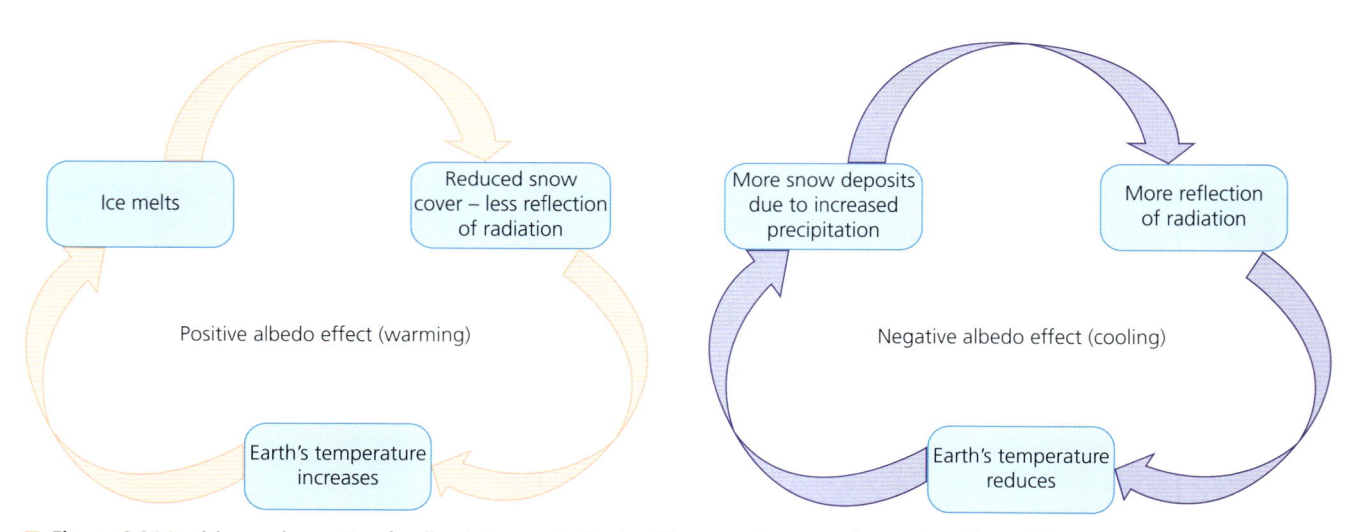

■ **Figure 6.24** Positive and negative feedback loops of albedo effect, causing warming and cooling of the Earth

> **ATL ACTIVITY**
>
> Using Figure 6.24 as a foundation, draw a systems diagram for one factor that can cause a positive feedback loop and one factor that can cause negative feedback loop in the climate system.

Sunlight received on the Earth is called insolation and it consists of X-rays, radio waves and the electromagnetic spectrum that includes infra-red, visible and ultra violet wavelengths. Variations in solar radiation can change the global energy balance, that is the difference between the amount of energy entering the Earth's system from the sun and the amount of energy leaving the system. If the overall flow of energy to and from the Earth does not balance, the average global temperature will change.

Figure 6.25 shows the complex balance between incoming and outgoing radiation. GHGs, clouds, aerosols and surface characteristics can influence this balance, leading to changes in global temperatures and climate patterns.

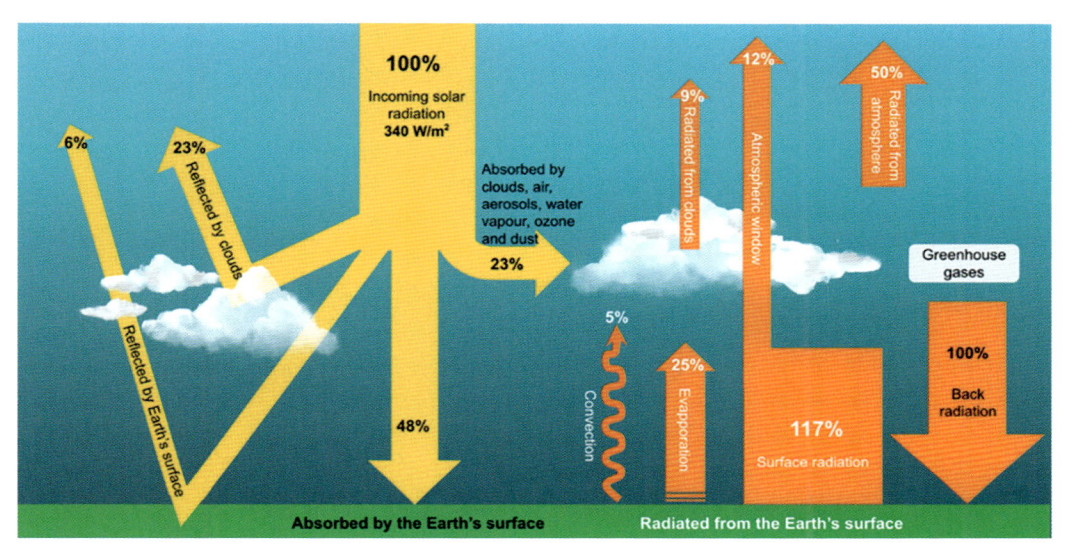

■ **Figure 6.25** The Earth's energy balance

Incoming solar radiation: Solar radiation reaching the Earth (insolation) is the main source of energy for the climate system. It covers the parts of the electromagnetic spectrum from radio waves through infrared and visible radiation to ultraviolet. A total of 341W/m^2 reaches the top of the Earth's atmosphere. Some of it is reflected by clouds, aerosols and the gases of the atmosphere. Another portion is reflected from the surface and returns to space through the atmosphere. The surface of the Earth absorbs around 48% of the incident solar radiation. The rest, around 23%, is absorbed by the atmosphere.

Outgoing terrestrial radiation: Of the incoming solar radiation that is absorbed by the Earth's surface, 25% is used for evaporation and 5% is conducted back to the atmosphere where it causes convection. The rest is emitted as infrared radiation, mostly heat (thermal infrared). The surface also absorbs energy reradiated by GHGs (back radiation) so the amount of radiation it emits is 396W/m^2, equivalent to 117% of the incoming solar radiation. Most of this heats the atmosphere which, in turn, radiates it into space. However, the gases in the atmosphere cannot absorb all wavelengths. Energy in some parts of the spectrum therefore passes straight through this 'atmospheric window' that allows direct radiative cooling of the Earth.

Terrestrial albedo changes can cause changes in the global energy balance (see Figure 6.24, page 544). If the overall flow of energy to and from the Earth does not balance, the average global temperature will change.

Figure 6.25 shows the complex balance between incoming and outgoing radiation. GHGs, clouds, aerosols and surface characteristics can influence this balance, leading to changes in global temperatures and climate patterns

6.2 Climate change – causes and impacts

Link

The production of methane by methanogenic bacteria present in the guts of ruminants is discussed in Topic 2.3.15 (page 153).

Methane, a GHG interferes with the global energy balance by trapping heat in the atmosphere, leading to warming (Figure 6.26). Agriculture (especially cattle farming), landfills, oil and natural gas are some of the main sources of anthropogenic methane. Natural sources of methane include wetlands, methane hydrates (calthrates), termites and anaerobic methanogenesis by bacteria in the oceans.

Positive feedback loop for methane:
- Global temperature rises due to increased emission of GHGs
- Permafrost begins to thaw as temperature rises
- Trapped CH_4 is released and dead organic matter is exposed for decomposition
- Decomposition releases further CH_4
- Concentration of CH_4 increases in the atmosphere

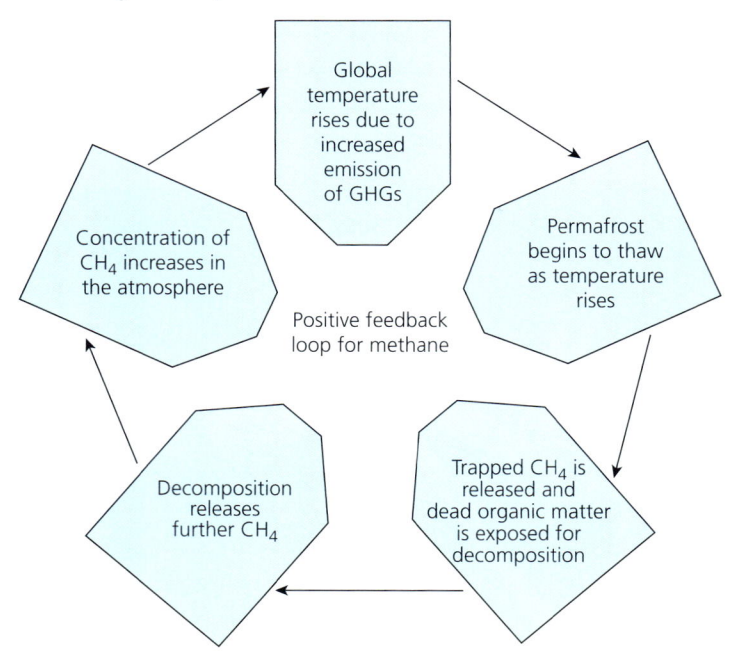

■ **Figure 6.26** Positive feedback loop of methane

ATL ACTIVITY

Depict the cause and effect relationship in climate change by designing a systems diagram for them.

6.2.8 Evidence suggests that the Earth has already passed the planetary boundary for climate change

The climate system is complex and a growing body of evidence suggests the Earth has already passed the planetary boundary for climate change (see Topic 1.3.19, page 72). The planetary boundaries for CO_2 and radiative forcing are set at 350 ppm and 1 W m_2 respectively. Currently, atmospheric CO_2 concentration is 417 ppm and the radiative forcing estimate for 2022 is 2.91 W m^2.

Link

There is more detail about planetary boundaries in Topic 1.3.19 (page 72) and the impacts of climate change on ecosystems are discussed in Topic 6.2.5 (page 537).

The 2021 IPCC report concluded that human-induced climate change is already affecting weather and climate in every region across the globe. Crossing the climate change boundary has led to an increase in incidents of heatwaves, heavy precipitation, droughts and tropical cyclones. Evidence that human influence on these has strengthened is given in IPCC reports. Each of the last three decades have shown successive increases in average temperatures, with oceans warming more than land. The Greenland and Antarctic ice sheets are melting at an alarming rate, and this is contributing to sea level rise. Sea levels have already risen by about 20 cm since 1880. In 2021, global sea level set a new record high – 97 mm above 1993 levels – and it is expected to rise by another 30 to 120 cm by the end of the century.

Potential human impacts of crossing the planetary boundary for climate change are increased displacement of people, increased poverty, food and water shortages and coastal flooding. Climate change challenges the food production system by shifting growing seasons. Many crops are lost due to drought, floods or extreme heat or have reduced yields. Lack of resources can make it difficult for vulnerable communities to move from high-risk areas like coastal regions. Forced migration, poverty and economic shocks due to climate change may also increase the risk of conflicts and social unrest.

6.2.9 Perspectives on climate change for both individuals and societies are influenced by many factors

Climate change affects everyone, irrespective of their socioeconomic and political status. The problems associated with climate change are so deep-rooted that policymakers find it difficult to address all issues equally. The perspectives that individuals and society have on climate change and its consequences also govern their actions. These perspectives develop due to factors such as lived experiences, economic interests, political agendas and cultural experiences.

The people of the Maldives and Bangladesh, for instance, face imminent threats from rising sea levels and thus their perspectives will be strongly influenced by the immediate impact of climate change on their homes and livelihoods. Having faced the harsh consequences themselves, these people will strongly believe that climate change is real and will fully support policies focused on mitigating and adapting to it. Indigenous peoples such as the Inuit and Saami in the Arctic, however, perceive climate change through a cultural lens. Climate change directly impacts their cultural traditions, such as herding reindeer, by changing weather patterns, distribution of vegetation and migration of animals.

Political leaders who have expressed scepticism about climate change not only influence public opinion, but also favour implementation of policies that are against combating climate change. The media has multiple biases and a strong influence on public opinion. For instance, the right-wing media may project the basic science behind anthropogenic global warming in a very different way than the left-wing media.

Companies and businesses, especially those involved with fossil fuels, might deny any involvement in anthropogenic enhancement of climate change. For instance, major oil and gas companies might have historically downplayed the effects of climate change for economic advantage. Such corporate perspectives influence public discourse and policies.

> **Link**
>
> Chapter 1.1 (page 3) discusses how perspectives shape our understanding of the environment.

> **ATL ACTIVITY**
>
> Environmental or climate refugees are people who migrate when their homes and communities are destroyed by environmental disasters. They are not protected by international law and so can be sent to refugee camps or back to their homes at any time.
> - Work as a class to design an environmental immigration policy that is fair and equitable for all stakeholders.
> - First, working individually, write down five points you would expect to see in an equitable policy.
> - Share ideas and come to a consensus on 10 points to be included in the policy.
> - Choose a community of climate refugees (for example, from Bangladesh, Maldives, Thailand, New Orleans).

- Discuss the policy in groups. Each group should take on the role of one of the following stakeholder groups:
  - climate refugees from the chosen community
  - the immigration ministry of the country where refuge is being sought
  - a climate activist NGO such as Greenpeace
  - a national trade and commerce organization.
- Share the responses of your group to each of the 10 agreed policy points with the whole class.

(HL) 6.2.10 Data collected over time by indirect and direct measurements provides opportunities for the study of climate change and land-use change

Observatories and weather stations all over the world help us record valuable data to understand how our climate is changing. For example, weather stations such as the Mauna Loa Observatory and the NOAA station at Wake Island harbor measure and transmit data on wind speed, atmospheric pressure, air temperature, tides, and CO_2 and CH_4 concentrations. Such stations and observatories are well equipped with instruments such as those shown in Figures 6.27 and 6.28. The data gathered is crucial for creating climate models.

■ **Figure 6.27** Humidity and temperature sensor

■ **Figure 6.28** An anemometer measures wind speed

Some data is collected by indirect methods using measurements of other quantities (proxies) from, for example, ice cores, dendrochronology and pollen from peat cores.
- Ice core analysis: Ice sheets are formed over several years of compression. In places that are always covered by snow, the ice sheets can be several centimetres thick. Ice cores from polar ice caps and glaciers are stratified and give an understanding of past climatic conditions. One method is to examine the ratio of oxygen isotopes in air bubbles trapped in the core (isotope analysis). Isotopes are atoms of the same element that have different numbers of neutrons. An oxygen molecule consists of 8 protons, 8 electrons, and 8 or 10 neutrons. 'Light' oxygen, ^{16}O, has 8 neutrons and 'heavy' oxygen, ^{18}O isotope is 'heavy'. Water rich in light oxygen evaporates more easily; water rich in heavy oxygen condenses more quickly. Moisture-rich air rising into the atmosphere or moving toward the polar regions cools and condenses. Water containing heavy oxygen falls out of the atmosphere faster than water containing light oxygen. Once an air mass reaches a high latitude, it contains fewer heavy oxygen isotopes. The fewer heavy isotopes in a polar ice core, the lower the ancient temperature.

Ice cores also contain CO_2 and dust deposited when the ice was formed. Changes in the concentration of and type of dust reveals information about land-use patterns and human activities at the time. Increased CO_2 levels in particular years indicate the loss of forest cover. As more land is cleared for agriculture or building human settlements, the soil loses its moisture. Dust from such bare soils can find its way into snow.

The core may also contain layers of ash recording a volcanic eruption.

■ **Figure 6.29** Scientist holding a piece of ice core drilled from ice sheets

- Dendrochronology: Tree ring dating is called dendrochronology. The growth rings of trees can provide detailed information about climate change. No trees have to be cut down to study the rings. An increment borer is used to extract a thin strip of wood from the trunk going from the bark to the centre of the tree. Trees are sensitive to local climate changes. The width, colour and pattern of tree rings reveal how the tree responded to the conditions it experienced. The width of the tree rings can be used to infer the amount of precipitation and the temperature during the year in which the ring was formed. Years with higher precipitation rates produce broader rings and dry seasons produce narrow rings. Changes in tree ring patterns can signal shifts in land use, such as deforestation or urbanization.

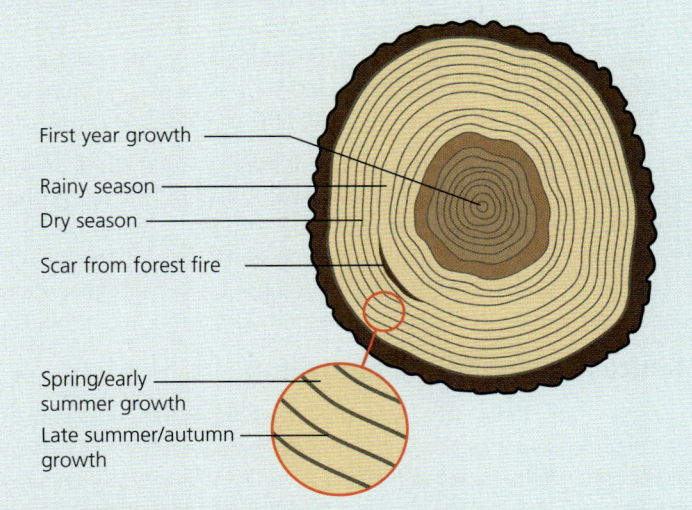

First year growth
Rainy season
Dry season
Scar from forest fire

Spring/early summer growth
Late summer/autumn growth

■ **Figure 6.30** Tree ring analysis

- Pollen analysis from peat cores and sedimentary rocks: Pollen grains trapped in sediment cores can be used to study past vegetation conditions. The pollen comes from plants that were growing in the soil when the sediment was deposited and wind-blown pollen from plants across a wider area. Changes in the types of pollen present can reveal shifts in land use, including deforestation, agriculture or urban development.
  Other direct methods of collecting data about climate and land-use include remote sensing techniques. These may use active or passive sensors.
- Active sensors have their own energy source for illumination and are used in instances when the desired wavelength is not emitted by the sun in sufficient quantity. Laser fluorosensors and synthetic aperture radar (SAR) are examples of active sensors. Radar sensors emit microwaves into the atmosphere. When these waves strike an object they are reflected back in the direction of the sensor and are detected.

6.2 Climate change – causes and impacts

- Passive sensors detect some of the sun's energy that is reflected by the Earth's surface. Most such measurements can therefore only take place during the day. However, thermal infrared radiation can be detected and recorded during both the day and night, depending on the intensity of the radiation.

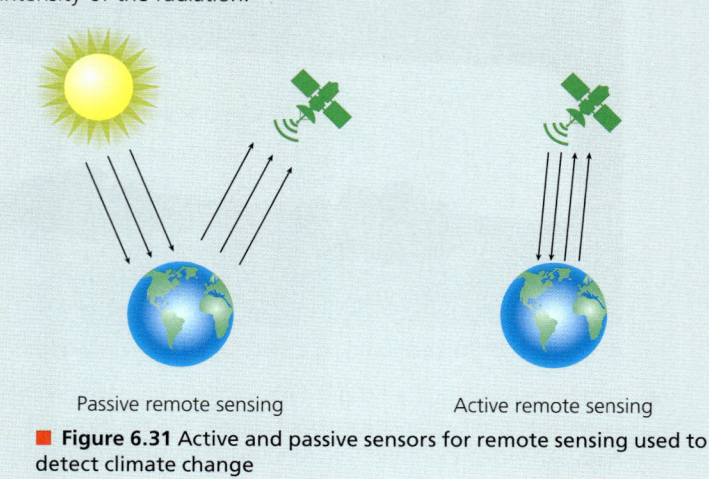

■ **Figure 6.31** Active and passive sensors for remote sensing used to detect climate change

6.2.11 Global climate models predict possible outputs or outcomes, and the validity of the models can be tested via a process known as hindcasting

Climate models use collected data and mathematical equations to quantify aspects of the Earth system and develop graphical representations of the climate. They provide useful information about changes in the climate today and in the future. There are two main types: general circulation models (GCMs) and Earth system models (ESMs). GCMs are used to study the atmosphere, ocean, land and ice. ESMs also include the biosphere and the cryosphere.

Computer simulations create a picture of the real world based on satellite imagery and real-time data collected by instruments on satellites, carried by balloons or fixed in particular areas. The information can be used to project future changes by making slight changes to the values of the inputs. The models are created by making some assumptions about how parts of the climate system interact and this may compromise the results. It is therefore important to check the validity of a model before using it for making scientific judgements and/or predictions.

Hindcasting in climate modelling

A process called hindcasting is used to check the validity of climate models. Through this process the computer-generated data is compared with actually recorded historical data to observe the degree of similarity between the two. A model that fails to compare well with historical data will be unable to produce realistic projections of the climate in future. In such cases, the parameters must be altered to fit the observations, making the outcome more reliable. These alterations are referred to as manipulations. Methods of hindcasting include:
- Model validation: Using a climate model to simulate past climate conditions and comparing the model outputs with observed data. This is a crucial step in validating the accuracy of a climate model. By simulating historical periods with known climate conditions, researchers can assess how well the model reproduces past climate variability.
Example: Simulating the climate of the last Ice Age using a model and comparing the results to geological evidence, such as ice core data and paleoclimate records. This allows researchers to assess whether the model accurately reproduces known temperature patterns, glacial advances, and atmospheric conditions during that historical period.

- Assessment of performance: Helps evaluate the overall performance of a climate model by comparing simulated data with actual observations. By assessing how well the model reproduces past climates, researchers can identify strengths and weaknesses in the model's ability to capture specific climate features or phenomena.
  Example: Using hindcasting, researchers can compare the modelled and observed temperature and precipitation patterns of the early twentieth century. If the model consistently overestimates rainfall in a specific region during this period, it indicates a performance issue that requires further investigation and refinement.
- Improvement of model projections: Provides insights into areas where a model may deviate from observed data, guiding adjustments to improve future projections. This allows researchers to refine model parameters or incorporate additional processes to enhance the model's predictive capabilities by identifying discrepancies between model outputs and actual data.
  Example: If a climate model consistently underestimates the frequency and intensity of tropical cyclones during historical periods, researchers can refine the model by adjusting parameters related to sea surface temperatures and atmospheric conditions. This improvement enhances the model's ability to project future tropical cyclone behaviour.
- Evaluation of the model based on data evidence: Involves comparing the modelled climate conditions with observational data to assess the reliability of the model. Researchers use hindcasting to ensure that a climate model aligns with historical data, providing evidence for the model's credibility and usefulness for future projections.
  Example: Hindcasting can compare simulated climate conditions of the mid-twentieth century with instrumental data. If the model accurately reproduces observed trends in global mean temperature, sea level rise and atmospheric carbon dioxide concentrations, it provides strong evidence of the model's reliability for simulating recent climate history.

Even though climate models have many physical parameters, there are some aspects of the atmosphere that are difficult to include. One such aspect is the impact of aerosols. These the amount of energy reflected, thus changing the climate sensitivity, and their composition is difficult to predict. Another important physical parameter is the volume of glacial meltwater. Since this continues to change every year, there is no possibility as yet of building this variable in to current climate models which are based on natural glacial melt rates. The excess melt that is taking place due to climate change has to be treated as an extra input that cannot be automatically calculated.

6.2.12 Climate models use different scenarios to predict possible impacts of climate change

Climate models consider the complexities and uncertainties associated with climate by using use a variety of scenarios to predict the potential impacts of climate change. These scenarios outline the different possible trends in emissions and atmospheric concentrations of GHGs. Climate models generate projections based on these concentrations to obtain a comprehensive understanding of the consequences of varying levels of anthropogenic actions on factors such as sea levels, precipitation and temperature. This can be useful for decision making and help policymakers to frame policies required to fight climate change. Representative Concentration Pathways (RCPs) were developed by the scientific community to support the work of the Intergovernmental Panel on Climate Change (IPCC). They describe several plausible ways in which anthropogenic emissions might change and are essential tools for assessing potential climate change impacts.

There are four main RCPs:
- RCP2.6 (Low Emissions): Represents a future where emissions peak around 2020 and decline rapidly, leading to a low concentration of GHGs by the end of the century.
- RCP4.5 (Intermediate Stabilization): Assumes that emissions peak around 2040 and then gradually decline, resulting in a mid-range concentration level by the end of the century.
- RCP6.0 (Intermediate Stabilization): Similar to RCP4.5, this scenario assumes a moderate stabilization of emissions. Emissions peak around 2080, and concentrations stabilize at a slightly higher level than RCP4.5 by the end of the century.

- **RCP8.5 (High Emissions):** Represents a future with continued high greenhouse gas emissions throughout the century. It assumes no significant efforts to mitigate emissions, resulting in a high concentration of GHGs by the end of the century.

Climate models based on these scenarios lead to the IPCC projecting the following:
- **Sea-level rise:** Sea levels could rise by 0.26 to 1.08 metres by 2100, relative to 1990 levels. This rise would be caused by the melting of glaciers and ice sheets, as well as the expansion of seawater due to warming.
- **Local temperature:** Global average temperatures could rise by 1.5 to 2.6 degrees Celsius by 2100, relative to 1990 levels. However, the actual temperature increase will vary from region to region. Some areas, such as the Arctic, are expected to warm more than others.

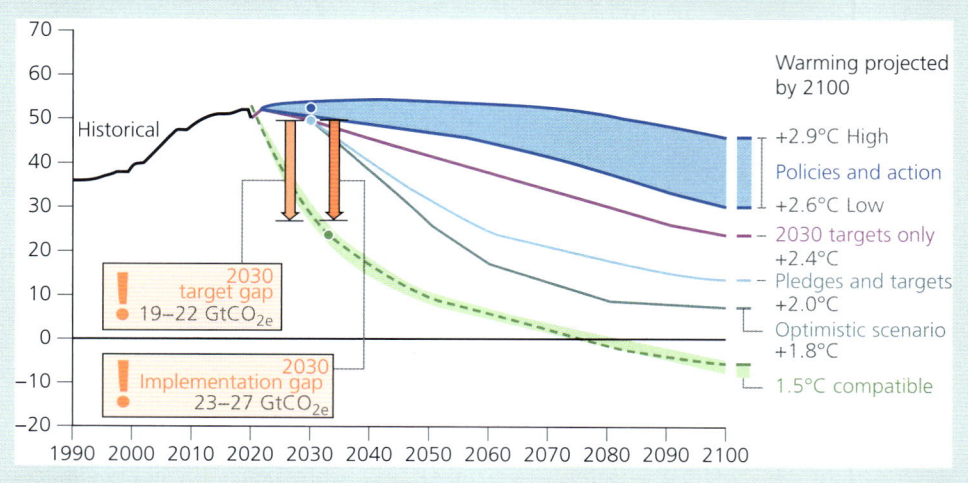

■ **Figure 6.32** Global warming projections for 2100

- **Precipitation patterns:** Precipitation patterns are likely to change in many parts of the world. Some areas are expected to become wetter, while others are expected to become drier. These changes are likely to be more pronounced in the tropics and subtropics.

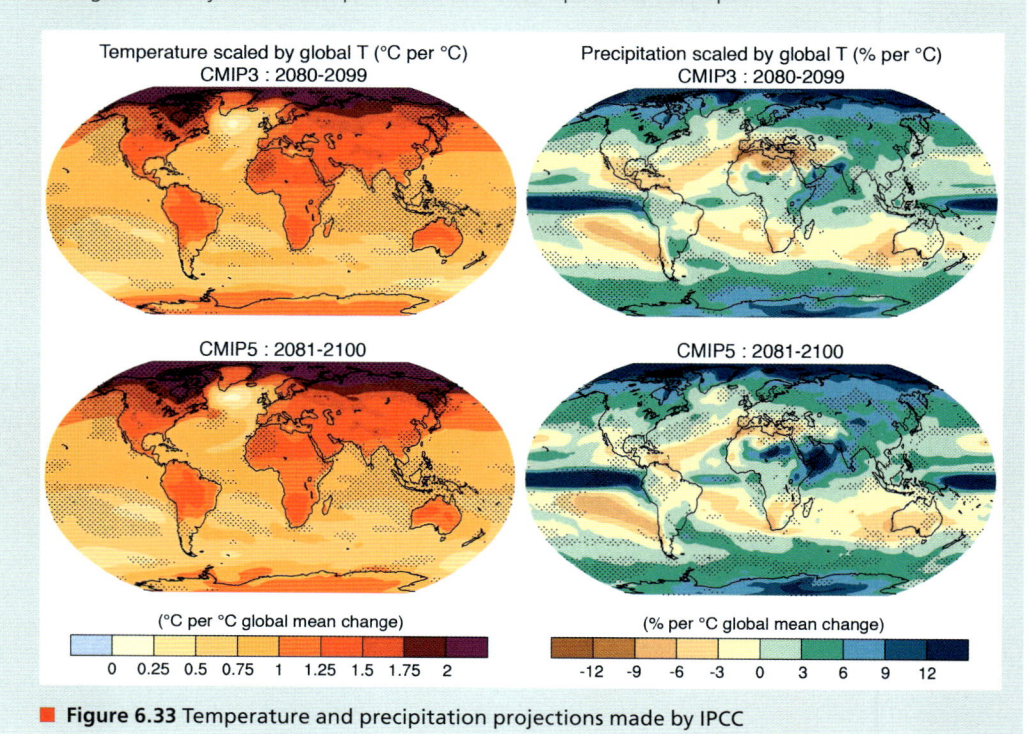

■ **Figure 6.33** Temperature and precipitation projections made by IPCC

> **Tool 4: Systems and models**
>
> **Correlation between carbon dioxide emission and temperature increases**
>
> Models can be used to simplify natural systems and help us understand their functioning. It has been clearly noted through research that there is a strong correlation between carbon dioxide emission and temperature increases. Use this online simulation to test this for yourself: **https://scied.ucar.edu/interactive/simple-climate-model**

> **ATL ACTIVITY**
>
> Conduct a debate on the following topic:
>
> "Climate models consider all possible perspectives to provide a realistic prediction of the climate."
>
> Debate protocol:
> - Opening statement – 3–4 minutes per side.
> - Response to opening statement – 3–4 minutes per side (only responses, no questions).
> - Main arguments – 15 minutes per side.
> - Open discussion and responses to main arguments – 25 minutes.
> - Closing statement – 3 minutes per side.

6.2.13 Climate models show the Earth may approach a critical threshold with changes to a new equilibrium. Local systems also have thresholds or tipping points

Link
Tipping points and their factors are discussed in detail in Topics 1.2.12 (page 40), 1.2.16 (page 43), 2.1.22 (page 107) and 2.1.25 (page 111).

Climate models show that the Earth may approach a critical threshold, beyond which the climate system could change rapidly and irreversibly. Factors such as increased melting of ice caps, release of carbon dioxide and methane from permafrost and changes to in albedo or cloud cover, will amplify changes to climatic conditions. These tipping points can be reached on both global and local scales, and have profound implications for the planet's climate and ecosystems.

Climate tipping points include:
- Melting of Antarctic ice sheet: One of the most significant global tipping points is the potential collapse of the Antarctic ice sheets. The loss of a substantial portion of the Antarctic ice could result in the rapid and irreversible sea-level rise, inundating coastal areas, displacing millions of people and causing widespread economic and ecological disruptions.
- Slowing of the Atlantic thermohaline circulation: The Atlantic Meridional Overturning Circulation (AMOC) is a crucial component of the Earth's climate system. Climate models suggest that changes in freshwater input from melting ice and increased precipitation due to global warming could disrupt the AMOC. This could lead to altered climate patterns, including cooler temperatures in parts of Europe and altered oceanic ecosystems.

■ Figure 6.34 Drift ice in the Antarctic

Link
There is more detail about how oceans distribute heat around the globe in Topic 2.4.6 (page 172)

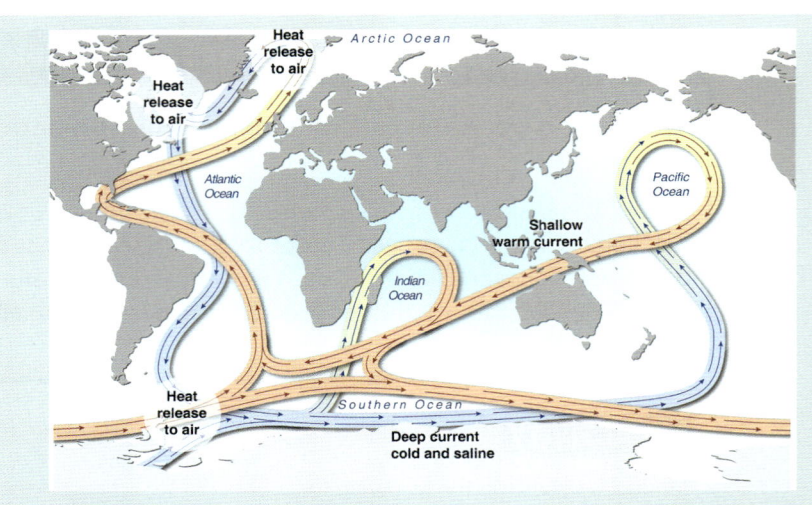

■ Figure 6.35 Thermohaline circulation in the North Atlantic ocean

- The Amazon rainforest–Cerrado transition (CAT): On a more local scale, the Amazon rainforest is approaching a tipping point known as the Amazon rainforest–Cerrado transition. The cerrado is a vast tropical and subtropical biome covering more than 20 per cent of Brazil. It includes a number of ecosystems including forests, to marshlands and open grassland and is a major source of water in Brazil. Deforestation in this region for the purpose of agriculture and pastures has altered rainfall patterns, leading to a shift from rainforest to savanna-like conditions. The temperature of air and soil in cleared areas are respectively 11 and 18 per cent higher than in forest areas and humidity levels are 14 per cent lower than in the forest areas. This transition could have severe consequences for biodiversity, local climates and the global carbon cycle.

6.2.14 Individual tipping points of the climate system may interact to create tipping cascades

As the Earth has crossed six of the nine planetary boundaries (see Topic 1.3.19, page 72), changes in the climate system have pushed it closer to several tipping points including:
- melting of ice sheets and glaciers in Greenland and the West Antarctic region
- changes in albedo, particularly in the Arctic
- thawing of permafrost and tundra, releasing methane into the atmosphere
- death of tropical coral reef due to warming of waters
- disruption of the AMOC
- dieback of Amazon rainforest and boreal forests
- dying of phytoplankton due to increasing ocean temperatures and acidification
- intensification of the El Niño–Southern Oscillation (ENSO).

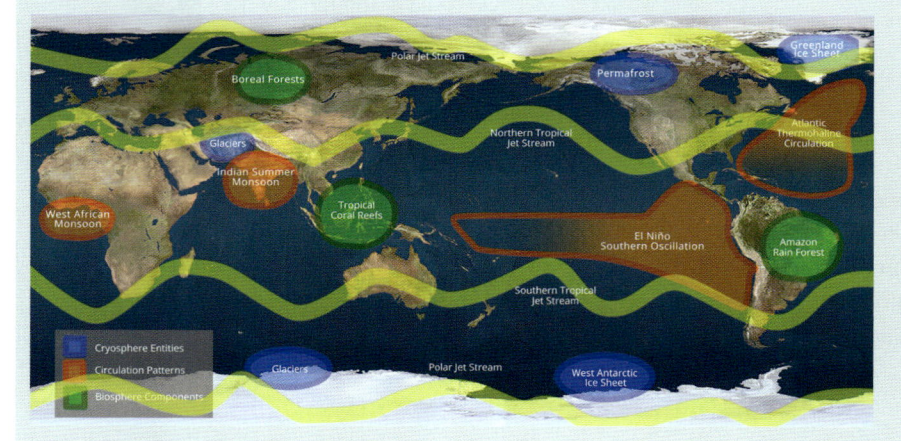

■ Figure 6.36 Global tipping points for climate systems

Theme 6: Atmosphere and climate change

◆ **Tipping cascade** – a sequence of tipping points, one leading to the other, which ultimately leads to system collapse.

◆ **Deepwater formation** – a process in which water in the ocean becomes denser and sinks to the ocean floor, creating deep ocean currents.

Individual tipping points within the climate system can interact and trigger **tipping cascades**. This domino effect can have devastating effects on large ecosystems, leading to forced migrations, socio-political instability and economic crisis.

Here is how tipping cascades work:

1 Interconnected feedback loops: A chain reaction can start when one component crosses its tipping point and causes changes that have an impact on other components. For example:
   - Greenland ice sheet melting: As global temperatures rise, the Greenland ice sheet undergoes accelerated melting. The ice sheet is a significant store of freshwater.
   - Impact on ocean currents: Freshwater from the melting ice sheet enters the North Atlantic. This influx reduces the salinity and density of seawater, particularly in the region where **deepwater formation** occurs.
   - Disruption of the Atlantic Meridional Overturning Circulation (AMOC): The altered density patterns weaken or disrupt the AMOC, a crucial component of global ocean circulation that plays a vital role in redistributing heat around the Earth.
   - Consequences for ecosystems: The weakened AMOC has cascading effects on marine ecosystems. Changes in ocean currents can influence the distribution of nutrients and impact the habitats of various marine species, causing populations to die back or forcing them to migrate. Fish that are sensitive to temperature and nutrient availability may change their migration patterns. These changes may reduce the biodiversity of an ecosystem, reducing its resilience.
   - Feedback to ice sheet melting: The changes in ocean currents, in turn, affect the melting rate of the Greenland ice sheet. Warmer ocean water accelerates the melting process, closing a feedback loop.

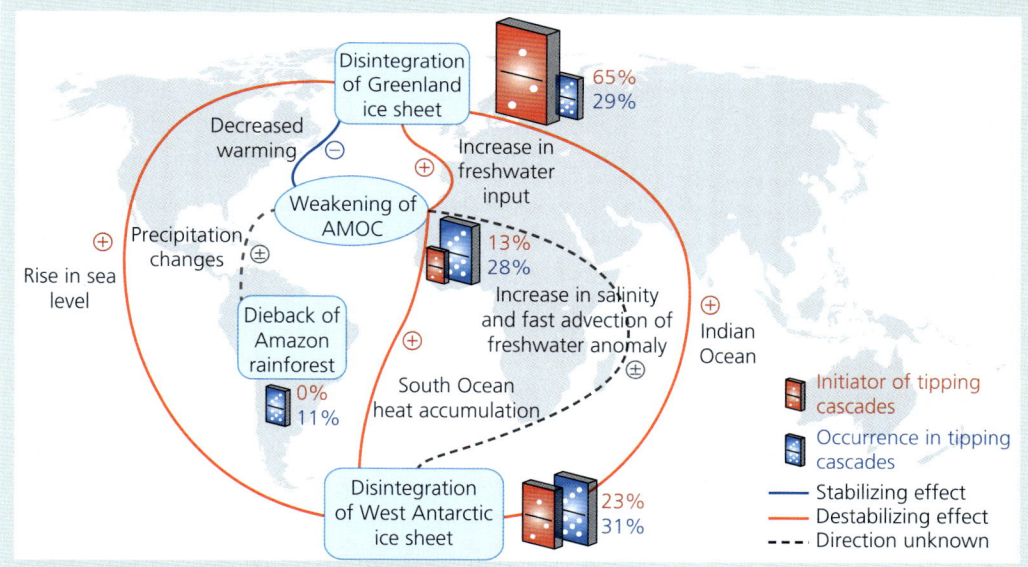

■ **Figure 6.37** Interaction of climate tipping points resulting in tipping cascades

2 Amplification of effects: Positive feedback loops amplify effects. For example, the melting of Arctic sea ice reduces the Earth's albedo, causing more sunlight to be absorbed rather than reflected back into space. This additional heat can further warm the surrounding ocean.

3 Cascading consequences: As one tipping point leads to another, a cascade of changes can occur. This can result in rapid and nonlinear shifts in the climate system, potentially leading to abrupt and severe impacts, such as accelerated sea level rise, disrupted weather patterns and ecosystem disruptions. For example, as global temperatures rise due to anthropogenic climate change, permafrost in Arctic regions begins to thaw. The organic matter within it decomposes, releasing methane, which amplifies the greenhouse effect, contributing to further global warming. The warming of the Arctic ocean may also lead to the release of methane hydrates from the seabed.

6.2 Climate change – causes and impacts

4. **Irreversibility**: Once a tipping cascade is initiated, it can be challenging or impossible to halt or reverse. The climate system may settle into a new less stable equilibrium state, with long-lasting consequences for ecosystems and human societies. For example, the disintegration of Antarctic ice shelves could lead to a substantial and permanent rise in global sea levels. Low-lying coastal areas and islands face the risk of submergence, leading to potential displacement of populations and loss of biodiversity.

Tipping cascades are a major concern for scientists and policymakers. They could have a devastating impact on the Earth's climate and ecosystems are difficult to predict, but they. Some potential consequences of tipping cascades include:

- Rapid and widespread changes in weather patterns, leading to more incidences of flash floods, droughts, forest fires, heatwaves.
- Collapse of major ecosystems, causing the release of high amounts of carbon dioxide into the atmosphere.
- Increased frequency of extinction as more and more species become vulnerable and are unable to cope with climate change.
- Enhanced human displacement from areas that are more prone to the effects of climate change. It has been predicted the number of climate migrants will increase with every passing year.

The exact impacts will depend on the specific tipping points that are crossed and the way in which the climate system responds.

Inquiry process

Inquiry 2: Collecting and processing data

- Investigate the effect of temperature on the bleaching of corals between the years 1990 to 2020.
  Use the website ourwoldindata.org to collect data on coral bleaching and climateanalyzer.org to collect data on mean sea temperature.
- Record the raw data using spreadsheets. Make use of appropriate titles, column and row headings, and use proper units wherever applicable.
- Process the data by calculating averages, means and relevant percentages to make a correlation between temperature and the percentage of coral bleaching every year.
- Present your findings in the form of processed charts, graphs (scatter plots, curve of best fit, for example) to demonstrate relevant patterns or correlations in the data, including error bars, where appropriate.
- Calculate, where appropriate, measures of dispersion, such as standard deviation and variance (minimum five values), interquartile range and coefficient of determination (R^2). Interpret values of the correlation coefficient (r) and identify correlations as positive or negative.

6.2.15 Global disparities in climate responsibility and vulnerability: political and economic implications and issues of equity

The impacts of the climate crisis are more pronounced in some regions of the world than in others, leaving humans and ecosystems unequally exposed to extreme weather patterns. Hurricanes, cyclones, floods, droughts and famines are pushing more people towards starvation and poverty. Usually, the countries that are most vulnerable are not the ones that

◆ **The Dry Corridor** – a term for a strip of land across El Salvador, Guatemala, Honduras and Nicaragua that is particularly at risk of extreme climate events such as long periods of drought.

are responsible for climate change. According to the United Nations World Food Programme, the countries most vulnerable to drought are South Sudan, Madagascar, Pakistan, Somalia, Sudan, Chad, and those in the Sahel and **the Dry Corridor**. Small island nations, such as Tuvalu, Kiribati and the Marshall Islands, have contributed minimally to global but sea level rise threatens their existence, with potential displacement of entire populations and loss of their cultural heritage.

Concentrations of CO_2 in the atmosphere first began to rise as coal and iron were exploited in the Industrial Revolution of the eighteenth century. European countries, which began to industrialize at this time, therefore have historically contributed to GHG levels. The major emitters of CO_2 today include China, the USA, India, Russia and Japan. These countries may also have high emissions per capita which is often linked to consumer-driven economies that and industrial activities lead to extensive energy consumption. Qatar and Kuwait have high per capita emissions despite their smaller populations.

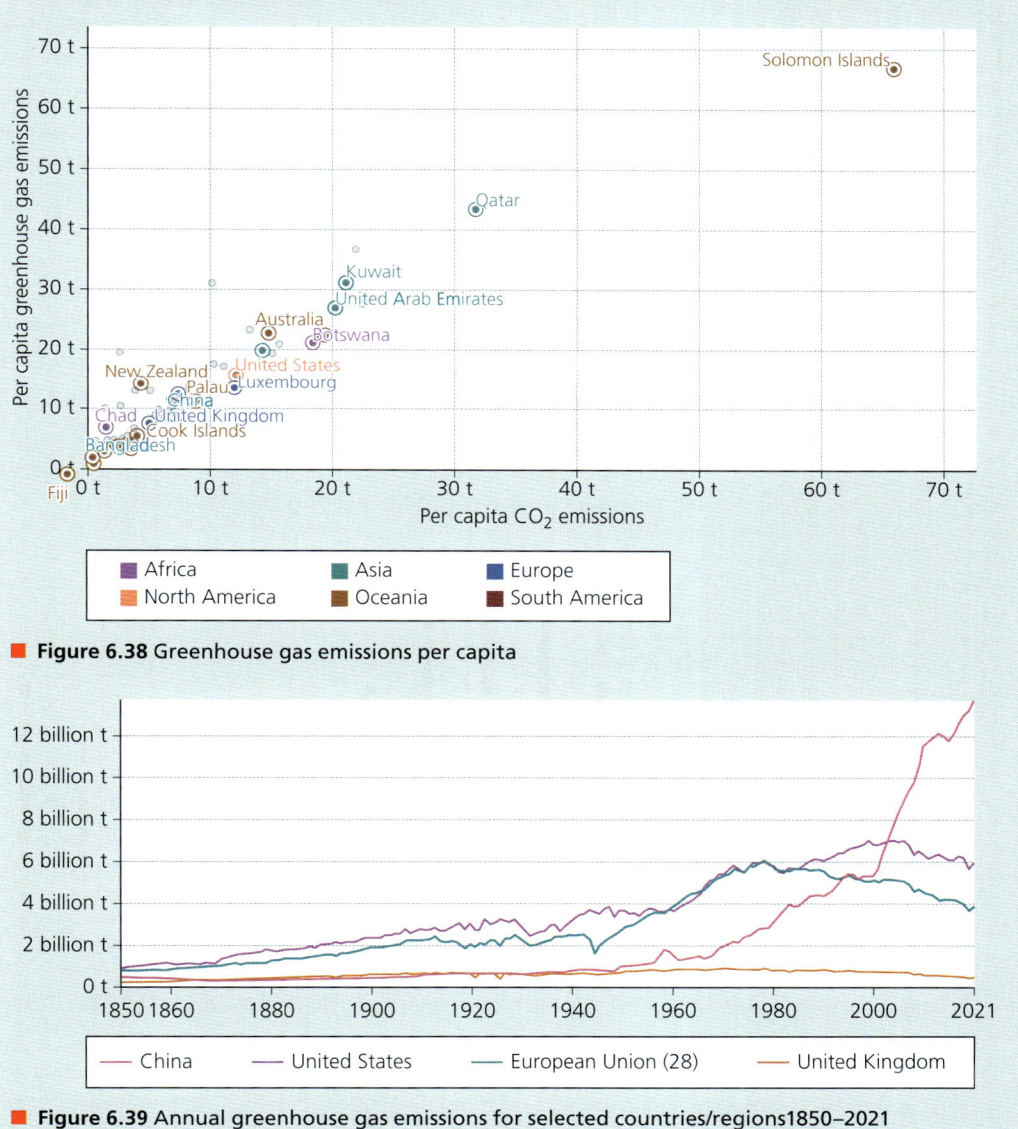

■ **Figure 6.38** Greenhouse gas emissions per capita

■ **Figure 6.39** Annual greenhouse gas emissions for selected countries/regions 1850–2021

6.2 Climate change – causes and impacts

The (Allied for Climate Transformation by 2025) (ACT2025) consortium aims to foster equity and sustainability and works to achieve ambitious outcomes at the UN FCCC climate negotiations. Discussions held at the 2023 COP28 summit revolved around the following four pillars of ACT2025:
1 Advancing just and equitable ambition.
2 Driving people-centred and livelihood-sustaining adaptation.
3 Designing fit-for-purpose institutions to address loss and damage.
4 Delivering finance to respond to the needs of the climate crises.

REAL-WORLD EXAMPLE

South Sudan

South Sudan is a landlocked country in eastern Central Africa. It is mostly covered by forests, swamps and grasslands. Sitting in the White Nile valley, South Sudan has some of Africa's most fertile soils. The majority of the population live in rural areas and are heavily dependent on the agricultural sector, focused on subsistence farming. Poverty rates are very high. South Sudan has an underdeveloped industrial and economic sector, and most of its goods and services are imported from Sudan. The country has been battling the impacts of climate change since the 1970s and the situation worsens every year.

Natural hazard risks prevalent in the country include floods, drought and famine. Sudden high-intensity rainfall leaves crops, animals and humans vulnerable to flooding and water-borne diseases. Harsh weather conditions have displaced many individuals and destroyed communities. Figure 6.41 shows actual and projected monthly average temperature anomalies for South Sudan. It shows a 2–3°C rise in the average temperature by the year 2050.

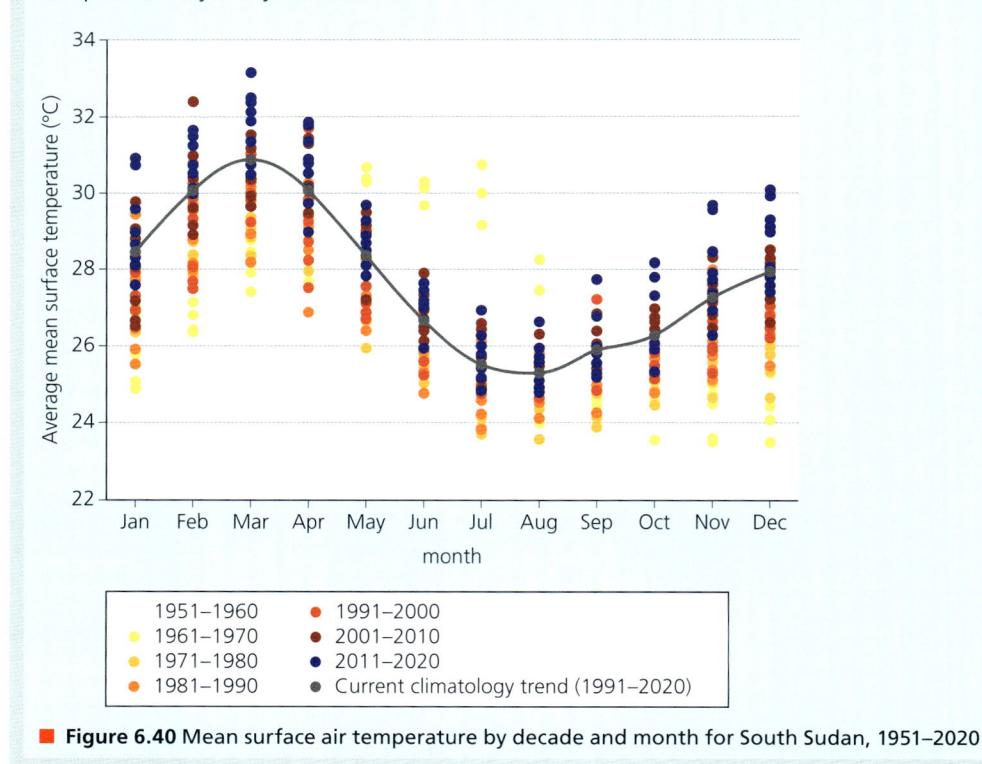

■ **Figure 6.40** Mean surface air temperature by decade and month for South Sudan, 1951–2020

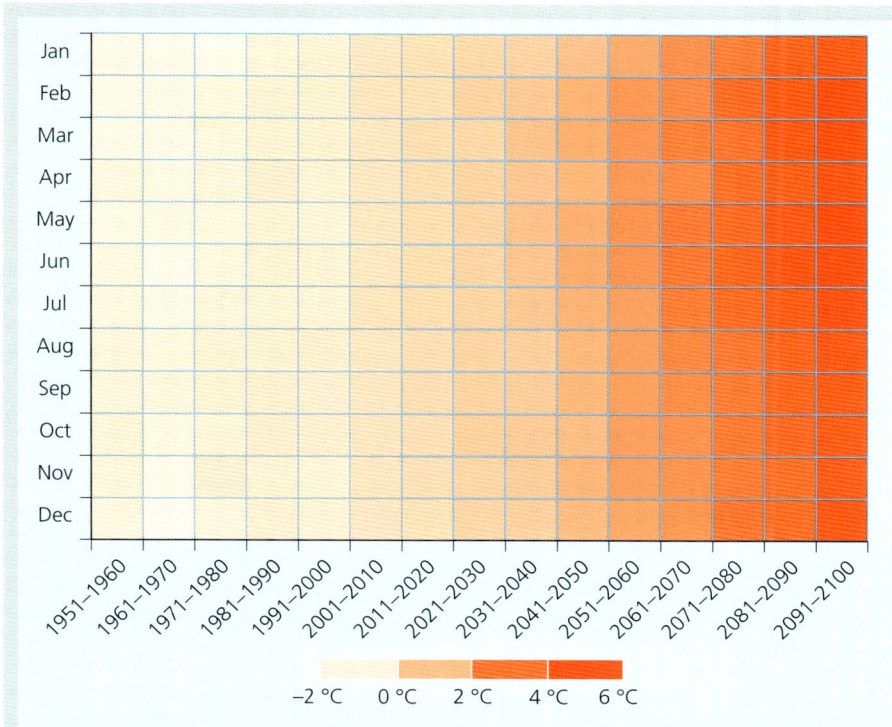

■ **Figure 6.41** Actual and projected mean surface air temperature anomalies for South Sudan, 1951–2100

● HL.a.8: Environmental law

Environmental law provides a legal framework for climate justice. It makes provisions for holding nations accountable for their contributions to climate change and ensuring equitable responses to its impacts. For example, the Paris Agreement establishes a framework for voluntary emission-reduction commitments by nations. It reflects a collective acknowledgment of historical responsibilities, with developed nations committing to support developing nations in mitigating and adapting to climate change. The implementation of the agreement is evaluated by accurate reporting. If a country that has signed the agreement fails to act to reduce GHG emissions or does not reach its targets, then it has to meet with a global committee of neutral researchers to evaluate and reflect on its climate status and plan new strategies. The country may experience diplomatic pressure, criticism and sanctions from other states. Although the Paris Agreement does not bind individual companies within a country to reduce their emissions, many businesses have voluntarily committed to participate in climate action against emissions.

ATL ACTIVITY

Emission-reduction strategies

The Paris Agreement covers climate change adaptation, mitigation and finance. Parties prepare plan to communicate the actions they will take to reduce GHG emissions.
- Working individually, develop one strategy that can be implemented at the local level and one at the national level to reduce GHG emissions.
- Working with a partner, share the strategies you have each developed. Discuss their strengths, weaknesses and limitations.
- Write a 250-word report to evaluating the suggested strategies.

6.2 Climate change – causes and impacts

HL.b.4: Environmental economics

Environmental economics has created solutions such as quotas, fines, taxes, tradable permits and carbon-neutral certification that ensure the polluter pays to limit the burden to society. Carbon-pricing mechanisms, such as carbon taxes or cap-and-trade systems, internalize the external cost of carbon emissions. This economic approach encourages polluters to take responsibility for their emissions by assigning a monetary value to carbon. European countries have operated a cap-and-trade programme since 2005. This system places a cap on the total emissions from industries. Companies can trade emission allowances, encouraging more efficient and responsible practices and emission reductions. Further, the economic valuation of of ecosystem services, such as the like storm protection mangroves offer coastal areas, supports the design of policies to protect these ecosystems and enhance resilience.

> **ATL ACTIVITY**
>
> **Analyse climate-mitigation policies**
> - Read the article, "The argument for a carbon price" (**https://ourworldindata.org/carbon-price**)
> - Write a reflective essay on implementing carbon pricing for environmental accounting.

HL.c.13: Environmental ethics

Environmental and social justice movements share overarching objectives in their advocacy efforts. These movements recognize the interconnectedness of environmental issues and social inequalities, understanding that achieving justice in one area often requires addressing concerns in the other.

- Environmental movements seek to ensure fair and equal treatment in environmental policies and the distribution of environmental benefits and burdens. All communities, regardless of socioeconomic factors or demographic characteristics, must have equal protection from environmental and health hazards.
- Social justice movements aim to address systemic inequalities and injustices within society, including issues related to race, class, gender, and other socioeconomic factors. Their objectives include creating a more equitable and just society where all individuals have equal opportunities, rights and access to resources.

For example, the Climate Justice Alliance is a coalition of environmental, indigenous and frontline community groups working toward a just transition away from fossil fuels. The alliance aims to address climate change while prioritizing the needs of communities that are disproportionately affected, emphasizing the connection between social and environmental justice.

Indigenous groups and environmentalists opposed the construction of the Belo Monte Dam in the Amazon rainforest, highlighting the impact on local ecosystems and the displacement of indigenous communities. Both environmental and social justice movements were central to the opposition, emphasizing the interconnectedness of protecting the environment and human rights.

> **TOK**
>
> To what extent do differing perspectives within environmental and social justice movements impact the formulation and effectiveness of their shared goals for creating societies where the costs and impacts of climate change are equitable and just?
>
> While approaching the above knowledge question, possible areas of exploration include how the language used in policies shapes the perception of environmental issues. You could also consider the role of emotions in motivating individuals to join environmental movements or take a stand for social justice.

> **ATL ACTIVITY**
>
> Raise awareness about the issue of climate change
>
> Create a presentation or display for the school to raise awareness about the issue of climate change.
> 1 Choose a topic: What do you want to focus on? Are you interested in the causes of climate change, the impacts of climate change or the solutions to climate change?
> 2 Do your research: Gather information about your chosen topic. This could include reading articles, watching documentaries or talking to experts.
> 3 Create your presentation or display: Once you have all of your materials, start creating your presentation or display. Be sure to include information that is accurate, relevant and engaging.
> 4 Deliver your presentation or display: Present your presentation or display to your chosen audience.

REVIEW QUESTIONS

1 Outline the differences between weather and climate.
2 What are some local and global impacts of climate change on marine ecosystems?
3 Discuss three factors that affect the perspectives of individuals and societies on climate change.

HL only

4 Evaluate the use of climate models to predict possible impacts of climate change.
5 With the help of named biotic and abiotic examples, explain how tipping points interact to form tipping cascades.

6.3 Climate change – mitigation and adaptation

> **Guiding question**
>
> - How can human societies address the causes and consequences of climate change?

SYLLABUS CONTENT

This chapter covers the following syllabus content:
- 6.3.1 To avoid the risk of catastrophic climate change, global action is required, rather than measures adopted only by certain states.
- 6.3.2 Decarbonization of the economy means reducing or ending the use of energy sources that result in CO_2 emissions, and their replacement with renewable energy sources.
- 6.3.3 A variety of mitigation strategies aim to address climate change.
- 6.3.4 Adaptation strategies aim to reduce adverse effects of climate change and maximize any positive consequences.
- 6.3.5 Individuals and societies on a range of scales are developing adaptation plans, such as National Adaptation Programmes of Action (NAPAs), and resilience and adaptation plans.

HL ONLY

- 6.3.6 Responses to climate change may be led by governments or a range of non-governmental stakeholders. Responses may include economic measures, legislation, goal-setting commitments and personal life changes.
- 6.3.7 The UN has played a key role in formulating global strategies to address climate change.
- 6.3.8 The IPCC has proposed a range of emissions scenarios with targets to reduce the risk of catastrophic climate change.
- 6.3.9 Technology is being developed and implemented to aid in the mitigation of climate change.
- 6.3.10 There are challenges to overcome in implementing climate management and intervention strategies.
- 6.3.11 Geoengineering is a mitigation strategy for climate change, treating the symptom not the cause.
- 6.3.12 A range of stakeholders play an important role in changing perspectives on climate change.
- 6.3.13 Perspectives on the necessity, practicality and urgency of action on climate change will vary between individuals and between societies.
- 6.3.14 The concept of the tragedy of the commons suggests that catastrophic climate change is likely unless there is international cooperation on an unprecedented scale.

6.3.1 To avoid the risk of catastrophic climate change, global action is required, rather than measures adopted only by certain states

Climate change is a global problem and international cooperation is needed to address it satisfactorily. Sovereign states have complete legal authority over their territory and can govern and regulate their decisions based on their own priorities and interests. It can, therefore, be challenging to have willing and equal participation in addressing the risk of catastrophic climate change. The United Nations serves as a common platform for states (countries) to collaborate and agree on global action directed towards pressing environmental issues. International law extends beyond the jurisdiction of individual nations and provides a robust framework and guidelines for negotiations, treaties, protocols and conventions to foster cooperation among nations.

- A negotiation is any form of direct or indirect communication whereby countries (parties) with opposing interests find common ground on issues that affect them. They use diplomatic channels to reach an agreement that best suits all member states. For example, each year climate negotiators meet to discuss possible ways to tackle the climate crisis. In December 2023, representatives of 197 countries met in the United Arab Emirates at COP 28 to discuss ending the use of fossil fuels.

- A treaty is a legally binding agreement between countries. It can be bilateral (between two countries) or multilateral (between three or more countries). The United Nations Framework Convention on Climate Change (UNFCCC), adopted in 1992, is an important treaty that recognizes the danger of climate change and the need for countries to work together. The formal agreementis usually between sovereign states but international organizations, individuals and businesses can also be a part of it.

- Protocols are less than treaties. They can be supplementary to a treaty, independent of a treaty or an optional instrument for a treaty. For example, the Kyoto Protocol, which was agreed by nations in 1997, is a subsidiary protocol to the UNFCCC. Similarly, the Montreal Protocol, signed by countries in 1987, is a result of the Vienna Convention for the Protection of the Ozone Layer, signed in 1985.

- Conventions are treaties that become legally binding when a state ratifies it. Ratification is the act that indicates a state consents to be bound to the treaty. Usually the instruments negotiated under international organizations such as the UN are entitled conventions, for example the Vienna Convention (1985) and Rotterdam Convention (2004).

> **Concept**
>
> **Systems and models**
>
> These international instruments provide a holistic approach to understanding and managing transboundary issues. In the context of climate change, systems thinking involves understanding the complex interactions between the atmosphere, oceans, ecosystems and human activities. Many of these are shared across borders. A systems approach allows analysis of cause-and-effect relationships and identification of appropriate solutions, keeping in mind the multiple factors that govern climate change.

Although highly industrialized countries bear a greater responsibility for emissions of greenhouse gases, combined efforts from multiple countries are the best means of fighting climate change.

◆ **Disposal cost** – the amount of money required to remove unwanted materials or pollutants (such as CO_2) from the environment.

◆ **Environmental degradation** – the deterioration of the environment due to overexploitation of resources, destruction of habitats and ecosystems, and accumulation of pollutants.

Policies may not always make producers accountable for the **disposal cost** of CO_2, and so they may continue to add it to the atmosphere. International cooperation allows participating nations to revisit their policies around resource use, environmental services and **environmental degradation**. Due to the global nature of climate change impacts, nations need to take action independently and collectively. The Paris Agreement, for instance, provides a pathway for developed nations to help developing nations with climate **mitigation** and **adaptation**. A tax on carbon emissions that are exported from one country to another could be implemented through a mix of policies, including regulations, market-based mechanisms and public investment. This would make it more expensive for countries to emit greenhouse gases, and would encourage them to reduce their emissions.

HL.a.7: Environmental law

Nations need to cooperate to effectively address transboundary pollution issues. For instance:
- Air pollution, including acid rain, global warming and damage to the ozone layer: Under the Geneva Convention on Long-Range Transboundary Air Pollution, participating nations commit to limit, to gradually prevent, and to reduce their discharges of air pollutants.
- International rivers and shared water bodies: These are governed by agreements that promote equitable sharing and sustainable management of rivers. For example, the Nile Basin Cooperative Framework Agreement obliges the basin states to conserve, manage and develop the river basin and its waters.

Bilateral and multilateral agreements facilitate joint efforts in resource management, disaster response and pollution control.

Disputes arising over transboundary pollution or resource management are resolved by international courts such as the International Court of Justice.

International organizations, like the United Nations Environment Programme (UNEP), work towards ensuring that nations adhere to their commitments.

ATL ACTIVITY

Read the research paper below and make notes individually:
- McIntyre, O. (2006) The Role of Customary Rules and Principles of International Environmental Law in the Protection of Shared International Freshwater Resources, *Natural Resources Journal*, Vol. 46, Issue 1.

Answer these questions:
- What are the main principles of international law concerning transboundary pollution and resource management?
- How do international legal frameworks address the challenges posed by pollution and resource exploitation that crosses national borders?
- Identify some case studies or examples where international law has been applied (or struggled to be applied) in addressing transboundary environmental issues.

Discuss your responses with your peers to understand their perspectives.

◆ **Mitigation** – actions that reduce the impacts of something, such as climate change.

◆ **Adaptation** – changes, such as the efforts made to live with the effects of climate change.

Mitigation and **adaptation** strategies can guide the effective implementation of climate agreements. Mitigation strategies (see Topic 6.3.3, page 567) aim to reduce GHG emissions and remove them from the atmosphere. For example, cap and trade to control pollution. Adaptation strategies (see Topic 6.3.4, page 570) help us to live with the impacts of climate change. For example, protecting cities from storm surges by building flood barriers.

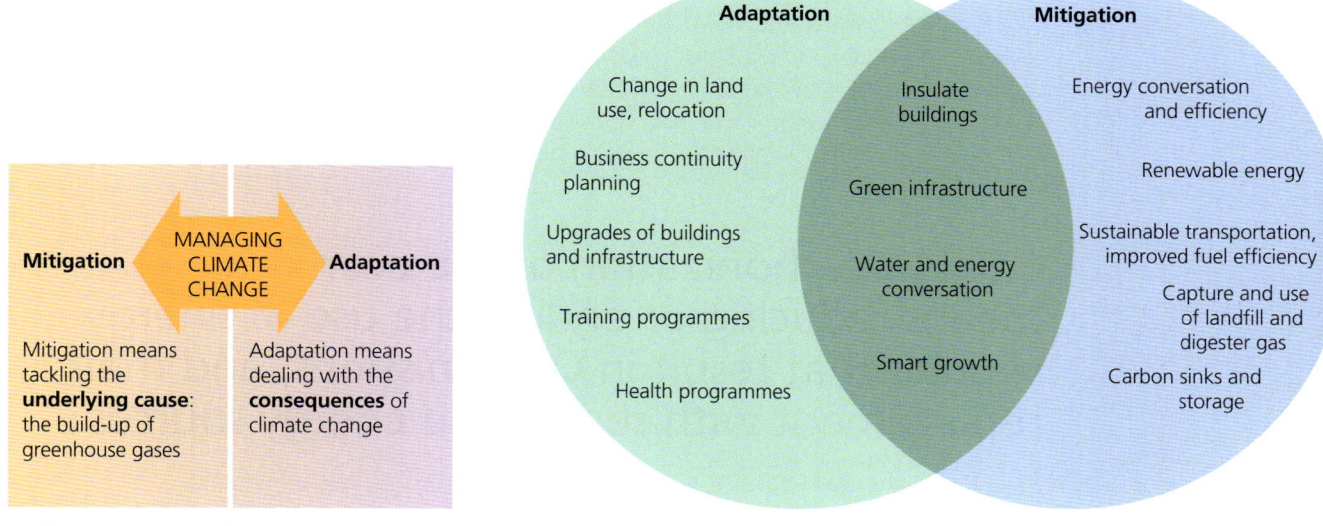

■ Figure 6.42 Ways of managing climate change   ■ Figure 6.43 Some examples of climate change mitigation and adaptation

HL.b.5: Environmental economics

Companies may spend time and money on marketing strategies, including promotion of product eco-friendliness and services, without actually changing their practices to become sustainable. Misleading claims about the environmental benefits of a product, service or corporate practice attracts the attention of environmentally conscious consumers. These claims include vague and generic environmental claims, incomplete information, false imagery and false certifications. This practice is called greenwashing.

ATL ACTIVITY

- Identify various types of greenwashing that companies may indulge in.
- Perform detailed research on at least two examples of greenwashing from the real world. Choose products that are most common or companies that people can easily recognize.
- Generate awareness about greenwashing by creating an infographic and putting it up in your school, or by creating a video.
- As a group, present your findings in an assembly.

HL.b.13: Environmental economics

The impact of economic growth on the environment can be positive and/or negative. The tendency to exploit natural resources increases with increased economic growth, leading to more waste and pollution. Sometimes it may even lead to overexploitation, loss of habitats and climate change. The positive impacts include advancement of technologies that can provide solutions to environmental issues such as pollution. Such technologies can help in decoupling environmental degradation from production and consumption. The overall well-being of society and environment may increase as more funds are diverted towards these sectors.

ATL ACTIVITY

Consider a hypothetical situation where the economy of a developing country is growing rapidly due to investments from foreign companies and rapid industrialization. A market for a variety of goods and services emerges, along with job opportunities, higher incomes

6.3 Climate change – mitigation and adaptation

> and a better standard of living. However, a large portion of the industries that support this economic growth are responsible for significant environmental degradation, causing pollution, deforestation and displacement of people.
>
> Write a reflective essay of 500–700 words entitled 'The ethical dilemma: A trade-off between economic growth, environmental impact and sustainable development'.

6.3.2 Decarbonization of the economy means reducing or ending the use of energy sources that result in CO_2 emissions, and their replacement with renewable energy sources

Combustion of fossil fuels, such as coal, oil and natural gas, is the principal source of anthropogenic carbon dioxide (CO_2) emissions. These emissions contribute significantly to climate change, which has a host of environmental and socioeconomic consequences. Decarbonization is the deliberate reduction or elimination of energy sources that release CO_2 and other greenhouse gases into the atmosphere, while replacing them with renewable and low-carbon alternatives. As such, decarbonization plays a central role in mitigating the impacts of climate change.

> **Concept**
>
> **Sustainability**
>
> Carbon neutrality is achieved when the amount of CO_2 added to the atmosphere by anthropogenic actions is equal to the amount of CO_2 removed from the atmosphere using all or some of the strategies discussed below. This balance between emissions and removal leads to net-zero emissions. Carbon neutrality will favour sustainability by slowing global warming and all the catastrophic events that follow it. It will also have a positive impact on the energy crisis, air pollution and land use changes.

Carbon neutrality roadmaps for different countries have a common focus on achieving net-zero emissions by a set target year (see Table 6.9 for examples). Each country decides these dates based on their current economic stability and availability of resources. Bhutan and Suriname are two countries that have already achieved carbon neutrality.

■ Table 6.9 Carbon neutrality status of selected countries (source: https://eciu.net/netzerotracker)

Country	Year of achieving carbon neutrality	Implementation level
Bhutan	2030	Achieved (self-declared)
Suriname	2030	Achieved (self-declared)
Finland	2035	In law
Sweden	2045	In law
Italy	2050	In policy documents
China	2060	In policy documents
Kuwait	2060	Declaration/pledge
Ghana	2070	Declaration/pledge
Bangladesh	2050	Proposed/in discussion
Malawi	2050	Proposed/in discussion

HL.b.14: Environmental economics

Eco-economic decoupling means breaking the link between economic growth and environmental degradation. It is possible to achieve the former without causing the latter. Efforts are being made across many countries, and through common platforms such as the UN, to achieve net-zero. Transitioning from fossil fuels to renewable energy sources alongside increasing carbon offsetting and use of carbon-capture technologies supports economic development. Carbon offsetting is when companies or individuals cancel out the impacts of some of their emissions by investing in projects that either reduce or store carbon to gain 'carbon credits'. These activities include preserving forests, planting trees, building wind farms and improving farming techniques to reduce the use of fossil fuels. Carbon credits allow emissions to be counterbalanced without altering current energy generation and usage. It enables economic activities to continue without an overall negative impact on the environment.

Inquiry process

Inquiry 2: Collecting and processing data

Collect and analyse data on choices supporting net-zero emissions to understand preferences and awareness.

- Create a simple survey with 10–15 questions about daily choices related to energy, transportation and lifestyle.
- Use Google Forms to create the survey.
- Include options that align with net-zero principles.
- Share the survey link with friends, family or classmates.
- Collect survey responses from at least 50 individuals over a designated period.
- Organize survey responses and input the data into a spreadsheet.
- Use analytical tools to analyse the data to identify common choices supporting net-zero emissions.
- Look for trends or patterns in participant responses.
- Create a presentation or infographic presenting key findings.
- Share the results with participants and discuss implications.

6.3.3 A variety of mitigation strategies aim to address climate change

Mitigation of climate change is essential as there is a limit beyond which adaptation strategies will not be effective. However, the success of mitigation strategies depends on how well they are implemented. Some actions can be implemented at individual, local levels, while others require policies to be implemented nationally or globally.

The first approach must be to reduce our consumption of energy. The world's per capita energy consumption for 2022 was 21,039 kWh and CO_2 emissions were 4.66 tonnes per person, compared to 12,978 kWh and 3.39 tonnes per person in 1965. Within approximately 60 years, per capita energy consumption has almost doubled. Since much energy is generated from fossil fuels, this increasing demand is a major factor in the increase in CO_2 emissions, which ultimately causes temperatures to rise and further increases the demand for energy.

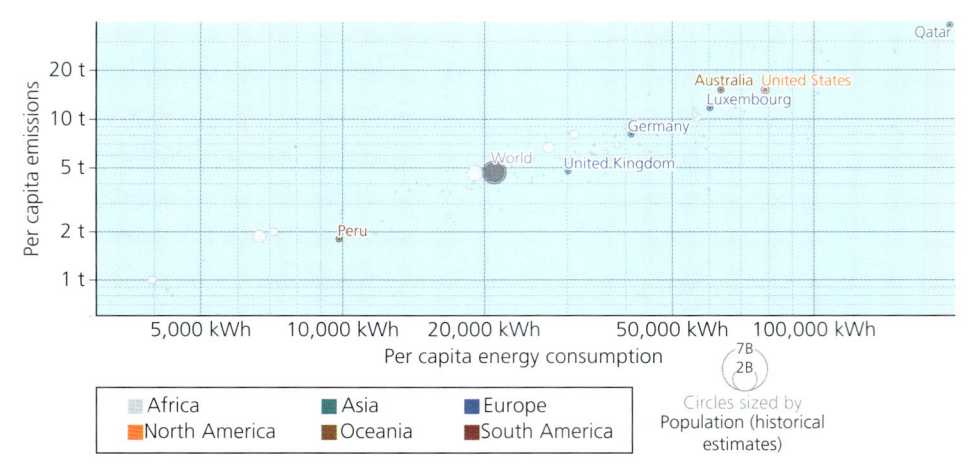

Figure 6.44 Energy consumption vs CO_2 emission for 2022

Mitigation strategies remove the cause of carbon dioxide emissions over a period of time. Investing in mitigation strategies can be both economically and ethically beneficial. There are three main categories:

1. Reducing the process of global warming.
   - Lifestyle changes. By making sustainable choices in our daily lives we can reduce our contribution to global warming. Examples include:
     - reducing consumption of meat, dairy and processed foods
     - using public transport or car-pooling whenever possible
     - using devices and appliances that reduce or optimize energy use such as thermostats and LED lighting
     - buying locally produced food and products that are not produced in an unstainable way
     - adapting reduce, reuse, recycle and repurpose strategies.
   - Geoengineering: Proposed large-scale interventions to remove significant amounts of CO_2 from the atmosphere (see below), increase the amount of radiation reflected back into space or reduce the amount reaching the surface by:
     - increasing surface albedo
     - injecting sulfur dioxide particles into the stratosphere
     - injecting tiny droplets of seawater into clouds
     - deploying tiny mirrors or reflectors into space.

2. Reducing production of GHGs.
   - Reducing the GHG emissions associated with energy production by:
     - using renewable or clean energy sources to generate electricity (see Figure 6.45)
     - improving urban design and building insulation to reduce the energy used for cooling and heating
     - upgrading to energy efficient appliances
     - giving industries and businesses a financial incentive to reduce their energy use through carbon taxation.
   - Changing agricultural practices. What is possible here is constrained by the need for food security. The major sources of emissions are:

Figure 6.45 Renewable energy in Germany: a biogas plant and solar array in Schleswig-Holstein

Theme 6: Atmosphere and climate change

- Livestock farming and cultivation of rice release large amounts of methane. Change here is linked to dietary choices.
- Clearing forests for agriculture releases CO_2 as well as removing a carbon sink.
- As well as requiring large amounts of energy to produce, fertilisers are responsible for nitrous oxide emissions. Precision agriculture techniques and use of varieties of crops that are better suited to conditions can reduce the amount of fertiliser (and water) required to ensure a good yield.

3 Removing CO_2 from the atmosphere using the techniques shown in Figure 6.46.

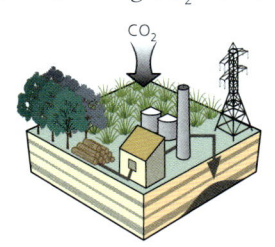

Bioenergy with carbon capture and storage (BECCS)
Fast-growing plants are harvested and burned to make energy. Exhaust carbon is captured and piped underground

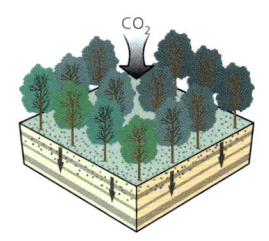

Forestation
Planted trees capture carbon dioxide as they grow. The carbon remains sequestered as long as forests are not cut down

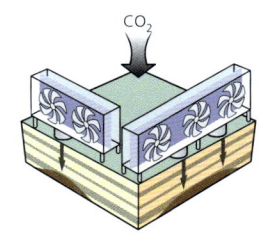

Direct air capture
Carbon dioxide in air selectively 'sticks' to chemicals in filters. Filters are reused after releasing pure carbon dioxide which can be stored underground

Biochar and soil sequestration
Charring biomass stores carbon in soil by making it resistant to decomposition. Altered tilling practices also enhance carbon storage

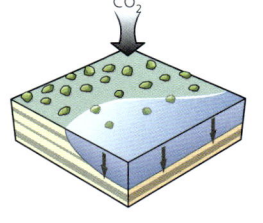

Enhanced weathering
When spread across fields or beaches and wetted, crushed silicate minerals like olivine naturally absorb carbon dioxide

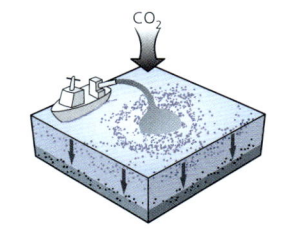

Ocean fertilisation
Injections of nutrients like iron stimulate phytoplankton blooms, which absorb carbon dioxide. When they die, they take the carbon to the sea floor

■ **Figure 6.46** The main ways to remove carbon dioxide from the atmosphere using the techniques shown in Figure 6.46.

Link
There is more detail about how some of these geoengineering techniques and methods of removing CO_2 from the atmosphere work in Topic 6.3.11 on page 584.

ATL ACTIVITY

Generate awareness about climate action
Make a poster to generate awareness about climate action in school.
- Identify and research in detail three personal behaviours that impact climate change.
- Include key facts and statistics.
- Use graphics to show the impact of recommended behaviours.
- Visualize concepts for easy understanding.
- Provide step-by-step plans for each behaviour.
- Use icons for clarity.
- Design visually engaging and motivational posters.
- Include quotes and emphasize collective impact.
- Display these posters where people can see them – in your classroom, corridors, the cafeteria or common area.

6.3.4 Adaptation strategies aim to reduce adverse effects of climate change and maximize any positive consequences

Adaptation strategies allow us to modify our existing practices to survive the effects of continuing climate change. Adaptation strategies can be grouped into two categories – structural adaptations, where we can build structures to help us cope with the effects, and non-structural adaptations, where we make changes to our choices and actions to better survive in the changing conditions.

1. Structural adaptations:
    - Flood defences: Structures such as seawalls, breakwaters and flood barriers help protect coastal communities and infrastructure in regions that are highly threatened by rising sea levels, storm surges and floods due to climate change. Some of these structures are permanent, for example the breakwaters, while some are movable, like the Thames Barrier in London, UK.
    - Desalination plants: These plants offer an opportunity to convert the salty sea water into freshwater. Erratic precipitation patterns due to climate change are leading to water scarcity, and this technology has proved useful to address water-shortage issues in many areas, such as the Ashkelon, Israel, which is one of the world's largest desalination plants.

■ **Figure 6.47** The Thames Barrier has been effective in preventing London from flooding

■ **Figure 6.48** Desalination plant in Hamburg port, Germany

2. Non-structural changes:
    - Agricultural practices: Changing weather patterns are affecting the sowing and growing seasons, making it difficult for farmers to grow their usual crops. Farmers can choose drought-resistant varieties and improved methods of irrigation, and can also align the sowing and growing seasons to the changing climate conditions. In parts of India, for example, farmers have successfully adopted drought-resistant millet varieties as a response to reduced rainfall.
    - Vaccination for new diseases: Climate change induced outbreaks of new diseases and their vectors have been on the rise for the past few decades. Developing vaccines to combat these diseases and stop their spread remains a big challenge. For example, warmer climates can cause a rise in the incidence of waterborne diseases such as cholera, malaria and salmonellosis.

> **REAL-WORLD EXAMPLE**
>
> ### The face of agriculture in 2080 due to climate change
>
> As the average temperature of the world increases, and seasonal weather patterns change, the possibility of growing conventional crops decreases. Climate catastrophes such as storms, hurricanes, long periods of drought and flash floods, have caused severe losses in the agricultural industry. Farmers are forced to adapt their crop choices to the new environment. By projecting the climate changes, we can get an idea of the possible agricultural productivity in the future. However, it is not completely possible to predict the relationship between such a dynamic system and human systems.
>
> ■ Figure 6.49 Projected change in agricultural productivity in 2080 (source: www.grida.no/resources/7299, Hugo Ahlenius, UNEP/GRID-Arendal)
>
> There are around 630 farms in the Ards Peninsula in Northern Ireland, where approximately 78 per cent of land is farmed. Due to its favourable climatic conditions, the area has a growing season of around 280 days. The fertile soils support mixed farming and pastures. The area receives less rainfall (less than 800 mm per year) than nearby upland regions (1600 mm per year). Climate predictions for 2080 suggest warm, dry summers and milder winters, with greater contrast between dry and wet periods and increased precipitation.
>
> A study conducted using this hypothetical situation asked farmers about a alternate crops, energy crops and their readiness to adapt to the situation overall. Low-cost and low maintenance energy crops are used for biofuel production. The responses fell into three categories: clear acceptance and adaptation, no adaptation, and uncertainty.
>
> Farmers willing to adapt were prepared to grow alternate crops such as forage maize and new grass species, and also energy crops such as willow. Forage maize, used as animal feed, is best grown in warm weather and harvested in winters, leaving the soils

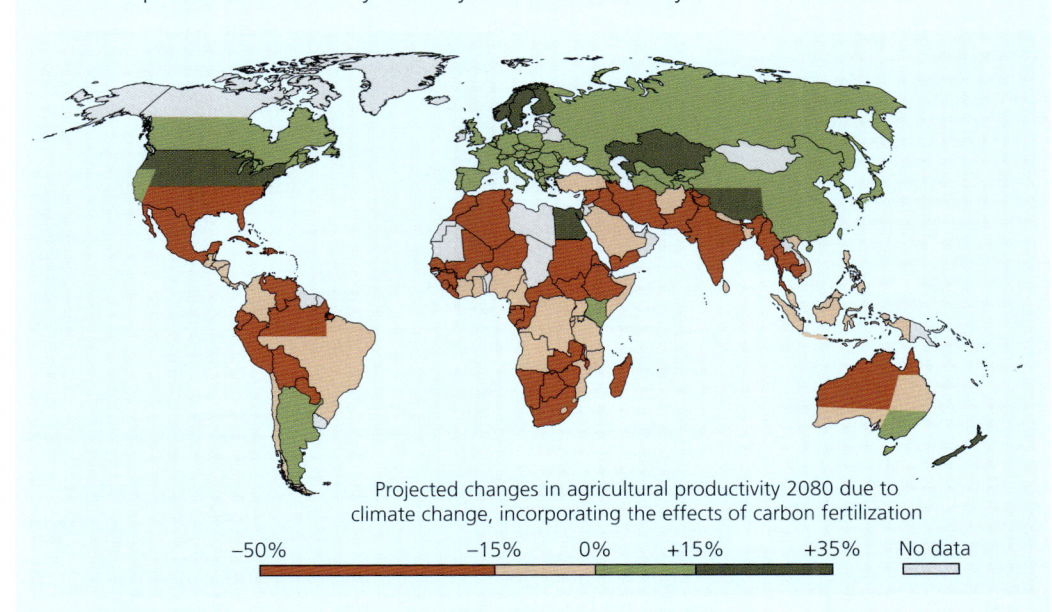

6.3 Climate change – mitigation and adaptation

barren and open to erosion. Storm surges and heavy rains will cause an erosion not just of the nutrient-rich topsoil, but also of the nitrate and phosphate reserves from the soil. Compaction of soil due to farm animals and use of heavy machinery will decrease infiltration, leading to increased runoff. Willow provides greater ground cover throughout the year. The roots hold the soil together, thus decreasing the chances of soil erosion due to heavy rainfall. The market for biomass (including willow) to be used as an energy source is increasing in Northern Ireland and will be more extensive in 2080.

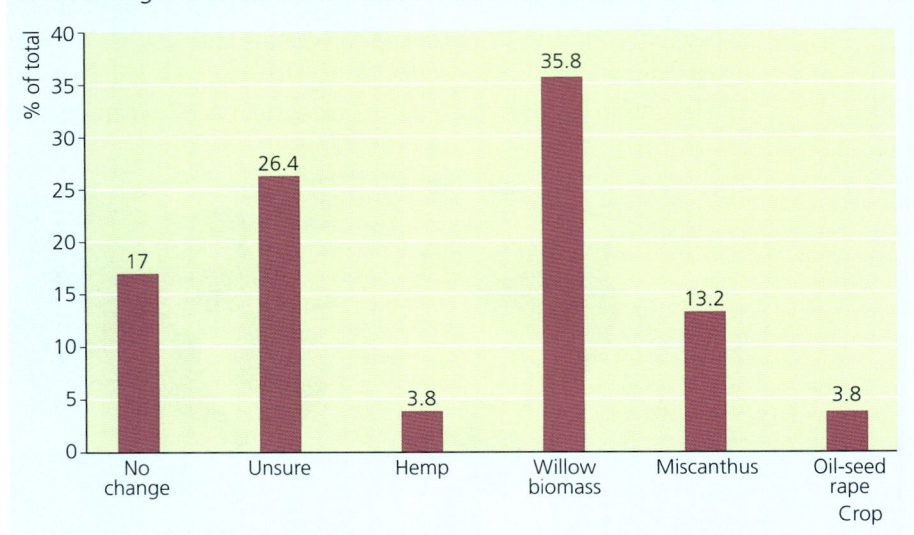

■ **Figure 6.50** Energy crops that farmers would most likely grow in 2080

Crop	% of total
No change	17
Unsure	26.4
Hemp	3.8
Willow biomass	35.8
Miscanthus	13.2
Oil-seed rape	3.8

Some farmers did not wish to adapt as they believed that the climate changes will not be so extreme to require them to change the choice of crop they grow.

Another group of farmers were uncertain about their willingness to adapt and change. They wanted to 'wait and see' what changes occur and then decide.

Tool 1: Experimental techniques

Questionnaires, Surveys, Interviews

A survey can help gather valuable insights into community attitudes, concerns, and willingness to participate in climate change mitigation initiatives like solar panel installation. The results can inform decision-making and communication strategies for implementing such projects.

- Design a survey consisting of 10–15 questions to understand the community's perspectives on installing solar panels on school buildings as a climate change mitigation strategy.
  - identify and justify your choice of an appropriate target audience
  - construct relevant open or closed questions with multiple-choice responses/ Likert scale, as appropriate
  - choose and justify an appropriate method and size of sample, i.e., random/ convenience/volunteer/purposive
  - show ethical awareness, i.e., anonymity/consent of respondents over the age of 12
  - pilot/trial the survey to gain feedback for modification.

> **Inquiry process**
>
> ### Inquiry 1: Exploring and designing
>
> Investigate the cause and effect relationship behind an environmental issue associated with climate change.
>
> - Formulate a relevant, concise research question to study the effect of ocean acidification on coral reefs.
> - Formulate a testable hypothesis, including a null and alternative hypothesis where appropriate.
> - Identify all appropriate variables, range of values and controls.
> - Design a method for collection of sufficient relevant data, repetitions and sample sizes for statistical tests.

6.3.5 Individuals and societies on a range of scales are developing adaptation plans, such as National Adaptation Programmes of Action, and resilience and adaptation plans

Adaptation to climate change is possible when there is awareness and willingness to take action. Awareness of the effects of climate change has been made possible by the combined efforts of a broad range of international organizations, scientific publications and reports, along with climate activists and media coverage. An adaptation strategy clearly outlines the vision and direction of action to adapt to the effects of climate change and the desired outcomes. An action plan includes steps that have to be taken to put the strategy into action. At the international level, organizations such as the UN Development Programme, UNFCCC and UNEP provide guidelines for developed and developing nations to make their own policy and legal frameworks to develop local priority activities to address the imminent consequences of climate change. These action plans can be developed at various scales.

■ National Adaptation Programmes of Action

In 2001, the Conference of Parties (COP) established the National Adaptation Programmes of Action (NAPA) to help the least economically developed countries (LDCs) to address the challenges of climate change. Provision was made for funding and advising these LDCs to create and implement their NAPAs.

The main objectives of NAPAs are:

- To identify and develop immediate and urgent NAPA activities to adapt to climate change and climate variability.
- To protect the life and livelihoods of the people, infrastructure, biodiversity and environment.
- To mainstream adaptation activities into national and sectoral development policies and strategies, development goals, visions and objectives.
- To increase public awareness of climate change impacts and adaptation activities in communities, civil society and government officials.

- To assist communities to improve and sustain human and technological capacity for environmentally friendly exploitation of natural resources in a more sustainable way in a changing climate.
- To complement national and community development activities that are hampered by adverse effects of climate change.
- To create long-term sustainable livelihood and development activities at both community and national level in changing climatic conditions.

The projects are finalized based on factors such as degree of risk factor, cost effectiveness, vulnerable groups in the community, and alignment to national goals and objectives.

> **REAL-WORLD EXAMPLE**
>
> ### NAPAs of Tanzania
>
> The Tanzania NAPA focuses on sectors such as agriculture, health, water and energy. It was informed by the National Development Vision 2025, which focused on quality livelihood, shared growth, peace, stability and unity, and more. Of the 72 projects suggested, 14 were prioritized based on Tanzania's conditions and local environment. The projects are categorized in sectors such as agriculture, water, health, energy and community as a whole.
>
> In order to bring in sustainable food production, the NAPA promotes drought-resistant food crops to eliminate hunger, food insecurity and malnutrition in drought-prone areas of Tanzania such as the Shinyanga, Dodoma and Singida regions. Maize is the staple crop grown in Tanzania. It requires a lot of water and has many negative effects on the soil ecosystem. Due to water scarcity, people in these regions have to walk miles to fetch drinking water. Focused actions include replacing maize with a relevant drought-tolerant crop, providing coastal communities with access to safe and clean drinking water, and generating awareness about the negative impacts of maize cultivation in relation to climate change.
>
> To promote protection of wildlife ecosystems and improve the livelihood of communities, the NAPA supports many nurseries and projects that involve replanting trees in degraded areas. It helps in generation of alternative sources of food and income as well as in reviving the ecosystem.
>
> Awareness projects related to prevalent diseases such as malaria, dysentery, cholera and meningitis are run throughout the villages and districts to make health opportunities available to all.
>
> > **ATL ACTIVITY**
> >
> > Answer the following questions based on the case study:
> > 1. What methods do you think the government of Tanzania adopted to create its NAPAs?
> > 2. Why does climate change make cultivation of maize in Tanzania difficult?
> > 3. Suggest two adaptation strategies that can be implemented at the local level to improve the existing health facilities.

■ Resilience and adaptation plans

Climate adaptation plans aim to reduce the vulnerability of natural and human systems to impacts of climate change. Climate resilience plans aim to anticipate, prepare for and plan responses to extreme climate events. Adaptation to climate change requires structural, social and physical approaches. Beyond the national level, many regions, cities and communities are developing resilience and adaptation plans to address the localized impacts of climate change. These plans may take one of the following approaches:

- Social approach: Adaptation planning with communities. Community adaptation action plans enable vulnerable communities to make collective decisions on priority actions towards climate change. They contain plans that can be implemented for different groups within the society. All participants agree to willingly cooperate towards the implementation of the plan. Depending on the vulnerability of a particular group, the intensity of action required may change. The whole procedure consists of four steps:

  1. Analysing the context and encouraging participation by stakeholders.
  2. Analysing climate vulnerability and adaptive capacity.
  3. Developing action plans.
  4. Implementation of action plans.

- Structural and physical approach: Identifying and implementing infrastructure improvements, such as building sea walls to prevent flooding, implementing green roofs and vertical gardens to decrease the urban heat island effect, and planting mangroves as natural barriers from storm surges and for water purification.

◆ **Adaptive capacity** – the ability of a region, community or system to adjust to the effects of climate change.

REAL-WORLD EXAMPLE

Resilience and adaptation plans in Malawi

Malawi, in southeastern Africa, is vulnerable to climate-related challenges, including droughts, floods and agricultural uncertainties. The country has initiated resilience and adaptation plans to address these issues and enhance the capacity of its communities to cope with climate change impacts. Building Resilience and Adaptation to Climate Change (BRACC) is a five-year programme in Malawi with the main aim of strengthening the resilience of vulnerable communities to withstand climate change.

Key initiatives are:

- Household- and community-level resilience-building activities – climate-smart agriculture: Malawi is predominantly agrarian, depending on subsistence farming. BRACC promotes less weather-dependent livelihoods and more sustainable crops such as millet, which consumes less water. It also educates farmers about crop diversification and post-harvest handling, and links them to the market so that they can sell their produce at a good price.
- Strengthening shock-sensitive protection systems – early warning systems: Malawi is highly susceptible to floods and droughts. These systems warn communities of any upcoming climate disaster. The members of the community are well informed and have time to prepare for the shock or to evacuate the area if needed.
- Contingency funding for shock response: In case of unavoidable shocks, there is a contingency plan that allows for repair of damage and rebuilding of the system.
- Natural resource management – renewable energy projects: Promotion of solar energy rather than the conventional dependence on biomass ensures that clean energy can be supplied to most places, including rural areas.
- Water-resource management: Irregular rainfall has led to water scarcity in Malawi. Teaching water-conservation techniques has empowered communities to prepare for the impacts climate change.

(HL) 6.3.6 Responses to climate change may be led by governments or a range of non-governmental stakeholders

Governmental and non-governmental stakeholders respond in different ways to issues related to climate change. Their responses may be predictable or unpredictable depending on their awareness of climate change and in the light of unprecedented extreme weather conditions.

Government responses to climate change

Governments are responsible for enacting laws and policies related to climate change, such as setting emissions standards, establishing international agreements and defining conservation measures. Factors such as political stance towards climate change and the stability of political leadership play a major role in shaping their responses to climate change.

Governmental stakeholders at the national level include departments, ministries and agencies responsible for environmental regulation, policy development and implementation. Regional and state governments, municipalities and local authorities control policy implementation and climate action at the local level. Intergovernmental organizations and bilateral partnerships bring governments of different countries together to take action on climate change by sharing best practices and providing technical assistance to each other.

Economic measures can include emissions trading schemes, carbon pricing, subsidies and tariffs. The European Union's Emissions Trading System sets a cap on the total amount of greenhouse gas that certain industries can emit. Companies receive or purchase emission allowances. Companies that are able to reduce emissions can sell their excess allowances to others. This scheme is called 'cap and trade' or 'allowance trading'. See Topic 6.3.6, page 576. Subsidies on renewable energy encourage people to shift away from conventional energy sources.

Legislative measures include carbon taxes. Usually, there is an undesirable trade-off between economic growth and carbon taxes as they reduce GDP. Therefore, many countries do not completely support this strategy. The impacts of carbon taxes may not be equitable, having more negative effects in low-income regions. In 2008, the Canadian province of British Columbia (BC) imposed a tax on the purchase and use of fossil fuels. This was initially set at $10 (Canadian) per tonne of CO_2eq emissions and has increased gradually since. The tax has led to decreased demand for fuel, increased use of public transport and a 5.2–10.9% improvement in air quality. Countries have set themselves targets to become carbon neutral, which requires certain policies to be implemented. Policies that govern the energy-production systems, such as increased production of solar cells and nuclear power plant regulation, form an integral part of legislative management focused on climate change.

Non-government stakeholder responses to climate change

Non-governmental stakeholders include international organizations, non-governmental organizations (NGOs), sustainable businesses, research and academic institutions, activist movements, community groups and individuals. They act by adhering to existing laws, demanding stricter measures and participating in climate actions at local levels.

Goal-setting commitments of industries include corporate social responsibility (CSR) setting standards to transition, committing to source a certain percentage of their energy requirements from renewable sources. Microsoft is working to reduce its fossil fuel demands and shift to sustainable aviation fuel by 2030. Accenture aims to achieve net-zero emissions by 2025. Companies that commit to reducing their carbon footprint support renewable energy businesses. In doing so, they gain a reputation of being environmentally friendly, which further promotes their own business.

Research institutes involved in climate research have a responsibility to find scientifically viable solutions to existing climate problems. For example, the National Oceanic and Atmospheric Administration conducts climate research, monitoring and modelling to enhance understanding of climate dynamics, extreme weather events and ocean–atmosphere interactions.

Individuals may feel motivated to make personal life changes to support climate action. Reducing their carbon footprint by reducing meat consumption, or adopting a vegan diet and minimalistic lifestyle, are examples. Awareness of the difference in the amount of water and GHG emissions associated with the production of 1 kg of beef or lamb compared to vegetables has encouraged many people to make different food choices. While such individual changes may be short-lived or insufficient to effectively contribute to reversing the effects of climate change, they do yield positive results at local levels.

Reactions to climate events can include economic risks, such as the changing cost of capital, more frequent and more successful litigation, policy change increasing the cost of doing business, a reduction in customers, or difficulty in sourcing suppliers who can comply with emissions targets.

Civil society in Kenya has developed its own response to climate change. Organizations such as Kenya Climate Change Working Group (KCCWG) and One Acre Fund act at different levels to promote climate action. KCCWG describes itself as a 'national network of Civil Society organisations uniting voices and action on climate change'. It was formed in Kenya in 2009, with support from partners such as CAFOD, Oxfam, UK Aid and WWF. It advocates climate change action and aims to bridge the gap between local communities and central government. It conducts 'climate hearings', for example, Kenya Vision 2030, which are ways to form policies for meaningful action to curb climate change. It is difficult to assess the success of such non-governmental organizations, therefore this is a case of lobbying. The government, however, published the 'Kenya National Adaptation Plan 2015–2030', which mentions the contributions of the civil societies.

The One Acre Fund in Kenya is a non-profit social enterprise that funds and trains people with small farms (smallholders) to assist them in moving out of poverty. Farmers are provided with basic necessities such as seeds, fertilizers and bags to transport crops, to ensure productive farming. One Acre also provides health and hygiene facilities to families of farmers.

ATL ACTIVITY

Investigate the mitigation and adaptation policies for climate change used by the regional or national government where you live.
- Conduct independent research into the climate context of the region/country where you live. Consider the geographic, climatic and socioeconomic factors influencing climate change in your region/country.
- Collaborate in small groups to identify and compile a list of mitigation and adaptation policies implemented in your region/country.
- Each member of the group can choose a different sector, such as agriculture, energy, health.
- Analyse the effectiveness of policies from the same sectors that are implemented in different regions/countries. Think of factors such as public awareness, stakeholder engagement, and the actual impact on reducing emissions or enhancing resilience. For example, different groups could speak about the mitigation or adaptation policies for the agriculture sector.
- Reflect on the perspectives of different stakeholders involved in making, or affected by, these policies.

(HL) 6.3.7 The UN has played a key role in formulating global strategies to address climate change

Global strategies to address climate changes have been extensively supported by the United Nations, with several key agreements resulting from UN summits. The UN Framework Convention on Climate Change (UNFCCC) was adopted in the 1992 Rio Earth Summit and was put into effect in 1994. The Convention states:

> The ultimate objective of this Convention and any related legal instruments that the Conference of the Parties may adopt is to achieve, in accordance with the relevant provisions of the Convention, stabilization of greenhouse gas concentrations in the atmosphere at a level that would prevent dangerous anthropogenic interference with the climate system.

The UNFCCC aims to achieve stabilization in such a way that ecosystems can adapt to climate change naturally and that there is no hindrance to food production, also allowing sustainable economic development to occur. It works through the Intergovernmental Panel on Climate Change (IPCC) and annual Conference of the Parties (COP) summits. However, it has not been able to slow down GHG emissions.

The Kyoto Protocol

The Kyoto Protocol was signed COP 3 which took place in Japan in 1997. The European Community and 37 MEDCs and economies in transition, committed to reduce their GHG emissions by an average of 5 per cent against 1990 levels over the period from 2008 to 2012. This would have led to an overall reduction of 20 per cent by 2012. The agreement was signed by 183 countries but some, including the USA, did not sign. They argued that industries in India and China, which had not agreed to reduce emissions, would an have unfair advantage. Although Canada and Australia signed the treaty, they did not implement it.

Under the 2012 Doha Amendment to Kyoto Protocol, countries committed to reduce their greenhouse emissions 18 per cent below 1990 levels by 2020.

The countries that signed the protocol monitor emissions and track all 'carbon credits' earned through:
1 Emissions trading: See Topic 6.3.6, page 576.
2 The clean development mechanism: Implementing an emission-reduction project (such as setting up a wind or solar energy harnessing system to provide electricity in a rural area) in a developing country.
3 Joint implementation: Investing in emissions-removal projects in another country.

The Kyoto Protocol also established an adaptation fund to support adaptation projects and programmes in developing countries.

The Kyoto Protocol proved beneficial to cutting down greenhouse emissions to some extent, it did not work well at the global level. Many governments put policies and legislation in place but implementation was a problem; very few businesses made environmentally conscious manufacturing choices. In 2015, the UNFCCC parties signed the Paris Climate Agreement, which effectively replaced the Kyoto Protocol.

COP

The Conference of Parties (COP) conventions are formal meetings of the 197 UNFCCC parties. COP reviews the measures each country has taken to reduce emissions and assesses the progress made towards the objective of the convention. The first COP was held in 2015 in Berlin, Germany. COP 28, held in Dubai, UAE, included the first global stocktake to assess progress towards the aims of the Paris Agreement. The agenda of COP 28 was to accelerate action across all areas by 2030. Transitioning away from fossil fuels is the most important target.

The IPCC

The Intergovernmental Panel on Climate Change (IPCC) was created in 1988 by the World Meteorological Organization (WMO) and the United Nations Environment Programme (UNEP) and now has 195 members. The IPCC evaluates scientific information related to things such as vulnerability to and impacts of climate change, projected climate predictions, and as mitigation and adaptation strategies. It reports its finding to policymakers to help them develop climate policies. International climate change negotiations rely on information published in the IPCC Assessment reports. Synthesis reports summarize key points from several more detailed reports on specific aspects of climate change. The synthesis of the sixth assessment report (AR6) was published in March 2023.

The main findings of AR6 are:
- Deforestation, urbanization, land use change and agriculture have put species at greater risk of extinction. This risk will increase dramatically if global warming exceeds 1.5°C–3°C in the future.
- Human-induced climate change is linked to potential loss and damage.
- The average global temperature in 2011–2020 was around 1.1°C higher than in 1850–1900.
- The frequency of extreme climate events is increasing. Heatwaves, rainstorms and droughts are becoming more prevalent.

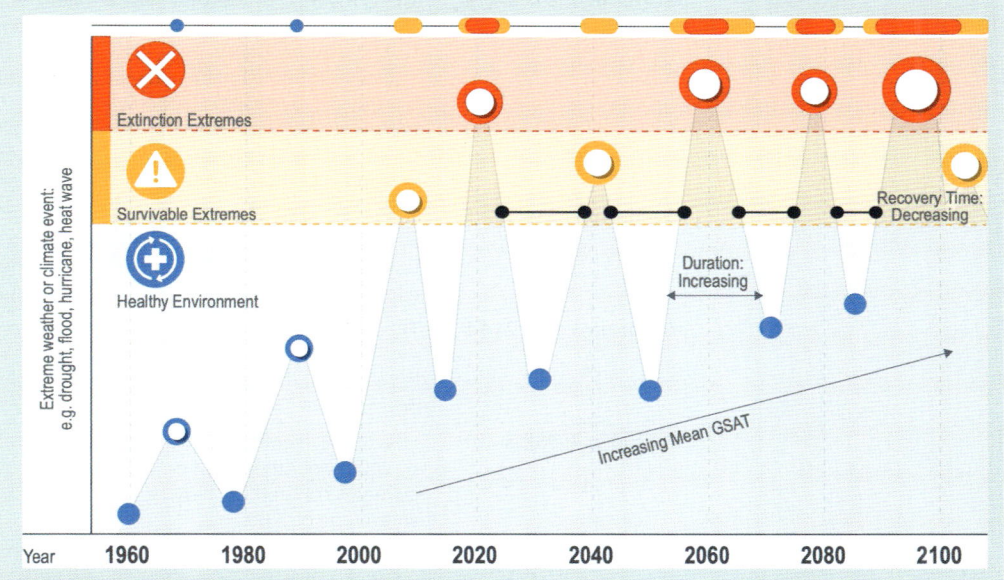

■ **Figure 6.51** How extinction risk is affected by changes in the frequency, duration and magnitude of extreme weather or climate events

- Climate change is causing ocean acidification, increased temperatures, changes to rainfall and runoff patterns, greater variability in river flow and water levels, and rising sea levels.
- The burden of climate change is unequally distributed.
- Increased extreme climate events are causing the spread of diseases.
- As temperature increases, the number and intensity of wildfires will increase, along with tree mortality, carbon loss, biodiversity loss and structural change.

The Kigali Amendment

The Kigali Amendment is the eighth amendment to the Montreal Protocol (the treaty for phasing out ozone depleting substances or ODSs), which came into force in 2019. 197 countries signed the agreement and committed to phasing down the production and consumption of hydrochlorofluorocarbons (H-FCs) by 80–85 per cent by 2045. HFCs replaced CFCs as they are less harmful to the ozone layer, but they are powerful greenhouse gases.

6.3 Climate change – mitigation and adaptation

(HL) 6.3.8 The IPCC has proposed a range of emissions scenarios with targets to reduce the risk of catastrophic climate change

GHG emissions have increased for many decades and efforts to reduce them have not been very successful so far. The level of GHG emissions in the future will depend on demographic changes, industrialization, urbanization, technological advancements and the level of fossil fuel use. The IPCC's sixth Assessment Report (AR6) gives five possible scenarios for the future of GHG emissions driven by anthropogenic actions. Each scenario describes a shared socioeconomic pathway (SSP). Numbers in the scenario title indicate how much extra energy GHG emissions add to the Earth's system. The greater the number, the higher the temperature rise. The five SSPs cover a range of challenges that arise in climate mitigation and adaptation, and indicate our ability to stabilize temperature increase due to GHG emissions.

■ Table 6.10 The five scenarios for shared socioeconomic pathways

Scenario with radiative forcing (W/m²)	Description
Scenario 1: SSP 1–1.9	This is the IPCC's most optimistic scenario, which aims to cut down CO_2 emissions to net-zero by 2050. This will be possible only if societies shift to sustainable practices in every area, the use of fossil fuels is drastically reduced, investments in the health and education sector increase, and overall well-being is improved. By keeping global warming to 1.5°C above the pre-industrial era, there is a possibility that slowly we will be able to bring global warming down to 1.4°C if we continue the sustainable practices. Even though the extreme weather conditions cannot be stopped, the ability to be prepared will increase.
Scenario 2: SSP 1–2.6	Global CO_2 emissions are greatly reduced but over a long period of time, reaching net-zero after 2050. Society shifts to sustainability as in Scenario 1, but temperatures stabilize around 1.8°C above the pre-industrial era by the end of the century.
Scenario 3: SSP 2–4.5	The CO_2 emissions remain mostly the same and begin to decline in the mid-century. However, the net-zero target is not reached by 2100. Societies have not adopted sustainability well and there is uneven income growth and development. Temperatures rise by 2.7°C by the end of the century.
Scenario 4: SSP 3–7.0	Steady rise in emissions and temperatures with almost double emissions of CO_2 by 2100. Countries focus on themselves for national and food security. This increases competition between them for resources. Temperatures stabilize at an average of 3.6°C rise.
Scenario 5: SSP 5–8.5	CO_2 emissions almost double by 2050. Fossil fuels thrust the global economy forward. Lifestyles are highly energy-intensive. The average global temperature has increased by 4.4°C by 2100.

It is difficult to predict which scenario is most probable because it all depends on how human societies choose to act. If we stopped all anthropogenic emissions at once, the Earth would take time to get back into balance and we would continue to experience extreme weather conditions for some time. The Arctic ice would continue to melt and add freshwater into the oceans, and sea level would continue to rise. The extent of the impact can only be assessed in the future.

(HL) 6.3.9 Technology is being developed and implemented to aid in the mitigation of climate change

Electric vehicles may reduce the use of fossil fuels, however there are several problems associated with them including, e-waste from their lithium batteries. Hydrogen fuel cells and metal hydride batteries may provide an eco-friendly alternative. These batteries have a long lifespan and high energy density.

Smart grids are electricity distribution networks that use digital technologies, software and sensors to match the supply of electricity to demand. They can reduce carbon emissions by transmitting electricity efficiently, reducing power losses in the grid and limiting the use of fossil-fuel power plants. Their use can make electricity cheaper for consumers by minimising the cost of producing and transmitting it. Smart grids help ensure a stable and reliable supply because they include components that can identify and address problems before they cause outages. In the United States, smart meters that can form part of a smart grid are found in more than 75% of households.

■ **Figure 6.52** A smart grid

Computers are an integral part of our life and are embedded in many elements that make up our cities, including buildings, streets, cars and electricity and water networks. They provide a variety of real-time data and allow us to understand how a city functions This, in turn, enables us to improve facilities and make the city more liveable, efficient and sustainable. A smart city uses various electronic devices and sensors to collect specific data and monitor changes. London, Amsterdam, Singapore, New York and Copenhagen are some well-known developing smart cities.

REAL-WORLD EXAMPLE

London: a smart city

London, UK, is a smart city. Technology controls and connects almost all aspects of the city. There are many 5G towers and electric vehicle (EV) charging points. Citizens can easily locate the nearest EV charging station using their smart phones. Its green infrastructure includes green roofs, living walls and easily accessible outdoor green spaces. Real-time control and sensors support payment for services, intelligent lighting and more efficient energy supply. Lampposts in London's streets are connected to the internet and many continuously monitor air pollution. Bunhill energy centre uses technology to harness residual heat from the underground train system to heat 700 homes, schools and a leisure centre. Smart phones or smartwatches can be used to plan and pay for journeys on public transport, see when the next bus or tube train is due and hire bikes and scooters. Free downloadable apps provide information on walking routes around the city. The Ultra Low Emission Zone tackles traffic congestion with the help of surveillance cameras. Cameras covering different locations in the city, allowing monitoring of emergencies. Locating recycling centres or bins has also become easier by the connection of the Internet of Things to mobile phones has made it easier for citizens and visitors to locate bins and recycling centres.

■ **Figure 6.53** Signage denoting the start of London's Ultra Low Emission Zone (ULEZ)

(HL) 6.3.10 There are challenges to overcome in implementing climate management and intervention strategies

The success of climate management and intervention strategies depends on collective acceptance, collaborative mitigation approaches and equitable partnership. There are some challenges that may result in a lack or incomplete implementation of management strategies, as detailed below.

Difference in perspectives

Climate scientists research and collect evidence that climate change is happening and is in part caused by anthropogenic actions. Nonetheless, a very small fraction of scientists suggest that the changes are a part of the Earth's evolution (see Topic 6.1.9, page 525) and have nothing to do with human actions. Other climate sceptics believe that the contribution of anthropogenic actions to global warming is negligible compared to the Earth's natural processes that emit GHGs.

There are generational differences in perspectives towards climate change. A study conducted by the University of Göttingen with young German citizens aged 15–29 years revealed that half of the young individuals were highly aware of climate change. These individuals also displayed a strong influence of this awareness on their diet choices.

People living in coastal and low-lying areas, who bear the harsh consequences of climate change, strongly believe that there is a drastic change in the Earth's environment and that it is becoming worse every year. People living inland and on uplands, who might not face the same severities of climate change, may not understand the need to act urgently (see Topic 6.2.9, page 547).

Theme 6: Atmosphere and climate change

Public opinion on climate change depends on a wide range of factors, including demographic changes, sociopolitical situations, economic stability, resource availability, social media groups and news reporting by the media. Most people look at climate change as an emergency and are willing to do something about it, but there are few people who actually take necessary actions at an individual level.

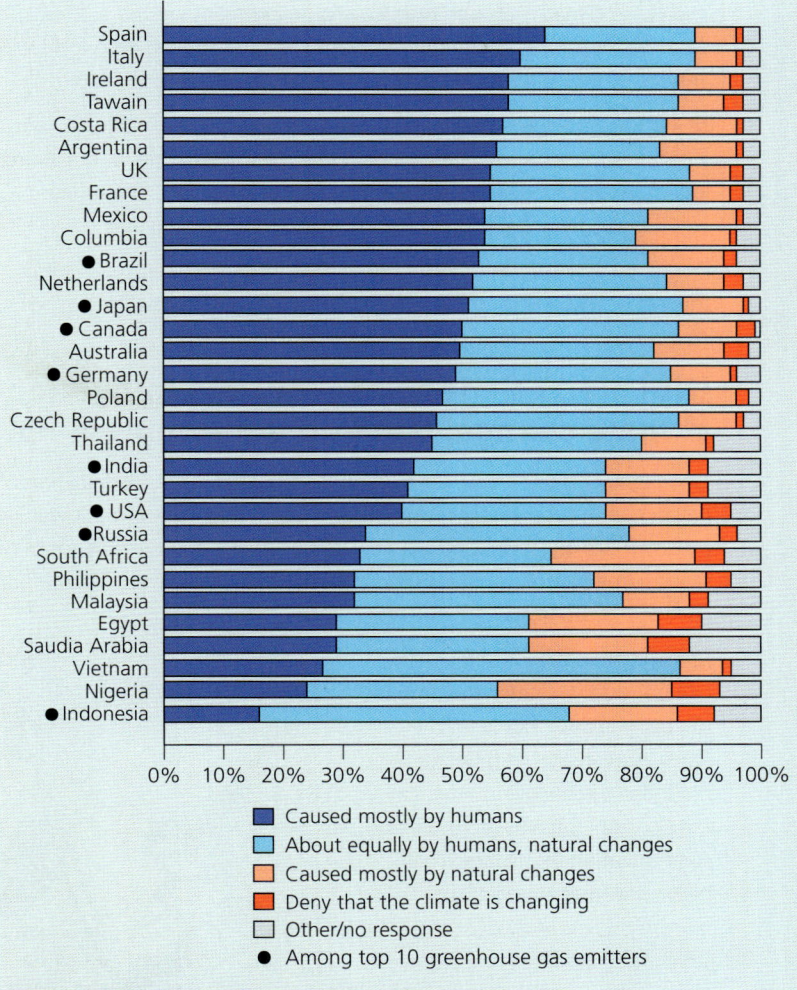

■ Figure 6.54 The world's perception of climate change

Financial constraints

Implementation of climate policies requires financial support, lack of which delays the process and sometimes even leads to complete failure to act. Anthropogenic climate change can be difficult to address, therefore solutions are needed that can be monitored and reported on. This requires the use of measurable variables. For instance, the increased use of renewable energy sources to replace fossil fuels requires technological knowhow and huge initial costs for set-up. Even though some poorer nations are aware of climate change consequences and are willing to take action, financial restraints do not permit them to transition to cleaner energy sources. Global collaboration, whereby developed nations support developing nations financially and technologically, and help them to meet their targets of reducing emissions, is key in such cases.

Lack of leadership from stakeholders

Policies that promote mitigation do not necessarily support fast economic growth and reduction of poverty. Institutional bottlenecks may act as barriers to successful planning and implementation of climate policies. Planning requires clear goal setting and a stepwise approach to prepare for the consequences or reduce the causes of climate change. The lack of well-informed personnel and weak climate research may leave a government lacking sufficient data to base its policies on. Additionally, a lack of commitment from national leaders to prioritize and advocate climate initiatives impedes the implementation of policies for climate management.

NGOs have their independent agendas and sometimes these may not be completely aligned with a nation's goals for climate change. Often NGOs struggle with resources and the financial and logistical support needed to create impactful initiatives. Sometimes NGOs do not have adequate scientific knowledge to convince people to join their initiatives.

Private companies, despite the growing emphasis on corporate social responsibility (CSR), may lack the will or leadership to implement sustainable business practices, which ultimately adds to the global carbon footprint.

(HL) 6.3.11 Geoengineering is a mitigation strategy for climate change, treating the symptom not the cause

Geoengineering refers to the deliberate and large-scale modification of the Earth's energy balance to reduce the rise in temperatures and stabilize it at a lower level than it would otherwise attain. There are several methods used to achieve this, such as space mirrors, ocean fertilization, stratospheric aerosols, cloud seeding, and burning biomass with carbon capture and storage (see Topic 6.3.3, page 567).

Space mirrors are satellites that can be used to reduce the amount of solar radiation reaching the Earth. Sun shields or reflectors made from reflective plastic film are positioned in space, to deflect solar radiation into space. They be deployed as a single large mirror or as an array of several small mirrors. Placing the right size of space mirror at the correct angle requires precision. Sending these mirrors into orbit and deploying them is challenging. This method requires a lot of energy for construction, transport and maintenance, including ground-support operations.

■ **Figure 6.55** The James Webb space telescope mirror

Historically, it has been observed that the summer temperatures of places with recent volcanic eruptions were cooler. This is attributed to the presence of sulfates in volcanic ash, which block sunlight. Stratospheric aerosol injection releases a range of tiny reflective particles in the stratosphere to scatter sunlight back into space. Climate-model simulation studies reveal that this technology holds potential to prevent further warming of the Earth and stabilize the temperature. Reflective sulfate aerosols are sprayed in the stratosphere with the help of high-altitude airplanes, tethered balloons, blimps or artillery. This technology is also considered environmentally friendly because the sulfate particles that are added into the atmosphere are temporary and can easily be

Theme 6: Atmosphere and climate change

removed from the system through the sulfur biogeochemical cycle. Besides sulfate, other reflective aerosols such as black carbon, metallic aluminium, aluminium oxide and barium titanate are also being considered. The main drawbacks of this technology are that it reduces cloud formation, thus affecting precipitation. Crop growth and replenishment of water bodies may be impacted, leading to drought and starvation. However, these problems can be handled with proper planning and management of food and water resources.

Clouds are made up of tiny water droplets or ice crystals that condense around dust or salt particles to form rain or snow. Precipitation cannot occur without these particles, which are called ice nuclei. Cloud seeding is a technique of weather modification by the addition of ice nuclei into subfreezing clouds in order to improve their ability to produce rain or snow. The nuclei provide a foundation for the formation of snowflakes. Silver iodide (AgI) is the most commonly used seeding agent. Other salts used for seeding are sodium chloride (NaCl) and calcium chloride (CaCl). Cloud seeding can be done using ground-based generators, which can be remotely controlled, or using aircraft.

Cloud seeding is regularly used in Russia and Thailand to fight heatwaves and wildfires. The USA, China and Australia, on the other hand, use the technology for drought mitigation. In the dry land of the United Arab Emirates, the technology is used extensively to improve agriculture. Although cloud seeding seems easy and is being used in many places to battle the consequences of climate change, there are environmental implications associated with it that cannot be ignored. It may cause excessive precipitation leading to floods in some areas, while induced rainfall in one place may lead to dry winds that cause drought in other places. It may also lead to potential health hazards due to accumulation of particulate matter in the atmosphere. Overuse or mishandling of silver iodide might trigger a chain reaction of environmental pollution.

Oceans are considered useful carbon sinks. There is approximately 35,000 gigatonnes of carbon in the deep oceans, compared to 750 gigatonnes in the atmosphere. Although the transfer of CO_2 into the deep ocean is a very slow process, it is efficient and has unlimited capacity. Phytoplankton are the microscopic producers in the marine ecosystems. Their population depends on the availability of nutrients such as nitrogen, phosphorus and iron. Ocean fertilization is the purposeful addition of these plant nutrients into the upper, sunlit layers of the ocean to increase photosynthesis in phytoplankton and other producers. This process utilizes dissolved CO_2 and increases its uptake from the atmosphere. The death of phytoplankton causes biomass to sink to the bottom and accumulate in the depths of the oceans, where it remains for several hundred years.

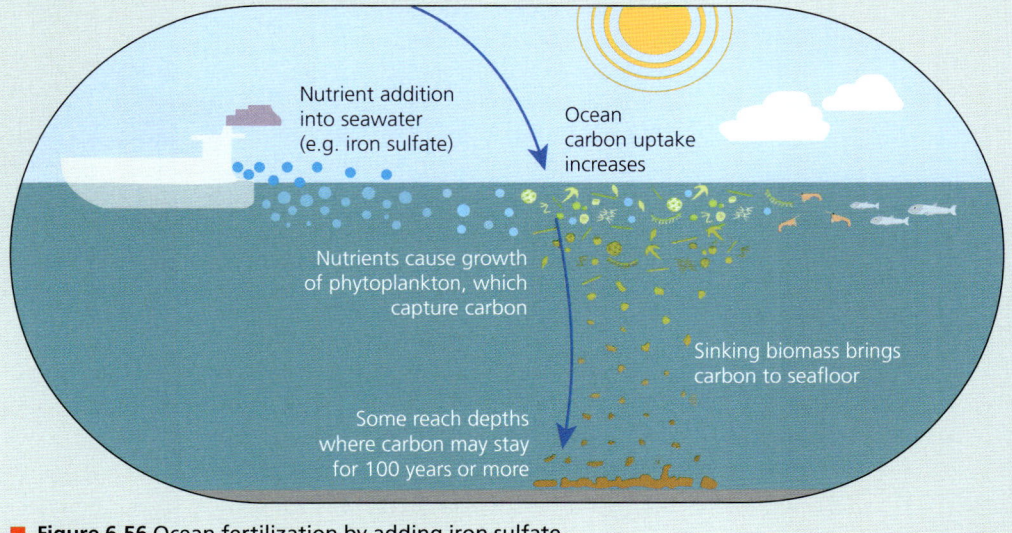

■ **Figure 6.56** Ocean fertilization by adding iron sulfate

6.3 Climate change – mitigation and adaptation

Research and development in climate-action strategies have come up with new technologies for the efficient reduction and removal of carbon emissions. Carbon dioxide removal (CDR) technologies are processes that manually remove CO_2 from the atmosphere and store it. These include:

- Direct air capture: This technology captures CO_2 directly from the atmosphere for storage or utilization. A fan draws air into the collector, from where it passes through a filter that selectively traps CO_2. When the filter is completely full of CO_2, the collector closes and the temperature is raised to 100°C. This causes the CO_2 to be released so it can be collected and permanently stored underground.

■ **Figure 6.57** Direct air capture

- Enhancing carbon sinks: Natural CO_2 sequesters such as soil, vegetation and the oceans are enhanced under programmes such as the UN's Reducing Emissions from Deforestation and Forest Degradation in Developing Countries (REDD+) created in 2008. The REDD+ programme safeguards existing forests in a relatively cost-effective manner. However, these strategies are effective only for the duration of the programme and may result in re-emissions once the project is over.

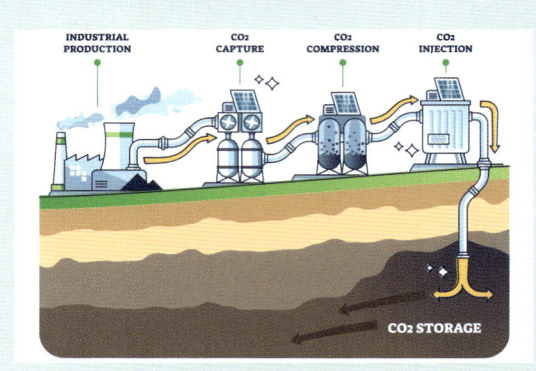

■ **Figure 6.58** Carbon capture and storage

- Carbon capture and storage (CCS): This technology involves capturing and separating CO_2 from other gases that are emitted by industries such as steel and cement production or from power plants where fossil fuels are burned to generate energy. It also captures gases that leak from geological formations. Captured CO_2 is then transported to a compressing unit, where it is compressed and stored by burying it deep underground, such as above the bedrock, deep inside the oceans or even under the ocean floor. The locations are chosen so that the stored CO_2 may not escape but remain there for millions of years. The Boundary Dam Power Station in Saskatchewan, Canada is the world's first CCS project. It is constructed on a coal-fired power plant and started functioning in 2014.

■ **Figure 6.59** The Boundary Dam Power Station in Saskatchewan, Canada

■ Table 6.11 Geoengineering techniques

Geoengineering technique	Examples	Chemical questions to research	Political questions to investigate
Carbon dioxide (or other GHG) removal. These techniques address the root cause of climate change.	Land-based carbon dioxide removal includes afforestation, air capture, and bio-energy with carbon sequestration.	How much CO_2 can be removed by these techniques, and over what time period?	Is it preferable to deal with the causes or the consequences of climate change?
	Ocean ecosystem methods include ocean afforestation and ocean fertilization.	How can the rate at which CO_2 is removed from the ocean surface be increased, and what are the likely environmental side effects of these techniques?	How much risk is too much risk?
Solar radiation management. These techniques attempt to offset the effects of increased GHG concentrations.	Surface albedo enhancement (by painting roofs white in urban areas, planting more reflective crop varieties, and covering deserts with a reflective surface) could make the Earth's surface brighter.	How can increasing the reflectiveness of the planet (albedo) cool the Earth?	Is it the responsibility of individuals, communities or governments to paint roofs?
	Shields or deflectors located in space could reduce the proportion of sunlight reaching the Earth.	What temperature drop can be achieved using this method (and how is this known)?	Who should fund and control these expensive shields and deflectors?
	Sulfate aerosols could be injected into the lower stratosphere to reflect sunlight before it reaches the Earth's surface, mimicking volcanic eruptions.	How much solar radiation can be reflected by sulfate aerosols?	How can decisions with global consequences be made fairly?

(HL) 6.3.12 A range of stakeholders play an important role in changing perspectives on climate change

Climate change affects everyone, although not equally. The stakeholders of climate change include people, businesses and organizations that contribute to climate change, and the policymakers, decision makers, media and people who are affected by the consequences of climate change. An individual's perspective on climate change can be influenced by many factors – media, culture, education, local community groups and NGOs, and/or the influence of charismatic personalities.

The media has considerable influence on public opinion, and the way climate change is reported leaves a strong mark on people's perception. People often resort to social media to bridge the gap between scientific consensus and public opinion. Very often, an individual's choice of media guides the perspective that they may develop, due to constant exposure to a certain type of information. Some media and social media platforms are known to focus on certain views of climate change, while some others are prejudiced towards a particular political party or cause. Some right-wing sources may deny man-made climate change or minimize the scale of it. Media that support this view may accentuate the idea of climate change being a natural process. Climate change is a threat that is difficult to perceive and this may cause humans to delay in taking action since the risks are not seen as immediate threats.

Culture has influenced our understanding of the environment since the very inception of humankind. Over the years, humans have left their impression on nature through the way they have used natural resources. Historically, humans have benefited from the environment, and in return they have maintained the ecosystems to be able to extract the benefits for longer durations. Culture provides a platform for individuals and communities to explore their relationship with nature and ingrain practices that support coexistence. Indigenous knowledge systems hold valuable information about ecosystems and can provide feasible solutions to many issues.

Education is a powerful tool that changes people's attitudes. It allows individuals to understand the complex nature of climate change and to take necessary actions to mitigate and adapt to the effects. Italy was the first country to make climate education compulsory in schools in 2019. This is beneficial because individuals who have this education will monitor their actions and make environmentally friendly decisions, thus contributing to the reduction of greenhouse gases. Awareness campaigns also aid in spreading relevant information about climate change and its effects and how to cope with it. Community-based initiatives involve entire local communities in taking action for climate change.

NGOs and local community groups acting at the local level are more closely linked to the needs of the community. Certain local communities are more impacted by the effects of climate change. NGOs working in these communities can offer timely help in adapting to climate change. For example, they can find funds to help the community in a low-lying area to build seawalls to protect itself from flooding. They may run awareness workshops educate people about the harms of climate change and how to prepare for a crisis. NGOs also help with faster resource mobilization in times of crisis. Individuals and groups in close contact with NGOs and local communities develop a reasoned understanding of climate change and, having experienced its harsh effects, they are also willing to take action and adopt a more sustainable lifestyle.

■ **Figure 6.60** Sir David Attenborough

Charismatic individuals and climate activists have influenced many people to accept that climate change is real and that action needs to be taken while there is still time. Documentaries featuring British presenter David Attenborough include *Extinction: The Facts*, which talks about existing dangers of mass extinction, and *The Truth About Climate Change*, which discusses global warming and its dangerous consequences. They evoke a sense of responsibility in individuals who view them. Attenborough has a unique and direct way of presenting facts without sugarcoating them. This approach is helpful in presenting the reality to people, who can then make informed choices. Al Gore, the former vice president of the United States of America who founded the Climate Reality Project, has also generated awareness about climate change. His speech on global warming was made into a film, *An Inconvenient Truth* (2006), which inspired many to take action against climate change. Greta Thunberg is a young climate activist from Sweden, famously known for her speech at the COP 24 United Nations climate change summit in 2018. She initiated the Fridays for Future movement, which mobilized millions of people around the world. Her motivating talks convince young people to take control of their future by forcing policymakers to prioritize action against climate change.

■ **Figure 6.61** Al Gore

■ **Figure 6.62** Greta Thunberg

(HL) 6.3.13 Perspectives on the necessity, practicality and urgency of action on climate change will vary between individuals and between societies

Individuals and groups react differently to climate change situations. Some might understand the urgency to act on it right away, while others might think that the drastic effects may not occur in their lifetime, so they can put it off until the last moment. The various dimensions in which these decisions exist is an indication of the complexity of this global issue.

Younger generations often express a sense of urgency as they understand that they will inherit this Earth with all its global issues if they don't do anything about it right now. Young climate activists like Greta Thunberg, Licypriya Kangujam, Txai Suruí, Russell Raymond and many more are demanding a safe future for the next generation. Other individuals, including older generations, exhibit a mix of beliefs ranging from conviction to scepticism and from resistance to immediate action. Individuals belonging to regions that are heavily reliant on traditional industries might resist sudden changes due to fear of economic loss.

Developed countries tend to have more resources to fight climate change than developing countries, although they may face challenges to transitioning efficiently in a short period of time. Nations like Sweden and Germany demonstrate their commitment to climate action by investing in renewable energy and emissions reduction. Developing countries may not be as willing to adopt net-zero emission strategies, arguing that industrialization is important for economic growth. Immediate development needs in some parts of Africa, India, China and Brazil may dissuade these countries from taking immediate steps to embrace decarbonization. They may however, follow a slow phasing out of fossil fuels and replace them with renewable energy sources.

Coastal areas experience direct impacts like sea-level rise and extreme weather events, fostering a stronger sense of urgency. Low-lying island nations like Tuvalu or Kiribati in the Pacific Islands, face imminent threats from rising sea levels, prompting urgent calls for global climate action. Locally managed adaptation plans identify vulnerable areas and gather funds to protect and monitor them. Environmental pressure is one of the main 'push' factors for migration of people from Tuvalu and Kiribati. Migration of people out of coastal areas has increased over the years, hanging the demography of these areas. The displaced community of Kiribati, for instance, find shelter in Australia or New Zealand. In contrast, in regions in Central Asia where the impacts of climate change are fewer, the urgency might be perceived differently, influencing local attitudes toward mitigation measures.

Fossil fuel-driven economies are reluctant to make the shift to decarbonization as it would hamper their economic growth. Fossil fuels are more cost effective than renewable sources such as solar and wind energy. They also do not require any special technical setup for energy production. On the flip side, economies that have transitioned to renewable energy emphasize the benefits of decarbonization.

HL.c.10: Environmental ethics

Consequentialist ethics is the view that the consequences of an action determine the morality of the action. The precautionary principle states that we should take action to prevent harm, even if there is uncertainty about the risks. This principle is often invoked in the context of climate change, as there is still uncertainty about the full extent of the risks.

It can be argued that there must be a fair distribution of resources between current and future generations. Given that current emissions are bound to have a significant impact on future generations, and that these emissions result from the depletion of existing resources, future generations should be left with sufficient resources to be able to deploy them as they would see fit. On the other hand, the natural resources also have interests that need protecting. How to protect those interests in the future is the preservation of future generations. It is not within the remit of the current generation to determine how future generations may or should value nature.

> **ATL ACTIVITY**
>
> **Balloon debate**
>
> Rate the following in ascending order of potential causes of climate change:
>
> - Deforestation
> - Industrial emissions
> - Cattle-generated methane gas
> - Overfishing
> - Plastic waste
> - Transport and vehicles
> - Agriculture
> - Manufacturing goods
> - Consuming too much
> - Packaged food industry
>
> - With a partner, discuss the rationale for any one rating you have each given.
> - Discuss with your partner an item from the list that you feel is not relevant. You may only discuss one item. If you both agree it is not relevant, discard the item and one other item chosen at random.
> - Repeat the process until you are left with only two items.
> - Write a 100–150-word reflection on how the two items left on your list are connected to consequentialist ethics.

Link

The tragedy of the commons is discussed in Topic 3.2.7 on page 231.

(HL) 6.3.14 The concept of the tragedy of the commons suggests that catastrophic climate change is likely unless there is international cooperation on an unprecedented scale

The tragedy of commons refers to a situation when the rational yet selfish actions of certain individuals or groups result in the depletion of a resource that belongs to everyone. According to Garrett Hardin (see Environmental Economics, below), 'the commons' are the finite shared resources such as forests, clean water, clean air and fisheries that are used collectively by many groups of people. For instance, the atmosphere is shared by all living beings. If one nation opts for extensive fossil fuel consumption for economic growth, it reaps immediate benefits in terms of energy and economic development. However, the atmospheric consequences, such as increased greenhouse gas concentrations, are shared globally and affect all nations. Similarly, when a nation takes the initiative to invest in CCS technologies for environmental restoration, it bears the costs and efforts individually. However, the benefits, such as reduced global carbon emissions and a stabilized climate, are enjoyed by all nations.

The tragedy arises when the benefits of conserving a shared resource are much lower than when the resource is owned individually. The distribution of benefits is governed in such a way that everyone gets an equal share irrespective of whether or not they have made efforts for its conservation. Because exploiters also receive the benefits without making any efforts, they continue to exploit. Conversely, nations that have taken necessary climate action still suffer due to the selfish interest of a few nations that continue to pollute the atmosphere and the oceans.

Climate change is a challenge that has consequences that cross borders and affect everyone. There is an urgency for nations to overcome individual interests and demonstrate an unparalleled level of international cooperation through agreements to stop polluters from continuing to pollute without any consequences, and to promote mitigation strategies to achieve net-zero as early as possible.

Global governance supports equitable sharing of burdens through agreements such as the Paris Agreement. Each party to this agreement establishes a Nationally Determined Contribution (NDC) to limit temperature rise. The NDC must be reviewed every five years. The success of the Paris Agreement depends on the continuous support and efforts from the majority of the parties, even if some fail to keep up.

REAL-WORLD EXAMPLE

Coffee consumption

Coffee consumption is an example of the tragedy of commons. Global demand for coffee is putting stress on the production sector. Some varieties of coffee beans have more acclaim than others. Wild coffee plants grow naturally and are commonly shared by the local society. However, overconsumption has led to traditional coffee-growing methods, giving way to coffee plantations. This requires clearing of forest land, which had led to habitat loss and 60 per cent of wild coffee species (including *Arabica*) facing extinction. Traditional methods grow coffee plants in the shade of forest trees, which ensures the berries get just the right amount of heat to ripen and the soil does not become too dry. Coffee plantations leave the plants exposed to the heat of sun, requiring more water for irrigation. According to a report released by the International Centre for Tropical Agriculture, almost 50 per cent of the land used for coffee cultivation will become unsuitable by 2050 due to over-farming.

Coffee beans have to go through various stages of processing (see Figure 6.63) before they can be packaged and sold as coffee. All these stages require water and energy, and contribute to GHG emissions. Dry processing of 1 kg green coffee releases approximately 15 kg CO_2 equivalent, while wet processing releases around 30 kg CO_2 equivalent.

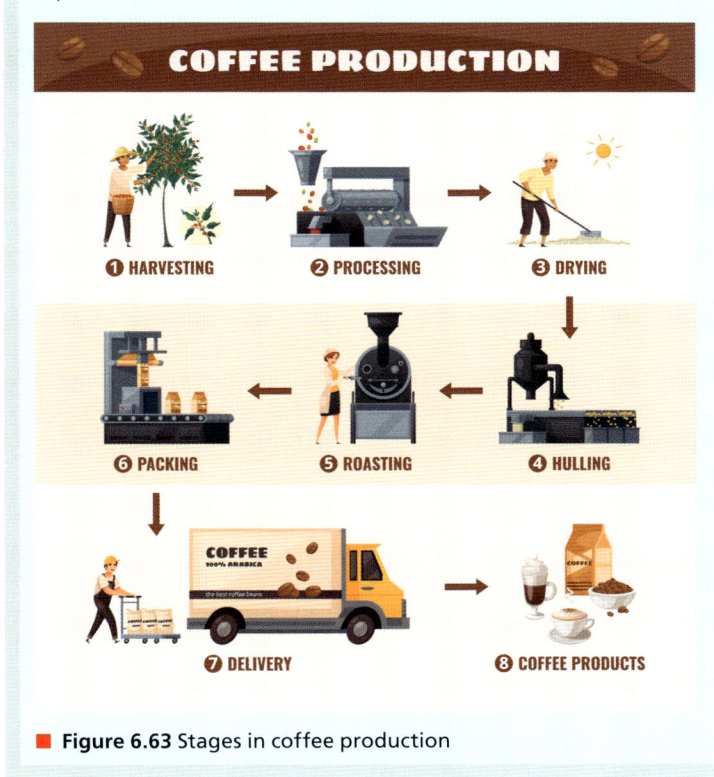

■ Figure 6.63 Stages in coffee production

6.3 Climate change – mitigation and adaptation

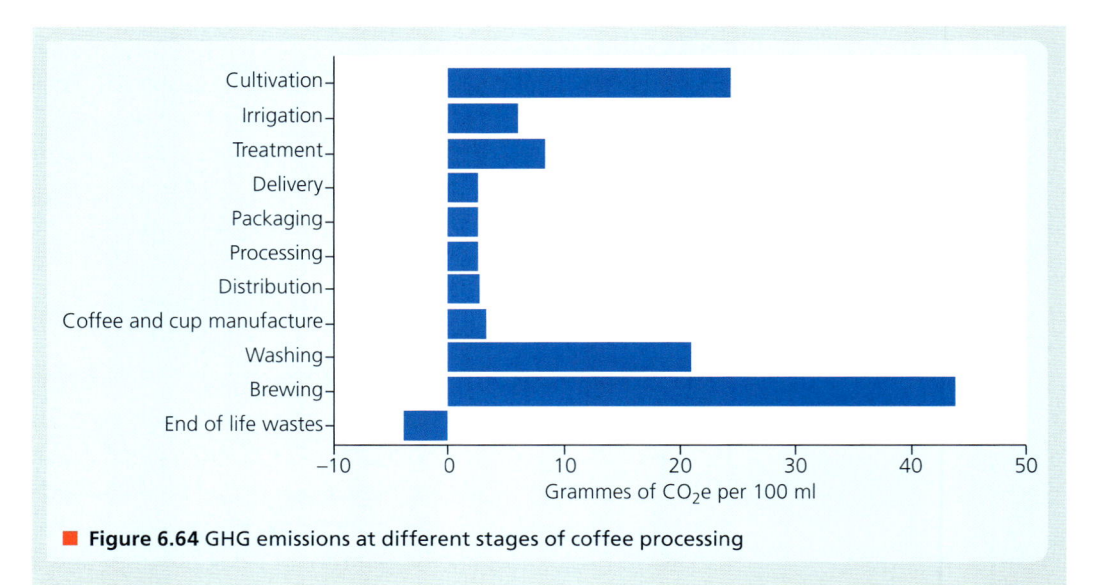

■ **Figure 6.64** GHG emissions at different stages of coffee processing

HL.b.6: Environmental economics

Overconsumption of a particular resource makes it unavailable for others, and leads to accumulation of waste that affects the society in general. This economic theory was proposed in 1833 by William Foster Lloyd. In 1968, Garett Hardin, an American economist and microbiologist, wrote an influential essay titled 'The Tragedy of Commons', which was published in the journal *Science*. This states that short-term interest and benefit override the unsustainable use of a resource and the impact it could have on the environment and public. Individuals exploiting the common resource may have a 'use it or lose it' mentality.

> **ATL ACTIVITY**
>
> Consider a shared resource such as water.
> - Identify how the exploitation of this can be an example of the tragedy of commons.
> - Write down some regulations that would help conserve the resource.
> - Share your ideas with your peers.

HL.c.13: Environmental ethics

The dominance of humans over nature and its exploitation results in some groups being deprived of a common resource. Garett Hardin argues that those who can use a resource for free will continue to consume more of it than those who have to pay for that resource. Although Hardin considered small-scale resources for which the cause-and-effect relationship could be seen immediately, the complex nature of climate change makes the tragedy of commons a temporal phenomenon stretching over a long period of time and making a stronger connection to the dynamic system.

The microeconomics of ethics proposed by preference utilitarians, including Peter Singer, Martha Nusbaum and Amartya Sen, culminated in the idea of nudge theory. This fundamentally argues that instead of trying to actively persuade people to behave a certain way, you gently nudge them towards accepting the behaviour that is in their best interest. Certain preferences count for more than others. To stimulate a certain behaviour, economic, social and cultural preferences should be taken into account and should weigh more heavily than some other preferences. Environmental ethics emphasizes the concept of stewardship, suggesting that humans have a moral duty to responsibly care for and preserve the environment. In the context of the tragedy of the commons, ethical considerations call for a shift from self-interest to collective responsibility to prevent resource depletion.

ATL ACTIVITY

Conduct a spectrum debate using prompts

Examples of prompts are given below:
- Nations should prioritize individual interests over collective action in managing common resources.
- Unilateral actions are more effective than global cooperation in resource management.
- Limited international cooperation might be more effective than comprehensive agreements.
- Global collaboration can be both promising and challenging.
- Certain challenges require collaborative efforts on a global scale.
- Nations can overcome potential challenges to effective international cooperation in resource management.

Respond by choosing any one of the following perspectives and give a reason for your choice. Where relevant, support your stance with examples.
- Strongly oppose
- Moderately oppose
- Neutral/undecided
- Moderately support
- Strongly support

REVIEW QUESTIONS

1 Outline the ways in which anthropogenic actions contribute to increased levels of greenhouse gas emissions.
2 Using appropriate examples, explain the difference between mitigation and adaptation for climate change.

HL ONLY

3 Outline the role of governmental and non-governmental stakeholders in climate action.
4 Discuss the challenges in implementing climate-management and possible strategies.

6.4 Stratospheric ozone

Guiding question

- How does the ozone layer maintain equilibrium?
- How does human activity change this equilibrium?

SYLLABUS CONTENT

This chapter covers the following syllabus content:
- ▶ 6.4.1 The sun emits electromagnetic radiation in a range of wavelengths, from low-frequency radio waves to high-frequency gamma radiation.
- ▶ 6.4.2 Shorter wavelengths of radiation (namely, UV radiation) have higher frequencies and therefore more energy, so pose an increased danger to life.
- ▶ 6.4.3 Stratospheric ozone absorbs UV radiation from the sun, reducing the amount that reaches the Earth's surface and therefore protecting living organisms from its harmful effects.
- ▶ 6.4.4 UV radiation reduces photosynthesis in phytoplankton and damages DNA by causing mutations and cancer. In humans, it causes sunburn, premature ageing of the skin and cataracts.
- ▶ 6.4.5 The relative concentration of ozone molecules has stayed constant over long periods of time due to a steady state of equilibrium between the concurrent processes of ozone formation and destruction.
- ▶ 6.4.6 Ozone-depleting substances (ODSs) destroy ozone molecules, augmenting the natural ozone breakdown process.
- ▶ 6.4.7 Ozone depletion allows increasing amounts of UVB radiation to reach the Earth's surface, which impacts ecosystems and human health.
- ▶ 6.4.8 The Montreal Protocol is an international treaty that regulates the production, trade and use of chlorofluorocarbons (CFCs) and other ODSs. It is regarded as the most successful example yet of international cooperation in management and intervention to resolve a significant environmental issue.
- ▶ 6.4.9 Actions taken in response to the Montreal Protocol have prevented the planetary boundary for stratospheric ozone depletion being crossed.

HL ONLY
- ▶ 6.4.10 ODSs release halogens, such as chlorine and fluorine, into the stratosphere, which break down ozone.
- ▶ 6.4.11 Polar stratospheric ozone depletion occurs in the spring due to the unique chemical and atmospheric conditions in the polar stratosphere.
- ▶ 6.4.12 Hydrofluorocarbons (HFCs) were developed to replace CFCs as they can be used in similar ways and cause much less ozone depletion, but they are potent GHGs. They have since been controlled by the Kigali Amendment to the Montreal Protocol.
- ▶ 6.4.13 Air conditioning units are energy-intensive, contribute to GHG emissions, and traditionally have contained ODSs.

6.4.1 The sun emits electromagnetic radiation in a range of wavelengths, from low-frequency radio waves to high-frequency gamma radiation

Like all other stars, the sun emits electromagnetic radiation. This radiation is emitted in a continuous spectrum, which means that it covers all wavelengths of light, from low-frequency radio waves to high-frequency gamma radiation. A major part of these radiations are invisible to human eyes. The different wavelengths of electromagnetic radiation are associated with different energies. Radio waves have the lowest energy, while gamma rays have the highest energy.

The sun's radiation is strongest in the visible light range, but it also emits significant amounts of radiation in other wavelengths, such as ultraviolet light, infrared light, and X-rays. This radiation is important for life on Earth. Infrared light is absorbed by the Earth's atmosphere and oceans, and it helps to keep the Earth warm. It also plays a role in photosynthesis, as it helps to transfer heat from the sun to the Earth's surface. Visible light is essential for photosynthesis in terrestrial and aquatic biomes.

6.4.2 Shorter wavelengths of radiation have higher frequencies and therefore more energy, so pose an increased danger to life

Short wavelength, high frequency radiation carries more energy and has greater penetrating power making it more harmful. The shortest-wavelength radiation from the sun (X-rays and gamma rays) is entirely absorbed by the atmosphere. Radiation in the UV range has wavelengths, between 100 and 400 nanometres (nm). UV radiation can affect ecosystems, disrupt photosynthesis, alter food chains, damage DNA and cause sunburn, skin cancer and cataracts. Radiation between 315 and 400 nm is called UVA. It makes up about 90 per cent of UV radiation and causes tanning. Radiation with wavelengths between 280 and 315 nm is called UVB and it penetrates deeper into the skin causing the damage to DNA that leads to skin cancer. Radiation between 100 and 280 nm is called UVC. Having the highest frequency and strongest energy, UVC is considered the most damaging and is known to cause cancer and cataracts. More details about the effects of UV radiation can be found in Topic 6.4.4 (see page 596).

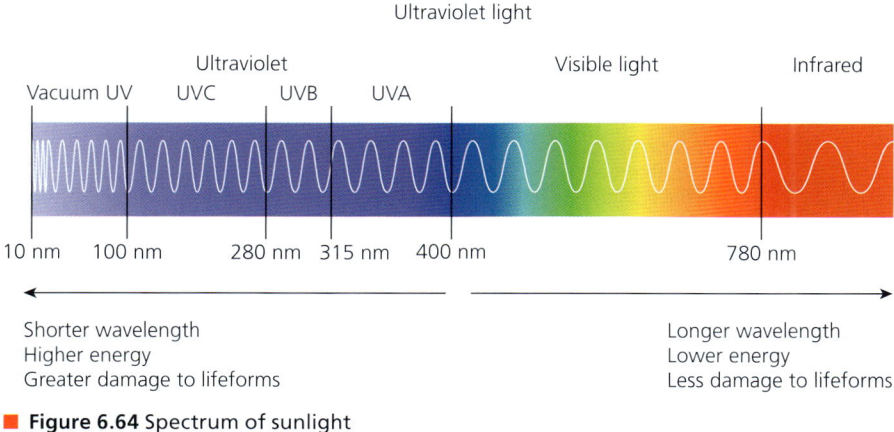

■ Figure 6.64 Spectrum of sunlight

6.4.3 Stratospheric ozone absorbs UV radiation from the sun, reducing the amount that reaches the Earth's surface and therefore protecting living organisms from its harmful effects

Stratospheric ozone envelops the Earth and absorbs harmful UV radiation, preventing it from reaching the Earth's surface. The thickness of the ozone layer varies across the planet and may fluctuate over short time scales. It completely absorbs UVC and most UVB. When ozone molecules (O_3) absorb UV radiation, they split into oxygen molecules (O_2) and oxygen atoms (O), which can combine to reform ozone molecules. The fast and continuous breaking and making of ozone molecules in the ozone layer protects life on Earth from the harmful effects of UV radiation.

6.4.4 UV radiation reduces photosynthesis in phytoplankton and damages DNA by causing mutations and cancer. In humans, it causes sunburn, premature ageing of the skin and cataracts

UVB radiations are harmful to photosynthetic organisms, including plants and phytoplankton. They cause the generation of reactive oxygen species (ROS), molecules that have at least one oxygen atom and one or more unpaired electrons. ROS include superoxide radicals, hydroxyl radicals, singlet oxygen, and free nitrogen radicals. They slow photosynthesis, especially that of phytoplankton in the ocean. This leads to reduced productivity that affects entire food webs.

Deoxyribonucleic acid (DNA) is the hereditary molecule that contains all information for transferring genes from one generation to the other. ROS oxidize DNA bases to cause mutations. This changes the shape of the DNA molecule, affecting protein synthesis or causing uncontrolled

Link

The ozone layer is being depleted by human activities, which release ozone-depleting substances (ODSs). This will be further discussed in Topic 6.4.6 (page 598).

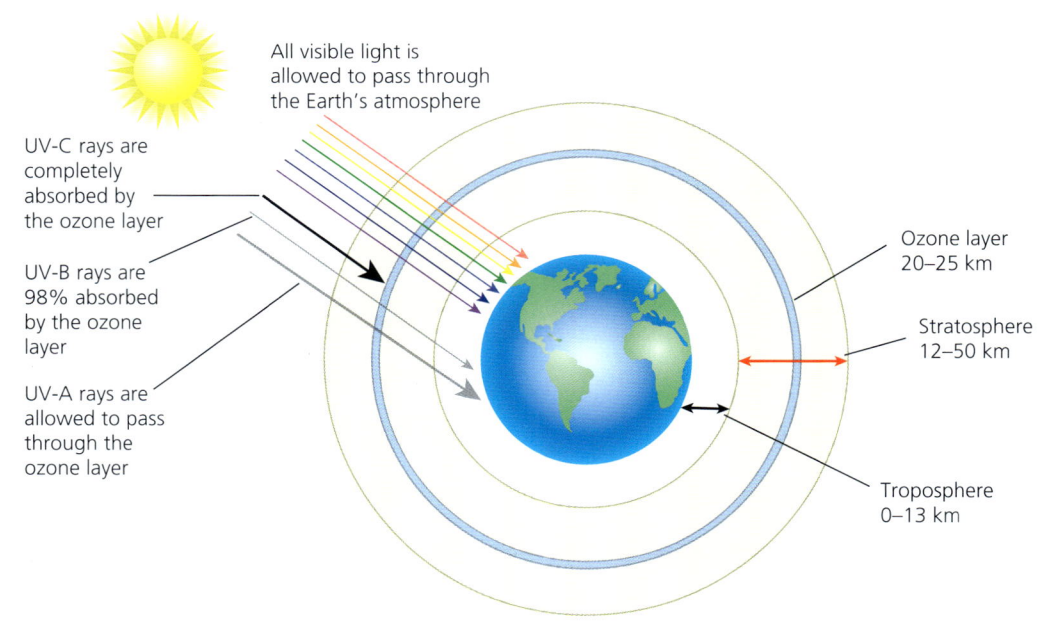

■ Figure 6.65 The ozone layer absorbs the harmful UV rays

Theme 6: Atmosphere and climate change

cell growth that leads to cancer. Overexposure to UV radiations can cause sunburns ROS damage collagen and elastic in the skin, causing premature ageing and sagging of the skin. This is called **photoageing**. Exposure of the eyes to UV radiation can lead to corneal damage, cataracts, and macular degeneration. **Cataracts** occur when the lens protein in the eye is damaged by UV radiation.

- ◆ **Photoageing** – when the sun causes premature ageing of the skin.
- ◆ **Cataract** – the formation of an opaque layer on the surface of the lens in the eye due to exposure to UV radiation.

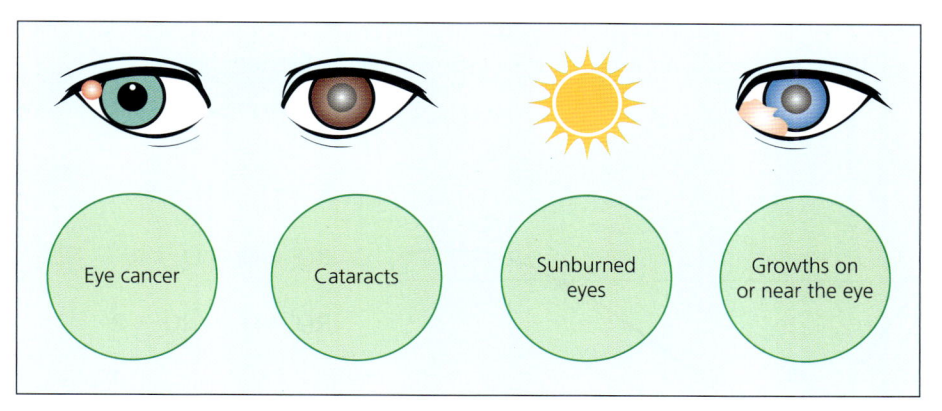

■ Figure 6.66 Ways in which UV radiation can harm the eyes

6.4.5 The relative concentration of ozone molecules has stayed constant over long periods of time due to a steady state of equilibrium between the concurrent processes of ozone formation and destruction

Ozone molecules in the stratospheric ozone layer are constantly being formed and destroyed. UV rays split oxygen molecules (O_2) into two oxygen atoms. Free oxygen atoms (O), are highly reactive and interact with other oxygen molecules (O_2) to form ozone molecules (O_3). This is a reversible reaction that naturally reaches a steady state equilibrium where the rates of formation and destruction are equal. This balance has kept the overall concentration of stratospheric ozone constant over long periods of time.

Common mistake

It is possible to confuse stratospheric ozone and tropospheric ozone.

Stratospheric ozone forms naturally in the upper atmosphere and protects the Earth from harmful UV radiation.

Tropospheric ozone is ground-level ozone formed by reactions between volatile organic compounds (emitted by vehicles) and nitrogen oxide in the presence of sunlight.

Link

The interaction of components of a system to maintain equilibrium is discussed in Chapter 1.2 (page 36), and the way the energy of a system remains constant is covered in Chapter 2.2 (page 120).

> **Common mistake**
>
> Some people think that the ozone hole is the cause of global warming. This is not true. The ozone layer blocks only UV radiation. It does not trap heat.

6.4.6 Ozone-depleting substances destroy ozone molecules, augmenting the natural ozone breakdown process

Ozone is destroyed by ozone-depleting substances (ODSs), which include chlorofluorocarbons (CFCs), hydrochlorofluorocarbons (HFCs), methyl bromide, and carbon tetrachloride. Ozone concentration is measured in Dobson units (DU). The excessive release of ODSs from the 1970s to the early 1990s disturbed the natural steady state equilibrium that kept the concentration of stratospheric ozone constant. This resulted in a rapid fall in levels of ozone and the formation of holes in the ozone layer, especially at the Earth's poles.

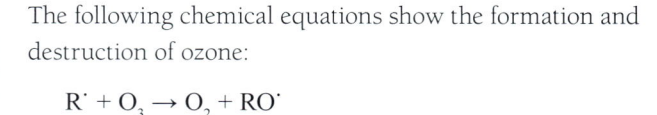

The following chemical equations show the formation and destruction of ozone:

$$R^{\cdot} + O_3 \rightarrow O_2 + RO^{\cdot}$$

$$RO^{\cdot} + O \rightarrow O_2 + R^{\cdot}$$

$$RO^{\cdot} + O_3 \rightarrow 2O_2 + R^{\cdot}$$

Where $R^{\cdot}$ indicates a halogen gas, Cl (chlorine), F (fluorine) or Br (bromine).

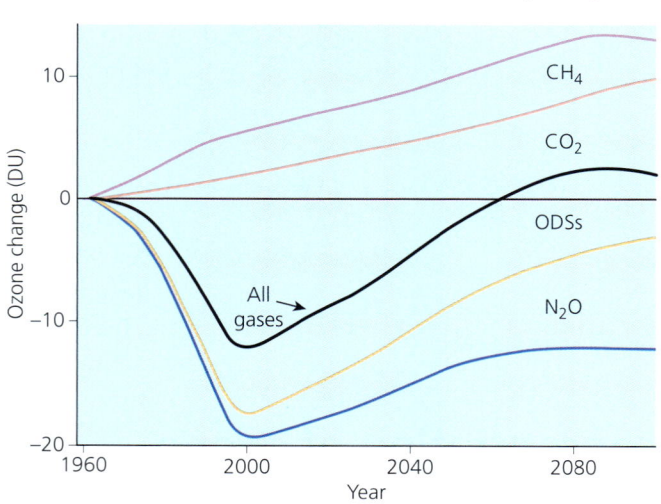

■ Figure 6.67 Changes in ozone in response to ODSs and greenhouse gases

Tool 3: Mathematics

Ozone-depleting substances cause the depletion of the ozone layer. The change in ozone concentration is recorded in Table 6.12.

■ Table 6.12 The change in ozone concentration

Year	Dobson units	Absolute change (DU)	Relative change (%)
1979	225.0	-97.0	-50
1984	163.6		
1989	127.0		
1994	93.3		
1999	102.9		
2004	123.5		
2009	107.9		
2014	128.6	-17.0	-15
2018	102.0		
2020	94.0		
2022	97.0		

1 Calculate the absolute change and relative change for each year.
   [Hint: Relative change % = (New value − Old value) / (Old value) × 100]
2 Plot the data on an appropriate graph using Excel.

6.4.7 Ozone depletion allows increasing amounts of UVB radiation to reach the Earth's surface, which impacts ecosystems and human health

The average level of stratospheric ozone across the globe is around 300 DU. The natural concentration is higher at the poles than at the equator and varies with the seasons. Unique climatic conditions in the Antarctic result in a seasonal decline in the concentration of ozone above the region from August to October each year (see Topic 6.4.11 page 602). This thinning of the ozone layer creates the so-called the ozone hole where ozone levels are limited to about 220 DU. The area, severity and duration of the ozone hole depends strongly on stratospheric weather conditions.

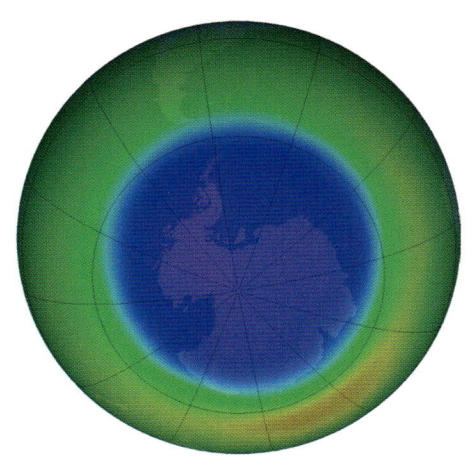

■ **Figure 6.68** The ozone hole above the Antarctic region in September 2023

Link
See Chapter 2.2 (page 120) for more detail on ecosystems.

◆ **Coloured dissolved organic matter** – the optically measurable component of dissolved organic matter. It absorbs light in the blue and UV region of the spectrum.

The extent of the largest ozone hole was 28.4 million square kilometers, recorded in September 2000. Thinning of the ozone layer allows harmful UVB radiation to penetrate the Earth's atmosphere and impact human health and ecosystems (see Topic 6.4.4, page 596). UV radiation causes damage to the ecosystem by inhibiting photosynthesis in plants, algae and phytoplankton, triggering the accumulation of ROS, and damaging the DNA of plants. As a result, the primary productivity of the ecosystem is reduced, leading to less energy flow in the ecosystem. Less oxygen is produced, endangering other aerobic species that depend on this oxygen. Overexposure to UV radiation kills nutrient cycling bacteria, thus affecting the nutrient cycle in ecosystems.

However, UV radiation has some positive impacts. It increases the rate at which leaves and other organic matter decay. In aquatic ecosystems UV radiation promotes the breakdown of **coloured dissolved organic matter**, making the water clear and allowing more sunlight to enter.

6.4.8 The Montreal Protocol is regarded as the most successful example yet of international cooperation in management and intervention to resolve a significant environmental issue

Chlorofluorocarbons (CFCs) that contributed to ozone depletion were widely found in everyday products such as aerosols, air conditioners and refrigerators from the 1970s until 1987, when a worldwide ban implemented under the Montreal Protocol resulted in a drop in their use. The Montreal Protocol is a multilateral agreement that regulates the production and consumption of ozone-depleting substances. It was adopted on 16 September 1987 and remains one of the rare treaties to have achieved universal ratification by 197 countries. It went into the implementation phase in 1989.

6.4 Stratospheric ozone

Link

The Kigali amendment of the Montreal Protocol is discussed in Topic 6.3.7 on page 578.

HL.a.3: Environmental law

The Montreal Protocol was adopted in 1987, but it took several years for it to come into force. This was because some countries, such as the United States, were lobbied by industries that were opposed to the treaty. These industries argued that the treaty would be too costly and that it would harm their businesses. Although the protocol was adopted and ratified by all major countries, lobbying efforts of these industries continued. For example, in the early 1990s, the United States tried to weaken the Montreal Protocol by allowing the production of ODSs for use in aerosol cans. This effort was unsuccessful, but it shows how difficult it can be to pass environmental laws in the face of lobbying from powerful industries. These laws can be successful but they need to be strong enough to withstand the pressure from lobbying.

HL.a.4: Environmental law

There are a number of factors that can affect the success of environmental law, including:
- The level of political commitment: Countries with strong political commitment to environmental protection are more likely to be successful in implementing environmental laws.
- The availability of resources: Countries with more resources are more likely to be able to implement environmental laws effectively.
- The level of public awareness: Countries with high levels of public awareness of environmental issues are more likely to be supportive of environmental laws.
- The strength of enforcement mechanisms: Countries with strong enforcement mechanisms are more likely to be able to ensure compliance with environmental laws.

HL.a.9: Environmental law

The Montreal Protocol established the Ozone Secretariat, which operates under the United Nations Environment Programme (UNEP). The Ozone Secretariat oversees the implementation of the protocol, facilitates meetings of the parties, and provides essential support for the treaty's objectives. It serves as a central hub for information and coordination among member nations. In 2016, the Kigali Amendment to the Montreal Protocol was implemented. This focuses on the phase-out of hydrofluorocarbons (HFCs) which were used as replacements for some ODSs eliminated by the original protocol. Although HFCs do not deplete the ozone layer, they are powerful greenhouse gases and, thus, contributors to climate change.

The Multilateral Fund for the Implementation of the Montreal Protocol was established in 1990 and is dedicated to reversing the damage to the Earth's ozone layer. This fund assists developing countries in meeting their commitments to phase out ODSs. It provides financial support and technical assistance for the transition to ozone-friendly technologies.

The protocol suggested that the production of ODSs be phased out over a period of time. The reasons why the Montreal Protocol has been so successful are:

- Early action: The Montreal Protocol was adopted in 1987, even though the scientific evidence of ozone depletion was not at that point fully clear. This early action gave countries time to phase out ODSs before they caused too much damage to the ozone layer.
- Flexibility: The Montreal Protocol has been amended several times since it was first adopted. These amendments have allowed the protocol to adapt to new scientific information and to address new challenges.
- Monitoring and enforcement: The Montreal Protocol has a number of mechanisms in place to monitor compliance with the treaty. These mechanisms have helped to ensure that countries are meeting their commitments to phase out ODSs.

The protocol provided a framework to companies that were producing ODSs so that they could plan their research in step with the targets of compliance.

6.4.9 Actions taken in response to the Montreal Protocol have prevented the planetary boundary for stratospheric ozone depletion being crossed

Since the Montreal Protocol was adopted, global production and consumption of ODSs has been reduced by more than 99 per cent.

Through a combination of regulations, restrictions, and phased reductions in the production and consumption of ODSs, the Montreal Protocol has led to a substantial reduction in their emissions. This deliberate and cooperative global effort has resulted in a significant recovery of the ozone layer. The decline in atmospheric concentrations of ODSs, notably CFCs, has been instrumental in allowing the stratospheric ozone layer to heal. Scientists expect the ozone layer to recover to its pre-1980 levels by 2050 in the mid latitudes and by 2065 over the Antarctic region. Had the Montreal Protocol not been implemented, the planetary boundary for stratospheric ozone depletion could have been breached, leading to severe consequences for human health and terrestrial and aquatic ecosystems.

Link

See Chapter 1.3 (page 48) to remind yourself about the concept of sustainability and the planetary boundaries model.

HL.a.7: Environmental law

The Montreal Protocol is an example of international cooperation and environmental governance. Parties to the protocol are legally bound to phase out the production and use of ozone-depleting substances such as CFCs. Environmental laws are instrumental in addressing global issues that cross borders. Transboundary pollution can be controlled by a flexible approach like that of the Montreal Protocol, where amendments and adjustments based on scientific assessments have enabled continuous progress in ozone-layer protection.

Various human activities, including industrial processes, transportation, and energy production, release pollutants such as sulfur dioxide (SO_2), nitrogen oxides (NO_x), and particulates into the atmosphere. Air pollutants emitted in one country can travel long distances, crossing international borders and affecting neighboring countries and regions. The Convention on Long-Range Transboundary Air Pollution (CLRTAP) addresses the transport of pollutants between countries and issues like acid rain. This convention was established to resolve the problem of acidification of lakes in Scandinavia and Canada due to SO_2 emitted by other countries.

ATL ACTIVITY

Research the background, objectives, key provisions, and outcomes of the Montreal Protocol. Prepare a slideshow detailing any one objective of the Protocol and present to your class.

(HL) 6.4.10 ODSs release halogens, such as chlorine and fluorine, into the stratosphere, which break down ozone

ODSs include CFCs, HFCs, hydrobromofluorocarbons (HBrFCs), halocarbons like methyl bromide methyl bromide, methyl chloroform, and chlorobromo methane. These compounds are stable in the troposphere but, in the stratosphere, they are degraded by UV light and release free radicals such as OH (hydroxide), O (oxygen), F (fluorine), Cl (chlorine) and ClO (chlorine monoxide). The chain reaction shown in Figure 6.69 occurs in the stratosphere. It causes a buildup of the free radicals that deplete much of the ozone layer.

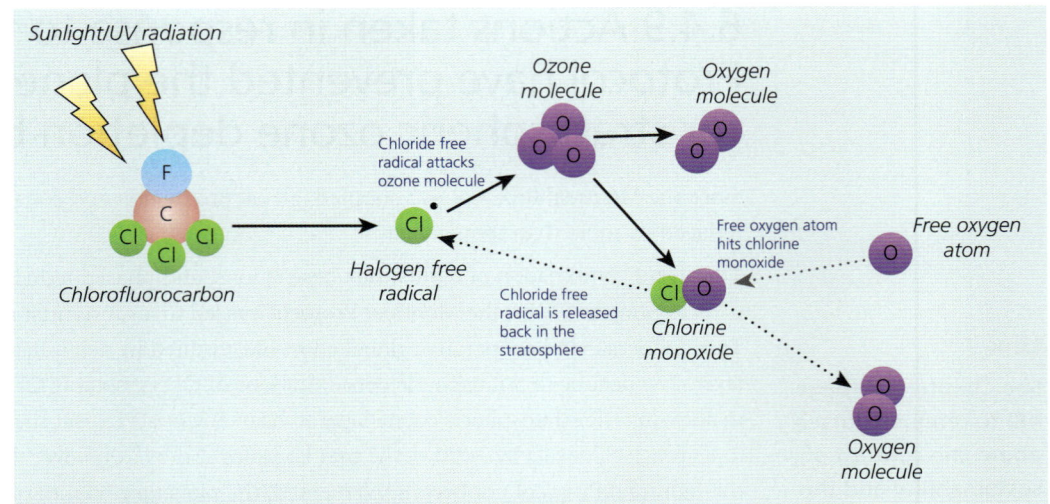

■ **Figure 6.69** How halocarbons can deplete atmospheric ozone

ODSs remain in the atmosphere for varying lengths of time. CFCs have an atmospheric lifespan of hundreds to thousands of years and continue to damage the ozone for as long as they are present. Methyl chloroform and carbon tetrachloride have shorter lifespans and decay quickly. Therefore, despite the phasing out of the ODSs through the Montreal Protocol, residues persist in the atmosphere.

> **ATL ACTIVITY**
>
> Conduct an awareness campaign for protection against UV exposure.
> - Research sun-safety campaigns such as Slip, Slop, Slap, Seek and Slide.
> - Design handmade or digital posters to generate awareness about the harmful effects of UVB.
> - Conduct a sun safety campaign in your school.

(HL) 6.4.11 Polar stratospheric ozone depletion occurs in the spring due to the unique chemical and atmospheric conditions in the polar stratosphere

In the mid-1970s, scientists noticed a significant decrease in ozone levels over Antarctica. They initially thought that the data was incorrect but by 1988 they had confirmed that a hole had formed in the ozone layer over the continent. The ozone hole is caused by a combination of unique atmospheric conditions over Antarctica. At both poles, low pressure creates a strong clockwise wind in the stratosphere, known as a polar vortex. The Antarctic polar vortex is more pronounced and persistent than the Arctic polar vortex.

The temperature difference between the poles and the mid-latitudes causes a change in the atmospheric pressure that results in atmospheric shifting pressure between the poles and the mid-latitudes. This is called the Arctic Oscillation (North Pole) and Antarctic Oscillation (South Pole). When the vortex is stable, the strong stratospheric winds circulate only in the polar regions. This is facilitated by the mid-latitude jet streams moving northwards (see Figure 6.70). However, when the vortex is unstable, for instance due to warming, the circulating winds become weaker. As a result, during the Antarctic spring in September, warmer winds and ODSs enter the polar vortex, causing the ozone layer over Antarctica to thin so much that it gives the appearance of a hole.

Antarctica has the coldest winter temperatures on Earth, which can reach −80°C. These cold temperatures allow the formation of polar stratospheric clouds (PSCs). PSCs are made of frozen water vapour and nitric acid. They attract chlorine and bromine molecules, which are released when the PSCs melt in the spring. The chlorine and bromine molecules then react with ozone, destroying it. The polar vortex prevents these chemicals from being diluted throughout the atmosphere, so they continue to destroy ozone over Antarctica. The strength of the polar vortex is directly correlated to the size of the ozone hole. In years with a strong polar vortex, the ozone hole is larger; in years with a weak polar vortex, the ozone hole is smaller.

Understanding the polar vortex

The Arctic polar vortex is a strong band of winds in the stratosphere, surrounding the North Pole 10–30 miles above the surface.

The polar vortex is far above and typically does not interact with the polar jet stream, the flow of winds in the troposphere 5–9 miles above the surface. But when the polar vortex is especially strong and stable, the jet stream stays farther north and has fewer "kinks." This keeps cold air contained over the Arctic and the mid-latitudes warmer than usual.

Every other year or so, the Arctic polar vortex dramatically weakens. The vortex can be pushed off the pole or split into two. Sometimes the polar jet stream mirrors this stratospheric upheaval, becoming weaker or wavy. At the surface, cold air is pushed southward to the mid-latitudes, and warm air is drawn up into the Arctic.

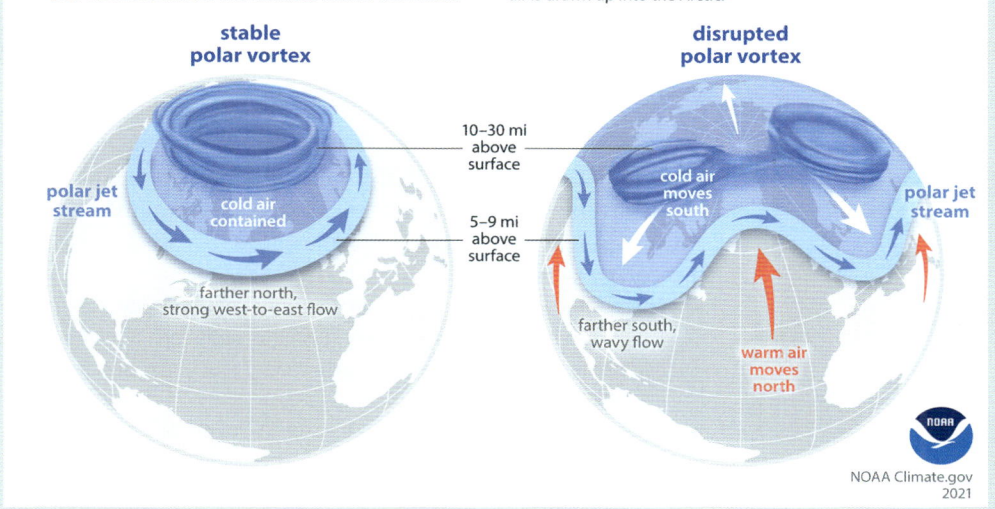

NOAA Climate.gov 2021

■ **Figure 6.70** How the Arctic polar vortex operates

(HL) 6.4.12 Hydrofluorocarbons were developed to replace CFCs, but they are potent GHGs and have since been controlled by the Kigali Amendment

During the phasing out of CFCs, it was thought that hydrofluorocarbons (HFCs) would be a good replacement for them. However, over the years, it has been observed that HFCs also have a negative impact on the atmosphere. Although their ozone-depleting capacity is much lower than that of CFCs, they do have an impact because HFCs have a longer lifespan in the atmosphere. They also have a high global warming potential (GWP), so they can trap more heat and raise the Earth's temperature.

In 2018, HFCs accounted for about 1 per cent of global GHG emissions and this number is expected to increase in the coming years if action is not taken to reduce HFC emissions.

The 2016 Kigali Amendment to the Montreal Protocol was a significant step towards reducing HFC emissions. It aims to reduce HFC emissions by up to 105 billion tonnes of carbon dioxide equivalent (CO_2e) and so help to avoid up to 0.5°C of global temperature rise by 2100. This mitigation approach is one of the most widespread in the world and also one that has received increased support over the years.

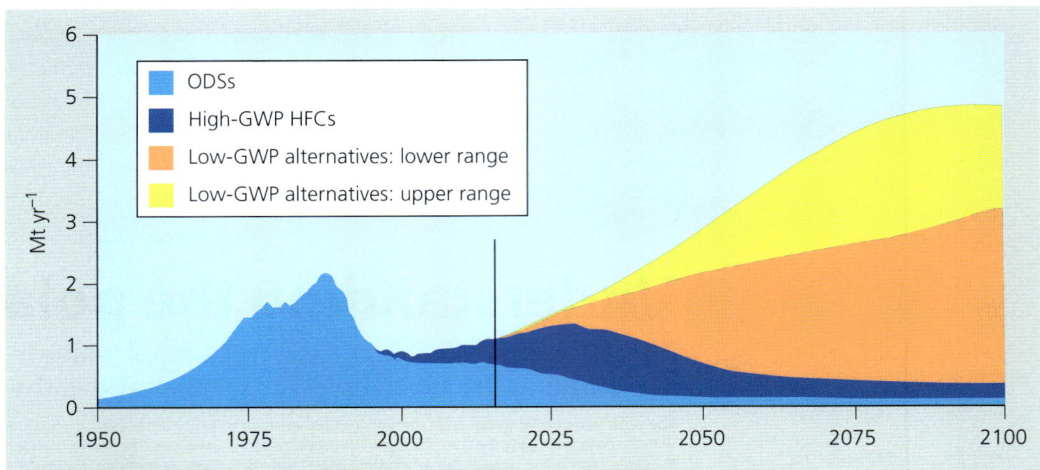

■ **Figure 6.71** Trends in the emissions of ODSs, HFCs and alternatives

(HL) 6.4.13 Air conditioning units are energy-intensive, contribute to GHG emissions, and traditionally have contained ODSs

Air conditioning units work by circulating refrigerant through a system of coils. The refrigerant, often an ODS, is a chemical that absorbs heat from the air inside the unit and then releases it to the air outside the unit. This process requires a lot of energy, which can contribute to GHG emissions. There are a number of things that can be done to make air conditioning units more environmentally friendly. These include:

- Energy-efficient units: Energy-efficient air conditioning units use less energy to cool a space, which can help to reduce greenhouse gas emissions. There are a number of ways to make an air conditioning unit more energy-efficient, such as using a variable-speed compressor and insulation.
- Solar panels: Solar panels can be used to power air conditioning units. This can help to reduce reliance on fossil fuels, since solar power does not produce greenhouse gases.
- Natural refrigerants: Natural refrigerants are substances that occur naturally in the environment and do not harm the ozone layer. Some examples of natural refrigerants include ammonia, propane, and CO_2. Natural refrigerants are often more expensive than HFCs, but they are becoming more affordable.
- Sealing leaks: Leaks in air conditioning units can release refrigerant into the atmosphere. It is important to seal any leaks in an air conditioning unit to prevent this from happening.
- Maintaining the unit regularly: Regular maintenance of an air conditioning unit can help to ensure that it is operating efficiently and that there are no leaks.
- Developing substitute refrigerants: Scientists are working to develop refrigerants that are less harmful to the environment than HFCs. Some promising alternatives include hydrofluoroolefins (HFOs) and carbon dioxide.
- Improving building design: Building design can have a significant impact on the energy efficiency of an air conditioning system. Features such as insulation, shading, and ventilation can help to reduce the amount of energy needed to cool a building.
- Greening and rewilding cities: Green spaces, such as trees and vegetation, can help to cool the air in cities. By increasing the amount of green space in cities, we can help to reduce the need for air conditioning.

Link

See Chapter 1.1 (page 3) to remind yourself about the importance of how perspectives drive our choices. See Chapter 6.2 (page 528) for details of the impacts of greenhouse gases on the atmosphere, and see Chapter 8.3 (page 740) for contributors to urban air pollution.

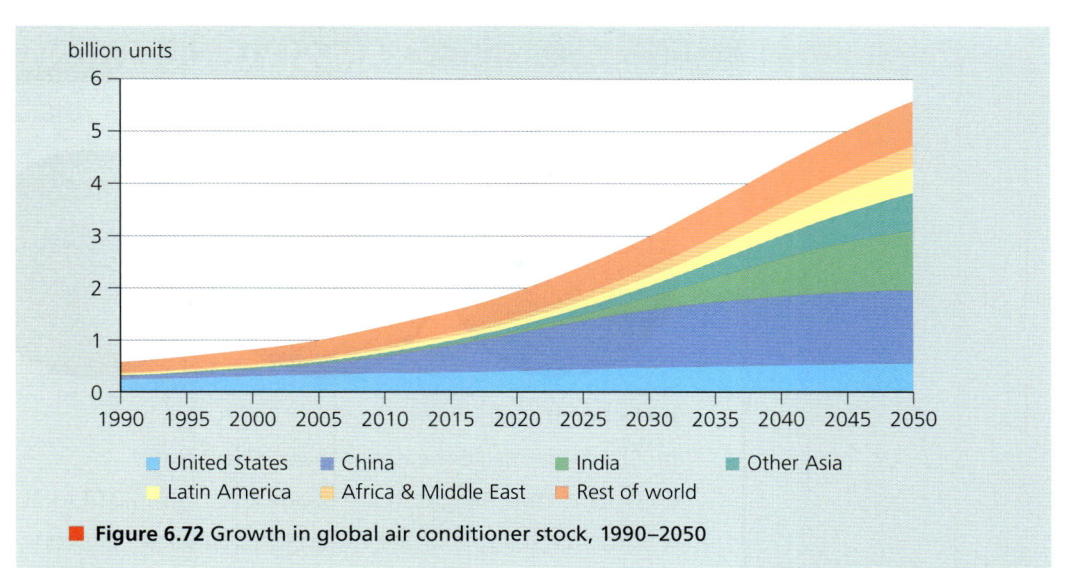

■ **Figure 6.72** Growth in global air conditioner stock, 1990–2050

HL.b.3: Environmental economics

Air conditioners contribute to greenhouse gas emissions, which impose cost on societies in the form of climate impacts. However, these costs are often not reflected in the price of air conditioning units, leading to market inefficiencies and externalities. There is a trade-off between the benefits of cooling and the costs associated with greenhouse gas emissions and ODSs. Economic analysis can help determine the most cost-effective strategies for reducing these negative impacts.

ATL ACTIVITY

Review the alternatives to air conditioning units

Using data available from the International Energy Agency (**www.iea.org**):
- Collect data on the use of air conditioning units in different societies.
- Present this data graphically, considering the reasons for the differences per capita.
- Present your findings to the school leadership.

REVIEW QUESTIONS

1. What are ozone depleting substances?
2. Outline the effects of UVB radiation on humans and ecosystems.
3. Discuss the role of the Montreal Protocol in regulating the trade in and use of ozone depleting substances.

HL ONLY

4. Explain the mechanism by which CFCs act as ozone depleting substances.
5. Discuss why the Kigali Amendment to the Montreal Protocol was made and what changes it has brought.
6. Explain how air conditioning units used throughout the world contribute to ozone depletion over the poles.

EXAM-STYLE QUESTIONS

1 The figure shows the ozone hole over Antarctica in 2018, 2019, 2020 and 2021.

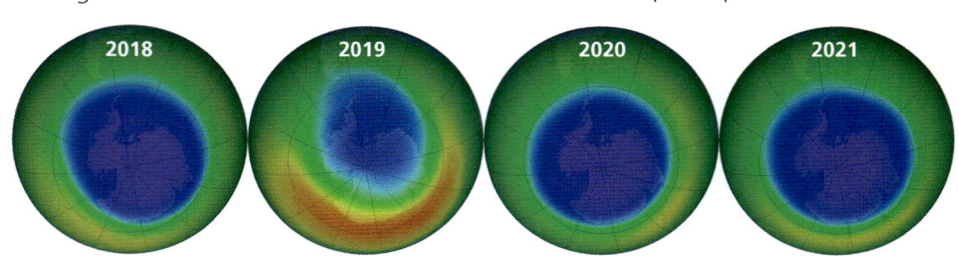

    a Define the term 'ozone hole'. [1 mark]
    b State the role of the ozone layer is necessary. [1 mark]
    c Outline the changes in the ozone layer from 2018 to 2021. [3 marks]
    d Describe the role of the Antarctic polar vortex in ozone depletion. [4 marks]

2 Discuss how Milankovitch cycles affect the amount of solar radiation reaching the Earth. [5 marks]

3 With the help of named examples, discuss how climate change affects ecosystems at a global scale causing shifting of biomes. [7 marks]

4 The figure shows the climate vulnerability of different areas around the world.

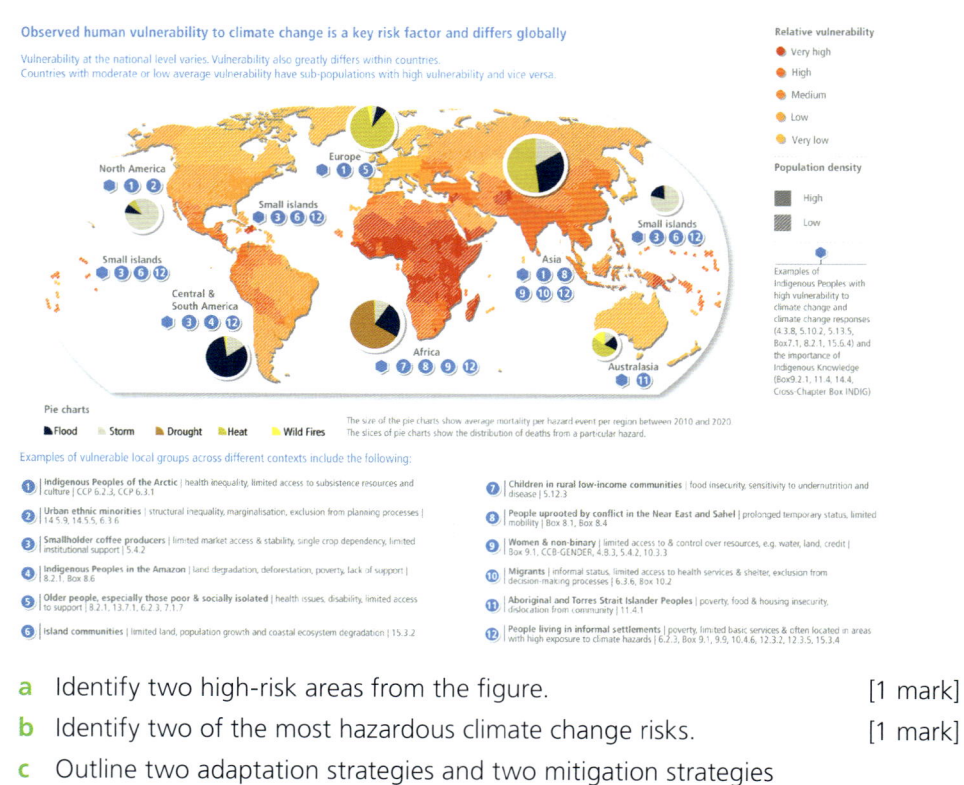

    a Identify two high-risk areas from the figure. [1 mark]
    b Identify two of the most hazardous climate change risks. [1 mark]
    c Outline two adaptation strategies and two mitigation strategies for overcoming climate vulnerability. [4 marks]
    d Discuss how climate mitigation strategies in high-income countries might differ from those used in low-income countries. [7 marks]
    e Comment on the impact of climate change on two named indigenous societies. [9 marks]

Theme 6: Atmosphere and climate change

Theme 7
Natural resources

7.1 Natural resources, uses and management

> **Guiding questions**
>
> - How does the renewability of natural capital have implications for its sustainable use?
> - How might societies reconcile competing perspectives on natural resource use?
> - To what extent can human societies use natural resources sustainably?

SYLLABUS CONTENT

This chapter covers the following syllabus content:
- ▶ 7.1.1 Natural resources are the raw materials and sources of energy used and consumed by society.
- ▶ 7.1.2 Natural capital is the stock of natural resources available on Earth.
- ▶ 7.1.3 Natural capital provides natural income in terms of goods and services.
- ▶ 7.1.4 The terms 'natural capital' and 'natural income' imply a particular perspective on nature.
- ▶ 7.1.5 Ecosystems provide life-supporting ecosystem services.
- ▶ 7.1.6 All resources are finite. Resources can be classified as either renewable or non-renewable.
- ▶ 7.1.7 Natural capital has aesthetic, cultural, economic, environmental, health, intrinsic, social, spiritual and technological value. The value of natural capital is influenced by these factors.
- ▶ 7.1.8 The value of natural capital is dynamic and can change over time.
- ▶ 7.1.9 The use of natural capital needs to be managed in order to ensure sustainability.
- ▶ 7.1.10 Resource security depends on the ability of societies to ensure the long-term availability of sufficient natural resources to meet demand.
- ▶ 7.1.11 The choices a society makes in using given natural resources are affected by many factors and reflect diverse perspectives.

HL ONLY
- ▶ 7.1.12 A range of different management and intervention strategies can be used to directly influence a society's use of natural capital.
- ▶ 7.1.13 The SDGs provide a framework for action by all countries in global partnership for natural resources use and management.
- ▶ 7.1.14 Sustainable resource management in development projects is addressed in an Environmental Impact Assessment (EIA).
- ▶ 7.1.15 Countries and regions have different guidance on the use of EIAs.
- ▶ 7.1.16 Making EIAs public allows local citizens to have a role as stakeholders in decision making.
- ▶ 7.1.17 While a given resource may be renewable, the associated means of extracting, harvesting, transporting, and processing it may be unsustainable.
- ▶ 7.1.18 Economic interests often favour short-term responses in production and consumption that undermine long-term sustainability.
- ▶ 7.1.19 Natural resource insecurity hinders socioeconomic development and can lead to environmental degradation and geopolitical tensions and conflicts.

- 7.1.20 Resource security can be brought about by reductions in demand, increases in supply or changing technologies.
- 7.1.21 Economic globalization can increase supply, making countries increasingly interdependent, but it may reduce national resource security.

7.1.1 Natural resources are the raw materials and sources of energy used and consumed by society

A resource is anything that is useful to humans. The Earth contains several resources that support its systems. The core and mantle of the Earth possess minerals and valuable rocks, the land has a variety of ecosystems such as forests that exist in a state of balance, the oceans and other freshwater bodies have life-sustaining water, and the atmosphere contains a mixture of gases required to continue life on Earth. Humans utilize these resources and in so doing create a relationship with nature. These raw materials and sources of energy that are used and consumed by society, such as air, water, land, minerals, soil, forests and living things, are called natural resources. Humans tend to have an anthropocentric approach towards these resources, and to overexploit them for their benefit.

> **Link**
> See Topic 1.1.8 (page 13) to remind yourself about anthropocentrism.

> **Common mistake**
> There is a tendency to confuse natural resources with natural capital. Keep in mind that natural resources are all those things that are available naturally, while natural capital comprises those natural resources that benefit human beings.

7.1.2 Natural capital is the stock of natural resources available on Earth

The natural resources, along with the ecosystem services they provide, that help in making the environment conducive for humans are called natural capital. This consists of all the natural stores of the Earth, including soil, rocks and minerals, water, air, fishing stocks, forests and biodiversity. The demands on the Earth made by the human population drastically increased from the onset of the Industrial Revolution. Since technologies for making life easier require more resources, this increasing demand has caused a decline of certain resources over the years.

Stocks of both renewable and non-renewable resources count as natural capital:

- Water is a renewable natural capital. It is a basic requirement of all living beings, and is also required in many production sectors such as agriculture and clothing manufacturing.
- Solar energy is also renewable natural capital. It is the sole source of light energy for plants to photosynthesize. Without the sun, ecosystems would not be able to sustain themselves.
- Non-renewable sources such as soil and minerals are also natural capital and are extensively used and consumed by humans and plants in the ecosystem.

> **HL.b.3: Environmental economics**
>
> Environmental goods and services are available to everyone. These goods are categorized as non-rival and non-excludable in consumption. This means that once a service such as water purification by reeds or mangroves has been done, it is accessible to all humans. There is no way of excluding some individuals from gaining the benefits of this.
>
> Markets often undervalue goods and services such as clean air, clean water, wildlife and forests. As a result, people might not realize their importance for the well-being of the plant. Some people may use these resources in a way that might harm the environment.

7.1 Natural resources, uses and management

◆ **Social cost** – the total cost borne by the society for the action of an industry.

For example, a factory that produces goods also pollutes the environment. Negative externalities occur because the cost associated with environmental pollution is not reflected in the factory's costs as it does not bear the **full social** cost of its activities. As a result, the market is inefficient, leading to net welfare loss for the society.

> **ATL ACTIVITY**
>
> **Analyse market failure and environmental impact**
> Explore real-world examples of market failure and its consequences on the environment. Create a five-point summary for each example.

7.1.3 Natural capital provides natural income in terms of goods and services

Economic ecologists define natural resources as things that can be sold for financial benefit and generate natural income. As a simple example, trees can provide many benefits and generate income in various ways. Obviously, the timber from trees can be sold for money. The trees also provide services: they absorb many tonnes of carbon dioxide via photosynthesis, lowering the concentration in the atmosphere. These trees, when present in an urban setting, also help in cooling the temperature. And when they are present near a river, they help in preventing soil erosion into the river. They also enable recreation as people can camp beneath them, climb them, or for certain trees, gather their fruit.

Natural capital that is well-managed can ensure steady natural income in the form of goods and services. Goods are marketable commodities (for example, food, fibre, timber, fuel, freshwater), while ecosystem services include all the essential services that we receive from nature (for example, carbon dioxide sequestration, water filtration, cooling, nutrient cycling, pollination, temperature regulation, flood prevention), as well as the non-material benefits we receive from nature, such as aesthetic value, and recreation.

■ **Figure 7.1** Tree habitats can be used by humans for recreation in a variety of ways

> **● Common mistake**
>
> It is incorrect to assume that natural income operates on the same principles as financial income. You must remember that natural income refers to the flow of benefits or goods and services that are obtained from natural resources and ecosystems. The mistake often arises when people try to quantify natural income solely in monetary terms, overlooking the diverse and intrinsic values of ecosystems.

7.1.4 The terms 'natural capital' and 'natural income' imply a particular perspective on nature

There are pros and cons to looking at nature as a source of goods and services for human benefit. On the positive side, this view can help us to appreciate the value of nature and to see it as an essential part of the economy. It can also lead to more sustainable resource management practices. For example, if we see forests as a source of natural capital, we are more likely to invest in sustainable forestry practices that ensure forests continue to provide us with valuable services such as carbon sequestration.

However, there are also some negative implications to considering nature as a collection of resources. One concern is that this view can lead to the exploitation of nature for human benefit without regard for the intrinsic value of other living things. This is particularly true in the context of more extreme manifestations of anthropocentrism, which put human requirements above all else.

Another concern is that seeing nature as natural capital might make people treat it like something to buy and sell. This means that nature is seen as a commodity that can be bought and sold, rather than as a public good that should be protected for the benefit of all. Such an approach can have negative consequences for biodiversity and ecosystem health. Resources such as minerals and oil can bring huge benefits to the countries that possess them. However, if the focus is only on extracting profit, then the exploitation becomes unsustainable. Despite being unsustainable, if the resource is available in considerable amount, the exploitation may continue for a long time, and countries may not worry about it getting depleted.

> **Concept**
>
> **Perspectives**
>
> The perspective taken towards nature plays a crucial role in deciding the ecological, economic, social and intrinsic value of natural resources. It influences how individuals, communities and policymakers perceive and value the ecosystem and its natural reserves. The terms 'natural capital' and 'natural income' indicate a perspective shift in economic thinking, acknowledging the intrinsic value of nature beyond mere commodities. 'Natural capital' reflects the view that assets provide a wide array of services to human societies, while 'natural income' reflects the perspective of deriving benefits from the sustainable use of natural capital.

> **● HL.b.2: Environmental economics**
>
> Perspective plays a crucial role in ecological economics as it influences how individuals, communities and policymakers perceive and value natural resources and ecosystems. Different perspectives can shape economic decisions and policies, impacting environmental sustainability.
>
> For instance, think of a local community that depends on a forest for sustenance, clean water and cultural practices. From the traditional economic perspective, the forest will be viewed simply as a source of revenue generation through timber. However, adopting an ecological economics perspective would involve recognizing the forest's multifaceted role in providing ecosystem services such as water purification, habitat support and climate regulation. Thus, the community might implement sustainable forest management practices that ensure the continued availability of these essential services. Even policymakers might explore economic instruments like payments for maintaining the forest's ecological integrity, acknowledging its non-market values.

> **ATL ACTIVITY**
>
> Select a natural resource (for example a mineral deposit, forest or water body).
> - Research the ecological, social and economic aspects of the chosen resource, considering its contributions to ecosystems, economic values and the well-being of society.
> - Identify and categorize different perspectives that stakeholders (such as environmentalists, industries, local communities) might have on the chosen resource.
> - Consider how these perspectives might influence the valuation and management of the chosen natural capital.
> - Create a visual representation (a mind map, infographic or poster) that illustrates the diverse perspectives on the valuation of the natural resource.
> - Write a short reflection on how each perspective shapes the understanding of natural capital.
> - Share your ideas with your peers in a class discussion.

HL.c.3: Environmental ethics

A variety of ethical frameworks and conflicting ethical values emerge from differing fundamental beliefs concerning the relationship between humans and nature.
- Technocentrism focuses on maximizing economic gains through technological innovation, even if it involves the depletion or alteration of natural resources. For example, intensive agriculture employs genetically modified crops to increase yields, disregarding potential ecological impacts.
- Ecocentrism focuses on the idea that all living organisms and ecosystems are interconnected, creating a balance and sustainable bond between humans and nature. For example, establishing protected natural reserves to conserve biodiversity and ecosystem integrity, prioritizes the preservation of natural processes.
- Anthropocentrism prioritizes fulfilling human needs. Laws and policies are put in place to avoid exploitation. For example, in India the National Fisheries Policy of 2020 helps to manage and regulate fisheries in a responsible and sustainable way.

■ Table 7.1 Differences in ethical frameworks

Criteria for ethical framework	Ecocentric	Technocentric	Anthropocentric
Approach to natural capital	Promotes the sustainable use of natural capital, emphasizing conservation and regeneration.	Seeks to exploit and commodify natural capital using cutting edge technology to extract maximum value.	Exploits natural resources for the betterment of humans.
View on natural income	Seeks to maintain a balance in natural income, ensuring the continued health and resilience of ecosystems.	May prioritize immediate gains from natural income, often leading to overexploitation.	Acknowledges that preserving natural capital is essential for securing a reliable natural income for future generations.

> **ATL ACTIVITY**
>
> Choose an environmental issue that interests you (such as deforestation, climate change or biodiversity).
> - Analyse the ethical considerations linked to the chosen environmental issue.
> - Explore diverse ethical frameworks (technocentric, ecocentric, anthropocentric) and their implications.

> **ATL ACTIVITY**
>
> Create a poster or presentation to generate awareness about natural capital and the need to use it sustainably.
> - Suggest why a society that is heavily inclined towards anthropocentrism may not be sustainable in the long run.
> - Present your poster to your friends and people in your community.

7.1.5 Ecosystems provide life-supporting ecosystem services

Ecosystems help us with many services. An ecosystem contains a variety of lifeforms that are interdependent on each other and also harmoniously interact with the abiotic factors of that given area.

Here are some examples of ecosystem services:

Ecosystem service	Description
Carbon sequestration	Trees and plants in forests absorb CO_2 from the atmosphere during photosynthesis, helping to mitigate climate change. An adult tree can sequester approximately 20 kg of CO_2 per year, contributing to cleaner air.
Soil erosion prevention	The roots of trees in forests hold the soil together, preventing erosion caused by wind and water. This helps to maintain soil fertility and stability.
Coastal protection	**Coral reefs** act as natural barriers, dissipating wave energy and protecting shorelines from storms, floods, and erosion.
Water purification	**Reeds** in **wetlands** transport oxygen to the soil, promoting the growth of aerobic microorganisms that decompose organic matter and clean the water. **Reed beds** are used for water treatment, improving water quality by acting as buffer zones for effluents and sewage before being released into the environment. **Mangroves** purify water by phytoremediation – filtering water and removing contaminants such as heavy metals.

◆ **Coral reef** – a huge ridge or rock in the sea formed by corals, which serves as an ecosystem, providing habitats to many marine organisms.
◆ **Reeds** – any species of tall grass found in wetlands throughout the temperate and tropical regions.
◆ **Reed beds** – area of water that is dominated by reeds.
◆ **Mangroves** – trees or shrubs that grow in coastal saline water.
◆ **Wetlands** – areas where water covers the soil.

> **Concept**
>
> **Systems and models**
>
> Ecosystems work in tandem to provide ecosystem services like pollination and water purification. In the interconnected nature of the systems, the output of one system may function as the input for another. Pollination is a good example, with the transfer of the pollen grains from the anthers (male part) to the stigma (female part). In water purification services, many harmful chemicals are absorbed and stored in the system.

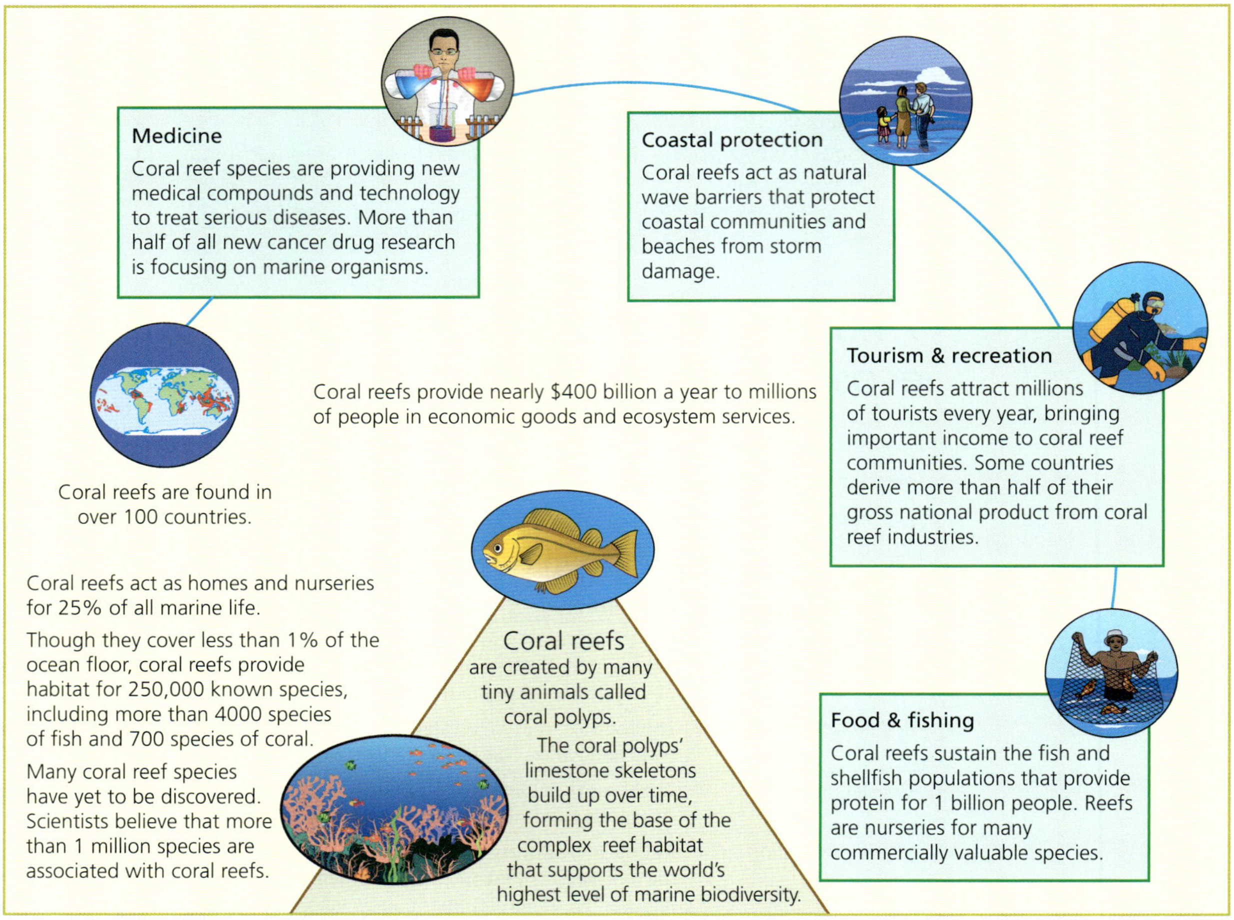

Medicine
Coral reef species are providing new medical compounds and technology to treat serious diseases. More than half of all new cancer drug research is focusing on marine organisms.

Coastal protection
Coral reefs act as natural wave barriers that protect coastal communities and beaches from storm damage.

Coral reefs provide nearly $400 billion a year to millions of people in economic goods and ecosystem services.

Tourism & recreation
Coral reefs attract millions of tourists every year, bringing important income to coral reef communities. Some countries derive more than half of their gross national product from coral reef industries.

Coral reefs are found in over 100 countries.

Coral reefs act as homes and nurseries for 25% of all marine life.

Though they cover less than 1% of the ocean floor, coral reefs provide habitat for 250,000 known species, including more than 4000 species of fish and 700 species of coral.

Many coral reef species have yet to be discovered. Scientists believe that more than 1 million species are associated with coral reefs.

Coral reefs are created by many tiny animals called coral polyps. The coral polyps' limestone skeletons build up over time, forming the base of the complex reef habitat that supports the world's highest level of marine biodiversity.

Food & fishing
Coral reefs sustain the fish and shellfish populations that provide protein for 1 billion people. Reefs are nurseries for many commercially valuable species.

■ **Figure 7.2** Coral reef ecosystem services

■ **Figure 7.3** Reed beds act as natural water filters

Link
See Topics 1.2.3 (page 27) and 1.2.4 (page 29) for more on how ecosystems provide services such as storage (e.g. removing heavy metal contamination), transformation (e.g. nutrient cycling), and transfer (e.g. a river bringing down minerals from the mountains).

7.1.6 All resources are finite. Resources can be classified as either renewable or non-renewable

Anything that can be used or adds value to life is termed as a resource. Naturally occurring resources are limited in quantity on Earth. These resources can be classified as renewable or non-renewable.

Theme 7: Natural resources

- Renewable resources are those natural resources that can be regenerated as quickly as they are consumed. For example, crops and timber are continuously used by humans and they are replaced by natural growth and reproduction. Water can be circulated and maintained through the water cycle.
- Non-renewable resources are those resources that take a very long time to regenerate and cannot be replenished as fast as they are consumed. Since they cannot re-enter the system within human lifetime, their use becomes unsustainable. For example, producing fossil fuels such as coal, crude oil, petroleum and natural gas takes millions of years and very specific conditions for their formation.

7.1.7 Natural capital has various different types of value. The value of natural capital is influenced by these factors

Natural capital, including ecosystems and organisms, has value. This value may or may not generate financial benefit. Natural capital can be classified based on the values in Table 7.2.

■ **Table 7.2** The values by which natural capital can be classified

Aesthetic value	The value that a resource possesses due to its appearance, for instance tourism at scenic spots like coasts and coral reefs.
Cultural value	Natural capital helps in shaping cultures and thus becomes an integral part of them. The local features of a given area may become culturally useful. For example, Mount Uluru in Australia has immense cultural value for the **Anangu people**. It serves as a site for practising sacred ceremonies (also giving it spiritual value) and rituals, and also as a foundation for passing traditional stories from one generation to another.
Economic value	A value that is determined by the market price of the natural capital. For example, timber can be sold to generate natural income.
Environmental value	The value that natural capital possesses due to its ability to maintain ecological balance, support biodiversity, and maintain the overall health of the ecosystem. For example, wetlands such as the Everglades of Florida, USA act as natural water filters, purifying water by trapping sediments and removing pollutants. They also help regulate water flow, preventing flooding during periods of heavy rainfall and providing a stable water supply during dry periods.
Health	The medicinal value that some natural capital holds, for example some plants such as the peace lily, bamboo palm and echinopsis are known to purify air by absorbing harmful carbon monoxide, formaldehyde and benzene vapours. The health value obtained from natural capital includes access to fresh water and nutrient-rich fruits, and mental well-being.
Intrinsic value	The inherent value of the natural capital, independent of its utility to humans. For example, the giant panda helps to maintain healthy bamboo forests by dispersing seeds that could not otherwise be spread. Even though the giant panda does not have any direct use value for humans, its actions of seed dispersal are beneficial to all humans as they help the forests to flourish.
Social value	Growth of a society relies on the natural capital that binds people together by providing better quality of life and well-being. For example, forests, natural parks and beaches are places people can relax and spend time together.
Spiritual value	Resources that reflect the spiritual significance of natural elements. For example, Mount Kailasa in Tibet is considered sacred in Hinduism, Buddhism and Jainism. It is believed to be the abode of the Hindu deity Shiva. Lakes (e.g. Titicaca in South America between Peru and Bolivia) and rivers (e.g. the Ganges) can also be very important places for people from various spiritual traditions.
Technological value	Some natural resources provide raw materials for technological advances, such as lithium used in batteries, silicon used in computer chips, and geothermal energy, wind energy, solar energy and tidal energy to produce electricity.

◆ **Anangu people** – members of several aboriginal Australian groups.

> **Concept**
>
> **Perspectives**
>
> People view and value the environment in diverse ways. Perspectives on natural capital can vary significantly, ranging from utilitarian and economic viewpoints, to cultural, spiritual and ecological considerations. Some may perceive natural capital primarily as a source of raw materials and ecosystem services with economic value, emphasizing its contribution to human well-being and development. The cultural and spiritual background of an individual provides a baseline for valuing nature and its role in shaping identity, traditions and worldview. The interconnectedness of all living organisms allows people to prioritize the environment over their personal needs.

HL.b.7: Environmental economics

Environmental accounting involves keeping track of natural resources, their use and the addition of new resources. Asset valuation can help in assigning economic values to natural assets such as forests and wetlands. Ecosystem service assessment provides detailed information about services such as pollination, flood protection, temperature regulation and water purification. These can be obtained from a variety of ecosystems. The cost of environmental degradation is an important aspect of environmental accounting and provides useful insights into the costs incurred due to environmental damage such as deforestation, soil degradation, loss of biodiversity and much more.

If a company uses natural resources to provide products that can generate revenue, this means the natural capital has economic value. However, the process of manufacturing products from these natural resources will also generate waste. This waste can be in the form of pollution, which can cause environmental degradation. This is a loss and needs to be set right. Money has to be spent in order to rectify this damage. This is known as an environmental cost. Proper environmental accounting ensures that we responsibly manage and sustainably use nature's resources, recognizing their value, preventing their overuse, and promoting equitable access for current and future generations.

◆ **Replacement cost** – the cost incurred to replace the existing system with a new or better system.

Economists use the **replacement cost** method to estimate the value of an asset. A wetland serves as a natural water purification system providing fresh water. Since it is not possible to calculate the value of 'purification' provided by the wetlands, we will look at the cost for replacing this wetland with a water purification plant. Assuming that the cost for this will be $50,000 and that the average lifespan of the wetlands would be around 50 years, the annual value of wetland can be calculated as:

Annual value of wetland = cost of replacement of wetland / average lifespan of wetland

= $50,000 / 50

= $1,000

Since the value of money does not remain the same over years, we assume that there is a discount rate of 5 per cent, the present value can be calculated as:

Present value = Annual value / (1 + discount rate)$^{\text{number of years}}$

$= 1,000 / (1 + 0.05)^{50}$

= $87.21

This is a very simple representation of environmental accounting that considers only a few aspects. In practice, it is a more sophisticated procedure because it involves uncertainties, ecosystem dynamics and stakeholder preferences. It must be remembered that the replacement cost method is one of many methods that can be used to assign monetary values to ecosystem services. Different methods yield different values and this is what makes the valuation process complicated.

> **ATL ACTIVITY**
>
> Research at least three ecosystems for the goods and services they provide. Discuss how environmental accounting can be done in these ecosystems.

HL.c.3: Environmental ethics

Guiding principles and moral considerations to foster responsible interactions with the environment are provided by ethical frameworks for the relationship between humans and nature. Environmental ethics emphasizes the intrinsic value of nature, and the moral responsibility to protect and preserve it. Aldo Leopold's land ethic, for instance, proposes a holistic view where humans see themselves as members of a broader ecological community, recognizing their role as stewards of the land. The Norwegian philosopher Arne Næss coined the term 'deep ecology' which offers a framework promoting a major shift in human consciousness, emphasizing the interconnectedness of all living beings and ecosystems. The environmental ethics of many indigenous peoples offer diverse perspectives rooted in cultural and spiritual connections to the land. For example, the 'seventh generation principle' which is based on the philosophy of the indigenous Haudenosaunee (pronounced as Hau-de-no-shaw-nee) people of North America reflects that the decisions we make today must result in a sustainable world, seven generations into the future. It is a commitment to making decisions that consider the well-being of future generations.

> **ATL ACTIVITY**
>
> Use the following links to access Aldo Leopold's land ethic. Read them and make notes. Share your understanding on the topic with your classmates.
> - www.aldoleopold.org/about/the-land-ethic
> - www.aldoleopold.org/about/land-ethics-and-social-justice

7.1.8 The value of natural capital is dynamic and can change over time

The value of natural capital is not fixed. It can change over time depending on how it is used and the increase or decrease of its demand. Changes in the social value, technological advancements and demand drive the valuation of natural capital over time. For example, in 1870 lithium was used in small quantities as an anticonvulsant and hypnotic drug. Its demand was limited and thus its value was very low. However, in the 1970s experiments began using lithium ion in making batteries, and products using the first lithium-ion batteries were developed in 1991. This was also a time when tremendous progress was being made in the field of electronics. This created a huge demand for lithium and as a result its value increased.

Similarly, whale oil, which is made out of whale blubber and bones, was very popular in the eighteenth and nineteenth centuries. It was used in lamps and candles, and as a lubricant. The whaling industry flourished during this time. However, the advancements in energy resources such as petroleum products and the discovery of the electric lightbulb meant the demand for whale oil declined, and so did its value.

Radium watches were trendy between 1920 and 1960. Awareness about the health hazards associated with the radioactivity resulted in a decline in demand for these products and thus a drop in the use of radium in watches.

HL.c.4: Environmental ethics

Instrumental value is the usefulness of an entity for humans, in the form of goods, services and opportunities for human development. Entities are valued for their contribution to human well-being or objectives rather than for their intrinsic worth. For example, trees have instrumental value as they provide timber for construction, oxygen for respiration and contribute to carbon sequestration. Similarly, bees have instrumental value due to their role in pollination, which is essential for agriculture and food production. The concept underscores the functional and practical importance of various entities in meeting human requirements and objectives.

ATL ACTIVITY

The utility of resources is shaped by cultural, societal and past experiences. Discuss with your peers the unique needs, practices and beliefs of various cultural perspectives that contribute to determining the instrumental value of an entity.
- Explore examples of how certain plants, animals or minerals hold distinct instrumental values across diverse cultural contexts.
- Explore the ethical dimensions of instrumental value by discussing the potential conflicts that arise when nature is primarily valued for its utility.

7.1.9 The use of natural capital needs to be managed in order to ensure sustainability

Long-term benefits can be harnessed from natural capital if it is not consumed more rapidly than it can be regenerated. At the same time, strategies for efficient waste management must be in place to ensure that the system remains stable for longer periods of time. Careful stewardship of natural capital can prove beneficial to society. For instance, harvesting fisheries beyond their natural replenishment rates can lead to the collapse of marine ecosystems, jeopardizing the livelihoods of fishing communities and threatening food security. Similarly, deforestation at a pace exceeding the natural regrowth of forests results in habitat loss, biodiversity decline and adverse effects on climate regulation.

REAL-WORLD EXAMPLE

Guano on Nauru

Nauru is a small island country to the east-northeast of Papua New Guinea in the Pacific Ocean. It is home to birds such as seagulls (brown noddy, black noddy), terns and frigatebirds. The excrement of these birds is called guano, and is rich in nitrogen and phosphate. Over several years, this guano accumulated in huge amounts over the entire plateau. In 1900, this phosphate reserve was discovered by a prospector, Albert Fuller Ellis. The Pacific Phosphate Company made its first sale of phosphate fertilizer in 1907. The country economically benefited from mining and exporting its phosphate reserves to the extent that most of the companies in the country worked in this industry. Most of the country's food and consumer goods was imported from other countries. The unsustainable extraction practices caused the resource to deplete faster than they could have imagined.

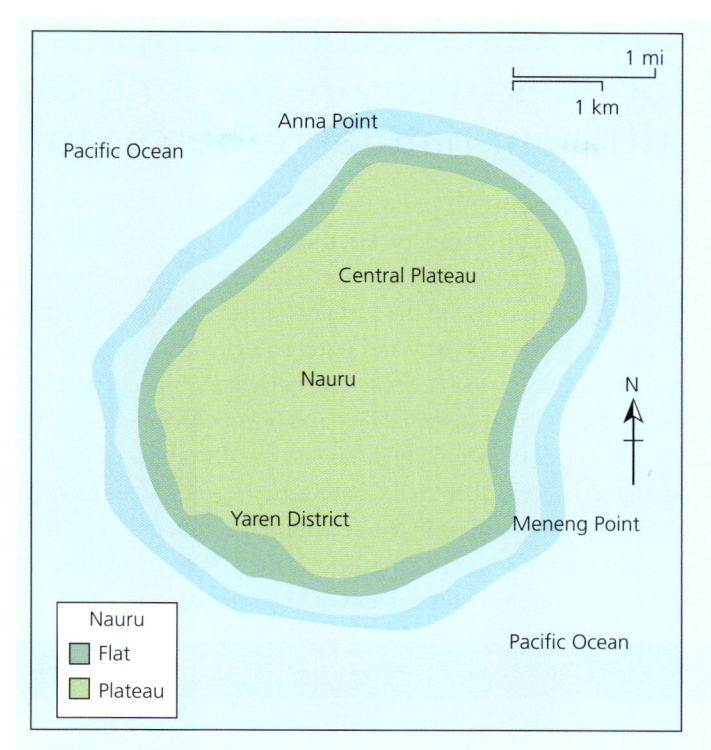

■ **Figure 7.4** Map of Nauru

The plateau on Nauru had wild almond and Pandanus trees and many shrubs that provided shelter to the birds. The clearing of land and topsoil for mining activity destroyed the habitats of the birds, forcing them to move away. This drastically brought down the bird population and further decreased the guano deposits. Nauru took a very big financial hit as there was no other business to support the economy. The country, which was for a time in the 1980s one of the richest in the world in terms of GDP per capita, was brought to its knees in 2004 with a badly broken economy.

■ **Figure 7.5** Phosphate extraction in the Plateau of Nauru

More details on Nauru can be found in this video: **https://youtu.be/IbMpaJuhHeM** and in this article: **https://journals.openedition.org/jso/7055**

7.1 Natural resources, uses and management

7.1.10 Resource security depends on the ability of societies to ensure the long-term availability of sufficient natural resources to meet demand

Resource security means the protection of important natural resources, such as water, soil, food and energy, to avoid their scarcity. Meeting resource security enables us to can protect ourselves from conflicts, hunger, war, and unwanted loss or pollution of natural resources. Some countries have designed policies around the sustainable use of resources. This can be better understood as food security, energy security and water security. For example, Sweden has a high level of food security, according to *The Economist* magazine's Global Food Security Index, with a strong emphasis on agricultural sustainability and organic farming practices. The country has a well-established tradition of valuing high-quality, locally produced food. Crisis and contingency policies such as the Swedish Civil Contingency Agency focus on increasing the productivity, viability and competitiveness in the agricultural sector, and the EU's Common Agricultural Policy (CAP) strategy aims at strengthening the supply chain. Sweden prioritizes environmental responsibility and contributes to a robust and secure food system.

Link
See Chapters 1.2 (page 24), 1.3 (page 47), 4.2 (page 291), 5.2 (page 456), 7.2 (page 634), 8.2 (page 719), 8.3 (page 740) and 9.2 (online) for more information on this topic.

> **Tool 2: Technology**
>
> **The Global Food Security Index**
>
> The following website provides information about the Global Food Security Index:
> **https://impact.economist.com/sustainability/project/food-security-index**
> - Choose 10 countries randomly and look at their Global Food Security Index.
> - Make a table to record the index.
> - Read the 'Key Findings' on the website.
> - Research the ability of the societies of your chosen countries to support food security. Record this in your table.
> - Discuss your findings with your peers.

According to the Global Water Security 2023 Assessment, Sweden is the most water-secure country. Sweden values water as a critical resource, and has developed water-conservation strategies that ensure easy and continuous access of freshwater for all its citizens. Policies like the Water Framework Directive help in the management of this useful resource in different sectors of society. Investment in water infrastructure, efficient irrigation practices, and public awareness campaigns contribute to Sweden's success in ensuring water security.

Ethiopia, on the other hand, is struggling for food and water security due to resource scarcity. It falls within an arid zone and thus receives very little precipitation, a situation that is worsening due to climate change. However, the government runs Productive Safety Net Programmes that ensure all households receive food and some money during a crisis in exchange for building or rehabilitating community assets such as gardens and dams. The Growth and Transformation Plan aims at steadily increasing agriculture production by 8 per cent annually. For this, the International Crops Research Institute for Semi-Arid Tropics (ICRISAT) has partnered with the Ethiopian government to help reclaim lost agricultural land. Soil treatment has been key to improving fertility and water-holding capacity, increasing food security.

Similarly, under the One WASH National Programme, the government of Ethiopia is able to provide increased access to safe drinking water to its rural and urban citizens. Efforts to align

Link

Water stress is covered in Chapter 4.2 (page 291).

Food security is covered in Topic 5.2.12 (page 489).

Urban planning for the best use of land is covered in Chapter 8.2 (page 719).

traditional practices with modern sustainability principles and to address environmental vulnerabilities are indicative of a shift towards a more positive environmental approach. However, challenges persist due to factors deeply rooted in historical practices.

Societies can contribute towards resource security by:

- taking care to avoid unsustainable rates of consumption of natural resources like oil, wood and gas
- resolving conflicts all around the world that could reduce the availability and accessibility of resources
- proper agreement on the use of shared resources
- taking action to reduce environmental degradation and climate change.

7.1.11 The choices a society makes in using given natural resources are affected by many factors and reflect diverse perspectives

Various factors affect the choices about the use and management of natural resources at the local. These are often influenced by economic, sociocultural, political, environmental, geographical, technological and historical considerations. Let's examine these factors in the context of a local forest.

◆ **Logging** – the felling of trees to cut and prepare for timber.

- The economic value of the forest can significantly impact local choices. If the region relies on timber and wood products, this may drive choices that favour **logging** and deforestation for economic gain. Local economies may be influenced by industries such as paper manufacturing or furniture production, impacting choices related to forest use. Alternatively, the forest might contribute to ecotourism, influencing choices towards conservation and responsible management. Local communities often have cultural ties to natural resources.

- Local communities often have cultural ties to natural resources. The community's traditions, values, and customary practices related to the forest may also guide choices. For example, if the forest is considered sacred or essential for cultural practices, preservation may be a priority.

- Political decisions regarding land-use and forest-conservation policies, or incentives for sustainable practices, can significantly influence the choices made. Choices may also be influenced by the need to protect endangered species, maintain ecosystem services, or mitigate the impacts of climate change through carbon sequestration by trees. Commitments made by governments in international agreements, such as the Paris Agreement to achieve net-zero carbon emissions, may influence local policies towards sustainable forest management.

- The location and geography of the forest influence its vulnerability to natural disasters, its accessibility and its potential for various uses. A forest located in a remote mountainous region will be less accessible to people, and its potential destruction by natural disasters such as landslides will prevail. If it is near a populated urban area with good access, people may want to conserve it for recreational purposes.

- Technological advancements greatly affect resource management. Modern equipment makes logging easier and more efficient, and also provides opportunities for sustainable management of logging, biodiversity monitoring and fire prevention.
- Past land-use practices and historical events can shape current choices. Understanding the history of the forest, including previous logging or conservation efforts, provides context for decision making.

> **Tool 4: Systems and models**
>
> **Perspectives towards natural resources**
>
> Choose two societies that differ in their perspectives towards natural resources.
> - Research and identify the influence of their diverse perspectives on the consumption of a particular named resource.
> - Consider economic, social, environmental and political factors.
> - Construct a systems diagram for resource-use choices by each society.
> - Indicate feedback loops in the system.
> - Show the flows using arrows.
>
> ■ Figure 7.6 Example of a systems diagram

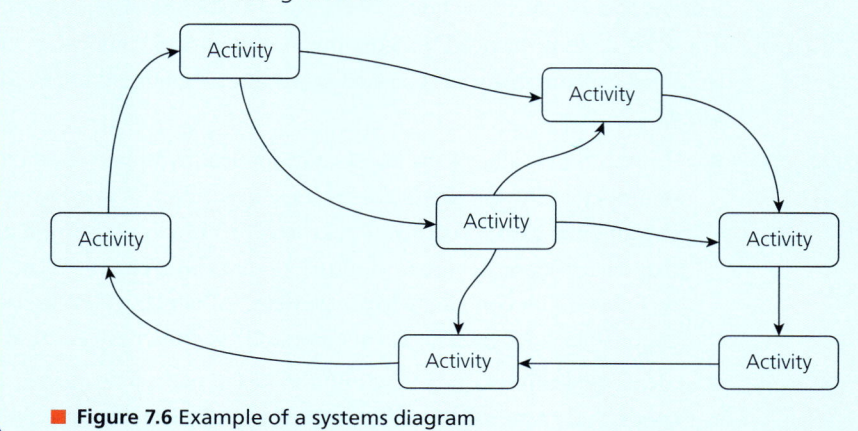

● **Common mistake**

It is easy to be mistaken that HFCs are safe to use, as they do not contribute to ozone depletion. HFCs, however, are greenhouse gases and therefore their release into the environment must also be curtailed.

HL.a.8: Environmental law

The Montreal Protocol (1989) is an international treaty designed to protect the ozone layer by phasing out the production and consumption of numerous substances that are responsible for ozone depletion, such as chlorofluorocarbons (CFCs) and halons. It has significantly influenced societal choices, particularly in industries using ozone-depleting substances. CFCs are used in the manufacture of aerosols, refrigerants and solvents used in the making of packing materials such as foams. As a legally binding agreement, the Montreal Protocol mandated the phase-out of these harmful substances, prompting industries to transition to ozone-friendly alternatives such as hydrochlorofluorocarbons and hydrofluorocarbons (HCFCs and HFCs, which are halogenated compounds usually associated with refrigerants). This has led to changes in manufacturing processes, technological innovations, and consumer choices to align with the goals of the protocol. The agreement demonstrates how international legal frameworks can drive transformative changes in societal practices related to a specific natural resource – in this case, the ozone layer.

HL.b.8: Environmental economics

Use value and non-use value significantly shape the choices people make in utilizing natural resources. While use value addresses the direct economic benefits, non-use value reflects the broader environmental, ethical and intrinsic values that influence decision making for both present and future generations. People are more likely to prioritize the utilization of natural resources that offer direct economic benefits, such as timber for construction, water for irrigation or medicinal plants for healthcare. People are willing to pay for the tangible goods and services derived from natural resources. Non-use value encompasses the value people place on a natural resource without directly using or consuming it. It includes both existence value (valuing the mere existence of a resource) and bequest value (valuing the resource for future generations).

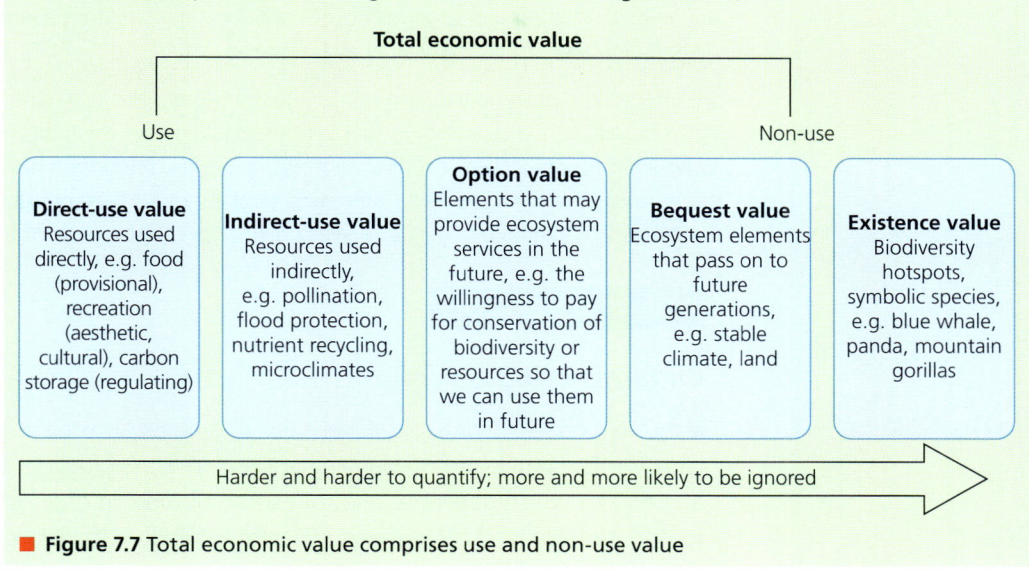

Total economic value

Use — Non-use

Direct-use value	Indirect-use value	Option value	Bequest value	Existence value
Resources used directly, e.g. food (provisional), recreation (aesthetic, cultural), carbon storage (regulating)	Resources used indirectly, e.g. pollination, flood protection, nutrient recycling, microclimates	Elements that may provide ecosystem services in the future, e.g. the willingness to pay for conservation of biodiversity or resources so that we can use them in future	Ecosystem elements that pass on to future generations, e.g. stable climate, land	Biodiversity hotspots, symbolic species, e.g. blue whale, panda, mountain gorillas

Harder and harder to quantify; more and more likely to be ignored →

■ Figure 7.7 Total economic value comprises use and non-use value

HL.c.8: Environmental ethics

There are three major approaches in traditional ethics – virtue ethics, deontological ethics and consequentialist ethics.
- Virtue ethics centres on virtue and character. The person who possesses virtue is considered to be good. It promotes sustainable resource management.
- Deontological ethics focuses on the morality of actions irrespective of their consequences. Choices related to resource use under deontology would be guided by ethical principles and rules. For example, if a society recognizes the duty to preserve biodiversity, deontological ethics might discourage actions that lead to the extinction of species, irrespective of the potential economic benefits.
- Consequentialist ethics evaluates the morality of action based on their outcomes or consequences. The ethical judgement depends on the overall well-being or utility generated. Consequentialist thinking would consider the overall impact on society, the environment and future generations. For instance, if sustainable logging practices lead to positive economic outcomes without severe environmental degradation, consequentialism might support such practices.

(HL) 7.1.12 A range of different management and intervention strategies can be used to directly influence a society's use of natural capital

A society's use of natural capital needs proper management to ensure sustainability. There are many management and intervention strategies that are used to this end and at different levels. At the government level a range of strategies and policies can be implemented. One effective approach involves the development and implementation of national action plans aligned with the Sustainable Development Goals (SDGs). These plans provide a comprehensive framework for achieving sustainable development and can guide policies related to natural resource use. For example, the involvement of municipalities and sub-national governments help in the implementation of the SDGs at the regional level and across sectors.

Government intervention becomes essential in steering society towards sustainable practices. This can include measures to reduce or halt the use of certain natural capital goods and services through the use of tools such as taxes, fines and legislation. For instance, increasing the price of fossil fuels and imposing carbon-emission restrictions are direct interventions aimed at curbing environmental degradation and promoting cleaner alternatives.

Simultaneously, governments can encourage the use of specific natural capital goods and services by employing subsidies, enacting supportive legislation, launching publicity campaigns, and investing in research and education. An illustrative example is the promotion of innovative technologies such as carbon-storing concrete, recyclable wind turbines and the biological production of ammonia, as alternatives to traditional, more environmentally impactful practices like the Haber process.

REAL-WORLD EXAMPLE

Carbon-storing concrete

Concrete is one the most prominent building materials used in the construction industry. The process of making cement requires fossil fuels to burn clay and limestone at 1,450°C, which generates a lot of carbon dioxide. In order to reduce the carbon footprint of Portland cement and concrete formation, the processes of carbon sequestration by carbonation of the concrete are employed. The carbon dioxide reacts with calcium ions present in the cement and forms tiny nano-sized particles of limestone that stay embedded in the concrete forever.

Carbonation is a reaction between carbon dioxide (CO_2) and calcium hydroxide ($Ca(OH)_2$) and hydrated calcium silicate in the concrete, to form insoluble carbonates ($CaCO_3$):

- Capture of the CO_2 produced during the cement making process
- Purification of the CO_2
- Injection of the CO_2 into pressure tanks
- Mixing of the liquefied CO_2 with concrete. The CO_2 reacts with the calcium present in the concrete to form calcium carbonate ($CaCO_3$)
- Embedding of the tiny $CaCO_3$ particles in the concrete.

Beyond government actions, the influence of NGOs, local communities and social movements is significant. Their engagement can shape societal behaviours through campaigns, social media initiatives, and practical actions like recycling programmes. These entities play a crucial role in raising awareness, mobilizing communities, and holding both governments and businesses accountable for their impact on natural capital. For example, Detox My Fashion by Greenpeace raises awareness about fast fashion and provides several solutions to the problem of waste generated due to the fashion industry.

ATL ACTIVITY

Get into groups of three or four.
- Pick a country and list the different management strategies used to influence its society's resource use.
- Present your group's findings to others.
- Collate all the findings and share them with students as a resource for this subtopic.

7.1.13 The SDGs provide a framework for action by all countries in global partnership for natural resources use and management

The Sustainable Development Goals (SDGs) serve as a global framework for collective action to address various challenges, including those related to natural resource use and management. The framework provides flexibility to countries to set an achievable target for the different SDGs. For example, one of the aims of SDG 15 is the sustainable management of forests. Forests provide a variety of natural goods and services, therefore their protection and conservation is of prime importance. Over the years around the world, the degree of deforestation and changes in land use have drastically impacted the ecosystem, biodiversity and the livelihood of people dependent on forests.

The Bonn Challenge is a global effort to restore 350 million hectares of degraded and deforested land by 2030. It is directly aligned with SDG 15, aiming to protect, restore and promote sustainable use of terrestrial ecosystems. Across the world, 61 countries have taken on the challenge and are working at regional and national levels to promote SDG 15. Participating countries commit to restoring large-scale landscapes through sustainable forest management, afforestation and reforestation projects. This initiative not only contributes to biodiversity conservation, it also supports community livelihoods and mitigates climate change by sequestering carbon.

Another example is SDG 6, access to clean water and sanitation. Water scarcity, pollution and inadequate access to clean water are global challenges impacting ecosystems and human well-being. The implementation of Integrated Water Resource Management (IWRM) strategies aligns with SDG 6, aiming to ensure the availability and sustainable management of water and sanitation for all. Countries adopting IWRM consider the entire water cycle, incorporating environmental, social and economic aspects. This includes measures such as watershed management, water-use efficiency, pollution control and community engagement. By integrating water management across sectors, IWRM supports sustainable development, ecosystem health, and the provision of clean water for communities.

Link

The SDGs are introduced in Topic 1.3.18 (page 70).

HL.b.17: Environmental economics

SDG 12 provides a framework for sustainable production and consumption. It encourages closed-loop systems where materials are reused, repaired, recycled and repurposed. SDG 12 and the circular economy framework both help in minimizing waste and maximizing the value extracted from resources. Circular economy practices, such as reducing reliance on virgin materials and minimizing carbon emissions through recycling, contribute to mitigating environmental impact, supporting the climate action objective. Practices such as product life extension contribute to job creation and economic growth.

ATL ACTIVITY

- Analyse your daily activities, habits and consumer choices. Identify areas where circular economy practices align with the SDGs, such as reducing single-use plastics, promoting recycling and supporting sustainable products.
- Set personal goals to minimize waste, choose eco-friendly products or participate in community initiatives fostering circular practices.
- Maintain a reflective journal to document your progress and challenges.

◆ **Environmental Impact Assessment** – a tool used to assess the environmental, social and economic impact of a project before it is sanctioned.

◆ **Impact** – the change from the baseline situation caused by an action.

7.1.14 Sustainable resource management in development projects is addressed in an Environmental Impact Assessment

The integration of sustainable resource management into development projects is a critical aspect addressed through the **Environmental Impact Assessment** (EIA) process. An EIA is a project-specific comprehensive tool used to evaluate the potential environmental, social and economic **impacts** of a development project before it commences. It ensures that the project aligns with the principles of sustainability. The process of an EIA ensures public participation.

The EIA process may differ between countries but it has the common components shown in Table 7.3.

■ Table 7.3 The common components of the EIA process

Component of the EIA process	Explanation
Screening	Looking at whether or not the project requires an EIA. Not all projects need to undergo the EIA process.
Scoping	Finding out what impacts or issues are associated with the project and what the current state of the environment is at the site of the project. Also takes into consideration the stakeholders involved in the entire project.
Predicting the scale of the potential impacts	Providing a clear overall picture of the damage that can occur to the environment.
Mitigation	Looking at strategies that can lower or manage the scale of potential impacts.
Evaluation	Deciding whether or not the impacts are acceptable.
Preparation of the Environmental Impact Statement (EIS) or Report	Documenting the proposal, impacts, impact-mitigation and management options, level of significance and concerns.
Review of the EIS	The EIS is open for public comment for a sufficient period of time.
Decision making	Public comments are considered and a decision made about whether to accept the proposal as is, or to modify the proposal or reject the proposal outright.
Monitoring and review	Developing an implementation plan, and beginning to monitor and review the project.
Environmental impact on ecosystems, biodiversity, air and water quality, and natural resources	Identifying and understanding these impacts is crucial for sustainable resource management.
Social aspect	Effects on local communities, cultural heritage and quality of life. This helps in addressing social concerns related to resource access, use and potential conflicts.
Economic aspect	Evaluating the project's contribution to local economies, job creation and overall economic sustainability. This ensures that economic benefits are balanced with the responsible use of resources.
Independent experts conduct detailed surveys to gather data on the existing environmental, social and economic conditions	This involves studying the project area, identifying potential impacts and engaging with local communities. The independence of these surveys enhances the credibility of the assessment. Audits during the project-implementation phase ensure that the planned measures to mitigate environmental and social impacts are being effectively implemented. Regular monitoring post-project completion helps assess the actual outcomes and identify any unforeseen issues.
Sustainability evaluation	Considering whether the project aligns with Sustainable Development Goals, balances economic growth with environmental conservation and ensures social equity. This evaluation guides decision-makers in adopting measures that promote sustainable resource use.
Public participation	Providing opportunities for public participation allows local communities and stakeholders to voice their concerns and contribute to decision making. This participatory approach ensures that the project's impacts on natural resources are well understood by those directly affected.

HL.a.2: Environmental law

Environmental laws provide the legal basis for regulating activities that may impact the environment. EIAs are often required by environmental laws and regulations before the approval of development projects. These laws set out the standards and requirements that must be met for sustainable resource management. They specify the circumstances under which an EIA is required, ensuring that sustainable resource management considerations are systematically integrated into project planning and decision making. Penalties are put in place for non-compliance with EIA requirements and other environmental standards.

7.1.15 Countries and regions have different guidance on the use of EIAs

Guidance on the use of EIAs varies between countries and regions due to differences in legal frameworks, regulatory structures and environmental priorities. Some nations have specific legislation mandating EIAs for certain types of projects, while others may rely on broader environmental laws. For example, in India the Environmental (Protection) Act of 1986 means that Environmental Clearance (EC) is mandatory for new projects, expansion or modernization.

The institutions responsible for overseeing EIAs vary. In some countries, there is a centralized environmental agency or department that manages and approves EIAs. Other regions may not have a particular agency responsible for managing and approving EIAs, rather they may have multiple agencies each with specific responsibilities. The scope and thresholds for when an EIA is required can differ significantly. Some countries have a defined list of projects that always necessitate an EIA, while others use criteria such as project size, potential impacts or location to determine whether an assessment is required.

◆ **Baseline study** – the analysis of the existing environmental conditions at the site of the project.

Baseline studies provide the groundwork for the EIA process. They are carried out to know what the physical and biological environment is like before the project begins. The following parameters might be considered:
- Habitat type and abundance: Total area of each habitat
- Species list: Number of species of flora and fauna
- Species diversity: Estimated abundance and calculated diversity
- List of endangered species
- Land use: Type and coverage
- Hydrology: Volume, discharge, flows and water quality
- Human population: Present population
- Soil: Quality, fertility, pH.

7.1.16 Making EIAs public allows local citizens to have a role as stakeholders in decision making

Making Environmental Impact Assessments (EIAs) public promotes transparency, accountability, and public participation in the decision-making process. There are many advantages to doing this:
- Local citizens become stakeholders with the ability to engage in discussions and contribute to decisions that may impact their communities.
- The transparency builds trust between the public and the project managers and regulatory authorities.
- Decision-making processes becomes more democratic as local people are allowed to voice their concerns, give feedback and assess the situation. This ensures reduction in biased decision making and favouritism.
- The report generated will be written in non-technical language so that ordinary people can understand it and contribute their feedback.
- It serves as an opportunity for public education and awareness about environmental issues.
- It helps individuals understand the connections between human activities and their environmental impacts, fostering a sense of responsibility and environmental stewardship.

7.1 Natural resources, uses and management

- Early engagement and addressing concerns in the assessment phase can lead to more collaborative and mutually beneficial outcomes.

> ### Inquiry process
>
> Inquiry 2: Collecting and processing data
>
> Inquiry 3: Concluding and evaluating
>
> Identify a big project in your local area, such as a flyover, shopping mall, road extension or special economic zone. Research its impact on biodiversity and local residents.
> - Interview local residents and other stakeholders, or collect information using a survey.
> - Collect past and present photographs of the site (if easily accessible).
> - Collect some real-time data, such as soil analysis, water analysis, noise-level measurements, socioeconomic changes or biodiversity measurements.
> - Construct relevant tables and graphs.
> - Make a presentation on the impacts of the project and present it to your class.

7.1.17 While a given resource may be renewable, the associated means of extracting, harvesting, transporting and processing it may be unsustainable

As well as a resource's inherent renewable nature, the methods and practices used in resource extraction, harvesting, transportation and processing all contribute to its sustainability. This idea emphasizes the importance of a resource's entire life cycle and is known as a sustainability footprint. Table 7.4 gives some important points to remember.

■ Table 7.4 Aspects of a sustainability footprint

Aspect of sustainability footprint	Explanation
Extraction practices	Timber from forests can be renewable, but unsustainable logging methods can lead to deforestation, soil erosion and loss of biodiversity, rendering the overall process unsustainable.
Harvesting techniques	Fishing is generally a renewable resource, but overfishing, destructive fishing methods (bottom trawling, dynamite fishing, cyanide fishing) and bycatch can deplete fish populations and harm marine ecosystems, making the overall practice unsustainable.
Transportation impact	Biofuels derived from crops may be renewable, but if they are transported over long distances using fossil fuels, the environmental impact and carbon footprint may negate the renewable aspect.
Processing and manufacturing	Certain metals like aluminium are infinitely recyclable. However, the energy-intensive process of extracting and refining aluminium from bauxite can be environmentally damaging, affecting the overall sustainability.
Waste generation	Biomass for energy production is renewable, but improper disposal of byproducts or waste from the processing stage can lead to pollution and environmental harm, affecting the overall sustainability of the resource.
Land use practices	Agriculture, while relying on renewable resources (plants and soil), can become unsustainable if it involves extensive deforestation, excessive use of chemical inputs or monoculture, leading to soil degradation and loss of biodiversity.
Water usage	Hydropower is a renewable energy source, but large-scale dams may alter ecosystems, disrupt aquatic habitats and displace communities, impacting the overall sustainability of the resource.
Social and cultural considerations	Traditional practices of using renewable resources may become unsustainable if modern extraction or harvesting methods disrupt cultural practices, indigenous rights or the well-being of the community.

Considering the entire life cycle of a resource allows for a more comprehensive understanding of its sustainability. This holistic approach is crucial for making informed decisions about resource management and utilization, promoting practices that minimize environmental impact, support social well-being and ensure long-term viability.

HL.b.13: Environmental economics

Economic growth has impacts on environmental welfare. The unsustainable means of extracting, harvesting, transporting and processing renewable resources may result in problems, such as environmental degradation, pollution or habitat destruction. This can be overcome by incorporating full-cost accounting, which includes the environmental and social costs of resource extraction and processing. Environmental decision-makers can use this information to make informed decisions. Unsustainable means of processing a resource may lead to a reduction in its availability for future generations. By introducing taxes such as cap-and-trade, the responsible and sustainable use of resources can be ensured.

ATL ACTIVITY

Use the links below to find out more about full-cost accounting and cap-and-trade methods.
- Full-cost accounting: **www.freshbooks.com/glossary/accounting/full-costing**
- Cap-and-trade method: **https://youtu.be/bxs6ZrxLvHg**

7.1.18 Economic interests often favour short-term responses in production and consumption that undermine long-term sustainability

The conflict between economic interests favouring short-term gains and the imperative for long-term sustainability is a significant challenge in various industries and economic activities. This conflict is often driven by immediate financial incentives, and these may not align with the more gradual and comprehensive benefits of sustainable practices.

Economic systems frequently prioritize short-term profit maximization over long-term sustainability. Companies may focus on immediate financial gains without fully accounting for the environmental and social costs associated with resource depletion, pollution or social inequality. Consumer demand often favours inexpensive and readily available goods, contributing to a culture of disposable and resource-intensive products. This consumer behaviour suits short-term preferences, but can lead to long-term consequences such as resource depletion and environmental degradation. Industries engaged in resource extraction, agriculture or manufacturing may prioritize rapid production methods and exploitation of natural resources for immediate economic returns. This approach can lead to overexploitation, habitat destruction and loss of biodiversity, undermining long-term ecological health.

All of this means that investors and financial institutions, driven by short-term return expectations, may be hesitant to support long-term sustainability initiatives that might have initial high costs but yield benefits over an extended period. This dynamic can discourage sustainable investment practices. Companies may feel compelled to meet quarterly or yearly financial targets, for instance, which can influence decision making to prioritize short-term gains over sustainable practices that may take years to yield returns. Businesses may **externalize** (shift) the costs associated with environmental degradation, social inequality or health impacts, placing the burden on society and ecosystems, rather than **internalizing** (bearing) these costs into their operations.

◆ **Externalizing cost** – the cost generated by the producers but carried by the society as a whole. For example, pollution of a water body by a factory dumping waste in it.

◆ **Internalizing cost** – the cost of pollution being borne by the polluters/producers. For example, by implementing taxes, fees or emissions trading systems.

7.1 Natural resources, uses and management

> **REAL-WORLD EXAMPLE**
>
> ### Whale hunting
>
> Although long hunted on a much smaller scale, whales were hunted commercially in the eighteenth and nineteenth centuries mainly for their blubber, which was in demand as a lubricant and fuel for lamps. Other products, such as soap and perfume from whale fat, clothing from whale skin and whale meat, can also be obtained from the giant marine mammal. Modern hunting tools such as exploding harpoons made whaling easier. The introduction of steam-powered ships empowered fisherfolk to exploit fresh and large fishing grounds for whales. The number of kills increased, which made more whale products available in the market and drove demand, which further pushed the industry to supply the requirements.
>
> Commercial whale hunting is the prime example of exploitation in the history of humankind. Blue whales were protected from commercial whaling in 1966, but illegal whaling continued. Antarctic blue whales were hunted to near extinction until the 1970s. Whaling operations reduced the Southern Ocean population from approximately 250,000 to a few hundred in less than 50 years. In 1982, the International Whaling Commission agreed to a global moratorium on commercial whaling, which came into effect in 1986. Although their population is slowly increasing, Antarctic blue whales are still classified as critically endangered by the International Union for Conservation of Nature (IUCN), with only a few thousand estimated to remain. Over the period, the overexploitation caused many whale species to become extinct. Including the environmental cost of whaling in the products derived from whales may help in reducing their demand, thereby curbing the whaling industry.

7.1.19 Natural resource insecurity hinders socioeconomic development and can lead to environmental degradation and geopolitical tensions and conflicts

Natural resource insecurity poses a significant threat to socioeconomic development, fostering environmental degradation, and fuelling geopolitical tensions and conflicts. Struggles over resource access and control not only hinder sustainable growth but also contribute to global power dynamics, emphasizing the urgency of responsible resource management for long-term stability.

> **REAL-WORLD EXAMPLE**
>
> ### Cobalt mining in the Democratic Republic of Congo, Africa
>
> The Democratic Republic of Congo is a major global supplier of cobalt, a critical component in the production of batteries for electric vehicles, mobile phones and other electronic devices. Cobalt mining has been driven by global demand for these technologies. The economy of DR Congo heavily relies on cobalt exports, and a significant portion of the country's GDP is linked to mining activities. However, this economic dependence on a single resource can create vulnerabilities when global demand fluctuates or when prices are subject to volatility.
>
> In the pursuit of meeting global demand for cobalt, DR Congo has faced challenges related to unregulated and environmentally damaging mining practices. Artisanal

mining, in particular, often involves manual extraction methods that lead to deforestation, soil erosion and water pollution. Mining has raised significant human rights concerns, including issues related to child labour, unsafe working conditions, and lack of social and environmental safeguards.

Although cobalt is mined in DR Congo, a substantial portion of the global processing capacity for cobalt is concentrated in China. This creates a geopolitical dynamic where China holds significant influence over the supply chain for a critical component of emerging technologies. China's dominance in cobalt processing provides it with a strategic advantage in the global supply chain for electric vehicles and electronics. This control over resources can influence diplomatic and economic relations with cobalt-producing countries, impacting geopolitical power dynamics.

The high economic value of cobalt, coupled with issues of governance, has led to conflicts and instability in DR Congo. Armed groups have exploited the cobalt trade to fund their activities, contributing to regional conflicts and humanitarian crises. Global demand for cobalt, driven by the transition to electric vehicles and renewable energy technologies, has heightened competition for control over cobalt resources. This can lead to geopolitical tensions among major consumer nations seeking to secure their access to these critical resources.

7.1.20 Resource security can be brought about by reductions in demand, increases in supply, or changing technologies

Ensuring resource security involves a multifaceted approach, achieved through reductions in demand, increases in supply and advancements in technology. Sustainable practices, such as improved resource efficiency and conservation, can alleviate demand pressures. Simultaneously, expanding supply through responsible extraction and exploration, coupled with the adoption of innovative technologies, contributes to a more resilient and secure resource landscape. Balancing these strategies is essential for promoting long-term resource sustainability and mitigating potential conflicts arising from scarcity.

Enhancing resource security involves strategic measures such as reducing demand through efficiency and conservation, and minimizing reliance on imported resources. For example, the three main realms of resource security are food, water and energy securities. In the sphere of food security, some novel technologies and advanced technologies can enhance the efficiency of farming practices. These technologies include precision agriculture, which includes the use of drones, sensors and data analytics. By precisely managing resources such as water, fertilizers and pesticides, precision agriculture reduces waste and increases yields, contributing to food security.

7.1.21 Economic globalization can increase supply, making countries increasingly interdependent, but it may reduce national resource security

◆ **Economic globalization** – the widespread international movement of goods, capital, services, technology and information.

◆ **LEDCs** – low economically developed countries.

Economic globalization means the increasing interdependence of countries because they trade goods, services and technology across borders and share money between their economies. On the one hand this has created a strong foundation for supporting **LEDCs** and has been well accepted by many economies around the world, both established and growing. However, the interconnectedness of global markets may lead to heightened competition for resources, potential vulnerabilities in **supply chains** and exposure to external shocks. Nations may find their resource security compromised as economic ties become more intricate, emphasizing the need for strategic resource management policies to navigate these complexities.

◆ **Global supply chain** – all the steps and procedures involved in manufacturing and delivering a product when those steps take place in more than one country.

◆ **Integrated Water Resources Management** – a process that promotes the development of water, land and other associated resources.

Advantages of economic globalization

Economic globalization facilitates the international trade of food products, allowing countries to access a diverse range of agricultural goods. For instance, Japan relies on global markets for a significant portion of its food supply, ensuring a stable and varied diet even with limited domestic agricultural resources. It also enables the transfer of renewable energy technologies across borders. Germany's expertise in solar energy and wind power has been shared globally, contributing to the adoption of sustainable energy practices and enhancing energy security in various nations such as Spain, India and China.

Water security is an important feature for every country. If a water body is shared between countries or states then there have to be set regulations for its use to ensure that the water body is used sustainably. Economic globalization facilitates the establishment of agreements and mechanisms for shared resource governance. The United Nations provides set protocols for managing shared water resources, such as the **Integrated Water Resources Management** process.

REAL-WORLD EXAMPLE

The Mekong River

The Mekong River runs through several countries including China, Laos and Vietnam. It benefits from collaborative efforts in water governance, ensuring sustainable water use and enhancing water security in the region. This governance has existed for more than 60 years, making the Mekong River one of the first transboundary rivers to be governed by an international river body and in accordance with the principles of equitable use.

Disadvantages of economic globalization

Globalization increases dependency on global supply chains. Countries that are heavily reliant on global food supply chains may face vulnerabilities. The 2007–08 food crisis highlighted the risks of dependency, impacting nations like Haiti, which experienced food shortages due to its reliance on imported rice. Globalization can expose countries to oil price fluctuations. For instance, during the 2014 oil price drop, oil-dependent economies like Nigeria faced economic challenges as revenues declined, impacting energy security and broader economic stability. Global trade in water-intensive products can contribute to water stress in exporting countries. For instance, Egypt's reliance on water-intensive wheat imports contributes to its virtual water footprint and places strain on water resources in exporting nations.

ATL ACTIVITY

How does the utilization of a specific product differ in contrasting societies?

Choose a specific product for investigation (for example, steel, concrete, inorganic fertilizer) based on relevance and availability of data.
- Select two contrasting societies or countries for comparison. Consider factors such as economic development, geographical location or cultural differences.
- Use reputable secondary data sources (such as Gapminder, Our World in Data and the World Bank) to gather relevant information on resource consumption, production and related indicators for your chosen societies.
- Compile the gathered data into organized datasets. Use spreadsheet software for this.
- Create graphical representations (for example line graphs, bar charts) to illustrate the trends and patterns of resource utilization over a specific period for both societies.
- Apply basic statistical tests to analyse the significance of differences or correlations in resource use between the selected societies. This may include t-tests, correlation analysis or other relevant statistical measures.

- Compare and contrast the resource utilization patterns between the two societies, and identify the potential factors influencing these patterns.

After completion of the task, answer the following questions:
- What economic, environmental or cultural factors may explain the observed differences in resource use?
- How might these patterns impact the sustainability and development of each society?

ATL ACTIVITY

Engage in citizen science initiatives

Engage in citizen science initiatives to develop scientific inquiry skills, foster a sense of environmental responsibility and contribute to real-world research.
- Visit citizen science platforms like Zooniverse or Project Noah.
- Select a project and actively contribute as instructed by the platform.
- Clearly document your observations and data according to the project's guidelines.
- Reflect on how your participation may contribute to scientific knowledge.
- Share your citizen science experience with a peer or teacher, discussing the challenges and insights.

ATL ACTIVITY

Investigate the sustainability of a local food-production system
- Choose a local farm (organic or conventional).
- Collect data by visiting the farm, interviewing farmers and researching online sources. Consider the types of crops grown, the water management practices and any certifications related to organic farming.
- Evaluate the sustainability practices employed by the farm. You may consider environmental conservation, biodiversity, well-being of local communities, and aspects like energy efficiency and waste reduction.
- Compare your findings with a contrasting farm (organic or conventional).
- Create a presentation or report summarizing your findings. If you visited an organic farm, discuss the benefits and challenges of its sustainability practices compared to conventional methods. Offer recommendations for consumers interested in supporting sustainable local food production.

REVIEW QUESTIONS

1. State the difference between natural capital and natural income. Give appropriate examples.
2. Discuss the advantages and disadvantages of using renewable sources in comparison to non-renewable sources.
3. Explain why Environmental Impact Assessments (EIAs) are essential.

EXAM-STYLE QUESTIONS

1. Outline the factors influencing resource security. [4 marks]
2. Discuss the potential ecosystem services and goods provided by a named ecosystem. [7 marks]
3. Comment on how the value of a named natural resource has changed over time. [7 marks]

7.1 Natural resources, uses and management

7.2 Energy sources, use and management

> **Guiding questions**
> - To what extent can energy consumption be equitable around the world?
> - How can energy production be sustainable?

SYLLABUS CONTENT

This chapter covers the following syllabus content:
- ▶ 7.2.1 Energy sources can be either renewable or non-renewable.
- ▶ 7.2.2 Global energy consumption is rising with increasing population and per capita demand.
- ▶ 7.2.3 The sustainability of energy sources varies significantly.
- ▶ 7.2.4 A variety of factors will affect the energy choices that a country makes.
- ▶ 7.2.5 Intermittent energy production from some renewable sources creates the need for energy-storage systems.
- ▶ 7.2.6 Energy conservation and energy efficiency may allow a country to be less dependent on importing a resource.

HL ONLY
- ▶ 7.2.7 Energy security for a country means access to affordable and reliable sources of energy.
- ▶ 7.2.8 The global economy mostly depends on finite reserves of fossil fuels as energy sources; these include coal, oil and natural gas.
- ▶ 7.2.9 Nuclear power is a non-renewable, low-carbon means of electricity production.
- ▶ 7.2.10 Battery storage is required on a large scale to meet global requirements for reduction of carbon emissions, but it requires mining, transporting, processing and construction, all of which produce emissions and pollution, and cause sociopolitical tensions.

7.2.1 Energy sources can be either renewable or non-renewable

Renewable energy sources, which include hydropower, wind, solar, tidal, wood and geothermal energy, use natural processes to produce energy on either a large or small scale. Renewable energy sources are considered to be more sustainable as they can be naturally replenished and have low environmental impact.

Non-renewable sources such as fossil fuels and nuclear power are also used to derive energy. A major part of the energy released by both types of resources is used to generate electricity. Nuclear power is considered a non-renewable source because the Earth has a finite source of the radioactive metal uranium, which is used in nuclear power generation.

In the past, renewable resources for energy production were not very popular due to a lack of technology to extract their full potential, as compared to the non-renewable resources. However, watermills have traditionally been used in various fields including agriculture and metallurgy. Their use for generation of electricity began in 1870 in England. Hydroelectricity is one of the oldest ways of producing electricity, followed by wind turbines, which were first used in 1887 in Scotland to produce electricity. Similarly, humans have always used firewood to produce energy. Development of technology provided better means of harnessing energy resources. Installation of solar and wind energy power plants involves high initial costs. The energy produced by wind turbines is inconsistent due to the change in wind speeds. This may be one of the factors affecting the slow acceptance of wind energy as an alternate energy source. However, with the diminishing reserves of non-renewable resources and their severe environmental impacts, the focus is rapidly turning towards renewable resources. These resources cause less damage to the environment and thus technology for their use is constantly being updated to improve its efficiency.

Common mistake

People often assume that renewable energy sources, such as solar or wind, are always environmentally friendly and sustainable, while non-renewable sources, like fossil fuels, are inherently harmful because their stores are limited and they pollute the atmosphere. This assumption may be due to oversimplifying the distinction between renewable or non-renewable energy sources solely based on the availability of the resource.

7.2.2 Global energy consumption is rising with increasing population and per capita demand

The increasing global population leads to increased resource consumption. Currently, most energy demand is being fulfilled by fossil fuels.

The availability and awareness of renewable sources of energy is gradually growing; however, it is not enough to meet the high demands of steel, concrete and fertilizer industries. Around 1.76 MJ of energy is required for the production of 1 kg of Portland **cement clinker**. The dry process of cement manufacturing requires a huge amount of energy at various steps. First, the raw materials (limestone, clay, shale, iron ore, gypsum) are mined and put into a crusher to make fine particles, which are then preheated to remove moisture. This process requires around 27 per cent energy. Further, the mixture is transferred into a kiln for clinker production, which itself consumes 25 per cent energy because the kiln has to be maintained at 1400°C –1500°C. The clinker is then cooled to stabilize its crystalline structure, then ground into fine cement particles, and finally packaged. This takes up almost 48 per cent of energy. Therefore, these industries may continue to depend on non-renewable energy resources for a long time to come.

Addressing the challenge of meeting the escalating energy demand requires a dual approach. This involves transitioning to alternative energy-production resources via the increasing adoption of renewables, and simultaneously focusing on reducing overall energy consumption. This transformation is driven by global and local changes, influenced by factors like climate change concerns, technological advancements, and shifting societal values towards sustainability. Globally there is an increased awareness around the use of alternative energy sources such as solar and wind energy. The Paris Agreement reflects a global commitment to reduce carbon emissions, influencing nations to transition towards cleaner energy sources and invest in sustainable practices. Technological advancements in the design of wind turbines has improved their efficiency. At the local level, governments are implementing policies that incentivize the adoption of renewable energy technologies, such as subsidies for installation of solar panels.

◆ **Cement clinker** – the solid material that is produced in the manufacture of Portland cement.

HL.b.12: Environmental economics

Global economic growth depends on the supply and demand of resources. Population growth is linked to an increasing per capita demand for energy resources. This creates unwanted pressure on the system for providing uninterrupted energy supply. A mix of fossil fuels, renewable energy sources and other sources of energy (waste to energy) are being used to meet the growing demand of industries, households, and transportation systems worldwide. However, this approach is based on linear economy, which does not account for waste generation and pollution.

Countries with larger populations experience a higher demand of energy to support industrial production, urbanization and the rising demand for consumer goods. The growing populations not only contribute to an overall increase in energy consumption, but also intensify per capita energy demands as people aspire to improve their quality of life. Policymakers, businesses and communities are faced with the intricate task of finding a balance between achieving economic prosperity and ensuring responsible energy usage.

> **ATL ACTIVITY**
>
> Choose an energy resource (for example coal, oil, solar energy) and research how its demand and supply have changed over the last 50 years.

HL.c.3: Environmental ethics

Resource stewardship refers to the responsible and sustainable management of resources to ensure that they are available for the future generations. The ethical judgement, here, is drawn in favour of natural resources on the grounds of compassion, respect and good stewardship. This includes judgements regarding fossil fuels and minerals that are used in the production of renewable technologies such as solar panels, wind turbines and batteries for e-vehicles. It requires collaboration between stakeholders including governments, industries, communities and individuals to balance competing interests and priorities, and to control pollution, habitat destruction, and climate change resulting from certain energy sources.

> **ATL ACTIVITY**
>
> Reflect on your personal habits and practices related to resource stewardship, and identify areas for improvement.

TOK

How can we discern and justify our moral obligation to leave behind a habitable and sustainable environment for future generations?

Approach this question through the key concepts. For example:
- Recognize that different individuals and communities may have varied perspectives on moral obligations to the environment, influenced by cultural, historical and personal factors.
- Reflect on the methods and reasoning used to justify moral obligations, considering ethical frameworks, scientific evidence and cultural values.

7.2.3 The sustainability of energy sources varies significantly

Energy sources have distinct ecological footprints including the direct emissions that they release and the environmental impacts associated with their extraction, processing, transportations and disposal. Fossil fuels are hydrocarbons, have a finite source, and are highly unsustainable as they produce huge amounts of carbon dioxide, a greenhouse gas. Furthermore, the extraction of fossil fuels such as coal and crude oil also involves significant environmental disruption, including habitat destruction and landscape alteration. Environmental restoration efforts may focus on reclamation and rehabilitation of mined areas, aiming to mitigate ecosystem impacts. Although nuclear energy generates low emissions, long-term nuclear waste disposal is a struggle. The mining of uranium for nuclear energy can lead to soil and water contamination, and so may require restoration efforts such as soil remediation and water purification to restore affected ecosystems. On the other hand, solar energy using photovoltaic cells, wind, and tidal energy using turbines, offer long term sustainability despite the challenges in manufacturing and end-of-life management.

Germany and other countries have made significant progress in incorporating renewable energy into their grids. France's commitment to nuclear energy has reduced its carbon footprint. Over 70 per cent of France's electricity is generated by nuclear power, demonstrating a low-carbon alternative that contributes to energy sustainability. Energy-efficiency technologies such as LED lighting, smart grids and energy-efficient appliances have reduced overall energy consumption in both countries.

> **REAL-WORLD EXAMPLE**
>
> ### Germany's energy transition
>
> Germany's power market has been transitioning from conventional to renewable energy sources for the past few decades. It now aims at reaching 80 per cent renewable-energy generation by the year 2030, phasing out the use of coal completely by this time. One notable example of Germany's commitment to sustainable energy is the Energiewende (energy transition) policy, introduced in 2011. The policy aims to reduce greenhouse gas emissions by 80 per cent by the year 2050. To achieve this, the country has significantly increased its renewable energy capacity, with a substantial portion of its electricity generated from wind turbines and solar panels, and is also looking at biomass and geothermal energy. In 2020, renewable energy sources accounted for over 45.2 per cent of Germany's electricity production (source: 'Renewable energy in Germany', Clean Energy Wire). Until March 2011, Germany was obtaining a quarter of its energy from nuclear energy generated in 17 nuclear plants.
>
> After Japan's Fukushima nuclear disaster in 2011, Germany decided to phase out its nuclear power plants and to focus on expanding onshore wind power. The last three nuclear reactors shut down in April 2023. This move clearly indicates the country's dedication to move away from non-renewable and hazardous sources for energy production. The National Action Plan on Energy Efficiency (NAPE) is aimed at promoting energy efficiency across different sectors of the economy. The focus is on expanding the use of renewables and efficient use of energy. There are three ways in which NAPE addresses the issue of energy efficiency:

- by stepping up energy efficiency in the building sector,
- by establishing energy efficiency as an investment and business model,
- by increasing individual responsibility for energy efficiency.

These efforts are reflected in Germany's commitment to the Paris Agreement and its aim to achieve a carbon-neutral economy by 2050.

Germany has developed many innovative technologies and business models. For example, the country has seen the rise of community-owned wind turbines, where local communities invest in and operate their own wind turbines, providing them with a sustainable source of energy and a share of the revenue generated. There are challenges associated with Germany's energy transition strategies, such as the intermittency of renewable energy sources, the need for grid expansion, and the associated costs have sparked discussions on the practicality and feasibility of such a transition.

Tables 7.5 and 7.6 list the advantages and disadvantages of renewable and non-renewable energy sources.

■ Table 7.5 Renewable energy sources

Advantages	Disadvantages
Renewable sources are naturally replenished.	Dependence on weather conditions for some sources like solar and wind, leading to fluctuations in energy production.
Reduced greenhouse gas emissions and minimal environmental degradation.	Installation of infrastructure can be expensive.
Growing renewable energy sectors contribute to employment opportunities.	May require significant (large) land areas.
Various options like solar, wind and hydropower offer flexibility.	Lower output of energy than some non-renewable sources.
Can be harnessed locally, reducing dependence on centralized power grids.	Technological advancements are needed to increase efficiency and storage capabilities.

■ Table 7.6 Non-renewable energy sources

Advantages	Disadvantages
Fossil fuels and nuclear sources provide concentrated energy.	They are a finite resource.
Offer consistent energy production, irrespective of weather conditions.	High level of greenhouse gas emissions, air and water pollution, and habitat disruption.
Well-established systems for extraction, refinement and distribution.	Dependence on specific regions for resource supply can lead to geopolitical tensions.
Steady power output, enabling continuous energy supply.	Nuclear energy carries risks associated with accidents and radioactive waste.
Accessible and ready for immediate use.	Resources have limited long-term viability.

You should also be able to evaluate the advantages and disadvantages of different energy sources.

■ Examples of non-renewable energy sources

Coal

Coal occurs naturally and is formed by the carbonation of prehistoric plants and animals over millions of years under very high temperature and pressure. As such, it is considered a non-renewable source as this is not quickly replaceable. As we move from the Earth's surface towards the mantle, there is a rise in temperature. The dead remains of plants and animals which were buried under layers and layers of rocks and dirt decay to form peat which ultimately becomes bituminous or lignite coal. The process of coal formation (called coalification) releases carbon dioxide and traps methane in small pockets in the coal. There are four main types of coals formed by this process: anthracite, bituminous, sub-bituminous and lignite. Each type differs in the amount of carbon content and the amount of heat energy it can produce. The burning of bituminous coal creates more pollution than anthracite and lignite since it produces greater particulate matter and has a higher sulfur content, resulting in acid rain.

■ Figure 7.8 The process of coal formation

During the Industrial Revolution (1760–1840), coal was one of the most extensively used fuels that drove production and transportation. Steam engines, a prominent invention of this era, were fuelled by coal, revolutionizing transportation, manufacturing and energy production.

The widespread availability of coal and its high energy density made it a favoured choice for powering steam engines and, later, electricity generation. Extensive coal reserves made it readily available and cheaper than other energy sources in the UK, the first industrialized country. This increased its popularity and it became the fuel of choice. The development of advanced mining technologies, such as steam-powered pumps and later electric machinery, facilitated deeper and more efficient extraction of coal. The construction of railways enabled the transportation of coal over longer distances, connecting the mines to the industrial centres.

The British coal industry played a crucial role in the Industrial Revolution. Coal mines proliferated across regions like Yorkshire and Lancashire in the north of England, powering factories and railways. Coal also fuelled the expansion of the American railway system and powered industries. As the world's largest coal producer and consumer, China heavily relies on coal for electricity generation and industrial processes. As such, the rapid industrialization of that country in recent decades has intensified coal usage.

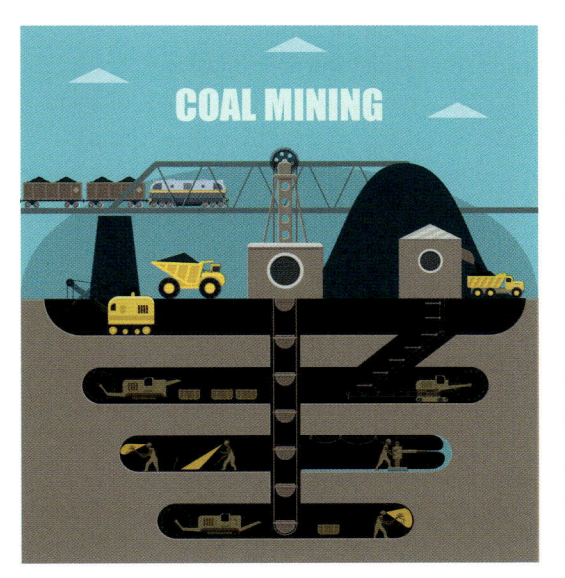

■ Figure 7.9 Mining of coal using technology

7.2 Energy sources, use and management

◆ **Exajoule (EJ)** – unit of energy equal to 10^18 joules.

China is, however, aiming to reduce its reliance on fossil fuels and promote renewable energy sources. Figure 7.10 shows a future projection of the current status of clean energy versus fossil fuel energy in China.

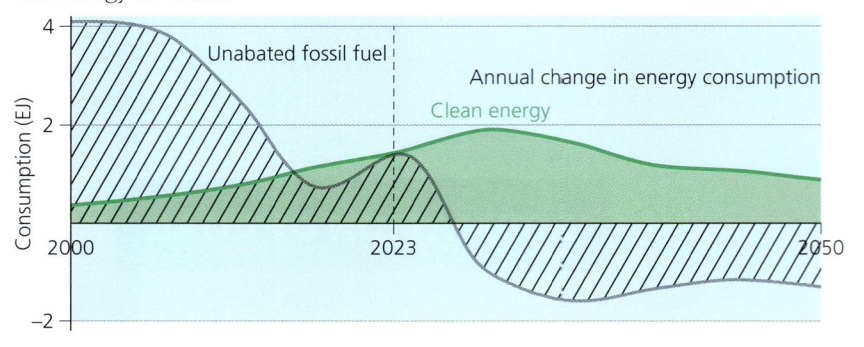

■ **Figure 7.10** World Energy Outlook 2023

Crude oil

Crude oil is also a fossil fuel. It gained popularity and value as petroleum, diesel, aircraft fuel, synthetic materials such as nylon, and other raw materials for petrochemical industries could be extracted from it. At that time, however, no one realized that fossil fuels could cause pollution and harm the environment.

Crude oil is formed from the remains of marine microorganisms that lived millions of years ago. Over time, these microscopic organisms accumulated in large quantities on the ocean floor, forming organic sediments. The organic sediments buried deep in the ocean beds were transformed into hydrocarbons due to heat and pressure. Biological matter breaks down into organic ketamine and inorganic bitumen matter, then further into different types of hydrocarbons depending on the pressure and heat generated by the layers of sediments. The resultant liquid and gaseous hydrocarbons are collectively known as crude oil. Underground reservoirs were created by the upward movement of the less dense crude oil through porous and permeable rock formations until it encountered an impermeable layer. Crude oil is considered to be a non-renewable energy source because it takes millions of years for the process to occur and crude oil to accumulate. These reservoirs became the primary targets for oil exploration. Crude oil reservoirs are mainly found in sedimentary rocks including **sandstone** and **limestone**. These porous reservoir rocks mainly contain carbonates, **quartz**, and **chert**. Crude oil can be extracted by drilling through these rocks and pumping the oil out.

◆ **Sandstone** – a type of sedimentary rock consisting of sand or quartz grains cemented together, typically red, yellow or brown in colour.

◆ **Limestone** – a type of sedimentary rock formed by the accumulation of organic remains (shells or corals) consisting mainly of calcium carbonate.

◆ **Quartz** – a hard mineral consisting of silica, typically occurring as colourless or white hexagonal prisms.

◆ **Chert** – a hard, dark, opaque rock composed of silica.

The Middle East, particularly countries like Saudi Arabia, Iran, and Iraq, holds some of the world's largest proven oil reserves. The discovery of vast oil fields in the early to mid-twentieth century transformed the economic landscape of the region. The discovery of the Spindletop oil field in Texas, USA, in 1901 marked a significant turning point, leading to a surge in oil production in the United States. Similarly, the Siberian oil fields, such as the Samotlor Field, contributed significantly to the oil output of the Soviet Union and later of Russia.

Crude oil emerged as another cornerstone of the Industrial Revolution, powering transportation, manufacturing, and various industries. The invention of the internal combustion engine, which ran on refined petroleum products, further

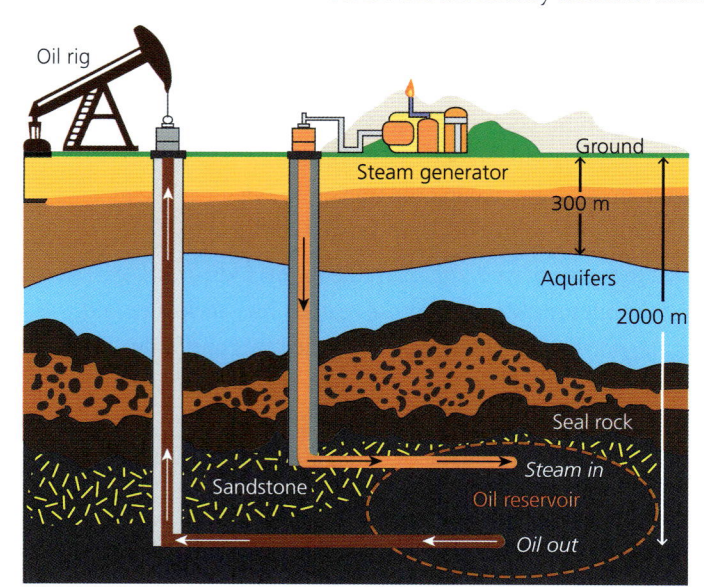

■ **Figure 7.11** Extraction of crude oil from reservoir rocks

640   **Theme 7:** Natural resources

◆ **Carcinogens** – substances that have the ability to trigger cancerous genes to be switched on.

◆ **Shale** – categorized as fine-grained sedimentary rock found deep inside the Earth.

increased the demand for crude oil. The extraction of crude oil from ocean beds affects marine animals through destruction of their habitat, and oil spills. Oil spills can also affect wetlands and oyster reefs, causing a disturbance in the lifecycle of the organisms and in their migration patterns. Since crude oil is a mixture of hydrocarbons, the volatile lighter portions of the mixture, containing benzene, toluene, xylene and ethylbenzene, are released into the air at the site of oil extraction and of oil spills. These chemicals are highly toxic to humans and other organisms as they are **carcinogens**. The use of crude oil negatively impacts the environment by increasing the percentage of greenhouse gases (CO_2, NO_2) in the atmosphere. SO_2 and NO_2 released from the combustion of crude oil may cause acid rain or acid deposits. It also causes the air quality index to decline due to release of particulate matter and soot.

REAL-WORLD EXAMPLE

Shale oil

Shale oil is a type of crude oil trapped in impermeable rocks. It is buried deep inside the Earth in sedimentary shale rocks, and it requires hydraulic fracturing for its extraction. Crude oil, on the other hand, is found in sandstone deposits that are easy to break. Shale oil can be used to extract diesel, gasoline, liquid petroleum gas (LPG), ammonia and sulfur. The spent rock can also be mixed with limestone to make cement, bricks, tile and pottery. Shale oil reserves are abundant in the United States, Russia, China, Argentina and Libya.

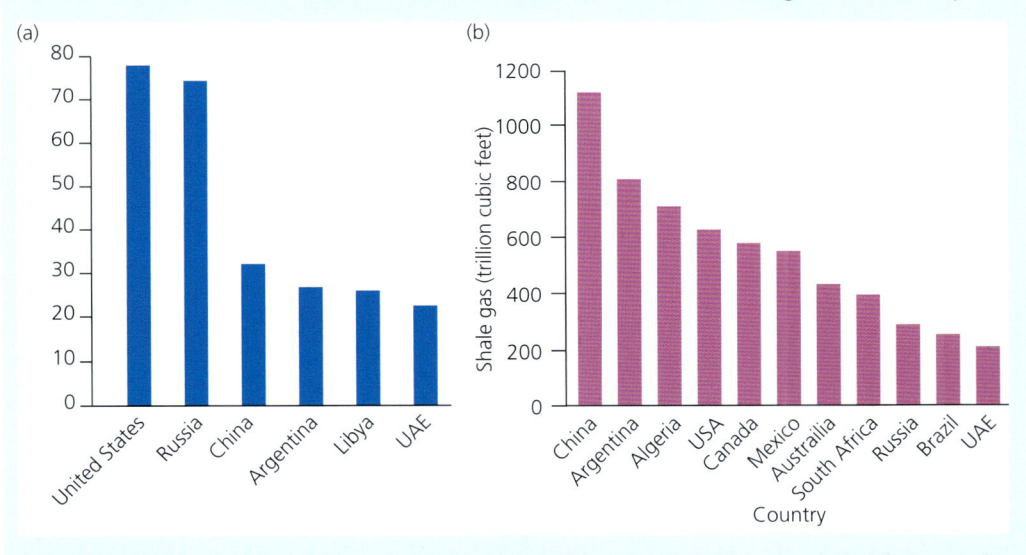

■ **Figure 7.12** (a) Shale oil reserves in various countries. (b) Countries with major shale oil reserves in the world

Countries with abundant shale resources may experience enhanced energy security, reducing vulnerability to disruptions in global oil markets. Shale oil production has allowed countries to diversify their sources of energy, reducing dependence on traditional oil-producing nations. Significant reserves have been found in China, the USA, Argentina and South Africa. The United States has gained greater energy independence, altering its geopolitical relationships with oil-exporting countries. Reserves are most abundant in Russia, Qatar and Algeria, thus these countries dominate the world markets for shale oil and gas. The increase in shale oil production, especially in the United States, has altered the dynamics of the global energy trade. Traditional oil-exporting countries, such as Russia and members of OPEC (the Organization of the Petroleum Exporting Countries), faced challenges as the USA became a major exporter of oil. China's total shale oil deposits have been estimated to be around 720 billion tonnes. Shale oil production has contributed to a more competitive market and challenged OPEC's ability to control oil prices and exert geopolitical influence.

Fracking, or hydraulic fracturing, is a method of extracting shale oil and natural gas from underground rock formations. In the USA, this technique has been widely utilized in shale oil extraction, particularly from formations such as the Bakken Formation in North Dakota and the Eagle Ford Shale in Texas. The process involves injecting a high-pressure fluid mixture (usually water, sand and chemicals) into the rock to fracture it, releasing the trapped oil and gas.

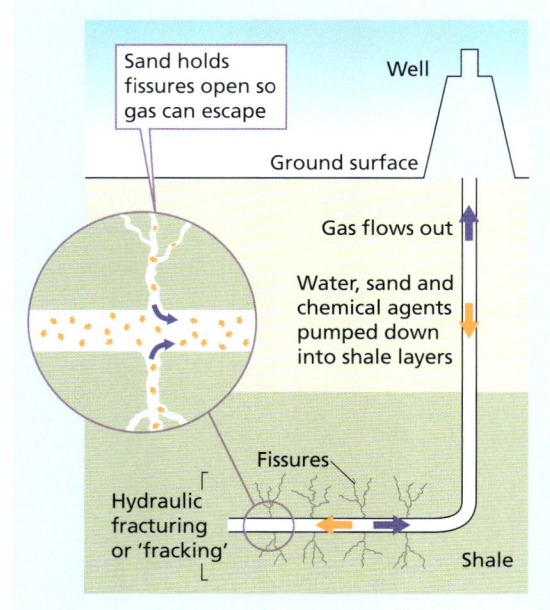

■ Figure 7.13 Fracking of shale oil

While fracking has significantly increased domestic oil and gas production, it is a controversial practice due to environmental concerns. Environmental implications include:

- The chemicals used in fracking fluid, as well as naturally occurring substances in the underground formations, can contaminate drinking water sources.
- Fracking operations release pollutants into the air, including volatile organic compounds (VOCs) and other hazardous air pollutants. VOCs are the major cause of smog. In the presence of sunlight, they react with NO_2 released by vehicles, industries and power plants to generate tropospheric ozone. This can cause respiratory problems in humans, such as difficulty in breathing, asthma, bronchitis, and inflammation of the lining of the lungs.
- Methane, a potent greenhouse gas, can also be released during extraction and transportation, contributing to climate change.
- Wastewater known as flowback, which is generated during the fracking process, contains chemicals and salts that can lead to soil and water contamination.
- The impact of hydraulic fracturing has been reported to generate some low-level seismic activity. In addition, the process of disposal of fracking wastewater by injecting it into deep Class II wells may be closely associated with larger earthquakes, especially in the United States.

(Source: National Research Council. 2013. Induced Seismicity Potential in Energy Technologies. Washington, DC: The National Academies Press.)

Tar sands or oil sands

Bitumen is a thick, black, sticky oil which clings to sand and clay. Tar sands are a mixture of sand, clay, water and bitumen. Since it is very thick it cannot be pumped out of the ground using conventional oil drilling methods. Instead, it has to be mined using the oil-pit mining or the in-situ mining method.

- Oil-pit mining: tar sands near the surface can be directly mined by digging open pits. They are then transferred to extraction plants where bitumen is separated from the sand, clay and water.

- In-situ mining: tar sands which are present at greater depths are extracted by in-situ mining. Two wells are drilled into the ground approximately 50 metres apart. One well serves as the injection well while the other serves as the production well. Hot steam or solvent is injected through. High-pressure steam heats up the bitumen, reducing its viscosity and allowing it to flow more easily. The bitumen then flows towards the production well, from where it can be pumped out. Various separation techniques are used to separate sand, clay and water from bitumen before it is sent for refining. This may involve distillation and upgrading process to improve the quality of bitumen.

Alberta in Canada and Venezuela have the largest tar sand deposits in the world.

Bitumen extracted from tar sands is used for the following purposes:

- Road construction: bitumen is used as a binding agent in asphalt concrete. Asphalt or black-top roads are made using this material. It makes the surface of the road durable and resistant to water and weathering.

- Roofing: bitumen is used in roofing materials which provides durability and protection against UV and harsh weather conditions.

- Waterproofing: bitumen is used as a waterproofing agent in basements, foundations, exterior walls, roofs, tunnels from water infiltration.

- Adhesives and sealants: bitumen is used in construction and manufacturing industries as bonding agent for materials like wood, concrete and metal. It is a very strong adhesive.

Natural gas and liquefaction

Liquefaction is a process that converts gaseous natural gas into a liquid state, known as liquefied natural gas (LNG). It involves cooling natural gas to extremely low temperatures, typically around -162°C. This causes the gas to condense into a liquid, reducing its volume significantly. The liquefied form is not only more economically viable for long-distance transportation, but also facilitates storage and distribution across different geographical locations.

Qatar, a pioneer in LNG production, has harnessed liquefaction technologies to emerge as the world's leading exporter of LNG. The country's investment in advanced liquefaction facilities, such as the Ras Laffan Industrial City complex, underscores the pivotal role of LNG in meeting global energy demands. Australia has also embraced LNG liquefaction, capitalizing on its abundant natural gas reserves. Projects like the Gorgon LNG facility contribute significantly to Australia's emergence as a major player in the global LNG market.

Nuclear energy/mining of uranium

Uranium is a radioactive element used in nuclear power plants to produce clean energy. The mining of uranium is a complex process that involves extracting the element from the ground and processing it into fuel for nuclear reactors. While there are environmental concerns associated with uranium mining, it is a necessary practice to meet the world's growing energy demands.

The McArthur River Mine is one of the largest uranium mines in the world, located in the Northern Territory of Australia. The mine has been operational since 1981 and has produced over 100,000 tons of uranium oxide concentrate. The mine has implemented various environmental measures, including rehabilitation of mined areas and management of groundwater resources. Uranium can be extracted from its ore by conventional mining from the rock, or by using chemicals to melt the uranium and then pumping it out from the ore while still inside the ground. In the former process, the ore is transported to mills where it is crushed and ground before uranium can be extracted from it by chemical leaching. The process leaves behind solid radioactive waste called **tailing**, and liquid radioactive waste called raffinates. These wastes are stored in specially designed ponds called impoundments. Uranium slowly decays to radium, which further decays to radon gas. Radon gas is highly radioactive.

◆ **Tailings** – any leftover materials from the process of mining ore.

Advantages of nuclear energy:

- The initial construction costs of nuclear power plants are high. However, once operational, nuclear power stations can produce electricity at a relatively low cost.
- It provides a constant and reliable source of energy, unlike some renewable sources that are intermittent.
- Nuclear power is considered a low-carbon energy source because the electricity-generation process does not produce carbon dioxide during operation, unlike fossil fuels. Compared to coal and natural gas, nuclear power plants also produce fewer toxic gases such as SO_2 and NO_2. It can therefore offer a good option to reduce greenhouse gas emissions and contribute to the fight against global warming.

Disadvantages of nuclear energy:

- The process of mining uranium can have environmental and health impacts, including habitat disruption and potential exposure to radioactive materials.
- Nuclear power plants release heated water into natural water bodies, which is called thermal pollution. This can alter the water chemistry and affect aquatic ecosystems.
- The potential for nuclear accidents is a significant concern, though these are relatively rare. Accidents like those at Chernobyl and Fukushima have underscored the serious consequences of such events. Exposure to nuclear radiation can cause radiation sickness or cancer at a later stage in life due to DNA mutations.
- Nuclear power generates radioactive waste, and the safe disposal of this waste is a long-term challenge. Storage containers are used to shield the environment from radiation, but concerns persist about the long-term safety and security of waste storage.

Common mistake

Nuclear energy is often wrongly thought to be renewable because it produces 'clean energy' (no air pollution or CO_2 is released). The source of nuclear energy (uranium or plutonium) is finite and will be completely depleted with regular use.

Examples of renewable energy sources

Renewable energy sources, such as solar energy produced using photovoltaic technologies, and wind energy harnessed through turbines, show promise in sustainability despite some challenges at present in manufacturing and end-of-life management. The financial costs of these technologies are borne by the companies, either directly or indirectly. Awareness about the impacts of renewable energy sources on the environment allows people to make the right choices.

Wind power

Wind energy is a source of renewable energy. Wind turbines convert kinetic energy of the wind into electrical energy. Modern wind turbines consist of blades mounted

Figure 7.14 A wind farm with a turbine system

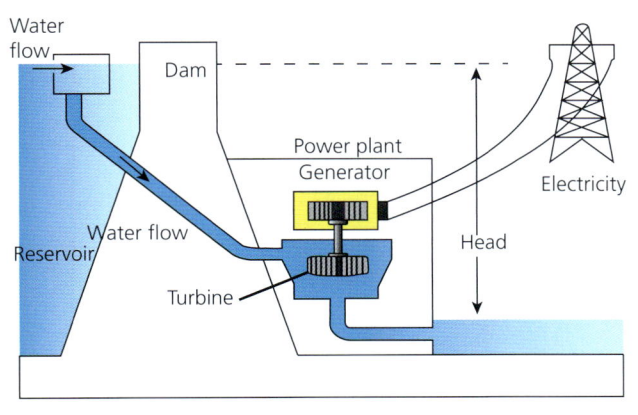

Figure 7.15 A hydroelectric power plant

on a rotor which is connected to a generator. Wind causes the rotor to spin, activating the generator and producing electricity.

Wind farms can be located either onshore or offshore to harness maximum wind current for energy production.

- Onshore wind farms: The Horse Hollow Wind Energy Center in Texas, USA, is one of the largest onshore wind farms. With over 700 wind turbines, it has a capacity to generate a significant amount of electricity, contributing to the state's renewable energy goals.
- Offshore wind farms: The Horns Rev Offshore Wind Farm off the coast of Denmark is a notable example.

Hydroelectric power

Hydroelectric power uses water to turn turbines that generate electricity. This is often within artificial dams built on rivers that have a good supply of water. The water is stored at a height (see Figure 7.15) and the gravitational force of the falling or flowing water turns turbines connected to generators. The kinetic energy of the moving water is converted into mechanical energy, which, in turn, generates electrical energy. These turbines can be switched on whenever required. The efficiency of the hydroelectric power plant depends on the volume of water stored in the dam and the height from which it falls or flows.

There are three types of dams:

- Reservoir-based: Large dams create reservoirs, providing a consistent water supply (for example, Hoover Dam).
- Run-of-river: These utilize naturally flowing water without large-scale reservoirs, maintaining the river's natural flow. For example, the Jiangxia Tidal Power Plant (Zhejiang Province, China) harnesses electricity from the Yangtze River without creating a large reservoir. It uses the natural ebb and flow of the river to generate electricity.
- Pumped storage: These employ two reservoirs at different elevations; excess energy is used to pump water to the higher reservoir for later use. For example, Dinorwig Power Station located in Wales, UK, is a pumped storage facility with lower and upper reservoirs. During periods of low energy demand, excess electricity from the grid is used to pump water from the lower to the upper reservoir. When energy demand is high, the stored water is released to generate electricity.

Advantages of hydroelectric power:

- It is renewable as it relies on the water cycle.
- It produces low greenhouse gas emissions and therefore has minimal carbon footprint compared to fossil fuels.
- It mostly provides consistent, reliable power.

Disadvantages of hydroelectric power:

- It alters aquatic and terrestrial ecosystems by affecting aquatic life and habitats.
- Dams reduce the flow of water downstream.
- Large dams may necessitate the relocation of communities.
- Initial construction expenses can be substantial.

Solar energy

Solar energy is a renewable energy source that harnesses the power of the sun to generate electricity. Solar panels, made of semiconductor materials like silicon, capture sunlight and convert it into electricity through the photovoltaic effect. The photovoltaic effect is the process that generates electric current in a photovoltaic cell when it is exposed to sunlight. Mirrors or lenses can also be used to focus sunlight on to a small area, producing heat that drives a turbine to generate electricity. This is called a concentrated solar power system. Inverters are used to convert direct current (DC) electricity from solar panels into alternating current (AC) for use in homes and the grid.

Advantages of solar power:

- Renewable: It is renewable as it relies on sunlight, which is generally abundant.
- Low environmental impact: There are minimal greenhouse gas emissions during operation.
- Scale: It can be deployed at various scales, from small residential installations to large solar farms.

Disadvantages of solar power:

- Inconsistent: It depends on sunlight availability (not all regions will have as much), thus requiring energy storage or backup sources for consistent supply of energy.
- Can be expensive: Installation can be expensive, although costs have been decreasing.
- Land area: Large-scale solar farms may require significant land area.
- Material used to make solar panels and photovoltaic cells, such as silicon, are limited (finite).

The installation of solar farms requires the clearing of larger land areas, which could adversely affect vegetation and wildlife, causing habitat loss, and disruption of rainfall and infiltration. The use of metals such as silicon, lead and chromium in the solar panels adds to the environmental hazards, especially in the end-of-life management of solar panels. As most of the solar panels will be sent to landfills, the possibility of these heavy metals leaching out into the soil and groundwater is high.

Geothermal power

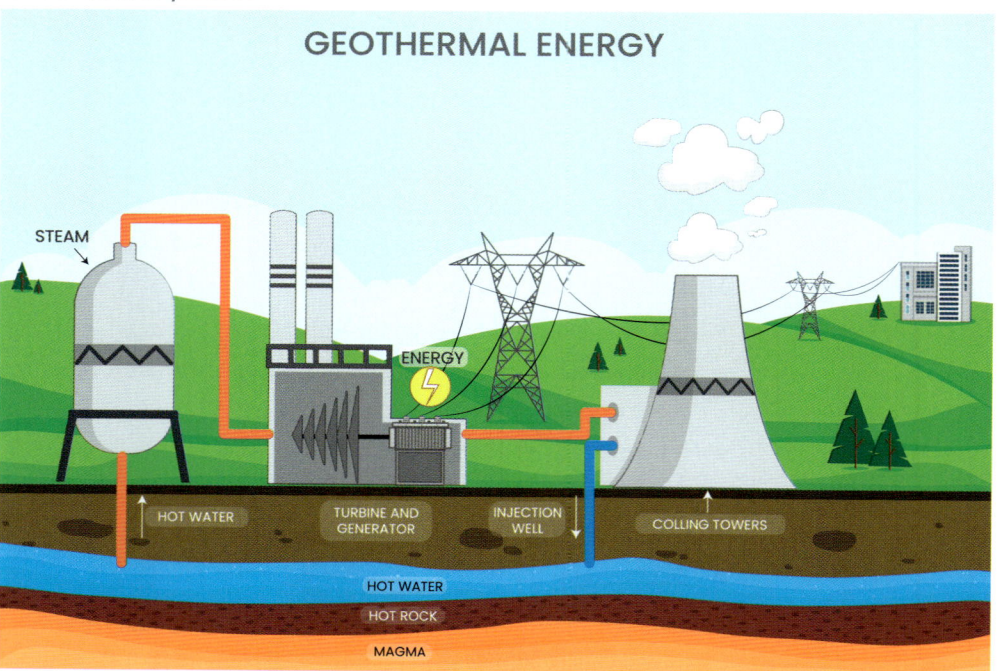

Figure 7.16 A geothermal power plant

Geothermal energy is a renewable energy source that harnesses the heat stored beneath the Earth's surface. There are three types of geothermal power plants:

- Dry steam plants extract high-pressure steam from underground reservoirs to directly drive turbines, generating electricity.
- Flash steam plants extract high-temperature water, convert it into steam, and use the steam to drive turbines.
- Binary cycle power plants transfer heat from geothermal hot water to another liquid with a lower boiling point, which is vaporized to drive turbines.

The Hellisheiði Geothermal Power Project in Iceland utilizes geothermal resources for both electricity generation and district heating.

Advantages of geothermal energy:

- It is environmentally friendly as it does not cause any pollution.
- It is a renewable form of energy as the heat reservoirs used to heat water are naturally present inside the Earth and will continue to heat water until the core of the Earth is hot.
- The energy produced is stable, efficient and sustainable. It does not fluctuate like wind energy. Approximately 70 per cent of the energy is converted to electricity.
- No fossil fuels are required to power the energy production.
- The lifespan of geothermal power plants is very long. Old power plants can be functional for 50–60 years, while newly built plants can work efficiently for 80–100 years.

Disadvantages or challenges of geothermal energy:

- Viable geothermal resources are mostly located in regions with tectonic activity.
- The areas can be remote and difficult to access to build a geothermal plant.
- Drilling and exploration expenses can be significant.
- Gases such as CO_2 which are sequestered and stored deep under several layers of the Earth can be released into the atmosphere during drilling.
- The drilling might set off minor earthquakes.
- The initial setting up of the geothermal plant requires a huge amount of money.
- It requires large amounts of water for cooling, which can be a challenge in areas with water scarcity.

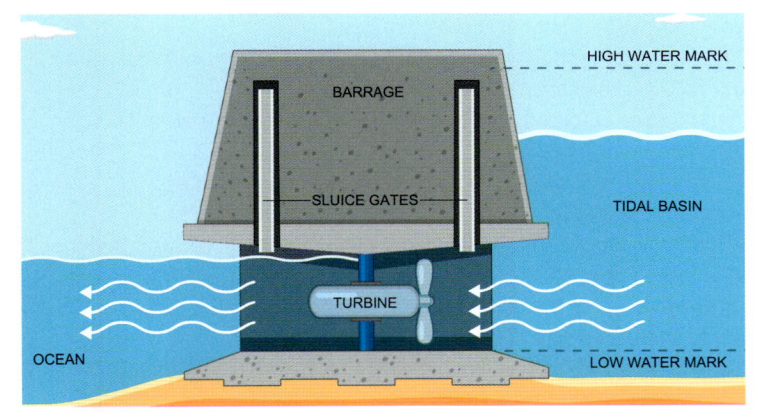

■ **Figure 7.17** Tidal energy generation

Tidal power

Tidal power uses the ebb and flow of tides to turn turbines which then generate electricity. Underwater turbines are submerged turbines placed in tidal streams, where the flow of water caused by the rising and falling tides turns the turbines to generate electricity. Tidal barrages are built across estuaries, also called tidal basins or lagoons. The rising tide fills the basin, and during the ebb tide the stored water is released through turbines to generate electricity. Similarly, tidal kites and tidal buoys can be used to harness the power of the tides to generate electricity.

Advantages of tidal power:

- Tidal patterns are predictable, allowing for accurate energy-generation forecasts.
- Tidal currents are dense and carry substantial energy, so a lot of power can be generated.
- It can work even at low speed of tides.
- Tidal power plants can have a lifespan of more than 80 years.

Disadvantages of tidal power:

- Construction of tidal barrages or underwater turbine installations can be expensive.
- The barrages disrupt the way in which marine lifeforms thrive and navigate in through the tides.
- Corrosion due to saltwater can be a major maintenance issue.
- As the strongest currents are close to the land, tidal power plants have to be built on coastlines.
- Electricity generated by tidal power is more expensive than that produced by wind and solar power.

HL.b.13: Environmental economics

As nations strive for economic development, the sources of energy they choose play a pivotal role in determining the long-term well-being both of their economies and of the environment. Economic growth often leads to increased energy demands. The sustainability of this growth depends on the types of energy sources employed. Countries that are heavily reliant on non-renewable energy, like coal and oil, may experience rapid economic growth. For instance, Saudi Arabia's economic boom has been largely driven by oil exports. However, this growth model poses environmental challenges, leading to air pollution and resource depletion, as seen in the case of countries with rapid industrialization. Costa Rica relies on renewable sources for over 99 per cent of its electricity, showcasing how a smaller nation can achieve economic growth while prioritizing environmental sustainability.

Developing nations often face the challenge of balancing economic growth with environmental considerations, leading to complex decision making. Collaborative efforts and international partnerships are crucial in supporting nations to transition towards sustainable energy practices, ensuring economic growth without compromising environmental well-being.

ATL ACTIVITY

Explore the interconnection between economic growth and environmental sustainability by analysing the energy choices of different countries and their impacts on both realms.
- Choose two contrasting countries with distinct economic profiles.
- Research and identify the primary energy sources each country relies on, such as fossil fuels or renewable energy. Consider factors like energy-efficiency initiatives.
- Investigate the policies and strategies employed by each country to ensure a sustainable energy mix.
- Examine economic growth indicators for both countries, such as GDP growth rates, employment trends and industry development.
- Analyse how each country's energy choices align with its economic growth goals, and their impact on employment and industry sectors.
- Evaluate the environmental implications of the chosen energy sources.
- Conduct a group discussion to share your findings with your peers.

7.2.4 A variety of factors will affect the energy choices that a country makes

Energy choice adopted by societies may depend on availability, accessibility, technological developments, energy efficiency, cultural attitudes, energy security, pollution and economic costs.

> **REAL-WORLD EXAMPLE**
>
> ### Denmark's wind energy
>
> Denmark is a global leader in wind energy. While Denmark has achieved a landmark in the use of wind power, there are certain challenges that remain associated with its energy choice.
>
> Advantages of wind energy:
>
> - Sustainability: Denmark is leveraging its abundant wind resources for sustainable power generation.
> - Economic cost: The increasing demand of wind power in Denmark has increased the competition among manufacturers which has resulted in the lowering of prices for wind turbines.
> - Energy efficiency: Wind turbines are relatively efficient, converting wind power into electricity with minimal environmental impact.
> - Availability: Denmark is located in the North Sea. This provides consistent and strong wind patterns, ensuring the availability of this renewable resource throughout the year.
> - Energy security: Diversification of energy sources, including wind, enhances Denmark's energy security.
>
> Disadvantages of wind energy:
>
> - Installation of wind turbines requires significant land area: Denmark being a small country (in terms of area) uses onshore, nearshore and offshore sites for wind farms.
> - Wind-power generation is dependent on weather conditions, resulting in intermittent and variable energy output: When the wind is not blowing, or is blowing too strongly, electricity production may decrease, posing challenges for maintaining a consistent and reliable energy supply.
> - Energy-storing challenges: Like any other alternative energy source, storing excess energy generated during windy periods for use during calm periods is a significant challenge. Current energy-storage technologies are not yet fully capable of efficiently storing large-scale wind-generated power for extended periods.
> - Cost of production: The overall cost of production makes wind energy expensive compared to coal or hydropower.
>
> ■ **Figure 7.18** Wind turbines near the Danish coast

> **REAL-WORLD EXAMPLE**
>
> ### Ukraine's nuclear energy
>
> Ukraine has huge reserves of uranium, so it uses nuclear energy as a significant contributor to meeting its energy needs. It has four power plants with 15 reactors that provide most of its electricity. Ukraine had been receiving most of its nuclear services and nuclear fuel from Russia, however after the Russian invasion of Ukraine on 24 February 2022, Russian troops took control of the Zaporizhzhia Nuclear Power Plant. This is now operational under the Russian Federation regulations. The nuclear power plant was cut off from the power grid during the war due to shelling. Widespread blackouts resulted, and the external power supply to all four of the country's nuclear plants was affected.
>
> The energy policy of Ukraine supports the use of uranium and coal to produce the major portion of its energy, as these are abundant in the region, with the remaining energy coming from oil and gas imported from Russia and Europe. Ukraine proposed to become part of the European energy market by 2017, and was finally connected to the Moldova energy grid in 2022. 'Energomost' was the energy bridge project which was agreed in March 2015 between Ukraine's Ukenergo distribution company and Polenergia of Poland. The aim of the project was to favour greater use of Ukraine's nuclear capacity and generate funds for expansion (source: 'Nuclear power in Ukraine', World Nuclear Association, 2023). Part of the project was to export electricity. However, the project did not proceed.
>
> Advantages of nuclear energy for Ukraine:
> - Sustainability: Since Ukraine has large reserves of uranium, it contributes to sustainable energy.
> - Energy efficiency: Ukraine produces 51 per cent of its electricity through nuclear power.
> - Availability: Uranium is abundantly available in Ukraine, making it easy to use this as an energy source.
> - Energy security: A diversified energy mix, including nuclear power, contributes to Ukraine's energy security, along with coal, oil and gas.
> - Management: Amid the Russian invasion, Ukraine is successfully able to manage and control its nuclear power plants.
>
> Disadvantages of nuclear energy:
> - Safety concerns: The Zaporizhzhia Nuclear Power Plant has become a constant threat due to the ongoing Russian attacks on it.
> - High initial costs: In the past, installation of nuclear reactors in Ukraine was funded by Russia. However, Ukraine has become increasingly financially independent to build its own nuclear power plants.
> - Radioactive waste: Proper disposal and management of radioactive waste remain challenges for the nuclear industry.
> - Limited domestic uranium resources: Ukraine relies on importing uranium for its nuclear reactors from Russia. The current war has hampered the supplies of uranium.
>
> Figure 7.19 shows how the source of electricity generation has changed over the years in Ukraine. In the early 1990s, coal was the primary source. This gave way to nuclear power around 1996. The amount of electricity generated by nuclear power has since remained almost steady, while the trend for coal has seen some major fluctuations. The use of oil as an energy source has reduced and become negligible. By contrast, natural gas and hydroelectricity have contributed almost a quarter of the total energy produced.

Figure 7.19 Electricity generation by source, Ukraine 1990–2020

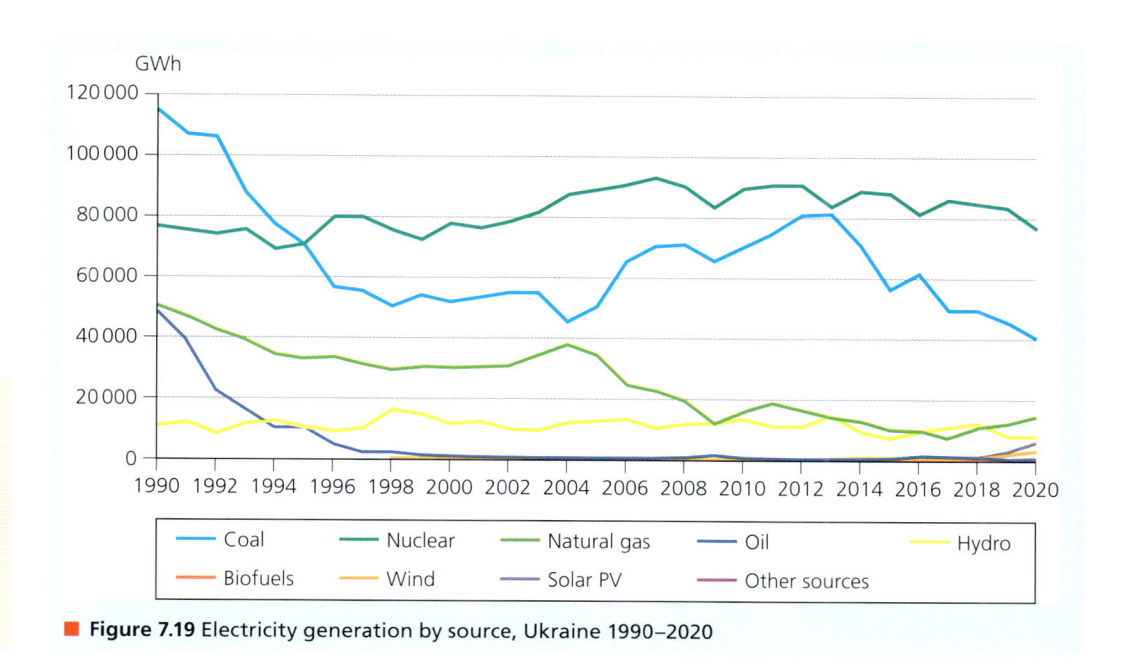

Link

Topic 1.1.7 (page 12) discusses how our inputs of information, perspectives, etc., shape our outputs such as the energy choices we make.

7.2.5 Intermittent energy production from some renewable sources creates the need for energy-storage systems

Weather conditions and daylight changes cause the flow of wind and the intensity of solar energy to fluctuate and therefore produce energy. This inconsistency creates a need for systems that can store excess energy when it's available, and release it when demand is high or renewable generation is low. Storage systems are necessary to maintain a stable power supply and make renewably sourced energy more viable.

Pumped hydroelectricity storage (PHS) is a well-established and widely employed method for storing and managing electrical energy. It stores energy by using two water reservoirs at different elevations. When the energy demand is low or the generation of electricity is in excess, the surplus electricity is used to pump water from the lower reservoir to the upper reservoir, effectively storing energy in the form of gravitational potential energy. When electricity demand is high, water from the upper reservoir is released, flowing downhill to the lower reservoir. This movement drives turbines, generating electricity to meet demand. PHS is known for its high efficiency, its quick response to changes in demand, and its contribution to grid stability by smoothing out peaks in energy demand.

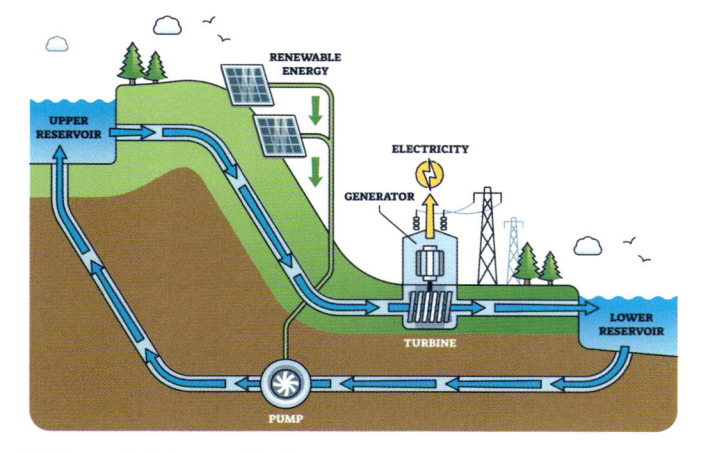

Figure 7.20 Pumped hydropower storage

Advantages of PHS:

- Energy storage: It allows for the storage of excess electricity generated during periods of high renewable energy production, such as when the wind is blowing strongly or during peak solar generation.
- Peak-shaving: The PHS system helps smooth out peaks in energy demand by releasing stored water to generate electricity during periods of high demand, providing a reliable and consistent power supply.

7.2 Energy sources, use and management

- Efficiency: It is efficient at maintaining grid stability because it can react quickly to variations in demand.
- Long lifespan: PHS infrastructure has a long lifespan, contributing to the sustainability and long-term reliability of the energy-storage solution.

There are certain challenges in the use of PHS:

- The storage system requires water to be pumped from a lower reservoir to a higher reservoir. This process requires energy, causing energy losses.
- Setting up PHS projects can be very cost-intensive.
- The applicability of this storage system is reduced since the reservoirs can only be built around water bodies (rivers, big lakes).
- The pumping of water causes alterations in its flow, which may disturb the natural course of aquatic animals. As a result these animals may die and cause a buildup of oxygen, leading to eutrophication. Deep water withdrawal from the upper reservoir also affects the water quality, eventually harming the aquatic ecosystem.

7.2.6 Energy conservation and energy efficiency may allow a country to be less dependent on importing a resource

Energy conservation and energy efficiency play an important role in reducing a country's dependency on importing external energy resources. Energy conservation involves altering behaviours to diminish energy consumption, encompassing practices like turning off lights, using heating and air conditioning more judiciously, and opting for less fuel-intensive modes of transportation such as trains. Meanwhile, energy efficiency entails the adoption of technologies designed to optimize energy usage, such as constructing buildings with improved insulation, utilizing low-energy intelligent lighting, and implementing sustainable product design principles within a circular economy framework. By promoting both behavioural changes and technological innovations, these measures contribute to the reduction of overall energy consumption.

■ Improved lighting technology

Improved lighting technology involves:

- Behavioural change: The adoption of energy-efficient lighting practices, such as turning off lights when not needed and using natural light during the day, reduces electricity consumption.
- Technological innovation: The implementation of low-energy intelligent lighting, such as LED bulbs, significantly lowers energy usage compared to traditional incandescent bulbs.

In countries like Japan, the widespread adoption of LED lighting in households and public spaces has led to substantial energy savings, contributing to reduced dependence on imported energy resources. Furthermore, there are many parts of Europe that use energy-saving street lighting. Norway exemplifies the implementation of intelligent lighting systems with sensor-based LED lights along roads, which turn on to 100 per cent brightness only when they sense movement indicating a vehicle is in close proximity. Their auto-dimming capabilities mean that at other times they reduce lighting to 20 per cent.

■ Circular economy and product design

Circular economy and product design involves:

- Behavioural change: Embracing a circular economy and designing products for easy recycling necessitates a shift in consumer behaviour towards responsible disposal and recycling practices. While this is a significant behavioural change, it contributes to long-term sustainability by reducing the environmental impact of resource extraction and waste.
- Technological innovation and impact: Innovations in product design that prioritize recyclability and reusability demonstrate effectiveness in technological advancements. For instance, designing goods to be easily recycled within a circular economy framework reflects a commitment to sustainable production practices.

Adding natural energy systems to a building, like green roofs and other similar systems, makes clean energy and makes the building more energy efficient. Using new packing materials that are recyclable and biodegradable reduces the need for producing petroleum derived polymers (polyethylene) for packaging.

Companies implementing circular economy principles, such as IKEA designing furniture for easy recycling, exemplify real-world impact. This approach contributes to energy conservation and a more sustainable resource-management system by reducing the environmental footprint of products.

● (HL) 7.2.7 Energy security for a country means access to affordable and reliable sources of energy

Energy security refers to the uninterrupted availability of energy that a country can provide for all its energy needs. Countries try to maintain a stock of resources to sustain themselves in periods of crisis. Energy choices will therefore depend to a very large extent on resources that can provide long-term energy security as well as on technologies that can provide alternative energy choices. Usually these reserves are present within the country itself. However, it is not always possible for countries to have such reserves, in which case they have to depend on other nearby countries to procure efficient and sustainable energy resources. A country can be energy secure or insecure, depending on whether it has sufficient energy resources. Energy insecurity arises either when people do not have access to energy resources, or when the access is unstable or unreliable. Sweden, Denmark and Finland are the most energy secure countries in the world and their energy choices are inclined towards renewable energy resources. They have the most well-planned energy systems to manage demand and supply efficiently.

REAL-WORLD EXAMPLE

Japan's energy security

Japan is considered to be one of the largest energy importers in the world, with almost 90 per cent of its primary energy supply coming from other countries (source: 'Japan Natural Gas Security Policy', International Energy Agency, 2022). Over the last 40 years, Japan has faced several harsh energy-supply conditions; the Great Tōhoku earthquake causing the Fukushima Daiichi nuclear disaster in 2011 being one of the prominent ones. This catastrophic event compelled Japan to reassess its energy mix and completely eliminate its dependence on nuclear power. The country diversified its energy sources, emphasizing renewables, natural gas and energy-efficiency measures. Japan's strategy was to increase reliance on liquified natural gas (LNG) and it emerged as the world's largest importer of LNG, enhancing its energy security by establishing

◆ **Climate diplomacy** – the practice and process of creating the international climate change regime and ensuring its effective operation.

diverse supply chains. This shift was instrumental in reducing the nation's vulnerability to the geopolitical uncertainties surrounding traditional fossil fuel suppliers.

Japan's commitment to energy security is also reflected in its **diplomatic** endeavours. The nation actively engages in international collaborations and agreements to secure stable energy supplies. For instance, it has established partnerships with resource-rich countries such as Australia, Malaysia and Qatar, among others, securing long-term agreements for the supply of LNG.

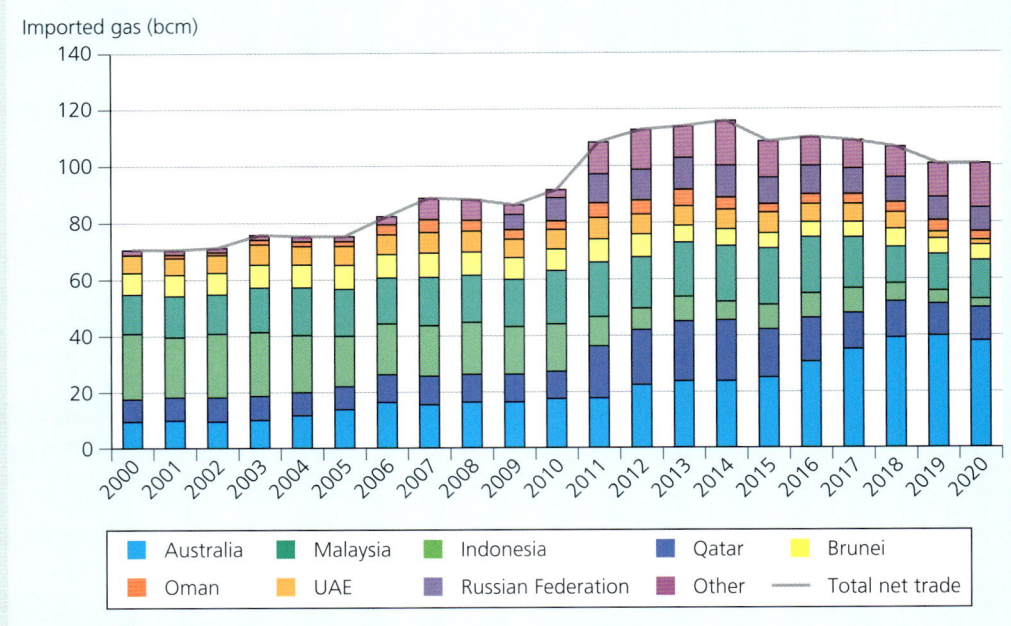

■ Figure 7.21 Net imports of natural gas in Japan, 2000–2020

Japan, being a member of the Paris Agreement, has intentions of becoming carbon neutral by 2050. Keeping these in mind, the country has invested heavily in harnessing solar power. However, its lack of land poses a challenge, thus Japan is one of the first countries to use floating solar panels for energy production.

■ Figure 7.22 Floating solar panels in Japan

◆ **International Energy Security Risk Index** – identifies the policies and other factors that contribute positively or negatively to international energy security.

REAL-WORLD EXAMPLE

Ukraine's energy security

Ukraine has an **International Energy Security Risk Index** of 1,463 according to the Global Energy Institute. The index itself is calculated by taking several energy-risk factors into consideration. A high score like 1,463 indicates that a country is less energy secure. Ukraine battles with energy insecurity due to geopolitical tensions and historical dependencies. The country has experienced disruptions in the supply of natural gas from Russia due to the ongoing conflicts, leading to concerns regarding energy reliability.

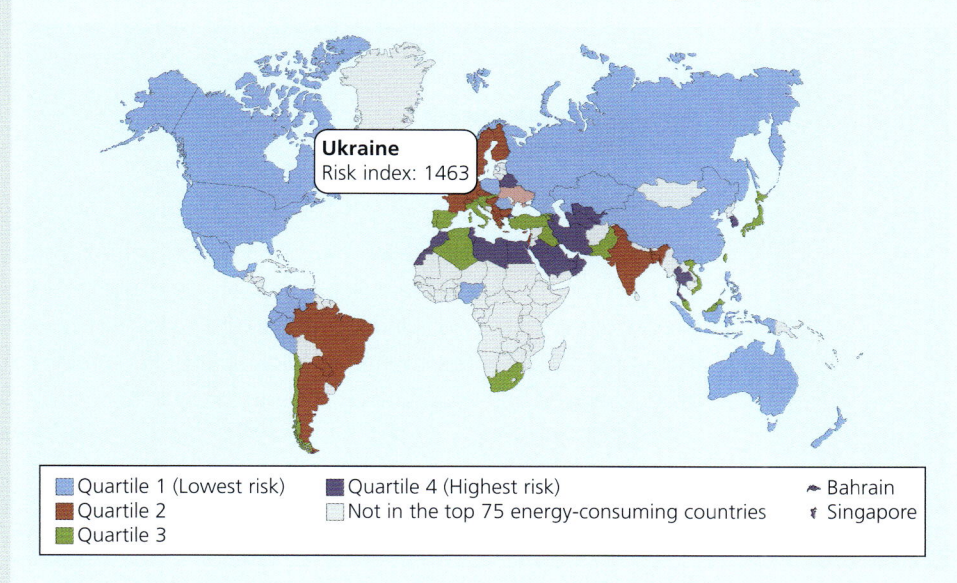

Ukraine
Risk index: 1463

- Quartile 1 (Lowest risk)
- Quartile 2
- Quartile 3
- Quartile 4 (Highest risk)
- Not in the top 75 energy-consuming countries
- Bahrain
- Singapore

■ **Figure 7.23** The International Energy Security Risk Index quartiles for all top 75 energy-consuming countries

Historically, Ukraine has been a producer of oil and gas, however this was not sufficient to meet its rising energy demands. Thus, it became heavily dependent on natural gas imports from Russia to meet its energy needs. Disputes over gas prices and transit fees between Ukraine and Russia led to interruptions in the supply from 2005 onwards, creating uncertainties in the country's energy security. Its energy security was seriously challenged during a prolonged gas-price negotiation with Russia in 2014, followed by the loss of governmental authority in Crimea, which is a prime location for significant offshore gas resources. Ukraine's energy infrastructure, including its ageing pipelines and outdated facilities, faces challenges in terms of maintenance and modernization. The inefficiencies in the energy sector contribute to energy losses, making the country more vulnerable to supply disruptions.

In response to these challenges, since 2015 Ukraine has been making efforts to diversify its energy mix and reduce dependence on Russian gas. As a result, nuclear power and coal have been exploited to meet its energy demands. Most of the energy requirement is fulfilled by nuclear power. While nuclear energy contributes to the energy mix, it also brings concerns about safety, especially given the proximity to the Chernobyl nuclear disaster site. Ukraine also has vast deposits of coalbed methane basins, including two portions of the Lublin Basin, which extends into Poland, and the Dnieper-Donets Basin in eastern Ukraine, where the majority of the country's coal is mined.

Sweden, Denmark and Finland are the most energy-secure countries in the world and their energy choices are inclined towards renewable energy resources. They have the most well-planned energy systems to manage demand and supply efficiently. Improving energy efficiency involves adopting technologies and practices that reduce energy consumption without compromising output or quality of services. This can include upgrading infrastructure, implementing energy-efficient technologies and promoting conservation. By using energy more efficiently, a country can meet its energy needs with less overall consumption, thereby enhancing energy security. It reduces dependence on external energy sources and enhances the resilience of the energy system.

REAL-WORLD EXAMPLE

Singapore's energy efficiency

Singapore, the small island city-state with limited natural resources, has prioritized energy efficiency in its sustainable development strategy. Singapore imports almost all of the natural gas from Malaysia, Indonesia and other parts of the world, which puts it at a high risk index of 2,211. The country is vulnerable to fluctuations in the prices of fuel due to global demand and supply shocks. Due to space limitations, Singapore cannot efficiently utilize solar and wind energy. It has, therefore, implemented several strategies for better management of energy by using technologies to design smart buildings and infrastructure. Intelligent building-management systems, efficient lighting and sensor technologies are integrated to optimize energy use. The Marina Bay Sands complex, an iconic integrated resort, is a testament to Singapore's commitment to energy-efficient design and operation.

The Green Mark Incentive Scheme allows building owners to use grants in order to upgrade older systems and make them more energy efficient. District cooling systems are put in place to reduce the use of air conditioning. Chilled water runs through the pipes in buildings, minimizing the need for individual cooling systems. A good example of this is the Jurong Lake District cooling network. Singapore also implements efficient waste-to-energy (WTE) practices, under which waste is used to generate energy. There are around four WTE plants in Singapore. The TuasOne WTE plant alone can incinerate around 3,600 tonnes of waste to produce 120 MW of electricity on a daily basis.

■ Figure 7.24 The Marina Bay Sands complex in Singapore

Reducing dependence on energy imports involves developing and utilizing domestic energy resources. This can include investing in renewable energy sources, increasing domestic production of fossil fuels, and implementing energy-conservation measures. For example, the United States has taken steps to decrease its reliance on imported energy supplies by increasing domestic production of oil and natural gas through technologies like hydraulic fracturing (fracking). These are of course non-renewable sources, but additionally there has been a push to expand renewable energy production, such as wind and solar, so reducing dependence on foreign oil and gas.

Diversifying energy sources involves expanding the mix of energy resources a country uses, including renewables (solar, wind, hydro), fossil fuels, nuclear and other sources. This reduces dependence on a single energy type. This diversified approach helps to mitigate risks associated with the variability of specific energy sources and provides a more stable energy supply.

Inquiry process

Inquiry 2: Collecting and processing data

Energy sources and their consumption change over time. Sometimes new sources of a particular energy resource are discovered and exploited, shifting the usage pattern. Both non-renewable and renewable resources are used by countries to varying degrees depending on the availability of a certain resource.

- Identify one renewable and one non-renewable energy source used in two different countries or societies.
- Choose reputable secondary data sources such as Gapminder, Our World in Data and the World Bank that provide comprehensive information on global energy consumption. Explore the pattern of energy source use in a given country and for the entire world.
- Collect data about the consumption of energy produced by the chosen resource.

  For example: Coal is extensively used in the USA. Look at the annual consumption of coal by the USA and then compare it to that of the world. Calculate what percentage of the world's coal consumption is contributed by the USA.
- Use spreadsheet software to record your findings, and process the data collected in the form of tables and graphs.
- Present your findings to the class.

7.2 Energy sources, use and management

7.2.8 The global economy mostly depends on finite reserves of fossil fuels as energy sources, including coal, oil and natural gas

The global economy has been heavily dependent on non-renewable energy resources. Fossil fuels have been the primary source of power for industries, transportation and electricity generation. However, this heavy reliance has led to environmental degradation and concerns about long-term sustainability.

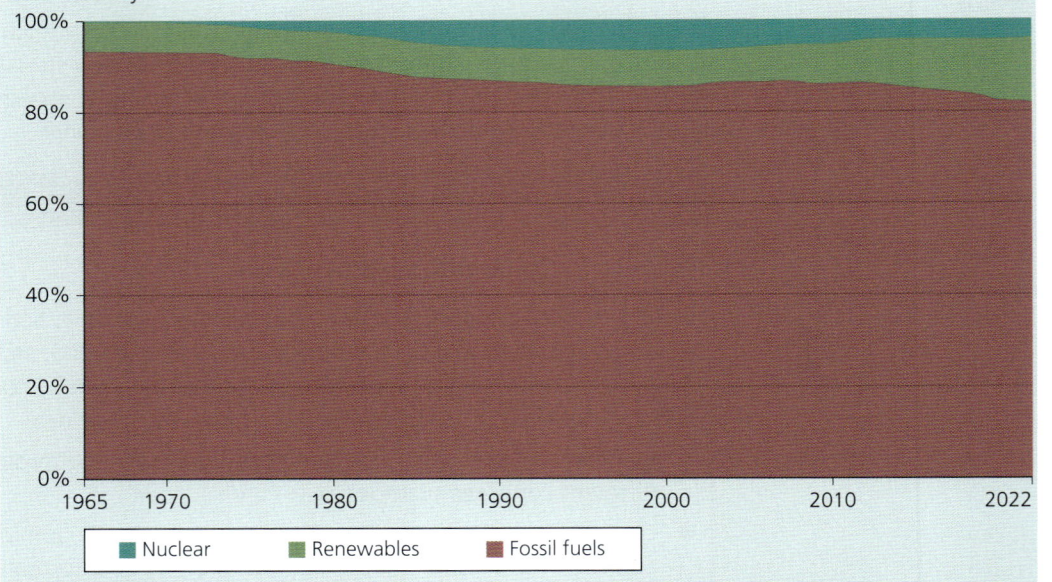

■ Figure 7.25 Primary energy consumption from fossil fuels, nuclear and renewables

There are several factors that influence the timeline for final depletion of these fuels:
- One factor is the rate of consumption, as increased demand for energy can mean the resource is depleted faster.
- Meanwhile, technological advancements in extraction methods can either prolong or speed up the depletion of these resources. For example, horizontal drilling and hydraulic fracturing (fracking) have revolutionized the oil and gas industries, allowing for the extraction of previously inaccessible reserves and increasing the overall global supply of these resources. These increased extraction rates can lead to faster depletion of the resources.
- Government policies and regulations, such as subsidies or carbon pricing (or choosing not to invest in renewable alternatives), can also play a significant role in influencing the timeline for depletion.
- The discovery of new reserves or alternative energy sources can impact the availability and longevity of non-renewable energy resources.

Overall, a combination of these factors will shape the future of global energy consumption and determine how long these resources will last.

7.2.9 Nuclear power is a non-renewable, low-carbon means of electricity production

Energy released by the fission of atoms is harnessed as nuclear energy. A chain reaction occurs when atoms of **radioactive** elements such as uranium and plutonium are forced to split by bombarding neutrons. This splitting releases a tremendous amount of energy all at once, causing an explosion. Many technologies have been developed to control the fission of radioactive elements inside a nuclear reactor and harness this energy for electricity.

◆ **Radioactivity** – the continuous release of particles and energy from the decay of the nuclei of certain atoms and their isotopes until it reaches a stable atom. For example, uranium (U_{235}) decays through several steps until it reaches lead (Pb_{206}), a stable atom.

Theme 7: Natural resources

Nuclear power plant

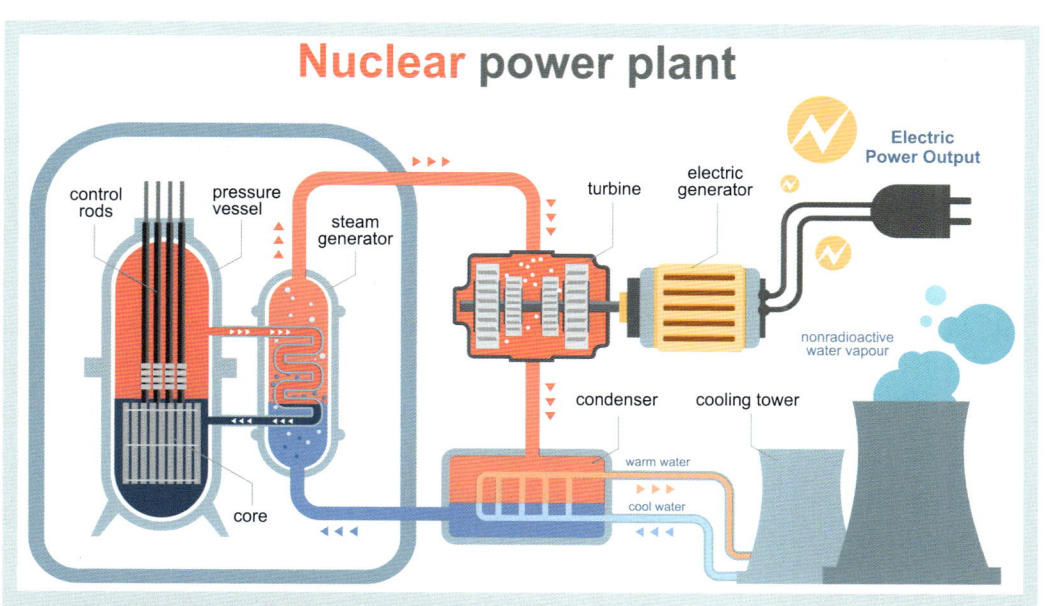

Figure 7.26 Diagram of a nuclear power plant

> ### Concept
>
> ### Perspectives
>
> Nuclear power has been met with mixed reactions from both the public and policymakers. While historical disasters have left an indelible negative mark on nuclear power, it has also proved to be one of the cleanest forms of energy. Policymakers look at the opportunity to use radioactive resources such as uranium to harness energy that can meet the needs of billions of homes for several years. However, the public, having heard the worst negative impacts of nuclear radiation, are often concerned about the safety risk of such a power plant. For example, a campaign group named Together Against Sizewell C challenged the plans for the Sizewell C power station to be built in Suffolk, UK, over environmental impacts. The UK government, however, aims to reduce its dependence on fossil fuels and energy imports, and considers Sizewell C to be an important part of the new UK energy strategy.

Nuclear power is a non-renewable energy source as it relies on finite radioactive resources like uranium and plutonium. It also generates huge amounts of nuclear waste, which is hazardous and difficult to discard. Some of this waste can continue to generate radiation for several thousand years after being disposed of. This is because, when a radioactive atom decays, it converts into a different element. The amount of time it takes for half of the original substance to decay is called its **half-life** ($t_{1/2}$).

◆ **Half-life** – the time taken for the radioactivity of a radioactive element to fall to half is original value.

Since it is highly radioactive, nuclear waste requires safe disposal methods to ensure that no form of life, water or air is exposed to it. Improper disposal of radioactive waste can cause water and soil pollution, and can have long-term consequences for ecosystems and wildlife. Prolonged exposure to radiation may cause radiation sickness, cancers, and can also affect the bone marrow. Radiation can also cause genetic mutations that can be transferred from one generation to the next.

7.2 Energy sources, use and management

The method used for the disposal of radioactive waste depends on the type of waste and its radioactive intensity:

- Geological deep disposal: This is the most widely used method and is employed for the long-term management of high-level nuclear waste, involving the placement of radioactive materials deep underground in stable geological formations, such as a salt dome or a layer of volcanic rock, which provides a natural barrier to prevent the escape of radioactive materials. The waste is isolated from the biosphere for thousands of years, allowing radioactive decay to reduce the radiation level to safe levels.
- Storage in salt mines: Radioactive waste can be stored in abandoned salt mines. Layers and beds of salt provide a natural barrier to prevent the escape of radioactive materials. The waste is placed in specially designed containers and buried in the salt deposits.
- Dry cask storage: This method involves storing the radioactive waste in dry casks made of steel or concrete. The casks are designed to provide a high level of security and protection against potential earthquakes and floods.
- Vitrification: This involves mixing the radioactive waste with glass or other materials to create a solid, stable form that can be stored for long periods of time. The resulting glass matrix can be buried in the ground.
- Encapsulation: This involves surrounding the radioactive waste with a protective material, such as concrete or metal, to reduce the risk of radiation exposure.

One notable example of disposal is the Onkalo spent nuclear fuel repository in Finland. Onkalo, situated on Olkiluoto Island, utilizes a combination of granite and clay formations for secure waste containment. The facility involves the construction of tunnels and deposition chambers deep underground. Similarly, the Waste Isolation Pilot Plant (WIPP) in New Mexico, USA, is dedicated to the disposal of transuranic radioactive waste. WIPP is located in a salt bed formation, providing geological stability for the isolation of long-lived radioactive materials.

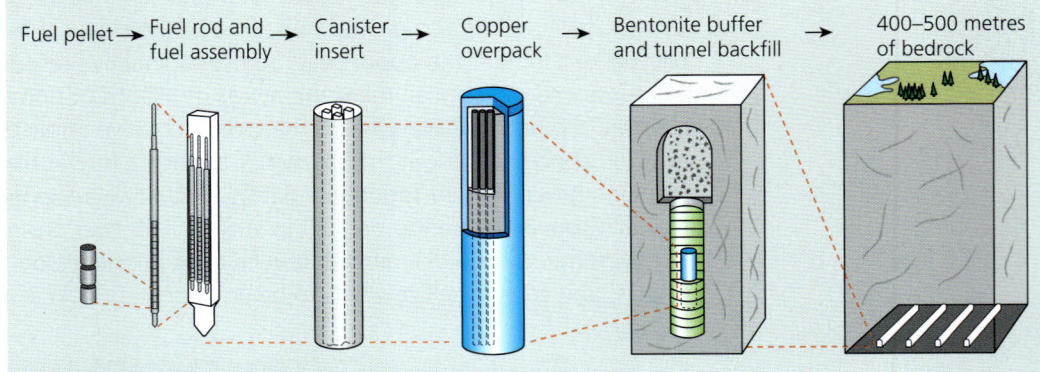

■ Figure 7.27 Geological deep disposal for nuclear waste

REAL-WORLD EXAMPLE

The history of the Navajo people and uranium mining

Navajo Nation occupies portions of northeastern Arizona, northwestern New Mexico, and southeastern Utah in the USA. The indigenous Navajo people live here. From World War II to 1997, uranium was mined in this area (mostly Arizona and New Mexico) and used by the government. Many Navajo people worked as miners in these mines. Even though the harmful effects of uranium exposure were known, the miners were not provided with sufficient protective measures. The mines were fully functional up until the 1970s and '80s, when uranium prices started dropping. Many mines were abandoned and as a result of this many Navajo people were left without a job. The aftermath of the mine closures included:

- Environmental contamination: Uranium mining involves the release of radioactive materials into the environment. Contaminated water sources became a significant issue, affecting both surface water and groundwater. Many Navajo communities rely on these local water sources for drinking water and agricultural purposes.
- Health impact: Miners and nearby communities were exposed to high levels of radiation, leading to health issues such as lung cancer and respiratory diseases. The radioactive dust from mining operations posed a severe threat to both miners and residents.
- Abandoned mines: Many uranium mines on Navajo lands were abandoned without proper reclamation. Open pits and tailings piles were left exposed, contributing to ongoing environmental degradation.
- Social and cultural impact: The extraction activities disrupted traditional Navajo ways of life. Moreover, many Navajo families were relocated for mining operations without adequate consideration for the social and cultural consequences.

In 2005, uranium mining was completely banned in Navajo Nation. The region still struggles with the negative impacts of mining. The United States Environmental Protection Act (EPA) has designated many sites for clean-up, however the challenge of exposure to radioactivity remains. According to the Radiation Exposure Compensation Act of 1990, the Navajo miners who were diagnosed with lung cancer and non-malignant respiratory diseases as a result of exposure to uranium were supposed to receive a 'compassionate payment' of $100,000, which either never reached the miners or was distributed in very small amounts due to bureaucracy.

REAL-WORLD EXAMPLE

Meltdown of huge nuclear reactors

In the history of nuclear power generation, a few incidents have left an indelible mark on the world. The partial meltdown of the Three Mile Island Nuclear Power Plant in Pennsylvania, USA in 1979 released radioactive gases and iodine (a poisonous liquid element) into the environment. Very low-level radionuclides were released from the meltdown and did not cause much harm to the people working in the plant, the environment or living beings. However, the accident raised serious safety concerns among the general public. New regulations for nuclear industries were set. The meltdown was categorized as Level 5 (Level 7 being the most severe – see Figure 7.28).

The International Nuclear and Radiological Event Scale

- 7 Major Accident
- 6 Serious Accident
- 5 Accident with Wider Consequences
- 4 Accident with Local Consequences
- 3 Serious Incident
- 2 Incident
- 1 Anomaly

Below Scale/Level 0 No Safety Significance

Accident / Incident

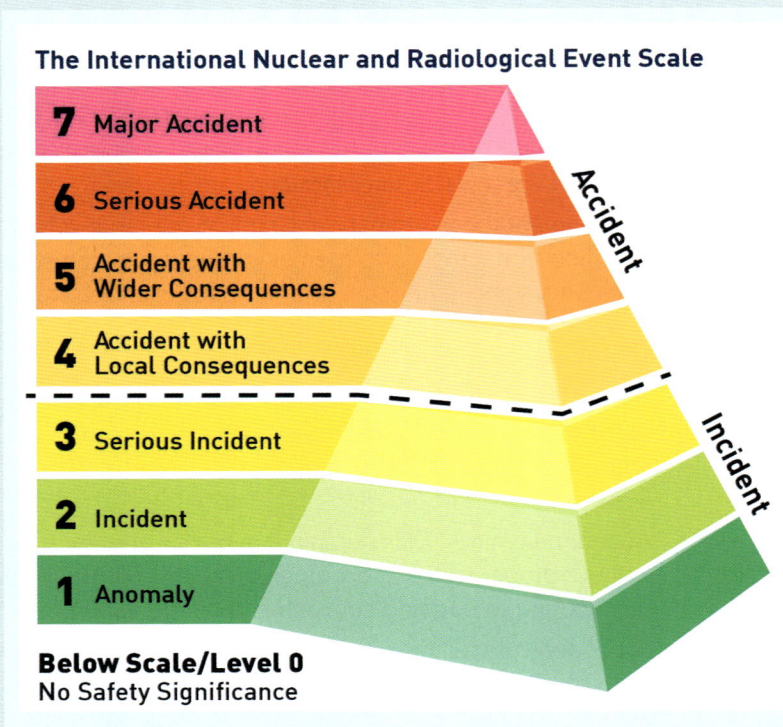

■ Figure 7.28 The scale for nuclear power accidents

Six years later, one of the worst disasters in the history of nuclear generation occurred in the Chernobyl nuclear power station in what is now Ukraine (then the USSR). During a test run, the nuclear reactor went out of control, causing an explosion that demolished the building, exposing the core and releasing large amounts of radiation into the atmosphere. This was a Level 7 accident as the exposure to high amounts of radiation was prolonged.

People had to be evacuated as the amount of radiation in the air was too high. Soil and water bodies became contaminated with exposure to radiation. Radionuclides were taken up by plants and then by animals. Immediately after the accident, the trees in the area surrounding the power plant became reddish-brown due to absorption of high-level radiation. This was called the 'Red Forest'. Radioactive material was deposited throughout the Northern Hemisphere, spreading over the whole of Europe and the rest of the world. Western Europe was the most affected after Ukraine, Belarus and Russia, resulting in 30,000 to 60,000 more deaths due to cancer. The local bodies of water still have low levels of Caesium-137 and Strontium-90.

The only other Level 7 accident in a nuclear power plant occurred in Fukushima Daiichi in 2011, where a major earthquake and a 15-metre tsunami disabled the power supply and cooling of three out of the four nuclear reactors, causing them to completely melt and releasing radiation into the atmosphere.

7.2.10 Battery storage is required for reduction of carbon emissions, but this brings challenges

Carbon emissions can be reduced by using energy-storage devices such as batteries. The increasing global demand for battery storage marks the transition towards sustainable energy. However, the production of batteries involves a complex chain of processes, from mining and transportation to processing and construction, each of which contributes to emissions and environmental challenges. The extraction and processing of elements like lithium, cobalt and rare earth elements for battery production release toxins into the environment, leading to pollution of both land and water bodies. This pollution can have detrimental effects on ecosystems and surrounding communities.

Lithium

The extraction of lithium involves resource-intensive mining operations, often leading to habitat disruption and landscape alteration. Processing lithium further generates toxins and pollutants, impacting both terrestrial and aquatic ecosystems. For example, lithium mining in the Salar de Atacama in Chile has been associated with water-scarcity issues and soil contamination. Massive brine evaporation ponds are used for lithium extraction, which alter the groundwater table and negatively impact wildlife, such as flamingo populations. Dam failures can result in the release of toxic materials into nearby rivers or land, causing severe environmental damage and posing risks to human health. The Brumadinho dam collapse in Brazil in 2019, which released mine tailings containing iron ore waste, resulted in a devastating environmental disaster, causing loss of lives and extensive ecological damage.

The distribution of lithium resources is concentrated in certain countries, like Australia and the lithium triangle of Chile, Bolivia and Argentina, yet the demand for lithium-ion batteries is global. This geographic mismatch creates a scenario where some countries with abundant resources become central to the supply chain, resulting in geopolitical tensions and conflicts over resource access.

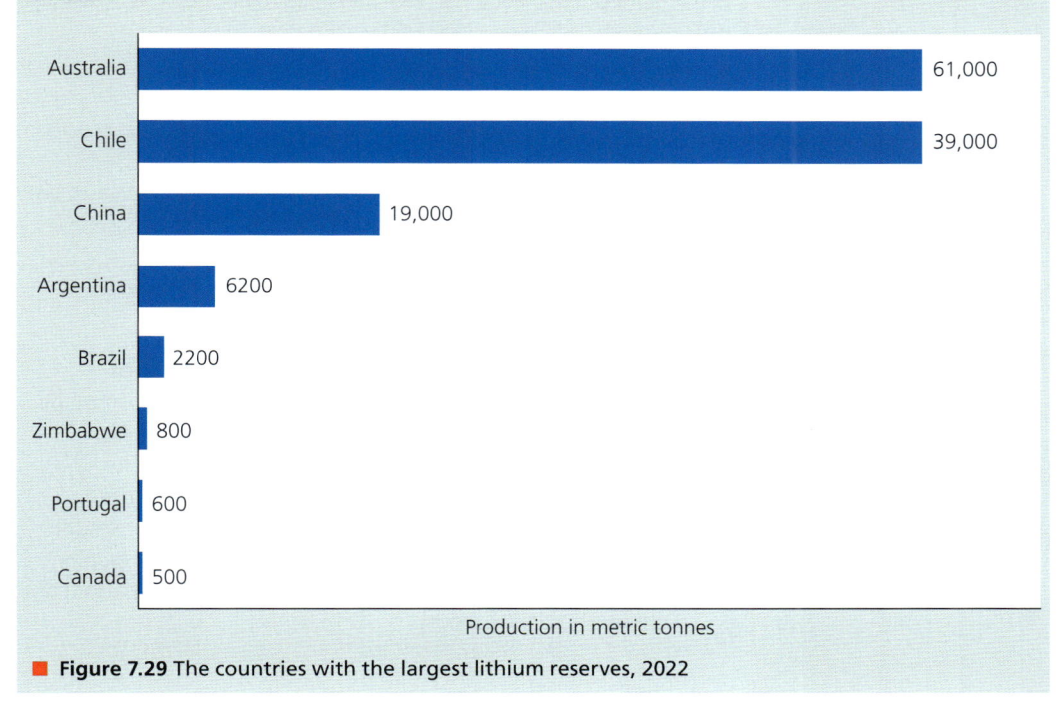

Country	Production in metric tonnes
Australia	61,000
Chile	39,000
China	19,000
Argentina	6200
Brazil	2200
Zimbabwe	800
Portugal	600
Canada	500

■ Figure 7.29 The countries with the largest lithium reserves, 2022

China is the leader in the production of lithium batteries, accounting for over 50 per cent of the world's lithium production. This dominance has significant geopolitical implications, particularly in the context of the ongoing electric vehicle revolution. As the demand for lithium-ion batteries, which rely on lithium as a key component, continues to grow, China's control over the market could give it a significant advantage in the electric vehicle market. In fact, many electric vehicle manufacturers have already begun to prioritize sourcing lithium from China, further solidifying the country's position in the market. However, this dominance also raises concerns about the potential for supply disruptions and the impact of geopolitical tensions on the global lithium supply chain.

Several factors govern China's dominance in the lithium-ion battery market. The first is the abundant reserves of lithium found in the country, mostly in the Tibet region. The country has cost effectively invested in the production and processing of lithium. China also has a vantage point of being on the crossroads of major trade routes. The rivalry between the USA and China in the electric vehicle and battery industries has far-reaching implications for economic dominance, technological leadership and national security. The USA is investing in strengthening the semiconductor industry and rebuilding the lithium supply chains to avoid overreliance on China's supply.

Cobalt

Cobalt mining, predominantly in the Democratic Republic of the Congo, has faced scrutiny due to hazardous conditions for miners and concerns over child labour. The extraction process itself can lead to environmental contamination.

The concentration of cobalt production in a limited number of countries, combined with the global demand for electric vehicles and electronic devices, has created challenges in ensuring a responsible and ethical supply chain.

Rare earth elements

Rare earth metals, such as neodymium, samarium, terbium and dysprosium, are a group of elements in the periodic table known as the lanthanide series. They have high electrical conductivity and are therefore used in batteries.
- Pollution from processing: China is a dominant player in rare earth element production, and the processing of these elements has led to environmental pollution in areas like Baotou, Inner Mongolia. Disposal of mining waste and processing byproducts poses challenges.
- Supply chain complexity: The global supply chain for rare earth elements involves several countries, with China controlling a significant portion.

> **ATL ACTIVITY**
>
> ### Conduct an energy-use survey in your school
> - Write a research question for investigating the energy consumption in a given society.
> - Identify the independent, dependent and control variables.
> - Formulate a testable hypothesis.
> - Prepare a questionnaire on Google Forms or SurveyMonkey.
> - Share the survey with the students and teachers of your school.
> - Collate the results and process the data into tables and graphs.
> - Write an academic report on the energy-consumption pattern of the chosen society. Include your findings in the report.

REVIEW QUESTIONS

1. What are renewable and non-renewable resources?
2. Write two advantages and two disadvantages of nuclear energy.
3. Evaluate the use of cobalt in battery storage devices.
4. Outline what is meant by energy security.
5. Discuss how the global demand for batteries can lead to geopolitical conflicts.

EXAM-STYLE QUESTIONS

1. Outline the challenges and benefits associated with the use of nuclear energy for electricity generation. [4 marks]
2. Discuss the effectiveness of government policies in promoting the transition from non-renewable to renewable energy sources in a specific region. [5 marks]
3. With the help of named examples, evaluate the environmental and economic implications of a country's reliance on non-renewable energy sources. [7 marks]
4. Discuss the role of technological advancements in enhancing the efficiency of energy production and consumption. [7 marks]

7.3 Solid waste

> **Guiding question**
> - How can societies sustainably manage waste?

> **SYLLABUS CONTENT**
>
> This chapter covers the following syllabus content:
> ▶ 7.3.1 Use of natural resources generates waste that can be classified by source or type.
> ▶ 7.3.2 Solid domestic waste (SDW) typically has diverse content.
> ▶ 7.3.3 The volume and composition of waste varies over time and between societies due to socioeconomic, political, environmental and technological factors.
> ▶ 7.3.4 The production, treatment and management of waste has environmental and social impacts, which may be experienced in a different location from where the waste was generated.
> ▶ 7.3.5 Ecosystems can absorb some waste, but pollution occurs when harmful substances are added to an environment at a rate faster than they are transformed into harmless substances.
> ▶ 7.3.6 Preventative strategies for waste management are more sustainable than restorative strategies.
> ▶ 7.3.7 Different waste-disposal options have different merits and demerits in terms of their impact on societies and ecosystems.
> ▶ 7.3.8 Sustainable options for management of SDW can be promoted in societies.
> ▶ 7.3.9 The principles of a circular economy provide a holistic perspective on sustainable waste management.
>
> *Note: There is no additional higher level content in 7.3.*

7.3.1 Use of natural resources generates waste that can be classified by source or type

Waste is the leftover, unwanted or unusable material generated by a system. In the process of extracting its usefulness, the resource might not be completely utilized or might generate some byproducts. Such byproducts therefore need to be removed and discarded. Solid waste includes food waste, packaging material, scrap material and microplastics.

Solid waste is produced in all areas of life. Households produce a diverse array of waste, ranging from everyday refuse to recyclables. Industrial activities contribute significantly to waste generation, producing solid, liquid and gaseous byproducts. Health sectors, through medical facilities, generate biomedical waste, necessitating specialized disposal methods. Additionally, agricultural practices contribute to the generation of organic and inorganic waste, such as crop residues and pesticide containers.

There are three main sources of waste:

- Domestic waste: Generated in a household as a result of day-to-day activities. It includes kitchen waste, paper, cardboard, glass, metal and e-waste.
- Industrial waste: It consists of waste generated as a result of production and manufacturing processes. It depends on the type of industry (timber industry would generate wood scraps, paint industry would generate chemical and solvent waste...). Common industrial waste includes metal scraps, oil, solvents, chemicals, dirt and gravel.
- Agricultural waste: It is the waste generated as a result of agricultural activities, such as growing of crops or rearing of animals (for example crop residue, unsaleable crops, animal excrement, left over fodder...).

Types of waste:

- Food waste: Organic waste generated during the production, processing, and consumption of food.
- Electronic waste (e-waste): Made up of unusable, damaged and outdated electronic items.
- Biohazardous waste: Waste contaminated with blood, bodily fluids, and other material that can be a threat to public health or the environment.
- Chemical waste: Any unused or excess of unwanted chemicals that can be harmful for humans and the environment. This includes expired chemicals, broken thermometers, chemical spills and material used to clean them, used oil (such as engine oil).

7.3.2 Solid domestic waste typically has diverse content

Solid domestic waste (SDW), often referred to as municipal solid waste (MSW), is the collective term for the different waste materials generated from households and businesses. SDW can include both recyclable and non-recyclable materials. It is a collection of a wide variety of waste ranging from food to metal to clothing to hazardous chemicals. Examples of SDW are shown in Table 7.7.

■ **Table 7.7** Examples of solid domestic waste

Type of waste	Example
Paper and cardboard	Newspapers, magazines, old notebooks and books, paper cups and plates, office paper, boxes, cartons and packing material
Glass	Empty bottles, jars, and other glass containers used for beverages and food
Metal	Aluminium cans, steel containers; often from food packaging
Plastic	Single-use plastic bottles, containers, packaging materials, pens and other plastic goods
Organic	Kitchen or garden waste, fruit and vegetable peels, scrap food, dried leaves and trimmings
Packaging	Bubble wrap, cling film, aluminium foil, Styrofoam
Construction debris	Concrete, cement, demolition material
Clothing	Discarded textiles, old clothing, accessories that are no longer in use
Hazardous	Batteries, paints and varnishes, light bulbs, other electric material
E-waste	Housekeeping goods, also called 'white goods' (washing machines, refrigerators, electric cookers, oven, grills, toasters and microwaves), electronic audiovisual goods, also called 'brown goods' (mobile phones, keyboards, monitors, laptops, tablets, televisions, printers, Wi-Fi routers, uninterruptible power supplies (UPS)

7.3.3 The volume and composition of waste varies over time and between societies due to various factors

Increased demand and supply of natural resources to produce material that has a short life span has caused a rapid increase in the amount of waste produced. The rate of waste production is much higher than the rate at which it is removed from the system, resulting in accumulation of unwanted waste in the form of landfills and floating plastic in water bodies. Waste-management techniques in many countries are not efficient enough to tackle the amount of waste generated and countries are therefore struggling with landfill issues.

The volume and composition of waste produced by different societies is dynamic and depends on factors including socioeconomic, political, environmental and technological elements. For instance, societies with higher disposable incomes might produce more packaging waste from increased consumption of convenience products. Figure 7.30 provides an overall picture of solid waste produced by different countries around the world. It is evident from the figure that MEDCs produce approximately 700 kg of solid waste per capita per year, while LEDCs or NICs produce around 200 kg.

■ Figure 7.30 The amount of annual waste generated by countries (source: World Bank What A Waste Global Database)

◆ **Fast fashion** – the production of inexpensive clothing by mass-market retailers in response to the latest trends.

◆ **Environmental constitutionalism** – a concept for providing constitutional protection to local and global environments.

■ Socioeconomic factors

Socioeconomic factors, such as income levels and consumer behaviour, can impact the types and amounts of products purchased, thereby influencing waste generation. The amount and type of waste produced by countries varies. In affluent societies, the prevalence of single-use products, excessive packaging, and a culture of convenience can contribute to higher volumes of waste compared to societies with lower disposable incomes, which may prioritize resourcefulness and sustainability. For example, the trend of **fast fashion** driven by consumer demand for inexpensive clothing contributes a significant volume of textile waste. The throwaway culture, influenced by the ability to regularly update wardrobes, household furnishings, appliances and utensils, results in discarded clothing items that contribute to landfill waste. In contrast, societies with limited financial resources may be more likely to exhibit a culture of thriftiness, reusing and repurposing items, thereby reducing overall waste generation.

Similarly, family size plays an important role in determining the volume of waste produced. Smaller families generate less waste than large families do. Environmental perspectives and the willingness to segregate waste can influence waste generation patterns. Societies which promote community awareness programmes and involvement of local governments are usually less wasteful than societies which lack community engagement. The volume of waste is much lower in countries which have adopted circular economy than in countries which still follow the linear economic model. Circular economies increase the life of products by repurposing items that would otherwise end up in landfill sites.

■ Political factors

Political factors that influence the volume and type of waste generated include policies, regulatory framework, **environmental constitutionalism**, funding and international agreements. Strict waste management policies such as landfill regulations, recycling policies and waste reduction targets directly impact waste generation and composition. Countries with strict recycling laws have less waste being sent to landfill and high rates of material recovery. In contrast, regions with weak or poorly enforced waste-management regulations may experience higher levels of pollution and waste mismanagement. In some countries like Germany the waste-management system requires citizens to sort their waste into multiple categories such as recycling and composting. The deposit refund system (also called the deposit return scheme, DRS) implemented in many countries allows citizens to take reusable containers such as beverage containers to return points and receive a 10-cent refund for each container. Prioritising the environment by including legally binding environmental policies in the constitution of a country is a primary way of uniting stakeholders to reduce waste generation. Implementation of waste reduction policies requires adequate funds, proper allocation of funds in research and development for innovative recycling technologies and novel sustainable materials, incentive programmes to encourage businesses and households to adopt waste reduction practices such as repurposing, education and outreach programmes focussed on waste reduction methods.

■ Environmental factors

Environmental factors such as geographical location, seasons, climate and population density play an important role in influencing the volume and composition of waste generated. Many of these factors are inter-linked. Geographical locations affect the type of waste generated. For example, waste accumulation may take place in residential areas located in the hills or mountains due to lesser frequency of waste collection. During peak tourist seasons, tourist destinations may experience high amounts of solid waste generation due to the increased number of people visiting restaurants or hotels and to the high amount of disposable consumer items being sold. In the agricultural sector, harvest seasons are usually followed by huge amounts of solid waste generated from crop remains in the fields. Hotter seasons may also contribute to over-ripening of fruits and vegetables, causing them to perish faster. Untimely rains, droughts and floods are some of the environmental factors that adversely affect the agricultural sector, increasing food loss. High population density and availability of natural resources in a region or country are also driving forces for waste generation. The production, consumption and trade patterns heavily rely on the availability of resources. Urban areas tend to produce more waste than rural areas due to higher population densities, increased consumption and commercial activity. Countries with high overall population densities have high per capita demand and thus a greater tendency of waste production. Discarded items such as single-use plastic, clothes, furniture, metal cans and food end up in landfills if not properly managed.

■ Technological factors

Electronic waste (e-waste) is a collection of electric and electronic equipment that has either stopped serving its original purpose due to breakage, or has lost its value due to replacement and redundancy. The rapid pace of technological advancements contribute to the growing issue of electronic waste (e-waste). The constant release of new models and features encourages consumers to upgrade their devices frequently, leading to the discarding of older ones, which become obsolete. We saw in Chapter 7.2 (page 634) how renewable energy sources can also produce e-waste, such as damaged solar panels, wind turbines and rotor blades. Additionally, technological advances in the nuclear energy sector have contributed to increased amounts of radioactive waste. On the other hand, technological advancements in new packaging materials can possibly contribute to a reduction in waste generation by replacing the non-biodegradable materials with biodegradable ones.

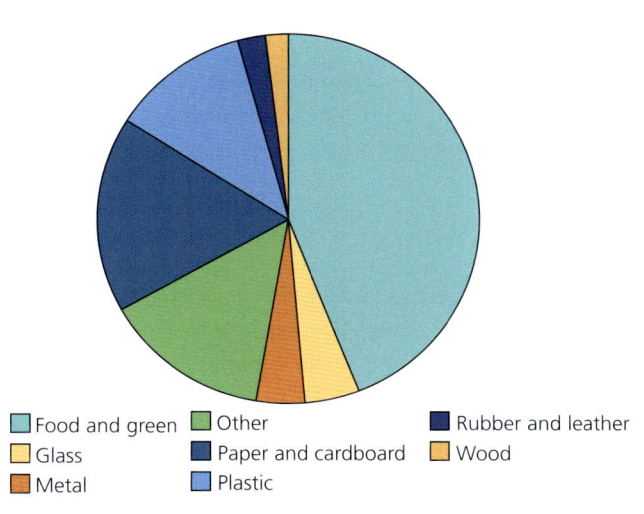

- Food and green
- Glass
- Metal
- Other
- Paper and cardboard
- Plastic
- Rubber and leather
- Wood

■ **Figure 7.31** The rubbish of the world – sorted

7.3.4 The production, treatment and management of waste has environmental and social impacts, which may be experienced in a different location from where the waste was generated

People think that when they throw something into a recycling bin, it actually goes for recycling. This may not always be the case. A portion or all of this waste ends up either in the suburbs of the cities or across borders in other countries. Often, there is a long chain of handling the recyclable waste from source to destination with no accountability.

> ### REAL-WORLD EXAMPLE
>
> ### Waste generation
>
> The global issue of waste generation is predominantly urban-centric in almost all countries around the world. Improper infrastructure, lack of waste-collection and disposal methods, and overpopulated cities contribute greatly to the huge volume of waste generated. According to the World Bank's report of 2018, 'What a Waste 2.0', the world was estimated to generate 2.24 billion tonnes of solid waste, which would amount to a footprint of 0.79 kg per person per day. It is estimated that with the rapid growth of population the waste generation will increase to 3.88 billion tonnes by 2050, which is a 73 per cent increase from 2020. The urban poor sector of societies experiences the major impact of unsustainably managed waste. This is because most of the unsegregated waste goes directly to landfills located on the outskirts of cities. In most countries land prices are low in the suburbs. These areas are mainly occupied by the urban poor who do not have other options and, in some cases, may be trying to escape conflict or environmental issues elsewhere. If the landfills are not well managed, then the residents of these areas may be continuously exposed to harmful conditions such as toxic fumes, gases, bad odours and contaminated water, causing serious health issues. This is a form of environmental injustice.

Waste-management strategies differ from country to country. Popular methods include landfill and incineration. In some low-income countries, the environmental regulations regarding waste disposal are lenient. Therefore, high-income countries such as the European Union, Japan, United Kingdom and the United States may export their electronic and plastic waste to countries such as Indonesia, Vietnam, Malaysia or Ghana. This waste travels long distances by ship to reach its destination. In the destination country, electronic waste is processed to recover usable material, while mixed-plastic waste is melted into plastic pellets, dumped in landfills or simply burned.

In regions like Agbogbloshie in Ghana, informal recycling operations contribute to soil and water pollution due to the improper disposal and burning of electronic components.

Dumping waste illegally, along with informal waste processing, harms local communities, exposed to hazardous materials and pollutants, which pose health risks. Unregulated import of plastic waste into a country may lead to bans on importing waste. This may affect the livelihood of informal waste pickers who rely on sorting and recycling these materials and selling them to recycling companies. The Basel Convention, an international treaty addressing the control of transboundary movements of hazardous wastes and their disposal, aims to prevent such environmental injustices. However, gaps in enforcement and compliance can occur, allowing the movement of hazardous waste to continue.

In March 2022, the United Nations Environment Assembly endorsed a resolution to End Plastic Pollution. A negotiating committee was established, due to complete its negotiations by December 2024. The resolution will focus on finding alternatives to using plastics and designing reusable and recyclable products and materials. It will also enhance international collaboration to share scientific and technological know-how of using alternative materials to plastic.

> **REAL-WORLD EXAMPLE**
>
> ### Pulau Indah recycling industries
>
> Pulau Indah is an island town and home to Malaysia's biggest port. Approximately 25–30 years ago, Pulau Indah was a natural swamp with a quiet island town feel feel with a population of approximately 25,000 people. It is now Malaysia's fastest-growing industrial hub. Malaysia started importing solid waste for recycling when China banned this practice in 2018. Pulau Indah was a perfect choice, as it had the largest port and a small population. Illegal plastic waste-recycling factories have flourished in Malaysia since 2017, making it one of the world's largest plastic importers.
>
> ■ Figure 7.32 The location of Pulau Indah
>
> Pulau Indah struggled as huge piles of plastic were dumped all over the island. The illegal recycling factories do not have the required machinery or protocols for waste disposal. The presence of these factories close to residential areas is a persistent source of air and water pollution. Open burning of plastic waste releases huge amounts of dioxins into the atmosphere, along with the prominent greenhouse gas carbon dioxide. These dioxins are called **persistent organic pollutants** (POPs), are very harmful and can remain in the body for a very long time. They may cause cancer, reproductive and developmental problems, and can interfere with hormones in humans. They remain in the body for a very long time.

◆ **Persistent organic pollutants** – organic compounds that are resistant to degradation chemically, biologically or physically.

● HL.b.16, HL.b.17: Environmental economics

Degrowth is a political and economic theory that focuses on reducing production and consumption to conserve natural resources and reduce environmental damage. It favours the stabilization of resource use, production and population remain stable over time. This model is directly related to the environmental and social impacts of the production, treatment and management of waste in several ways. It focuses on anti-capitalism, zero growth and anti-consumerism(reducing the inclination to buy unnecessary things). The theory lays greater emphasis on sustainability and resource efficiency directly relates to the environmental and social impacts of waste. For example, implementing policies that prioritize waste reduction and recycling initiatives. Similarly, the circular economy model aims at reducing waste by maximising product lifespan. It also minimizes resource extraction by encouraging people to reuse and recycle products.

The European Union has been at the forefront of adopting policies related to sustainable development and circular economies. Strategies such as the European Green Deal aim to make the EU climate neutral by 2050. One of the key focus areas of the strategy is circular economy. Under the Green Deal, there are provisions for degrowth as well as green growth. The Circular Economy Action plan focuses primarily on reduction of waste by implementing the following:

- designing sustainable products: durable, reusable, recyclable
- empowering consumers: providing information about product lifespan, repair services, spare parts, repair manuals, sustainability labels, protection from greenwashing
- promoting circularity in production processes: avoiding new material, reusing existing material
- enhancing product value chain by introducing right to repair electronic goods, common charger, durable cables, improved collection of e-waste, return to seller schemes, phasing out non-rechargeable batteries, end-of-life treatment of electronic vehicles
- packaging material: reducing over-packaging, using reusable and recyclable packing material, reducing the complexity of packaging material
- restrictions on plastics: avoiding the intentional use of microplastics, providing relevant data and information on the existing microplastic levels.

ATL ACTIVITY

Write down five ways in which the degrowth model can help reduce solid waste. Discuss with your peers how the degrowth model can be successfully implemented at the local, national and international levels.

HL.c.13: Environmental ethics

Equitable and just societies aim to ensure that no particular section of the population bears an unreasonable burden of the environmental and social impacts associated with the production, treatment and management of waste. Policies that explicitly address environmental justice concerns ensure that waste-management facilities and activities are distributed equitably across different communities. For example, regulations restrict the location of landfills or waste-treatment plants in areas with vulnerable populations, prevent the concentration of environmental burdens in marginalized communities, ensure that community voices are heard and that their concerns are addressed by engaging them in decision-making processes, improve the accessibility of waste-management infrastructure to all sections of the societyand create more jobs in the waste-management sector, such as litter collectors and street sweepers. These regulations thereby distribute the work and improve the efficiency of the system.

ATL ACTIVITY

Choose a country and list the equitable practices it follows for waste management.

TOK

On what basis can we choose to perform an action without knowing the potential harm it may cause to other non-human lives on Earth?

Consider the challenges of predicting global consequences and the ethical implications of our interconnected actions on a global scale.

◆ **Biodegradable** – substances that are capable of being degraded by microorganisms into smaller, simpler non-toxic molecules.

◆ **Non-biodegradable** – substances that cannot be degraded by microorganisms into smaller molecules.

◆ **Pollution** – the introduction of substances into the environment which can have harmful or poisonous effects.

7.3.5 Ecosystems can absorb some waste, but pollution occurs when harmful substances are added to an environment at a rate faster than they are transformed into harmless substances

Ecosystems have a natural ability to absorb and process certain levels of waste. This is achieved by certain biological, chemical and physical processes. Wetlands, for example, act as natural filters, absorbing and breaking down pollutants in water. Ecosystems are capable of taking care of **biodegradable** waste. For instance, bacteria and other living organisms (particularly fungi) feed on organic waste and return the nutrients back to the environment. Some wastes like plastic, concrete, heavy metals and radioactive waste cannot be assimilated by microorganisms. Such wastes are called **non-biodegradable** wastes.

Pollution is the introduction of harmful substances into the environment at a faster rate than they can be removed or converted into harmless substances. It can contaminate water sources, impact air quality and disrupt the natural food chain. Pollution of any kind can have severe impacts on biodiversity due to accumulation of harmful substances that affect the health of plants, animals and microorganisms in ecosystems. For example, the decline in amphibian populations, for example, linked to water pollution and habitat degradation due to climate crisis, is a global concern, with over 40 per cent of amphibian species at risk.

Biodegradability plays a crucial role in waste transformation, influencing pollution levels in ecosystems. Plastics, known for their low biodegradability, contribute significantly to environmental pollution and stay in the ecosystem for a very long time. It may take around 20 to 500 years for plastic to degrade. According to IUCN, at least 14 million tons of plastic enter the oceans each year and contribute to 80 per cent of the marine debris from surface waters to deep-sea sediments. The production of plastic has experienced a significant increase, from 2 million tonnes in 1950 to 348 million tonnes in 2017, resulting in the establishment of a global plastic-producing industry valued at US$522.6 billion. Projections indicate that this industry will double its capacity by the year 2040. The removal of plastic from the ecosystem greatly depends on its type. There are seven common types of plastics, as shown in Figure 7.33.

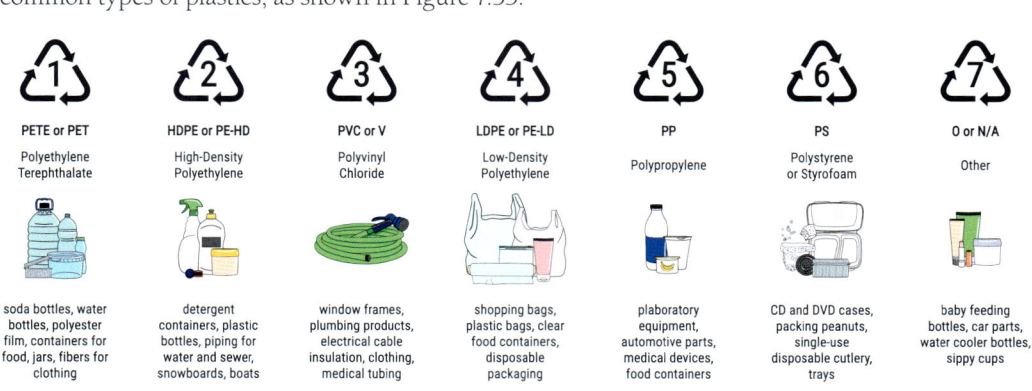

1 PETE or PET	2 HDPE or PE-HD	3 PVC or V	4 LDPE or PE-LD	5 PP	6 PS	7 O or N/A
Polyethylene Terephthalate	High-Density Polyethylene	Polyvinyl Chloride	Low-Density Polyethylene	Polypropylene	Polystyrene or Styrofoam	Other
soda bottles, water bottles, polyester film, containers for food, jars, fibers for clothing	detergent containers, plastic bottles, piping for water and sewer, snowboards, boats	window frames, plumbing products, electrical cable insulation, clothing, medical tubing	shopping bags, plastic bags, clear food containers, disposable packaging	plaboratory equipment, automotive parts, medical devices, food containers	CD and DVD cases, packing peanuts, single-use disposable cutlery, trays	baby feeding bottles, car parts, water cooler bottles, sippy cups

■ **Figure 7.33** The seven common types of plastics (source: Greenpeace)

◆ **Gyre** – a giant circular oceanic surface current.

> **REAL-WORLD EXAMPLE**
>
> ### The Great Pacific Garbage Patch
>
> The Great Pacific Garbage Patch (GPGP), located in the Pacific Ocean, is the largest of the five offshore garbage-accumulation sites in the world's oceans. GPGP is actually two distinct collections of debris, bounded by the massive North Pacific Subtropical **Gyre**. The western garbage patch is located near Japan, and the eastern garbage patch is located near Hawaii and California. Four currents, namely the California current, the North Equatorial current, the Kuroshio current and the North Pacific current, make up the gyre. These rotate clockwise in an area of approximately 20 million square kilometres.
>
> ■ **Figure 7.34** Location of the Great Pacific Garbage Patch
>
> Measurements made by the Ocean Cleanup team estimate the total mass of the patch to be 100,000 tonnes. Vertically, the GPGP is distributed within the top few metres of the ocean. It consists of dispersed particles, largely microplastics, and other debris suspended in the water column. The majority of this pollution originates from land-based sources, with approximately 80 per cent coming from activities in North America and Asia. The centre of the patch is relatively calm and stable. The vortex pulls the debris into the centre, where it stays and continues to rotate because all of the plastic waste is non-biodegradable.
>
> The major part of the GPGP is made up of tiny plastic particles called **microplastics**, formed by **photodegradation**. They are not always visible to the naked eye and therefore make the water cloudy and appear like a soup. These are the main sources of marine pollution. Marine debris consists of heavy plastic materials, and 70 per cent of it sinks to the bottom of the ocean. Fishing nets and other fishing equipment are some of the major components of the GPGP. Marine animals are exposed to grave danger due to the presence of this floating plastic graveyard. For example, the loggerhead turtle confuses plastic bags with jellyfish, which are its preferred food, and feeds on them. Some birds confuse the microplastics for fish eggs and feed them to their chicks. As a result, the chicks die due to starvation or organ rupture due to accumulation of plastic.

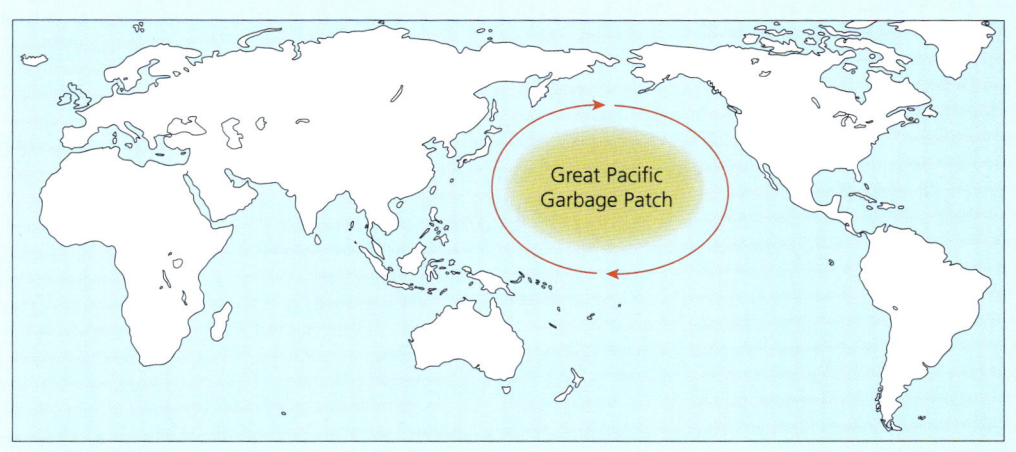

◆ **Microplastics** – extremely small (less than 5 mm) particles of plastic.

◆ **Photodegradation** – the degradation of plastic by exposure to sunlight.

Link

Topic 1.3.19 (page 72) discusses the planetary boundary of novel entities, which has been crossed. Microplastics are a type of novel entity that is damaging the marine ecosystem.

Despite increasing efforts in some places to recycle it, electronic waste is still mostly sent to landfill. The heavy metals such as chromium present in electronic devices leach out over time and infiltrate into the soil, eventually contaminating the groundwater.

7.3 Solid waste

7.3.6 Preventative strategies for waste management are more sustainable than restorative strategies

Restorative strategies for waste management like recycling, require energy. Therefore, efforts are being made to focus on reducing and preventing waste creation. This approach is widely seen as being more environmentally beneficial. The primary emphasis lies in modifying human behaviour, regulating waste-disposal practices, and promoting the conscientious use of resources. Preventative strategies for waste management are inherently more sustainable than restorative strategies as they focus on addressing the root causes of waste generation.

According to the Green Economy Report by the United Nations Environment Programme (UNEP), 2011, the order of preference for waste-management strategies indicates prevention as the best method, followed by reduction, recycling, recovery and disposal.

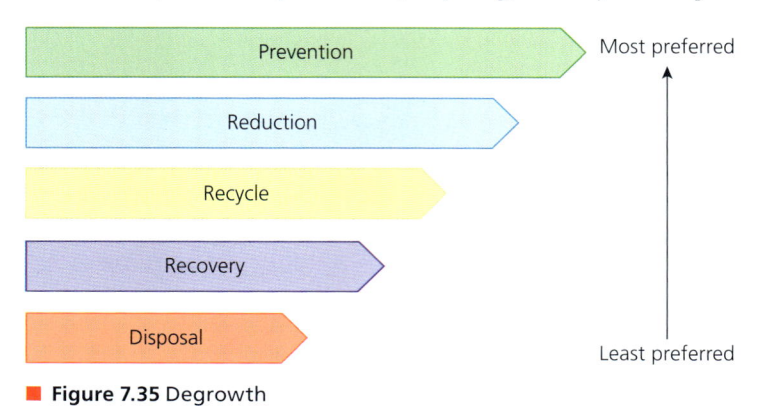

■ **Figure 7.35** Degrowth

■ Altering human behaviour

Education, incentive and regulation can be used to alter human behaviour. Awareness programmes, subsidies, refunds, fines and taxes are some of the ways in which change in human outlook towards waste creation can be changed:

- Education: Campaigns informing people about the negative impacts of waste generation and the benefits of recycling have proved helpful in slowly conditioning people's thoughts towards the environmental benefit. Introducing waste management lessons in school curricula promotes children to make environmentally friendly choices early in their lives. Workshops conducted to teach waste reduction techniques in real-life settings encourage people to take steps towards embracing conscious consumerism and minimalism. Individuals learn the importance of buying only what is necessary. For instance, campaigns like 'Buy Nothing' Day encourage people to abstain from non-essential purchases on a designated day, spreading awareness about the impact of overconsumption.

- Incentives and regulations: Community-based recycling programmes, such as those encouraging citizens to separate recyclables from general waste, actively promote the reuse of materials. In cities like San Francisco, where stringent recycling regulations are enforced, there has been a notable reduction in waste sent to landfills. Incentivizing public behaviour to return empty containers for a refund encourages recycling and helps in reduction of waste creation. Fines and penalties for littering discourage people from littering public places. In many countries, violation of waste management procedures may incur a hefty fine.

■ Regulating waste disposal practices

- Responsible waste disposal: Waste-disposal initiatives, like the 'Pay-As-You-Throw' programmes in cities such as Taipei, Taiwan, incentivize responsible waste disposal by charging households based on the amount of waste generated. This encourages residents to reduce their waste and recycle more.

- Regulatory measures: Countries like Sweden implement policies that impose a tax on incineration, creating financial incentives for businesses to explore alternative waste-management solutions. Such regulations stimulate industries to adopt environmentally friendly practices.

Even with these initiatives, some waste will still be produced. Restorative strategies are therefore necessary alongside prevention.

> **REAL-WORLD EXAMPLE**
>
> #### Reducing the production of waste
>
> Roubaix is one of poorest towns in France, with 46 per cent of people living below poverty line. Lacking waste collection and treatment practices, Roubaix decided to find its own way to reduce waste at the source. Families, schools and businesses were urged to join the zero-waste challenge. Around 100 families volunteered to participate in a year-long challenge and were called 'zero-waste families'. They received training in 14 different workshops held over the period of one year. The workshops taught volunteers to halve their waste by making small changes in their daily choices. The training included food waste reduction and composting, homemade cleaning products and cosmetics. Volunteer families were given a weighing scale to measure non-residual waste generation. They had to periodically weigh their trash and report it to the city. This gave a sense of awareness among the volunteers about the actual 'weight' of the trash they generate. This information-based approach allowed people to understand how waste generation takes place from the choices we make in our daily lives.
>
> France was one of the first countries in the world to implement the 'extended producer responsibility' law, which holds producers financially responsible for the waste they create even after their products are sold. In Roubaix, the zero-waste families could significantly cut down on their waste. On average, volunteers could save $1,088 by cutting down their waste generation by almost 50 per cent. Almost one-third of the waste generated belonged to organic waste from the kitchen and could easily be composted, while recycling would help to sort another third, which comprised glass and metal. Ten per cent of the waste was plastic. This can easily be avoided by using reusable alternatives. Volunteers deliberately chose sustainable options for shopping bags and containers, ate very few takeaway meals, and switched to homemade laundry detergents. Roubaix continues to focus on promoting the zero-waste programme in schools.
>
> *(Source: www.theguardian.com/environment/2023/dec/07/its-kind-of-gross-but-we-can-do-it-how-a-community-learned-to-go-zero-waste)*

■ Clean-up and restoration

Restorative efforts to clean up plastic pollution from rivers involve projects like The Ocean Cleanup, initiated by the Dutch entrepreneur Boyan Slat, which uses floating barriers called 'interceptors' and a conveyor that make use of the natural flow of water to collect and concentrate plastic debris. The catamaran design of the interceptor allows the plastic debris to freely flow into the device where it is removed. The design also reduces any risk of entanglement or disturbance to aquatic organisms. The project installs its cleaning systems in strategically chosen locations to maximize efficacy of debris capture. Collected plastic debris is processed and recycled whenever possible, contributing to a circular economy approach. Efforts are also made to explore innovative ways to repurpose collected plastic for constructive applications.

The Ocean Cleanup collaborates with governments, industries and environmental organizations globally to generate public awareness and mobilize support for ocean conservation. International efforts to address the transboundary nature of oceanic plastic pollution.

There are several challenges within the Ocean Cleanup project that make it difficult to achieve sustainability. For instance, cleaning up oceanic garbage patches is an immense undertaking due to the vastness of the affected areas and the minute size of plastic particles. Improvement of technologies can help capture microplastics effectively. At a given time, only a small area can be cleaned up and there is a lack of precision in the process. Large-scale cleanup initiatives require substantial energy, resources and financial investments. The use of cleanup technologies, particularly those involving nets and barriers, carries the risk of harming marine life. Attempts to capture plastic debris may inadvertently catch and harm aquatic organisms, disrupting fragile ecosystems. The continuous input of plastic garbage is one of the main challenges, as debris continues to enter the marine ecosystem.

■ **Figure 7.36** An Ocean Cleanup interceptor collecting rubbish from Klang River, Malaysia

REAL-WORLD EXAMPLE

Landfill reclamation at Freshkills Park, USA

Located on Staten Island, New York, Freshkills Park stands as a unique example of urban restoration, transforming what was once the world's largest landfill into a vast green space. The project is operated by New York City Department of Parks and Recreation. The park opened partially in February 2012 and is scheduled to be in full operation by 2035–37. The project revitalized the landscape, created recreational opportunities and restored ecosystems, demonstrating the potential for repurposing degraded areas for public benefit. Freshkills was historically a landfill, and municipal waste from New York City was dumped in this landfill from 1947 to 2001. While restoring the land, this approach does not eliminate the environmental impact caused by the landfill's active use.

Challenges associated with the restoration project include substantial financial investments, technological infrastructure and human resources. The sustainability of such efforts depends on ongoing financial and community support. Restoration also involves the reintroduction of plant and animal species to establish the ecosystem. There could be a possibility of introducing a non-native species, which could cause havoc in the ecosystem. Shifts in demographics, land use or urban planning may influence the park's viability and relevance over time.

■ Thoughtful use of resources

The idea that reducing consumption is better for the Earth makes a lot of sense. First, it saves important natural resources like trees and energy, because we don't need as much to make new material. For example, buying fewer clothes and choosing ones that last longer helps cut down on the huge amount of waste created by fast fashion. The idea of a circular economy promotes increasing the life of a commodity by recycling and repurposing it. As a result, things get reused instead of being thrown away. Companies like Loop from Quebec (see Figure 7.37) facilitate reuse. The company collaborates with brands, retailers and manufacturers to help in the sale and collection of reusable and returnable items. This is called a reverse supply chain, where products are delivered to the customer and at the same time the empty package is collected, ensuring that less packaging material is thrown away. The packaging material is hygienically cleaned and returned to the manufacturer for them to use to pack products for other customers.

How does Loop work?

1. Buy your Loop products in reusable packaging and enjoy
2. Return your empty containers
3. We hygienically clean and refill
4. All ready for reuse!

■ **Figure 7.37** Loop is an example of circular economy

In terms of transport, car-pooling, ride sharing and choosing cleaner ways to reach our destination (public transport, cycling, walking) are some of the easy ways in which we can reduce the need for manufacturing new cars. With respect to technological advances, choosing not to replace their phone so often is a conscious choice that customers can make.

7.3.7 Different waste-disposal options have different merits and demerits in terms of their impact on societies and ecosystems

There are advantages and disadvantages of different disposal methods, influencing both societies and ecosystems differently. The choice of how we handle waste has wide-ranging effects on the environment and human well-being.

> **Concept**
>
> **Sustainability**
>
> Sustainability needs to be built in to the backbone of waste disposal. Sweden's efficient waste-to-energy plants and Germany's comprehensive recycling systems contribute significantly to reducing the environmental footprint by minimizing the extraction of raw materials and mitigating pollution. Germany has a five-level waste hierarchy, comprising waste prevention, reuse, recycling, energy recovery and waste disposal.

> Different bins are used for different types of waste. It has blue bins for non-greasy paper, newspaper and cardboard, yellow or orange bins for plastic and metal containers, green or brown bins for biodegradable material, grey or black bins for things that you can't sell, donate or recycle. Waste containing glass is collected separately, in three different bins: brown bins for brown coloured glass, white bins for transparent glass and green bins for any other coloured glass. These are called bottle banks.
>
> Even waste that is dumped can be put to more sustainable use. The largest waste dump in the world, Apex Regional Landfill in Las Vegas, Nevada, covers approximately 2,200 acres of land. It has an on-site gas-to-energy plant that generates approximately 11 megawatts of renewable electricity.

■ Landfill

Landfill sites are designated areas where solid waste is disposed of and managed. These sites are engineered to contain and isolate waste from the surrounding environment to prevent contamination and adverse impacts on public health. The process of landfilling involves depositing waste into specifically designed pits or trenches within the landfill area. Landfills are the most common method of waste disposal in many countries.

The key features of a landfill are listed in Table 7.8. The layout of a typical landfill is shown in Figure 7.38.

■ **Table 7.8** The key features of a landfill

Liner system	Modern landfills are equipped with liner systems made of materials like clay and synthetic liners to prevent the leakage of leachate (liquid formed as waste breaks down) into the soil and groundwater.
Compacted waste	Waste is compacted and spread in layers to maximize the use of available space. Compaction reduces the volume of waste and minimizes the potential for settling.
Cover material	Once a section of the landfill is filled, it is covered with a layer of soil or other materials to control odours, reduce windblown debris and discourage pests.
Gas collection	Landfills generate gases, primarily methane, as a byproduct of the decomposition of organic waste. Gas collection systems are installed to capture these gases, and in some cases utilize them for energy.
Leachate collection	Leachate, formed as rainwater percolates through the waste, is collected and treated to prevent contamination of groundwater.
Monitoring systems	Landfills are equipped with monitoring systems to track environmental indicators, including groundwater quality, gas emissions and the stability of the landfill structure.
Closure and post-closure care	Once a landfill reaches its capacity, it undergoes closure procedures, including the installation of a final cover. Post-closure care involves continued monitoring and maintenance to address potential environmental issues.

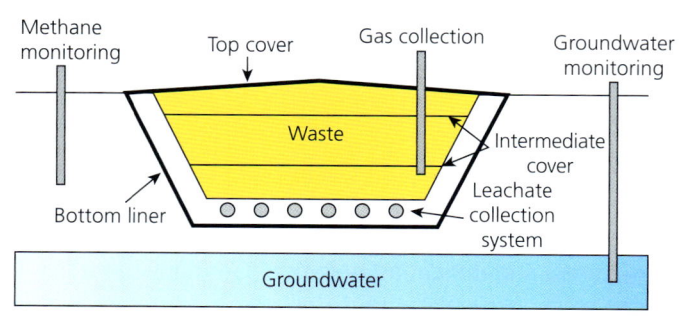

■ **Figure 7.38** The layout of a typical landfill

Theme 7: Natural resources

Merits and demerits of landfill

Merits:

- Landfill is efficient for large-scale waste disposal, especially in countries with large open spaces.
- Landfill is a cost-effective method in the short term (30–50 years), and does not require complicated technology.
- Landfills can serve as an energy source, as the methane and carbon dioxide gas can be upgraded and used as fuel. The byproducts of landfills can also be used to produce energy.
- The land used for landfill can be reclaimed at any time. When the landfill becomes old enough and can hold no more, the top can be built over.
- It produces fewer emissions than incineration.
- Dedicated landfills can serve to reduce the littering of waste everywhere. If managed properly, the landfills can help to keep cities, towns and villages clean.

Demerits:

- Landfill has environmental risks, such as leachate and gas (methane) emissions.
- There is limited space availability for new landfills in areas with space constraints. Landfills usually require large spaces to accommodate all the trash that they receive. Smaller cities or countries do not have much area to dedicate for landfills.
- Landfills have potential for soil and water contamination due to leachate, leading to groundwater and aquifer contamination.
- The lining of landfills has a limited lifespan, even though it may be 100 years.
- Pests such as rodents, flies and mosquitoes are attracted to the landfill and may cause the spread of diseases in humans and other animals.

Incineration and waste-to-energy

Incinerators are facilities designed to burn waste materials at high temperatures, typically for the purpose of reducing the volume of waste, minimizing its environmental impact and generating energy.

Waste-to-energy (WTE) is a process that harnesses the energy present in the waste material by burning it under controlled conditions. This energy would otherwise be wasted in the landfills. Waste-to-energy facilities utilize advanced technologies to combust waste at high temperatures, producing heat that is then used to generate steam. This steam, in turn, drives turbines connected to generators, producing electricity.

Merits and demerits of incineration

Merits:

- Incineration reduces the volume of waste, so is particularly useful in densely populated areas.
- Incineration can effectively destroy certain hazardous waste materials, reducing their environmental impact.
- Incinerators operate within controlled conditions, minimizing the release of pollutants when properly designed and managed.
- Facilities like the Spittelau waste-to-energy plant in Vienna, Austria, can generate both heat and electricity from waste incineration.

Demerits:

- Incineration can release pollutants such as particulate matter, dioxins and heavy metals into the air. Advanced air pollution-control devices are required to address this issue.
- Public perception and concerns about air quality can lead to opposition to the establishment of incineration facilities in certain communities.
- Establishing and operating incineration facilities can involve significant financial investment.
- Incineration produces ash, which may contain non-combustible materials and residues from pollution-control devices. Proper disposal of ash is crucial.
- While WTE facilities can reduce methane emissions from landfills, incineration may emit carbon dioxide and contribute to climate change.
- Plans for waste incinerators in various locations may face opposition due to environmental and health concerns.

Exporting waste

Global waste trade is the legal buying and selling of waste between two or more countries for recycling or disposal by other means. Many MEDCs export their waste to LEDCs, where it may be sorted, treated, recycled, and used to produce energy by a waste-to-energy technique, or simply be dumped in open dumps, creating an environmental issue in the LEDCs. From 1988 to 2016, the global waste-trading industry had a total value of $98.3 billion. Mostly plastic and e-waste is exported to LEDCs. According to a report published by Statista in May 2023, the major exporters of plastic waste include Japan and wealthy European Union countries. Countries and regions which import plastic waste include Turkey, India, Eastern Europe and Latin America. In 2022, the major exporters of e-waste were France and Germany while South Korea was the major importer of e-waste.

Merits and demerits of exporting waste

Merits:

- Exporting waste can facilitate resource recovery, allowing materials to be recycled and reused in other countries' manufacturing processes.
- Waste exports contribute to the global trade in recyclable materials, supporting economic activities and job creation in both exporting and importing countries.
- Exporting waste can provide economic benefits for the importing country, generating revenue through the sale of recyclables.
- Some countries may lack specialized recycling facilities. Exporting waste allows access to advanced facilities in other nations for proper processing.
- By exporting waste, countries may divert materials from landfills, contributing to waste reduction and more sustainable waste-management practices.

Demerits:

- The transportation of waste over long distances results in greater carbon emissions, contributing further to environmental pollution and climate change.
- Improper disposal practices in receiving countries may lead to environmental and health hazards, impacting local communities and ecosystems.
- Exporting waste to developing countries, where regulations and infrastructure may be inadequate, raises ethical concerns regarding responsible waste management.

- Relying on waste export markets may hinder the development of domestic recycling and waste-management infrastructure, creating a dependency on external solutions.
- Exporting waste means relinquishing control over the waste-management process, potentially leading to unintended consequences and improper handling.

Recycling

Recycling is a process through which materials or products are collected, processed and transformed into new items, thereby diverting them from being disposed of as waste. The goal of recycling is to recover valuable resources, conserve energy, and reduce environmental impacts associated with the extraction and processing of raw materials to make new products.

The recycling process typically involves the following steps:

- Collection: Recyclable materials, such as paper, glass, plastic and metal, are collected from households, businesses or designated collection points.
- Sorting: The collected materials are sorted into categories to separate different types of recyclables. This step is crucial to ensure that materials can be processed efficiently.
- Processing: The sorted materials undergo processing, which may include cleaning, shredding or melting, depending on the material type. Processing prepares the materials for the next stage of manufacturing.
- Manufacturing: The processed materials are used as raw materials in the manufacturing of new products. For example, recycled plastic can be used to produce new plastic items.
- Distribution: The newly manufactured products are distributed for sale, completing the recycling loop.

Merits and demerits of recycling

Merits:

- Recycling conserves natural resources by reducing the need for extracting and processing raw materials.
- The manufacturing of products from recycled materials often requires less energy, lowering emissions.
- Recycling diverts materials from landfill, reducing the volume of waste and associated environmental impact.
- Recycling contributes to lower greenhouse gas emissions associated with the production of new materials.
- It creates jobs, leading to economic benefits.

Demerits:

- Establishing and maintaining recycling programmes can be costly, and economic viability depends on market conditions.
- Products from the recycling process may not always be as durable as new ones, and may not always be in demand, thus affecting the economic benefits.
- The logistics of collecting, sorting and transporting recyclables can be complex, especially in regions with limited infrastructure.
- Not all materials can be effectively recycled, and certain items may have limited recycling options.
- Recycling products can require some new materials, for example paper requires some virgin wood, so it is not a full substitute for reducing consumption.

Even though the merits of recycling outweigh the demerits, recycling of solid waste has not been the most popular choice around the world (see Figure 7.39).

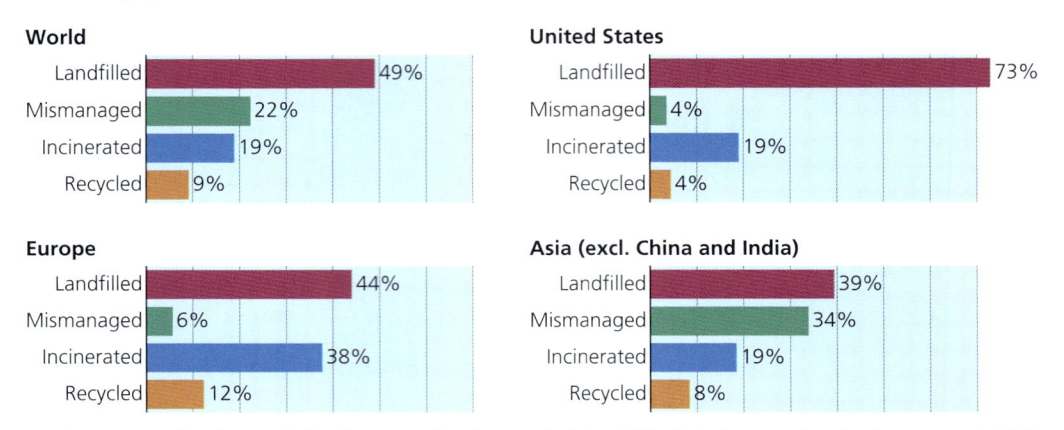

World
- Landfilled: 49%
- Mismanaged: 22%
- Incinerated: 19%
- Recycled: 9%

United States
- Landfilled: 73%
- Mismanaged: 4%
- Incinerated: 19%
- Recycled: 4%

Europe
- Landfilled: 44%
- Mismanaged: 6%
- Incinerated: 38%
- Recycled: 12%

Asia (excl. China and India)
- Landfilled: 39%
- Mismanaged: 34%
- Incinerated: 19%
- Recycled: 8%

■ **Figure 7.39** The share of plastic waste that is recycled, landfilled, incinerated and mismanaged, 2019

Composting

Composting is the natural process of recycling organic matter into fertilizer. Dead leaves, twigs, and kitchen waste such as vegetable peels, egg shells and leftover food, can be decomposed by microorganisms to provide rich manure that can enrich soil fertility. Composting is widely practised for both household and large-scale waste management and has several benefits for the environment and soil health.

Compost Life Cycle

Food → Scraps → Compost → Fertilizer → Grow

■ **Figure 7.40** Composting has several benefits for the environment and soil health

The key steps in the composting process are:

- Collection of organic waste: Organic materials such as kitchen scraps (fruit and vegetable peels, coffee grounds), garden waste (grass clippings, leaves) and other biodegradable materials are collected.
- Layering and balancing: The collected organic materials are layered in a compost bin or pile, aiming for a balance between 'green' (nitrogen-rich) and 'brown' (carbon-rich) materials. Green materials provide nitrogen, while brown materials provide carbon.
- Aeration and moisture: Regular turning or mixing of the compost pile introduces oxygen, promoting aerobic decomposition. Adequate moisture is also required to support microbial activity.
- Microbial decomposition: Microorganisms such as bacteria, fungi and other decomposers break down the organic matter into simpler compounds. This microbial activity generates heat, accelerating the composting process.
- Maturation and curing: After the initial decomposition phase, the compost undergoes a maturation or curing period, allowing the composting process to complete, and stabilizing the final product.
- Use of compost: Once matured, the compost becomes nutrient-rich and can be used to enhance soil fertility, structure and water retention in gardens, landscaping and agriculture.

Link
More about the nitrogen cycle is covered in Topic 2.3.17 (page 154).

Merits and demerits of composting

Merits:

- Composting diverts organic waste from landfills, reducing the volume of waste and minimizing methane emissions.
- Compost is a valuable soil conditioner, enhancing soil fertility, structure and water retention in gardens and agriculture.
- It promotes sustainable waste-management practices at a variety of levels, both industrial and household.
- Composting is a cost-effective and natural process that reduces dependence on synthetic fertilizers and chemicals.
- Community composting initiatives and educational programmes promote environmental awareness and community engagement.

Demerits:

- Improperly managed compost piles may produce odours, and the aesthetic impact may be a concern in certain settings.
- The process may be time consuming as it is a biochemical process and involves microorganisms.
- Organic materials may attract pests, such as rodents or insects, if the compost pile is not properly managed.
- It requires dedicated space for compost pits or bins, which may be a limitation for large-scale composting in urban areas.
- Educating the community may present challenges in some contexts.

◆ **Environmental consciousness** – the characteristics or factors that determine the behaviour of humans towards the environment.

⬤ HL.c.9, HL.c.10: Environmental ethics

Human actions towards responsible waste management can be viewed through the lenses of virtue ethics and consequentialist ethics. **Environmental consciousness** means being concerned about the consequences of human actions on the environment. Individuals with a strong sense of empathy may consider the potential impacts of their actions on fellow citizens, communities and the environment when throwing away, burning, or buying and selling waste. They are likely to make decisions that contribute positively to waste-management practices and may choose alternative methods of waste disposal in order to minimize the negative impacts.

The actions of people who are concerned about the outcomes of human actions, and who for this reason may choose methods that minimize harm to the environment, public health and social well-being, fall under the umbrella of consequential ethics.

ATL ACTIVITY

Reflect on your personal waste management
- Take a moment to reflect on your own waste-management practices.
- Consider instances of waste disposal, burning or purchasing/selling items related to waste (for example, selling old books, magazines, newspapers, clothes).
- Identify any virtues (such as empathy, responsibility, environmental consciousness) that you believe guide your decisions in relation to your waste-management practices.

7.3.8 Sustainable options for management of SDW can be promoted in societies

Approximately 3 billion people around the world lack proper waste-disposal facilities. It has been estimated by the United Nations that the total solid waste generated by 2050 will be around 4 billion tonnes, which is double the waste generated in 2016. Waste is generated by households, businesses and shops, and disposed of into streets, water bodies and on to open land. This not only causes pollution and negatively affects the health of the citizens, it also impacts the aesthetics of the place and disrupts its regular functioning.

It is the urban poor who have to bear the most negative impacts of improper waste disposal. Until recently, the budget for waste management was insufficient to procure proper disposal machinery. The importance of managing waste to achieve sustainability is being recognized globally. This has resulted in the identification of strategies that can be implemented at the local, national and global levels.

■ Taxes and incentives

Implementing taxes on non-recyclable or non-compostable materials can discourage their use, while providing incentives for businesses and individuals to choose sustainable alternatives. In countries like Denmark and Sweden, there are high taxes on landfill disposal, incentivizing businesses and individuals to reduce waste and adopt recycling practices.

Container deposit schemes that offer incentives to customers for returning their used containers (plastic and glass) are becoming popular in many countries. The refund amount can vary. In Scotland, the customer has to pay 20 pence when they purchase a drink that comes in a single-use plastic bottle. They then get a refund when they return the empty bottle.

Figure 7.41 People in Australia returning used containers for a refund

■ Social policies

Some countries have implemented strict social policies encouraging responsible waste behaviour and promoting a culture of waste reduction, reuse and recycling. South Korea implemented a policy called 'pay-as-you-throw', where households are charged based on the amount of waste generated, encouraging citizens to minimize waste.

■ Legislation

Some countries enforce legislation that regulates waste-management practices, sets recycling targets, and establishes penalties for improper disposal. The European Union has stringent waste-management legislation, including directives to reduce landfilling and increase recycling rates among member countries. Plastic bags are banned or regulated in many countries and they are not freely provided to customers in shopping markets; customers have to pay for a plastic bag or carry their own shopping bags. For example, in Singapore customers are charged 5 cents per single-use plastic carry bag. Supermarkets are encouraged to use alternatives, such as paper bags or cloth bags, to deliver products to customers.

■ Education and awareness

Citizens must be made aware with respect to the common facilities that they use. The tendency to litter in public areas is one of the major concerns in many cities around the world. Copenhagen has been able to overcome this problem by not only generating awareness among the citizens, but also employing a smart-sweeping programme across the city. There are around 200 workers using sweeping machines to keep the streets litter-free. The Too Good to Waste programme in Calgary, Canada, aims at reducing the amount of waste sent to landfills by separating and recycling certain materials. A website called Green Calgary is part of this programme, and provides citizens with advice on making environmentally ethical choices. Programmes like Keep America Beautiful and NGOs like Waste Warriors are dedicated to spreading awareness and promoting litter prevention, recycling and community beautification.

■ Campaigns

Public campaigns help in emphasizing the importance of reducing, reusing and recycling waste.

The United Nations–Habitat launched the Waste Wise Cities campaign to encourage cities around the world to become waste-wise by following 12 key principles of the solid waste-management strategy:

1. Assess the quantity and type of waste generated.
2. Improve the collection and transportation of waste.
3. Ensure the environmentally safe disposal of waste.
4. Promote the 5Rs – Rethink, Refuse, Reduce, Reuse, Recycle.
5. Empower and work with all waste stakeholders.
6. Establish better working conditions for waste workers.
7. Implement innovative technological alternatives.
8. Make long-term strategic plans for urbanization, which fully consider solid waste generation and treatment.
9. Design incentives promoting a circular economy.
10. Encourage 'rethinking on waste' through public education.
11. Regularly review progress on waste management.
12. Strive to achieve the SDGs (Sustainable Development Goals) and NUA (New Urban Agenda).

The Waste Wise Cities Tool allows cities to assess their performance on municipal solid domestic waste management using SDG indicator 11.6.1. It has been successfully field tested in Nairobi and Mombasa in Kenya and Mahé Island in Seychelles, and has resulted in the creation of waste-management protocols. Many cities around the world have adopted this campaign for proper waste management. Other examples of campaigns:

- UN–Habitat's initiative introduced in 2019, My Waste, Our Wealth, aims at sensitizing children and youth, and enforcing behavioural change at the community level.
- Every year, 30 March is marked as the International Day of Zero Waste. Indore, a city in India, has been successfully implementing the campaign for six consecutive years under the Swatch Bharat Mission (Clean India Mission), which focuses on improving solid waste management in urban areas.
- Cagayan de Oro on the island of Mindanao in the Philippines has become the first city in the world to successfully complete a Waste Wise Cities Tool assessment on its own.

Improved access to disposal facilities

The major problem with waste disposal is the accessibility of waste-disposal systems. A lack of proper dustbins or a dedicated place for waste disposal forces the public to be thoughtless. Different dustbins for different types of wastes ensures proper segregation of waste and makes sorting less time consuming at the waste-collection centre. Facilities for recycling, composting and drop-off for hazardous wastes are also important for local authorities to provide.

San Francisco, California, USA, has implemented a comprehensive waste-management system with a focus on improved access to disposal facilities. There is an extensive network of recycling bins and composting facilities throughout the city, making it easy for residents and businesses to dispose of recyclables and organic waste conveniently. It uses the three-bin waste-collection system, which includes separate bins for recyclables, compostables and landfill-bound waste. It has opened community drop-off centres where residents can bring hazardous materials, electronics and bulky items for proper disposal. Awareness and education about the type of waste and which bins to use for what type of waste is an integral part of its solid waste management plan, which has a multilingual approach to reach out to maximum citizens.

> ### HL.a.12: Environmental law
>
> The integration of legal and economic strategies is crucial for maintaining sustainable options in the management of solid domestic waste. Governments can establish and enforce regulations that mandate proper waste disposal, recycling, and the reduction of hazardous materials. Legal frameworks can incorporate Extended Producer Responsibility (EPR) policies, making producers responsible for the entire lifecycle of their products, including proper disposal and recycling. In the European Union, EPR is applied to various products, including packaging and electronic waste. Governments can use economic instruments such as financial incentives and subsidies to encourage businesses and individuals to adopt sustainable waste-management practices. For instance, tax credits may be offered for companies that invest in recycling technologies. Collaborative efforts between the public and private sectors can lead to more effective waste management. Public–private partnerships can leverage the strengths of both sectors to implement sustainable waste solutions.
>
> **ATL ACTIVITY**
>
> Research and explore successful examples of Extended Producer Responsibility (EPR) implementations in the European Union. Prepare a concise presentation outlining the key elements of your findings.

7.3.9 The principles of a circular economy provide a holistic perspective on sustainable waste management

The principles of a circular economy offer a comprehensive framework for reimagining and optimizing the entire lifecycle of products, emphasizing the reduction of waste and the continual use and regeneration of resources. In the context of sustainable waste management, these principles guide practices that minimize environmental impact and promote a more sustainable and resilient economy. Circular economies prioritize the design of products with a focus on longevity, repairability and recyclability. Products are created to withstand extended use, be easily repairable, and have components that can be efficiently recycled at the end of their life.

This emphasis on product life extension mitigates premature disposal, reducing the overall volume of waste generated and promoting a culture of responsible consumption. Resource inputs are carefully managed, promoting the use of renewable materials, reducing raw material extraction and lowering overall resource consumption. Circular economies advocate closed-loop systems, where materials are continually circulated through recycling, remanufacturing and recovery processes. Unnecessary waste creation is reduced by restricting the amount of packaging and encouraging the use of sustainable materials.

> **REAL-WORLD EXAMPLE**
>
> ### Circular path of a plastic bottle
>
> 1. Manufacturing: A plastic bottle is manufactured using circular design principles. Recycled PET (polyethylene terephthalate) is used to make the bottle. The impact of this is reduced demand for virgin plastic.
>
> 2. Distribution and use: Efficient supply chains are a key to reducing emissions due to transportation of products. Unnecessary packaging is minimized and materials with low environmental impact are used. Reusable containers are used wherever possible. By advertising and promoting responsible consumption, unwanted demand can be reduced.
>
> 3. End of life: Appropriate recycling systems for the plastic bottle are accessible once it reaches the end of its life. The closed-loop system ensures that the plastic bottle is recycled rather than discarded.
>
> 4. Recycling: The plastic bottle is sent to a recycling facility, where it is processed into raw materials. These materials are then used to manufacture new plastic products, continuing the cycle.
>
> 5. Product recovery: The bottle is sent through the product-recovery process to extract whatever material can be reused to make another bottle or some other product. Traditional recycling involves collecting plastic waste, sorting it based on resin types, and then melting and reforming the plastic into new products. However, the limitation is that as the product passes through each recycling stage, it loses some of its vitality and thus there is only a certain number of recycles that any given product can undergo. Chemical recycling could be a solution to this, as it breaks down plastic into its molecular components. **Pyrolysis**, **depolymerization**, and other modern chemical processescould be a solution to this, as they break down plastic into its molecular components.

◆ **Pyrolysis** – the thermo-chemical decomposition of organic material into liquid, gases and solid residue at high temperatures in the absence of oxygen.

◆ **Depolymerization** – breaks down polymers into monomers using chemical reactions.

Link

The circular economy model is explained in Topic 1.3.21 (page 77).

> **ATL ACTIVITY**
>
> Visit or contact a local recycling centre, if you have one, to learn about how waste is handled locally.
> - Prepare a set of questions that you would ask the people in charge of different sections at the recycling centre.
> - Record their responses in your notebook or on your tablet. You could even interview them and record an audio/video of their response.
> - Collect photos (if permissible) of the different stages involved in recycling.
> - Collate the responses and write a magazine-style article on recycling using the information gathered.

Inquiry process

Inquiry 1: Exploring and designing

Inquiry 2: Collecting and processing data

Inquiry 3: Concluding and evaluating

Imagine you are an environmentalist working on the environmental issue of improper waste management and the factors affecting waste generation and management.

- Formulate a research question that will allow you to investigate the reasons for improper waste management. You may think on the following lines:
  - Find out what happens to waste in your society – how much is recycled, reused, remade, goes to landfill or incineration, or is shipped to another country.
  - Compare the issue of improper waste management in two contrasting societies.
- Write down a hypothesis and identify the variables that you will work with for this investigation.
- Create a survey with at least 10–15 relevant questions that will allow you to respond to the research question you have formulated.
- Share the survey with the target audience and collect responses through the most appropriate process. You could use Google Forms or other survey software for this.
- Collect and process the data – use a spreadsheet for this.
- Use a statistical analysis tool such as a chi square or t-test.
- Represent your processed data in the form of processed tables and/or graphs.
- Analyse and evaluate your findings.
- Provide a valid conclusion for your investigation.

ATL ACTIVITY

Raise awareness of circular economy options in your community
- Work collaboratively and design posters with slogans.
- Write a script for a short street play portraying the importance of adopting a circular economy.
- Present your street play to the local community.

ATL ACTIVITY

Either in your school:
- Discuss with your classmates and start a repurposing club.

Or in your society:
- Discuss with your residential manager about how you might start a library of things.
- Request that people donate household items they do not use but that may be of use to others.
- Things can be loaned from the library and returned when the use is over.

REVIEW QUESTIONS

1. Outline the types of solid domestic waste.
2. Explain how landfills can be successfully used as a waste management strategy.
3. State the disadvantages of incineration.
4. Differentiate between landfills and composting.

EXAM-STYLE QUESTIONS

1. Figure 7.42 shows the recycling rate of solid domestic waste in the UK.

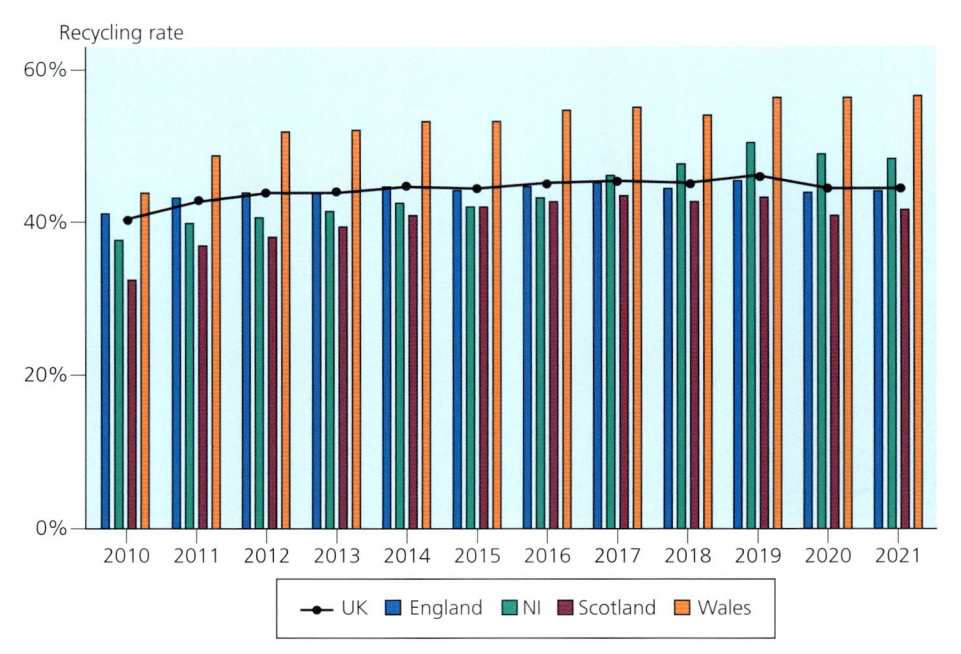

Figure 7.42 The recycling rate of solid domestic waste in the UK (source: WasteDataFlow, Defra Statistics)

Discuss how the recycling rate in the UK changed from 2010 to 2021. [7 marks]

2. Differentiate between any two solid waste disposal methods, giving their advantages and disadvantages. [7 marks]

3. With the help of named examples, suggest valid reasons for the hierarchy of waste-management strategies shown in Figure 7.43. [9 marks]

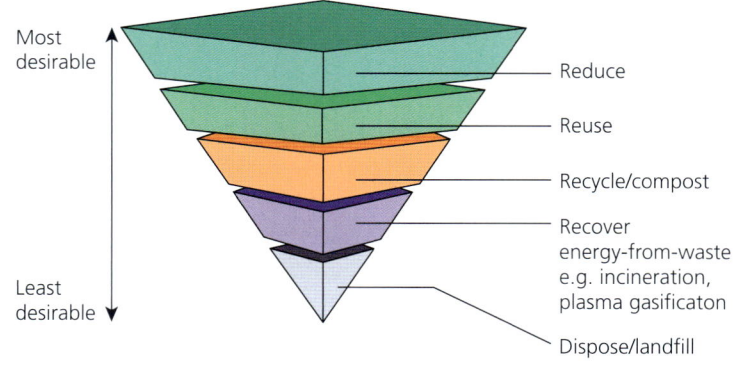

Figure 7.43 The hierarchy of waste-management strategies

4. State two advantages and two disadvantages of a named method of 'disposal' in the management of urban waste. [4 marks]

5. Outline three strategies for the management of a named example of industrial waste. [3 marks]

Theme 7: Natural resources

Theme 8
Human populations and urban systems

8.1 Human populations

Guiding questions

- How can the dynamics of human populations be measured and compared?
- To what extent can the future growth of the human population be accurately predicted?

SYLLABUS CONTENT

This chapter covers the following syllabus content:
- 8.1.1 Births and immigration are inputs to a human population.
- 8.1.2 Deaths and emigration are outputs from a human population.
- 8.1.3 Population dynamics can be quantified and analysed by calculating total fertility rate, life expectancy, doubling time and natural increase.
- 8.1.4 Global human population has followed a rapid growth curve. Models are used to predict the growth of the future global human population.
- 8.1.5 Population and migration policies can be employed to directly manage growth rates of human populations.
- 8.1.6 Human population growth can also be managed indirectly through economic, social, health, development and other policies that impact births, deaths and migration.
- 8.1.7 The composition of human populations can be modelled and compared using age–sex pyramids.
- 8.1.8 The Demographic Transition Model (DTM) describes the changing levels of births and deaths in a human population through different stages of development over time.

HL ONLY
- 8.1.9 Rapid human population growth has increased stress on the Earth's systems.
- 8.1.10 Age–sex pyramids can be used to determine the dependency ratio and population momentum.
- 8.1.11 Using examples of two countries in different stages of the DTM, consider the reasons for the patterns and trends in population structure and growth.
- 8.1.12 Environmental issues such as climate change, drought and land degradation are causing environmental migration.

8.1.1 Births and immigration are inputs to a human population

Early human populations were stable and slow growing, reaching 1 billion in the early 1800s and 8 billion in 2022 (see Figure 8.1). The inputs that result in population growth are births and immigration rates. The Industrial Revolution of the nineteenth century saw dramatic increases in the population due to advancements in agriculture and technology, and improvements in standards of living, which resulted in increased life expectancy. Since international travel has become more accessible, living and working in different parts of the world has become easier. This led to an increase in the level of immigration around the world, which meant immigration began to have a marked impact on population growth in specific areas.

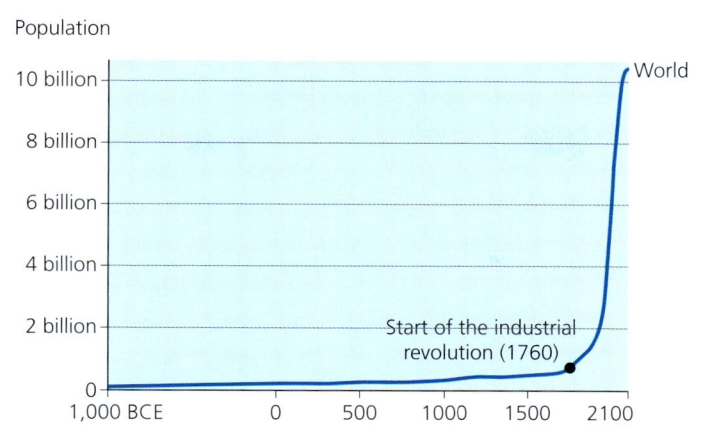

■ **Figure 8.1** Human population growth from BCE to 2100. Future projections are based on the UN's medium-fertility scenario

8.1.2 Deaths and emigration are outputs from a human population

The natural output of the human population system is death, and as standards of living, access to healthcare and education all increase, this output level natural reduces as life expectancy increases. The population boom of the twentieth century saw the population rise to 6 billion by the year 2000 due to ever-increasing medical advancements. These resulted in reductions in death rates, for example the crude death rate per 1,000 people reduced from 13.4 in 1980 to 7.3 in 2022.

8.1.3 Population dynamics can be quantified and analysed

The human population is a complex, dynamic system that is influenced by many internal and external factors. Population dynamics allow these complexities to be measured and compared in a standardized way. Birth rates and death rates are both elements of population dynamics that can impact the size of a population at any given time. This allows these measurements to be scaled to allow specific populations to be measured and compared. Therefore, measurements could be used at a country, city or town level to allow comparisons of different scales of systems. For instance, capital cities could be compared in terms of the immigration and emigration rates over time, or compared with one another. This could be correlated with development measures, such as Gross Domestic Product (GDP). This can help drive decisions regarding the development of services and infrastructure to support the growing populations.

Birth and death rates measure natural fluctuations in populations already living within an area. Immigration and emigration rates show the flow of people to and from other countries or cities who come to live in the area. This allows both internal and external growth factors to be assessed and the demographics of the population to be monitored. For instance, a country like Japan has a population with many people who are outside the potential earning bracket, putting strain on the need for services and support for the growing elderly population, and a reducing GDP.

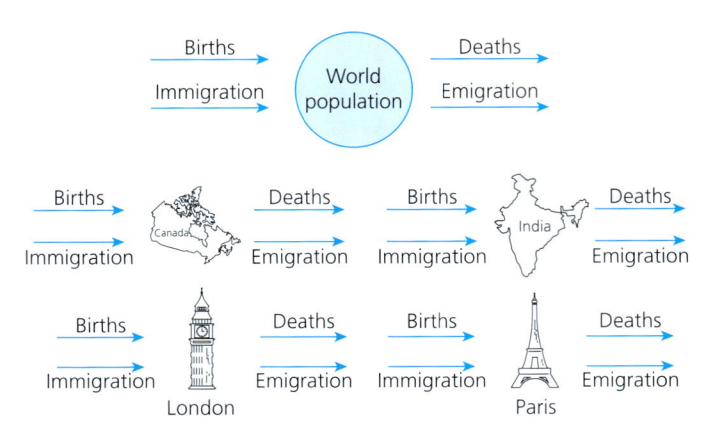

■ **Figure 8.2** Simplified population change systems models showing the scalability of this technique

Births and deaths are two ways that a population can increase and decrease from within. The rates given are per 1,000 people, so if a country has a birth rate of 11, this means that for every 1,000 people in the country 11 children were born in that year. The death rate measures the number per 1,000 of any age group who die.

The total fertility rate is a hypothetical value, so it is based on a series of assumptions, such as that all women within the designated fertility range will have children. It also defines the fertility period as ages 15 to 50, even though it is possible for women, girls and other people with wombs to have children outside this age bracket. This therefore means that this value is less reliable than the ones that are measured from information such as census data. However, this statistic allows predictions to be made regarding how much the population has the potential to expand within a given period of time.

Life expectancy is an average value that relates to the age from birth that a person can expect to live to. This is based on an average case scenario and can obviously change throughout an individual's life.

■ **Table 8.1** Quantitative human population measurements showing what each value represents

Measurement	Meaning	Additional information
Crude birth rate*	Number of live births per 1,000 people, per year	
Immigration rate*	Number of immigrants per 1,000 people, per year	
Crude death rate*	Number of deaths per 1,000 people, per year	
Emigration rate*	Number of emigrants per 1,000 people, per year	
Total fertility rate	Average number of births per woman of childbearing age	Childbearing age range is typically considered to be 15–50 years old
Life expectancy	Average number of years a person can be expected to live	
Natural increase rate (NIR)	The difference between the number of births and deaths over a given time	This is calculated for given values: Birth rate – Death rate
Doubling time	The numbers of years for a population to double in size, given the current rate of natural increase	This is calculated using the NIR and the Rule of 70: 70 / Natural increase rate

* These rates can be applied to any scale, for example to compare a town, city, country or region.

Tool 3: Mathematics

Calculate population statistics

You will need to use basic mathematical skills to calculate population statistics.

- The NIR is calculated from taking the death rate from the birth rate. This gives you the net number of individuals that the given population changes by in that year.
- The doubling time is calculated using the Rule of 70. This is the number of years needed for a variable to double. To determine the doubling time of anything, 70 is divided by the growth of the variable, in this case the natural increase rate (NIR).
  - For example: In 2021, Australia had a birth rate of 11.54 people per 1,000 and a death rate of 6.42 people per 1,000, resulting in a natural increase rate of 5.12 per 1,000.

    Divide 70 by 5.12 to give a doubling time of 13.68 years.
  - For example: In 2021, Gabon had a birth rate of 27.14 people per 1,000 and a death rate of only 7.26 people per 1,000, giving this country a much higher NIR of 19.88.

    Divide 70 by 19.88 to show that Gabon's population will double in 3.52 years based on the dynamics of 2021's population.

Tool 1: Experimental techniques

Secondary data collection

To investigate population dynamics you need to access secondary data. The most important considerations with secondary data are the reliability of the source and the amount of data you have access to. In order to ensure that the data are reliable, you need to use a suitable source. Government sites and UN statistics such as Our World in Data (**https://ourworldindata.org**) are highly reliable and provide information on how to best cite the source to ensure you are complying with plagiarism-avoidance guidelines.

How much data you need depends on what your research question is asking and the analysis you intend to carry out. If you are looking for patterns within a country over time, you need to make sure that you have a sufficient number of years of available data to show patterns. This should be at least 30 data points and would allow a correlation to be calculated. If you are comparing different countries then you need to consider either comparing two locations (or groups – such as development levels), with at least 20 data points per location, allowing you to do a simple t-test; or at least five different groups, which would allow you to conduct an ANOVA test on the data.

If you do not collect enough data then your analysis will not take on all relevant factors and may result in a false conclusion.

Population dynamics such as life expectancy and fertility rate are dependent upon many social and political factors, such as the government's population policies and traditional views of the 'value' of particular genders in some cultures. The fertility rate and life expectancy in fact fluctuate all the time in relation to many social and political factors, and can both significantly impact the size of the population overall.

8.1 Human populations

The size of the population to begin with is also a very important factor when analysing a country's population demographics. If the country already has a high population then even a low fertility rate will result in a more rapid overall population increase than for a country with a relatively small population. These elements can also be shown by looking at the natural increase rate (NIR) and doubling time, which are calculated from the raw data (see Table 8.1 on page 696). To get a clearer picture of a population, all of these data need to be available (see Table 8.2 on page 699).

Common mistake

Don't forget to describe the direction of differences or relationships. If your processed data show a significant difference or relationship, you need to go back and look at what the pattern is. For example, when looking at the relationship shown in Figure 8.3 (page 700), the direction of the relationship is a negative one. Countries with high fertility rates also had lower life expectancy rates. Lower fertility rate countries tended to also have higher life expectancy.

It is not enough just to state that there is a significant relationship between two variables.

Inquiry process

Inquiry 2: Collecting and processing data

When interpreting data, the important things to include are:
- Finding patterns, differences and relationships, and being able to describe and explain what they show.
- Identifying any outliers that do not fit the same 'pattern' as the others, and determining if that is a data collection error or a real anomaly.
- Understanding the source of the data, if it is reliable and if there is any possibility of bias in the collection of those data.

Start by looking closely at the data. Can some be grouped together as similar or different? Are any of the data related to each other? Is there a pattern? Is that pattern showing a positive or negative relationship?

Concept

Perspectives

In countries where the fertility rates are low, women are often choosing to either have no, or fewer, children for financial reasons or to focus more on their careers.

Some countries, such as Nigeria, have high fertility rates due to the high number of children (5.24) per childbearing female. In this situation improvements in medical care have resulted in a larger proportion of children that are born surviving to adulthood, leading to an increase in the overall population of that country.

In many ways there is a trend towards countries with lower levels of development having higher fertility rates than countries with well-developed social systems. Therefore, populations in economically developing countries are expanding more rapidly than those in more developed countries, causing increased stress on resources that are already being unsustainably used.

■ **Table 8.2** Data showing population dynamic measurements for a range of countries for 2021 (source: https://ourworldindata.org/world-population-growth)

Country	Birth rate	Death rate	Population size	Fertility rate	Life expectancy*	Natural increase %	Doubling time (years)
Australia	11.54	6.422	25,921,094	1.61	83.20	5.12	13.68
Bangladesh	17.82	5.68	169,356,240	1.98	72.87	12.14	5.77
Bhutan	12.48	6.477	777,500	1.41	72.08	6.00	11.66
Brazil	12.88	8.326	214,326,220	1.64	76.08	4.55	15.37
Cambodia	19.33	6.838	16,589,031	2.3	70.05	12.49	5.60
Canada	9.82	7.81	38,155,012	1.46	81.75	2.01	34.83
China	7.63	7.448	1,425,893,500	1.16	77.10	0.18	384.62
Denmark	10.81	9.998	5,854,246	1.72	81.55	0.81	86.21
Gabon	27.14	7.258	2,341,185	3.49	66.69	19.88	3.52
Iceland	12.36	6.79	370,338	1.73	83.07	5.57	12.57
Japan	6.57	12.64	124,612,530	1.30	84.62	-6.07	-11.53
Myanmar	17.10	9.769	53,798,090	2.15	67.36	7.33	9.55
New Zealand	12.44	6.627	5,129,730	1.77	82.06	5.81	12.04
Nigeria	37.12	13.083	213,401,330	5.24	55.02	24.04	2.91
Norway	9.98	7.757	5,403,021	1.50	83.21	2.22	31.49
United States	11.06	9.743	336,997,630	1.66	79.82	1.32	53.15

* 2020

ATL ACTIVITY

Table 8.2 shows a selection of countries and their population statistics. Many different things can be determined from data such as this. To get the most out of data tables you need to learn how to read them. Use the Inquiry process box (page 698) for guidance and practice.

Use the data in the table to determine if there is a correlation between fertility rate and life expectancy.

- Go to www.socscistatistics.com/tests/spearman/default.aspx.
- Click on 'Take me to the calculator!' and two boxes will appear.
- Copy the fertility data from the table into the 'X Values' box.
- Copy the life expectancy into the 'Y Values' box. These are your raw data.
- Click 'Calculate R'. The statistical results will be displayed in blue and show r_s and p.
- If the p value is less than or equal to 0.05 then there is a significant relationship; if the p value is greater than 0.05 then the relationship is not significant.
- In order to see what that relationship looks like, copy the raw data into a spreadsheet program (for example, Microsoft Excel or Google Sheets).
- Make a scatter plot with fertility rate on the x-axis and life expectancy on the y-axis. This will help you to see the relationship.
- If it is a positive relationship, then one factor increases as the other does. If it is a negative relationship, then an increase in one factor results in a reduction in the other factor. One way to check your interpretation is that if the r_s value is negative then so is the relationship.

When you create a statement about the statistical results, there needs to be a description of the direction of the relationship. For example, the results from the data in Table 8.2 show that there is a significant negative relationship between fertility rate and life expectancy ($r_s = -0.62$, $p = 0.009$). As the fertility rate increases, the life expectancy significantly decreases.

Discuss reasons for this relationship and apply these skills to the next Inquiry process.

a)

X Values	Y Values
1.61	83.2
1.98	72.87
1.64	76.08
1.41	72.08
1.16	77.1
2.3	70.05
1.46	81.75
1.72	81.55
3.49	66.69
1.73	83.07
1.3	84.62
2.15	67.36
1.77	82.06
5.24	55.02
1.5	83.21
1.66	79.82

b) $r_s = -0.62353$, p (2-tailed) = 0.00986
By normal standards, the association between the two variables would be considered statistically significant.

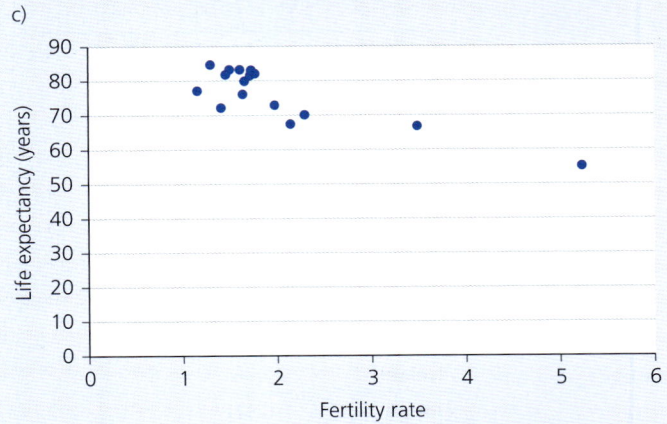

c) [Scatter plot: Life expectancy (years) vs Fertility rate]

■ **Figure 8.3** Results of a Spearman's rank correlation calculation using **www.socscistatistics.com**, showing (a) raw data entry, (b) calculated statistical results, and (c) a scatter plot (from Microsoft Excel) of the raw data visualizing the results

● Common mistake

Don't forget to look at the big picture. All calculated values of population dynamics tell a part of the whole story.

● Common mistake

Don't assume that a statistical value of p that is not significant means there is no relationship (or difference). There could still be differences in the averages, but if the variation in the data is large there may be data that overlap between groups, making the relationship or difference not as clear as if there is no overlap in the level of variation (shown by standard error or standard deviation bars on graphs).

Inquiry process

Inquiry 2: Collecting and processing data

Secondary data collection is an important aspect of ESS. Where data are sourced from is very important as it impacts the reliability of the data. To ensure there is no bias, reliable collection and robust information, it is necessary to use trustworthy sites – anything that is .gov, .edu or .org indicates that it is likely to be from a reputable source. Be careful when using data from organizations that may have some vested interest in the data being shown.

Data collection

Determine the data to be collected; this could be for your Internal Assessment or an Extended Essay. In this case, you will look for human population dynamics for different countries.

- Find a reliable source of data, for example **https://ourworldindata.org/world-population-growth**. Using one source can be good for data as they should all have been collected in a comparable way.
- Browse through the settings to decide which data you will request. This will include the required data time range and specific statistics, for example 2010–30, birth and death rates, fertility rate, life expectancy or infant mortality. Using this site you can also select which countries will be included in the data search. Once you have completed the search, the website gives you the option of viewing or downloading the results as data, a graph or a map (if relevant to the search).

Theme 8: Human populations and urban systems

- Select the countries you want to download the data for (this will depend on what your question is). For example, randomly select 30 countries from the list and request the birth rate, death rate, fertility rate and life expectancy; these can then be used to calculate the natural increase rate and doubling time (see Table 8.2 for guidance). Download the .csv data file. This can be opened in most spreadsheets.

Data processing

How you process your data will depend on your question. You could look at relationships between variables, for example using Spearman's rank to look at whether fertility rates are correlated with life expectancy:

- Use an online statistics package to calculate basic statistics, such as **www.socscistatistics.com/tests/spearman/default.aspx**.
- Click on 'Take me to the calculator!' and add the fertility rates into the 'X Values' box and the life expectancy in the 'Y Values' box (see Figure 8.3a, using data from Table 8.2).
- Click 'Calculate R' and the program runs the analysis for you.
- The values you are interested in are the r_s and the p value. The p value determines if the relationship is significant or not; a p value of less than 0.05 is considered to be significant. The r_s value shows the direction of the relationship and the strength of the relationship. A negative value indicates a negative relationship, and a positive value shows a positive relationship. A value close to 1 or -1 indicates a strong relationship between the two factors (see Figure 8.3b, using data from Table 8.2).
- To show this visually, plot a scatter plot with the fertility rate as the x-axis and the life expectancy as the y-axis (see Figure 8.3c, using data from Table 8.2).

8.1.4 Global human population has followed a rapid growth curve. Models are used to predict future growth

The ability to calculate doubling time from known data allows predictions to be made in terms of what can be expected in the future. Many predictions simply look to determine what happens when conditions remain as they currently are, while some create a scenario that is applied to determine the impact that that scenario would have on the current situation.

The United Nations (UN) has made three main predictions on the development of the human population. These projections relate to different scenarios that could take place in the future, and they focus on changes in fertility rates. These are low, medium and high variant predictions that base projections on differences in fertility levels.

◼ Low projection

The low variant projection assumes that there is a gradual decline in fertility rates over time, thus slowing population growth. This prediction relates to improved family planning and education, and an increase in urbanization, which all tend to lead to smaller family sizes. The low variant projections are likely to be presented in conjunction with the implementation of anti-natal policies by the government to achieve these changes.

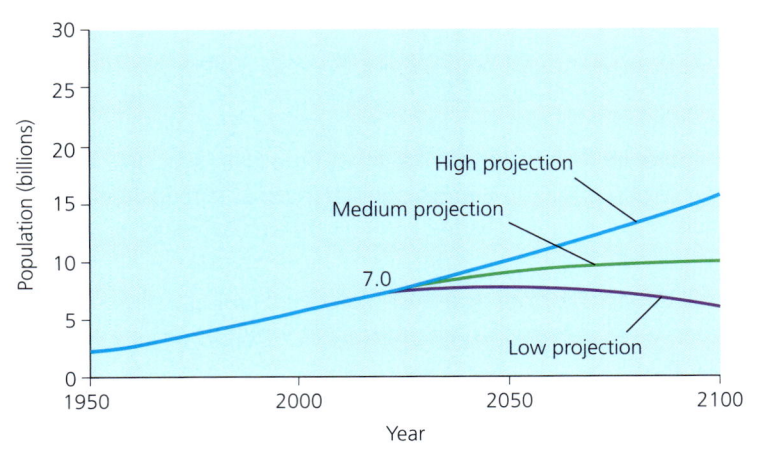

Figure 8.4 The three UN population projections up to 2100

Medium projection

The medium projection is considered to be the most likely scenario as it reflects current trends in fertility, mortality and migration. This scenario reflects what would occur if there were no major changes in population policies of different countries. Yet, this projection takes into account the ongoing improvements and developments in healthcare and economic development.

High projection

The high projection assumes that fertility rates will remain high overall, leading to a rapidly growing population. In this situation the availability of family planning services, medical care, etc., will reduce as the population rapidly expands. This projection is also expected to result in quick depletion of resources due to the rapid increases in some regions of the world.

8.1.5 Population and migration policies can be employed to directly manage growth rates of human populations

Population policies are developed at a governmental level and address the aims of the country in relation to how it wants the population to change over time. These policies relate to either increasing or decreasing the birth rate in a population, or developing migration policies focusing on increasing (immigration) or decreasing (emigration) numbers. **Pro-natalist** policies and practices are those that encourage people to have more children. **Anti-natalist** policies and practices are those that discourage people from having children. Migration policies can be targeted towards specific demographics to fill skills gaps or population numbers in the country's population.

◆ **Pro-natalist** – relating to a policy or practice that encourages people to have more children.

◆ **Anti-natalist** – relating to a policy or practice that discourages people from having children.

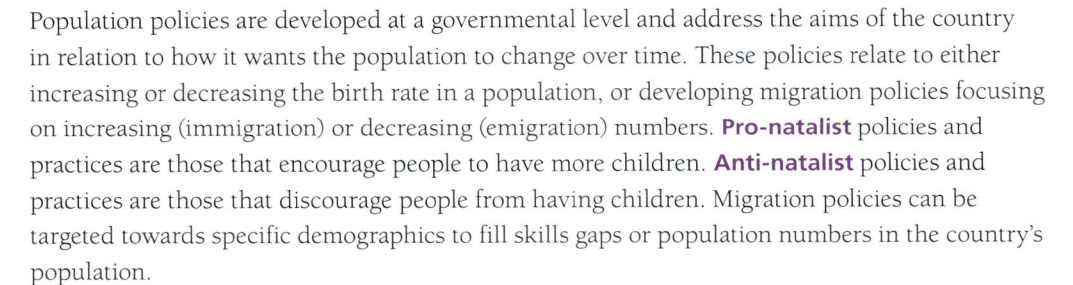

Table 8.3 Examples of countries adopting one of the three population development strategies

Pro-natalist	Anti-natalist	Migration
Japan	Ethiopia	Australia
Finland	China	Sweden
Singapore	India	UAE

Government policies

One way that the population of a country can be directly influenced is through specific policies developed by a government to either increase or decrease its population. There is a complex history behind birth policies, but in the majority of countries the focus has shifted more towards promoting reproductive rights, gender equality, and voluntary family planning as means of addressing population issues, rather than the coercive measures of the past.

A pro-natalist policy aims to increase birth rates and population growth through encouraging families to have more children. This is achieved through a number of direct policies relating to pro-family and financial incentives.

Pro-family incentives

Pro-family incentives aim at making parenthood less of a financial and logistical burden for families. This includes highly beneficial extended maternity and paternity leave, tax incentives that get larger with each child added to the family, free childcare to allow parents to work, and free healthcare and education. France has a policy that provides tax incentives, financial bonuses, and parental leave provision aimed at increasing the fertility rate within the country. These policies promote the importance of family values and are aimed at the working population, who may currently be choosing to focus more on careers over family.

Financial incentives

Other policies are aimed at providing financial bonuses to encourage population growth. This is a strategy that aims at rewarding those who have large families to allow them to have a higher standard of living. These policies include tax benefits, subsidized childcare and education grants. While similar to the pro-family policies, these simply focus on the financial aspect. Singapore has gone one step further by providing sizable 'Baby Bonuses' to parents (see Real-world example box below for more details).

HL.a.6: Environmental law

Similar to many environmental laws, population laws are mostly developed on a national level by government agencies. International law can drive the overall focus of specific national strategies. Countries struggling to control population growth due to poverty, lack of education, resources and infrastructure will receive international focus through the UN.

> **ATL ACTIVITY**
>
> Discuss with a partner if you think the law should be allowed to decide how many children someone can have.

Link

More information about the tragedy of the commons is available in Topic 3.2.7 (page 231), and about the changes in demand for aquatic food sources in Chapter 4.3 (page 331). Make a mind map of the connections between these areas of the syllabus.

HL.b.6: Environmental economics

As populations increase so does their demand for products and services, which can lead to demand exceeding sustainable production. This is particularly the case in terms of many of our natural resources that are considered 'common' property, such as fish and the oceans. It is hard to monitor the common resources, and these can become overexploited in countries that are experiencing rapid population growth, as they are essentially free resources.

HL.b.11: Environmental economics

Countries can develop targeted immigration policies that are directed towards capacity-building in those areas where expertise is lacking, such as medical expertise and financial experts. This helps the country to boost its population with individuals who will improve the overall profile and productivity of the country.

> **ATL ACTIVITY**
>
> Consider the differences between the impacts on the economy of attracting unskilled or highly skilled workers. What impacts might these have on the infrastructure of a country?

HL.b.13: Environmental economics

Developments that focus on economic growth and GDP tend to prioritize these policies over those protecting the environment within the country. Bhutan is a country that focuses on Gross National Happiness (GNH) and prioritizes the maintenance of the environment over the rapid development of its economy. Tourism is controlled to minimize its impact on a handful of parts of the country, and much of the country's electricity comes from renewable energy sources, mostly hydropower.

> **ATL ACTIVITY**
>
> **Investigate Gross National Happiness**
>
> Investigate what Gross National Happiness is, and consider how this might impact the environmental focus for Bhutan.

HL.c.10: Environmental ethics

Consequentialist ethics states that if the outcome of an act is good then it does not matter if the act itself was good or bad. If the outcome of implementing strict anti-natalist policies will benefit the country as a whole, then under this view it does not matter if the law results in unlawful practices that favour male babies due to their status as an heir.

> **ATL ACTIVITY**
>
> Discuss these questions as a class:
> - How would you feel if your human rights were restricted for the good of all?
> - Is it morally right to impact individuals for a greater good?

> **REAL-WORLD EXAMPLE**
>
> **Pro-natalist policy: Singapore**
>
> Singapore has recently been experiencing a significant reduction in fertility and birth rates. This has led to a decrease in the working population and greater proportions of elderly people needing support. This led to the development and implementation of a pro-natalist policy aimed at encouraging couples to have multiple children.
>
> The policy includes many incentives that will increasingly benefit parents. The 'Made for Families' campaign aims to make Singapore a place that encourages harmony in work/life balance. This campaign starts all the way from developing social dating networks to aid single people in finding a partner. Support is then preferentially given to families, for things such as housing support and loans, to help selling properties, and tax breaks relating to the number of children they have. Considerable support and pregnancy packages and leave are available, along with a new Baby Bonus Scheme that started from February 2023 and gives $300USD per child born to a family.
>
> This is on top of the already generous range of tax incentives and partner programme discounts for items such food, clothing and food. These are all designed to encourage young couples to have larger families without it negatively impacting their standard of living (source: information collected from **www.madeforfamilies.gov.sg/about-us/made-for-families**).

◆ **Exponential** – an increase, or decrease, that changes more and more rapidly.

Anti-natalist policy: Ethiopia

Ethiopia introduced a National Population Policy in 1993 aimed at reducing the population, which was growing at an **exponential** rate. The policy was initially highly successful, with a drop in the population growth rate from 2.9 in 1993 to 2.6 in 2005, and a fertility rate drop from 6.4 in 1994 to 5.4 in 2005.

Despite this, the population of Ethiopia has continued to grow, with a growth rate of 2.54 in 2022. This can be partly attributed to continued improvements in medical care and contraception, and a drop in infant mortality rates, from over 130 deaths per 1,000 live births in 1993, to 39 per 1,000 live births in 2022. These improvements in medical care, contraception and maternity care were also coupled with improvements in sanitation and living conditions for many.

However, the majority of the population of Ethiopia still live in rural communities where fertility rates and infant mortality are considerably higher than the national average and the availability of care and support is not as widespread as in the urban areas.

Concept

Sustainability

Maintaining the same policy without reviewing the level of population growth at regular points can result in dramatic changes in the population structure that can take time to recover from. South Korea found this when it introduced an anti-natalist policy to reduce population growth in the 1960s and 1970s. However, its policy was too effective, and by 2020 South Korea had one of the lowest fertility rates in the world. As a result of this, the country has now reversed its stance and is offering tax incentives and educational grants to people who have children.

Concept

Perspectives

While pro-natal policies that offer incentives are beneficial for the child-bearing population of a country, the money for these tax breaks has to come from somewhere. The economic burden for the rest of the country's taxpayers can be vast. These incentives dramatically increase government spending over time. This then requires governments to either increase other taxes or to create new taxes to help cover the increased spending. Increased taxes can be harmful to local economic development as smaller businesses may not be able to survive these adjustments.

If a country has introduced these policies in reaction to the impact of the growing ageing population in the country, these incentives might result in a rebalance in the demographic issues. Increased population size also has the potential to improve the economic standing of the country by increasing demand and spending, which can stimulate further development and growth.

Migration policies

One way that a country can enhance its population is through the implementation of migration policies to control the influx of people from other countries. Some countries, such as Australia, have very strict immigration policies that restrict certain age groups and impose a cap on the number of immigrants allowed to enter each year; other countries, like Sweden, adopted a very open immigration policy that was later modified to reduce the free influx of people after a huge peak in immigration in 2014 due to an influx of refugees. This resulted in Sweden developing tighter regulations on the immigration of refugees.

Australia has a long history of immigration and its complex immigration policy favours young, skilled workers. Its policy has annual caps on the number of people who can enter the country for residence purposes. The United Arab Emirates (UAE) is a country that relies on migrant workers to provide the labour force needed to deal with the huge amount of construction taking place in areas like Dubai. These workers mostly come from countries such as Bangladesh and India and are mostly male. These people leave their families to be able to work and earn enough money to send home. This has resulted in a population in the UAE that is not well distributed.

When people immigrate into a country in huge numbers, they have come from somewhere. Make sure you think about the big picture and where these people have come from and the impact of their emigration.

8.1.6 Human population growth can also be managed indirectly through economic, social, health, development and other policies that impact births, deaths and migration

There are other ways that population levels can be impacted. Countries' policies and development can have an indirect impact on birth rates, death rates, and the level of migration by focusing on improving sanitary and medical conditions, standards of living and levels of education. Although these are not directly related to birth and death rates, they have a significant impact on them as they change conditions that impact people's needs and decisions.

Many of the SDGs that were developed by the United Nations deal with areas relating to human populations, such as Quality education (SDG 4), Gender equality (SDG 5), and Clean water and sanitation (SDG 6). The introduction of the SDGs has driven policy development in these areas in numerous countries, resulting in indirect impacts on the birth and death rates within those countries.

> **Link**
> More information about the scope and limitations of the SDGs is available in Topic 1.3.18 (page 70).

Globally, improvements in the availability of education for females has led to greater empowerment and a reduction in the fertility rate (see Figure 8.5). Providing education for females can help them understand the impacts of having a large family on their own economics and their ability to work. Education needs to also include health issues such as the correct use and availability of contraception, nutrition and pregnancy help. In Nigeria, the Women's Economic Empowerment (WEE) policy is beginning to be implemented in order to empower women to enter the workforce, be able to compete in the economic market, and help boost the overall economy of the country. This has been introduced in an attempt to reduce fertility rates and population expansion in the country.

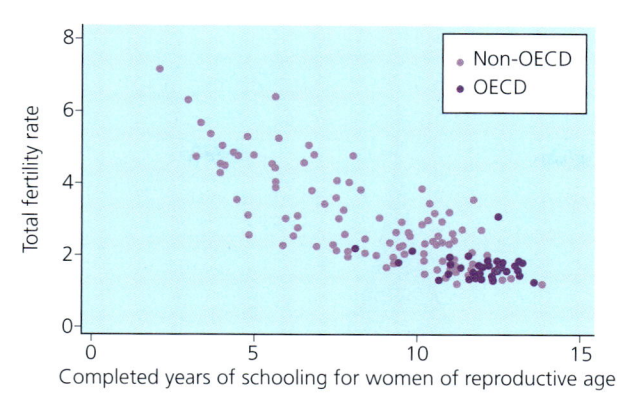

■ **Figure 8.5** Impact of female education on fertility rates (source: Kim, J. Female education and its impact on fertility. *IZA World of Labor* 2023: 228. https://doi.org.10.15185/izawol.228.v2)

As clean water and adequate sanitation become more and more available in some of the poorest parts of the world, we can see their impact on infant mortality rates (down by 15 per cent) and drops in many intestinal-based illnesses. In turn, more children live through infancy, thus reducing the need for families to have more children to 'replace' those who did not survive. Healthier children will also learn and thrive with the improved education that is often part of aid in countries that need it.

8.1.7 The composition of human populations can be modelled and compared using age–sex pyramids

Human populations can be mapped at different scales to show the distribution of different age groups for males and females. One way of looking closely at the population is through population pyramids. These display the number of males and females in each age bracket. This provides a visualization of the structure of the population and allows predictions to be made as to what the population age distribution will look like in the future. Population pyramids can also show the impact of major disasters, mass migration, or the impact of major events such as war. Equally, when a country implements a natal policy, its progress can be seen in future population pyramids.

The use of population pyramids as a tool to map population structure and change has developed alongside advances in data collection and technology. This has allowed models to be created that have demonstrated patterns in the shapes of pyramid that can identify the level of development of a country. This allows assumptions to be made in relation to these shapes. For instance, a pyramid that is characterized by a very wide base that rapidly reduces (see Figure 8.6) indicates a country in the earliest stages of development, where both birth and death rates are very high. A narrow base with a wider band higher up the pyramid represents a country that has a large elderly population with few children being born, indicating that this population will soon start to decline rapidly (see Figure 8.6).

Developing populations will have rapid drops in the size of the population as age increases, and will often have a low life expectancy, resulting in a typical pyramid shape such as that of Gabon (see Figure 8.6). A stationary population is one that has no real changes in population numbers over time, with birth and death rates being relatively similar. A pyramid for a country in this stage is characterized by few changes in the populations from birth rates towards life expectancy levels, resulting in a population pyramid with straight sides (see Canada in Figure 8.6). Once a population encounters high levels of improvement in healthcare, etc., the population increase often starts to decline as the fertility rates drop.

> **● Common mistake**
>
> Remember that these pyramids can show a snapshot in time, but they are also very powerful predictive models if you link them to the events that are taking place within that country.

Gabon population by age and sex: 2023

Annotations:
- Low life expectancy due to limited medical treatments and low standards of living
- upper limit (64 years)
- lower limit (15 years)
- High infant mortality rate

Canada population by age and sex: 2023

Annotations:
- Product of the baby boom and immigration (1945–1960s)
- upper limit (64 years)
- lower limit (15 years)
- Canada's immigration policy encouraged immigration, increasing the working population

Japan population by age and sex: 2023

Annotations:
- Post WWII birth rates dropped dramatically due to the number of people of childbearing age who lost their lives
- Japan has one of the highest life expectancy rates – leaving a large elderly population who need supporting
- upper limit (64 years)
- Women working in tech industry; prioritize their career and have children later, if at all, due to the high cost of childcare
- lower limit (15 years)

Legend: Median — 80% prediction interval — 90% prediction interval

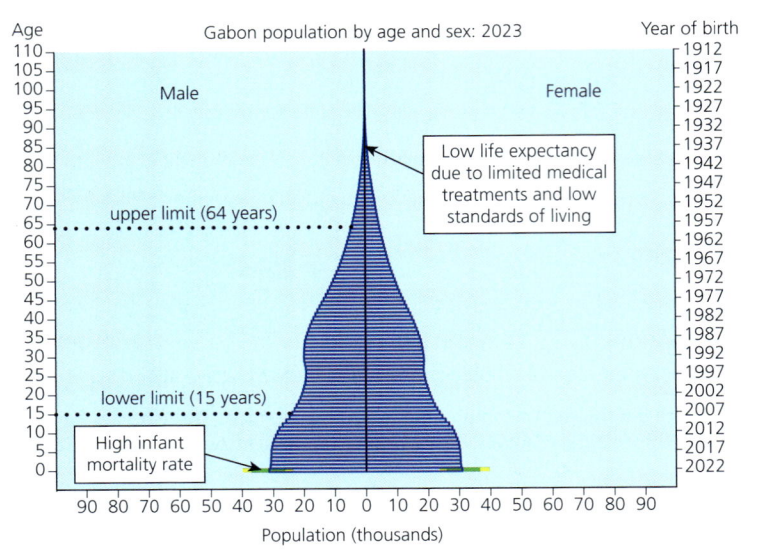

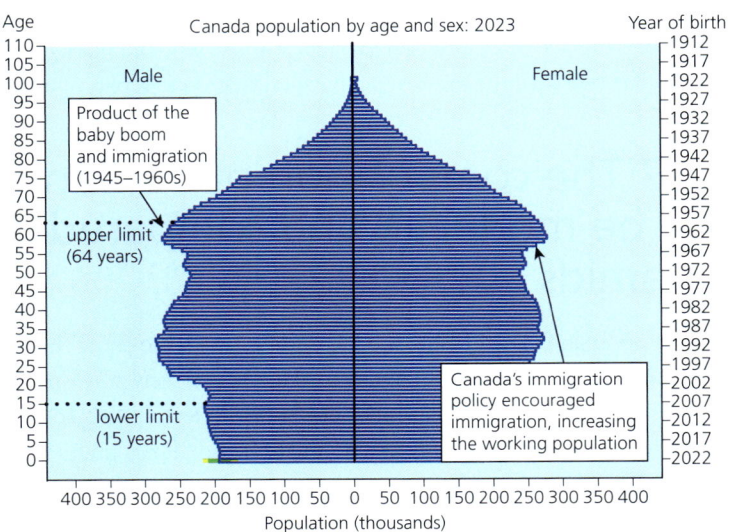

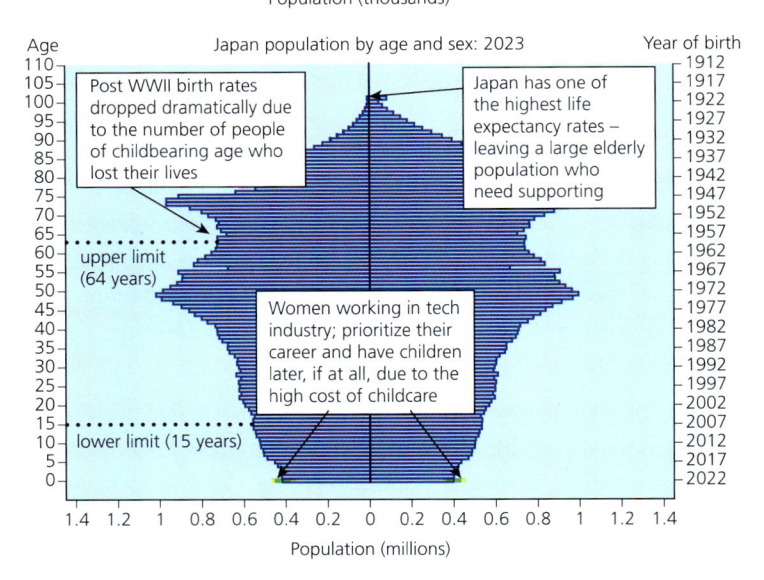

■ **Figure 8.6** Age–sex population pyramids for Gabon, Canada and Japan

> **Concept**
>
> **Systems and models**
>
> Modelling of human population growth takes place via age–sex population pyramids. Interpreting these is an important skill for ESS. Models allow projected data to be applied and predictions to be made based on those projections. A country might apply increased birth rates to determine the potential impact of pro-natal policy.
>
> While modelling is a good tool to use in ESS, there are some aspects that need to be understood when using models. Models are oversimplified projections of what might happen, but they don't take into account any country-specific issues, such as preferential treatment of one sex of baby over another. Models are also developed on recorded data only, so the strength of the models is only as strong as the data that have been collected and used. Data collection can be variable in different countries, with some taking very detailed accounts of population dynamics while others may have a large number of undocumented births.

8.1.8 The Demographic Transition Model describes the changing levels of births and deaths in a human population through different stages of development over time

Another method of measuring and modelling population is through the use of the Demographic Transition Model (DTM). The Demographic Transition Model is a theory that was developed in the late 1920s to map the changes in population dynamics as a country undergoes economic development and industrialization. It provides a framework for understanding the cause and consequences of population dynamics changes over time. There are five stages to the model, representing different stages in development. Stage 1 represents the lowest stage of development.

- Stage 1: Characterized by high birth and high death rates, likely to be present in pre-industrialized countries with small populations that are not increasing very rapidly. There are now no countries that fit this stage. A population pyramid for this stage would have a wide base that rapidly reduces due to high infant mortality. The pyramid would also be shorter than in later stages due to the lower levels of life expectancy (see Figure 8.7).
- Stage 2: Death rates start to drop as development increases with the possible start of industrialization of the country. The birth rate remains high at this stage, representing a period of rapid population growth. Countries including Gabon, Yemen and Uganda are some of the countries that are still in Stage 2 of the DTM.
- Stage 3: As the country settles into being industrialized, the birth rate starts to slow down and the death rate remains low, resulting in a slower increase in comparison to Stage 2. Mexico, Thailand and Columbia are examples of countries currently at this stage of development. A population pyramid at this stage would begin to adopt a dome shape as the fertility rate stabilizes and the birth and death rates become similar (see Figure 8.7).
- Stage 4: Both birth and death rates are low and the population increase dramatically slows. This is commonly post-industrialization where the population is relatively stable, as is the case in Australia, Canada, Singapore, South Korea and much of Europe.

- Stage 5: Declining population as the death rates rise with an increase in the size of the elderly population. This is coupled with continuing reductions in the fertility rate, resulting in a reduction in the population growth and size. Examples of countries in this stage are Germany, Japan, Greece and Croatia.

Age–sex pyramids and the DTM work in conjunction with one another, and each stage of the DTM represents a subsequent pyramid shape (see Figure 8.7). This reinforces the interpretation of the model outcomes. You can also use one method and simply infer either the developmental stage or the type of pyramid the country will have based on this understanding.

> **Common mistake**
>
> Always consider scale. Some countries were in Stage 1 hundreds of years ago, and some have just left. There were very different issues happening while countries were going through these changes.

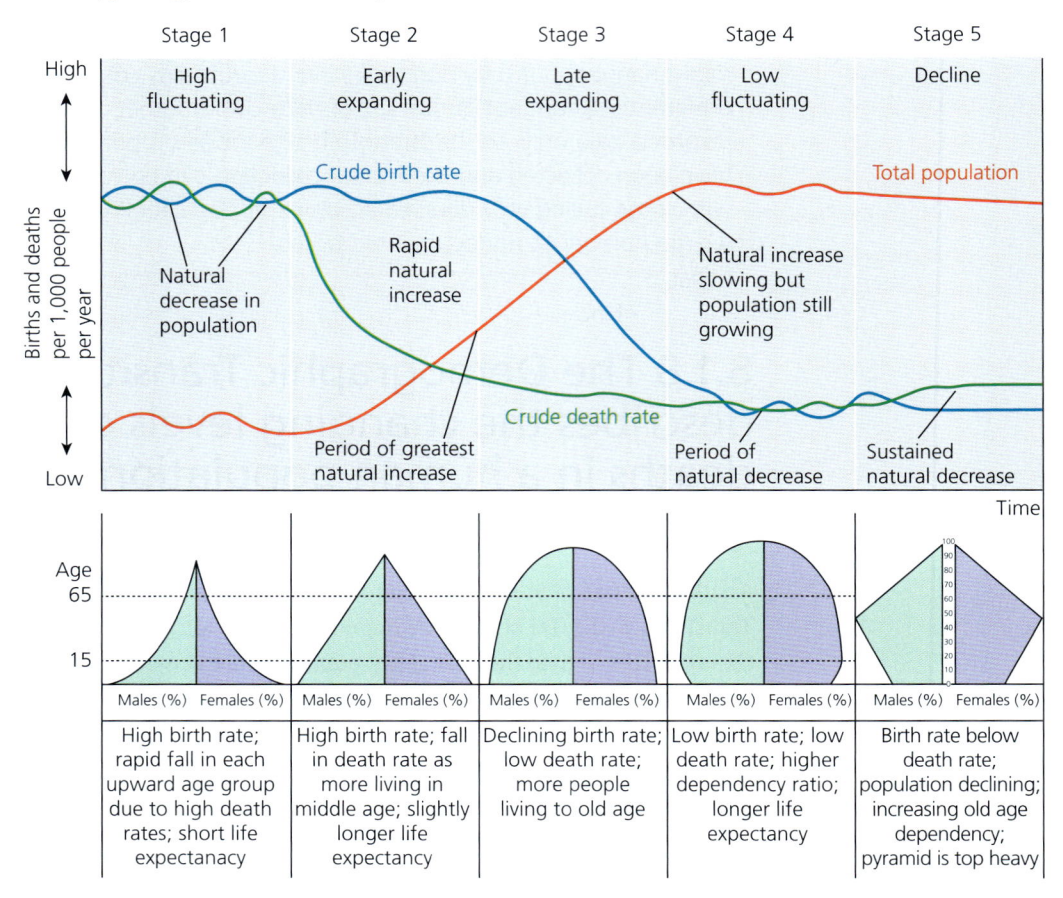

■ **Figure 8.7** Demographic Transition Model mapping the stages, and representative age–sex population pyramids relevant to each stage

Concept

Perspectives

Clear classification of countries into DTM stages is made more complex due to the migration of individuals around the world. For instance, Germany should simply fit in Stage 5 with a natural increase rate that is lower than the **replacement rate** and a population that is beginning to decline. However, the overall population of the country is rising due mostly to an increased immigration rate.

◆ **Replacement rate** – a fertility rate that results in the population exactly replacing itself in one generation. This is an average of 2.1 births per female of reproductive age in developed nations, such as the USA.

> ### ● TOK
>
> Can a country's level of development be judged solely on the basis of the status of its population demographics?

● (HL) 8.1.9 Rapid human population growth has increased stress on the Earth's systems

As the world's population has grown, the volume of resources we have taken from the planet has also grown. Not only has the sheer number of people grown to over 8 billion, but our level of needs and wants has also increased. Development has brought many changes in the way we live that require us to consume more resources to keep up with ever-increasing technologies and rapidly changing trends. As these levels of development spread throughout the world, we are likely to see further increases in the global population and their impact on all aspects of the Earth's systems.

The UN has created predictions that future growth, based on current trends, will reach 9.7 billion by 2050 and around 10.4 billion by the end of the century. This represents a significant reduction in the rate of increase in the second half of this century. Currently some of the rapidly developing countries, such as China and India, represent a huge proportion of the global population. China has been the country with the largest population since around 1950, but this was surpassed by India in 2023, marking a new shift in population dynamics. Predictions include that India's population will peak in around 2064. As India continues to grow, the pressure on the natural environment will continue to worsen. However, once both India's and China's populations begin to decline, the overall population growth will slow down and eventually begin to decrease.

Despite long-term population projections showing a reduction in the growth rate, it is hard to predict how the Earth's systems will survive. Current population numbers are already depleting many of our resources, for which demand will continue to grow, unless some widespread changes in behaviour are actioned. As consumption and population growth continue it is fair to assume that the ecological footprints of different countries will also begin to increase, particularly that of India.

◆ **Biocapacity** – the ability of a biological area to provide ongoing natural resources, while naturally absorbing the waste and pollution.

There is a huge disparity in the distribution of **biocapacity** around the world, as with the distribution of the human population. Some countries (for example, Sweden and Brazil) have a large biocapacity and can support increased population growth, but many (for example, UAE, China and Gabon) cannot and are already working in a deficit in some areas. The distributions of biocapacity and human population, although linked, do not necessarily correlate with each other due to many geographic, environmental and population factors. One way to visualize this is using the doughnut economics model.

Doughnut economics, and population growth and projections

Link

More information about the doughnut economics model is available in Topic 1.3.20 (page 75).

The doughnut economics model is a new way of looking at economic, environmental and social development as a collective, dynamic model. The older models of economic growth that centred around raising GDP need to be modified to centre around the doughnut economy. The model is dynamic and it is possible to easily identify where there are undershoots in the social foundation and where there are overshoots in the ecological ceilings. This allows population growth projections to be modelled in order to determine areas where balance needs to be resumed. For instance, if there is an overshoot in the biodiversity loss section for a specific country, policy can be developed to address these factors. Equally, areas can be addressed that are essential for developing a social safe-zone in the doughnut, such as availability of clean water and reliable electricity.

8.1 Human populations

Gabon | Japan

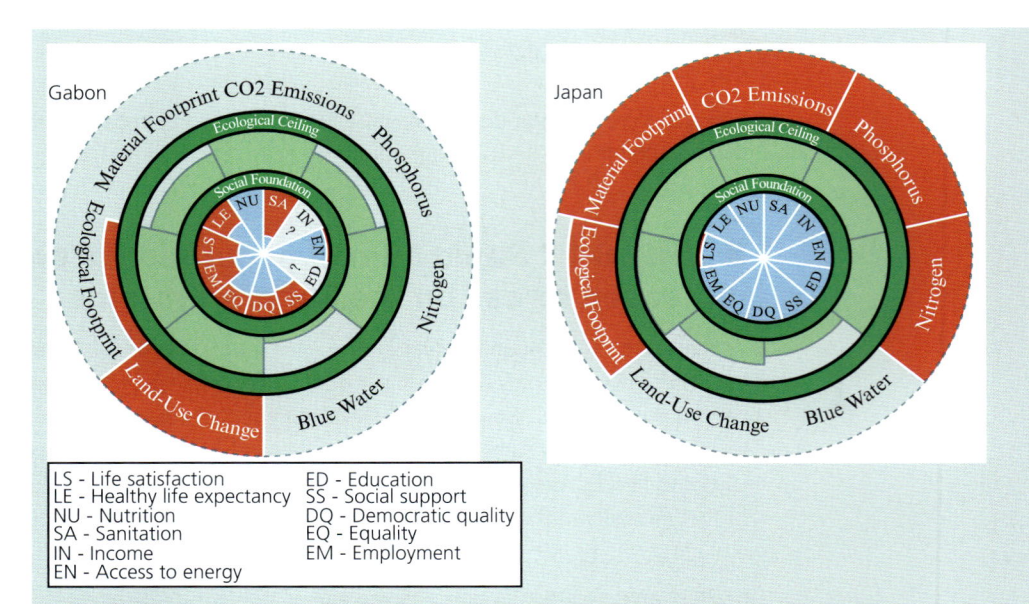

LS - Life satisfaction	ED - Education
LE - Healthy life expectancy	SS - Social support
NU - Nutrition	DQ - Democratic quality
SA - Sanitation	EQ - Equality
IN - Income	EM - Employment
EN - Access to energy	

■ **Figure 8.8** The doughnut economics models for Gabon and Japan (data from 2011)

The doughnut economics model can be used to compare the positive and negative aspects of each country's impact on the planetary boundaries and how well it is addressing the resilience of its social structure. These two aspects are clearly linked, and countries that have overshot on the planetary boundaries often also have resilience in their social foundation. Japan clearly has a highly developed social foundation, with only life satisfaction experiencing any limitations (see Figure 8.8), whereas Gabon has many areas of the social foundation that are underdeveloped, some to the point of not being assessed.

When comparing their impact on the planetary boundary, Japan's impact is much greater than that of Gabon, in all areas. Japan has an ageing population and is highly developed with well-established industries, such as car manufacturing. Gabon, on the other hand, is an underdeveloped African country that is still in the second stage of the DTM. The biggest environmental issue for the country is the level of deforestation, which has removed nearly 80 per cent of the original forest due to urban expansion, infrastructure development and logging.

8.1.10 Age–sex pyramids can be used to determine the dependency ratio and population momentum

Age–sex pyramids can show a lot about the economic potential of a population. In most countries, people begin work from 15 and the average age of retirement is around 64. This is the proportion of the population who have the earning power in that country, and who are therefore supporting the other parts of the population. The age groups that are outside this earning bracket are children and the elders. The larger these portions are, the higher the proportion of people that the earning population are supporting.

Populations that have high birth rates will have more young dependents to support, but these countries also often have lower life expectancy. Therefore the elders portion of the population is small (refer back to Figure 8.6). Countries at the highest stage of the DTM, Stage 5, are experiencing the opposite issues. Japan, for instance, has a decreasing birth rate as the population is choosing to have fewer children, however due to their lifestyle and development most people live to an old age. This gives the country a high dependency ratio who need medical care and support (see Figure 8.6). Canada's population pyramid (see Figure 8.6) shows a low- to mid-level of dependency as there is a relatively large number of people in the working age group, but also an increasingly large old population. As the birth rates in Canada have begun to drop, the dependency ration will gradually increase over time.

Population momentum can also be determined by looking at the age–sex pyramids. Despite the fact that throughout the world fertility rates are decreasing, the population continues to increase. This is due to population momentum and the continuous increase in the childbearing population. Therefore, even if the number of children per woman does not increase, the number of people able to have children is increasing. This is most obvious in countries where the death rate is beginning to decrease but the birth rates are still high (DTM Stage 2). As a country moves through the DTM stages the momentum reduces, but it does not begin to decrease until the base of the pyramid is significantly reduced over a number of years.

Tool 3: Mathematics

Dependency ratio

$$\text{Dependency ratio} = 100 \times \frac{(\text{Population } 0-14) + (\text{Population } 65+)}{\text{Population } 15-64}$$

Concept

Systems and models

The dependency ratio calculation and age–sex population pyramids can be used together to predict how the ratio is likely to change over time. As with all models there are limitations to using standard calculations and assumptions to create realistic predictions. Some of the limitations and weaknesses of this way of looking at economic dependency in a country include the fact that the calculation assumes that every person within the earning age group will be working. This does not take into account the rate of unemployment within that country, or the proportion of parents on maternity or paternity leave within the population. It also does not take into account people within that group who are not able to work due to illness, incarceration or other issues, which will increase the economic dependency.

Furthermore, the age bracket for employment may not be relevant in some countries. Some people are earning money before the age of 15 and some after 64. Final limitations of the calculations relate to the fact that earning potential is not equal for all individuals, and neither is their need for support outside the working age group.

8.1.11 Using examples of two countries in different stages of the DTM, consider the reasons for the patterns and trends in population structure and growth

The level of development within a country is often responsible for the population dynamics within that country. There are other factors that can have a greater or equal impact on population development, resulting in countries sometimes remaining within a certain development stage for a long time. These countries will only progress once specific issues have been addressed. Afghanistan, for example, is one of the poorest Asian countries. It is in Stage 2 of the DTM and has been struggling to advance in its development due to political, social and economic issues.

By the time a country reaches Stage 4 of the DTM, it will have a large population from growing through Stages 1–3. By Stage 4 the population begins to stabilize, often resulting in more balance in economic and social issues. Despite being the leader in population expansion for many years, China is currently in Stage 4 of the DTM and its population increase is starting to stabilize, resulting in India taking over as the most rapidly growing population in the world. Despite being at Stage 4 of the DTM, China does not fit the pattern of many of the other countries in this stage, such as Canada, the USA and Spain.

■ Table 8.4 Historical, cultural, religious, economic, social and political reasons for Afghanistan and China currently residing in Stages 2 and 4 of the DTM, respectively

	Afghanistan – Stage 2	China – Stage 4
Historical	Instability began in 1973 and the country has been involved in global and national political struggles ever since.	Post World War II China experienced a population boom, which gave rise to massive increases in population and resulted in the one-child policy coming into effect.
Cultural	Family is a very important part of the Afghan culture, and much of the country is rural farming communities so children are also seen as free labour.	Birth rates were artificially brought down through the anti-natal policy. Poverty in the rural communities is still very high. Males are seen as being more valuable in society and therefore the sex ratio in the country is highly skewed.
Religious	Although contraception is not banned, many do not use it for religious reasons.	There are no specific religious reasons for the demographic transition stage.
Economic	Previously dependent on foreign financial aid, the economy of the country suffered greatly when troops and aid were withdrawn from the country. The country's economic growth is slow.	Economically China has developed significantly over a very short period of time. There is a very large workforce helping to drive economic development.
Social	75 per cent of the poorest communities are illiterate. Sanitation and electricity are still lacking in many rural communities.	Post the one-child policy the population of young people is highly skewed, resulting in many men remaining unmarried. This has led to an artificial drop in the birth rate. This is further worsened as young males migrate away to find someone to marry.
Political	Internal and external political turmoil is resulting in increased migration rates. Afghanistan has high levels of valuable natural resources, resulting in political issues relating to how much it should exploit these resources to alleviate the poverty in the country.	Government developed the one-child policy to control population growth, and reversed this in recent years to encourage couples to have more than one child, due to issues with a reducing workforce and growing ageing population.

Each country has its own specific issues that have historically impacted, and are currently impacting, development. The improvements in medical availability, sanitation and food availability mean that populations might develop at a faster rate than the economic stability of the country can increase. This means there are now countries that are economically underdeveloped, but their populations match the patterns seen for Stage 4 of the DTM.

HL.a.2: Environmental law

The development of goods and services to enhance the living conditions within developing countries needs to be carefully managed to ensure that these elements are not developed in a way that negatively impacts the natural environment. Environmental Impact Assessments are a legal requirement in many countries, and a large part of the planning is this process of investigating the potential impact of any development on the natural systems and environment.

ATL ACTIVITY

As a group, look at the UNEP review of Impact Assessments (**www.unep.org/resources/assessment/assessing-environmental-impacts-global-review-legislation**). How variable are they around the world? Consider what this might mean for the environment in these areas.

HL.b.4: Environmental economics

The rapid development of countries such as China has resulted in many negative impacts on the air and water quality. Many organizations have outsourced manufacturing processes to these developing countries to avoid increasing the environmental footprint within their own countries. Many countries have tight regulations on pollution and may adopt a polluter-pays policy to reduce the burden on the community and force organizations to take responsibility for their waste. However, the outsourcing of manufacturing to other countries can bypass these rules and regulations. In addition, these organizations are able to create products for a fraction of the cost in relation to employees, fair wages, medical benefits, and so on.

ATL ACTIVITY

Investigate the impact of outsourcing

What are the pros and cons of this business practice for the company, the environment and the employees?

HL.c.10: Environmental ethics

The one-child policy in China was designed to help reduce the population growth and pressure on resources. However, there have been a number of unanticipated outcomes from the implementation of this strategy. For instance, having a male heir in Chinese culture can be of great importance. This has driven a preference for one sex of baby, resulting in a population that has many more males than females. Due to the lack of marriage opportunities, many individuals are leaving China in search of a partner.

ATL ACTIVITY

Consider if there are any other countries with populations that are skewed to one gender. What are the factors involved in these differences? Hint: Look at Qatar.

TOK

Are traditional knowledge and behaviours more important than making changes for the greater good?

Concept

Sustainability

To be sustainable, a country needs to ensure that the growth in population does not overtake the rate of development of economic and social issues. If population growth rises too quickly, resources may be used at a rate that is not sustainable. This will result in a depletion of resources. Balancing the increase in population and the availability of sufficient resources is key to improving sustainability.

8.1 Human populations

8.1.12 Environmental issues such as climate change, drought and land degradation are causing environmental migration

Factors that cause people to migrate were addressed earlier in the chapter, but these are not the only reasons why people migrate. Increasingly, environmental issues have been responsible for some considerable migration events. These are often the result of the direct impact of anthropogenically influenced climate change. Coastal areas have been most frequently hit due to the direct impact of climate change on sea-level rise. In many countries coastal areas are hubs for human communities due to the access to resources. Therefore, coastal flooding can displace large portions of populations in some areas. Some of the most vulnerable communities are therefore those in small island locations that could easily be devastated by coastal flooding.

Drought is another major climate issue that drives people to seek refuge elsewhere. Drought impacts all aspects of agriculture and therefore impacts the habitat from the foundations. Without plants, animals will migrate or die, leading to the collapse of the ecosystem. As we are reliant upon plants and animals for our food, this also dramatically impacts the suitability of an area to live in.

REAL-WORLD EXAMPLE

Tuvalu, South Pacific

The Tuvaluan people are migrating from the islands as climate change threatens to remove the islands from existence. The community is spread throughout some small coral and atoll islands in the South Pacific Ocean, which are rapidly being over-washed by the rising ocean. Due to the small size of the islands there are times when the land is inundated with sea water, making growing crops almost impossible due to the salinity levels. Around one-fifth of the population of 12,000 people have already left the island due to health, lack of education and regular flooding. Coastal reclamation is taking place through the deposition of large volumes of sand, at great cost to the government and without guarantee of long-term success. However, in November 2023 Australia set up a residence and citizenship programme for residents of Tuvalu.

Concept

Systems and models

Models are extensively used to look at the impact of flooding and this allows predictions to be made as to how long it will take for islands in the Pacific Ocean to disappear. These predictions are used by climate scientists to develop action plans to address potential risk. The same is taking place in relation to desertification through the use of satellite imagery and mapping, to allow the annual fluctuations to be recorded. This kind of information can inform drastic measures, such as moving a community, if the risk is too great. Following a large storm off the coast of France in 2010, and the rising sea levels in the affected coastal area, the government destroyed numerous apartment buildings, forcing hundreds of people to move out of the community due to the risk of continuous and worsening flooding.

HL.a.5: Environmental law

Environmental law has been introduced into the constitution of many countries. Switzerland developed the first addition of protection of the environment to its constitution in 1971, and since then many countries have followed suit, but the implementation and effectiveness of these additions are highly variable around the world.

ATL ACTIVITY

Either individually, as groups or as a class:
- Search the internet for the constitution of a specific country (some are easier to find than others).
- Use the Find function in your document reader to search for 'environment' in the document.
- How many times is it mentioned in the constitution?
- Scroll to each reference to the environment and have a look at the level of detail in the statements. Are they clear and explained, or do they just mention 'the environment'?

Using the internet, search for information about the success of the environmental policies of the country you have chosen.
- How much is your country following the environmental elements of its constitution?
- If you are working in small groups, compare your results. Are your countries similar or different?
- Consider the possible social, political, economic, religious, cultural and historical reasons for differences or similarities in the development and implementation of environmental laws in your different countries.

HL.c.10, HL.c.11: Environmental ethics

Consequentialist ethics and rights-based ethics are both related to the migration of communities due to negative environmental conditions. Rapid and sustained change in weather patterns is resulting in some areas being impacted by regular, repeated and significant flooding. This can force people to leave these areas due to the continual disruption of their homes. These are consequences of our continual overuse of resources and burning of fossil fuels without considering the rights of those people living in these continually disrupted parts of the world.

ATL ACTIVITY

Consider how ethical it is that the communities who do not create the environmental problem are often the ones who have to deal with the consequences. Discuss this as a class.

REVIEW QUESTIONS

1. Explain, using the data in Table 8.2 (page 699), which country has a reducing population.
2. List three countries in Table 8.2 that have high fertility rates and low doubling time.
3. Describe the population trends in the Scandinavian countries in Table 8.2.
4. Identify the country in Table 8.2 with the longest doubling time.
5. Australia and New Zealand have relatively high rates of natural increase for developed countries. Given that their fertility rates are low, determine what might be the main reason for the high rate of population increase.

8.1 Human populations

EXAM-STYLE QUESTIONS

1. List one reason why a country might need to develop an anti-natalist policy. [1 mark]
2. Explain direct and indirect ways a government might persuade residents to decrease the number of children they have. [5 marks]
3. Explain why there are now no countries in Stage 1 of the Demographic Transition Model. [2 marks]

8.2 Urban systems and urban planning

> **Guiding questions**
>
> - To what extent are urban systems similar to natural ecosystems?
> - How can reimagining urban systems create a more sustainable future?

SYLLABUS CONTENT

This chapter covers the following syllabus content:
- ▶ 8.2.1 Urban ecosystems, like all ecosystems, are composed of biotic components (plants, animals and other forms of life) and abiotic components (soil, water, air, climate and topography).
- ▶ 8.2.2 An urban area is a built-up area with a high population density, buildings and infrastructure.
- ▶ 8.2.3 An urban area works as a system.
- ▶ 8.2.4 Urbanization is the population shift from rural to urban areas.
- ▶ 8.2.5 Due to rural–urban migration, a greater proportion of the human population now lives in urban rather than rural systems, and this proportion is increasing.
- ▶ 8.2.6 Suburbanization is due to the movement of people from dense central urban areas to lower-density peripheral areas.
- ▶ 8.2.7 The expansion of the urban and suburban systems results in changes in the environment.
- ▶ 8.2.8 Urban planning helps decide the best way to use land and buildings.
- ▶ 8.2.9 Modern urban planning may involve considering the sustainability of the urban system.
- ▶ 8.2.10 Ecological urban planning is a more holistic approach that treats the urban system as an ecosystem, understanding the complex relationships between its biotic and abiotic components.

HL ONLY
- ▶ 8.2.11 Ecological urban planning will follow principles of urban compactness, mixed land use and social mix practice.
- ▶ 8.2.12 Societies are developing systems that address urban sustainability by using models such as the circular economy or doughnut economics to promote sustainability within urban systems.
- ▶ 8.2.13 Green architecture minimizes the harmful effects of construction projects on human health and the environment, and aims to safeguard air, water and earth by choosing environmentally friendly building materials and construction practices.

8.2.1 Urban areas contain urban ecosystems

An urban ecosystem is very much the same as every other ecosystem, with a combination of biotic and abiotic elements that interact together to make the ecosystem function correctly. An urban system has many elements not necessarily present within other ecosystems due to the combination of the built and natural environments.

Abiotic factors can be split into four main areas, in addition to the regular ecological abiotic aspects (see Chapter 2.1, page 84). These four areas are the physical environment, infrastructure, economic opportunities and technological advancements.

1. The physical environment: This is the combination of regular abiotic factors like climate, topography, soil composition and availability of water resources. The way that the built environment is designed can result in changes in all of these factors. For example, the temperature in a city is related to the density and height of buildings, amount of air conditioning use, and the topography of the area surrounding the city.
2. Infrastructure: The presence of infrastructure like housing, transportation systems, utilities (for example, water supply and electricity), and waste management services affects the growth and sustainability of urban populations.
3. Economic opportunities: The availability of jobs, income levels, and economic prospects within urban areas are critical factors that attract people to migrate to and settle in cities.
4. Technological advancements: Access to modern technology, including communication networks, internet connectivity and digital infrastructure, influences the urban population's growth and development.

The biotic factors in an urban system are also somewhat more complex than in other, non-anthropogenic, habitats. The four major areas of biotic elements are the human interactions, the flora and fauna, health and disease, and finally the food systems.

1. Human interactions: Urban areas are dominated by the humans who live and interact within them, and their cultural dynamics. This includes social networks, migration patterns, cultural diversity and community interactions, all of which influence the growth and composition of urban populations.
2. Flora and fauna: Urban areas have unique ecosystems that support a variety of plant and animal species. The presence of green spaces, parks and urban forests contributes to the quality of life in cities and supports biodiversity.
3. Health and disease: Biotic factors like the spread of diseases, access to healthcare facilities and sanitation conditions impact the well-being and health of urban populations.
4. Food systems: Urban areas have their own food systems, including agriculture, food production, distribution networks, and access to fresh and nutritious food. These factors affect the nutritional status and overall health of urban populations.

These abiotic and biotic elements are interconnected, and each element influences the others. In order to develop sustainability in urban systems, planners need to take all of these elements into account.

● Common mistake

Try not to limit your view to your own experiences. Many people around the world live in very different conditions, climates, political situations and levels of poverty than you do. Watching a variety of documentaries, for instance, will help you gain these perspectives.

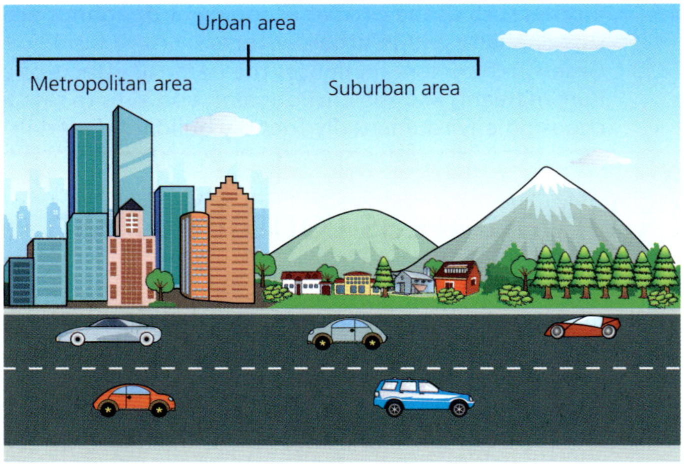

■ **Figure 8.9** Diagram representing the density of metropolitan and suburban parts of the urban environment

Theme 8: Human populations and urban systems

8.2.2 An urban area is a built-up area with a high population density, buildings and infrastructure

The urban environment consists of different zones with differing functions. Although the urban system covers all urban areas, there are very different conditions present in different zones due to the density and dispersal of buildings, transport networks, residential and commercial zones. The metropolitan area represents that central part of a city, where the density of businesses, transportation hubs, and more recently residential areas is much greater than in the suburban areas (see Figure 8.9). In larger cities this is usually where buildings expand vertically due to limited space to increase the density without having to increase the footprint of the buildings. While this allows greater volumes of residents and businesses to be present in a small area, there are advantages and disadvantages to high-rise living.

Greater volumes of people living in metropolitan areas significantly reduces the use of private transport to commute to work and decreases the ownership of vehicles, resulting in a reduction in transport-related air pollution. While the use of public transport is higher in urban areas, this remains a more sustainable approach to reducing emissions. As building increases vertically, this creates an area where air does not flow freely, as well as increasing heat generated from air conditioning units. These two factors both contribute to increased urban air pollution and an increased **urban heat island effect**, respectively.

◆ **Urban heat island effect** – an issue that takes place when cities replace the majority of natural areas with pavements, buildings and roads, which all absorb and trap heat within the city. This results in the temperature in urban areas being higher than in surrounding natural and less densely populated areas.

8.2.3 An urban area works as a system

An urban system is a complex web of interconnected elements and factors that contribute to the functioning of the system. For an urban environment to function correctly, a range of factors need to be considered. The availability of power and the creation of energy are paramount to the efficiency of the area. The use of water and the creation, removal and treatment of sewage, along with all other domestic and retail waste, needs to be balanced if sustainability is to be achieved. Transport systems allow for ease of connectivity within an urban area, but also contribute significantly to the direct and indirect air pollution issues.

Like any system, there are inputs, outputs and processes that take place to either pull people to that area or push them away. In a city, the inputs include immigration and births for the human population; water, air, nutrients, animals and plants for the natural elements; and energy, pollutants and goods are some of the anthropogenic inputs. Outputs include deaths and emigration for the human population. The outputs for the natural and anthropogenic aspects are likely to be very similar to inputs, just in different volumes. For instance, goods come in, and although some goods will leave the city system, goods that come in will likely be turned into waste and that is the form by which they leave the system.

> **Tool 4: Systems and models**
>
> **Create a systems diagram**
>
> One way of visualizing a system is to create a systems diagram. In general, boxes are used to represent a storage of a component and arrows are used to represent the flow of something within the system. Systems diagrams can take many forms, from highly simplified spider diagrams that show the links within the system, to complex webs that provide detailed information relating to the relative inputs and outputs from all parts of the system. Using a systems diagram allows complex interaction to be visualized. These can then allow modelling to take place by either increasing or decreasing inputs and outputs to determine the potential impacts of future change in the population and its requirements.

8.2.4 Urbanization is the population shift from rural to urban areas

◆ **Urbanization** – an increase in the number of people living within urban systems, generally due to migration from rural areas.

Urbanization refers to the process by which an increasing proportion of a country's population resides in urban areas or cities. It involves the movement of people from rural or suburban areas to urban centres, leading to the growth and expansion of city populations. Urbanization is typically accompanied by changes in social, economic and physical environments.

As the population of an urban area increases, the landscape also changes. Housing is needed to accommodate more people and, with the limited space available in some urban areas, this often results in buildings expanding vertically, increasing housing availability without increasing the space needed. As this movement takes place, all aspects of the infrastructure also need to grow to allow the population growth to be maintained and sustainable. In order to keep people in the city there needs to be sufficient availability of food, clean water, sanitation, electricity and transport. The increasing volume of people within urban areas results in higher volumes of waste and pollution. Solid domestic waste from residential and business premises will increase as these areas expand and develop. Therefore, the systems required to manage this also need to develop to keep up with demand. As these aspects all develop, the natural ecosystem often shrinks, resulting in less efficient natural systems as the urban environment grows.

8.2.5 Due to rural–urban migration, a greater proportion of the human population now lives in urban rather than rural systems, and this proportion is increasing

The migration of people to urban settlements is not a new phenomenon and some major European cities, like Rome and Paris, developed in Roman times as hubs for trade and the government. Since this, the Industrial Revolution in Europe drove the industrialization of cities such as Manchester, Liverpool and London in the UK. These cities were the centre for many industries that significantly helped to drive development in the UK.

Since this period, there has been a steady increase in industrialization worldwide, resulting in urban environments developing globally (see Figure 8.14 on page 739).

Inquiry process

Inquiry 2: Collecting and processing data

Global maps can be excellent visual representations of many different types of data. These allow patterns to be easily identified. Sometimes you might only have the map form and therefore need to interpret patterns directly from the image (see Figure 8.14 in the Review questions on page 739). Use this figure to answer the Review questions and develop an understanding of how to extract information from secondary sources. Be aware of the key for any figure when you are interpreting information.

Many parts of the world are heavily urbanized and these areas are, for the most part, continuing to expand. Some places such as Singapore and Kuwait have urban populations of 100 per cent. Most of the higher levels of urbanization are found in developed countries in Europe. Many of the countries that are still developing, such as Niger and Burundi, have the lowest urban percentages with 16.9 and 14.4, respectively. Gabon is an African country that does not follow the same pattern and has 90.7 per cent of the population living in urban areas. This is due to Gabon being the fourth-largest oil producer in Sub-Saharan Africa, resulting in significant economic development in relation to other African countries. This business is all concentrated in the major cities that act as hubs for the offshore oil mining industry. This increase in work has led to large-scale migration from rural to urban areas.

The trend to urbanization is not universal, and there are some countries where populations are starting to migrate back to rural areas. Japan, Bulgaria and Poland all have negative urban migration values, showing these areas are decreasing in size.

Each country has its own set of factors that drive people to move; these can be things that attract them about the destination or deter them from staying in the rural location. These are known as push and pull factors.

Push factors are aspects that have driven people from rural to urban areas. These include lack of resources and facilities, limited job availability, poverty, lack of diversity and lack of higher education, among other things. Most rural areas provide education up to a certain level, but a lot of further and higher education establishments tend to be based in urban areas. This causes people in their late teens to leave to pursue better education. These migrants do not return to the rural area once their education has finished, instead they find jobs within urban areas. The reason for this is that the job opportunities in rural areas are often very limited, especially in careers relating to business and commerce, and government, example. Other factors that drive people away from rural areas are the lack of local facilities such as shopping centres, entertainment venues and medical support.

Conversely, the pull factors that encourage people to move to urban areas include the availability of work and wider higher education opportunities, stable communities, access to services and infrastructure, and a perceived improvement in the standard of living. These factors are similar and often complement each other, so the push of the lack of job opportunities in rural areas is coupled with the increased level of opportunities in urban areas. These therefore reinforce each other, resulting in the dramatic increase in populations in these areas.

Some migration is voluntary and is due to specific decisions made by individuals to change their life situation. Forced migration also takes place, however. For instance, in many developing countries adults will leave rural areas to earn money to send home to support the growing family. This often results in families being separated due to financial requirements. This pull towards the urban environment can be perceived to be beneficial to lower-income rural households. However, on arrival in urban areas it is rapidly apparent that earning potential is soon balanced out by the increased cost of living within urban areas, thus leaving both parts of these divided families in poverty. Forced migration can be a result of external factors such as natural disasters, civil unrest, persecution of minority groups and human trafficking. These people do not leave by choice, but move to urban areas with dreams of a better life, which may not realistically exist.

As urban centres try to expand, this can result in developments encroaching on suburban areas, increasing urban sprawl. These suburban areas are characteristically less densely populated and built-up than urban centres. They are areas of lower-density buildings and fewer businesses, resulting in increased residential areas. This expansion of cities is not as detrimental to the natural environment as densely populated city centre areas, but the issues of pollution, reduction of green space, and impacts on the water cycle still take place in suburban areas.

In some parts of the world the trend is starting to reverse. Developments in communication capabilities, flexible working, and attempts to reduce company costs have resulted in many sectors allowing workers to work remotely, eliminating the need to live in the city or to commute to work. Many middle-income workers living in cities have opted for a slower pace of life and better work–life balance, and have taken advantage of this to move back to the quiet rural areas. This, however, is causing the cost of living in rural areas to increase as businesses respond to the changing wants and needs of this influx of people.

REAL-WORLD EXAMPLE

Urban–rural migration patterns in the United Kingdom

Currently over 80 per cent of the population of the United Kingdom (UK) live in urban or suburban communities. Industrialization in the nineteenth century was one of the major triggers that started the mass migration of people from rural to urban areas. As jobs became more freely available at the many factories and industries in the cities, many families left rural England to seek a better standard of living.

In order to accommodate the influx of people, urban planning was used to develop a network of water-treatment and sewage-disposal facilities, transportation networks and public buildings. Major cities later became associated with specific industries, with northern cities in the UK such as Manchester and Liverpool focusing on textiles and import and export, respectively. These growing industries needed a steadily increasing workforce, encouraging people in rural areas to migrate with the promise of good wages and housing.

In the 1980s in the UK many of the longstanding industries began to decline or ceased operating, leaving many derelict areas in inner cities. This led to a shift towards the promotion of cultural centres to encourage continued growth. Urbanization continued into the early 2000s as regeneration projects in derelict industrial areas, such as Manchester, began to gain funding and support, resulting in more modern and attractive city living options that attracted many young people to stay in the city after university.

Recently there has been a reversal in this trend as people begin to return to rural areas. The increased cost of city living, increased congestion, parking prices, noise and air pollution are driving some to leave for the rural areas. Figure 8.10(a) shows the difference in age groups living in rural and urban areas. Rural areas are clearly dominated by people over 45 and under 19, and that those between 20 and 45 years old are more likely to live in urban areas. This age group includes that age when teenagers traditionally leave home for work or university in the UK. Availability of both education and work is greater in the cities, attracting this age group to move from the rural communities to the urban ones.

(Source: www.gov.uk/government/statistics/rural-population-and-migration/rural-population-and-migration#mid-year-population-2020)

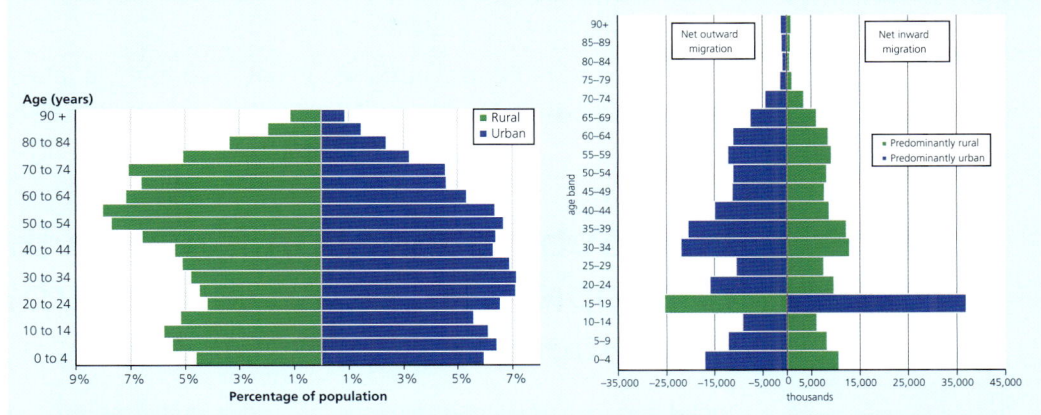

■ **Figure 8.10** (a) Percentage of the population (within England) residing in rural and urban areas by age bands, mid-year 2011 to mid-year 2020. (b) Net internal migration (within England) to predominantly rural areas by age bands, mid-year 2011 to mid-year 2020

Concept

Perspectives

The reasons for migration might be similar around the world, however the reasons for and strength of the push or pull factor is very much dependent on the status of the country. Around 80 per cent of populations in more developed countries were living in urban communities in 2021, while the value ranged from 35 to 50 per cent in developing countries. There are many different factors impacting these trends, but urbanization is increasing in countries that are still developing their industries and infrastructures, whereas some countries that have well-developed business and industries are experiencing a return of residents from urban to rural areas.

Countries still in lower development stages are still experiencing significant urbanization as the overall conditions in rural areas are often somewhat lower than in the cities.

Concept

Sustainability

The development of megacities around the world is not sustainable. While it is necessary for countries to develop their cities in line with population growth and migration, this is causing many social, environmental and economic issues in the regions where the city is located.

8.2 Urban systems and urban planning

> While development comes with a clear development in cities and cultural, economic and business hubs, there is a point where the shift of people is no longer beneficial for either the rural or urban habitats. Many people are driven from rural areas by poverty and a lack of opportunities, but as the cities continue to grow the availability of those opportunities drops, leaving some migrants homeless and in poverty. Once job opportunities become less available it can be difficult for families to financially be able to choose to return to their rural home.
>
> These factors can result in increased crime and low standards of living, and this negatively impacts the environment. As migration to these megacities continues the level of air, water and ground pollution can also increase, if careful urban planning and management are not present.
>
> In order to be more sustainable, countries need to try to retain some of the population in rural areas so as not to ruin the economic and social development.

8.2.6 Suburbanization is due to the movement of people from dense central urban areas to lower-density peripheral areas

Suburbanization refers to the movement of people from densely populated urban centres to lower-density residential areas located on the outskirts of cities or in surrounding suburban regions. This phenomenon leads to the expansion of urban areas beyond their central cores.

As people move out of the central urban areas, the population density in the city centre decreases. At the same time, suburban areas experience an increase in population due to the influx of people seeking more affordable living, quieter environments, and sometimes better schools. This involves the conversion of more agricultural or undeveloped land into residential neighbourhoods to accommodate the spread of the population. These areas also then require infrastructure such as roads, schools and healthcare facilities to accommodate the growing population, resulting in increased construction and development in these areas, coupled with the loss of natural habitats.

The development of comprehensive and reliable transportation networks can help with the expansion of suburban areas. However, people commuting to work in the city centre might contribute to increased traffic congestion and subsequent pollution on roads leading into the city. Also, unplanned or poorly managed suburbanization can lead to urban sprawl, where development occurs in a scattered and disorganized manner. This can have negative environmental and social consequences.

Suburbanization has certain benefits, such as improved living standards for some, but it also presents challenges related to increased reliance on cars, infrastructure demands and potential loss of open spaces. Proper urban planning and sustainable development practices are essential to ensure that suburbanization contributes positively to the overall urban environment, and the well-being both of city centres and suburban communities.

Some countries also make attempts to limit the level of urban sprawl by protecting the natural habitats between cities and neighbouring towns. This stops towns from growing too large and connecting with nearby cities. These areas are known as greenbelts and help to maintain biodiversity in suburban and rural habitats.

8.2.7 The expansion of the urban and suburban systems results in changes in the environment

As populations have moved into urban centres and these have begun to expand, there have been many significant impacts on the environment. Most notable is the level of habitat loss that has occurred in order to create space for the urban areas to expand into. Along with this loss of habitat also comes the inevitable loss of all of the species of plants and animals in areas associated with this expansion. Many urban expansions result in the loss of farmland previously used for crops or livestock. However, these crops and animals will just be moved as the population growth requires them for its food supply.

The conversion of any natural landscape into an urban one results in the development of an environment dominated by concrete. When it rains in urban areas the water is no longer able to be absorbed into the soil as it is covered by an artificial surface. This can lead to the underlying soil lacking in water and nutrients, which over time can lead to the loss of urban trees.

Link

See Chapter 8.3, page 740, for more detail relating to urban air pollution

This removal of vegetation also removes the natural filtering of the atmosphere by plants. During the process of photosynthesis, plants absorb CO_2 and release oxygen, helping to improve urban air quality. Once they are removed, CO_2 can increase, along with other air pollutants such as SO_2 and particulate matter.

As the area is converted from natural land to an urban environment, the albedo is increased, leading to the urban heat island effect that increases temperatures in cities. This potentially results in increased use of air conditioning and energy to cool buildings.

Land use changes also impact balance in the water cycle and the air quality in these areas. Rivers that run through urban environments are often exposed to higher volumes of pollution from urban traffic, runoff and solid domestic waste. Some are forced to flow underground in concrete tunnels to allow construction to take place over the river. This changes the level of groundwater in urban areas, and pollution will impact downstream habitats. Air quality is directly reduced by the increase in vehicles producing primary air pollutants, which eventually interact with sunlight to create ozone and photochemical smog.

REAL-WORLD EXAMPLE

Land use change in Beijing

Figure 8.11 shows a series of land use change maps as Beijing expanded between 1984 and 2018. Prior to major development, Beijing was surrounded by arable land and forest. There are several clear patterns of land use change that took place as the urban centre expanded and developed. The most notable was the dramatic loss of arable land in the immediate vicinity of the expanding city. Secondly, as the expansion developed, the density of forest gradually reduced.

One other major impact is related to the water provision. To the northeast of the city is the largest manmade reservoir, Miyun Reservoir. As the expansion began, the reservoir expanded as the water needs of the area dramatically changed from irrigation to urban water needs. However, as the development continued there was a distinct reduction in the footprint of the reservoir, suggesting that the demand for water in the areas was increasing, resulting in the overextraction of water. In addition to concerns regarding future water availability in this wet land, the reservoir became a very important site due to its location on the migration pathway of many bird species.

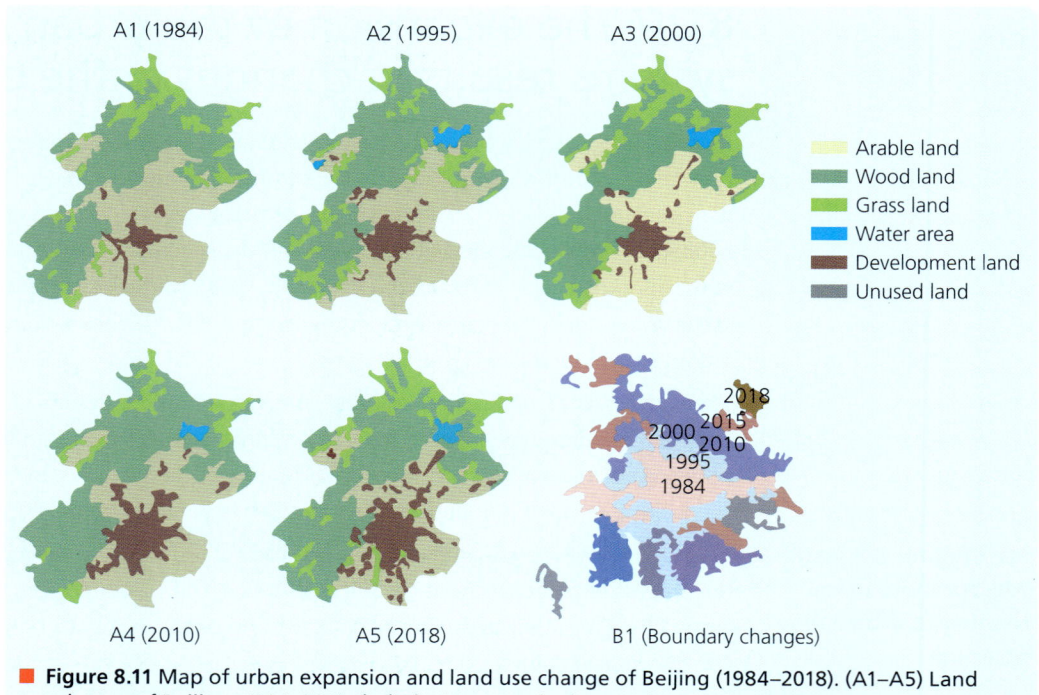

Figure 8.11 Map of urban expansion and land use change of Beijing (1984–2018). (A1–A5) Land use change of Beijing, 1984–2018. (B1) The process of urban expansion, 1984–2018

8.2.8 Urban planning helps decide the best way to use land and buildings

In order to ensure that a new town functions effectively and efficiently, successful urban planning has become increasingly important. As with any major development, the impacts on the environment need to be minimized. There are five main phases in the process: the exploration phase; the planning phase; the zoning phase; the design and implementation phase; and the final operation phase where the building of the urban area can commence.

- The exploration phase: The exploration phase involves gathering information, conducting research, and analysing data to understand the existing conditions and challenges of the urban area. It includes activities such as assessing demographics, studying land use patterns, identifying infrastructure needs, and engaging with stakeholders to understand their concerns and aspirations.

- The planning phase: In the planning phase factors like population projections, transportation needs, environmental impact and economic feasibility are considered. The plan may include goals, objectives, policies and strategies to guide decision making in subsequent phases.

- The zoning phase: In the zoning phase planners work on creating zoning ordinances, development codes, and design guidelines that determine how different areas of the city can be used and developed. It includes assigning zones for residential, commercial, industrial and recreational purposes, as well as establishing rules for elements such as building height.

- The design and implementation phase: The design phase includes designing the streetscapes, public spaces, parks, infrastructure systems, and other elements that contribute to the physical form and character of the city. The implementation phase involves coordinating with architects, engineers and developers to ensure that the plans are carried out correctly and in line with the established guidelines.

Theme 8: Human populations and urban systems

- The final operational phase: The final operational phase comes in post development, when the urban space is complete and functioning. This involves managing and maintaining the infrastructure, public spaces and services of the city. It includes activities such as ongoing monitoring, evaluation and adaptation of the urban plan as conditions and needs change over time.

8.2.9 Modern urban planning considers the sustainability of the urban system

Modern urban planning for sustainability focuses on creating cities and urban areas that are environmentally responsible, socially equitable, and economically viable in the long term. This focuses more on the efficiency of land use, provision of integrated public transport, renewable energy supply, and the use of smart technologies to monitor and control specific aspects of the environment. Green buildings and the incorporation of the natural environment are also key elements of sustainable urban planning. The social sustainability of urban areas is also key to ongoing sustainable development of community, society, equity and all aspects of culture. Economic sustainability is also essential to maintain a level of goods and services that will continually attract residents to move to and stay in urban communities.

REAL-WORLD EXAMPLE

The Cerdá Plan, Barcelona

Sustainable city planning has taken place for many years in Barcelona. The Cerdá Plan in 1860 created a system that distributed individuals evenly throughout the city and eliminated the potential of areas developing as specific 'rich' or 'poor' areas of the city. Buildings were designed in block layouts with green space incorporated into the design. The careful designing of the network of streets has aided the flow of air in the city, and as the population continued to increase this design made navigating the city easier as it aided the flow of traffic. This has resulted in an innovative block design to the city that Barcelona is now grouping together into 'Superblocks'. These areas will provide private spaces for residents only to drive in, thus alleviating the issue of limited parking for residents due to visitor numbers. This makes these areas safer and the air cleaner due to the reduced level of transport passing through.

In 2015, the Sustainable Development Goals were created, with SDG 11 focusing on sustainable cities and communities. This particular goal aims to improve inclusivity, equality, safety and resilience in urban environments, and focuses on providing adequate housing to replace slum areas in developing countries. Improved and safe transport networks with accessibility for people with disabilities is also a main focus of SDG 11, as is the protection of world heritage buildings and green spaces. Management and maintenance of safe and clean water is another main factor in the development of a sustainable urban area. One element of sustainability within the goal is to provide the least developed countries with financial support and education to provide them with the skills to develop new settlements using local sustainable products.

Some of the main aspects of developing sustainable cities are related to the decarbonization of urban travel, an increase in energy efficiency, and community involvement and development. These will tackle the aspects of the environmental, and the social and economic issues currently present in major cities. This can be achieved through either redevelopment of existing cities or the creation of new, smart cities.

> **REAL-WORLD EXAMPLE**
>
> ### Vancouver, Canada
>
> Vancouver has developed an innovative water-management model that has not only reduced pressure on municipal water supplies, but is also beginning to slowly repair the natural watershed area. Vancouver's Rain City Strategy includes government policies designed to increase green architecture, water permeable surfaces, and effective rainwater capture and management within the city. It includes leakage-reporting programmes and the redevelopment of wastewater treatment in the city, along with citywide projects to switch to water-saving devices and management of personal water use. Alongside this, a new tiered water-pricing system that charges the heaviest users the most has encouraged businesses and individuals to engage in water-saving behaviours in order to reduce their water bills. Together, these strategies have together resulted in Vancouver being rated as one of the top cities in the world for water conservation.

■ Redeveloping existing cities

Many older cities were not originally developed with the natural environment in mind. Inner-city industrial areas common in part of Europe have left large abandoned buildings in some major cities, which are sometimes protected for their heritage. This poses the problem of how to regenerate these areas while maintaining historical elements of the city's heritage. Repurposing large factory or industrial buildings has been a key part of regenerating many of the industrial parts of northern England. Community involvement in redeveloping these inner-city areas to provide integrated housing, office and lifestyle facilities has transformed these long derelict areas into bright, vibrant community areas.

Link
A real-world example of a smart city (London) is given in Topic 6.3.9 (page 580).

■ Creating smart cities

Some of the countries that have developed in recent years have taken advantage of the development of urban intelligence. This is a key part of any new city development as it allows the easy monitoring and control of lighting, water availability, traffic flow, security and other such services in these urban areas.

Having the ability to remotely control many aspects of the city is an attractive proposal that has proved to be a difficult aim to achieve. There have been a number of attempts to create smart cities, Masdar City in Abu Dhabi being one example of a city that is designed to be fully sustainable. The city was designed to be a sustainable oasis in the desert, with smart control of energy use, water management, waste reduction and monitoring, alongside the use of traditional building-design features. This was aimed at reducing the temperature within the city, and therefore the need for large amounts of energy to keep homes cold. Plans included the creation of new natural materials, and an artificial reef system on the coast to include the development of the coastal ecosystem. Despite these highly ambitious plans, the development of Masdar City has yet to be completed, with few residents choosing to move into the city.

■ **Figure 8.12** Masdar's central courtyard with a wind tower designed to keep the streets cooler

8.2.10 Ecological urban planning is a more holistic approach that treats the urban system as an ecosystem

Ecological urban planning addresses the triple bottom line of environment, economics and equity. This means that, in order to be more sustainable, new developments need to equally take into account the financial, social and environmental impacts and costs of the development. New developments need to provide accessibility for all, use renewable energies, and employ careful design that can reduce energy needs and increase connectivity with the environment.

One of the biggest changes in urban environments is the need for harmonious and simultaneous development of the natural and urban environments to provide safe and sustainable spaces. Harmony and connectivity can be achieved in a number of different ways, including urban farms, **biophilic design**, urban ecology and the development of **regenerative architecture**. Through the use of these sustainable initiatives it is possible to create more balance between the abiotic and biotic elements of urban systems. Apple Park in California, USA has a biophilic design to maximize engagement and connectivity with nature. The doughnut-shaped building does not require heating or cooling for more than nine months of the year and runs on 100 per cent renewable energy. The design of the building allows maximum connection with nature and it is surrounded by forest. This design seamlessly links the internal work environment with the natural environment, allowing reduced energy needs and a lower environmental impact of running and maintaining the building.

Urban farming and urban ecology focus on increasing biodiversity and productivity of the habitats. Urban farms can provide fresh, organic produce in areas that have traditionally had a lack of access to fresh ingredients. Improving the natural environment also impacts the mood and well-being of residents in the area and can help reconnect people in the city with the natural environment.

◆ **Biophilic design** – building design that aims to create seamless connection with nature through the direct and indirect use of nature, space and place in design.

◆ **Regenerative architecture** – incorporating building features, such as green roofs, wind turbines and photovoltaic cells, that harness or clean the natural environment.

> **Concept**
>
> ### Perspectives
>
> The idea of creating a more ecologically friendly location can be achieved on all different scales and does not need to be on a whole city level. It can be confined to pockets of biodiversity such as urban parks, parks, allotments and river banks. The ideal environmental scenario is to have cities either created to be sustainable, or regenerated to maximize environmental benefits. Singapore is an example of a city that has been successfully developed to incorporate many sustainable features, such as its urban farm that has adopted vertical farming technology to produce fresh produce in the city. Singapore also has around 50 per cent of the city's footprint covered in green spaces. Paris is a densely populated, old city that does not have many spaces available to return to nature, but the city boasts the largest urban farm in Europe, Agripolis, which is situated on top of one of the large buildings in the central area of the city.
>
> It is not always possible to create something on a large scale, therefore individual building developers need to focus on making ecological planning a core part of any project. Equally, we can all make changes to our own residential space to engage in different aspects of ecological planning. Using plants in the residence, maximizing natural light and air flow, and planting urban gardens with productive plants to create some free fresh food.

> **REAL-WORLD EXAMPLE**
>
> **Singapore: sustainable city**
>
> Singapore is a global import, business and port city that is heavily constrained due to its size and geography. Despite this, Singapore is one of the most sustainable cities in the world. It has a well-designed urban planning strategy that integrates green spaces, parks and nature reserves into its cityscape. The city is known for its lush greenery, with abundant trees and plants, which improves air quality, reduces the urban heat island effect, and enhances overall quality of life.
>
> The limited ground space in Singapore is used efficiently with vertical residences using vertical farming technology to reduce the need to import all fresh produce, while increasing biodiversity. Large parts of the city have been pedestrianized to encourage walking or cycling, and there is a comprehensive and affordable public transport system encouraging car owners not to drive in the city, thus significantly reducing emissions levels.
>
> Being an island, Singapore has limited freshwater and limited space to deal with waste. Efficient rainwater collection and recycling programmes have been developed to address these issues. These schemes, and many more, make Singapore a leader in sustainable urban development.

Biodiversity in urban areas is another aspect that is significantly improved through careful implementation of sustainable practices. For instance, living walls and green roofs can improve the diversity of plants as well as attracting pollinators such as bees and butterflies to the area. Regenerating canal areas and derelict sites can also attract significant bird and small mammal species to these areas. As biodiversity increases, resilient communities that can withstand changes and counteract some aspects of urban air pollution can be found in the urban environment.

> **Concept**
>
> **Sustainability**
>
> Urban planning and development for sustainability is a delicate balancing act that needs to be developed on the basis of a comprehensive needs and viability study. An Environmental Impact Assessment (EIA) is carefully conducted to consider how the population is going to continue to grow to allow for further expansion without compromising the green space provision or reducing sustainable initiatives. Local targets for reducing energy usage, developing more circular practices and community involvement are all needed in order to accommodate change.

Link

More information about the development of Environmental Impact Assessments (EIAs) is available in Topics 7.1.14–7.1.16 (pages 627 and 629).

(HL) 8.2.11 Ecological urban planning will follow principles of urban compactness, mixed land use, and social mix practice

The aim of ecological urban planning is to create sustainable, inclusive and resilient urban environments that benefit both people and the natural environment. This can be achieved by following a number of principles and practices that enhance the connection between humans and their environment. Careful planning can create areas with a design to minimize urban sprawl through the use of dense developments that are designed to help preserve natural habitats and green spaces outside the city. This reduces the fragmentation of ecosystems around the city and promotes biodiversity conservation. This can be achieved by building vertically and integrating essential commercial services and green space into the compact area, bringing essential goods and services closer to residents and reducing the need for transportation.

Further to this, the development of green public transport networks can reduce the number of vehicles on the road and the reliance on fossil fuels. Social connectivity is also enhanced in these areas, particularly if common recreational green spaces are part of the design. These can help to improve mental and physical health, as well as improving the local air quality. Sustainable water management and green infrastructures also help improve the overall resilience of the area.

Energy usage in urban areas is a major problem that not only causes localized pollution, but also uses unsustainable amounts of natural resources. Shifting towards the use of renewable energy alongside planning to reduce the absorption and trapping of heat can reduce the need for energy for cooling and minimize the urban heat island effect. Simple measures that change the albedo of the cityscape can reduce the urban heat island effect. Green spaces help with this, but some cities have gone further and have been adopting light-coloured roads, car parks and roofs to reflect more heat away from the area. New paint technology has produced super white coatings that have been reported to have decreased the temperature in some Californian cities by 10–15°C and could help eliminate the need for such extensive use of energy to cool our homes. One city in the USA, Culdesac in Phoenix, has gone one step further and is a city with no parking, designed for residents to no longer need a personal vehicle. Easy access to reliable transport away from the town is provided, but within the town everything is within walking distance. This dramatically reduces the amount of land needed to develop this town, as well as reducing congestion and air pollution, and minimizing the land use conversion that would be required to house the same number of residents, as well as parking.

Employing a combination of these elements offers a holistic approach to urban development, balancing the needs of people, the environment, and economic growth that can lead to more sustainable, vibrant and liveable citie.

HL.a.2: Environmental law

All construction and development projects in some countries are subject to strict Environmental Impact Assessments (EIAs) that carefully consider the impact of all stages of development, clean-up and long-term management of the environment. These are legally binding and can use legislation relating to protected species or habitats to halt construction and force it to move locations. However, bribery and corruption are common in some countries so the reliability of these is not guaranteed.

ATL ACTIVITY

Consider if this might be a factor in companies moving their factories to other countries where Environmental Impact Assessments and regulations are not as tight.

HL.c.1–13: Environmental ethics

As part of the environmental system we, as humans, should be more aware of the space that we are converting to urbanized structures and systems. As we continue to expand, we are systematically removing natural habitats to make space for urban environments. However, what is the impact of the loss of habitat or species and are these actions disregarding the intrinsic rights of all beings? In the past, ignorance was a valid excuse for the damage we did, but now we know the impact and can trace it directly back to our actions.

> **ATL ACTIVITY**
>
> Search for any local examples of habitat destruction. Consider taking action through raising awareness and supporting the protection of a habitat. Join a local conservation organization and volunteer some time to help raise money and awareness, as well as considering helping with habitat restoration and management.

8.2.12 Societies are developing systems that address urban sustainability by using models such as the circular economy or doughnut economics to promote sustainability

As urban sustainability gains momentum, planners and governments are looking to concepts such as the circular economy and models like doughnut economics. Both have similar features but address the issues in different ways. In reality, implementing circular economy practices could be used to address the shortfalls in social foundations.

The circular economy focuses on eliminating waste, creating circular systems that reduce the need for new resources and waste management, and regenerating natural systems. The system moves away from our usual pattern of taking, using and wasting resources through consumption, to a system that is more mindful of our resource needs and use, prioritizing the reuse, refurbishment and repurposing of goods and materials to eliminate waste from the cycle (see Figure 8.13). This works in harmony with the regeneration of our natural systems and harnessing renewable energy sources to move towards carbon-free energy.

Link

More information about the development of sustainability and the new models that are currently being developed is available in Chapters 1.2 and 1.3 (pages 24 and 47).

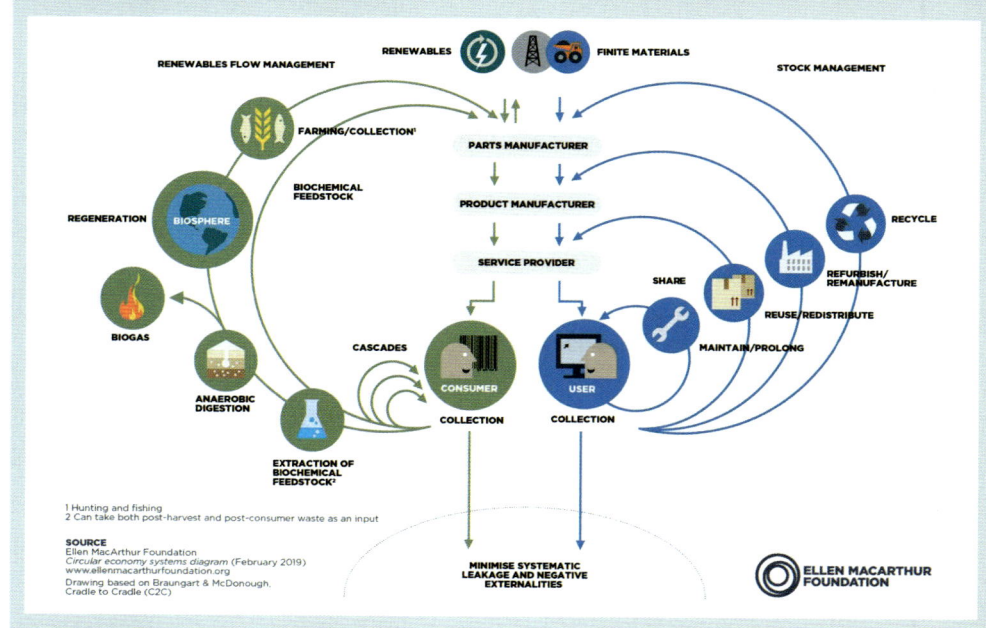

■ Figure 8.13 Diagrammatic representation of the circular economy model

Implementing a circular economy system within an urban area can be achieved in numerous ways, including:
- Incorporating aspects of green architecture to increase the biodiversity of the urban ecosystem.
- Increasing green spaces to provide a service for the urban residents and also to help improve the air quality in the local area.
- Creating sharing networks for goods that are not used frequently, such as tools or a vehicle. This reduces costs for residents, creates community networks and provides employment opportunities.
- Rideshare, food delivery, car rental and accommodation-letting apps have all been designed to create a society where there is no need for individuals to own expensive items themselves, instead opting to rent those goods or services when they are needed. This reduces the need for resources to provide each household with the goods solely for personal use, and reduces the eventual need to assimilate them as waste.

REAL-WORLD EXAMPLE

Schoonschip, Amsterdam and Kamikatsu, Japan

Amsterdam has created a circular, floating community (Schoonschip) within the city that is based around the principles of the circular economy. The project, which was completed in 2021, is home to around 100 residents who have created a close-knit community, where sharing goods, using renewable energy and encouraging the wild environment are all commonplace. Due to the fact that this community has been developed as a floating platform, it is also resistant to the potential impacts of sea level rise on Amsterdam's low-lying topography.

Kamikatsu in Japan is a community that has been committed to being zero waste for decades. Everything that is classified as waste in the city has to be sorted into one of the 38 different categories of waste. These are then all either reused, refurbished, repurposed, or used in another way that means they are no longer be classified as waste.

ATL ACTIVITY

Find out if you can create a system in your school where plastic that comes on to the site is ground into pellets and remelted into panels of hard plastic that can then be used to make furniture, clipboards and many other things. Precious Plastics gives free access to the blueprints of its machines so you can have them made for use at school.

This is a great service activity for all ages and can even provide income to allow you to create your own plastic-recycling enterprise.

Link
See Figure 8.13 (page 734), which represents the circular economy model.

TOK
To what extent are sharing-economy focused apps such as Uber really discouraging consumption?

Concept

Sustainability

Apps such as Grab and Uber have been branching out to make anything available to be delivered to your door, whenever you want. This increased convenience does not really discourage or reduce consumption.

Delivery providers are required to provide good packaging to stop spillage of food or possible tampering. This is not always recyclable, and it results in increased waste of packaging, a lot of which is made from ether plastics or Styrofoam and is therefore difficult to dispose of.

8.2 Urban systems and urban planning

HL.b.5: Environmental economics

Just because companies provide services that are linked to sharing goods or services, it does not mean that they are environmentally friendly or ethical employers. Greenwashing can persuade customers that a service is environmentally friendly and therefore that the company also is. This puts focus on practices that might be unsustainable. For instance, many delivery-based apps pay minimal wages, demand long hours and often penalize workers for late delivery or other issues, encouraging them to drive dangerously to ensure they are not charged for the late delivery. Claiming that a company is environmentally friendly can encourage more people to use its services. Lack of trust in these claims impacts the work of those organizations that are doing genuine environmental work.

HL.b.17: Environmental economics

The problem with the circular economy is that it relies on all stakeholders being responsible for their portion of the business in terms of sustainability practices. Due to greenwashing, this can be hard to take at face value. This often means that this is not investigated until someone pushes the agenda.

> **ATL ACTIVITY**
>
> Are there any major companies that used to greenwash and have positively changed their practices? You could start your search in the media to find reports of greenwashing.

HL.c.1: Environmental ethics

Greenwashing brings in the issue of ethics and what is right and wrong. It is hard to prove that the lack of taking responsibility is truly done on purpose. Understanding the complexities in the way the environment functions is not common knowledge, so it could be argued that organizations are not aware of the cascade of potential impacts of their practices. The environmental integrity of business practices simply isn't everyone's priority.

> **ATL ACTIVITY**
>
> Consider any organizations that have been identified as using poor environmental practices. How has this impacted their business and reputation? Does it stop people using them?

> **REAL-WORLD EXAMPLE**
>
> ### Prague, a circular city
>
> Prague is one of the most developed European cities in terms of circular practices and sustainability. The city has developed schemes for reducing compostable kitchen waste, as well as collection and redistribution of food waste for use as fertilizer or the generation of biogas, which is used to power the vehicles that collect the city's waste. In order to address the land use surrounding the city, all agriculture must be organic to reduce chemical pollution and provide local fresh produce for city residents. The city also provides options for renting bicycles or cars to encourage limited ownership and discourage overuse of personal vehicles. The success of this programme is reliant upon the residents taking these practices on board and changing their behaviours. The city addresses this with citywide and school programmes designed to educate residents about the importance of their part in the community.

> **Concept**
>
> ### Systems and models
>
> Modelling is hugely important in the planning of an urban system based around the circular economy. In order to ensure that the system will withstand the future growth and pressures of the systems, careful planning needs to ensure continual reduction of waste, providing sufficient 'clean' energy to support the needs of future generations. Using modelling allows planners to determine where stresses might be within the system and to test the limits of the model. For instance, the floating community in Amsterdam is designed to support a certain-sized population; if that threshold is surpassed then the community will cease to be truly sustainable. Modelling would have allowed the designers to work this into the development to ensure that the community can support the potential expansion of the families living there over the long term.

8.2.13 Green architecture minimizes the harmful effects of construction projects

When considering the development of urban areas, the materials that are used need to be carefully considered to ensure the durability and longevity of the structures. This has led to the development of many innovations designed to reuse waste materials to create new construction materials. This is being developed on all scales and levels of economics, from poorer communities using plastic bottles to create structures, to waste plastic being recycled to create 3D-printed buildings. This not only eliminates waste, but also means new resources are not needed for new construction projects. Some of the most innovative techniques have come from embracing local traditions, materials and knowledge alongside new technological innovation to create more sustainable urban environments.

■ Table 8.5 Examples of green architecture techniques

Bio-based 3D-printed homes made from wood residuals.	Wind tower (barajeel) buildings originating back to ancient Egypt and Persia.	Homes are made from plastic bottles filled with wet, compressed sand and other materials.
Relatively cheap, recyclable, durable. Very fast to construct.	The structure is designed to draw the heat out of the building while directing cooler air back in. This system can reduce the temperature inside a building by up to 10°C, reducing reliance on air conditioning.	This is a novel way of creating homes directly from waste and has been employed to provide housing for some of the most vulnerable communities.

Vernacular architecture harnesses local traditions, materials and knowledge alongside modern understanding of sustainable design. These building styles have evolved to fit the particular climate, location and availability of abundant resources. Vernacular architecture in Thailand is heavily influenced by the abundance, strength and versatility of bamboo, with buildings often being raised on stilts to reduce the impact of flash flooding, which is common in the wet season.

Through a combination of ancient knowledge and new technological innovations, there are numerous ways in which building design can improve the sustainability of a building, and the surrounding community. For instance, combining traditional design with photovoltaic roof panels or windows, rainwater capture and natural elements can improve building energy efficiency and reduce the overall environmental impact of urban areas.

HL.b.1: Environment economics

Creating structures from waste materials or natural local and replenishable materials makes them affordable for more people to create and repair their own home. Investigate a community that uses recycled materials to create structures. Consider what their main reasons for this are.

REVIEW QUESTIONS

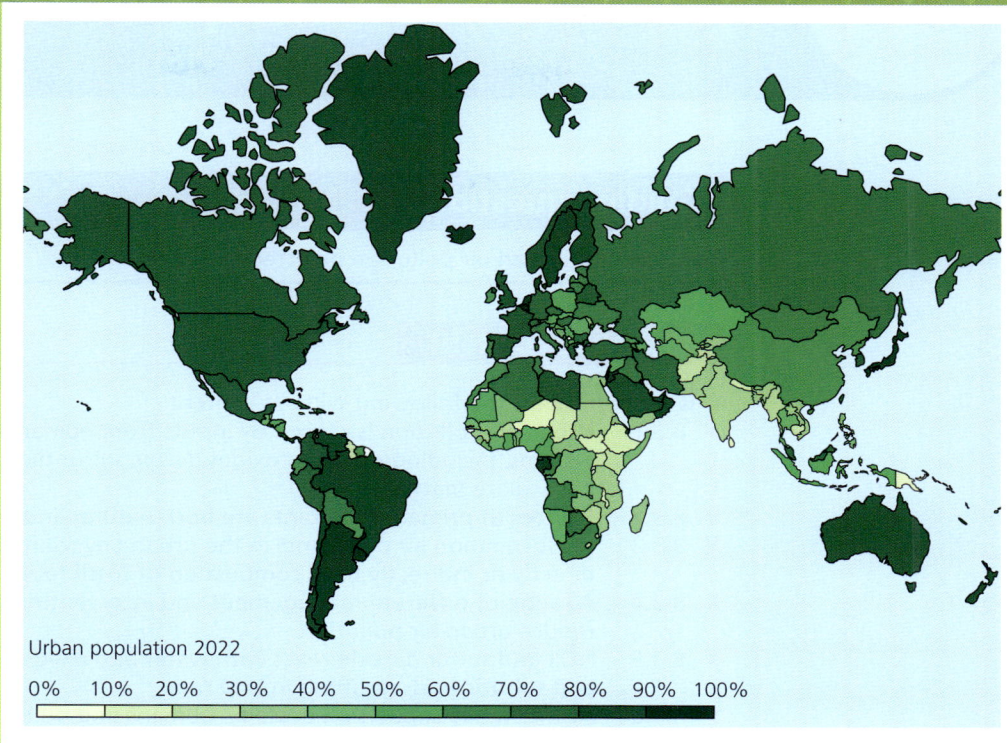

■ **Figure 8.14** The percentage urbanization for each country in 2022

1. Identify the continent with the lowest level of urbanization in most countries.
2. Compare and contrast the levels of urbanization in North and South America.
3. Identify a potential reason for low urban populations in India.
4. Suggest reasons for Australia having such low rural populations.
5. Name one push and one pull factor that result in migration from rural to urban environments.
6. Explain one way that the water management of a city can improve sustainability.
7. Name one way that established buildings can be made more sustainable.
8. List two biotic and two abiotic elements of an urban habitat.
9. Explain why people migrate from rural to urban areas.
10. Describe the impact of urban sprawl on natural environments.
11. Explain, using examples, how green architecture could improve our cities.

8.3 Urban air pollution

> **Guiding question**
> - How can urban air pollution be effectively managed?

SYLLABUS CONTENT

This chapter covers the following syllabus content:
- 8.3.1 Urban air pollution is caused by inputs from human activities to atmospheric systems, including nitrogen oxides (NOx), sulfur dioxide, carbon monoxide and particulate matter.
- 8.3.2 Sources of primary pollutants are both natural and anthropogenic.
- 8.3.3 Most common air pollutants in the urban environment are derived either directly or indirectly from combustion of fossil fuels.
- 8.3.4 A range of different management and intervention strategies can be used to reduce urban air pollution.
- 8.3.5 NOx and sulfur dioxide react with water and oxygen in the air to produce nitric and sulfuric acid, resulting in acid rain.
- 8.3.6 Acid rain has impacts on ecology, humans and buildings.
- 8.3.7 Management and intervention strategies are used to reduce the impact of sulfur dioxide and NOx on ecosystems, and to minimize their effects.

HL ONLY
- 8.3.8 Photochemical smog is formed when sunlight acts on primary pollutants, causing their chemical transformation into secondary pollutants.
- 8.3.9 Meteorological and topographical factors can intensify processes that cause photochemical smog formation.
- 8.3.10 Direct impacts of tropospheric ozone are both biological and physical.
- 8.3.11 Indirect impacts of tropospheric ozone include societal costs and lost economic output.

8.3.1 Urban air pollution is caused by inputs from human activities to atmospheric systems

> **Concept**
>
> **Systems and models**
>
> Like in any system, the atmospheric system has many inputs, outputs and processes involved. Urban systems have large volumes of pollutants being added to the system. This has created secondary interactions, and further pollutants that are harmful to plants, animals and our environments. A systems model can show a simplification of how all of these interact to create the complex pollution issues commonly observed in urban areas.

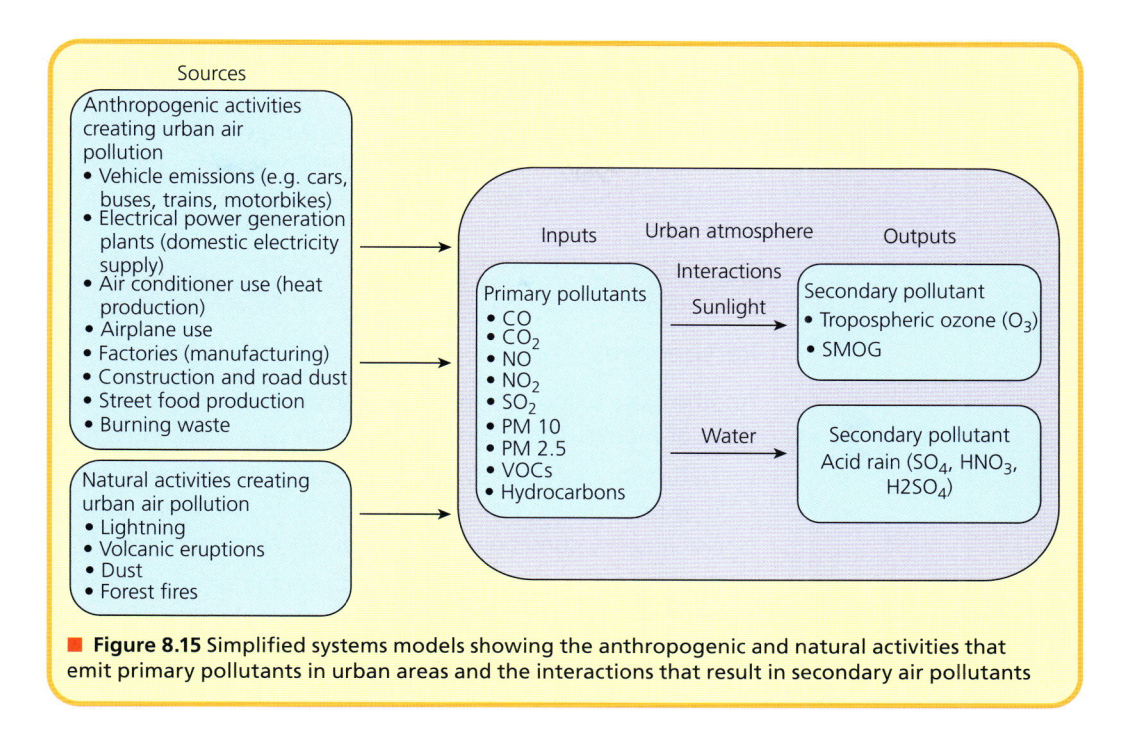

Figure 8.15 Simplified systems models showing the anthropogenic and natural activities that emit primary pollutants in urban areas and the interactions that result in secondary air pollutants

Sources

Anthropogenic activities creating urban air pollution
- Vehicle emissions (e.g. cars, buses, trains, motorbikes)
- Electrical power generation plants (domestic electricity supply)
- Air conditioner use (heat production)
- Airplane use
- Factories (manufacturing)
- Construction and road dust
- Street food production
- Burning waste

Natural activities creating urban air pollution
- Lightning
- Volcanic eruptions
- Dust
- Forest fires

Urban atmosphere

Inputs — Primary pollutants
- CO
- CO_2
- NO
- NO_2
- SO_2
- PM 10
- PM 2.5
- VOCs
- Hydrocarbons

Interactions: Sunlight → Outputs — Secondary pollutant
- Tropospheric ozone (O_3)
- SMOG

Water → Secondary pollutant
- Acid rain (SO_4, HNO_3, $H2SO_4$)

Air pollution in urban areas is a highly complex issue as the specific causes and their relative contributions to urban air pollution can vary depending on the city, its geographical location, level of industrialization, transportation patterns and regulatory measures in place. Urban environments are not only subjected to widespread atmospheric issues, such as the depletion of the ozone layer, but they are also subject to many localized issues that are exacerbated by the built environment and volume of activities in cities. There are some air pollutants that are immediately active upon release, while others develop over time as interactions in the atmosphere take place. Nitrogen oxides (NOx), carbon monoxide (CO) and sulfur dioxide (SO_2) gases are common air pollutants in urban areas. Particulate matter (PM) is also produced in abundance in these areas.

8.3.2 Sources of primary pollutants are both natural and anthropogenic

While many primary air pollutants present in the urban environment are created anthropogenically, there are still a number that are naturally produced and further worsen the air pollution situation (see Figure 8.15). Some pollutants, such as particulate matter (PM10 or PM2.5), are produced both naturally and through human activities. Natural sources of PM10 and PM2.5 come from natural dust created when soil is left uncovered and lacking in water. This dust then gets picked up by winds and becomes part of the air, and this is also the case with large desert storms. Humans create PM10 and PM2.5 from construction, industry, roads, burning fossil fuels, vehicle brakes and tyres, and the open burning of waste.

◆ **Primary pollutant** – a pollutant that is active as soon as it is emitted.

Other natural sources of **primary pollutants** are forest fires and volcanic eruptions, although the occurrence and regularity of these is highly variable and dependent on location. Forest fires and other wildfires produce a range of primary pollutants such as smoke, PM, CO, NOx and volatile organic compounds; volcanic eruptions create SO_2, volcanic ash and volcanic gases such as CO and CO_2. Anthropogenic primary pollutants include NO_2, SO_2, NOx, PM, CO and CO_2 generated by the wide range of activities conducted throughout the city, such as transportation and construction.

8.3 Urban air pollution

● Common mistake

People are often confused by the difference between ozone in the troposphere and in the stratosphere. In the stratosphere, the ozone layer protects us from harmful UV rays; in the troposphere (where we live), it is regarded as pollution.

Concept

Perspectives

Another way to look at primary urban air pollutants is to categorize them as natural, static or mobile. This way the impact can be assessed in a more dynamic way.

- Natural primary pollutants only occur under certain conditions, for example dust will only become an issue when it is dry and windy, natural forest fires tend to occur in hot and dry conditions, and volcanoes erupt infrequently.
- Static anthropogenic pollutants stay in the same place and mostly create similar levels of pollution at most times. These include heat from air conditioners and heaters, industrial pollutants, power stations, home cooking and heating.
- Mobile pollutants can move and can cause spikes in waste. Emissions from transportation tend to be worse during the morning and evening periods when many people are commuting to or from work or school.

It is likely that the level of pollution will be worse in developing countries. Many of these countries have poorly constructed housing and rely on direct burning of fossil fuels in the house for cooking and heating. This will increase the level of exposure to pollutants produced in the home. Vehicle standards may also be lower, with limited testing of emissions or implementation of regulations. Combine these factors with the already lower standard of living and healthcare, and the human populations in these communities are more greatly impacted.

Concept

Perspectives

Direct, real-time data on air quality are readily available and give a snapshot of what is happening worldwide. Longer-term trends and impacts can be determined through an understanding of species of lichens and their distribution in specific areas. Lichens are symbiotic mutualistic organisms made up of algae and fungi. The algae provide food for the fungi and in return the fungi provide protection for the algae. As a result, lichens do not have roots. While this means they can attach to many different surfaces, it also means that they derive all their nutrients and water from the air. Lichens are sensitive to nitrogen (N) and sulfur (S) pollution, and each group of lichens has different tolerances to levels of these elements in the atmosphere.

There are three main groups of lichens. Each group has different tolerances to pollution, with the leaf foliose species having the lowest tolerance to pollution and only being found in areas of clean air. Crustose species have the highest tolerance to pollution, and some species even thrive in polluted environments. Therefore, changes in air quality result in changes in lichen community diversity and sizes. These changes take place over time, so they give a longer perspective on air quality.

(Source: www.nps.gov/articles/lichens-as-bioindicators.htm)

Link

More information about the complex relationships and interactions between different organisms within ecosystems is available in Topic 2.1.9 (page 91).

Tool 2: Technology

AQI/PM readers

There are many handheld devices available on the market that record either the air quality index (AQI) or particulate matter (PM). Their accuracy is variable, but as long as the type and accuracy are noted down and the same device is used throughout the experiment, then this does not matter, as each value will be recorded to the same degree of accuracy. These measurements will provide real-time data that can be compared between different times of the year or day. Many cities around the world collect real-time data on air quality indicators, so secondary data can easily be sourced for this. However, when using secondary data remember the factors such as weather conditions that will significantly impact pollution levels between different countries on the same day.

Tool 1: Experimental techniques

Lichens as bioindicators of air pollution

Field data collection is critical when looking at lichen species diversity and distribution as **bioindicators**. Lichens can often be found on tree trunks. In urban areas, graveyards can be a good place to collect data, but permission must be obtained before carrying this out.

Data collection could be conducted in two similar habitats in different locations, one inside a city and one outside. This would allow direct comparisons to be made between a rural and an urban environment. Replication in each location would need to be around 20–30 data points per location. This could represent assessing 30 trees or 30 gravestones in each location. Data can be collected from a selected portion of the substrate being tested. For example, samples can be taken between x and y centimetres up the trunk. Number, size and type of lichen can be measured.

Limitations

It can be hard to find reliable sources for identifying lichens, so it is difficult to know exactly which group they fit into and so whether they are pollution-tolerant or not. Additionally, similar tree species of similar sizes need to be sampled to allow direct comparisons to be made.

◆ **Bioindicators** – species that are able to indirectly indicate certain conditions, usually in relation to pollution. Lichens are bioindicators for air pollution.

Tool 2: Technology

Map lichen distribution to measure pollution in Dortmund, Germany

Mapping software has been used by scientists in Dortmund, Germany to create choropleth maps of the distribution of lichens in the city. The lichens for the entire city have been mapped three times, in 1989, 1997 and 2006. The distribution of different

Clean air is made of mostly oxygen (O₂) and nitrogen (N₂). There's four times as much nitrogen as oxygen.

But nitrogen oxides (NOₓ) and volatile organic compounds (VOCs) get into air as fuel burns.

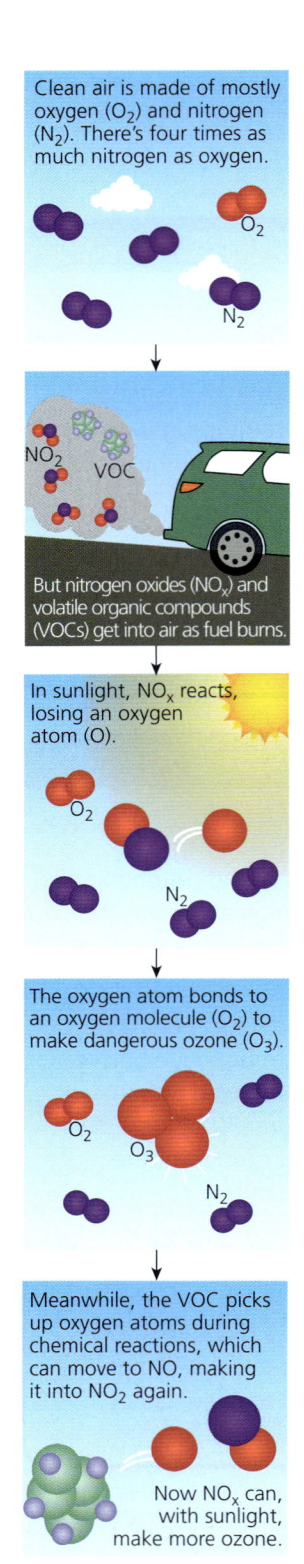

In sunlight, NOₓ reacts, losing an oxygen atom (O).

The oxygen atom bonds to an oxygen molecule (O₂) to make dangerous ozone (O₃).

Meanwhile, the VOC picks up oxygen atoms during chemical reactions, which can move to NO, making it into NO₂ again.

Now NOₓ can, with sunlight, make more ozone.

■ **Figure 8.17** Diagrammatic representation of the formation of ground level (tropospheric) ozone

types was then applied to maps of the city. This has allowed a clear indication of how the air quality has changed over time. The map for 1989 (see Figure 8.16) shows some considerable hotspots of pollution, based on the lichens found.

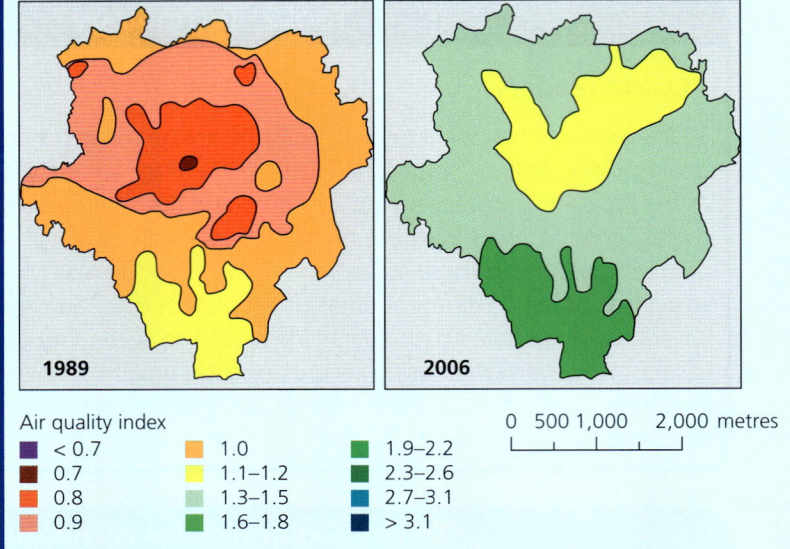

Air quality index
- < 0.7
- 0.7
- 0.8
- 0.9
- 1.0
- 1.1–1.2
- 1.3–1.5
- 1.6–1.8
- 1.9–2.2
- 2.3–2.6
- 2.7–3.1
- > 3.1

■ **Figure 8.16** Distribution of urban air pollutants in Dortmund, Germany in 1989 and 2006; distribution is determined by mapping the distribution of lichen species

Limitations

Only some pollutants can be assessed using lichens and therefore to create a full picture the use of indicator species should be paired with the collection of real-time data.

8.3.3 Most common air pollutants in the urban environment are derived either directly or indirectly from combustion of fossil fuels

Fossil fuels such as coal, oil and natural gas are used in urban environments for power generation, domestic cooking and heating, and transportation and industrial processes. Transportation accounts for a large proportion of these pollutants due to the volume of public transport systems, personal transportation and business transport that use fossil fuels and produce emissions.

When fossil fuels are burned they create a number of different pollutants. Volatile organic compounds (VOCs) and carbon monoxide (CO) are both released when combustion of fossil fuel is incomplete. VOCs include compounds such as benzenes and formaldehyde, and they contribute to the formation of tropospheric (ground-level) ozone. Complete combustion of coal and oil releases the sulfur and nitrogen present in these fuels, which then bond with oxygen to create nitrogen oxides (NOx) and sulfur dioxide (SO_2) (see Figure 8.17).

Particulate matter (PM) is also created when these fuels combust, in the form of either liquid or solid particles of substances like soot, metals, and particles of nitrates and sulphates. Larger particles, PM10, can be inhaled and will be deposited in the upper parts of the lung surface due to their size, where they can cause tissue damage and inflammation. PM2.5 are referred to as fine particles and these are able to reach much deeper into the lungs. Therefore, regardless of the source of PM, it is detrimental to health.

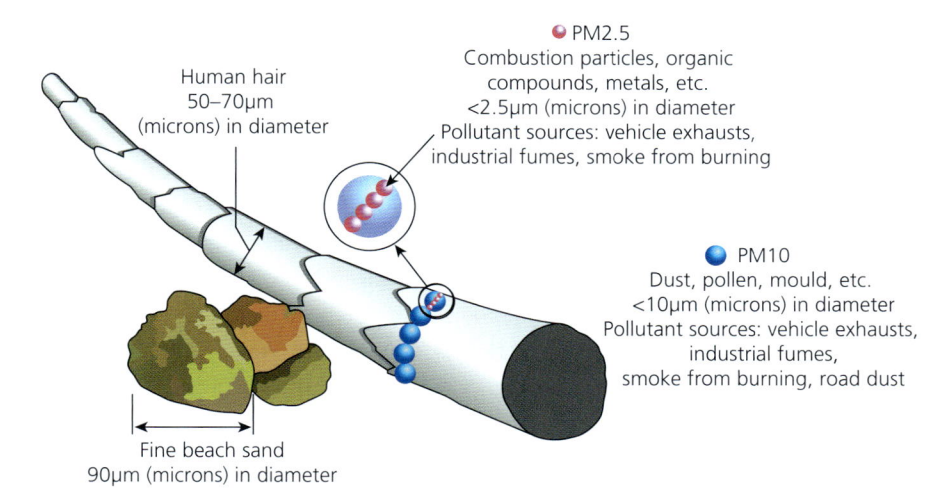

■ **Figure 8.18** Diagram showing particulate matter size comparisons for PM10 and PM2.5

8.3.4 A range of different management and intervention strategies can be used to reduce urban air pollution

As the rapid growth of cities continues and the release of GHG's continues to be of concern, countries are developing innovative strategies to address the issue of urban air pollution. Strategies that combine the adoption of public transportation that uses clean fuels and electric vehicles, alongside the implementation of pedestrianized central areas, are able to reduce air pollution in urban areas.

REAL-WORLD EXAMPLE

Amsterdam: bicycle-friendly city

Amsterdam has been the most bicycle-friendly city in the world since the 1960s, when increased traffic volumes, speeds and pedestrian deaths led residents to campaign against the development of the road systems in the city. In the 1970s oil prices increased and the government implemented occasional car-free days in the city, and by the 1980s the Dutch government began to develop comprehensive cycleways throughout Holland. Amsterdam's network of cycle paths are well-lit, clearly signposted and separated from pedestrians and other road users to provide safe alternative transport options that are used by all parts of the community, from toddlers to the elderly (see Figure 8.19).

■ **Figure 8.19** Separated bicycle and pedestrian pathway in Amsterdam, providing safe transportation alternatives in the city

Incorporating parks, trees and plant abundance in cities significantly improves air quality due to the natural process of photosynthesis and the conversion of CO_2 into O_2. Plants can be incorporated into cities in a variety of ways including planting the central strip of roads, planting trees, green roofs, green walls and productive areas such as urban farms (see Figure 8.20). Green walls are walls of plants that are attached to either the inside or outside of a building. Hydroponics or soil systems can be used to grow these walls, which have their own irrigation system. One of the first examples of green-wall technology was in Paris in 2004. Singapore, Dusseldorf (Germany) and Mexico have all got notable examples of this technology being used on various scales. Green roofs have also been implemented worldwide, with some US cities such as Philadelphia, New York City, Chicago and Washington D.C. offering financial incentives to encourage more buildings to develop green-roof technology.

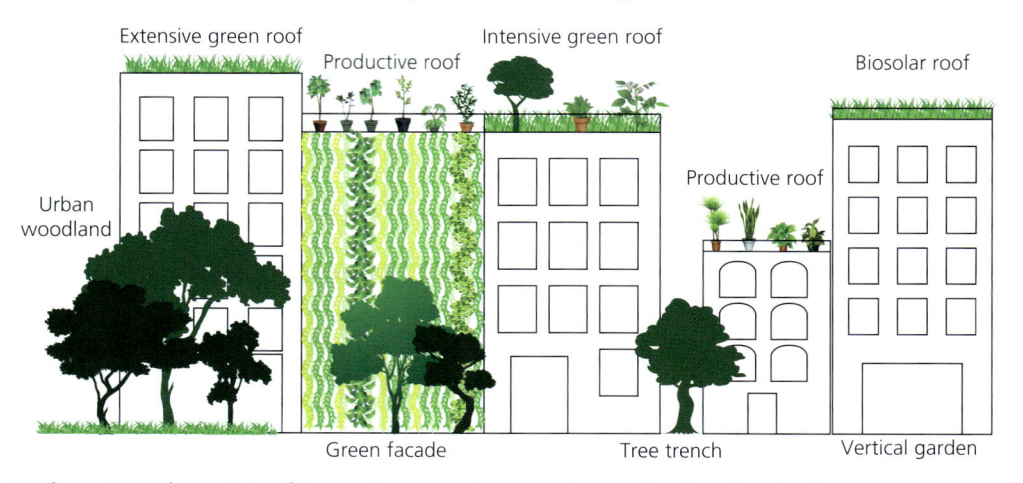

■ **Figure 8.20** The variety of innovative ways plants can be used in the urban landscape

ATL ACTIVITY

Select a major city close to your school and investigate the ways the city is reducing air pollution. Take a look at transport strategies that discourage personal vehicles in the city, city centre planning to reduce vehicles in specific areas, and any innovative ways of incorporating plants into the environment. This activity can be completed as a class project with different groups selecting major cities in different countries to also investigate the diversity of strategies around the world. Present your findings to the rest of the class.

8.3.5 NOx and sulfur dioxide react with water and oxygen in the air to produce nitric and sulfuric acid, resulting in acid rain

Some of the primary pollutants released in the urban environment are the precursors of acid rain. Nitrogen oxides and sulfur dioxide are released and combine with water vapour in the atmosphere to form nitric acid and sulfuric acid solutions. These then return to the surface as acid deposition in the form of either wet or dry deposition.

■ **Table 8.6** The chemical equations for the formation of acid deposition

Formation of nitric acid	Formation of sulfuric acid
$NO_2 + H_2O \rightarrow HNO_3$	$SO_2 + H_2O + O_2 \rightarrow H_2SO_4$

Theme 8: Human populations and urban systems

8.3.6 Acid rain has impacts on ecology, humans and buildings

Acid rain is any kind of precipitation that has a pH lower than the normal rainwater pH range of around between 5.6 and 6.5, which is slightly acidic. The addition of nitric and/or sulfuric acid from primary pollution interactions can lead to an increase in acidity up to a pH of around 4.3 (see Figure 8.21). Acid rain has significant impacts on many different parts of the biotic and abiotic environment.

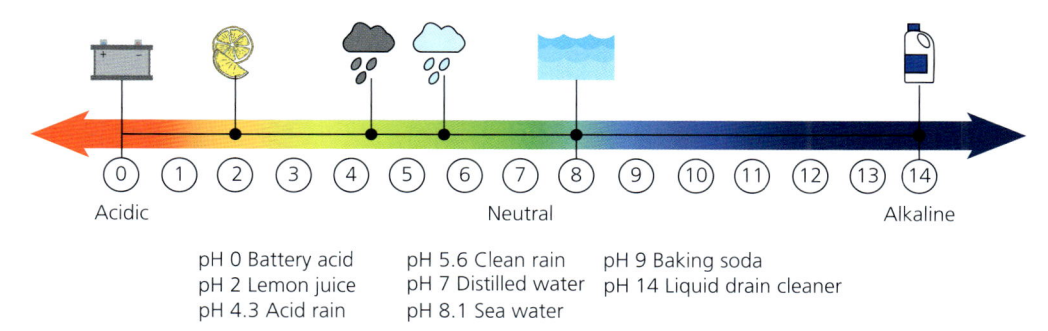

pH 0 Battery acid
pH 2 Lemon juice
pH 4.3 Acid rain
pH 5.6 Clean rain
pH 7 Distilled water
pH 8.1 Sea water
pH 9 Baking soda
pH 14 Liquid drain cleaner

■ **Figure 8.21** The position of 'normal' rain and acid rain of the pH scale

The ecological impacts of acid rain on the aquatic system, the soil system, and the flora and fauna are varied and complex as different plants and animals have different tolerances to levels of acidity in the environment (see Theme 4, page 267). Acid rain penetrates soil and is consequently taken up by plants. Beyond the tolerance range of each plant species, it will begin to produce lower-quality yields, have limited growth, and eventually die if the pH remains far outside its tolerance (see Figure 8.22). Seed germination can also be impacted by acid rain, resulting in limited plant growth in areas of high acidity. Further to this, acid rain impacts the soils as it binds with or dissolves essential soil minerals such as nitrogen and phosphorus, removing them from the soil through the process of leaching. Nitrogen and phosphorus are essential to plant growth and a loss of these nutrients will dramatically stunt the growth of any plants. Some plants are highly sensitive to acidic soil, such as species of evergreen trees that shed their needle leaves, dramatically impacting the whole ecosystem.

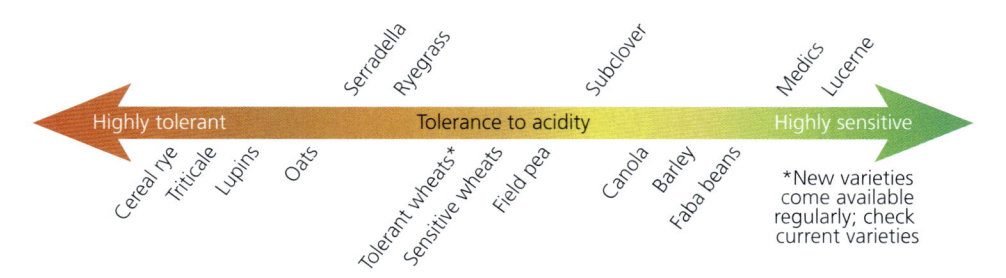

■ **Figure 8.22** The relative tolerance of different crops in Australia to acidity

> ## Tool 1: Experimental techniques
>
> ### Laboratory work
>
> Acid rain can easily be simulated in a laboratory setting and it can be used to determine impacts on seed germination and plant growth. Things to consider before conducting the experiment:
>
> - Replication: To be fully replicated for an ESS Internal Assessment, you ideally want to have five independent treatments and these should each be repeated at least five times. For a pH experiment looking at the impact of different levels on seed germination, select five different pH solutions that are within and outside the tolerance range for the seeds being investigated.
>
> - Justification for techniques and strategies: You need to say why you have chosen the methods you are using in your experiment. For a seed germination experiment, you would need to justify why you have chosen specific pH values, the type of seed(s) used, the length of the exposure to the pH solution, and the length of time for observations.
>
> - Appropriate quantitative measurements: This will generate the data that you will later process in some way to answer your specific research question. In the case of germination, only recording the number germinated will lead to very limited data. Further information such as root length could also be measured.
>
> - Appropriate procedure: This is not a new experiment so it is fine to search for relevant techniques, as long as you credit the source and justify why it was chosen.
>
> - Safety and ethics: All experimental work comes with some risks, and these need to be addressed and relevant precautions need to be taken. The ethics of actions also need to be considered for each experiment. For a seed germination experiment, there are risks associated with using different acid solutions and regular laboratory working safety must be followed. Ethically, the impact of disposal of the solutions, plants and soils after the experiment needs to be considered to eliminate any harm.

■ Figure 8.23 The impact of acid rain on ancient stonework

The deposition of acidic pollutants in the atmosphere may differ depending on specific conditions in the area. Areas with high levels of precipitation will experience wet deposition via rain, sleet or snow, whereas areas with low precipitation will experience dry deposition where the acidic particles are present as dust and gases. Humans are not directly impacted by wet deposition of acid pollution, but dry deposition of nitrate and sulphate particles is a component of PM2.5, and is therefore responsible for lung damage and respiratory issues. High levels of PM2.5 in the atmosphere can result in limited visibility, respiratory illness and eye irritation.

Building materials, such as marble, limestone and steel, are all impacted by acid rain, causing stone structures to dissolve and discolour, and metal structures to rust. Probably the most obvious impact is seen on old stone buildings in Europe and, famously, the Taj Mahal. The higher level of acidity in the precipitation slowly erodes and discolours buildings. While this is not always a structural issue, many of these are historic buildings and monuments.

> **Link**
>
> More information about the three different levels of pollution management applied in relation to water pollution is available in Topic 4.4.8 (page 393).

8.3.7 Management and intervention strategies are used to reduce the impact of sulfur dioxide and NOx on ecosystems and to minimize their effects

The most successful strategies are those that eliminate the production of the pollutant altogether, as these mean there is no need for costly clean-up. Some pollutants cannot currently be totally eliminated, and therefore interventions can only reduce the level of the pollutant that is released into the environment. The least favourable strategy is related to clean-up, as this takes place post release and can be costly and lengthy.

■ Changing behaviour

Many global, national and local government agencies are working to drive better and more reliable provision of renewable, non-carbon energy production so as to reduce the need to burn fossil fuels. The UN Sustainable Development Goals (SDGs) include targets to both directly and indirectly address these issues. For instance:

- SDG 7: Affordable and clean energy that directly focuses on the switch from fossil fuels to cleaner, renewable energy production.
- SDG 9: Industry, innovation and infrastructure.
- SDG 12: Responsible consumption and production.
- SDG 13: Climate action.

These goals have led to national governments developing considerable fossil fuel reduction targets within their urban habitats. These are focused on decarbonizing the public transport systems, combining on-site renewable energy and offshore agreements to power residential, commercial and business needs, developing waste-to-energy schemes, and much more. Some cities have set targets to become carbon neutral, with some like Sydney, Australia already achieving this status. These targets are being met by developing integrated energy and fuel alternative initiatives.

There are tax incentives in Sydney to buy electric vehicles, as well as exemption from paying certain road tolls and taxes, coupled with the provision of easy access to fast and reliable EV charging stations. This makes switching personal vehicles to electric ones a more attractive prospect for consumers. In 2019 the San Francisco, USA government developed a strategy, the EV Roadmap, to reach 100 per cent reduction in transportation emissions by 2040. This looks at developing a fully electric public transportation system, as well as developing a network of over 750 public EV charging stations, a number of which are free to use and others that are Tesla specific. This has significantly encouraged the population to invest in EV technology due to the ease of charging within the city.

8.3 Urban air pollution

> **ATL ACTIVITY**
>
> ### Engagement
>
> Create a challenge for your class, grade level or even school to make pledges to change their behaviour.
> - Individuals can promise to change behaviours to reduce consumption of electricity and fuel. This can be as simple as turning off the water when you are cleaning your teeth, to car-pooling with other families to reduce vehicle use.
> - Create infographics to be displayed in the school of the ways we can all contribute to reducing waste. Include shopping local, supporting ethical small businesses, and the importance of thinking before acting and being mindful of your daily choices.
> - Support this with messages in assemblies and community communications to get everyone involved. Furthermore, consider including sustainable initiatives such as banning single-use plastic, plastic straws, etc., from your school campus or events.

Reducing the release of pollutants

While the overall aim should be to eliminate fossil fuel burning, this will happen at different rates around the world, depending on the economy and level of development of each country. While in the majority of developed countries many of their targets are related to regeneration and sustainability in all areas, developing countries, particularly in parts of Africa, are still focusing on developing functioning infrastructures to support their growing populations.

In order to reduce the release of pollutants, there are a few innovations that can be implemented to break down primary pollutants prior to emission. The catalytic convertor (see Figure 8.24) is a device that is put into the exhaust system of a vehicle that still uses fossil fuels, in order to reduce NOx emissions from exhaust fumes. The exhaust fumes pass through the catalytic converter, which has an internal design that significantly increases the volume of exhaust that can be cleaned at any one time. This breaks down the harmful NOx, hydrocarbons and CO into O_2, CO_2, N_2 and H_2O, and reduces the impact on air pollution.

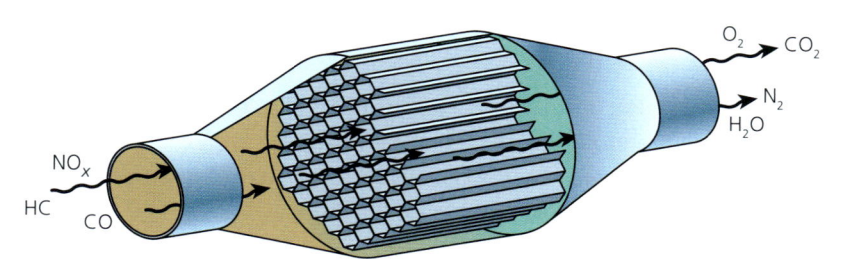

■ Figure 8.24 A catalytic converter

When newly added to an exhaust, a catalytic converter has been shown to be around 99 per cent effective at removing primary pollutants from exhaust emissions. Large industrial emissions can be dealt with in a similar way with the inclusion of scrubbers that remove particulate matter and gases from industrial exhaust and flue gas streams in the chimney stacks to clean the air prior to release. These are around 95 per cent effective at removing air pollutants prior to release. While highly effective, neither of these devices is mandatory and therefore these pollutants are still being released into the atmosphere.

■ Restoring damaged systems

The worst-case scenario for pollution management is the need for clean-up and post-pollution restoration of the habitat. However, even though there are many innovations to reduce emissions, acid rain still occurs and restoration strategies are still required. Soil restoration is relatively simple and economically viable, whereby the application of limestone to the soil effectively neutralizes the acid and returns the soil to a more neutral pH. Further additions of leached calcium help the plants to recover and begin to repair the damage from acid rain. This can be applied directly to soil in either powder or pellet form and is activated by rain. Pellets are the more effective option as they are less likely to be lost due to wind erosion. The addition of limestone is also an effective way of treating acidic water systems. Stone and metal damage to buildings can be minimized by coating the surface in a water-resistant solution to stop the acid from being able to dissolve or corrode the structure.

● ((HL) 8.3.8 Photochemical smog is formed when sunlight acts on primary pollutants, causing their chemical transformation into secondary pollutants

Primary pollutants are directly emitted into the atmosphere via vehicles, industrial processes and power plants. Nitrogen oxides (NOx) and volatile organic compounds (VOCs) are primary pollutants that undergo complex chemical reactions when they interact with sunlight. The chemical reactions triggered by sunlight result in the formation of **secondary pollutants**, including tropospheric or ground-level ozone (O_3), peroxyacyl nitrates (PANs), and secondary organic aerosols (SOAs). The accumulation of these secondary pollutants, particularly ozone, along with the existing primary pollutants, leads to the formation of photochemical smog.

◆ **Secondary pollutant** – pollutants that form within the atmosphere as a result of chemical reactions taking place in the atmosphere.

> **Link**
>
> See Chapter 6.3 (page 562) for more detail on the legislation around general air pollution, and Chapter 6.4 (page 594) for links to stratospheric ozone depletion. This will give you a clear overview of all aspects of air pollution.

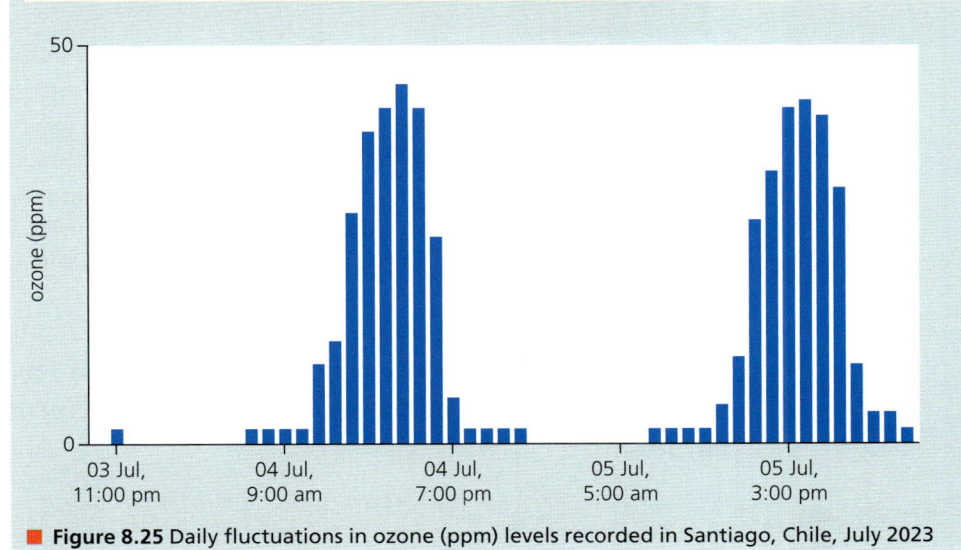

■ **Figure 8.25** Daily fluctuations in ozone (ppm) levels recorded in Santiago, Chile, July 2023

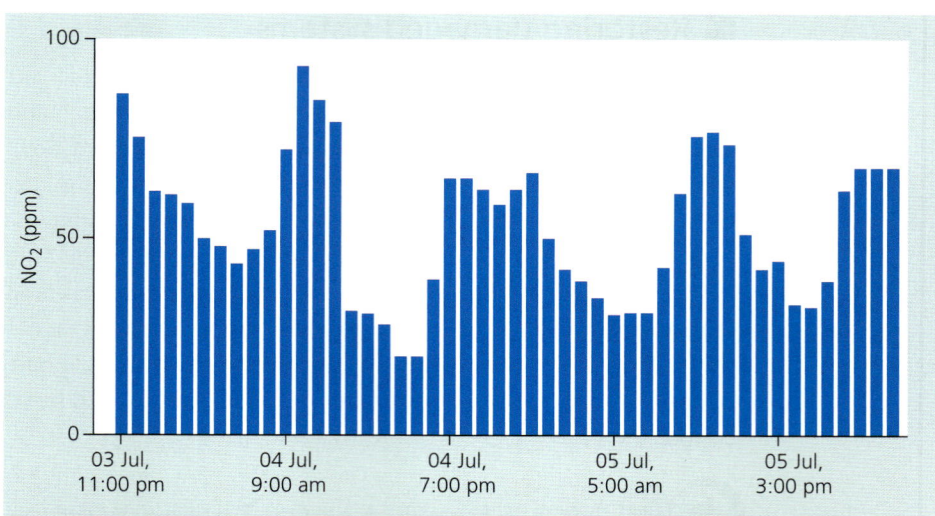

■ Figure 8.26 Daily fluctuations in NO_2 (ppm) recorded in Santiago, Chile, July 2023

> ### Concept
>
> **Systems and models**
>
> Mathematical models and simulations are frequently used to predict potential pollution events in urban areas. Scientists have constructed models that can use the forecasted weather patterns to calculate when urban air pollution is likely to build up and create health concerns within cities. These models enable urban design to be mindful of the potential hotspots where pollution levels could build up. Pedestrian-only zones in potential hotspot areas can help to eliminate the build-up of the primary pollutants responsible for the eventual creation of photochemical smog.

> **ATL ACTIVITY**
>
> **Application**
>
> Try creating a city that addresses the needs of the community while ensuring there is enough green space to absorb the pollution created by the human population. Can it really be done?
>
> This can be a group activity where you are able to discuss what additions need to be made for the growing urban population, where it should be placed and why. Discuss choices with other groups and compare the final results (source: **https://games4sustainability.org/gamepedia/urbanclimatearchitect**).

8.3.9 Meteorological and topographical factors can intensify processes that cause photochemical smog formation

Photochemical smog is worse in some parts of the world than others, and at different times of the year. This is because there are numerous factors relating to the local weather conditions, and some topographical features that can worsen the build-up of photochemical smog. There are numerous cities around the world that experience a number of these contributing factors at the same time, resulting in regular major pollution events in some cities, such as Santiago, Chile, San Fransisco, USA, and Chiang Mai, Thailand.

Meteorological factors

Sunlight is the biggest climatic contributor to the development of photochemical smog. Therefore, the closer the location is to the equator, the more likely it is that smog will regularly build up within urban areas. These areas not only have more intense sunlight, but the day length can also be a factor in how long each day primary pollutants will be able to transform into secondary pollutants. Cities that are also at a high elevation will be exposed to great sunlight, and therefore have more potential to develop smog.

Air movement is another major factor in worsening the development of photochemical smog. Stagnant air, which is common in areas that experience monsoon rains, reduces the movement of the pollution created within the city, allowing primary pollutants to build up, allowing more time for conversion to secondary pollutants to take place. This can be made even worse in cities that are located within a topographical bowl, as the lack of air movement and the warmer air at the top of the valley or bowl will result in a nightly temperature inversion that effectively traps the air and pollutants in the low-lying areas. These factors once again result in the build-up of primary pollutants, especially overnight, providing the perfect conditions for the development of photochemical smog once the sunlight levels increase throughout the day.

Topographical factors

The local topography, as shown in the previous section, can also impact the severity of the build-up of photochemical smog. This also applies to the landscape of the city itself, as buildings that are clustered together and have more than a few storeys will create limits to air movement at ground level, where many of the primary pollutants are released. This can result in pollution that is particularly dangerous to human health, due to where it is located.

There are many cities in the world that frequently experience severe pollution events due to a combination of the worsening factors. Primary pollutants from vehicle emissions in cities build up most during the morning and afternoon rush hour, leading to their availability to transform into ozone as the day progresses. Santiago, Chile and Los Angeles, USA are two cities that commonly experience bad air pollution due to these topographical and meteorological factors (source: **https://data.footprintnetwork.org/#**).

Tool 1: Experimental techniques

Secondary data collection

There are many sites that will give you access to real-time and historic air quality data. Ensure that the site you use is an official site; many governments have their own air quality networks that can be accessed.

When accessing secondary data, ensure you know how the values are collected, for instance are they real-time, hourly average or some other timescale? This is important if comparisons between areas are going to be made. Remember to collect enough data to suitably replicate measurements to allow an analysis to be conducted. If daily patterns are being tested then ensure that enough days are used to provide sufficient data to allow the statistical analysis to be reliable. You will need around one month's worth of data to allow a real pattern to be presented.

Questionnaires

A survey will help to collect data about the impact of pollution on humans. Significant and regular pollution has impacts on human behaviour and activities. This can allow assumptions to be made regarding the potential economic impacts of the pollution.

> Questionnaires can determine if pollution causes people to migrate to avoid areas of pollution build-up, or any modifications are made to their behaviour during periods of poor air quality. Questions could also relate to access and use of personal and home protection, such as PM2.5 masks and air purifiers, to determine how the community deals with the pollution.

8.3.10 Direct impacts of tropospheric ozone are both biological and physical

The impacts of tropospheric ozone on the environment and human health are wide reaching. Human health is greatly impacted by exposure to tropospheric ozone, and the increased rates of respiratory illnesses in areas where this pollution is frequent. Those who are already susceptible to respiratory illnesses, such as those with asthma, may be impacted at a lower pollution level than individuals who do not have already-compromised systems. Lung capacity and function can be impaired by long-term exposure, and even short-term exposure can lead to coughing, shortness of breath, tiredness and eye irritation. The effects can lead to permanent, chronic lung damage and a weakened immune system.

Plants are also negatively affected by tropospheric ozone as they take in ozone through their stomata. This ozone disrupts plant tissues during respiration and results in damage to those cells, reducing photosynthetic potential and eventually growth. Leaves become discoloured, losing the ability to harness light and convert it into energy for the plant to grow. Over the long term, this will result in smaller individuals that do not produce fruit at the same rate is typical for that species. Therefore, tropospheric ozone can significantly lower crop yields, which can also reduce the plant's ability to repel pests.

Exposure to tropospheric ozone can damage natural and synthetic materials such as fabrics, elastic, rubber, fabric dyes and some types of paint. Items made with rubber become brittle and crack, elastic becomes brittle and loses its ability to stretch. Many dyes that are used in paints and fabrics also degrade and fade with tropospheric ozone exposure.

8.3.11 Indirect impacts of tropospheric ozone include societal costs and lost economic output

All of the direct impacts on humans, materials and plants have implications for society and the economy. Not only are there the direct costs related to replacing materials such as tyres, rubber seals and clothing once they have degraded, but there are also the indirect economic stresses relating to the lack of tourism in areas severely impacted by tropospheric ozone and photochemical smog. Tourism can be significantly impacted by urban air pollution, and visitor numbers dramatically drop during these pollution events. Records of hospitalizations for respiratory issues show increases in admittance during these events, putting strain on local facilities, insurers and the local government.

REAL-WORLD EXAMPLE

Chiang Mai, Thailand

The northern Thai city of Chiang Mai has been impacted by what locals refer to as 'Smoke Season', annually for many years, with dramatic increases in severity observed in recent years. The city is located in a bowl and has a cooler dry season that results in nightly temperature inversions. This is made worse by the lack of wind and a drop in the height of the mixing layer, creating a temperature inversion that traps pollution in the city. These factors create the perfect topographical and climatic conditions for the build-up of tropospheric ozone.

This time of the year is when local, regional and neighbouring rice and corn farmers burn the stubble from their crops. This is also the time when a mushroom considered as a delicacy can be collected in the extensive forests surrounding the city. Controlled burning and creation of fire breaks is employed to attempt to eliminate wildfires that can quickly get out of control.

These two elements, along with an increase in car ownership in Northern Thailand over the last 10 years, have created a perfect storm of events that has resulted in air quality index (AQI) levels reaching up to, and exceeding, 500 on occasion (the WHO upper limit of the No Risk category of AQI is 50).

During this time visibility is significantly impacted, sometimes grounding aircraft. Residents are advised to stay inside, wear AQI-filtering masks, seal and purify houses, and refrain from exercise. Many businesses have invested in purifying systems to keep their offices, shops and restaurants safe. However, during this time the level of tourism, which focuses on outdoor activities, significantly drops, causing businesses to close annually due to the decrease in customers during this time. Large proportions of the extensive digital nomad population who live in Chiang Mai leave in favour of the clean air of the south of Thailand.

Government initiatives are in place that ban burning during this time of year and impose heavy fines for those caught deliberately setting fires on their land or in the forests. However, due to the size of the region and limited easy access to some areas, few are prosecuted annually.

Significant to this issue is the fact that a portion of the air pollution during this period also comes from bordering countries, which do not have the same burning laws and bans in place. This is a common issue in a number of Southeast Asian countries, such as Malaysia, Myanmar, Thailand and Laos.

REVIEW QUESTIONS

1. List three primary urban air pollutants.
2. List two natural and two anthropogenic sources of urban air pollution.
3. Explain how pedestrianizing urban areas can reduce air pollution.
4. Explain how a catalytic converter eliminates most primary pollutants from vehicle exhausts.
5. Explain how topography can worsen photochemical smog in urban habitats.
6. Describe which human activity is responsible for acid rain.
7. State an impact of acid rain on plant growth.
8. What time of the day is ground-level ozone likely to be at its highest?

EXAM-STYLE QUESTIONS

1. Explain why changing behaviour is the best strategy to reduce the production of the primary pollutants responsible for the development of acid rain. [3 marks]
2. Explain ways that local communities can protect against tropospheric ozone. [6 marks]
3. Describe, with the use of equations, how acid rain forms. [3 marks]
4. Compare and contrast the development of new sustainable cities and the change of old cities into sustainable areas. [5 marks]

Glossary

Abiotic factor – an ecological property associated with the non-living part of an environment, such as temperature, pH and humidity.

Abstraction – the process of taking water out of surface or groundwater for human use.

Adaptation – changes, such as the efforts made to live with the effects of climate change.

Adaptive capacity – the ability of a region, community or system to adjust to the effects of climate change.

Adhesion – the tendency of molecules to stick to other substances. Water molecules adhere to many substances, such as cellulose.

Advection – the horizontal movement of air.

Aerobic – involving or using oxygen. When aerobic respiration occurs, carbon from the organic molecules is combined with oxygen, and carbon dioxide is produced.

Aerobic decomposition – the breakdown of dead organic matter by decomposers such as bacteria and fungi, using oxygen. The process usually produces substances containing oxygen, such as carbon dioxide, nitrates and phosphates.

Aggregate – a clump of soil particles, usually held together by organic matter, or humus.

Agriculture – the production of living organisms for human use, including cultivation of crops and rearing of livestock, mostly for food.

Agrochemical – any chemical substance that is used in farming, including fertilizers, pesticides, hormones and antibiotics.

Albedo – the level of light reflected away from a surface. Dark colours have a low albedo as most of the heat is absorbed, whereas light colours have a high albedo as there is more heat reflected than absorbed.

Algal bloom – the rapid growth of algae, including phytoplankton and macroalgae. The algae often form a dense layer covering the surface of the water, or make the water appear green.

Anaerobic – occurring without oxygen. When anaerobic respiration occurs, carbon from the organic molecules may be released as methane.

Anaerobic decomposition – the breakdown of dead organic matter by special anaerobic decomposers, including some bacteria and fungi. The process usually produces substances containing no oxygen, such as methane, ammonia and hydrogen sulphide.

Animal husbandry – the practice of farming and caring for animals.

Anoxia – the complete absence of dissolved oxygen in water, due to extreme oxygen depletion.

Anthropogenic – produced by, or originating from, human action. Organic matter, in general, refers to substances that come from living organisms. This might include dead parts of plants and animals, or substances released from their bodies, such as faeces or other waste products. Organic matter is usually made up of carbon-rich compounds such as carbohydrates, proteins or lipids.

Antifouling – a substance used to prevent biofouling, such as certain types of paint.

Anti-natalist – relating to a policy or practice that discourages people from having children.

Apex predator – the top-level predator in a food chain or ecosystem, not preyed upon by other organisms.

Aphotic zone – the water below 1,000 m, where no light is present. Sometimes called the midnight zone.

Aquaculture – the production of aquatic plants or animals in water, usually for food. It is sometimes called fish farming.

Aquifer – an area of rock or sediment that holds groundwater.

Arable – related to the growth of crop plants.

Artisanal fishing – small-scale, non-commercial fishing, typically using traditional fishing methods. Relatively small numbers of fish are usually caught, to be used in the home or to sell for household income.

Aspect – the direction an area of land, such as a slope or mountainside, faces. The aspect may be towards or away from the sun, wind or rain, for instance. This may affect the processes of weathering and erosion.

Atmosphere – the layer of gases that surrounds the Earth. It's composed of various gases like nitrogen, oxygen, carbon dioxide, and others, and contains variable amounts of water vapour.

Baseline study – the analysis of the existing environmental conditions at the site of the project.

Benthic – related to the bottom of the sea or other body of water. The term can be used to describe the habitat or the species that live there.

Bioaccumulation – the movement of a pollutant from the environment into living organisms, resulting in higher concentrations of the pollutant in organisms than in the environment. Also a process that takes place within an organism, involving the faster absorption and buildup of substance concentration in tissues compared to removal. This accumulation commonly happens through the consumption of contaminated food and through direct absorption from water, often occurring simultaneously.

Biocapacity – the ability of a biological area to provide on-going natural resources, while naturally absorbing the waste and pollution.

Biodegradable – substances that are capable of being degraded by microorganisms into smaller, simpler non-toxic molecules.

Biodiesel – a fuel produced by chemical modification of vegetable oil, which can be used as a substitute for diesel fuel.

Biodiversity – the variety and variability of living organisms.

Biofouling – the growth of living organisms such as algae, oysters and barnacles on surfaces underwater, such as the hulls or propellers of ships.

Biofuel – a fuel produced from living organisms, such as ethanol or biodiesel.

Bioindicators – species that are able to indirectly indicate certain conditions, usually in relation to pollution. Lichens are bioindicators for air pollution.

Biomagnification – the increase in concentration of pollutants in organisms at higher trophic levels, typically attributed to the persistence of pollutants that resist natural breakdown processes, leading to their faster transfer up the food chain compared to breakdown or elimination. The highest concentrations are found at the top of the food chain.

Biomass – the matter that makes up living organisms.

Biome shift – a change in the distribution of plant species causing a change in the biome.

Biophilic design – building design that aims to create seamless connection with nature through the direct and indirect use of nature, space and place in design.

Bioremediation – the use of living organisms, particularly bacteria, to break down or remove pollutants from the environment.

Biosphere – refers to the region of the Earth that encompasses all living organisms and their interactions with each other and their environment.

Biosphere integrity – the overall health and functioning of ecosystems and the biodiversity they support. It is a fundamental component of the Planetary Boundaries model.

Biotic factor – an ecological property associated with the living part of an environment, such as competition or predation.

Bioturbation – the disturbance or mixing of soils by living organisms, such as burrowing animals.

Body Mass Index (BMI) – a widely used, simple measure, which relates mass to height. It is calculated by dividing a person's mass (in kilograms) by their height (in metres), squared. Depending on their BMI, a person may be classified as underweight, healthy weight, overweight, or obese.

Buffer – a substance that resists changes in pH. When acid or alkali is added to a buffer, the pH remains relatively constant.

Bycatch – fish or other organisms caught unintentionally, and usually thrown back in the water. These may be species with no commercial value, endangered or protected species, or individuals that are too small.

Calcareous – describing a rock or soil that contains a high percentage of calcium carbonate.

Calcium carbonate – $CaCO_3$. A compound found in the shells of molluscs and crustaceans, and in rocks such as limestone.

Carbon dating – a technique used to date fossil records that looks at the amount that the Carbon-14 molecules have broken down.

Carbon sequestration – the removal and storage of carbon dioxide from the atmosphere by natural or artificial means.

Carbon sink – a process that takes carbon dioxide out of the atmosphere.

Carcinogen – a substance that causes cancer. Many pollutants are suspected of being carcinogenic, but it is very difficult to prove this.

Carrying capacity – the maximum population size that can be supported by a particular environment. If the population rises above the carrying capacity, resources will become depleted and the population will fall.

Cash crop – a crop that is grown, typically in large quantities, primarily for sale, often on the international market. Cash crops are intended to provide income, but not necessarily food. Where a country or region focuses on producing cash crops for export, it may need to import most of its food from elsewhere.

Cataract – the formation of an opaque layer on the surface of the lens in the eye due to exposure to UV radiation.

Cement clinker – the solid material that is produced in the manufacture of Portland cement.

Clay – soil made up mostly of tiny particles, less than 0.002 mm in diameter.

Climate – long-term weather patterns, including temperatures, winds and precipitation. Climate varies around the globe, and can change over time. Some soils may have experienced different climates over the course of their formation.

Climate diplomacy – the practice and process of creating the international climate change regime and ensuring its effective operation.

Coevolution – the evolution of two different species in response to changes in the other species.

Cohesion – the tendency of molecules to stick together.

Coloured dissolved organic matter – the optically measurable component of dissolved organic matter. It absorbs light in the blue and UV region of the spectrum.

Commercial farming – generally large-scale farming as a business, aimed at earning a profit.

Commercial fishing – fishing for profit, typically using large-scale, modern vessels and methods, and catching large numbers of fish.

Community – all the interacting populations of organisms living in the same area at the same time.

Compost – partially decomposed plant matter, which can be produced by collecting farm, garden or kitchen waste in a suitable container, and allowing it to decompose. Compost can be used as organic fertilizer.

Coral polyp – tiny, individual animals that coral colonies are made up of. Each polyp looks a little like a small sea anemone, with tentacles around its mouth.

Coral reef – a huge ridge or rock in the sea formed by corals, which serves as an ecosystem, providing habitats to many marine organisms.

Cornucopian view – the belief that considers that nature is there to be made use of by humanity.

Covalent bond – a chemical bond between two atoms, where the atoms share a pair of electrons.

Crop residue – the leftover remains of crops after a harvest. Typically parts of the crop plant remain on the land after the useful parts (e.g. fruits or grain) are collected.

Crude oil – petroleum as it is found in the Earth. It is a mixture of many different substances called hydrocarbons. Crude oil must be refined, or separated into its components, before it is useful.

Dead zone – an area of water where oxygen levels have been severely depleted, and few animals can survive. These often occur in enclosed parts of the ocean where pollutants such as nitrates and phosphates accumulate.

Decomposer – a microorganism that break down organic matter by secreting enzymes on to it, and absorbing the products of decomposition.

Deepwater formation – a process in which water in the ocean becomes denser and sinks to the ocean floor, creating deep ocean currents.

Denitrifying bacteria – bacteria found in soils that convert nitrate and nitrite ions to nitrous oxide and nitrogen gas. An excess of denitrifying bacteria reduces the nitrogen content of the soil, and increases the emission of nitrous oxide.

Density-dependent factors – factors that influence population growth, and that change as the population increases in size, such as stress, disease, predation and competition.

Deontological ethics – an approach to ethics that considers the rights of different people or things. In this system, an action may be ethically wrong if it infringes on the rights of others.

Depolymerization – breaks down polymers into monomers using chemical reactions.

Deposition – the accumulation of sediments in a particular place, after they have been transported from elsewhere. Deposition may occur in low-lying areas, at the bottoms of slopes, or in the lower course of a river.

Desertification – a form of soil degradation that may occur in dry climates. The loss of vegetation cover and organic matter, and the accumulation of salt, may result in the soil becoming desert-like, and being unable to support vegetation.

Detritivore – an animal that eats detritus.

Detritivorous food web – a food web that is based around the actions of detritivores, which feed on dead and decaying plant material and are responsible for the breakdown of nutrients and eventual incorporation into the soil.

Detritus – partially decomposed organic matter.

Disinfection – a process intended to kill microorganisms, and therefore prevent the spread of diseases.

Disposal cost – the amount of money required to remove unwanted materials or pollutants (such as CO_2) from the environment.

Dormant – in a state similar to sleep. Dormant seeds have very little biological activity going on inside them.

Duration of cropping – how long a particular piece of agricultural land is used for growing crops before it is either rotated to grow a different crop or left empty for a period of time. It is essentially the time period between planting and harvesting the crops in a specific field.

Dysphotic zone – the water from about 200 m to 1,000 m deep, where light is rapidly absorbed. The deeper the water, the less light is present. Sometimes called the twilight zone.

Ecological footprint per capita – the total ecological footprint of a region, divided by the population of that region.

Ecological overshoot – the point at which human demand exceeds the regenerative capacity of the natural ecosystems. Humanity's ecological footprint exceeds what the planet can regenerate.

Economic globalization – the widespread international movement of goods, capital, services, technology and information.

Economic status – an individual's income and occupation. For example, a person with a high income may have a different perspective on the importance of economic growth than someone with a lower income.

Economic water scarcity – limited availability of water because a society lacks the ability to store and transport water.

Ecosystem – an ecological unit that includes all the organisms living in a particular area (the community) and the physical and chemical components with which they interact.

Ecotourism – a type of tourism that is conscious of the environmental footprint of human populations and focused around the local environment.

Electronegativity – the tendency of an atom to draw the electrons in its covalent bonds towards itself. In a water molecule, oxygen is more electronegative than hydrogen.

Eluviation – the process of washing substances downwards, away from a soil layer, into a layer below.

Emergent properties – these are unpredictable outcomes from the interactions between different components within a system.

Endemic – a species that is only found in one geographical location and is not naturally found in any other area. These species are usually found in isolated areas.

Environmental consciousness – the characteristics or factors that determine the behaviour of humans towards the environment.

Environmental degradation – the deterioration of the environment due to overexploitation of resources, destruction of habitats and ecosystems, and accumulation of pollutants.

Environmental Impact Assessment – a tool used to assess the environmental, social and economic impact of a project before it is sanctioned.

Equilibrium – balance between inputs and outputs in a system. In an ecosystem, this could be the balance between the number of organisms and their needs.

Ethical beliefs – the moral principles and values that an individual holds. For example, a person with strong religious beliefs may have a different perspective on the morality of certain actions than someone without strong religious beliefs.

Euphotic zone – the water near the surface of a lake or the ocean and down to about 200 m deep, where light can easily penetrate. Sometimes called the sunlight zone.

Eutrophication – a process of excessive nutrient enrichment, particularly nitrogen and phosphorus, in water bodies, leading to rapid and dense growth of algae and other aquatic plants. This excessive growth can result in oxygen depletion, harming aquatic life and disrupting the ecological balance of the ecosystem.

Evolution – the change in heritable traits passed from generation to generation.

Ex situ **conservation** – conservation that is carried out outside the natural habitat, such as in a zoo, sanctuary, or botanical garden.

Exponential – an increase, or decrease, that changes more and more rapidly.

Externality – a cost or benefit for a business that is not included in the normal accounts of the business. External costs and benefits are felt by people or environments outside of the business.

Externalizing cost – the cost generated by the producers but carried by the society as a whole. For example, pollution of a water body by a factory dumping waste in it.

Exudate – a fluid secreted by a living organism.

Fallow – left uncultivated in order to recover fertility. In conventional farming, land may be left fallow for a time after it has been ploughed. In slash-and-burn farming, land must be left fallow for long enough for the ecosystem to recover. This may take decades.

Famine – extreme, widespread scarcity of food. Multiple factors may contribute to famine, including adverse weather or natural disasters, which may cause crop failure, and poverty, economic downturns, and war, which may limit a society's ability to import food.

Farmers' market – a market where local farmers can sell their produce directly to consumers.

Fertilizer – a substance added to the soil or water to promote the growth of plants such as crops. They often contain large amounts of useful minerals such as nitrates, phosphates and potassium.

Fishing effort – a measure of the level of fishing activity. It takes into account how many vessels are active, how much time they spend fishing, and what sort of fishing gear they use.

Flagship species – a species that is selected to be the symbol or image associated with a conservation programme. For instance, the giant panda is the flagship for the World Wildlife Fund (WWF).

Forage – living plant matter, such as grasses or leaves, which livestock can feed on.

Forestry – the management of forests for human use. This can include the cultivation of forests for timber or paper production, and the management of forests for conservation.

Fracking – a technique for extracting natural gas from shale rock formations, involving injecting high-pressure water, sand and chemicals underground.

Freshwater – water with very low salinity, usually found in rivers, lakes, glaciers and ice caps, and groundwater.

Furrow – a trench or trough created by ploughing a field. Furrows are the lower part of the soil, which is typically used for irrigation. The raised soil between furrows, where the crops are normally planted, are called ridges.

Gene – an individual stretch of DNA, usually found in the nucleus of an organism's cells. Genes influence the characteristics of the organism. Each gene typically influences one particular characteristic.

Genetic drift – a change in the gene pool of a small population.

Genetic engineering – a laboratory technique for extracting individual genes from one organism and inserting them into another organism. Also called genetic modification or genetic manipulation. When successful, the organism that receives the transferred gene will have the selected characteristic from the organism that provided the gene. The organism that receives the genes is described as genetically modified, or transgenic.

Genetic modification – a technique that takes DNA from one organism and inserts it into another the DNA of another species to change the characteristics of a living organism.

Geomorphology – the shape of the surface of the Earth, including mountains, valleys, plains, etc.

Glacial ice – ice found on land, near the poles, or on top of tall mountains. Glaciers slowly slide over the surface of the land, towards the sea.

Glaciation – the effects of glaciers on the rocks and soils. Glaciers flow slowly over the land, scouring the soil away from where it first forms, and depositing it elsewhere.

Global hectare – a unit that measures the average productivity of the world's biologically productive land and sea areas in a given year.

Global supply chain – all the steps and procedures involved in manufacturing and delivering a product when those steps take place in more than one country.

Goods – commodities that can be sold to gain profits.

Gravel – stony or rocky particles in soil, which are bigger than 2.0 mm in diameter.

Green infrastructure – eco-infrastructure/nature-based infrastructure involves using natural elements like plants and ecosystems to help with things like water management and environmental sustainability.

Greenwashing – a deceptive practice of conveying a false impression about how environmentally sound products or actions are.

Groundwater – water found in porous rocks underground.

Gyre – a giant circular oceanic surface current.

Habitat fragmentation – the breaking up of areas of continuous habitat into unconnected, smaller parts.

Hazardous waste – garbage that contains specific, dangerous substances or items, such as toxic, carcinogenic or explosive materials.

Heavy metal – a metal element with relatively high density. They are mostly found in the lower central part of the Periodic Table.

Herder – a pastoralist farmer who raises herds of livestock.

Holistic – describes how the parts of a system are connected and can only be fully understood by including reference to their part in the whole system.

Holocene epoch – the current geological epoch (era) in which we are living. It began after the Pleistocene epoch.

Humus – fully decomposed organic matter, made up of a sticky, dark substance, which often coats the mineral soil particles.

Hunger – in general, the need for food. When discussing the extent of world hunger, it refers to the percentage of people who are undernourished.

Hydrocarbons – substances made up mostly of carbon and hydrogen. These are the main constituents of all fossil fuels.

Hydrogen bond – a relatively strong force of attraction between two molecules, which happens when there is a hydrogen atom in one molecule, and a strongly electronegative atom, like oxygen, in the other molecule.

Hydrological cycle – also known as the water cycle, this is the movement, distribution and transformation of water on Earth. The continuous flow from low to high areas, and then back to low areas.

Hydrophilic – attracted to water. Hydrophilic substances are usually soluble in water.

Hydrophobic – repelled by, or not attracted to, water. Hydrophobic substances are usually insoluble in water.

Hydrosphere – the total sum of the Earth's water in all its forms, including water on the surface (in oceans, seas, lakes, rivers, and groundwater) and water vapour in the atmosphere.

Hyphae – thin, branching, thread-like structures of fungi.

Hypoxia – the condition of having a very low oxygen concentration in water due to oxygen depletion.

Illuviation – the process of accumulation of substances in a soil layer, which have been washed away from a layer above.

Impact – the change from the baseline situation caused by an action.

***In situ* conservation** – conservation that is carried out in the habitat of the species.

Industrial Revolution – defined as the shift or change from an agrarian economy to one that is dominated by machine manufacturing and industries.

Infiltration – the entry of water into the soil.

Inorganic matter – substances that do not come from living organisms. In the soil, inorganic matter mostly comes from the underlying bedrock, or from the air or water. It can include a wide range of substances, containing many different elements and compounds.

Insoluble – unable to dissolve in a particular solvent, such as water.

Integrated Water Resources Management – a process that promotes the development of water, land and other associated resources.

Intercropping – an agricultural practice in which different crops are planted in close proximity to one another, in order to make efficient use of the many resources in the soil, and to capitalize on mutually beneficial relationships between crops.

Internalizing cost – the cost of pollution being borne by the polluters/producers. For example, by implementing taxes, fees or emissions trading systems.

International Energy Security Risk Index – identifies the policies and other factors that contribute positively or negatively to international energy security.

Intertidal zone – the area of the shoreline that lies between the high tide and low tide marks.

Intrinsic value – the value that a thing has in and of itself. Intrinsic value does not depend on whether a thing is useful or beneficial to people.

Introgression – the transfer of genetic material from one species to the genetic pool of another one.

Invasive species – harmful, non-native species that are rapidly able to spread and outcompete local species.

Inversely proportional – if something is inversely proportional to something else, it increases as the other thing decreases, or vice versa.

Ion – a charged particle, which is formed when an atom or molecule gains or loses an electron.

Irrigation – the application of water to soil in order to help plants grow. This is usually done where there is not enough natural rainfall to supply plants.

Keeling curve – a graph that represents the concentration of carbon dioxide (CO_2) in the Earth's atmosphere since 1958, recorded at the Mauna Loa Observatory.

Keystone species – a species that has a disproportionately large impact on a habitat. For instance, the African elephant is a keystone species due to the impact of its presence on all other aspects of the environment.

Leachate – a liquid that washes away as water flows over something, usually into the soil.

Leaching – the process where nitrates in waterlogged soils are washed away and lost due to the movement of water through the soil.

LEDCs – low economically developed countries.

Legume – a member of the plant family Fabaceae, which possesses root nodules (swellings) containing symbiotic nitrogen-fixing bacteria. Legumes include peas and beans, as well as a variety of herbs and trees that are not known for producing food. Well-known, non-food examples include clover, acacia and poinciana trees, and the sensitive plant Mimosa pudica.

Ley – an area of land on which grasses are grown for a limited period of time.

Limiting factor – an environmental factor that determines the rate of a process, often because it is lacking (e.g. in dim light, the rate of photosynthesis is slow; when the light gets brighter, the rate increases).

Lithosphere – the solid outermost shell of the Earth, consisting of the crust and the uppermost part of the mantle.

Livestock – animals grown on farms, for food or other products such as wool.

Loam – soil made up of a mixture of sand, silt and clay, and containing a significant amount of organic matter.

Logging – the felling of trees to cut and prepare for timber.

Logistic growth – growth of a population that shows a typical S-shaped or sigmoid curve. Growth starts slow, then speeds up, becoming exponential, then slows down, approaching zero.

Glossary

Macronutrient – a substance that a plant needs in large amounts. Nitrogen, phosphorus and potassium (N, P and K) are macronutrients for all plants, and these are the main constituents of plant fertilizer.

Malnourished – poorly fed; eating a diet that does not provide the right nutrients for a healthy life. Malnourishment can include both undernourishment and overnourishment, as well as nutrient deficiency.

Mangroves – trees or shrubs that grow in coastal saline water.

Manure – faecal matter of animals, usually livestock, or humans, which can be used as an organic fertilizer.

Marginal land – land that is considered to be unsuitable for agriculture. The term is not well-defined, however, and can sometimes be used to justify converting land used for grazing or small-scale farming to other uses.

Market failure – the concept in economics by which the free market fails to allocate resources efficiently. In environmental economics, many cases of environmental harm can be described as market failures.

Maximum sustainable yield – the highest yield that it is possible to achieve without depleting the fish stock.

Methane clathrate – a solid, ice-like structure formed by methane hydrate, where a molecule of methane is enclosed within a cage-like structure formed of water molecules.

Methane hydrate – a compound formed by methane associated with water molecules.

Methanotroph – a microorganism that uses methane as a source of energy and carbon to carry out primary production. Methanotrophs convert methane into more complex, energy-rich organic molecules, which generally remain in the soil.

Micronutrient – a substance that a plant needs in relatively small quantities.

Microplastics – extremely small (less than 5 mm) particles of plastic.

Mitigation – actions that reduce the impacts of something, such as climate change.

Mollisol – a rich, soft, dark soil, typical of temperate grasslands, with high organic matter content in the A horizon, or topsoil. Mollisols often make fertile farmland.

Moral standing – deserving of, or entitled to, ethical consideration. If something has moral standing, we should consider the effect of our actions on it.

Mutualism – a relationship between two species in which both species benefit.

Mycorrhizae – a large group of fungi that have a strong mutualistic relationship with many plants, exchanging water and minerals for organic molecules, and possibly forming network-like connections between plants. The name mycorrhizae comes from words meaning fungus (myco) and root (rhiza).

Natural capital – the total value of the natural resources in a place (e.g. all the fish in the fishery) that can produce sustainable natural income.

Natural income – the yield obtained from the natural resources, and the natural increase in the value of natural capital over time, due to the natural growth of the resource (e.g. the reproduction of fish increases the size of the fish population).

Natural selection – the change in genetic composition of a species over time as a result of the natural pressures within that species.

Negative feedback – a regulatory mechanism in population management where an increase in a particular factor, such as population density, leads to a response that reduces or counteracts that increase, thereby maintaining balance in the system.

Nitrates – compounds containing the nitrate ion, NO_3^-. They are an essential nutrient for plants and algae, and often included in synthetic fertilizers.

Nitrifying bacteria – bacteria found in soils that convert ammonium ions to nitrite and nitrate ions.

Nitrogen-fixing bacteria – bacteria found in soils, and sometimes living in plants, which take nitrogen from the air and convert it into soluble compounds that add fertility to the soil. A lack of nitrogen-fixing bacteria means that soils lack a natural source of nitrogen.

Nitrogenous fertilizer – a fertilizer that is rich in nitrogen compounds, such as ammonium salts and nitrates.

Nitrous oxide – N_2O, a greenhouse gas emitted from soils. Use of nitrogenous fertilizers increases the emission of nitrous oxide from soil.

Nomadic – related to people who move from place to place as a way of life. Nomadic people do not have one settled, permanent home.

Non-governmental organization (NGO) – a non-profit organization that is independent of any government. Its main purpose is to address a social, environmental or political issue.

Nutrient deficiency – illness caused by lack of particular nutrients, such as proteins, or specific vitamins or minerals. Deficiency of each nutrient leads to specific health conditions (e.g. vitamin C deficiency causes scurvy).

Obesity – the condition of having significantly more body fat than average, often defined as having a Body Mass Index of over 30. Being obese significantly increases the risk of several health problems.

Ocean acidification – a decrease in the ocean's pH that happens when carbon dioxide dissolves in ocean water.

Oligotrophic – having low concentrations of nutrients. Most naturally occurring waterways are oligotrophic.

Open dump – a site where garbage is deposited, but not covered.

Organic – in the context of agriculture, organic means originating in nature, and not including anything man-made. Organic fertilizers include manure and compost. Organic pesticides include natural products from plants that repel pests.

Organic matter – substances from living organisms, which can decompose. It can come from natural sources, such as falling leaves, but can also include waste from toilets, farms and kitchens.

Overgrazed – the state of a field or other habitat that has had too much of its vegetation eaten by herbivorous livestock. The vegetation on overgrazed land is depleted to the extent that it cannot grow back.

Overnourished – eating a diet that contains too much food overall, or too much of particular nutrients, leading to poor health.

Overshoot day – the day in the year by which the annual resource budget of our planet has been exhausted. For a country, it is the day Earth Overshoot Day would be if everyone on the planet consumed at their rate.

Overweight – the condition of having more than an average amount of body fat, generally considered a risk to health. Overweight is often defined as having a Body Mass Index of more than 25, but less than 30.

Oxygen depletion – a reduction in the concentration of oxygen that is dissolved in water. It is often caused by nutrient or organic matter pollution, and causes the death of many animals.

Parasite – a close, symbiotic relationship between two species in which one species benefits, and the other suffers.

Parent rock – the rock that provides the mineral components of a particular soil.

Pastoralism – a form of animal husbandry in which animals are grazed in open fields.

Pastoralist – a farmer who breeds animals such as cattle and sheep.

Pathogen – a disease-causing organism, such as viruses, some bacteria, and some other single-celled organisms.

Pedosphere – the layer of soil that covers most of the Earth's terrestrial surface.

Per capita GDP – a measurement of the economic output of a nation per person.

Percolate – the downward movement of water through the soil.

Permafrost – an area of permanently frozen ground, where all the water in the soil is frozen throughout the year. Permafrost is mostly found in the Arctic tundra.

Persistent – remaining for a long time.

Personal assumptions – made up of our beliefs and preconceptions, which are based on our previous experiences as well as our cultural background and the education with which we have been provided. For instance, someone who has grown up in a rural area may have a different perspective on the importance of protecting natural resources, or on how necessary it is to drive a car, than someone who has grown up in an urban area.

Personal values – can include beliefs about the importance of family, community, and personal responsibility. For example, an individual with a strong sense of community may have a different perspective on the importance of protecting the environment than someone who places a higher emphasis on individual rights.

Perspective – a point of view, a particular way of seeing or considering something.

Pesticide – a substance that is used on farms (e.g. sprayed on to crops, or into the soil), to kill (or sometimes repel) pests such as insects or fungi that would otherwise harm the crops.

pH – a measure of how acidic or alkaline a solution is. A pH of 7 is neutral; less than 7 is acidic, and more than 7 is alkaline.

Phosphates – compounds containing the phosphate ion, PO_4^{3-}. They are an essential nutrient for plants and algae, and often included in synthetic fertilizers.

Photoageing – when the sun causes premature ageing of the skin.

Photodegradation – the degradation of plastic by exposure to sunlight.

Physical water scarcity – limited availability of water because there is insufficient water in an area.

Pick-your-own farm – a farm that allows paying customers to pick the crops they want directly from the fields.

Planetary boundary – a concept to define the environmental limits within which humans can survive and develop.

Ploughing – using heavy cutting or digging tools to break up the surface of the soil, removing any vegetation, and mixing (or turning over) the top few centimetres of the soil. Also spelled plowing.

Poaching – the illegal hunting and harvesting of animals.

Polar ice caps – large masses of ice found near the poles. At the North Pole, the ice mostly floats on the Arctic Ocean. At the South Pole, the ice sits on the continent of Antarctica.

Polar molecule – a molecule, like water, that has one end with a slight positive charge, and one end with a slight negative charge.

Polluter-pays principle – the principle that the person or business that causes pollution (or other forms of environmental harm) should bear the cost of the damage they cause.

Population – all the members of a single species that live in the same area at the same time (and are capable of interbreeding).

Porous rock – rock with interconnected open spaces or pores that allow fluids (such as water, oil or gas) to pass through and be stored within the rock structure.

Pragmatism – a philosophical approach that relates to utility and practicality.

Precipitation – any form of water, liquid or solid, that falls from the atmosphere to the ground.

Primary pollutant – a pollutant that is active as soon as it is emitted.

Production – the process of accumulation of energy in living organisms. The term can also refer to the amount of energy stored in living organisms. Usually expressed in kilojoules (kJ).

Productivity – the rate of accumulation of energy in living organisms. Usually expressed in kilojoules per square metre per year (kJ m-2 y-1).

Pro-natalist – relating to a policy or practice that encourages people to have more children.

Pyrolysis – the heating of plastic waste in the absence of oxygen, leading to the breakdown of complex polymers into simpler hydrocarbons.

Quota – a limit placed on the amount of a resource that can be extracted (e.g. the amount of fish that can be caught). It might be applied to individuals, companies, or countries.

Radioactivity – the continuous release of particles and energy from the decay of the nuclei of certain atoms and their isotopes until it reaches a stable atom. For example, uranium (U_{235}) decays through several steps until it reaches lead (Pb_{206}), a stable atom.

Radiosondes – instrument packages carried aloft by a helium-filled 'weather balloon' that measure vertical profiles of air temperature, relative humidity, and pressure from the ground up to 30 km.

Ramsar site – a wetland designated as having international importance.

Glossary

Rangeland – areas of land where wild or domesticated herbivores graze. Rangelands are often covered by grasses or shrubs, and herbivores are able to roam over wide areas, unless their movements are restricted, for instance by fencing.

Reeds – any species of tall grass found in wetlands throughout the temperate and tropical regions.

Regenerative architecture – incorporating building features, such as green roofs, wind turbines and photovoltaic cells, that harness or clean the natural environment.

Relief – variation in height or elevation of the land. A large change in elevation over a short distance produces a steep slope or gradient, which might be prone to erosion, due to gravity or fast-flowing water.

Replacement cost – the cost incurred to replace the existing system with a new or better system.

Replacement rate – a fertility rate that results in the population exactly replacing itself in one generation. This is an average of 2.1 births per female of reproductive age in developed nations, such as the USA.

Residence time – the average length of time that water remains in one location, such as a lake, aquifer or enclosed area of the sea. A high residence time means that pollutants in that area tend to accumulate more than they would if the water flowed freely away.

Resilience – in an environmental context, the inherent ability to absorb various disturbances and reorganize while undergoing state changes to maintain critical functions.

Respiration – the breakdown of organic molecules by living cells, to produce energy. Respiration may be aerobic or anaerobic.

Ruminant – one of several species of herbivorous mammals, including cattle, sheep and goats, which have a special chamber in their stomachs called a rumen where some digestion of plant matter occurs. Ruminants pass food back and forth between their mouths and their rumens, repeatedly chewing and swallowing. Bacteria in the rumen break down plant matter and produce methane, which is released from the animal's mouth.

Salinity – a measure of how much salt is dissolved in water (the dissolved NaCl content of water), usually measured in parts per thousand (‰).

Salinization – the process of addition of salt to soil, usually as a result of excessive irrigation.

Saltwater – water with a high salinity, usually found in the ocean, but also in some inland seas.

Sand – soil made up mostly of relatively large particles, from 0.05 to 2.0 mm in diameter.

Sanitary landfill – a carefully constructed site designed for the disposal of garbage. The ground is usually excavated to form a large pit, and the bottom of the pit is lined with an impermeable material. Garbage is evenly spread within the pit, and covered with soil.

Saturated – having the maximum amount of a solute in a solution. When a solution is saturated, no more of the solute can dissolve.

Scientific understanding – another factor that determines our perspectives. It includes our knowledge and understanding of scientific concepts and principles, as well as our knowledge of the natural world, social and physical sciences. For example, if you have a strong understanding of ecology then you may have a different perspective on how human activities impact the environment than a student with limited knowledge in this area.

Secondary pollutant – pollutants that form within the atmosphere as a result of chemical reactions taking place in the atmosphere.

Sediment – small pieces of insoluble material that can be carried by water. Sediments may come from natural erosion of rocks and soils, but can also be released in sewage, and in agricultural runoff.

Seed bank – a store of living, but usually dormant, seeds, which can grow into new plants. Artificial seed banks are widely used in conservation of plant biodiversity. Soil can act as a natural seed bank.

Selective breeding – a process of cross-breeding individual plants or animals with desirable characteristics or traits, to produce offspring with those favourable traits. The method has been used since the very beginnings of agriculture, but was improved by developments in the modern field of genetics.

Seral community – intermediate stage in ecological succession, representing a transitional group of species in an area undergoing environmental change.

Services – the direct and indirect contributions ecosystems provide for human well-being and quality of life.

Sewage – any form of wastewater from human activities.

Silt – soil made up mostly of intermediate-sized particles, from 0.002 to 0.05 mm in diameter:

Silviculture – the practice of managing and controlling the establishment, growth and maintenance of forests and woodlands. It is a branch of forestry that involves applying scientific principles and techniques to ensure the sustained production of various forest resources, such as timber, fuelwood and non-timber forest products, while also considering ecological and environmental objectives.

Soil compaction – a result of heavy loads pressing down on soil. The spaces between mineral particles and organic matter are lost, meaning that air and water cannot move through the soil easily, and plants' roots cannot grow easily.

Soil contamination – the addition of harmful substances or pollutants to soil, either through deliberate disposal or by accidental or unintentional leakage of wastes.

Soluble – able to dissolve in a particular solvent, such as water.

Solute – the substance that is dissolved in a solution.

Solution – a mixture formed when small particles of one substance are dissolved in another substance.

Solvent – the substance that dissolves the solute in a solution.

Speciation – the gradual change of individuals of the same species, resulting in distinct species that are no longer able to breed.

Specific heat capacity – the amount of energy needed to raise the temperature of 1 g of a substance by 1°C. Water has an unusually high specific heat capacity.

Specific latent heat – the amount of energy needed to change the state of 1 g of a substance.

Stakeholder – an individual, group or organization with an interest in or concern with an issue or a project.

Stratification – the formation of layers in a liquid or gas.

Subsidy – a payment made by governments to producers (e.g. farmers) to lower the cost of production of an item. Effectively, the government pays part of the cost of production, so that the farmer's costs are lowered. Subsidies have often been used to encourage farmers to produce goods that the government sees as essential for the country.

Subsistence farming – small-scale farming primarily to provide food for the farmer's family, sometimes with a little excess sold for income.

Surface tension – a force of attraction between molecules on the surface of a liquid, which holds the molecules together. Surface tension can resist weaker forces that push or pull against the surface.

Synthesize – to combine or produce something by bringing together different elements or components to form a coherent whole.

Synthetic – made by humans; artificial, manufactured or man-made.

Synthetic fertilizer – a fertilizer that does not occur in nature, but is manufactured by humans from natural raw materials. They usually contain specified amounts of nutrients, particularly nitrogen (N), phosphorus (P), and potassium (K).

Tax – a payment made to the government by various parties within a country. Governments generally use taxes to raise funds to cover the cost of running the country, providing services to citizens, etc., but they can also use taxes to manipulate demand. High taxes on alcohol and tobacco, for instance, have both effects: they raise money for the government, and they raise the prices of those products so that fewer people might buy them.

Taxonomy – the science of classifying and organizing in a hierarchical manner.

Temperature anomaly – the difference between a measured temperature and a reference value or long-term average. It is positive if the measured temperature is above the reference value, and negative if the measured temperature is below the reference value.

The Dry Corridor – a term for a strip of land across El Salvador, Guatemala, Honduras and Nicaragua that is particularly at risk of extreme climate events such as long periods of drought.

Thermal stratification – the formation of layers with different temperatures in water. Warmer water is less dense, and rises above cooler water.

Thermocline – a layer of water, beginning a few metres below the surface, where the temperature drops rapidly with depth. The thermocline separates the warm surface water from the deeper, colder water.

Tillage – the practice of ploughing the land.

Tipping cascade – a sequence of tipping points, one leading to the other, which ultimately leads to system collapse.

Tipping point – the point that represents the critical threshold beyond which an environmental system undergoes rapid and often irreversible changes.

Toxic – causing harmful effects on health. How significant the effects are may depend on the chemical nature of a substance, the chemical form it is in, its concentration, and the duration of exposure.

Transparent – allows light to pass through.

Trophic – refers to the different levels in a food chain or web, indicating an organism's position and role in the transfer of energy and nutrients within an ecosystem.

Undernourished – eating a diet that provides less than the required amount of nutrients. This could refer to low energy content, or lack of particular nutrients such as protein or vitamins.

Urban heat island effect – an issue that takes place when cities replace the majority of natural areas with pavements, buildings and roads, which all absorb and trap heat within the city. This results in the temperature in urban areas being higher than in surrounding natural and less densely populated areas.

Urbanization – an increase in the number of people living within urban systems, generally due to migration from rural areas.

Vegan – a person who does not eat any products made from animals. Some vegans also prefer not to use non-food animal products, such as leather. 'Vegan' is more consistently used than 'vegetarian', but there are still variations from person to person.

Vegetarian – generally, a person who does not eat meat. The term is used inconsistently, however, as some vegetarians may eat some animal products, such as eggs, while others do not.

Vertical integration – a business model where one party carries out multiple functions in the supply chain. A farmer may package and process the food on site, before selling it to wholesalers, retailers, or consumers.

Water accessibility – a measure of how easy it is for people to obtain water. Accessibility may be affected by the location of the water, or the cost of obtaining it.

Water quality – a measure of the properties of water relative to what the water is needed for. For instance, drinking water should be clean, clear, colourless and odourless, and have no toxic chemicals or disease-causing organisms in it.

Water scarcity – limited availability of water. Scarcity may be physical or economic.

Water security – having access to enough safe drinking water to meet people's needs.

Water stress – difficulty meeting the water needs of people or ecosystems, due to lack of availability, poor water quality, or inability to access water.

Water-borne disease – a disease that is contracted by drinking or sometimes bathing in contaminated water. Most water-borne diseases are caused by pathogens in the water.

Weathering – The process of rocks breaking up into smaller and smaller fragments, due the action of wind, water and living organisms.

Wetland – an area of land that is regularly or permanently flooded by water, such as a swamp, marsh or bog.

World Heritage Site – a site designated by UNESCO as being culturally, historically or scientifically significant.

Worldview – a broad and comprehensive framework that shapes people's perceptions and understandings of their surroundings as well as their actions.

Xerophyte – a plant that is adapted to survive in a habitat that provides limited access to free water, such as a desert or frozen soil.

Yield – the total production of a product. In fisheries and agriculture, it is usually measured in mass of product (e.g. tonnes of fish or tomatoes).

Zooxanthellae – single-celled algae that live inside the tissues of coral polyps, giving the coral their colour. Zooxanthellae and corals have a mutualistic relationship, where the algae use the corals' waste products, and in turn provide food to the corals.

Acknowledgements

The Publishers would like to thank Bhavya Prabu for her contribution towards Theme 5.
IB material has been reproduced with kind permission from the International Baccalaureate Organization.

Photo credits

p. 1 © marina_larina/stock.adobe.com; **p. 4** © Davivd/stock.adobe.com; **p. 9** © Gallup/Statista; **p. 12** © Frank Chuang/UCL CASA; **p. 13** © Steffen Lehmann; **p. 16** © MediaWorldImages/Alamy Stock Photo; **p. 20** © Associated Press/Alamy Stock Photo; **p. 28** © Anna/stock.adobe.com; **p. 31** © Pat Eyre/Alamy Stock Photo; **p. 34** © Paul Atkinson/stock.adobe.com; **p. 45** © Dusan Kostic/stock.adobe.com; **p. 48 tl** © Curioso.Photography/stock.adobe.com, **tr** © noon@photo/stock.adobe.com, **b** © marina_larina/stock.adobe.com; **p. 53** © Global Footprint Network 2024, www.overshootday.org; **p. 73 t** © Azote for Stockholm Resilience Centre, based on analysis in Richardson et al 2023, **b** © Azote for Stockholm Resilience Centre, Stockholm University. Based on Richardson et al. 2023, Steffen et al. 2015, and Rockström et al. 2009; **p. 74** © Bruce Thomson/www.naturepl.com; **p. 75** © Kate Raworth and Christian Guthier. CC-BY-SA 4.0; **p. 78** © Ellen MacArthur Foundation (Circular economy system diagram, 2019); **p. 83** © Andrew Lichtenstein/Corbis – Getty Images; **p. 86 l** © Nicolette Wollentin/stock.adobe.com, **r** © Fanfo/stock.adobe.com; **p. 91** © Marshal Hedin/Wikipedia; **p. 98** © The Photolibrary Wales/Alamy Stock Photo; **p. 104 t** © Irina K./stock.adobe.com, **b** © Georgette Apol/Alamy Stock Photo; **p. 108** © Amazon Conservation's Monitoring of the Andean Amazon Program (MAAP); **p. 109 t** © wollertz/stock.adobe.com, **b** © Lori Labrecque/stock.adobe.com; **p. 110** © Azote for Stockholm Resilience Centre, Stockholm University. Based on Richardson et al. 2023, Steffen et al. 2015, and Rockström et al. 2009; **p. 115 t** © blickwinkel/Alamy Stock Photo, **b** © Dr. Steve A. Johnson, University of Florida, IFAS; **p. 121** © NASA's Goddard Space Flight Center; **p. 124** © NOAA MESA Project/https://creativecommons.org/licenses/by/2.0; micro_photo, Witold Krasowski, Sergey Belov/stock.adobe.com; **p. 140 l** © martin33/Shutterstock, **r** © noon@photo/stock.adobe.com; **p. 141 tl** © Mike Mareen/stock.adobe.com, **tr** © geogphotos/Alamy Stock Photo; **p. 142** © Julio Etchart/Alamy Stock Photo; **p. 145 t** © Photocolorsteph/stock.adobe.com, **b** © Oregon Wild; **p. 146 t** © Andrew Lichtenstein/Corbis – Getty Images, **b** © VectorMine/Shutterstock; **p. 147 t** © Matauw/stock.adobe.com, **m** © kamilpetran/stock.adobe.com, **b** © weisesicht/stock.adobe.com; **p. 150 t** © Science Photo Library/Alamy Stock Photo, **b** © Climate Central; **p. 151** © romaset/stock.adobe.com; **p. 155** © European Environment Agency (EEA); **p. 156** © Julie McMahon, University of Illinois Urbana-Champaign; **p. 157** © VectorMine/stock.adobe.com; **p. 169** © a7880ss/stock.adobe.com; **p. 171** © J. Marini/Shutterstock; **p. 172** © heavypong/stock.adobe.com; **p. 173** © BlueRingMedia/Shutterstock; **p. 175** © zombiu26/Shutterstock; **p. 176** © Denis Onyodi/IFRC via ZUMA Wire/Shutterstock; **p. 179** © blueringmedia/stock.adobe.com; **p. 184 t** © Aerial Film Studio/stock.adobe.com, **b** © Matthew R McClure/Shutterstock; **p. 187** https://earthjustice.org; **p. 191** © hammonphoto/stock.adobe.com; **p. 208** © Conservation International/https://zenodo.org/records/4311850#.X8_jMdhKg2w; **p. 210 tl to br** © egorxfi/stock.adobe.com, © Nik_Merkulov/stock.adobe.com, © monticellllo/stock.adobe.com, © denira/stock.adobe.com, © MIGUEL GARCIA SAAVED/stock.adobe.com; **p. 211** © hammonphoto/stock.adobe.com; **p. 213** © Aldona/stock.adobe.com; **p. 215** © Nature Picture Library/Alamy Stock Photo; **p. 216** © Dr. R. Tim Patterson; **p. 229** © IUCN (2012). IUCN Red List Categories and Criteria: Version 3.1. IUCN, Gland Switzerland and Cambridge, UK.; **p. 230 l** © Thierry Costa/Wirestock Creators/stock.adobe.com, **m** © slowmotiongli/stock.adobe.com, **r** © Goinyk/stock.adobe.com; **p. 242** © 2024 Planet Labs PBC; **p. 243** © Azote for Stockholm Resilience Centre, Stockholm University. Based on Richardson et al. 2023, Steffen et al. 2015, and Rockström et al. 2009; **p. 255** © Secretariat of the Convention on Biological Diversity; **p. 264** © Stephen Elliott; **p. 265** © steve bly/Alamy Stock Photo; **p. 267** © Newscom/Alamy Stock Photo; **p. 276 t** © Viiviien/stock.adobe.com, **b** © Kalyakan/stock.adobe.com; **p. 277** © FEMA/Alamy Stock Photo; **p. 278 l** © Justin Kase zsixz/Alamy Stock Photo, **r** © Serhii Chrucky/Alamy Stock Photo; **p. 281 t** © Daniel Jędzura/stock.adobe.com, **b** © leomalsam/stock.adobe.com; **p. 293** © Martha Thierry/Detroit Free Press via ZUMA Wire; **p. 296 t** © Dave Willman/stock.adobe.com, **b** © Myjourney/stock.adobe.com; **p. 299** © parkerspics/stock.adobe.com; **p. 301** © PHOTOSTOCK-ISRAEL/SCIENCE PHOTO LIBRARY; **p. 304** © Porcupen/stock.adobe.com; **p. 305** © siriboon/stock.adobe.com; **p. 306** © Sean Sprague/Alamy Stock Photo; **p. 307** ©Jens Thomas/Farm Urban; **p. 309** © Ros Drinkwater/Alamy Stock Photo; **p. 312** © Azote for Stockholm Resilience Centre, Stockholm University. Based on Richardson et al. 2023, Steffen et al. 2015, and Rockström et al. 2009; **p. 315** © Jose Luis Stephens/stock.adobe.com; **p. 316** © dpa picture alliance/Alamy Stock Photo; **p. 322** © The Asahi Shimbun/Getty Images; **p. 323 t** © Quirky China/Shutterstock, **b** © VectorMine/stock.adobe.com; **p. 332** © WIM VAN EGMOND/SCIENCE PHOTO LIBRARY, **b** © Nature Picture Library/Alamy Stock Photo; **p. 333 tl** © cherylvb/stock.adobe.com, **tr** © dam/stock.adobe.com, **b** © zilvergolf/stock.adobe.com; **p. 334** © dam/stock.adobe.com; **p. 335 t** © pikumin/stock.adobe.com, **b** © Vankad/stock.adobe.com; **p. 337** © Walter/stock.adobe.com; **p. 338** © Holland-PhotostockNL/stock.adobe.com; **p. 339 t** © whitcomberd/stock.adobe.com, **m** © WILDLIFE GmbH/Alamy Stock Photo, **b** © WILDLIFE GmbH/Alamy Stock Photo; **p. 342** © Design Pics Inc/Alamy Stock Photo; **p. 346** © Newscom/Alamy Stock Photo; **p. 347** © Auscape International Pty Ltd/Alamy Stock Photo; **p. 349 t** © SCUBAZOO/SCIENCE PHOTO LIBRARY, **b** © Marine Stewardship Council; **p. 351** map supplied by Geospatial Services, Great Barrier Reef Marine Park Authority, © Commonwealth of Australia (GBRMPA); **p. 354 t** © Gabriel/stock.adobe.com, **b** © Angela/stock.adobe.com; **p. 355** © Associated Press/Alamy Stock Photo; **p. 356** © PiLensPhoto/stock.adobe.com; **p. 376 t** © waechter-media.de/stock.adobe.com, **b** © michaeljung/stock.adobe.com; **p. 380 l** © Jouni/stock.adobe.com, **r** © Sipa US/Alamy Stock Photo; **p. 383** © WILDLIFE GmbH/Alamy Stock Photo; **p. 389** © Roman Bjuty/stock.adobe.com; **p. 392** © Guajillo studio/stock.adobe.com; **p. 398 t** © blickwinkel/Alamy Stock Photo, **ml** © Ekky/stock.adobe.com, **mr** © Macroscopic Solution/stock.adobe.com, **b** © Michael de Nysschen/stock.adobe.com; **p. 402 t** © Maurizio/stock.adobe.com, **b** © David Pimborough/stock.adobe.com; **p. 404 t** © troutnut/stock.adobe.com, **b** © Víctor/stock.adobe.com; **p. 405 l** © Rostislav/stock.adobe.com, **r** © blickwinkel/Alamy Stock Photo; **p. 406** © MShieldsPhotos/Alamy Stock Photo; **p. 412** © PA Images/Alamy Stock Photo; **p. 415** © murasal/stock.adobe.com; **p. 420** © murasal/stock.adobe.com; **p. 425** © maykal/stock.adobe.com; **p. 426** © Jared Quentin/stock.adobe.com; **p. 429 tl** © Mediscan/Alamy Stock Photo, **tr** © Nigel Cattlin /Alamy Stock Photo, **bl** © tum2282/stock.adobe.com, **br** © patila/stock.adobe.com; **p. 431 t** © DR JEREMY BURGESS/SCIENCE PHOTO LIBRARY, **m** © Grant Heilman Photography/Alamy Stock Photo; **p. 433 t** © Tomasz/stock.adobe.com, **b** © Becris/stock.adobe.com; **p. 435** © Pixel-Shot/stock.adobe.com; **p. 436 t** © Riccardo Pravettoni, UNEP/GRID-Arendal/www.grida.no/resources/6658, **b** © Danita Delimont Creative/Alamy Stock Photo; **p. 437 t** © Martyn Williams Photography/Alamy

Stock Photo, **b** © GRID-Arendal/Nunataryuk/www.grida.no/resources/13519; **p. 440** © Sandra A. Dunlap/Shutterstock; **p. 441** © Cranfield University and for the Controller of HMSO 2024 used with permission; **p. 442 t** © Vector Tradition/stock.adobe.com, **b** © SoilPaparazzi/stock.adobe.com; **p. 443** © Stockfotos/stock.adobe.com; **p. 445** © Silvan Wick-Ecology / Alamy Stock Photo; **p. 446 t** © Matthew J. Thomas/stock.adobe.com, **b** © Anthony Gallagher/Shutterstock; **p. 447** © MarkGodden/stock.adobe.com; **p. 451** Public domain/https://commons.wikimedia.org/wiki/File:Single_ring.JPG; **p. 452** © Isya' Geodafit/Shutterstock; **p. 455** © MIKKEL JUUL JENSEN/SCIENCE PHOTO LIBRARY; **p. 458** © Baifran I LOVE U/stock.adobe.com; **p. 461** © Bernd Ege/stock.adobe.com; **p. 462** © Marcin Rogozinski/stock.adobe.com; **p. 463** © t4nkyong/stock.adobe.com; **p. 464** © Sly/stock.adobe.com; **p. 465 t** © RethaAretha/Shutterstock, **m** © laura/stock.adobe.com; **p. 465 b** © ponsulak/stock.adobe.com; **p. 466** © Science History Images/Alamy Stock Photo; **p. 470 t** © Aleksandar Todorovic/stock.adobe.com, **b** © Julio Etchart/Alamy Stock Photo; **p. 471** © Joseph Sorrentino/Alamy Stock Photo; **p. 475** © henrique ferrera/Shutterstock; **p. 476** © Hemis/Alamy Stock Photo; **p. 477** © wakr10/stock.adobe.com; **p. 478** © GiangHai/stock.adobe.com; **p. 479** © Юлия Пархоменко/stock.adobe.com; **p. 480 t** © jacquimartin/stock.adobe.com, **b** © torwaiphoto/stock.adobe.com; **p. 484** © Tee11/stock.adobe.com; **p. 485 m** © MemoryMan/stock.adobe.com, **b** © paul/stock.adobe.com; **p. 487 l** © Lalita/stock.adobe.com, **r** © CRISTINA PEDRAZZINI/SCIENCE PHOTO LIBRARY; **p. 488** © PA Images/Alamy Stock Photo; **p. 495** © Margaret Burlingham/stock.adobe.com; **p. 498 t** © creativenature.nl/stock.adobe.com, **b** © josefkubes/stock.adobe.com; **p. 500** © alisonhancock/stock.adobe.com; **p. 502** This graphic was prepared by EAT and is included in an adapted summary of the Commission Food in The Anthropocene: the EAT-*Lancet* Commission on Healthy Diets from Sustainable Food Systems. The entire Commission can be found online at eatforum.org/eat-lancet-commission.; **p. 503 t** © Imago Photo/stock.adobe.com, **b** © Sofia Royo/stock.adobe.com; **p. 505** © Asim Patel/stock.adobe.com; **p. 507** © Maggie Sully/Alamy Stock Photo; **p. 513** © Mamunur Rashid/Shutterstock; **p. 515** © bigmouse108/stock.adobe.com; **p. 518** © Dimitrios/stock.adobe.com; **p. 525** © farbkombinat/stock.adobe.com; **p. 530 t** © ESA, **b** © NOAA, measured at the Mauna Loa Observatory; **p. 531** © Sebastian Noethlichs/Shutterstock; **p. 540** © picture alliance/TopFoto; **p. 541** © Shutterstock / Mamunur Rashid; **p. 542** © Neil/stock.adobe.com; **p. 548 l** © Nick Beer/stock.adobe.com, **r** © Rob Byron/stock.adobe.com; **p. 549** © Daniel J. Cox/Alamy Stock Photo; **p. 552** Figure 1 in Stocker, T.F., D. Qin, G.-K. Plattner, L.V. Alexander, S.K. Allen, N.L. Bindoff, F.-M. Bréon, J.A. Church, U. Cubasch, S. Emori, P. Forster, P. Friedlingstein, N. Gillett, J.M. Gregory, D.L. Hartmann, E. Jansen, B. Kirtman, R. Knutti, K. Krishna Kumar, P. Lemke, J. Marotzke, V. Masson-Delmotte, G.A. Meehl, I.I. Mokhov, S. Piao, V. Ramaswamy, D. Randall, M. Rhein, M. Rojas, C. Sabine, D. Shindell, L.D. Talley, D.G. Vaughan and S.-P. Xie, 2013: Technical Summary. In: Climate Change 2013: The Physical Science Basis. Contribution of Working Group I to the Fifth Assessment Report of the Intergovernmental Panel on Climate Change [Stocker, T.F., D. Qin, G.-K. Plattner, M. Tignor, S.K. Allen, J. Boschung, A. Nauels, Y. Xia, V. Bex and P.M. Midgley (eds.)]. Cambridge University Press, Cambridge, UK and New York, NY, USA; **p. 553** © blickwinkel/Alamy Stock Photo; **p. 554 t** © Hugo Ahlenius, UNEP/GRID-Arendal/www.grida.no/resources/5228, **b** © KARSTEN SCHNEIDER/SCIENCE PHOTO LIBRARY; **p. 568** © Christian Schwier/stock.adobe.com; **p. 570 l** © Marc Pinter/stock.adobe.com, **r** © Andrea Izzotti/stock.adobe.com; **p. 579** Figure Cross-Chapter Box EXTREMES.1 in Parmesan, C., M.D. Morecroft, Y. Trisurat, R. Adrian, G.Z. Anshari, A. Arneth, Q. Gao, P. Gonzalez, R. Harris, J. Price, N. Stevens, and G.H. Talukdar, 2022: Terrestrial and Freshwater Ecosystems and their Services. In: Climate Change 2022: Impacts, Adaptation, and Vulnerability. Contribution of Working Group II to the Sixth Assessment Report of the Intergovernmental Panel on Climate Change [H.-O. Pörtner, D.C. Roberts, M. Tignor, E.S. Poloczanska, K. Mintenbeck, A. Alegría, M. Craig, S. Langsdorf, S. Löschke, V. Möller, A. Okem, B. Rama (eds.)]. Cambridge University Press, Cambridge, UK and New York, NY, USA, pp. 197-377, doi:10.1017/9781009325844.004; **p. 581** © metamorworks/stock.adobe.com; **p. 582** © Jansos/Alamy Stock Photo; **p. 584** © NASA/Desiree Stover; **p. 586 t** © Orjan Ellingvag/Alamy Stock Photo, **m** © VectorMine/stock.adobe.com, **b** © Design Pics Inc/Alamy Stock Photo; **p. 588 t** © PA Images/Alamy Stock Photo, **m** © Associated Press/Alamy Stock Photo, **b** © PA Images/Alamy Stock Photo; **p. 591** © Shutterstock / Macrovector; **p. 599** © NASA Goddard Space Flight Center; **p. 603** © NOAA Climate.gov; **p. 606 t** © NASA Goddard Space Flight Center, **b** Figure 8.6 in Birkmann, J., E. Liwenga, R. Pandey, E. Boyd, R. Djalante, F. Gemenne, W. Leal Filho, P.F. Pinho, L. Stringer, and D. Wrathall, 2022: Poverty, Livelihoods and Sustainable Development. In: Climate Change 2022: Impacts, Adaptation and Vulnerability. Contribution of Working Group II to the Sixth Assessment Report of the Intergovernmental Panel on Climate Change [H.-O. Pörtner, D.C. Roberts, M. Tignor, E.S. Poloczanska, K. Mintenbeck, A. Alegría, M. Craig, S. Langsdorf, S. Löschke, V. Möller, A. Okem, B. Rama (eds.)]. Cambridge University Press, Cambridge, UK and New York, NY, USA, pp. 1171–1274, doi:10.1017/9781009325844.010; **p. 607** © Shutterstock/Robb1037; **p. 610** © smuki/stock.adobe.com; **p. 614** © alagz/stock.adobe.com; **p. 619** © Shutterstock/Robert Szymanski; **p. 639 t** © VectorMine/stock.adobe.com, **b** © Shutterstock/piscari; **p. 645** © Design Pics Inc/Alamy Stock Photo; **p. 646** © Nandalal/stock.adobe.com; **p. 647** © blueringmedia/stock.adobe.com; **p. 649** © Michele Ursi/stock.adobe.com; **p. 651** © VectorMine/stock.adobe.com; **p. 654** © show999/stock.adobe.com; **p. 656** © DirkDaniel/stock.adobe.com; **p. 659** © AllahFoto/stock.adobe.com; **p. 662** © U.S. Nuclear Regulatory Commission/flickr.com/photos/nrcgov; **p. 668** © The World Bank/Statista; **p. 674** © m.malinika/stock.adobe.com; **p. 678** © TAUFIK ART/Shutterstock; **p. 679** © TerraCycle, Inc. & Loop Global; **p. 684** © Vikivector/stock.adobe.com; **p. 687** © TAUFIK ART/Shutterstock; **p. 693** © Iain Masterton/Alamy Stock Photo; **p. 730** © Iain Masterton/Alamy Stock Photo; **p. 734** © Ellen MacArthur Foundation (Circular economy system diagram, 2019) ; **p. 737 l** © University of Maine Advanced Structures and Composites Center (ASCC), **m** © gumbao/stock.adobe.com, **r** © Aminu Abubakar/Getty Images; **p. 745** © Jochen Tack/Alamy Stock Photo; **p. 748** © adrian davies/Alamy Stock Photo

■ Text and artwork credits

p. 18 adapted from Dave, Dhaval & Dench, Daniel & Kenkel, Donald & Mathios, Alan & Wang, Hua. (2020). News that Takes Your Breath Away: Risk Perceptions During an Outbreak of Vaping-related Lung Injuries. Journal of Risk and Uncertainty. 60.10.1007/s11166-020-09329-2/Springer/Nature; **p. 71** SDG 17 icons from United Nations Sustainable Development Goals/www.un.org/sustainabledevelopment. The content of this publication has not been approved by the United Nations and does not reflect the views of the United Nations or its officials or Member States; **p. 153** reproduced with permission from energyeducation.ca; p.194 data from International Union for Conservation of Nature (IUCN) Red List (2022)/Our World In Data/https://creativecommons.org/licenses/by/4.0; **p. 297** data from 'Water withdrawals and consumption', Aquastat – processed by Our World in Data/https://creativecommons.org/licenses/by/4.0; **p. 532** data from NASA/NOAA Global Monitoring Laboratory; **p. 571** adapted from Hugo Ahlenius, UNEP/GRID-Arendal/www.grida.no/resources/7299; **p. 605** IEA (2021), Growth in global air conditioner stock, 1990-2050, IEA, Paris www.iea.org/data-and-statistics/charts/growth-in-global-air-conditioner-stock-1990-2050, Licence: CC BY 4.0

Index

A

abiotic factors   88–9, 91, 94, 105, 164, 179, 721–2
abstraction   275–6
acidification   72, 148, 150, 285, 347
acid rain   749–50
adenosine triphosphate (ATP)   122–3
adhesion   282
aerosols   517–18, 599–600
African elephants   91, 109–10
agriculture   130, 133, 135, 158, 159, 443, 458v–507
   drought-resistant crops   308
   harmful practices   476
   impact on carbon cycle   147
   impact on water   275–6, 296–8, 374, 481
   livestock   308–9, 458, 469, 476, 479–81, 497
   sustainability   54–5, 467–8, 498–9
agrochemicals   466
agroforestry   475, 477
air conditioning units   604–5
air pollution   159, 327, 743–57
air pressure   171
albedo   32, 37
algae   194, 389, 392–3, 397–9
alternate wetting and drying (AWD)   485
Amazon forest biome   107–8
ammonia   158
ammonification   157
Amsterdam   76–7, 747
Anangu people   617
animal husbandry   469
animal rights   371–2
Anthropocene   215–16
Anthropocene Working Group (AWG)   216
anthropocentrism   14–15, 482
anthroposphere   32
aphotic zone   283

aquaculture   158, 275, 307, 353–7
aquaponics   307
aquatic ecosystems   332
aquatic foods   335–6
aquifers   271, 326
arable land   458
Aral Sea   275–6
architecture   733, 739–40
   see also urban planning
Arctic sea ice   40–1
Arctic tundra   168, 437
artificial selection   210
Atlantic Meridional Overturning Circulation (AMOC)   289, 555
atmosphere   32, 140, 514–23, 529
atmospheric aerosol loading   72
atmospheric carbon dioxide   531–2, 535–6, 557
atmospheric circulation   170–1, 516, 521–2, 529
atmospheric systems   34–5
autotrophs   122, 131

B

Baka people   238–9
bananas   461, 487
bats   92, 200
battery storage   664–5
beaches   179
behavioural isolation   207
behaviour-over-time graph   17
Bhutan   56
bioaccumulation   118, 129, 374
biocapacity   64, 68–9
biodiesel   458
biodiversity   51, 102, 109, 193, 196
   and evolution   197–201, 209
   hotspots   207–8, 234–7
   key biodiversity areas (KBAs)   236–7
   loss   73, 110–11, 159, 173, 241
biofuels   458–9

biogeochemical cycles   140–2, 431
biogeochemical flows   72
biomagnification   118, 129
biomass   45, 50, 126, 128, 137, 285
biomes   164–74
   atmospheric circulation   170–1
   boreal forest   167
   hot desert   166–7
   shifts   173, 537
   temperate forest   168
   tropical rainforests   166, 438–9
   tundra   168, 437
biosphere   31, 72, 85, 140
Biosphere 2   30–1
biosphere integrity   110
biotic factors   91, 93, 179, 722
birds of paradise   207
birth rates   697–8
bitumen   645
black carbon aerosols   517
blast fishing   338, 339–40
blue water   67
body mass index (BMI)   508
boreal forest   167
bottom trawling   338
Brazilian bromeliads   34
brazil nuts   503
Brockovich, Erin   378
bromeliads   34
Brundtland Report 1987   57
bushmeat   505
butterfly diagram   78
bycatch   338

C

calcium carbonate   285
Canada, key biodiversity areas (KBAs)   237
cane toad   223–5
carbon, organic and inorganic forms   142
carbon capture   151

carbon capture and storage (CCS) 151
carbon capture and utilization (CCU) 151
carbon cycle 144, 146–8, 150–2, 436–8, 454–5
carbon dating 212–13
carbon dioxide 45, 65, 130, 141–2, 145, 148, 150, 517, 521, 531
carbon footprint 65–7
carbon monoxide 746
carbon neutrality 566
carbon sequestration 145–6, 151, 284–5
carbon storage 45
carbon synthesis 131
carcinogens 378, 643
carnivores 131, 189, 430
Carson, Rachel 20
cash crop 458, 509
cellular respiration *see* respiration
cement clinker 637
cheetahs 193, 254
chemoautotrophs 132
Chengdu Research Base 250
chert 642
Chinese medicine 220
chlorofluorocarbons (CFCs) 599–600, 624
chromium VI 378
circular economy model 77–80, 655, 681, 691
citizen science 69–70, 317
cladistic classification 112–13, 196
cladograms 213–14
classification
    organisms and species 86–8, 112–14, 117, 193–4, 196
    soil 433
climate 164, 174, 529
climate change 20, 73, 117, 220, 546–7
    adaptation strategies 569–75
    and fish populations 346
    global warming 173
    impacts on ecosystems 537–9
    impacts on human societies 540–4
    and indigenous populations 55
    mitigation strategies 567–9, 584–7
    over time 523–5
    perceptions of 4–5, 583

responses to 576–90
    tipping point 40
climate modelling 550–6
climate zones 164–8
closed systems 30–1, 137
clownfish 104
coal 640–2
coastal erosion 4
coastal zonation 179
cobalt mining 632–3, 666
cod fishing 232–3, 342, 365–6
coevolution 200
coffee consumption 591–2
cohesion 281
commercial fishing 337
    see also fish farming
common resources 46, 231–4
    see also tragedy of the commons
community 101, 179–88
community management 234
compost 475, 686
composting 686–7
concrete 626
condensation 273
Connel, Joseph 114–15
consequentialist ethics 625
conservation 6, 226, 246–53
consumers 123–4, 156
contamination 118
Convention on Biological Diversity (CBD) 251
Convention on the International Trade in Endangered Species (CITES) 59, 205, 228, 334
COP 28 (Conference of Parties) 2023 20, 578
coral polyp 346
coral reef ecosystems 41, 150, 152, 346–8, 538, 616
corn 487
cornucopian view 13
covalently bonds 280
cover cropping 147, 478
Critical Ecosystems Partnership Fund (CEPF) 236
critical thinking 6
crop residue 485

crop rotation 147, 477
crude oil 380–1, 642–3
crustaceans 333
cryosphere 31
cultural identity 62
cultural relativism 5
cultural sustainability 55
cultural value 617
cyanide fishing 338, 339

D

DaisyWorld Model 32–3, 36
dams 45, 299, 321
Darwin, Charles 197–9
death, and decomposition 157
death rates 697–8
decarbonization 566
deciduous forests 440
decomposers 124, 125, 131, 144, 157, 430
decomposition 157
Deepwater Horizon oil spill 61, 380–1
defecation 144
deforestation 13, 45, 48, 51, 107–8, 130, 140, 142, 146, 151, 158, 217, 276–7
Demographic Transition Model (DTM) 711–12
dendrochronology 549
denitrification 155
deontological ethics 625
desalination 301, 324–5
desertification 476, 538
deserts 166–7
detergents 377
detritivores 34, 123, 430
dichlorodiphenyltrichloroethane (DDT) 129, 375–6
disasters 20, 214, 510, 564, 639, 655, 657, 664–5, 709
disease 21, 91–4, 129, 193, 262, 293, 295, 303, 318, 375, 377–9, 396, 399, 543
dissolved oxygen (DO) 89, 383–4, 388
dolphins 369
domestic waste 669
doughnut economics model 75–7, 262, 713–14

Environmental Systems and Societies

drip irrigation 308
drought-resistant crops 308
dumps 380
dysphotic zone 283

E

earthquakes 32
ecocentrism 13–14, 482
ecological footprints 64–9, 136, 316–17, 630–1
ecological overshoot 52–3
ecological pyramids 128
ecological succession 38, 102–3
ecologists 13–14
economics 46, 56, 65, 72, 75–7, 97, 148, 152, 211, 240, 618, 713–14
economic sustainability 49, 55–7, 467
ecosystems 25, 29, 34, 58–9, 85, 93, 111–12, 124–6, 179–88, 332
ecotourism 264
effluents 377–9
electric vehicles 580–1
electronegativity 280
elephants 91, 109–10, 209, 220
El Niño–Southern Oscillation (ENSO) 174–5
emergent properties 42–3
employment rate 59
endangered species *see* species: endangered and threatened
energy balance 545–6
energy conservation 654
energy cycle 121–3
energy flow 105, 124
energy security 655–9
energy sources *see* non-renewable energy sources; renewable energy sources
energy storage systems 653–4, 664–5
energy transfer efficiency 136
environmental ethics 6, 9, 14–15, 33, 211, 234, 240, 300–1, 369, 482, 625
Environmental Impact Assessments (EIAs) 152, 628–9
environmental law 13, 17, 72, 74, 233, 234, 239–40, 348–9, 600
environmental movement 19–20
environmental sustainability 49–54, 468

see also sustainability
environmental value system 12–13
equalities 62
equilibrium 36–41, 107, 141–2, 185
equity 54
Estonia 256
estuary barrages 321–2
ethical beliefs 3, 6, 14
ethics *see* environmental ethics
euphotic zone 283
eutrophication 130, 158–9, 388–93, 399
evaporation 272, 529
evolution 197–201, 209, 211–13
Exclusive Economic Zone (EEZ) 368–9
excretion 157
exhaust fumes 752
exosphere 515, 516
ex situ conservation 246–50
externality 379
extinction events 214–15
extinction rates 111, 217
extinction risk 226, 247
extinct species 73, 229

F

famine 509–11
farmers' markets 499–500
farming *see* agriculture
fast fashion 671
feedback loops 43, 106
  negative 36–8, 43, 94, 544
  positive 38–40, 261, 341, 544, 546
fertilizers 158, 376, 388–9, 466, 474–5, 488–9
fisheries 341–2, 348
fish farming 158, 333
see also aquaculture
fishing 337–8
  marine protected areas (MPAs) 350–2
  maximum sustainable yields (MSY) 344–6, 362–4
  monitoring stocks 361–2
  overfishing 45, 232–4, 340–3
flagship species 250
flooding 4, 277–8, 485, 540–1
flow diagrams 106–7
flows 26–7, 29–30

food
  aquatic 335–6
  production 66, 459
  security 489–91
  supply chains 499
  waste 484
food chain 124–6
food miles 499
food web 124, 125, 127, 196, 207, 430
Foreign Direct Investment (FDI) 59
Forest Restoration Unit (FORRU) 263
forests 54, 145, 165, 167–8, 438–40
  see also deforestation; reforestation
fossil fuels 65–6, 130, 140, 145–6, 148, 151, 152–3, 209, 535–6, 640–1, 659–60, 746
fossil records 212–17
fracking 16, 644
freshwater 72, 271, 292–5, 311
freshwater pond ecosystem 38
frog populations 34
fungi 194, 431
future generations 57

G

Gaia hypothesis 32–3
Galápagos Islands 197–8, 229
Ganges River 295
gases, solubility 282
Gatheru, Wawa 19–20
gender inequalities 54, 62
genetic diversity 193, 210
genetic drift 86
genetic modification 209, 471, 486–8
genetic mutation 197–8, 206
geographical isolation 201, 206
geological timeframe 212–15
geosphere 32
geothermal power 648–9
ghost fishing 338, 339
giant panda 230, 250–1
glacial ice 271
Global Boundary Stratotype Section and Points (GSSPs) 215
global geochemical cycles 30
global hectares 64
globalization 21, 634

Index

Global Partnership on Nutrient Management (GPNM) 162
global temperatures *see* temperatures: global
global warming 173, 176, 286, 437–8
glucose 122
Goat Island Marine Reserve 352
goods 51, 148
Gore, Al 20
grasslands 438, 492–3
Great Barrier Reef 347–8, 351, 538
Great Pacific Garbage Patch (GPGP) 677
green architecture 739–40
Green GDP 60
 see also gross domestic product (GDP)
greenhouse effect 32, 519, 529, 535–6
greenhouse gases 65, 145, 151, 153–4, 173, 216, 326, 484, 517–19, 580, 643
greenhouses 497–8
green infrastructure 160
Greenland ice sheet 41
Green Revolution 20, 471–3
greenwashing 57
green water 67
grey water 67, 305–6
gross domestic product (GDP) 56, 59–60, 531, 697
gross national happiness (GNH) 56
gross national product (GNP) 59
gross productivity (GP) 126–7, 187
groundwater 271, 302
Gulf Stream 289

H

Haber, Fritz 158
Haber process 158
habitats 103–4, 117, 193–4, 196, 222–3, 248–53, 227
Hawaiian Islands 184
hazardous wastes 380
heat capacity 283
herbal leys 475
herbivores 91, 123, 131, 430
herders 469
heterotrophs 131
hierarchy of emergence 43
hindcasting 550–1

Holocene epoch 72
homeostasis 32
hot desert 166–7
Human Development Index (HDI) 59
human impacts 45–6, 107–8, 110–11, 118, 130, 140, 150–1, 158, 189, 209–10, 216–17, 220–43, 274–8
 see also anthroposphere
human populations 95–7, 696–718
hunger 459–60, 490–1
hunting 371–2, 505–6, 632
hurricanes 176
hydraulic fracturing *see* fracking
hydroelectricity 637, 647
hydrofluorocarbons (HFCs) 603–4, 624
hydrogen 158
hydrogen bonds 280
hydrological cycle 269–77
hydroponics 307
hydrosphere 31, 140

I

ice 271, 284
ice core analysis 548–9
ice melt 40–1, 55, 437–8
iguanas 197, 199
incineration 683–4
indigenous communities 55, 238–40, 370, 617
industrial effluents 377–9
Industrial Revolution 530, 641, 696
inequalities 54, 62
infrared radiation 517, 519
insects 194
in situ conservation 246–50
insolation 545
intercropping 471
Intergovernmental Oceanographic Commission (IOC) 290
Intergovernmental Panel on Climate Change (IPCC) 579–80
International Convention for the Regulation of Whaling (ICRW) 369
International Union for Conservation of Nature (IUCN) 226–9, 238, 253–4
International Whaling Commission (IWC) 369–70

intertidal community 109
intertidal zone 114
intrinsic value 6, 617
introgression 114
invasive species 46, 221, 223–5
invertebrates 333
ionic substances 281
Irish potato famine 510
irrigation 275, 308
isolation 30, 201, 206–7
Isthmus of Panama 201

J

James Webb space telescope 584
Japanese knotweed 223
justice, environmental 60–3, 241

K

Kaua'i 184
Keeling curve 530
key biodiversity areas (KBAs) 236–7
keystone species 109, 250
Kigali Amendment 579, 603
Knepp Estate Rewilding Project 262
Kyoto Protocol 1997 563, 578

L

Lake Victoria 221
land, global availability 457
landfill 54, 380, 677, 680, 682–3
La Niña 174–5
latent heat 283
law *see* environmental law
leachate 380
leaching 156
legumes 475
Leopold, Aldo 33, 211
life cycles 116–17
life expectancy 698
lighting, energy-efficient 654
lime 478
limestone 152–3, 642
limiting factor 358
linear economic model 79
linguistic diversity 55
Linnaean taxonomy 86
Linnaeus, Carl 86

liquefaction   645
lithium   665–6
lithosphere   32, 140, 152
littering   21
livestock   308–9, 458, 469, 476, 479–81, 497
lizards   115
Lovelock, James   20, 32–3
low-carbon technologies   150

M

Maasai communities   61, 241–2, 469–70
macrophytes   332–3
Maldives   4
malnourishment   508
mammals   334
mangrove trees   333
manmade systems   26–7
manure   475, 478
Māori language   55
marginal land   458
Margulis, Lynn   33
mariculture   356
marine iguanas   197, 199
marine protected areas (MPAs)   350
Marine Stewardship Council (MSC)   349
market failure   46, 379
mass-extinction events   214–15
maximum sustainable yields (MSY)   135–6, 344–6, 362–4
McKibben, Bill   20
meat consumption   20–1, 479–82, 500–1, 505
Mediterranean Basin Biodiversity Hotspot   234–5
Mekong River   634
melting   273
mercury   129
mesosphere   515
methane   65, 153–4, 485–6, 517, 536, 546
microcosms   28–9
microplastics   129–30, 677
migration policies   708
Milankovitch cycles   522–3
milpa   471

Minamata Disease   20, 129, 378–9
mob grazing   497
molluscs   333
Montreal Protocol 1987   563, 599–601, 624
moths   200
Murray–Darling basin   538
mussels   109
mutualists   430
mycorrhizae   431, 475

N

Nagoya Protocol   251
National Action Plan on Energy Efficiency (NAPE)   639–40
National Adaptation Programmes of Action (NAPA)   573–4
National Oceanic and Atmospheric Administration (NOAA)   176
national parks   254–5
   see also Yellowstone National Park
natural capital   50–2, 611–20, 625
natural gas   645
natural resources   50–2, 58–9, 334, 611–34
natural selection   197–9
Navajo language   55
negative feedback loops   36–8, 43, 94, 544
Netherlands, coastal zonation   179
net primary productivity (NPP)   126–7, 133–4, 135
net productivity (NP)   126–7
nitrogen   106, 141, 154–60, 161, 484
nitrogen cycle   106, 154–60, 432
nitrous oxide   65, 474, 484
noise pollution   327
nomadic farming   464–5, 469–70
non-biodegradable pollutants   129
non-governmental organizations (NGOs)   17
non-renewable energy sources   636, 640–6
non-renewable resources   617
nori   335
North Pole   35

no-till farming   147, 495–6
nuclear energy   639, 645–6, 652–3, 660–4
nutrient cycling   105–6, 107, 109–10, 124

O

obesity   508–9
oceanic absorption   148
oceans
   acidification   72, 148, 150, 285, 347
   and carbon cycle   148
   carbon sink   284
   currents   172–3
   geographic patterns   360–1
   intertidal community   109
   intertidal zone   114
oil spills   61
overfishing   45
solar radiation   172–3
stratification   285–7, 358–60
temperatures   50, 176, 285–6, 360
thermohaline circulation   287–9
oil spills   61, 380–1
ontological anthropocentrism   14
open systems   30, 105
organic farming   466
organic fertilizers   475
organic matter   153, 432, 436, 478
organisms   85–6
   abiotic and biotic conditions   88–91
   biodiversity   193–4
   classification   86–8, 112–14, 193–4
   competition   92
   parasitic relationships   91–2
   populations   88
Ostrom, Elinor   234
overconsumption   591–2
overcropping   476
overexploitation   58–9, 341–3
overfishing   45, 232–3, 340–1
overgrazing   469, 476
overnourishment   508
overshoot days   52–3
oxygen depletion   377

Index

ozone-depleting substances (ODSs) 598–601, 624
ozone layer *see* stratospheric ozone

P

palm oil industry 238
pandas 230, 250–1
pangolins 504–5
parasites 91–2, 123, 430
Paris Agreement 2015 437, 637
pastoral farming 458
pastoralism 61, 469–70, 479
pathogens 377
peatlands 436–7
pentadactyl limb 213
per capita income 59
permaculture 494
permafrost 168, 437–8
persistent organic pollutants (POPs) 674
perspectives 3–11, 13–15, 21
pesticides 130, 374–5, 466
petroleum 141
phosphorus 161
photoautotrophs 132
photodegradation 677
photosynthesis 107, 121–4, 132, 144–5, 156, 346, 358
pH scale 150, 284–5, 347, 384, 749
phytoplankton 124, 332
pick-your-own farms 499
Pinta Island tortoise 229
pioneer species 183
Planetary Boundaries model 72–5, 110, 159, 243–4, 311–12, 546–7
plankton 332, 346
plant-based diet 481–2, 500–1
plastic pollution 21, 48, 72, 129–30, 233, 382–3, 676
plate tectonics 32
poaching 220
polar climates 174
polar ice caps 271
polar molecules 280
polar substances 282
polar vortex 602–3
pollutants 129–30, 743, 746, 748, 752
polluter-pays principle 379

pollution 46, 72, 118, 161, 216, 220, 302, 676
  air 159, 327, 743–57
  noise 346
  plastic 21, 48, 72, 129–30, 233, 382–3, 676
  water 141, 374–412
polychlorinated biphenyls (PCBs) 129, 397
polyculture 477
population dynamics 697–701
populations 93–100, 179, 696–718
positive feedback loops 38–40, 261, 341, 544, 546
potato crops 487
poverty 54, 72, 235, 293, 318, 509–10, 557–8, 577, 679, 716, 726–8
poverty rate 59
prairies 45
precipitation 269, 273, 529
  see also rainfall
predation 91
predator–prey relationships 36, 44, 104, 106, 124
predators 123, 186, 257
preservation 246
primary productivity 132–3, 207–8
production 358
productivity 358
pro-family incentives 705–6
property rights 46
purple sea stars 109
pyramids, ecological 128

Q

quality of life 54
quartz 642
Quaternary period 524
queen conch 334

R

race 62
radioactivity 660–1
rainfall 107, 165–9, 174, 379–80
rainforests 41, 107, 165–6, 438–9
  see also deforestation
rainwater-catchment systems 301, 306–7

random sampling 97–8
Raworth, Kate 75
realized niche 115
recycling 79, 685–6
Red List of Threatened Species 226–7
reforestation 151, 263
refurbishing 79
regenerative architecture 733
remanufacturing 79
renewable energy sources 150, 636, 640, 646–53
renewable resources 50, 617
reservoirs 299
resilience 43–5, 48, 193, 196, 537
respiration 121–3, 126, 141
reusing 79
rewilding 256–7, 261–2, 494
rhinos 220, 230
rice 485, 487
Rio+20 conference 2012 20
Rio Earth Summit 1992 20
rivers 278, 295, 634
Rockström, Johan 72
Rotterdam Convention 2004 563
ruminants 480, 486
rural–urban migration 724–6

S

salinity 88, 286
salinization 427
salmon farming 356–7
saltwater 4, 271, 322, 326
sampling methods 97–9
San Diego Zoo 248–9
sandstone 642
saprotrophs 124
Sargassum bloom 392
savannah grasslands 109–10
scavengers 124
sea anemone 104
seafood 335–6
seagrass restoration 145
sea levels, rising 4, 540–1
seals 369
seaweed 335
secondary productivity 133, 135
selective breeding 471–2

self-sustaining system   28–9
sewage   376–7, 400
shale oil   643
shrimp farming   353–4
silvicultural systems   133, 135
Simpson's reciprocal index   203
sinks   141, 145–6
smart cities   581–2, 732
smoking   17–18
social justice   300–1
social sustainability   49, 54–5, 467–8
societal systems   26–7
soil   142, 151, 217
   biodiversity   429–31
   and carbon cycle   436–8, 454–5
   classification   433
   compaction   476
   composition   417–20
   conservation   476, 477
   enhancement   477–8
   food web   430
   functional roles of organisms   430
   macronutrients and micronutrients   428
   maps   441
   organic matter   432, 436
   profiles   439–43
   properties   449–54
   regeneration   493–4
   texture   89, 433–5
solar cell system   25
solar distillation   324–5
solar power   648
solar radiation   34–5, 172–3, 516, 529, 545, 595
solvents   281
South Pole   35
soybeans   487
speciation   201
species   85–6
   biodiversity   193, 202–3
   classification   86–8, 112–14, 117, 193–4, 196
   conservation   247–53
   endangered and threatened   226–9, 230, 504–5
   flagship   250

interactions   101, 104
invasive   46, 221, 223–5
keystone   109, 250
pioneer   183
realized niche   115
trophic tiers   102
see also organisms
stores   26, 44–6, 141, 145–6
stratification   285–7, 358–60
stratosphere   72, 515
stratospheric ozone   596–605
strip cultivation   477
sublimation   273
subsistence farming   458, 464
suburbanization   728
succession   102, 182–8
Sumatran rhino   230
sunlight   545
   distribution   516
   spectrum   595
surface tension   281
sustainability   7–8, 29, 48–9, 106, 161, 196
   agricultural systems   467–8, 498–9
   biocapacity   64, 68–9
   circular economy model   77–80
   citizen science   69–70
   and diet   479–81
   doughnut economics model   75–7
   ecological footprints   64–9, 136, 316–17, 630–1
   economic   49, 55–7, 467
   energy sources   639
   environmental   49–54, 468
   indicators   63–8
   models of   49
   natural resources   50–2, 58–9
   social   49, 54–5, 467–8
   sustainable development   57–8
   Sustainable Development Goals (SDGs)   70–2, 205, 460, 627, 708, 731
   urban planning   731–4
sustainable development   57–8
Svalbard global seed vault   248
synthetic fertilizers   474
systematic sampling   98
system resilience   43–5, 48

systems approach   26–30, 34
systems diagrams   25, 27–8, 30, 78, 106–7, 544

T

tar sands   645
taxonomy   86–8, 113–14, 193–4, 196
technocentrism   13–14
temperate climates   174
temperate forest   168
temperate grasslands   438
temperatures
   boreal forest   167
   by climate type   174
   global   40–1, 50–1, 73, 176, 516
   hot desert   167
   temperate forest   168
   tropical rainforests   166
   tundra   168
   of water   384
terrariums   28–9
terrestrial radiation   545
thermal pollution   379, 399
thermocline   285–6
thermodynamics
   first law   120
   second law   123, 137–8
thermohaline circulation   287–9, 359
thermosphere   515–16
Third Agricultural Revolution   472
Three Gorges Dam   299–300
Thunberg, Greta   19
tidal power   649–50
tilapia   333
tillage   147, 151, 477, 495–6
tin   397
tipping points   40–1, 107–8, 111, 554–6
toads   223–5
tortoises   229
tourism   252, 264
trade winds   171
tragedy of the commons   46, 231–2, 590
transects   179–80
transect sampling   98
transfers   29–30
transformations   29–30
transparency   283

transpiration   272
transport, carbon footprint   66
trawling   338
tree planting   145, 151
trees
   boreal forest   167
   dendrochronology   549
   systems diagram   25
   temperate forest   168
   tropical rainforests   166, 438–9
tributyltin (TBT)   397
tricellular atmospheric circulation   516–17
trophic mismatch   117
trophic tiers   102, 122, 125–6, 127, 136–7
tropical climates   174
tropical cyclones   176
tropical rainforests   166, 438–9
troposphere   515
truffles   503–4
tuna fishing   366–7
tundra   168, 437
turbidity   385
turnover   358
typhoons   176

U

undernourishment   459–60
United Nations
   Climate Change Conferences   20
   Convention on the Law of the Sea (UNCLOS)   368–9
   Educational, Scientific and Cultural Organization (UNESCO)   254
   Exclusive Economic Zone (EEZ)   368–9
   Framework Convention on Climate Change (UNFCCC)   563, 578
   Sustainable Development Goals (SDGs)   70–2, 205, 460, 627, 708, 731
unsustainable development   60
uranium   645–6, 662–3
urban ecosystems   721–2
urban environment   723–40
urban heat island (UHI) effect   541–2
urbanization   130, 158, 277–8, 541, 724
urban planning   730–4
urban runoff   379–80
urban–rural migration   726–7

V

values   3, 6–11, 17–18
vegan diet   481, 500
vegetarian diet   481, 500
vertical farms   498
Vienna Convention 1985   563
virtue ethics   234, 625
volatile organic compounds (VOCs)   746
volcanic activity   32
volcanic gases   525

W

Wallace-Wells, David   20
waste   668–92
   biodegradable   676
   composting   686–7
   degrowth   674, 678
   domestic   669
   exporting   684–5
   food   484
   hazardous   380
   incineration   683–4
   landfill   54, 380, 677, 680, 682–3
   management   673–5, 679–91
   recycling   79, 685–6
   strategies   678–88
waste-to-energy (WTE)   683
wastewater treatment   160, 400–4
water   67, 269–71
   access to   317–18
   conservation   304–9
   cycle   34–5
   flows   272–3
   footprint   67–8
   footprints   316–17
   hydrological cycle   269–77
   hydrosphere   31
   impact of agriculture   275–6, 296–8, 374, 481
   impact of deforestation   276–7
   impact of urbanization   277–8
   physical and chemical properties   280–4
   pollution   141, 374–412
   quality   317, 323–4, 386, 408
   regulations   313–16, 409–10
   scarcity   303–4, 309–10, 317
   security   292–5
   stores   271, 302
   stress   317–22
   supply   298–303
   transformations   272–3
   turbidity   385
   use   296–8, 311
   wastewater treatment   160, 400–4
watercress   335
Watson, Andrew   32
weather   164, 170–1, 174–6, 529
   rainfall   107, 165–9, 174
   tropical cyclones   176
   see also climate
weathering   417, 426
wetlands   302, 436
whales   369–72, 632
wheat   462, 472
White, Lynn Jr.   15
white-nose syndrome   92
wildfires   542
wind power   646–7, 651
winds   171
wolves   186, 261
Women's Economic Empowerment (WEE) policy   708–9
woodlands, ancient   247
   see also forests; trees
World Heritage sites   55
worldviews   5, 12

X

xerophytic plants   167

Y

Yangtze River   299–300
Yellowstone National Park   186–7, 249, 256, 261

Z

Zealandia Te Māra a Tāne   253
zonation   179, 182
zooplankton   124, 332, 347
zoos   247, 248–9
zooxanthellae   346